CRC HANDBOOK OF BIOSOLAR RESOURCES

Editor-in-Chief

Oskar R. Zaborsky, Ph.D.

VOLUME I: BASIC PRINCIPLES

Akira Mitsui, Ph.D.
Professor
Division of Biology and Living Resources
School of Marine and Atmospheric Science
University of Miami
Miami, Florida

Clanton C. Black, Jr., Ph.D.
Professor
Boyd Graduate Studies Research Center
University of Georgia
Athens, Georgia

VOLUME II: RESOURCE MATERIALS

Thomas A. McClure, Ph.D.
Senior Economist
Battelle Columbus Laboratories
Columbus, Ohio

Edward S. Lipinsky
Senior Research Leader
Battelle Columbus Laboratories
Columbus, Ohio

VOLUME III: RESOURCE PRODUCTS/APPLICATIONS

Fred Shafizadeh, Ph.D., D.Sc.
Professor
Department of Chemistry
University of Montana
Missoula, Montana

CRC Handbook of Biosolar Resources

Editor-in-Chief

Oskar R. Zaborsky
National Science Foundation
Washington, D.C.

Volume I
Part 1
Basic Principles

Editors

Akira Mitsui
Professor
Rosenstiel School of Marine
and Atmospheric Science
University of Miami
Miami, Florida

Clanton C. Black
Professor
Boyd Graduate Studies Research
Center
University of Georgia
Athens, Georgia

CRC Press, Inc.
Boca Raton, Florida

Library of Congress Cataloging in Publication Data
Main entry under title:

CRC handbook of biosolar resources.

Includes bibliographies and indexes.
CONTENTS: v. 1. Fundamental principles.--v. 2. Resource materials.
1. Biomass energy--Handbooks, manuals, etc.
2. Agriculture--Handbooks, manuals, etc. I. Zaborsky, Oskar.
TP360.C18 333.95'3 79-27371
ISBN 0-8493-3470-5 (set)

Direct all inquiries to CRC Press, Inc., 2000 N.W. 24th Street, Boca Raton, Florida, 33431.

International Standard Book Number 0-8493-3470-5 (set)
International Standard Book Number 0-8493-3471-3 (Volume I, Part I)
International Standard Book Number 0-8493-3472-1 (Volume I, Part II)

Library of Congress Card Number 79-27371
Printed in the United States

HANDBOOK OF BIOSOLAR RESOURCES

SERIES PREFACE

The purpose of this multivolume handbook is to serve as a reference source of the most pertinent information on biosolar resources ultimately intended for utilization. The term "biosolar" refers to biological systems (especially plant and microbial systems) which are directly dependent on solar energy for their life processes and growth. The series represents an integrated approach to critical information on the fundamental principles of photosynthetic organisms, the more significant biomass resources, and major products and applications. The series focuses on renewable resources for ultimate utility as energy, fuels, and other materials.

Since antiquity, biosolar resources have been essential to man's well-being and quality of life. Presently, this somewhat neglected and unappreciated resource (except for foods) is becoming a priority topic of study, discussion, and debate in many circles of society both at national and international levels. However, for the interested individual, there was not yet available a single source of information which presented the more relevant data in a convienient and understandable style. It is our hope that this three-volume series will fill this need. We also hope that the series will serve as a vehicle for organizing and disseminating additional information as it is generated.

From the outset of this endeavor, we recognized the immense diversity and complexity of biosolar resources as well as the delicate and subtle interdependency of the comprising elements. During the course of assembling the data, the generalization that the economic viability of biosolar resources is dictated by the fullest utility became even more appreciated. Regrettably we also recognized from the outset that certain difficult choices had to be made with regard to the subject matter to be included and the scope of the information presented to arrive at a manageable activity. The topics chosen were largely determined by their relevancy to biosolar resources, the availability of information, or the need to provide available data in an integrated manner suitable for use by individuals interested in biomass. A major objective was to provide data of present or potential commercial significance and not to be dominated by transitory research interests or subjective viewpoints. Yet, we equally recognized that new knowledge and directions must be found for the future and that biosolar resources represent an alternative option, but one which needs to be assessed fully and critically.

The initial series of three volumes seeks to present data at several levels for various intended users. The entire series is based on the natural flow of events from producing to using biosolar resources. At times, there may be some overlap between the volumes but we view this more as a benefit than as a detriment. The authors may have a different viewpoint or purpose for the particular resource described and individual volumes are somewhat self-consistent. We also recognized the existence of other CRC handbooks and other reference books and have avoided extensive duplication of readily available data.

The Editor-in-Chief, the Editors, and CRC Press, Inc. invite comments and criticisms from users on the selection of topics, accuracy and scope of the information presented. We particularly desire to hear about errors and superior data as they are generated.

The Editor-in-Chief expresses his gratitude to the hundreds of individuals who have contributed their time and expertise to this undertaking. In particular, I wish to thank my colleague Volume Editors and members of the Advisory Board for their outstanding assistance, constant objectivity, and critical review. I also thank the many authors for their able contributions prepared in a very short time. I give special thanks to Mrs.

Linda Pilkington for her able assistance. Lastly, I particularly wish to note the patience and understanding of my wife, Marcia Lee, during this endeavor and the moral stimulation from my children, Erik Christian and Claudia Lee.

Oskar R. Zaborsky
McLean, Virginia
August, 1980

PREFACE: VOLUME I

Biosolar energy is a widely recognized renewable resource upon which all mankind is dependent. Volume I, Parts 1 and 2, presents the current state of knowledge on the basic biological processes for converting sunlight into chemical bond energy by photosynthetic organisms. This knowledge is collected with the hope that it will assist in more effectively utilizing photosynthetic organisms as renewable resources.

In preparing these volumes, skilled authors from around the world were asked to contribute concise articles containing up-to-date and pertinent information in several major areas. Section 1 presents the fundamentals of light absorption processes, photosynthetic pigments, and electron transport systems. Section 2 provides a description of assimilation and dissimilation of the major elements of photosynthetic organisms. Section 3 of Part 1 presents the major biosynthetic pathways essential for the production of biosolar resources. Section 4 presents a general classification of photosynthetic organisms and information on sources and collections.

Section 5 of Part 2 provides a general description of the characteristics of selected organisms. The response of photosynthetic organisms to major environmental factors is given in Section 6 along with information on photosynthetic organisms from underexploited environments. Section 7 presents information on the primary productivity of biosolar organisms. Section 8 presents the physical resources and inputs needed for organisms to live, grow, and reproduce.

These categories were created with the intention of assisting users in finding information on related subjects. An attempt was made to deal completely with the broad subject of biosolar resources. However, restrictions on the time available for the formulation of the manuscripts and limitations on the realistic size of the volume have led us to omit some areas. The relative emphasis placed on different subjects could be debatable. For example, because of the Handbook's theme of solar energy and related resources, special emphasis has been placed on the physical, physiological, and metabolic aspects of light as an environmental resource.

Throughout these volumes, emphasis has been placed on providing information in a format which would provide understandable and useful data to persons from a wide variety of disciplines. For this reason, the contributors were asked to use tables and figures whenever possible and to provide concise statements of the current state of knowledge on their subject.

Even though an attempt was made to present information in as uniform a manner as possible, there is some variability among the presentations by each author. This is almost inevitable considering the number of authors involved. A case in point is the designation of photosynthetic bacteria and blue-green algae. Some contemporary systematists no longer use the term photosynthetic bacteria and refer to blue-green algae as cyanobacteria. However, since the former terms are still common in the literature their use has not been restricted. Another example of disparity in approach is the variability of units used in the measurement of light intensity. While current research favors the use of quantum units, a wide diversity of other units remains in use, including lux, foot-candles, and langley per minute. While the editors tried to make the units uniform, this was impossible in some cases.

In an effort such as this, editors encounter delinquent authors, and some subjects were covered by the editors and their laboratory staff. Though we planned contributions on subjects such as light and temperature responses of terrestrial photosynthetic organisms, nutrients in terrestrial environments, and other topics, the promised manuscripts were not prepared. Thus, in fairness to contributing authors and due to a time constraint, the editors hesitantly dropped these topics.

With all the inherent difficulties which go along with the formulation of a handbook aimed at broad goals, we believe the information will be a valuable aid to students, researchers, and administrators interested in the expanding field of biosolar resource utilization.

Sincerest thanks go to the skilled contributors who donated their valuable time and expertise in an effort to make this an authoritative and timely contribution to the field of biosolar resources. In addition, the volume editors would like to thank Dr. Oskar R. Zaborsky, the Editor-in-Chief, for his constant encouragement and enthusiasm during the innumerable hours spent putting these volumes together.

We also would like to express appreciation and thanks for the editorial assistance of Ms. Cecelia Langley, Ms. Gay Ingram, Ms. Tia Maria, Ms. Susana Barciela, Ms. Hedy Mattson, and Ms. Margaret Ahearn at the University of Miami.

Akira Mitsui
Clanton C. Black

EDITOR-IN-CHIEF

Oskar R. Zaborsky, Ph.D., is a Program Manager at the National Science Foundation in Washington, D.C.

Dr. Zaborsky attended the Philadelphia College of Pharmacy and Science (B.Sc. in Chemistry, 1964), the University of Chicago (Ph.D. in Chemistry, 1968), and Harvard University (postdoctoral fellow, 1968-69). Prior to his present position with the Foundation, Dr. Zaborsky was with the Corporate Research Laboratories at the Exxon Research and Engineering Company in Linden, New Jersey. At Exxon, his research dealt with the chemical modification of proteins and the immobilization of enzymes. At the Foundation, Dr. Zaborsky has been involved in programs dealing with enzyme technology, renewable resources, and substitute materials.

Dr. Zaborsky's present scientific and administrative interests include enzyme technology, biotechnology, catalysis, renewable resources, and biosolar resources. He has published numerous articles on these topics and is the author of the book, *Immobilized Enzymes,* which was the first comprehensive treatise on this subject. Dr. Zaborsky's research on immobilized enzymes at Exxon also led to patents, and he is a founding editor of the journal, *Enzyme and Microbial Technology.*

Dr. Zaborsky is a member of several professional societies, including the American Chemical Society and the American Association for the Advancement of Science.

THE EDITORS

Akira Mitsui, Ph.D., is Professor of Marine Biochemistry and Bioenergetics, Division of Biology and Living Resources, School of Marine and Atmospheric Science at the University of Miami.

Dr. Mitsui received his B.S. in Biology from the Faculty of Science at the University of Tokyo in 1953, his M.S. in 1955, and his Ph.D. in 1958 in plant physiology from the University of Tokyo. From 1958 to 1964 he was a teaching and research faculty member at the Institute of Applied Microbiology, University of Tokyo. From 1960 to 1963, on leave from the University of Tokyo, he was an exchange scientist between the U.S. and the Japanese governments at the Department of Cell Physiology, University of California, at Berkeley, California. From 1964 to 1970 he was an Associate Professor at the Department of Biochemistry, Yokahama Medical School, Yokahama, Japan. He was a visiting scientist at the Department of Plant Sciences, Indiana University at Bloomington, Indiana from 1970 to 1971, and at the Kettering Research Laboratory at Yellow Springs, Ohio, from 1971 to 1972. Since 1972 he has been a professor at the University of Miami.

Dr. Mitsui has published many papers and articles dealing with hydrogen photoproduction, nitrogen fixation, photophosphorylation, photosynthetic electron carriers, and utilization of marine photosynthetic organisms.

Throughout the years he has actively participated in many international biological solar energy conversion and hydrogen energy conferences as chairman, co-chairman, organizer, advisory member, and invited speaker. He is co-editor of several books related to the subject of solar energy bioconversion.

Clanton C. Black, Jr., Ph.D., is a Professor of Biochemistry at the University of Georgia. He received a Ph.D. from the University of Florida in 1960. He then spent 2 years at Cornell University, Ithaca, as a National Institutes of Health Postdoctoral Fellow in the Biochemistry Department and 1 year as a Charles F. Kettering Foundation Fellow in Yellow Springs, Ohio. From 1963 to 1967 he was a Staff Scientist at the C. F. Kettering Research Laboratory and an Assistant Professor at Antioch College in Yellow Springs, Ohio.

Dr. Black is a Fellow of the American Association for the Advancement of Sciences. He is a member of the Research Advisory Committee of the Agency for International Development. He has served as President and in other positions in the American Society of Plant Physiologists. He is an Editor of *Plant Physiology*.

Dr. Black's research work centers on the biochemistry of photosynthetic organisms, with emphasis upon the variations and efficiencies at which higher plants assimilate their essential elements. He has lectured in many countries on these topics and published over 100 research papers on the biochemistry of photosynthesis.

ADVISORY BOARD

CONTRIBUTORS

Part 1

Takashi Akazawa
Professor
School of Agriculture
Nagoya University
Chikusa, Nagoya, Japan

Louise E. Anderson
Professor
Department of Biological Sciences
University of Illinois at Chicago Circle
Chicago, Illinois

Sumio Asami
Postdoctoral Fellow
Research Institute for Biochemical Regulation
School of Agriculture
Nagoya University
Chikusa, Nagoya, Japan

Reinhard Bachofen
Professor
Institut für Pflanzenbiologie
Universität Zurich
Zürich, Switzerland

James A. Bassham
Senior Scientist
Lawrence Berkeley Laboratory
University of California
Berkeley, California

D. S. Bendall
Lecturer
Department of Biochemistry
University of Cambridge
Cambridge, England

Peter Böger
Professor Chair
Lehrstuhl fur Physiologie und Biochemie der Pflanzen
Universität Konstanz
Konstanz, West Germany

Herbert Böhme
Fakultät für Biologie
Fachbereich Biologie
Universität Konstanz
Konstanz, West Germany

Jeanette S. Brown
Research Biologist
Carnegie Institution of Washington
Stanford, California

Bob B. Buchanan
Professor
Section of Cell Physiology
Department of Plant and Soil Biology
University of California
Berkeley, California

David J. Chapman
Professor
Department of Biology
University of California
Los Angeles, California

Mitsuo Chihara
Professor
Institute of Biological Sciences
University of Tsukuba
Sakura-mura, Ibaraki-ken
Japan

Brian H. Davies
Senior Lecturer
Department of Biochemistry and Agricultural Biochemistry
The University College of Wales
Penglais, Aberystwyth
United Kingdom

Deborah Delmer
Associate Professor
MSU-DOE Plant Research Laboratory
Michigan State University
East Lansing, Michigan

Richard A. Dilley
Professor
Department of Biological Sciences
Purdue University
West Lafayette, Indiana

Eirik O. Duerr
Post Doctoral Associate
School of Marine and Atmospheric Science
University of Miami
Miami, Florida

L. N. M. Duysens
Professor
Department of Biophysics
Huygens Laboratory of the State University
Leiden, The Netherlands

Gerald E. Edwards
Professor
Department of Botany
Washington State University
Pullman, Washington

Waldemar Eichenberger
Associate Professor
Department of Biochemistry
University of Bern
Bern, Switzerland

William Fenical
Associate Research Chemist and Lecturer
Institute of Marine Resources
Scripps Institution of Oceanography
La Jolla, California

W. R. Finnerty
Professor and Chairman
Department of Microbiology
University of Georgia
Athens, Georgia

Howard Gest
Distinguished Professor
Photosynthetic Bacteria Group
Department of Biology
Indiana University
Bloomington, Indiana

Douglas Graham
Senior Principal Research Scientist
Leader, Plant Physiology Group
CSIRO Division of Food Research
North Ryde, N.S.W., Australia

Jürg R. Gysi
Postdoctoral Fellow
Department of Biology
University of California
Los Angeles, California

M. D. Hatch
Chief Research Scientist
Division of Plant Industry
Commonwealth Scientific and Industrial Research Organization
Canberra City, Australia

Takayoshi Higuchi
Professor and Director
Wood Research Institute
Kyoto University
Uji, Kyoto, Japan

Takekazu Horio
Professor
Division of Enzymology
Institute for Protein Research
Osaka University
Osaka, Japan

Johannes F. Imhoff
Department of Microbiology
University of Bonn
Bonn, West Germany

S. Izawa
Professor
Department of Biological Sciences
Wayne State University
Detroit, Michigan

K. W. Joy
Professor
Department of Biology
Carleton University
Ottawa, Canada

Tomisaburo Kakuno
Instructor
Division of Enzymology
Institute for Protein Research
Osaka University
Osaka, Japan

Sakae Katoh
Professor
Department of Pure and Applied Sciences
College of General Education
University of Tokyo
Tokyo, Japan

Bacon Ke
Senior Investigator
Charles F. Kettering Research Laboratory
Yellow Springs, Ohio

Donald L. Keister
Senior Investigator
Charles F. Kettering Research Laboratory
Yellow Springs, Ohio

S. B. Ku
Plant Physiologist
Department of Biochemistry
University of Georgia and
R. B. Russell Agricultural Research Center
USDA/SEA
Athens, Georgia

Shuzo Kumazawa
Post Doctoral Associate
School of Marine and Atmospheric Science
University of Miami
Miami, Florida

H. K. Lichtenthaler
Professor
Botanical Institute
University of Karlsruhe
Karlsruhe, West Germany

George H. Lorimer
Research Scientist
Central Research and Development Department
E. I. DuPont de Nemours and Company Experimental Station
Wilmington, Delaware

Michael Madigan
Assistant Professor
Department of Microbiology
Southern Illinois University
Carbondale, Illinois

Kazutosi Nisizawa
Professor
Department of Fisheries
College of Agriculture and Veterinary Medicine
Nihon University
Tokyo, Japan

Masayuki Ohmori
Assistant Professor
Ocean Research Institute
University of Tokyo
Tokyo, Japan

Glenn W. Patterson
Professor and Chairman
Department of Botany
University of Maryland
College Park, Maryland

Norbert Pfennig
Professor
Fakultät für Biologie
Universität Konstanz
Konstanz, West Germany

P. Pohl
Professor
Institut für Pharmazeutische Biologie
Universität Kiel
Kiel, West Germany

Jack Preiss
Department of Biochemistry and Biophysics
University of California
Davis, California

Karin Schmidt
Institut für Mikrobiologie der Universität
Göttingen, West Germany

Karel R. Schubert
Associate Professor
Department of Biochemistry
Michigan State University
East Lansing, Michigan

David S. Seigler
Professor
Department of Botany
University of Illinois
Urbana, Illinois

Horst Senger
Professor
Fachbereich Biologie der Philipps Universitatät
Marburg, West Germany

Masateru Shin
Associate Professor
Department of Biology, Faculty of Science
Kobe University
Kobe, Japan

Paul C. Silva
Research Botanist
Department of Botany
University of California
Berkeley, California

Mario Snozzi
Lecturer
Institut für Pflanzenbiologie
Universität Zürich
Zürich, Switzerland

Harry E. Sommer
Assistant Professor
School of Forest Resources
University of Georgia
Athens, Georgia

Tatsuo Sugiyama
Associate Professor
Department of Agricultural Chemistry
School of Agriculture
Nagoya University
Nagoya, Japan

Ian W. Sutherland
Reader
Department of Microbiology
University of Edinburgh
Edinburgh, Scotland

Tetsuko Takabe
Assistant Professor
Research Institute for Biochemical Regulation
School of Agriculture
Nagoya University
Chikusa, Nagoya, Japan

Manfred Tevini
Professor
Department of Botany
University of Karlsruhe
Karlsruhe, West Germany

John F. Thompson
Plant Physiologist
U.S. Plant, Soil and Nutrition Laboratory
U.S. Department of Agriculture
Ithaca, New York

Achim Trebst
Professor
Department of Biology
Ruhr-University
Bochum, West Germany

Hans G. Truper
Professor
Department of Microbiology
Rheinische Friedrich-Wilhelms-Universitat
Bonn, West Germany

Rienk van Grondelle
Postdoctoral Fellow
Department of Biophysics
Huygens Laboratory of the State University
Leiden, The Netherlands

George A. White
Plant Introduction Officer
Germplasm Resources Laboratory
USDA Agricultural Research Center
Beltsville, Maryland

David A. Young
Assistant Professor
Department of Botany
University of Illinois
Urbana, Illinois

Jinpei Yamashita
Associate Professor
Division of Enzymology
Institute for Protein Research
Osaka University
Osaka, Japan

CONTRIBUTORS
Part 2

Yusho Aruga
Associate Professor
Laboratory of Phycology
Tokyo University of Fisheries
Tokyo, Japan

Reinhard Bachofen
Professor
Institut für Pflanzenbiologie
Universität Zürich
Zürich, Switzerland

Melvin S. Brown
Research Assistant
Division of Biology and Living Resources
Rosenstiel School of Marine and Atmospheric Science
University of Miami
Miami, Florida

R. Harold Brown
Professor
Department of Agronomy
University of Georgia
Athens, Georgia

J. S. Bunt
Director
Australian Institute of Marine Science
Townsville, Australia

James H. Carpenter
Professor and Chairman
Marine and Atmospheric Chemistry
Rosenstiel School of Marine and Atmospheric Science
University of Miami
Miami, Florida

David Coon
Associate Research Specialist
Marine Science Institute
University of California
Santa Barbara, California

T. P. Croughan
Professor
Rice Experiment Station
Louisiana State University
Crowley, Louisiana

W. Marshall Darley
Associate Professor
Department of Botany
University of Georgia
Athens, Georgia

Clinton J. Dawes
Professor
Department of Biology
University of South Florida
Tampa, Florida

Eirik O. Duerr
Post Doctoral Associate
School of Marine and Atmospheric Science
University of Miami
Miami, Florida

Hubert Durand-Chastel
General Manager
Texaco Company
Mexico City, Mexico

Rana A. Fine
Research Associate and Professor
Rosenstiel School of Marine and Atmospheric Science
University of Miami
Miami, Florida

Leon A. Garrard
Associate Professor
Department of Agronomy
University of Florida
Gainesville, Florida

B. Clifford Gerwick
Senior Research Biologist
Dow Chemical Company
Walnut Creek, California

Howard R. Gordon
Professor
Department of Physics
University of Miami
Coral Gables, Florida

Yoshio Hasegawa
Research Adviser
Marine Ecology Research Institute
Tokyo, Japan

Jean François Henry
Energy Planning and Design Corporation
Herndon, Virginia

Ulrich Horstmann
Research Scientist
Institut für Meereskunde
Universitat Kiel
Kiel, West Germany

Roland L. Hulstrom
Branch Chief
Renewable Resource Assessment
Solar Energy Research Institute
Golden, Colorado

Shun-ei Ichimura
Professor and Director
Institute of Biological Sciences
University of Tsukuba
Ibaraki, Japan

Johannes F. Imhoff
Department of Microbiology
University of Bonn
Bonn, West Germany

Akio Kamiya
Research Associate
Laboratory of Chemistry
Faculty of Pharmaceutical Sciences
Teikyo University
Kanagawa, Japan

Alfred R. Loeblich
Director
Marine Science Program and Associate Professor
Department of Biology
University of Houston
Houston, Texas

Jack R. Mauney
Plant Physiologist
Agricultural Research Service
U.S. Department of Agriculture
Phoenix, Arizona

Frank J. Millero
Professor
Rosenstiel School of Marine and Atmospheric Science
University of Miami
Miami, Florida

Akio Miura
Associate Professor
Laboratory of Algal Cultivation
Tokyo University of Fisheries
Tokyo, Japan

Shigetoh Miyachi
Professor
Institute of Applied Microbiology
University of Tokyo
Tokyo, Japan

Sigehiro Morita
Professor
Department of Environmental Science and Conservation
Tokyo University of Agriculture and Technology
Tokyo, Japan

John W. Morse
Associate Professor
Rosenstiel School of Marine and Atmospheric Science
University of Miami
Miami, Florida

Wheeler J. North
Professor
W. M. Keck Laboratory
California Institute of Technology
Corona del Mar, California

Edward J. Phlips
Post Doctoral Associate
Department of Biology and Living Resources
Rosenstiel School of Marine and Atmospheric Science
University of Miami
Miami, Florida

D. W. Rains
Professor
Department of Agronomy and Range Science and
Director
Plant Growth Laboratory
University ofCalifornia
Davis, California

Ferdinand Schanz
First Assistant
Hydrobiologcal-Limnological Station
University of Zurich
Kilchberg, Switzerland

Kurt Schneider
Institut für Pflanzenbiologie
Universität Zürich
Zürich, Switzerland

Horst Senger
Professor
Fachbereich Biologie der Universität
Lahnberge, West Germany

Ivan Show
IDS Associates
Encinitas, California

Bruce N. Smith
Professor
Department of Botany and Range Science
Brigham Young University
Provo, Utah

Samuel C. Snedaker
Associate Professor
Division of Biology and Living Resources
Rosenstiel School of Marine and Atmospheric Science
University of Miami
Miami, Florida

Mario Snozzi
Institut für Pflanzenbiologie
Universität Zürich
Zürich, Switzerland

Stan R. Szarek
Associate Professor
Department of Botany and Microbiology
Arizona State University
Tempe, Arizona

Masayuki Takahashi
Associate Professor
Institute of Biological Sciences
University of Tsukuba
Ibaraki, Japan

Howard J. Teas
Professor
Department of Biology
University of Miami
Coral Gables, Florida

Anitra Thorhaug
Professor
Department of Biology
Florida International University
Miami, Florida

Thai K. Van
Plant Physiologist
Science and Education Administration—Agricultural Research
United States Department of Agriculture
Fort Lauderdale, Florida

Kurt Wälti
Scientific Assistant
Hydrobiological Station
University of Zurich
Kilchberg, Switzerland

William N. Wheeler
Research Associate
Department of Biology
Simon Fraser University
Burnaby, British Columbia
Canada

Bernt Zeitzschel
Professor and Executive Director
Institut für Meereskunde
Universität Kiel
Kiel, West Germany

Rod G. Zika
Assistant Professor
Division of Marine and Atmospheric Chemistry
Rosenstiel School of Marine and Atmospheric Science
University of Miami
Miami, Florida

TABLE OF CONTENTS

Part 1

SECTION 1: LIGHT ABSORPTION PROCESSES AND ELECTRON TRANSPORT SYSTEMS

SECTION 2: ASSIMILATION AND DISSIMILATION OF MAJOR ELEMENTS

SECTION 3: MAJOR BIOSYNTHETIC PATHWAYS

SECTION 4: GENERAL CLASSIFICATION OF PHOTOSYNTHETIC ORGANISMS

Part 2

SECTION 5: GENERAL CHARACTERISTICS OF PHOTOSYNTHETIC ORGANISMS

SECTION 6: RESPONSE OF PHOTOSYNTHETIC ORGANISMS TO MAJOR ENVIRONMENTAL FACTORS

SECTION 7: BIOLOGICAL RESOURCES: PRIMARY PRODUCTIVITY

SECTION 8: PHYSICAL RESOURCES AND INPUTS

Section 1
Light Absorption Processes and Electron Transport Systems

BIOSOLAR RESOURCES: FUNDAMENTAL BIOLOGICAL PROCESSES

Clanton C. Black, Jr.

INTRODUCTION

Plant, animal, and most bacterial life as we know it today exists because of photosynthesis. Biologically, photosynthetic organisms are unique in their ability to repeatedly convert light energy into chemical bond energy in sufficient quantities to live, grow, and reproduce. Photosynthesis has likely been active for approximately 3×10^9 years on earth, and the storage of photosynthetic products began perhaps about 3×10^8 years ago. The stored products of photosynthesis comprise our present-day fossil fuels.

Photosynthesis results in the production of approximately 2×10^{11} tons/year of dry organic matter, with terrestrial plants producing about 1.2×10^{11} tons/year and oceans producing 0.8×10^{11} tons/year.[1-3] Plant photosynthesis can be presented as:

$$H_2O + CO_2 \xrightarrow[\text{sunlight}]{\text{plants}} \text{organic matter} + O_2 \quad (1)$$

The reduction of carbon via photosynthesis is the central process in the production of organic matter, since total plant matter usually is near 45% carbon. However, photosynthesis also results in the reduction of nitrogen and sulfur along with carbon. From these light-driven reduction processes, plus the assimilation of the other essential elements needed for photosynthetic organisms to live, earth obtains an almost endless variety of products. Even our very existence is absolutely dependent upon photosynthesis for food and the O_2 needed for us to obtain energy from our food. Thus, the necessity of photosynthesis for life on earth and its continuous renewable nature are unquestionable. Photosynthesis is our dependable biosolar resource.

A TIME SCALE FOR PHOTOSYNTHESIS

The scope of photosynthesis can be imagined by thinking about the diversity of photosynthetic organisms known to man or about the removal of 10^{11} tons of CO_2 each year from our atmosphere. However, perhaps one of the most informative ways to visualize the grandeur of photosynthesis was presented by Kamen,[2] who divided the process of photosynthesis into a logarithmic time scale. Figure 1 is a modified representation of his ideas. The time scale of photosynthesis is in seconds, and one must think in terms of time in orders of magnitude in seconds to grasp Figure 1.[2]

The initial absorption of light occurs in 10^{-15} sec, followed by photon stabilization as excitation energy in one of the photosynthetic pigments such as chlorophyll. Photon stabilization must be followed by photochemical reactions resulting in the formation of oxidizing and reducing components, or the photon will be reemitted as fluorescence, usually in about 10^{-9} sec. Formation of oxidizing and reducing components results in the conversion of the photon energy into components which can be used in enzyme-catalyzed reactions.

The enzyme-catalyzed reactions of biochemistry require energy and ultimately result in the storage of energy in a variety of chemical substances. During these biochemical conversions, we find the first products of photosynthesis most of us recognize, such as O_2, or we first observe the assimilation of CO_2.

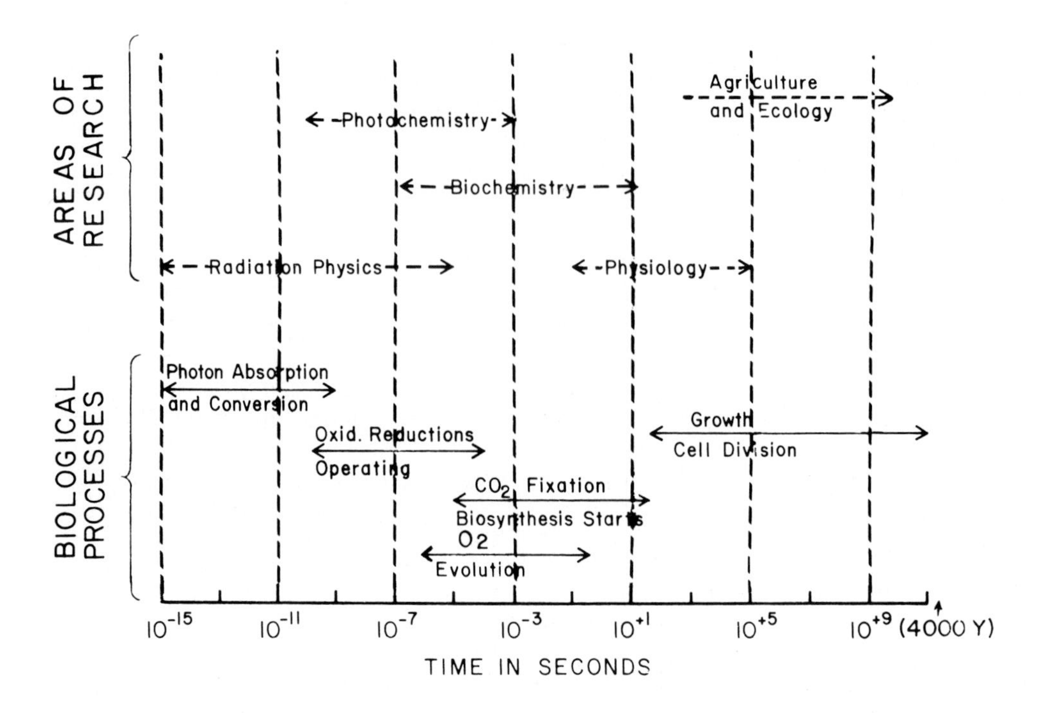

FIGURE 1. The time in seconds of the photosynthetic process beginning with light absorption in 10^{-15} sec and ending with the growth of a tree in about 10^{11} sec or 4000 years.[2]

As biochemical processes proceed, other biological processes blend in, such as cell division and growth. Environmental constraints and genetic information also begin to influence the products of photosynthesis. From about 10^1 sec to about 10^9 sec, photosynthesis is studied as agriculture and ecology. Here we recognize higher plants, photosynthetic bacteria, and algae as primary products of photosynthesis. Indeed, the oldest living plants, such as bristle cone pines, take about 10^{11} sec approximately (4000 years) to produce (Figure 1).

From the primary light absorption at 10^{-15} sec to the production of the oldest plants at 10^{11} sec, a time lapse of 26 orders of magnitude has occurred!

THE MECHANISM OF PHOTOSYNTHESIS

Nearly 50 years ago, van Niel compared the metabolism of anaerobically growing photosynthetic bacteria with green plants.[4] From his comparative viewpoint and research, he formulated bacterial and plant photosynthesis as follows:

Bacterial photosynthesis

$$2H_2A + CO_2 \xrightarrow[\text{light}]{\text{bacteriochlorophyll}} CH_2O + H_2O + 2A \qquad (2)$$

Green plant photosynthesis

$$2H_2O + CO_2 \xrightarrow[\text{light}]{\text{chlorophyll}} CH_2O + H_2O + O_2 \qquad (3)$$

In the equations, H_2A and H_2O play similar roles. H_2A stands for a reduced substance which photosynthetic bacteria require for photosynthesis, and this can be H_2S, H_2, acetone, isopropanol, or a variety of other oxidizable substrates.[5] Thus, van Niel thought of bacterial and plant photosynthesis as being quite similar, even though they have different requirements, and O_2 is produced only by green plants.

van Niel also formulated a scheme for photosynthesis which has been the dominant research model in photosynthesis for nearly five decades. In brief, the green plant photosynthesis scheme is

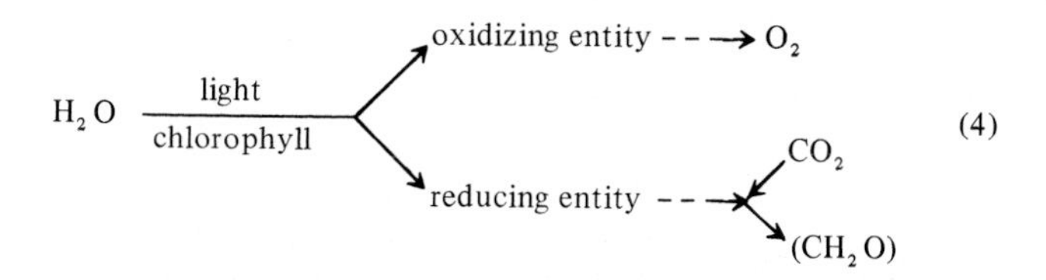

A similar scheme was presented for bacterial photosynthesis such that the primary reaction in photosynthesis is the formation of a reducing entity and an oxidizing entity. In the last four decades, much careful research work has allowed scientists to identify many of the reducing entities. However, much less is known even today about the identity of oxidizing entities, particularly in green plants. Today we would likely say that the primary reaction in photosynthesis is a charge separation. As a side note, at the same time, Shibata formulated similar comparative ideas about plant and bacterial photosynthesis, but his work in Japan was not widely known by other scientists until recently.[6]

REQUIREMENTS OF PHOTOSYNTHESIS AS A RENEWABLE RESOURCE

Photosynthesis not only requires light, CO_2, and water (or an oxidizable substrate in the case of photosynthetic bacteria), but also mineral elements. Today, it is commonly accepted that 16 elements are required as nutrients for plants to live, grow, and complete their life cycle.[7] All plants require the essential elements of C, H, O, N, P, S, K, Ca, Mg, B, Cl, Cu, Fe, Mn, Mo, and Zn. Other elements, e.g., Co, Se, Si, and Na, also are essential for a few plants, but not for all.

For a plant to grow and reproduce, it must assimilate these elements which it extracts from its environment. Some of the essential elements, such as K, Ca, Mg, B, Cl, Cu, Fe, Mn, Mo, and Zn, are assimilated fairly directly by plants through the formation of coordinate or electrostatic bonds. Much of the H and O in the organic components of plants comes from the aqueous milieu of cells, and some O_2 fixation occurs through a variety of plant oxygenases such as RuBP oxygenase. The assimilation of C, N, S, and P into organic matter are major processes involving photosynthesis.

PHOTOSYNTHETIC ASSIMILATORY ACTIVITIES

Generally, the assimilation of essential elements into organic matter involves photosynthesis. Some assimilatory processes only occur during photosynthesis, while others utilize products of photosynthesis. For example, the net fixation and reduction of CO_2 only occurs during photosynthesis, and photosynthesis is needed for plants to reduce quantities of nitrogen and sulfur. An element such as phosphate may be assimilated either nonphotosynthetically, as during oxidative phosphorylation, or photosynthetically, as during photophosphorylation. However, some essential elements, such as metals in metalloenzymes, can be assimilated quite independently of photosynthesis.

If we consider C, N, S, and P assimilation in greater detail, we find that these elements generally are not directly assimilated, but rather are transformed prior to, or during, assimilation into organic molecules. In the assimilation of these four major elements, photosynthetic organisms obtain them from their environments where the elements commonly exist as oxides, namely, CO_2, NO^-_3, SO^{2-}_4, and PO^{3-}_4.

Most CO_2 is assimilated from the atmosphere via a carboxylation reaction and the carboxyl group ultimately is reduced to an aldehyde in a two electron-requiring process. With plants, nitrate is the principal form of inorganic nitrogen available to roots, and it is swept up through the transpiration stream to leaves, where it is reduced in green cells to NH_3 in an eight-electron reduction process. Likewise, SO^{2-}_4 is the principal form of sulfur in soils for roots, and it is brought up to leaves via transpiration, reduced, and incorporated into cysteine, also in an eight-electron reduction process. Phosphate, however, is not reduced during assimilation. In soil, roots generally take up phosphoric acid which also is swept up in the transpiration stream to the leaves, where it is mostly incorporated into ATP via mitochondrical respiration, photophosphorylation, or glycolysis. Phosphate from ATP is rapidly moved into a variety of metabolites, such as sugar phosphates, phospholipids, and nucleotides. In intact plants, C, N, S, and P assimilation also occurs in nongreen tissue, but at reduced rates compared to photosynthetic tissues. Thus, C, N, and S are markedly transformed during photosynthetic assimilation into organic molecules, whereas P is directly incorporated.

The free energy changes for these assimilatory processes are

	ΔG kJ mol^{-1}	
$CO_2 + H_2O \rightarrow (CH_2O) + O_2$	+478	(5)
$H_2SO_4 \rightarrow H_2S + 2O_2$	+715	(6)
$HNO_3 + H_2O \rightarrow NH_3 + 2O_2$	+347	(7)
$ADP + PO_4^{3-} \rightarrow ATP$	+ 31	(8)

All of these reactions are positive and thus require energy input to proceed. The primary energy input process is through photosynthesis. Indeed, Equation 1 for plant photosynthesis could be modified in the following general fashion:

$$H_2O + SO_4^{2-} \xrightarrow[\text{sunlight}]{\text{plants}} \text{organic matter} + O_2 \quad (9)$$

$$H_2O + NO_3^{-} \xrightarrow[\text{sunlight}]{\text{plants}} \text{organic matter} + O_2 \quad (10)$$

$$H_2O + ADP + PO_4^{3-} \xrightarrow[\text{sunlight}]{\text{plants}} \text{organic matter} + O_2 \quad (11)$$

The assimilation of either C, N, S, or P could be thought of as photosynthesis! All of these photosynthetic processes must occur for plants to exist, but CO_2 assimilation is the dominant process which results in the synthesis of the major storage products of photosynthesis, such as cellulose.

The biochemistry of CO_2 assimilation has attracted much research since $^{14}CO_2$ became available after World War II, and we now recognize at least seven biochemical pathways for carbon flow during photosynthesis. These biochemical pathways are the C_3 cycle which is present in plants and bacteria, three variations within C_4 photosynthesis in higher plants, two variations within Crassulacean acid metabolism in plants, and the reductive tricarboxylic acid cycle in photosynthetic bacteria. Clearly, photosynthetic organisms have evolved considerable diversity in the biochemical pathways employed by specific organisms for the central process of CO_2 assimilation and reduction.

A similar biochemical diversity is not known for the assimilation of sulfur or phosphate. However, plants plus bacteria have developed the ability to assimilate gaseous nitrogen through the process of N_2 fixation.

For the assimilation of CO_2, NO^-_3, and SO^{2-}_4 to proceed, a strong reductant and ATP is needed. The strong reductant is produced from photosynthesis, as is the ATP. The strong reductant and ATP are produced from light-driven electron transport reactions and then used in the assimilation of C, N, and S.

DISSIMILATORY ACTIVITIES IN PHOTOSYNTHETIC ORGANISMS

In the dark, photosynthetic organisms live by using essentially the same biochemical reactions to obtain energy as nonphotosynthetic organisms. Thus, for photosynthetic organisms to have a net increase in organic matter, production in the light must exceed consumption in the dark. Usually this is a five- to tenfold excess in the light, but the excess varies with factors, such as age of the organism and the environment. In addition, some processes seem to work as "dissimilatory" activities.

The most intensively studied dissimilatory activity is the loss of CO_2 through photorespiration. It can result in the loss of CO_2 from green plants equal to nearly 35% of the rate of photosynthetic CO_2 assimilation. Hence, photorespiration is a major CO_2 loss process in green leaves and may be present in all organisms utilizing RuBP carboxylase because this enzyme from all organisms also has oxygenase activity. A major research effort around the world is to find ways to modify the contribution of photorespiration to net CO_2 assimilation. Studies are centered upon using genetics, chemicals and the environment to modify photorespiration or upon finding plants in nature which have modified the loss of photorespiratory CO_2.

In addition to the loss of CO_2, some plants are known to lose reduced carbon such as the light-dependent loss of isoprene by certain trees.[8] Plants also lose reduced sulfur in a light-dependent process as H_2S[9], and NH_3 is reportedly lost by *Zea mays*[10] in a light-dependent process. These losses of C, N, and S through light-dependent dissimilatory processes have not been studied as intensively as photorespiration. Hopefully we will soon know the contributions or detrimental features of such activities to the net organic matter production of photosynthetic organisms.

PHOTOSYNTHESIS AND PRODUCTIVITY

In agriculture, crop breeders have tried to select higher-yield crop varieties using the rate of leaf photosynthesis as a selection criterion. At first thought, one would assume that a higher rate of photosynthesis would equal a higher yield. However, in fact, few positive correlations have been detected between photosynthesis and yield,[11] particularly economic yield, such as grain or fruit production.

Commonly accepted knowledge in biology states that net organic matter production occurs in photosynthetic organisms in which the extra energy input is furnished by sunlight. Why have crop breeders and crop physiologists found little relationship between yield and photosynthesis? One portion of the answer is in the measurement of photosynthesis. Much of the literature data are single leaf measurements, some are at a single time, and some are at only one plant age. These values are used directly to

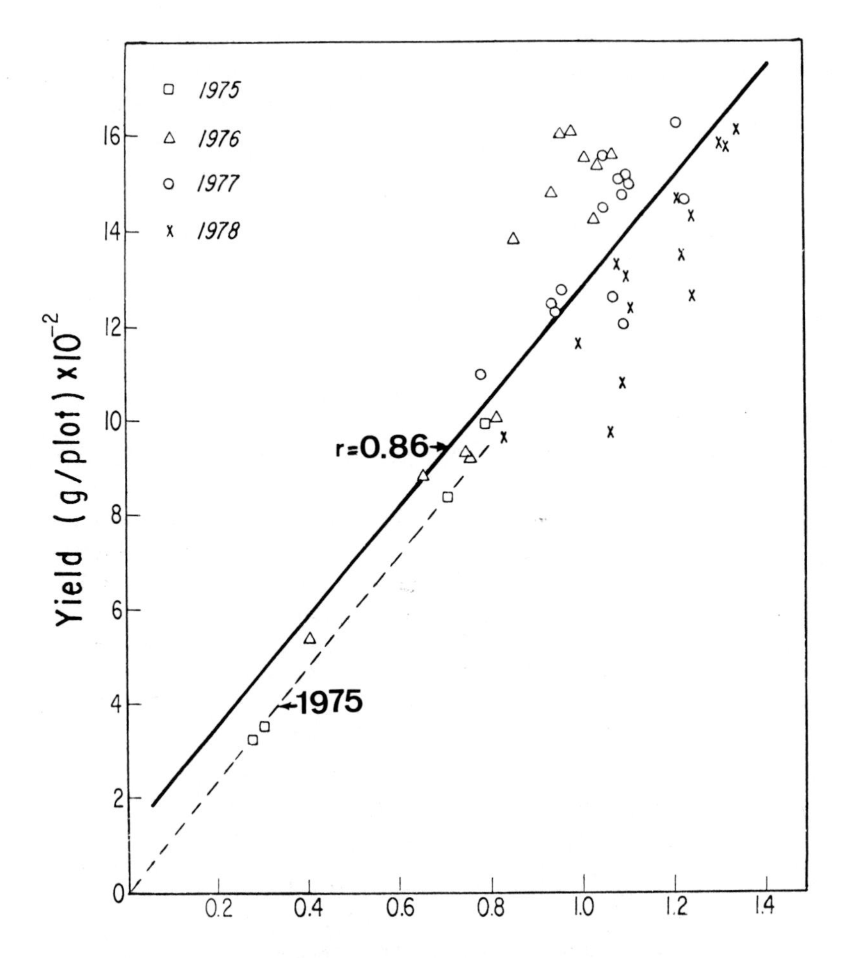

FIGURE 2. Seasonal net photosynthesis estimated from weekly measurements of soybean canopy photosynthesis. The yield is all aboveground organic matter. The solid line is a computer-fitted line showing a regression coefficient of r = 0.86. The dashed line connects only the 1975 data. Courtesy of Dr. A. L. Christy of the Monsanto Company, St. Louis, Mo.

compare with yield, or correction calculations are made to obtain seasonal photosynthesis. With the variations associated with changes in metabolism with plant age and the seasonal changes in environment, it seems unlikely that single measurements of photosynthesis are useful relative to yield, particularly since yield is a very complex integration by a plant of its genetic information and the environment it has experienced.

Recently, however, measurements have been obtained of seasonal net photosynthesis as shown in Figure 2 which relate quite strongly with yield. In these studies, canopy photosynthesis was measured regularly throughout the growing season, and data were collected on solar radiation. By obtaining seasonal net photosynthesis and plotting it against soybean yield, positive correlations between photosynthesis and yield were obtained. Both lines in Figure 2 show a good linear relationship between yield and photosynthesis. Though these results are satisfying, obtaining seasonal photosynthesis is a laborous task, and yield is an extremely complex subject which demands much future research.

Finally, if we convert the 2×10^{11} tons of organic matter produced each year by photosynthetic organisms into energy units, we find that currently photosynthesis stores about eight to ten times more energy than mankind consumes. Global photosynthetic productivity equals 3×10^{21} J stored annually, and in 1970 man expended 3×10^{20}J.[3] Clearly man's consumption is increasing, but global photosynthesis is not changing in concert!

REFERENCES

1. **Rabinowitch, E. I.,** *Photosynthesis,* Vol. 1, Interscience, New York, 1945.
2. **Kamen, M.,** *Primary Processes in Photosynthesis,* Academic Press, New York, 1963.
3. **Boardman, N. K.,** The energy budget in solar energy conversion in ecological and agricultural systems, in *Living Systems as Energy Converters,* Buvet, R., Allen M. J., and Massue, J. P., Eds., Elsevier, Amsterdam, 307, 1977.
4. **van Niel, C. B.,** The bacterial photosynthesis and their importance for the general problem of photosynthesis, *Adv. Enzymol.,* 1, 263, 1941.
5. **Gest, H., San Pietro, A., and Vernon, L. P., Eds.,** *Bacterial Photosynthesis,* Antioch Press, Yellow Springs, Ohio, 1963.
6. **Shibata, K.,** *Carbon and Nitrogen Assimilation,* (transl. by Gest, H., and Togaski, R. K.,) Japan Science Press, Tokyo, 1975.
7. **Epstein, E.,** *Mineral Nutrition of Plants: Principles and Perspectives,* John Wiley & Sons, New York, 1972.
8. **Sanadze, G. A. and Dzhaiani, G. I.,** On the incorporation of carbon into an isoprene molecule from $^{13}CO_2$ assimilated during photosynthesis, in *Proc. 2nd Int. Congr. on Photosynthesis Research,* Forti, G., Avron, M., and Melandri, A., Eds., N.V. Publishers, The Hague, 1972, 1958.
9. **Wilson, L. G., Bressan, R. A. and Filner, P.,** Light-dependent emission of hydrogen sulfide from plants, *Plant Physiol.,* 61, 184, 1978.
10. **Faraquot, G.,** Ammonia emission from corn leaves, *Science,* 204, 1937, 1979.
11. **Wallace, D. H., Peet, M. M., and Ozbun, J. L.,** Studies of CO_2 metabolism in *Phaseolus vulgaris* L. and applications in breeding, in CO_2 Metabolism and Plant Productivity, Burris, R. H. and Black, C. C., Eds University Park Press, Baltimore, 1976, 43.

PRIMARY LIGHT ABSORPTION PROCESSES AND THERMODYNAMICS IN PHOTOSYNTHESIS*

Rienk van Grondelle and Louis N. M. Duysens

INTRODUCTION

Even in strong sunlight, the number of photons absorbed per second per pigment molecule (e.g., Chl *a*) (Chl *a*) is only about ten. The subsequent electron transfer reactions are much more rapid (10^3 to 10^{11}/sec). For this reason in all photosynthetic systems, extensive transfer of electronic excitation energy occurs, concentrating the absorbed excitation energy on a small number of so-called RCs in which the electron transport reaction is initiated. The occurrence of excitation energy transfer results in an appreciable economy of specialized structures which are necessary for electron transfer and subsequent electron transport. This chapter covers the energy transferring and converting structures and introduces the mechanisms involved.

The RC of all photosynthetic organisms contains a specialized chlorophyll dimer, the so-called RC Chl, called P. The excited form, P*, is able to transfer an electron to a nearby electron acceptor I, resulting in the formation of a strong oxidant, P^+, and a strong reductant I^- by the process of charge separation. The excitation of P may occur via direct light absorption by the RC pigment or, with much greater probability, via the transfer of excitation energy captured by one of the light-harvesting pigments to the RC. The ratio of light-harvesting or antenna molecules to RCs is of the order of 200 in oxygen-evolving organisms and varies between 40 and 200 in photosynthetic purple bacteria. The latter contain Bchl instead of Chl *a* as the major antenna and RC pigment. The RC, together with its associated antenna pigments, is often called a photosynthetic unit (PSU), but, as shall be discussed later, energy transfer may occur between units. In this case, we may consider a PSU associated with a RC to consist of those antenna molecules that transfer excitation energy to this center with a probability greater than to other RCs.

In oxygen-evolving organisms, two photosystems, PS1 and PS2, operate in series, each with its own RC, RC1 and RC2 (with P_1, I_1, and P_2, I_2 as primary donors and acceptors) to drive electrons from H_2O to CO_2 (see Figure 1, Scheme 1). The identities of I_1 and I_2 are unknown, but I_2 might be pheophytin. In photosynthetic purple bacteria, only one photosystem occurs which drives electrons from a reduced sulfur compound or other "hydrogen donor" to CO_2 (Figure 1, Scheme 2). The identity of I in photosynthetic purple bacteria is now firmly established as being a Bph. Both in photosynthetic purple bacteria and oxygen-evolving organisms, the electron transport is coupled to the formation of ATP from ADP and P_i; the ATP is necessary to drive the reduction of CO_2 in the C_3 cycle.

This chapter only deals with the initial rapid processes involved in photosynthesis which occur in timescales between 10^{-15} and 10^{-6} sec and which include light absorption, energy transfer, and charge separation.

* Abbreviations: ATP, adenosine triphosphate; Bchl, bacteriochlorophyll; Bph, bacteriopheophytin; *C. vinosum*, *Chromatium vinosum*; Chl, chlorophyll; cyt, cytochrome; ENDOR, electron nuclear double resonance; ESR, electron spin resonance; NAD, nicotinamide adenine dinucleotide; NADP, nicotinamide adenine dinucleotide phosphate; P_i, inorganic phosphate; RC, reaction center; *R. sphaeroides*, *Rhodopseudomonas sphaeroides*; *R. rubrum*, *Rhodospirillum rubrum*; *R. viridis*, *Rhodopseudomonas viridis*.

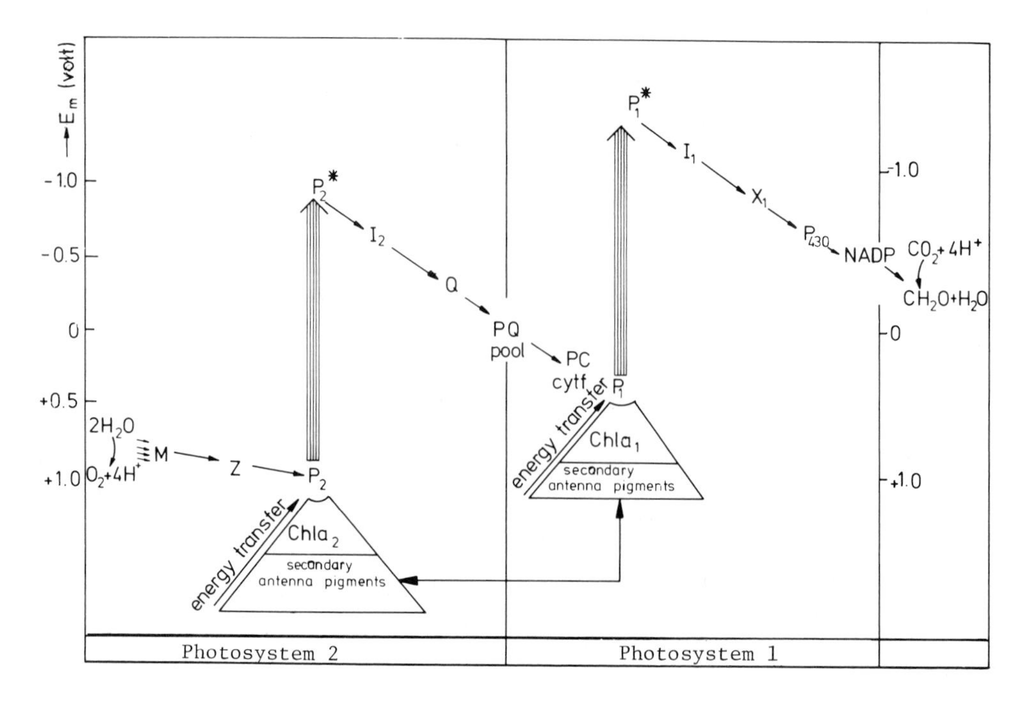

Scheme 1

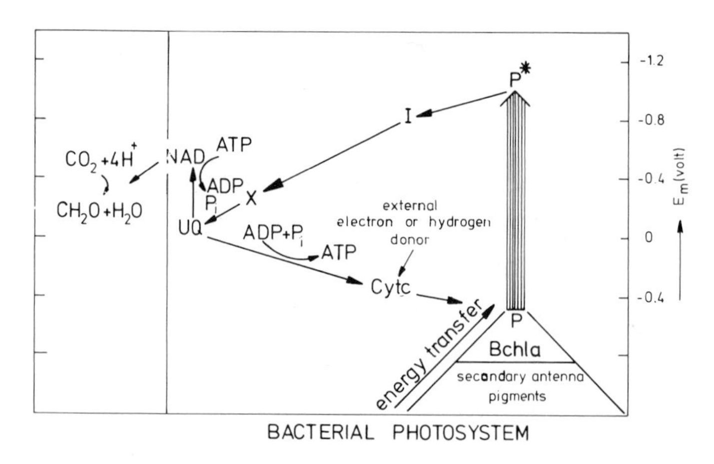

Scheme 2

FIGURE 1. Scheme 1. Schematic representation of electron transport in higher plants. The thick vertical arrows represent light reactions 1 and 2. See text for a discussion of the identity of P_1,I_1,X_1, P_{430}, P_2,I_2, Q, Z, and M. Other electron transport components are PC = plastocyanin; cyt *f* = cytochrome *f*; PQ = plastiquinone. The vertical scale represents the approximate oxidation/reduction potentials of the components involved. Scheme 2. Schematic representation of electron transport in photosynthetic purple bacteria. The thick vertical arrow represents the light reaction, the identity of P, I, X, UQ, and cyt *c* is discussed in the text. The cyclic electron transport drives the formation of ATP which is necessary for the reduction of NAD^+ by the reduced UQ by a process called "reversed electron flow". The vertical scale represents the approximate oxidation/reduction potential of the components involved.

PHOTOSYNTHETIC PIGMENTS IN VITRO AND IN VIVO

The Essential Pigments in Photosynthesis

The most important pigment molecules in photosynthetic organisms are the chlorophylls: Chl *a* in plants and algae and Bchl *a* and *b* in photosynthetic purple bacteria. Figure 2A shows the molecular structure of Chl *a*; Figure 2B shows that of Bchl *a*. Also the absorbance spectra of these pigments are shown in vitro (see Figure 2C) and in vivo (see Figure 2D). In addition many organisms contain additional antenna pigments, i.e., most higher plants contain Chl *b* (see Figures 2A and 2C) and carotenoids (see Figure 2D); red and blue algae contain the phycobilins; purple bacteria contain carotenoids (see Figure 2D), and green bacteria have Bchl *c*. The presence of these pigments, which absorb light at different wavelengths, guarantees a maximal utilization of incident sunlight. Quanta collected by these accessory pigments are transferred to Chl *a* or Bchl *a* (or *b*).

Light Absorption by Pigment Molecules and Possible Decay Pathways for the Excitation

Upon absorption of a quantum of light, an electron of the pigment molecule is exited from the ground state of the molecule, often denoted by S_o (S is singlet), to a higher unfilled orbital or excited state, often denoted by S_n, (where n $\geqslant$ 1). The molecular orbitals involved are associated with the system of conjugated bands and have π symmetry;[1] these transitions are called $\pi - \pi^*$ transitions. This process of excitation is very rapid (10^{15} s^{-1}), and because of the "discrete" energy differences between S_n and S_o, shows a marked dependence on the exciting wavelength which is characterized by the absorption spectrum of the molecule. Figure 2C shows the absorption spectra of Chl *a*,*b*, and Bchl *a* in vitro. In large organic molecules like Chl and Bchl, strong absorption bands occur in the visible and adjacent spectral regions because of the extensive conjugated bonding. Moreover an electronic transition has more than one band because vibrational quantum states also can be excited. For example, in Chl *a* the main band of the lowest excited singlet state in vivo is found around 678 nm (14.750 cm^{-1}), a minor band is found at 620 nm (16.130 cm^{-1}), and a higher excited state is found at 438 nm (22.830 cm^{-1}) (see Figure 2d which shows the absorption spectrum for the green algae *Chlorella* and for the purple bacterium *R. rubrum*). In comparison to the in vitro absorption spectra, the in vivo absorption peaks are often shifted to longer wavelengths and the bands are broadened or even multiple (especially in purple bacteria) because the pigments are associated with different proteins (or, more generally, experience different environments).

Because of rapid vibrational relaxation, a molecule like Chl *a* will, when promoted to a higher excited state (S_n(n $\geqslant$2), rapidly decay (10^{-13} to 10^{-12} s) to its lowest excited state (S_1). Chl *a* in dilute solution, 5, has a lifetime τ of 5×10^{-9} sec. In the following, S_1 of a pigment molecule A will be denoted by A*, then Chl *b** has a lifetime τ = 3.9 ns and for Bchl *a** τ = 4.5 nsec.

In most dilute solutions, the decay of Chl *a** to the groundstate is governed by three processes.[2] Part (30%) of the excitation energy may be lost by the emission of light (fluorescence), part (<10%) by vibrational relaxation, and part (>60%) by the decay to the triplet state (intersystem crossing). In the latter transition, the total spin $\bar{S}$ of the electrons of the conjugated band system changes from $\bar{S}$ = 0 (singlet state) to $\bar{S}$ = 1 (triplet state). In large porphyrins like Chl *a*, the process of intersystem crossing, although formally forbidden because of the conservation of spin angular momentum, is often allowed because of a small admixture of the triplet eigin functions with singlet states because of spin orbit coupling.

From the fluorescence yield ϕ and the observed lifetime τ, one can calculate that the natural lifetime τ_n of Chl *a** is τ_n = k_f^{-1} = 15 nsec, where τ_n is the lifetime if the excited singlet state would only decay by fluorescence (τ_n = τ/ϕ). The fluorescence yield of Chl *b** in dilute solution is 12%, while the yield of Bchl *a** is about 25%. Fluorescence of excited carotenoids has only been observed with picosecond fluorime-

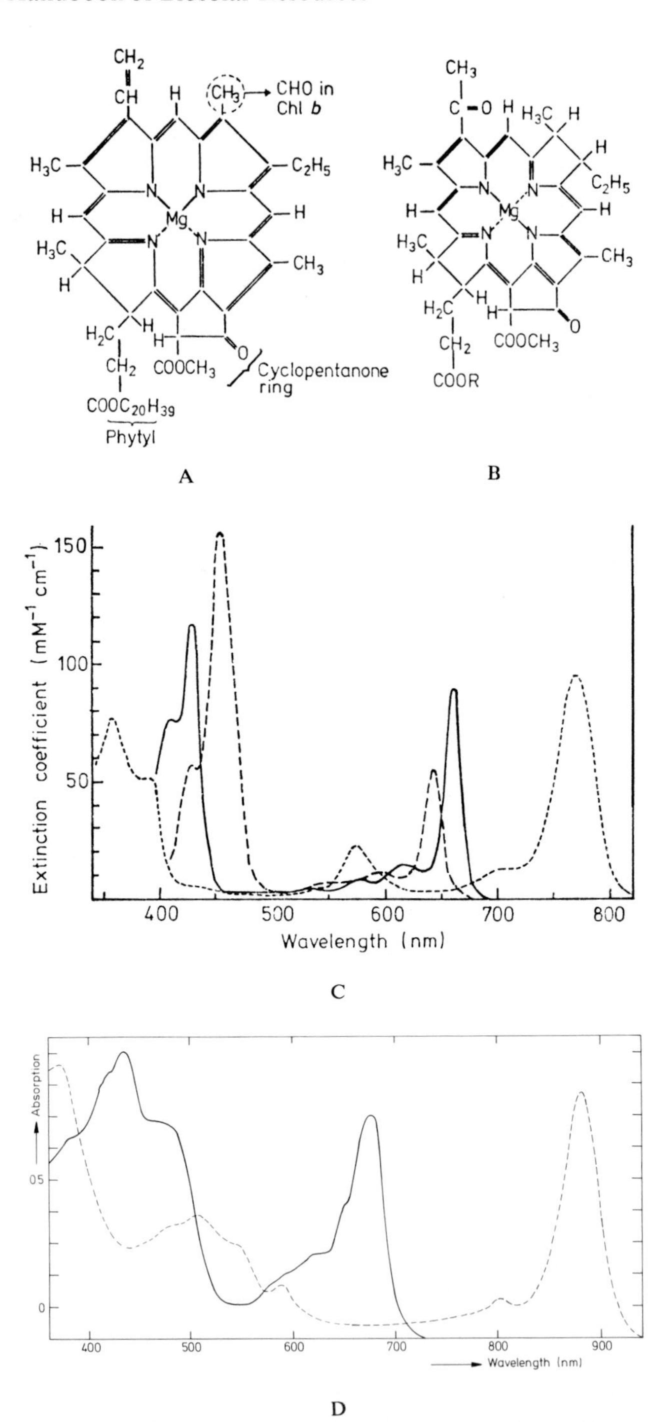

FIGURE 2. A. The structure of Chl *a* and *b*. B. structure of Bchl *a*. The group R shown in the structure of Bchl *a* has not been identified with certainty, but is probably a phytyl similar to that in CHl *a*. C. Absorption spectra for Chl *a* (————), Chl *b* (-------), and Bchl *a* (-------) dissolved in ether. D. In vivo absorption spectra (optical pathlength is 1 cm) of the green alga *Chlorella vulgaris* (————) and of the nonsulfur bacterium *R. rubrum* (-------). *C. vulgaris* contains as pigments Chl *a* (678 nm), Chl *b* (650 nm), and β-carotene (475 nm); *R. rubrum* contains Bchl *a* (898 nm, 800 nm, the latter band is probably a RC pigment) and spirilloxanthin (480, 510, and 545 nm).

try,[3] and yields are very low. Carotenoid triplet yields in solution are also very low[4] (between 0.2% and 2%). Thus, decay of excited carotenoids in solution occurs almost exclusively via internal conversion.

ENERGY TRANSFER IN PHOTOSYNTHESIS

Energy Transfer Between Photosynthetic Pigments In Vivo

The measurement of the fluorescence action spectra[5] in green plants has demonstrated that the light energy absorbed by carotenoids and Chl *b* is transferred to Chl *a* at efficiencies of 40 and 90%, respectively. As a model for quanta collection and distribution by both photosystems of higher plants the system is thought to consist of a PS1 antenna, Chl a_1, a PS2 antenna, Chl a_2, and the light-harvesting pigment protein complexes (LHPCs) containing Chl *a*, Chl *b*, and carotenoids in the approximate ratio of 3:3:1.[6] The Chl a_1 antenna transfers excitation energy to the RC of PS1, and the Chl a_2 antenna transfers to RC2. The quanta collected by the LHPCs are distributed among Chl a_1 and Chl a_2, depending on the wavelength, intensity, and duration of the exciting actinic light.[7-10] In cell-free chloroplast preparations, this distribution is affected by the presence or absence of ions,[11] specifically magnesium (Mg^{++}). In red and blue-green algae, light absorbed by the phycobilins is also transferred efficiently to Chl *a*. For example, the red algae *Porphyridium cruentum* contains the phycobilins, such as phycoerythrin (PE), phycocyanin (PC), and allophycocyanin (APC). Energy is transferred in the sequence PE → PC → APC → Chl *a*, with an overall efficiency of more than 90%.

In photosynthetic bacteria, light absorbed by carotenoids is transferred to Bchl *a* with variable efficiency[5] (see Table 1). Many photosynthetic purple bacteria contain several spectral forms of Bchl *a*. For example, B800, B850, and B870 (the number reflects the wavelength of maximum absorption) are found in *R. sphaeroides*, B800, B820, B850, and B890 are found in *C. vinosum*, and B880 is found in *R. Rubrum* (see Figure 2D). In all cases, quanta absorbed by pigments at the shorter wavelengths are transferred to the Bchl *a* absorbing at the longest wavelength.

As there are many Chl *a* or Bchl *a* molecules per RC, excitation energy is transferred between them until it finally reaches the RC. The Chl *a* → Chl *a* or Bchl *a* → Bchl *a* energy transfer also is very efficient. Under normal conditions, total losses do not exceed 10% before the excitation energy is trapped in the RC.

The Molecular Mechanism of Energy Transfer

The transfer of excitation energy may be governed by the mechanism of "inductive resonance",[12-15] a theory proposed by Förster.[16,17] The basic physical hypothesis in this theory is that a weak (≤ 1 cm^{-1}) interaction exists between neighboring pigment molecules, which is caused by electrostatic, at first approximation, electric dipole-dipole interaction between two states of the system, the initial state in which the donor D is excited and the acceptor A is in the ground state, and the final state in which A is excited and D is in the ground state. The initial state of the donor-acceptor pair is described by a wave function $\psi_D{}^*\psi_A$ (where $\psi_D{}^*$ is the wave function of the excited donor of D*, ψ_A of A), and that of the final state is represented by $\psi_D\psi_A{}^*$ (D is in the ground state; A* is in the excited state). The rate of transfer is governed by the magnitude of the interaction energy U,

$$U = \langle \psi_{D*}\psi_A | V | \psi_D \psi_{A*} \rangle \quad (1)$$

Table 1
FÖRSTER CRITICAL DISTANCE (R_o), ESTIMATED DISTANCE (R), AND EFFICIENCY FOR ENERGY TRANSFER BETWEEN A SET OF PIGMENT PAIRS[a]

D	A	R_o(Å)	R(Å)	ϕ_D	τ_D(ns)	In vivo estimated transfer efficiency (%)	In vivo donor fluorescence lifetime	In vivo estimated donor fluorescence yield[c] (%)
Chl *b*	Chl *a*	70	≥11[b]	0.12	3.9	>99	—	—
Chl *a*	Chl *a*	69	17	0.32	5.1	>99	0.3—0.65 nsec;2 nsec [k]	2,8
		65[22]	≥12[b]					
		92[23]						
β Carotene	Chl *a* *Chlorella*	50[f]	15	<10[-4e]	0.055[27]	40	—	—
Fucoxanthol	Chl *a* (*Navicula minima*)	—	—	—	—	>90	—	—
Bchl *a*	Bchl *a*	90[25]	>31[25]	0.25[26]	4.5[26]	>99	0.1—0.4 nsec;0.5—2 nsec[k]	2,6—15[d]
Spirilloxanthin	Bchl *a* (*R. rubrum*)	56[f]	16	≤10[-4e]	—	40	—	—
Neurosporene	Bchl *a* (*R. sphaeroides*)	56[f]	10	≤10[-4e]	—	80—90	—	—
Phycoerythrin	Phycocyanin (*Porphyridium cruentum*)	63[20],64[20]	<44[h],<39[h]	0.98[20],0.48[20]	3.2[20],2.9[20]	94[21] > 99[28]	280 psec[21] 70 psec[28]	0.4[28]
Phycoerythrin	Phycocyanin (*Nostoc.* sp.)	59[20],60[20]	<36[h],<39[h]	0.45[20],0.67[20]	2.8[20],2.6[20]	98[21]	—	—
Phycocyanin	Allophycocyanin (*P. cruentum*)	64[20]	<41[h]	0.59[20]	2.0[20]	>90,>99	90 psec[28]	0.3[28]
Phycocyanin	Allophycocyanin (*Nostoc.*sp.)	65[20]	<40[h]	0.52[20]	2.2[20]	>90	—	—
Allophycocyanin	Chl *a*	70[i]	<46[h]	0.68[20]	2.7[20],4[24]	>90,>99	120 psec[28]	0.2[28]

Note: Donor fluorescence yields in vivo and in vitro; fluorescence lifetimes in vitro and in vivo.

[a] The numbers mentioned in Table 1 which have no reference number assigned to them have been taken from Reference 12.

[b] Calculated in Reference 12, assuming a maximal transfer rate $k_D{}^*{}_A \leq 10^{13}$ sec^{-1}.

[c] If in the columns two numbers are given, these apply to photochemically active or inactive traps.
[d] The maximal antenna fluorescence yield shows much variation in photosynthetic purple bacteria.[19]
[e] Estimated fluorescence yield.
[f] Estimated R_o values assuming perfect overlap between carotenoid fluorescence emission and Chl *a* (or Bchl *a*) absorption spectra.[5]
[g] For the species *P. cruentum* and *Nostoc.* sp., two types of phycoerythrin have been isolated which have different properties.[20,21]
[h] Calculated, assuming a transfer efficiency $\geq 90\%$.
[i] Estimated by the authors, assuming perfect overlap between allophycocyanin fluorescence emission and Chl *a* absorption spectra.
[k] See for fluorescence lifetimes section.

where V is the electric dipole-dipole operator,

$$V = \frac{1}{k^2 R^3} \left\{ \vec{P}_D \cdot \vec{P}_A - \frac{3(\vec{p}_D \cdot \vec{R})(\vec{p}_A \cdot \vec{R})}{R^2} \right\} \quad (2)$$

where $\vec{P}_D$ and $\vec{P}_A$ are the electric dipole moment operators for D and A, $\vec{R}$ is the distance vector between D and A (R is $|\vec{R}|$), and k is the dielectric constant of the medium.

For weak interactions, the rate of energy transfer from D* to A, or k_{DA}, can be calculated from first-order time-dependent perturbation theory from which it follows that k_{DA} is related to U by:

$$k_{D*A} \sim U^2 \quad (3)$$

If one assumes that the pigment molecules are randomly oriented and the average is taken over all possible donor orientations, k_{D^*A} can be written as:

$$k_{D*A} = \frac{\phi_D}{\tau} \cdot \frac{1}{R^6} \left[\frac{3}{4\pi} \int_0^\infty \lambda\!\!\!^{-*4} F_D(\omega)\sigma_A(\omega)d\omega \right] \quad (4)$$

in which $\lambda\!\!\!^- = \lambda/2\pi n$; λ is the wavelength, n is the index of refraction of the medium, τ is the donor lifetime in the absence of energy transfer, $F_D(\omega)$ is the normalized fluorescence emission spectrum of D on an angular frequency scale ($\omega = 2\pi\nu$, where ν is the frequency), $\sigma_A(\omega)$ is the absorption cross section of the acceptor, and ϕ_D is the fluorescence yield of the donor in the absence of the acceptor. The expression between the brackets in Equation 4 is called the overlap or resonance integral because it depends on the degree of overlap or resonance between the emission spectrum of the energy donor and the absorption spectrum of the energy acceptor. If we put this expression equal to R_o^6, Equation 4 can be written as[12]

$$k_{D*A} = \frac{\phi_D}{\tau} \left(\frac{R_o}{R} \right)^6 \quad (5)$$

If $R = R_o$, there are equal probabilities for transfer and for internal deactivation of the primarily excited molecule. R_o can be precisely calculated from the experimental data: the excited state lifetime, the fluorescence yield, and the absorption and fluorescence emission spectra. Table 1 summarizes the values for R_o for energy transfer between a number of pigments.

Models for Pigment Systems Transferring Excitation Energy to Reaction Centers

Through the mechanism of energy transfer, the absorbed light energy is rapidly transferred to the antenna pigment absorbing at the longest wavelength. From there the energy is transferred between more or less identical pigment molecules until it finally reaches a RC where the energy may be trapped and the primary photochemical reaction is initiated. Under the conditions where the Förster formula is applicable ($U \leqslant 1$ cm^{-1}, $R \geqslant 10$ A°, these two conditions are complimentary), the motion of the excitation can be described as incoherent and diffusive and leads to a random walk formulation of energy transfer.[15] If we assume that at each moment in time the excitation is localized on a certain pigment molecule, the vibrational relaxation takes place before the next transfer occurs, then the transfer can be described as a hopping process in which the excitation jumps from one Chl *a* molecule to another, and the probability for transfer (P_{tf}) is just determined by the transfer rate (k_{D^*A}), the number of closest neighbors (c) of a donor molecule, and the rate of internal deactivation (k_l). For example, in a regular lattice the probability of transfer is given by

$$P_{tf} = \frac{c\, k_{D^*A}}{k_l + c\, k_{D^*A}} \qquad (6)$$

In a two-dimensional regularly spaced pigment matrix with N_o antenna molecules per RC, the number of jumps, $\bar{n}$, required to reach a RC is approximately:[29]

$$\bar{n} = \frac{(0.318\, N_o \ln N_o + 0.195\, N_o)}{(N_o - 1)} N_o \qquad (7)$$

which comes out as about 40 for N = 30, the size of the Bchl *a* antenna in, for example, *R. sphaeroides* R-26. The observed fluorescence lifetime after a weak picosecond excitation pulse is about 300 psec in *R. sphaeroides* R-26.[25] If the excitation energy is trapped by the RC on the first encounter, this would mean that the average transfer rate, k_{tf} ($k_{tf} = c\, k_{D^*A}$), in the antenna of *R. sphaeroides* R-26 is of the order of 0.12 $psec^{-1}$; for two other mutants of *R. sphaeroides* (see Table 1), these transfer rates might be a factor 3 larger, as suggested by the reported short fluorescence decay times (in the order of 100 psec). For the PS2 antenna, a minimal value $k_{tf} = 1.5\ psec^{-1}$ was calculated from a fluorescence yield of 3%, when all the RCs were in the active state P_2I_2Q (where Q is the secondary acceptor of PS2, a quinone), a rate of fluorescence $k_f = 6.7 * 10^7 sec^{-1}$ and $N_o = 200$.[30]

The above considerations apply in case the RCs are all in a trapping state and the excitation is trapped by the RC on the first encounter. Therefore, $\bar{n}$ and k_{tf} are lower limits. If the above assumptions are not made, the question arises what will happen to the excitation energy if an inactive RC is encountered or if an active RC transfers the excitation back to the surrounding antenna pigments.

To describe this process, we have to make an assumption about the interaction between different PSUs. If they are separated, an excitation in a certain unit will only be able to visit the RC in that unit. If the RC of the unit happens to be closed, the excitation will be lost. If, however, the units are connected, an excitation in a certain PSU may migrate to a neighboring PSU, the rate of inter unit migration would then depend on the degree of connection.

The Relationship Between the Antenna Fluorescence Yield and the State of the RCs

In the case where the photoreaction units are connected, the antenna fluorescence yield ϕ can be related to the fraction of weakly or nontrapping RCs, q^-, by the following equation:[19,30]

$$\phi = \frac{a}{1 - pq^-} \qquad (8)$$

$$a = \frac{N_o k_f}{k_t + N_o k_l} \qquad (9)$$

$$p = \frac{k_t - k_t'}{k_t + N_o k_l} \qquad (10)$$

where k_t, k_t' are trapping rates of a RC in state P* I Q, respectively, P* I Q^-, and k_l is the rate of deactivation of Chl *a** in the absence of energy transfer. For the derivation of Equations 8 to 10, it was assumed that

$$k_l << k_t, k_t' < k_{tf} \quad (11)$$

The excitation passes the RC several times before being trapped. For PS2 a titration of q^- vs. ϕ gives p = 0.74.[31] In the purple bacterium *R. rubrum* where the weakly trapping state P^+I X (X is a short notation for the secondary acceptor in photosynthetic purple bacteria) is formed, p = 0.5[32]

The validity of the inequality (Equation 11) depends on the values of the different rate constants involved. In purple bacteria, direct measurement of k_t yields a value[33] $k_t = 2.5 \times 10^{11}$ s^{-1}. For PS2 k_t is estimated[30] to be 4×10^{11} s^{-1}. These numbers are in the same range as the minimal values obtained for k_{tf} in the antenna of PS2 and most purple bacteria.

Time Dependence of the Fluorescence Decay

In the last few years, fluorimetry using picosecond laser pulses has frequently been applied to photosynthetic systems.[3,27] Often pulses of too high energies were used which excited the PSU more than once during the lifetime of the excitation and induced effects that will be discussed later. Fluorescence decay times in photosynthetic purple bacteria after low-intensity picosecond pulses range between 100 psec for several mutants of *R. sphaeroides*[25] containing the B800/B850 antenna and carotenoids (Strain 2.4.1 which is similar to the wild type and strain GA which contains only neurosporene as a carotenoid), 200 psec for *R. sphaeroides* strain 1760-1,[34] 300 psec for *R. sphaeroides* R-26[25] (contains only B870 and no carotenoid), and 400 psec for *R. rubrum.*[35] Paschenko et al.[34] and Govindjee et al.[35] also observed an increase in the fluorescence lifetime upon the oxidation of P870 for the *R. sphaeroides* Strain 1760-1 (to 500 to 550 psec) and for *R. rubrum* to about 1 nsec). Kinetic analysis of the fluorescence decays indicates a near exponential time course at low excitation energies.[25]

For higher plants and algae, the situation is less straightforward. Most authors seem to agree on a lifetime of several hundred picoseconds for PS2; in *Chlorella*, 650 psec,[36] 450 psec,[37] 400 psec,[38] and 490 psec;[39] in spinach chloroplasts, 410 psec[39] (with a 3.6% component of 1.5 nsec) and 200 to 320 psec;[38] in pea chloroplasts, 300 psec;[40] and in PS2 fractions 500 psec,[41] and 340 psec.[39] These numbers agree reasonably well with those obtained by the phase shift method.[42-45] Closing of PS2 RCs increases the lifetime to about 2 nsec.[37] For PS1 the few data reported point to a lifetime equal to or in the order of 100 psec (80 psec,[40] 100 psec,[41] 110 psec[39]); upon cooling to 100°K, this increases to 1.5 nsec in spinach chloroplasts[46] and 1.9 nsec in PS1 particles.[41]

Kinetic analysis of the PS2 fluorescence decay curves has indicated a nonexponential decay when the RCs are in the active state.[37,41] These fluorescence decays have been claimed to be in agreement with a model where the excitation is trapped by the RC on the first encounter.[47,48] In the model described previously, the fluorescence decay would be monophasic and exponential if only one kind of photosystem would be the cause of the fluorescence which is not the case in higher plants.[49] The lifetime of the fluorescence would be linear with yield. Such relations have been established, using a phase shift method.[50,51] A complete theory of energy migration within a PSU will have to account for the relations in Equation 8, especially those relating ϕ to q^- and for ϕ as a function of time. It will be important to consider the problem of the inhomogeneity of the fluorescing and energy transferring systems in the analysis of $\phi(q^-)$ and $\phi(t)$.

Singlet-Singlet and Singlet-Triplet Fusion in the Antenna Pigments of Photosynthetic Organisms

If the energy of the fluorescence exciting pulse is increased, it is observed that the yield of the fluorescence and its lifetime decreases. The effect has been observed for picosecond pulses,[25,37,52-55] nsec pulses,[56,57] μsec pulses,[58-61] and using multiple pulse excitation.[46,53,62] Two possible mechanisms to explain these effects are given. For a more complete discussion, the reader is referred to References 3, 27, and 63.

Singlet-Singlet Fusion

Singlet-singlet fusion is described by the following equation:

$$S_1 + S_1 \xrightarrow{\gamma} S_o + S_n \rightarrow S_o + S_1 \tag{12}$$

The observed decrease in fluorescence yield and the kinetics could be described by

$$\frac{d[S_1]}{dt} = -k[S_1] - \gamma[S_1]^2 + c\,I(t) \tag{13}$$

where $k = 1/\tau$ is the decay rate of the excited state at low intensities of excitation, γ is the bimolecular rate constant for S_1-S_1-annihilation, c is a constant, and I(t) is the photon density of the excitation pulse as a function of time. Equation 13 is based on the matrix model for the antenna pigment system.[53] If I(t) can be approximated by $I(t) = I_o\delta(t)$ (for picosecond pulse excitation, where $\delta(t)$ is dirac delta function), then Equation 13 can be solved and after integration yields:

$$\frac{\phi}{\phi_o} = \frac{k}{\gamma[S_1(o)]} \ln\left(1 + \frac{\gamma[S_1(o)]}{k}\right) \tag{14}$$

where ϕ_o is the fluorescence yield at low excitation energies, and $S_1(o)$ is the number of excited states generated by the picosecond pulse at t = 0. Equation 14 has been found to describe the decrease in fluorescence quantum yield at high pulse energies in photosynthetic purple bacteria,[25] chloroplast,[53,55] and algae.[47,64] The parameter γ is related with the Förster transfer rate.[55] Calculation of $k_D{}^*{}_A$ for chloroplasts[55] from this type of experiment gives $k_D{}^*{}_A \geqslant 3 \times 10^{11}\ s^{-1}$.

Singlet-Triplet Fusion

The process of singlet-triplet fusion[52,56] is described by

$$S_1 + T_1 \xrightarrow{\beta} S_o + T_n \rightarrow S_o + T_1 \tag{15}$$

and can be included in Equation 13 by adding an additional term, $-\beta[S_1]\,[T_1]$, on the right hand side. β is the bimolecular rate constant for singlet-triplet fusion, $[T_1]$ is the concentration of triplet states. If the generation of the triplet state by the excitation light also is included, two differential equations are obtained for which the solution can be found numerically.[47] In photosynthetic purple bacteria, the antenna Bchl triplet ($Bchl^{TR}$) is formed with a low yield[65,66] (2%) and decays if no carotenoid is present, as is the case for certain mutants, with a $t_{1/2} = 70$ μsec.[65] In carotenoid-containing strains, $Bchl^{TR}$ is rapidly ($t_{1/2} = 20$ nsec) converted into Car^{TR} which then has a decay time of $t_{1/2} = 2$ to 8 μsec.[65] Both $Bchl^{TR}$ and Car^{TR} are strong fluorescence quenchers.[57,61,67] $Bchl^{TR}$, quenches about five times stronger than Car^{TR}. $Bchl^{TR}$, if present in a concentration of approximately one per RC, quenches the fluorescence about tenfold, as compared to photochemical active traps which quench the fluorescence about four times.

In algae and chloroplasts, a Car^{TR} is formed via Chl^{TR} in PS2[58,60,68] when the PS2 RCs are in state $P_2I_2Q^-$. The yield has been calculated to be about 25%.[69] The decay of the fluorescence quenching state has a rate constant $k_c = k_c' + k_c'' [O_2]$, where $k_c' = 0.12 \times 10^6$ sec^{-1}, $k_c'' = 0.9 \times 10^9$ M^{-1} sec^{-1} and $[O_2]$ is the oxygen concentration in the medium in moles per liter.

The triplet state of Chl*a* or Bchl*a* reacts with oxygen, resulting in a harmful reaction by producing singlet oxygen.[61,65,69] Carotenoids may protect against damage by capturing the triplet excitation from Chl *a* (see Reference 70 for an alternative view), while Car^{TR} does not react with oxygen.

Using single picosecond pulses, it has been found that at low temperatures (100°K) the PS1 fluorescence yield decreases with increasing flash intensity, although the lifetime is not affected.[71] The dependence of ϕ on flash intensity was similar for PS1 and PS2 fluorescence.[55] From this it has been concluded that the main process of excitation annihilation takes place in the LPHCs coupled to PS1 and PS2.[55] In addition a rise time of 140 psec for the PS1 fluorescence is observed[72] upon excitation at 530 nm, suggesting that the larger part of the quanta arriving in PS1 have been collected by LHPCs. Upon excitation with a train of flashes, a fluorescence quencher is formed in PS1 which is a long-lived relative to the pulse time separation (6 to 7 nsec). The quenching may be because of triplet states.[46]

CHARGE SEPARATION

This part reviews the process of charge separation occurring in the RC, discusses the status in photosynthetic purple bacteria, and assumes that this can also be used as the basis for what occurs in RCs of PS1 and PS2. Data on the primary photochemical reactions in photosynthetic green bacteria are in References 73 to 76.

Photosynthetic Purple Bacteria

Many of the results mentioned in this section are extensively described and reviewed elsewhere.[77-80] The enormous progress which has been made during the last years in the field of primary electron transfer reactions in bacterial photosynthesis is, for a large part, based on the availability of purified RC preparations of several species of photosynthetic purple bacteria which still retain their full photochemical activity.[79,81,82] Purified RCs from *R. sphaeroides, R. viridis,* and *R. rubrum* have four molecules of Bchl and two molecules of Bph per functional unit. Two of these Bchl molecules form the Bchl dimer, P. The other two molecules absorb around 800 nm and are therefore called the P800 (in *R. sphaeroides* and *R. rubrum*) Bchls. RCs purified from strains that synthesize carotenoids normally contain one carotenoid molecule per RC.[83,84] In addition, purified RCs usually contain one or more quinones (which can either be ubiquinone or menaquinone) and one atom of nonheme iron.[85-92] After excitation of the RC Bchl dimer, an electron is transferred to an electron acceptor, Bph[93-97] (I), with a rate constant[33] $k_t = 2.5 \times 10^{11}$ sec^{-1}. The electron transfer process may be described as a vibronically coupled tunneling process[98,99] over a distance of 5 to 10 A. If the secondary acceptor X (a quinone-iron complex) is in the oxidized state before illumination, the electron is further transported[100] and the state $P^+I X^-$ ($t_{1/2}$ = 200 psec) is formed which has a recombination half time, $t_{1/2} \geqslant 20$ msec.[79] The free energy changes involved in this process are related to the midpoint potentials of the primary reactants which are +470 mV for P^+/P in most species[79] and about −180 mV[102] for X/X^-.

In the photosynthetic purple bacterium *C. vinosum*, the formation of P^+ is accompanied[79] by a bleaching of absorption bands at 890 and 605 nm and the generation of a band around 1250 nm. The complicated changes around 800 nm are most likely because of several processes.[103-105] The much smaller absorption band of P at 810 nm, the second transition of the Bchl dimer, bleaches (the 870- and the 810-nm transition moments make a 90° angle and can most likely be ascribed to the two exciton components of the Bchl dimer). At the same time, the absorption band of one of the two P800 Bchl molecules shifts to shorter wavelengths. The changes in the 380- to 440-nm region of the spectrum are because of shifting of the blue absorption bands of RC Bchls upon oxidation of P.

The reduction of I to I^- induces a bleaching of the Bph bands at 758 and 542 nm.[93-97] In addition, the blue absorption band shifts to longer wavelengths. The reduction of I is also accompanied by changes in the 800-nm region because of an electrochromic shift of the second of the two P800 Bchl molecules, probably induced by the nearby negative charge on the RC Bph.

The results of the reduction of $X \rightarrow X^-$ appear mainly in the UV part of the spectrum. The changes in *Chromatium* are characteristic for the reduction of a menaquinone to its semiquinone anion[91] (in *R. sphaeroides* and *R. rubrum* X most likely is a ubiquinone).[85-92]

P^+870, I^-, and X^- are radicals and at least theoretically detectable by ESR. The g values and line halfwidths are found: for P^+, $g = 2.0025$, $\Delta H = 9.4$ G;[106,107] for I^-, $g = 2.0035$, $\Delta H = 13$ G;[90,94,96] for X^-, $g = 2.05$, 1.83, and 1.68.[82,90] The ESR spectrum which appears upon the reduction of X suggests the involvement of a transition metal, probably the iron in the RC.[92,100,108] However, the iron does not become reduced upon illumination,[109] and RCs depleted of iron still show a primary photochemical reaction occurring with high efficiency, and under these conditions an ESR signal is observed, which can be ascribed to the UQ^- radical.[110,111] Thus, it seems that the line shape of the X^- radical is in fact because of a reduced quinone (UQ^-) in which the unpaired electron has a strong exchange interaction with the d electrons of the iron which is in close proximity to the UQ.[112] A strong indication of the dimeric structure of P comes from ENDOR experiments which showed that the unpaired electron remaining on P^+ after oxidation is delocalized over two Bchl molecules.[107] In the state $P\,I^-X^-$, the electron spin on I^- and X^- are magnetically coupled, giving rise to a doublet.[94] For example, in *C. vinosum,* there is a splitting of 60 G around $g = 2.0$. Magnetic coupling between electron spins on different radicals are a powerful tool for the investigation of structural aspects of the primary photochemical process, although the energies involved are small.[90]

In intact cells, which by light are brought into the state $P^+I\,X^-$, P^+ oxidizes Cyt *c* and X^- reduces a quinone. The reoxidation of reduced quinone by oxidized Cyt leads to the formation of ATP.

If X is reduced before illumination the excitation of P still produces a charge separation,[113-115] $P^*I\,X^- \rightarrow P^+I^-X^-$, but now the state $P^+I^-X^-$ has a lifetime of 10 nsec at 293 K and decays to form the excited singlet state $P^*\,I\,X^-$, the ground state $P\,I\,X^-$ or the triplet state $P^{TR}I\,X^-$. It has been shown[115] that the increased antenna Bchl emission, observed when the RCs are the state $P\,I\,X^-$, is in fact because of recombination from $P^+I^-X^-$ to $P^*I\,X^-$ and from P^* back to antenna Bchl. From the decrease of this emission with decreasing temperature, the energy difference between $P^+I^-X^-$ and $P^*I\,X^-$ is between 0.11 and 0.15 eV.

The formation of the RC triplet state can be explained within the framework of the so-called radical pair theory[90] as follows. Upon formation of the pair P^+I^-, the spins of the two unpaired electrons are opposite, giving a total spin $\vec{S} = \vec{S}_{P+} + \vec{S}_{I-} = 0$. The radicals are so far apart that the interaction between the spins (exchange, dipolar) is supposed to be small. However, each spin sees a different magnetic environment (e.g., because of hyperfine interaction or spin orbit coupling resulting in a different g value for the electron on P^+ and the electron on I^-), and thus they will precess around some local magnetic axis with different Larmor frequency, and dephasing between the spins occurs. Upon recombination there exists a finite probability that the total spin $\vec{S} = \vec{S}_{P+} + \vec{S}_{I-} = 1$, so that then the triplet state $P^{TR}I\ X^-$ is formed. These hypotheses are consistent with the effect of an external magnetic field on triplet formation and antenna Bchl fluorescence.[116-118] The population probability for the three triplet sublevels (which are close together) in zero magnetic field is not much different;[119] however, at high magnetic field, the $T_{\pm 1}$ levels are $\pm g\beta H$ apart from the T_o level (g is the electron g-factor, β is the Bohr magneton, and H the strength of the magnetic field) so that the singlet of $P^+I^-X^-$ can mix only with the T_o level, giving rise to a strong spin polarization. This pattern cannot be explained in terms of intersystem crossing in a single molecule, as first noted by Schaafsma (as quoted in Reference 120), and a decrease occurs in the flash-induced triplet yield.

At room temperature, the triplet yield in RCs in state P I X^- is about 15% for isolated RCs[113] and about 30% for intact systems.[115] This increases to about 100% at 77 K.[113,115] The reason for this is that at low temperature, the rate of the back reaction $P^+I^- \rightarrow P^*I$ strongly diminishes. $P^{TR}I\ X^-$ decays with a $t_{1/2} = 60\ \mu s$ at 293 K and with a $t_{1/2} = 120\ \mu s$ at 77 K. If a carotenoid is present, P^{TR} is rapidly converted into Car^{TR} ($t_{1/2} = 20$ ns at 293 K). This process shows a temperature-dependent rate;[121] below 50 K only P^{TR} can be observed, and the kinetics are no longer influenced by the presence of the carotenoid in the RC.

Photosystem 2

The primary and subsequent electron transfer reactions of the oxygen evolving photosystem 2 (PS2) may be described by the equations:[122]

$$\text{M Z P}_2\text{I}_2\text{Q(1)} + h\nu \rightarrow \text{M Z P}_2^*\text{I}_2\text{Q(2)} \rightarrow \text{M Z P}_2^+\text{I}_2^-\text{Q(3)}$$

$$\rightarrow \text{M Z P}_2^+\text{I}_2\text{Q}^-\text{(4)} \rightarrow \text{M Z}^+\text{P}_2\text{I}_2\text{Q}^-\text{(5)} \qquad (16)$$

As suggested by the difference spectrum $(P^+_2 - P_2)$[123], P_2 may be a Chl *a* dimer. P_2 (sometimes called P_{680}) is, upon transfer of an electron, transformed into a cation radical P^+_2. The E_m of P^+_2/P is supposed to be above +850 mV because PS2 still operates at this potential.[124] I is a hypothetical intermediate electron acceptor, Q is a quinone, probably plastoquinone, which is reduced to the semiquinone anion, and M and Z are unknown electron donors. The electron is transported from A^- via the plastoquinone pool and light reaction 1 to carbon dioxide, which is the final acceptor.

For the formation of an oxygen molecule, four electrons have to be transported by the reaction as shown in Equation 16 or:

$$M^{n+}\ Z^+ \rightarrow M^{(n+1)^+}\ Z \quad \text{(4 times)} \qquad (17)$$

$$M^{4+} + 2H_2O \rightarrow M + O_2 + 4H^+ \qquad (18)$$

Evidence concerning reactions of State (4) in Equation 16 was first obtained by measuring rapid luminescence and fluorescence yield kinetics[125,126] following flash excitation, where the luminescence is emitted by the reaction:

$$P_2O^- \rightarrow P_2^*Q \rightarrow P_2Q + h\nu_L \qquad (19)$$

Because P^+ is more rapidly reduced than Q^- is oxidized and the amplitude of the luminescence is a measure for the number of RC2s remaining in the state $P^+_2I_2Q^-$, the luminescence decay reflects the reduction of P^+_2. Similarly the fluorescence yield is an inverse measure of the concentration of $[P_2{}^+I_2Q^-]$, since P^+ is a quencher of the Chl *a* fluorescence,[125] so that the fluorescence yield in state $P^+_2I_2Q^-$ is low and that in state $P_2I_2Q^-$ is high.[127] This is provided no other fluorescence quenchers are present. It was found that the rate of reduction of P^+_2 depended among other things on the charge on MZ.[30] These and other conclusions have been confirmed by measurements of changes in light absorption presumably caused by the redox reaction of P_2.[128-130] The methods used were not rapid enough to measure the nsec reduction of P^+_2 occurring in a large part of the RC2s. Only recently, this more rapid reduction (half time 35 nsec) of P^+_2 has been observed[131] by measuring the absorption increase at 820 nm. A slower P^+_2 reduction has also been observed in PS2 particles,[132] in chloroplasts at low pH,[133] and in chloroplasts treated with hydroxylamine.[126,131] In addition a transient ESR signal has been ascribed to P^+_2, centering around g = 2.0025.[134,135]

The primary acceptor of PS2 is thought to be a quinone, Q.[136,137] In PS2 particles, where the secondary electron transport is blocked, the difference spectrum of the Q → Q^- transition is ascribed to the formation of a semiquinone anion. The formation of Q^- is coupled to a three- to fourfold increase in the antenna fluorescence.[127] This was concluded from the diminished trapping rate of the RC2 with Q^- present. Another possibility is that the increased Chl *a* emission is caused by luminescence,[138,139] resulting from the recombination reaction $P^+_2I^-_2 \rightarrow P_2I_2 + h\nu_L$.

Evidence for the presence of an intermediary acceptor I_2 is indirect. A rapid (1 μsec) luminescence component has been observed[140] when the RC2s are excited in state $P_2I_2Q^-$. This luminescence might be because of a recombination between I^-_2, formed by a flash and a positive charge on the donor side of PS2. In PS2 particles, but not in chloroplasts, the reduction of pheophytin[138] has been observed in state Q^-. Recently the experiment on a magnetic field effect on the antenna fluorescence also suggested the presence of an intermediary.[118]

Photosystem 1

Through the use of conventional spectroscopic methods, the primary donor of PS1 in plants and algae has been identified as P_{700} (P_1).[141] The oxidation of P_1 to P^+_1 is accompanied with absorbance decreases at 700, 680, and 430 nm[141,142] and an absorbance increase at 815 nm.[143,144] The latter change is characteristic of the formation of a Chl a^+ radical.[145] P_1 probably is a Chl *a* dimer.[146-148] P^+_1 shows an ESR signal with a g value of 2.0035 and ΔH of 7.2 G.[149] Some controversy exists on the oxidation/reduction midpoint potential. E_m values have been reported to be between +375 and +520 mV.[141,150-153] The reduction of P^+_1 after a light flash occurs in several phases at half times of about 10 μs and a few hundred microseconds.[154,155] Kinetic studies of P_1 in PS1 particles and in chloroplasts have shown the existence of a number of acceptors as described by the following model for PS1 electron transport:[156]

$$P_1 \rightarrow I_1 \rightarrow X_1 \rightarrow P_{430} \tag{20}$$

The arrows indicate the direction of electron transport. Low-temperature and psec-spectroscopic experiments have indicated that I_1 may involve a Chl *a* dimer having absorbance bands at 420, 450, and 700 nm.[157,158] X_1 is probably an iron-sulfur protein, judging from the ESR spectra,[159,160] with principal g-value components at 1.78, 1.88, and 2.08.[161,162] X_1 has a broad absorption band in the 400- to 500-nm region.[157] Denaturation of PS1 particles and their bound iron-sulfur proteins induces a loss of changes which could be attributed to $P^+_1X^-_1$ formation;[163] X_1/X^-_1 may have a midpoint potential $E_m = -730$ mV.[164] The recombination between P^+_{700} and X^-_1 gives rise to luminescence.[165]

P_{430}, which shows an absorbance decrease upon reduction around 430 nm,[166,167] is currently believed to consist of two iron-sulfur proteins[168-171] (may be 4Fe-4S centers), Center A and Center B which can be reduced irreversibly by light at low temperatures. The centers have an E_m of −540 (Center A) and −590 mV (Center B) at pH 10, and their reduced minus oxidized ESR spectra show principal g values at g = 1.86, 1.94, and 2.05 (Center A) and g = 1.82, 1.92, and 2.05 (Center B). At room temperature, P^+_1 P^-_{430} has a recombination time of 30 msec both for Center A and Center B.[156] If P_{430} is reduced before illumination. $P^+_1X^-_1$ is formed and has a recombination time of 250 μs at 293 K.[156] This slows down to 130 msec[157] at 5 K. If X_1 is also reduced before the excitation of $P_1^+I^-$, recombination occurs with a half time of 10 nsec (at 293 K)[158] (in part to a state which may be identified as the P_1 triplet state). The P_1 triplet state then has a lifetime $t_{1/2} = 3$ μsec at 293 K.[156,158] The latter rate slows down to 1.3 msec[157] at 5 K, which is also the case for the kinetically corresponding ESR change. If all iron-sulfur proteins are in the oxidized state before illumination, forward electron transport occurs and I^-_1 is reoxidized with $t_{1/2} = 200$ psec,[158] a rate comparable to that found in photosynthetic purple bacteria.

The spin polarization[172,173] of the ESR signal attributed to the primary donor of PS1 is related to the radical pair mechanism between P^+_1 and I^-_1 and X^-_1.

It is not known why so many electron transporting intermediates are present. The reason is probably not thermodynamic (there are no extra losses), but rather a kinetic one. It is possible that they help in transporting the electron rapidly across the membrane and thereby minimize the rate of back reactions.[174]

THERMODYNAMICS OF PHOTOSYNTHETIC PROCESSES

As has been described in the preceding sections, the process of photosynthesis carried out by green plants, algae, and some bacteria can be described as the conversion of light energy into (chemical) Gibbs' free energy. This section demonstrates by means of a few simple examples how the laws of thermodynamics can be applied to estimate the efficiency of this energy storage process and the stability of the products. The presentation here closely follows that of Ross and Calvin,[175] Knox,[176] and Almgren.[177]

Free Energy of a System of Excited Pigment Molecules in a Stationary State and Storage Efficiency

Consider a set of pigment molecules, say Chl *a*. Chl has a lowest excited singlet state Chl* at an energy $h\nu_o$ above the ground state. For simplicity we assume that the density of states of Chl* is the same as that of Chl. Then one can write for the difference in chemical potential between the ground state and the excited state:[178]

$$\Delta\mu = \mu(\mathrm{Chl}^*) - \mu(\mathrm{Chl}) = \mu^o(\mathrm{Chl}^*) - \mu^o(\mathrm{Chl}) + kT \ln [\mathrm{Chl}^*] - kT \ln [\mathrm{Chl}]$$

$$= h\nu_o + kT \ln \frac{[\mathrm{Chl}^*]}{[\mathrm{Chl}]} \tag{21}$$

In the dark, at thermal equilibrium, $\Delta\mu = 0$, and thus, the well-known Boltzmann equation is obtained.

$$\left\{\frac{[\mathrm{Chl}^*]}{[\mathrm{Chl}]}\right\}_{eq} = e^{-h\nu_o/kT} \tag{22}$$

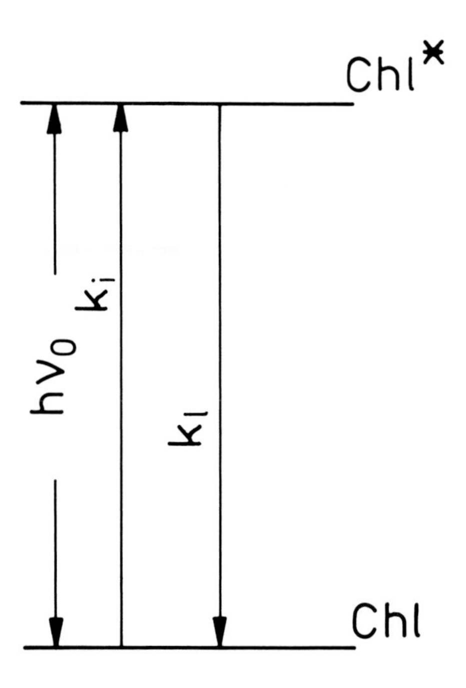

FIGURE 3. Kinetic scheme showing the excitation of chlorophyll to its lowest excited singlet state at a rate k_i, and subsequent loss processes represented by $k_l = k_f + k_{ic}$, where k_f is the rate of fluorescence and k_{ic} is the rate of internal conversion.

If the pigment molecules are excited at a rate k_i (where k_i in sunlight does not exceed 10 sec^{-1} for Chl), the steady-state concentration of Chl* is given by (see Figure 3):

$$k_i\,[\mathrm{Chl}] = k_l\,[\mathrm{Chl}^*]$$

or

$$\left\{\frac{[\mathrm{Chl}^*]}{[\mathrm{Chl}]}\right\}_{max} = \frac{k_i}{k_l} \tag{23}$$

where $k_l = k_f + k_{ic}$ is the rate of all loss processes such as fluorescence (rate constant k_f) and internal conversion (k_{ic}). The formation of a metastable triplet state is not considered. Equation 23 gives the maximum concentration of excited states at a certain light intensity. Any other process, which would occur from the excited state, e.g., a storage process or an additional loss process, would decrease the concentration of excited states. Table 2 shows the fraction of excited states at a number of light intensities and the associated changes in free energy calculated from Equations 21 and 23.

For light intensities of one tenth to full sunlight (Rows 2 and 3 of Table 2), the maximum chemical potential obtainable is 73 to 77% of the incident photon energy. Thus, the maximum amount of useful work which can be extracted from such a system is about 73 to 77% of the absorbed photon energy.

Earlier estimates were obtained by Duysens,[179] who compared the photoconversion process with a Carnot cycle. For such a cycle, the maximum obtainable efficiency ε, if energy flows from a bath at temperature T_1 to a bath at T_2 is given by:

$$\epsilon = \frac{T_1 - T_2}{T_1} \tag{24}$$

Table 2
FRACTION OF EXCITED STATES AND ASSOCIATED CHARGES IN FREE ENERGY, CALCULATED FOR A NUMBER OF LIGHT INTENSITIES (k_i)

k_i (s^{-1})	$\frac{[chl^*]}{[chl]}$	$\Delta\mu$ (eV)
k_i is taken as the intensity of black body radiation at T = 300 K	2.4×10^{-31} (the dark equilibrium value)	0
$k_i = 1\ s^{-1}$, normal sunlight	10^{-8}	1.34
$k_i = 10\ s^{-1}$, high-intensity sunlight	10^{-7}	1.41
$k_i = 10^8 s^{-1}$, a high-intensity picosecond flash	1	1.82

Note: The other parameters are: $k_l = 10^8\ s^{-1}$, T = 300 K, and $h\nu_o = 1.82$ eV.

By applying these arguments to algal photosynthesis, it has been proposed that in weak light the maximum efficiency of photosynthesis could be considered as a flow of black body radiation from a temperature of approximately 1100 K to a bath at 300 K. The maximum obtainable efficiency is (Equation 24) $\varepsilon = 73\%$ which is similar to the result obtained above.

The next section considers the problem of storage of chemical free energy using a kinetic approach, assuming the storage reaction is fast and quasi-irreversible (in the next paragraph this latter condition will be relaxed) and has a rate k_{st} (see Figure 4).

The steady-state concentration of excited states in the light is given by:

$$k_i\,[Chl] = (k_l + k_{st})\,[Chl^*]$$

or

$$\frac{[Chl^*]}{[Chl]} = \frac{k_i}{k_l + k_{st}} \qquad (25)$$

Moreover the quantum yield of the storage process in the absence of a back reaction, ϕ^o_{st}, is given by:

$$\phi^o_{st} = \frac{k_{st}}{k_l + k_{st}} \qquad (26)$$

Inserting Equations 25 and 26 into 21, we can calculate the quantity:

$$\phi^o_{st} \cdot \Delta\mu = \phi^o_{st} \left\{ h\nu_o + kT \ln \left[\frac{k_i}{k_{st}} \phi^o_{st} \right] \right\} \qquad (27)$$

Equation 27 now represents the amount of stored free energy per photon absorbed. Maximization of $\phi^o_{st} \cdot \Delta\mu$ yields optimal values for $\Delta\mu$ and ϕ^o_{st} that lead to a maximum conversion efficiency.[175] In Table 3, an example is given; for this specific case a maximum conversion efficiency is obtained if loss processes do not exceed 2%. It should be emphasized that this example is not realistic. If the back reaction is discarded, we in fact have a system where no storage of energy takes place. Clarification will follow in the next section.

Storage Process in the Presence of a Back Reaction and Accumulated Product

So far we have not considered the back reactions; however, for thermodynamic reasons, a back reaction always occurs. We now make the assumption that the storage reaction can be described by the following simple reaction scheme (see Figure 5): a substrate S is transformed by the energy in the excited chlorophyll molecules into a product P. To symbolize the activity of the light, we define the presence of the nonexcited state of chlorophyll as Chl S and the excited state as Chl*S. Only from the latter state is the product P formed spontaneously, whereby Chl* returns to the ground state. At "normal" light intensities, the fraction of excited states is extremely small, and therefore the concentration of pigment molecules in the ground state is assumed to be constant. The substrate can be thought of as a redox pair D A, (i.e., D is the electron donor, and A is the electron acceptor) which by the light changes into D^+A^-; the light is thus stored as redox energy. Of course, the product P can decay to the substrate S by a reaction in which the stored energy is used to perform work or even by a loss reaction. However, the highest efficiency is achieved if the decay of P occurs via the excited state of the pigment molecules and the unavoidable loss processes, represented by k_l. Furthermore, it seems realistic to assume that the forward reaction occurs with a high quantum yield in the absence of a back reaction, or ϕ^o_{st} approaches one, which condition is met if $k_{st} >> k_l$.

After an induction period when the light is switched on, the substrate and the product will be at more or less constant potentials given as:

$$\mu(S) = \mu(\text{Chl S}) = \mu^o(\text{Chl S}) + kT\ \ln\ [\text{Chl S}]$$

$$\mu(P) = \mu^o(P) + kT\ \ln\ [P]$$

and the chemical potential of the excited state is:

$$\begin{aligned}\mu(\text{Chl}^*\text{S}) &= \mu^o(\text{Chl}^*\text{S}) + kT\ \ln\ [\text{Chl}^*\text{S}] \\ &= \mu^o(\text{Chl S}) + h\nu_o + kT\ \ln\ [\text{Chl}^*\text{S}] \end{aligned} \qquad (28)$$

If after the accumulation of products P at a potential $\mu(P)$ the light is switched off, the amount of excited pigment molecules will be kept appreciably above the equilibrium value because of the back reaction. Thus, in the dark, assuming a steady state:

$$k_b\ P = (k_{st} + k_l)\ [\text{Chl}^*\text{S}]_{dark} \qquad (29)$$

If now $k_{st} >> k_l$, then the difference in free energy between Chl*S and P will be almost zero:

$$\Delta\mu(\text{Chl}^*\text{S, P})_{dark} = \mu(\text{Chl}^*\text{S})_{dark} - \mu(P) = 0$$

$$\mu(\text{Chl}^*\text{S})_{dark} = \mu(P) \qquad (30)$$

This can be interpreted by saying that the concentration of excited chlorophyll molecules in the dark is only determined by the chemical potential of the stored products.

Moreover, in the light we have the steady-state equation:

$$k_b[P] + k_i[\text{Chl S}] = (k_{st} + k_l)\ [\text{Chl}^*\text{s}]_{ss} \qquad (31)$$

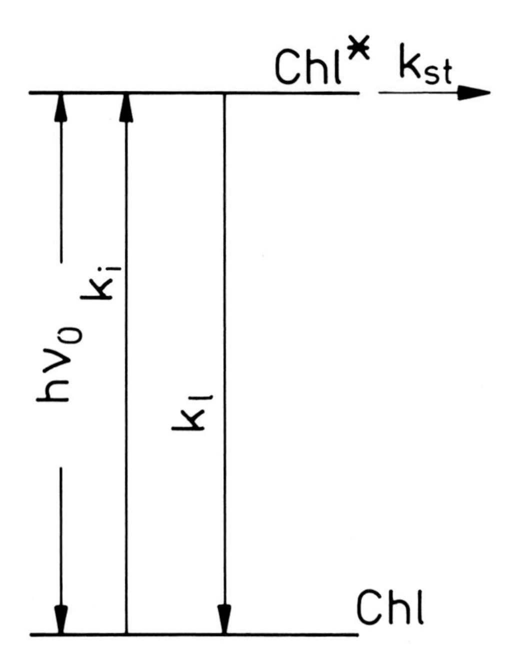

FIGURE 4. As Figure 3, but now including a quasi-irreversible storage process from the lowest excited singlet state at a rate k_{st}.

Table 3
FREE ENERGY STORED PER PHOTON ABSORBED CALCULATED FOR A NUMBER OF STORAGE RATES FROM EQUATION 27 FOR k_i = 10 s^{-1}, k_l = 10^8 s^{-1}, T = 300 K, AND $h\nu_o$ = 1.81 eV

k_{st} (s^{-1})	$Ø^{o}_{st}$	$\cdot\ \Delta\mu$ (eV)	$Ø^{o}_{st} \cdot \Delta\mu$ (eV)
0	0	1.82	0
10^8	0.5	1.79	0.90
10^9	0.91	1.75	1.59
5×10^9	0.98	1.71	1.68
10^{11}	1.00	1.63	1.63

and for the quantum efficiency of storage we find under these steady-state conditions:

$$\phi^{ss}_{st} = \frac{k_i[\text{Chl S}] - k_l\,[\text{Chl}^*\text{S}]_{ss}}{k_i[\text{Chl S}]} = 1 - \frac{k_l}{k_i}\,\frac{[\text{Chl}^*\text{S}]_{ss}}{[\text{Chl S}]} \qquad (32)$$

Using Equation 31, we obtain:

$$\phi^{ss}_{st} = 1 - \frac{k_l}{k_i}\left\{\frac{k_b}{(k_{st} + k_l)}\;\frac{[P]}{[\text{Chl S}]} + \frac{k_i}{(k_{st} + k_l)}\right\} \qquad (33)$$

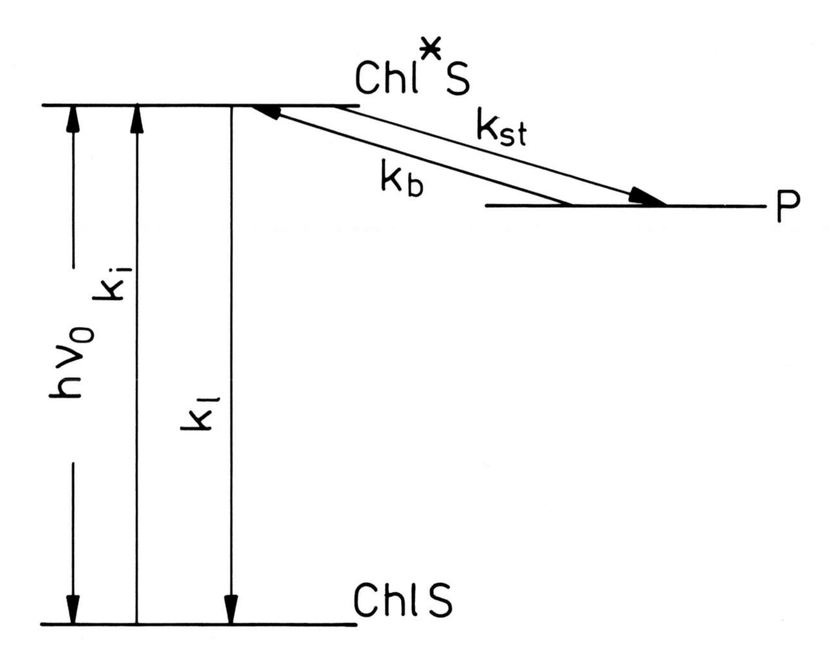

FIGURE 5. A substrate molecule S in the presence of pigment molecules in the ground state, represented by Chl S, is excited by a rate k_i to a state Chl*S which is a notation of substrate molecules S in the presence of excited pigment molecules Chl*. The energy in this excited state can be used to turn a molecule S into a product molecule P; the excited pigment molecule returns to the ground state. The meaning of k_l and k_{st} is as in Figure 4; k_b is the rate of the back reaction in which the product P again produces S in the presence of an excited pigment molecule Chl*.

Rewriting Equation 33 and inserting Equation 29, we find:

$$\phi_{st}^{ss} = \frac{k_{st}}{(k_{st} + k_l)} - \frac{k_l\, k_b}{k_i(k_{st} + k_l)} \frac{[P]}{[Chl\ S]} = \phi_{st}^{o} - \frac{k_l}{k_i} \left\{ \frac{[Chl^*S]_{dark}}{[Chl\ S]} \right\} \tag{34}$$

where ϕ^o_{st} is the storage quantum yield in the absence of a back reaction (see Equation 26) and $[Chl^*S]_{dark}$ is given by Equations 28 and 30.

$$\ln \left\{ \frac{[Chl^*S]_{dark}}{[Chl\ S]} \right\} = \frac{\mu(P) - \mu(S) - h\nu_o}{kT} \tag{35}$$

For illustration we shall discuss these equations for the case of the photosynthetic bacterium *R. rubrum* which has a main absorption peak at 880 nm ($h\nu_o$ = 1.41 eV). The product of the primary photochemical act is the oxidation of a *c*-type cyt at a potential of +0.3 V and the reduction of a quinone at a potential of −0.2 V (this is probably on the low side).[174] From these numbers, we can calculate the concentration of excited states in the dark (Equation 35), using $\mu(P) - \mu(S) = 0.5$ eV:

$$\frac{[Chl^*S]_{dark}}{[Chl\ S]} = 6.3 \times 10^{-16}$$

We now calculate the storage quantum efficiency under steady-state conditions, ϕ_{st}^{ss}

(Equation 34) for $k_i = 1\ sec^{-1}$, normal sunlight, and $k_i \sim 10^{-6}\ s^{-1}$ which may be a realistic approximation of the local light intensity at the place where a bacterium like *R. rubrum* lives (in mud!).

Assuming $\phi^o_{st} = 1$ and $k_l = 10^8\ sec^{-1}$, it follows that:

$$\text{for} \quad k_i = 1\ s^{-1}, \phi^{ss}_{st} = 1 - (6.3 \times 10^{-8}) \approx 1$$

$$\text{and for} \quad k_i = 10^{-6}\ s^{-1}, \phi^{ss}_{st} = 1 - (6.3 \times 10^{-2}) \approx 0.94$$

Thus, even at an extremely low light intensity of $10^{-6}\ sec^{-1}$, a bacterium like *R. rubrum* is able to store light energy with a quantum efficiency, which is larger than 90%.

From Equations 34 and 35, we conclude that if a photochemical system is to operate at a low light intensity and the quantum efficiency is to remain high, the potential difference $\mu(P) - \mu(S)$ between the product and the substrate must be smaller than for a system which operates with a similar efficiency, but at a much higher light intensity. Thus, if in our example above, *R. rubrum* was living at a light intensity $k_1 = 1\ sec^{-1}$, $\mu(P) - \mu(S)$ might increase to 0.86 eV before ϕ^{ss}_{st} would drop below 0.94.

Stability of the Products

This section calculates the stability of products formed in light after it has been switched off.

In the dark the rate of loss of free energy is given by:

$$R_l = k_l\ [Chl^*S]_{dark}\ [\mu(P) - \mu(S)] \qquad (36)$$

In a light period Δt and intensity k_i, the amount of free energy stored would equal:

$$G_{st} = k_i\ [Chl\ S]\ \phi^{ss}_{st}\ \Delta t\ [\mu(P) - \mu(S)] \qquad (37)$$

This amount of free energy will be lost in a time t_l, using Equations 36 and 37 given as:

$$t_l = \frac{G_{st}}{R_l} = \frac{k_i\ \phi^{ss}_{st}\ \Delta t\ [Chl\ S]}{k_l\ [Chl^*S]_{dark}} \qquad (38)$$

For example in *R. rubrum* under "normal" sunlight ($k_i = 1\ sec^{-1}$), we find:

$$t_l = \frac{10^{-8} \times 1}{6.3 \times 10^{16}}\ \Delta t = 1.6 \times 10^7\ \Delta t$$

The stored energy is lost in a period 1.6×10^7 times longer than the illumination time. The stability of the products thus depends on the potential difference $\mu(P) - \mu(S)$ which determines $[Chl^*S]_{dark}$ and the rate of illumination, k_i.

ACKNOWLEDGMENTS

The authors are grateful to the members of the Department of Biophysics of the

State University of Leiden for helpful discussions on several of the subjects, Dr. A. J. Hoff for critically reading the manuscript, Mr. A. H. M. de Wit for recording the spectra in Figure 2D, and to Mrs. J. M. Philips for typing the manuscript. Rienk van Grondelle acknowledges a grant from the EEG Solar Energy Program, Section D.

REFERENCES

1. **Gouterman, M.**,Optical spectra and electronic structure of porphyrins, in *The Porphyrins,* Vol. 3 (Part A), Dolphin, D., Ed., Academic Press, New York, 1978, 1.
2. **Bowers, P. G. and Porter, G.**, *Proc. R. Soc. London,* 296, 435, 1967.
3. **Holten, D. and Windsor, M. W.**, *Annu. Rev. Biophys. Bioeng.,* 7, 189, 1978.
4. **Bensasson, R., Land, E. J., and Maudinas, B.**, *Photochem. Photobiol.,* 23, 189, 1976.
5. **Duysens, L. N. M.**, Thesis, University of Utrecht, 1952.
6. **Thornber, J. P., Alberte, R. S., Hunter, F. A., Shiozowa, J. A., and Kan, K.-S.**, in *Chlorophyll-Proteins, Reaction Centers and Photosynthetic Membranes,* Brookhaven Symposium in Biology No. 28, Olson, J. M. and Hind, G., Eds., Brookhaven National Laboratory, Upton, N.Y., 1977, 132.
7. **Duysens, L. N. M., Amesz, J., and Kamp, B. M.**, *Nature (London),* 190, 510, 1961.
8. **Wang, R. T. and Myers, J.**, *Biochim. Biophys. Acta,* 347, 134, 1974.
9. **Wang, R. T. and Myers, J.**, *Photochem. Photobiol.,* 23, 405, 1976.
10. **Butler, W. L.**, *Annu. Rev. Plant Physiol.,* 29, 345, 1979.
11. **Barber, J.**, Ionic regulation in intact chloroplasts and its effect on primary photosynthetic processes, in *The Intact Chloroplast,* Barber, J., Ed., Elsevier, Amsterdam, 1976, 84.
12. **Duysens, L. N. M.**, Photosynthesis, *Progress in Biophysics,* 14, Pergamon Press, Oxford, 1964, 1.
13. **Borisov, A. Yu. and Godik, V. I.**, *Biochim. Biophys. Acta,* 301, 227, 1973.
14. **Knox, R. S.**, in *Bioenergetics of Photosynthesis,* Govindjee, S. R., Ed., Academic Press, New York, 1975, 183.
15. **Knox, R. S.**, in *The Intact Chloroplast,* Vol. 2, Barber, J., Ed., Elsevier, Amsterdam, 1977, 55.
16. **Förster, T.**, *Ann. Phys.,* 2, 55, 1948.
17. **Förster, T.**, Delocalized excitation and excitation transfer, in *Modern Quantum Chemistry, Part III,* Sinanoglu, O., Ed., Academic Press, New York, 1965, 93.
18. **Dexter, D. L.**, *J. Chem. Phys.,* 21, 836, 1953.
19. **Van Grondelle, R.**, Thesis, University of Leiden, The Netherlands, 1978.
20. **Grabowski, J. and Gantt, E.**, *Photochem Photobiol.,* 28, 39, 1978.
21. **Grabowski, J. and Gantt, E.**, *Photochem. Photobiol.,* 28, 47, 1978.
22. **Colbow, K.**, *Biochim. Biophys. Acta,* 314, 320, 1973.
23. **Knox, R. S.**, *Physica,* 39, 361, 1968.
24. **Searle, G. F. W., Barber, J., Porter, G., and Tredwell, C. J.**, *Biochim. Biophys. Acta,* 501, 246, 1978.
25. **Campillo, A. J., Hyer, R. C., Monger, T. G., Parson, W. W., and Shapiro, S.**, *Proc. Natl. Acad. Sci. U.S.A.,* 74, 1997, 1977.
26. **Zankel, K. L., Reed, D. W., and Clayton, R. K.**, *Proc. Natl. Acad. Sci. U.S.A.,* 61, 1248, 1968.
27. **Campillo, A. J. and Shapiro, S. L.**, in *Ultrashort Light Pulses,* Shapiro, S. L., Ed., Springer-Verlag, Heidelberg, 1977, 317.
28. **Porter, G., Tredwell, G. J., Searle, G. F. W., and Barber, J.**, *Biochim. Biophys. Acta,* 501, 232, 1978.
29. **Montroll, E. W.**, *J. Math. Phys. (N.Y.),* 10, 753, 1969.
30. **Duysens, L. N. M.**, *Proc. of CIBA Foundation Symp.,* Fitz Simons, D.W., Ed., in press.
31. **Van Gorkom, H. J., Pulles, M. P. J., and Etienne, A. L.**, *Proc. Symp. on Photosynthetic Oxygen Evolution at Tübingen,* Metzner, H., Ed., Academic Press, New York, 1978, 135.
32. **Vredenberg, W. J. and Duysens, L. N. M.**, *Nature (London),* 197, 355, 1963.
33. **Parson, W. W.**, 6th Congr. Biophysical Society, Kyoto, Japan, personal communication, 1978.
34. **Paschenko, V. Z., Kononenko, A. A., Protasov, S. P., Rubin, A. B., Rubin, L. B., and Uspenskaya, N. Ya.**, *Biochim. Biophys. Acta,* 461, 403, 1977.
35. **Govindjee, S. R., Hammond, J. H., and Merkelo, H.**, *Biophys. J.,* 12, 809, 1972.
36. **Campillo, A. J., Kollman, V. H., and Shapiro, S. L.**, *Science,* 193, 227, 1976.

37. Porter, G., Synowiec, J. A., and Tredwell, C. J., *Biochim. Biophys. Acta,* 459, 329, 1977.
38. Sauer, K. and Brewington, G. T., *Proc. of 4th Int. Congr. Photosynthesis,* Hall, D. O., Coombs, J., and Goodwin T. W., Eds., The Biochemical Society, 1978, 409.
39. Beddard G. S., Fleming, G. R., Porter, G., Searle, G. F. W., and Synowiec, J. A., *Biochim. Biophys. Acta,* 545, 165, 1979.
40. Paschenko, V. Z., Protasov, S. P., Rubin, A. B., Timofeev, K. N., Zamazova, L. M., and Rubin, L. B., *Biochim. Biophys. Acta,* 408, 143, 1975.
41. Searle, G. F. W., Barber, J., Harris, L., Porter, G., and Tredwell, C. J., *Biochim. Biophys. Acta,* 459, 390, 1977.
42. Butler, W. L. and Norris, K. H., *Biochim. Biophys. Acta,* 66, 72, 1963.
43. Muller, A., Lumry, R., and Walker, M. S., *Photochem. Photobiol.,* 9, 113, 1969.
44. Mar, T., Govindjee, S. R., Singhal, G. S., and Merkelo, H., *Biophys. J.,* 12, 797, 1972.
45. Hervo, G., Paillotin, G., and Thiery, J., *J. Chem. Phys.,* 72, 761, 1975.
46. Geacintov, N. E., Swenberg, A. J., Campillo, A., Hyer, R. C., Shapiro, S. L., and Winn, K. R., *Biophys. J.,* 24, 347, 1978.
47. Beddard, G. S. and Porter, G., *Biochim. Biophys. Acta,* 462, 63, 1977.
48. Altmann, J. A., Beddard, G. S., and Porter, G., *Chem. Phys. Lett.,* 58, 54, 1978.
49. Melis, A. and Homann, P., *Photochem. Photobiol.,* 23, 343, 1976.
50. Moya, I., *Biochim. Biophys. Acta,* 368, 214, 1974.
51. Moya, I., Govindjee, S. R., Vernotte, C., and Briantais, J.-M., *FEBS Lett.,* 75, 13, 1977.
52. Campillo, A. J., Shapiro, S. L., Kollman, V. H., Winn, K. R., and Hyer, R. C., *Biophys. J.,* 16, 93, 1976.
53. Swenberg, C. E., Geacintov, N. E., and Pope, M., *Biophys. J.,* 16, 1447, 1976.
54. Harris, L., Porter, G., Synowiec, J. A., Tredwell, C. J., and Barber, J., *Biochim. Biophys. Acta,* 449, 329, 1976.
55. Geacintov, N. E., Barber, J., Swenberg, C. E., and Paillotin, G., *Photochem. Photobiol.,* 26, 629, 1977.
56. Mauzerall, D., *Biophys. J.,* 16, 87, 1976.
57. Monger, T. G. and Parson, W. W., *Biochim. Biophys. Acta,* 460, 393, 1977.
58. Duysens, L. N. M., van der Schatte Olivier, T. E., and den Haan, G. A., Abstr. Int. Congr. on Photobiology, Abstr. VI, Bochum no. 277, 1972.
59. Zankel, K. L., Biochim. Biophys. Acta, 325, 138, 1973.
60. Den Haan, G. A., Duysens, L. N. M., and Egberts, D. M. J., *Biochim. Biophys. Acta,* 368, 409, 1974.
61. Renger, G. and Wolff, C., *Biochim. Biophys. Acta,* 460, 47, 1977.
62. Geacintov, N. E. and Breton, J., *Biophys. J.,* 17, 1, 1977.
63. Campillo, A. J. and Shapiro, S. L., *Photochem. Photobiol.,* 28, 975, 1978.
64. Campillo, A. J., Hyer, R. C., Shapiro, S. L., and Swenberg, C. E., *Chem. Phys. Lett.,* 48, 495, 1977.
65. Monger, T. G., Cogdell, R. J., and Parson, W. W., *Biochim. Biophys. Acta,* 449, 136, 1976.
66. Kung, M. C. and Devault, D., *Photochem. Photobiol.,* 24, 87, 1976.
67. Rahman, T. S. and Knox, R. S., *Phys. Status Solidi B,* 58, 715, 1973.
68. Breton, J. and Mathis, P., *C.R. Acad. Sci. Paris,* 271, 1094, 1970.
69. Den Haan, G. A., Thesis, University of Leiden, 1977.
70. Koka, P. and Song, P. S., *Photochem. Photobiol.,* 28, 509, 1978.
71. Geacintov, N. E., Breton, J., Swenberg, C. E., Campillo, A. J., Hyer, R. C., and Shapiro, S. L., *Biochim. Biophys. Acta,* 461, 306, 1977.
72. Campillo, A. J., Shapiro, S. L., Geacintov, N. E., and Swenberg, C. E., *FEBS Lett.,* 83, 316, 1977.
73. Fowler, C. F., Nugent, N. A., and Fuller, R. C., *Proc. Natl. Acad. Sci. U.S.A.,* 68, 2278, 1971.
74. Prince, R. C. and Olson, J. M., *Biochim. Biophys. Acta,* 423, 357, 1976.
75. Knaff, D. B. and Malkin, R., *Biochim. Biophys. Acta,* 430, 244, 1976.
76. Jennings, J. V. and Evans, M. C. W., *FEBS Lett.,* 75, 33, 1977.
77. Clayton, R. K., *Annu. Rev. Biophys. Bioeng.,* 2, 131, 1973.
78. Parson, W. W., *Annu. Rev. Microbiol.,* 28, 41, 1978.
79. Parson, W. W. and Cogdell, R. J., *Biochim. Biophys. Acta,* 416, 105, 1975.
80. Blankenship, R. E. and Parson, W. W., *Annu. Rev. Biochem.,* 47, 635, 1978.
81. Olson, J. M. and Thornber, J. P., Photosynthetic reaction centers, in *Membrane Proteins in Energy Transduction,* Capaldi, R. A., Ed., Marcel Dekker, New York, in press.
82. Feher, G. and Okamura, M. Y., Chemical composition and properties of reaction centers, in *The Photosynthetic Bacteria,* Clayton, R. K. and Sistrom, W. R., Eds., Plenum Press, New York, in press.
83. Van der Rest, M. and Gingras, G., *J. Biol. Chem.,* 249, 6446, 1974.

84. Cogdell, R. J., Parson, W. W., and Kerr, M. A., *Biochim. Biophys. Acta,* 430, 83, 1976.
85. Clayton, R. K. and Straley, S. C., *Biochem. Biophys. Res. Commun.,* 39, 1114, 1970.
86. Clayton, R. K. and Straley, S. C., *Biophys J.,* 12, 221, 1972.
87. Slooten, L., *Biochim. Biophys.,* 275, 208, 1972.
88. Slooten, L., Thesis, University of Leiden, The Netherlands, 1973.
89. Okumura, M. Y., Ackerson, L. C., Isaacson, R. A., Parson, W. W., and Feher, G., *Biophys. Soc. Abstr.,* 16, 67, 1976.
90. Hoff, A. J., Application of ESR in photosynthesis, *Physics Reports,* 54, 126, 1979.
91. Romijn, J. C. and Amesz, J., *Biochim. Biophys. Acta,* 461, 327, 1977.
92. Feher, G., *Photochem. Photobiol.,* 14, 373, 1971.
93. Shuvalov, V. A. and Klimov, V. V., *Biochim. Biophys. Acta,* 440, 587, 1976.
94. Tiede, D. M., Prince, R. C., and Dutton, P. L., *Biochim. Biophys. Acta,* 449, 447, 1976.
95. Van Grondelle, R., Romijn, J. C., and Holmes, N. G., *FEBS Lett.,* 72, 187, 1976.
96. Prince, R. C., Tiede, D. M., Thornber, J. P., and Dutton, P. L., *Biochim. Biophys. Acta,* 462, 467, 1977.
97. Fajer, J., Brune, D. C., Davis, M. S., Forman, A., and Spaulding, L. D., *Proc. Natl. Acad. Sci. U.S.A.,* 72, 4956, 1975.
98. Jortner, J., *J. Chem. Phys.,* 64 (12), 4860, 1976.
99. Hopfield, J. J., *Proc. Natl. Acad. Sci. U.S.A.,* 71, 3640, 1974.
100. Rockley, M. G., Windsor, M. W., Cogdell, R. J., and Parson, W. W., *Proc. Natl. Acad. Sci. U.S.A.,* 72, 2251, 1975.
101. Kaufmann, K. J., Dutton, P. L., Netzel, T. L., Leigh, J. S., and Rentzepis, P. M., *Science,* 188, 1301, 1975.
102. Prince, R. C. and Dutton, P. L., *Arch. Biochem. Biophys.,* 172, 329, 1976.
103. Vermeglio, A., Breton, J., Paillotin, G., and Cogdell, R. J., *Biochim. Biophys. Acta,* 501, 514, 1978.
104. Vermeglio, A. and Clayton, R. K., *Biochim. Biophys. Acta,* 449, 500, 1976.
105. Shuvalov, V. A., Asadov, A. A., and Krakhmaleva, I. N., *FEBS Lett.,* 76, 240, 1977.
106. McElroy, J. D., Feher, G., and Mauzerall, D. C., *Biochim. Biophys. Acta,* 267, 363, 1972.
107. Feher, G., Hoff, A. J., Isaacson, R. A., and Ackerson, L. C., *Ann. N.Y. Acad. Sci.,* 244, 239, 1975.
108. Leigh, J. S. and Dutton, P. L., *Biochem. Biophys. Res. Commun.,* 46, 414, 1972.
109. Debunner, P. G., Schultz, G. E., Feher, G., and Okamura, M. Y., *Biophys. J.,* 15, 226, 1975.
110. Feher, G., Okamura, M. Y., and McElroy, J. D., *Biochim. Biophys. Acta,* 267, 222, 1972.
111. Loach, P. A. and Hall, R. L., *Proc. Natl. Acad. Sci. U.S.A.,* 69, 786, 1972.
112. Butler, W. L., Johnston, D. C., Okamura, M. Y., Shore, H. B., and Feher, G., *Biophys. Soc. and Am. Phys. Soc. Abstr.,* M-AM-A3, *Biophys. J.,* 21, 8a, 1978.
113. Parson, W. W., Clayton, R. K., and Cogdell, R. J., *Biochim. Biophys. Acta,* 387, 265, 1975.
114. Holmes, N. G., van Grondelle, R., Hoff, A. J., and Duysens, L. N. M., *FEBS Lett.,* 70, 185, 1976.
115. Van Grondelle, R., Holmes, N. G., Rademaker, H., and Duysens, L. N. M., *Biochim. Biophys. Acta,* 503, 10, 1978.
116. Blankenship, R. E., Schaafsma, T. J., and Parson, W. W., *Biochim. Biophys. Acta,* 461, 297, 1977.
117. Hoff, A. J., Rademaker, H., van Grondelle, R., and Duysens, L. N. M., *Biochim. Biophys. Acta,* 460, 547, 1977.
118. Rademaker, H., Hoff, A. J., and Duysens, L. N. M., *Biochim. Biophys. Acta,* in the press.
119. Hoff, A. J. and Gorter de Vries, H., *Biochim. Biophys. Acta,* 503, 94, 1978.
120. Norris, J. R., Uphaus, R. A., and Katz, J. J., *Chem. Phys. Lett.,* 31, 157, 1975.
121. Parson, W. W. and Monger, T. G., *Brookhaven Symp. Biol.,* 28, 195, 1977.
122. Amess, J. and Duysens, L. N. M., in *Primary Processes in Photosynthesis,* Barber, J., Ed., Elsevier, Amsterdam, 1977, 149.
123. van Gorkom, H. J., Thesis, University of Leiden, The Netherlands, 1976.
124. Bearden, A. J. and Malkin, R., *Biochim. Biophys. Acta,* 325, 266, 1973.
125. Duysens, L. N. M., den Haan, G. A., and van Best, J. A., in *Proc. 3rd Int. Congr. on Photosynthesis,* Vol. 1, Avron, M., Ed., Elsevier Scientific, New York, 1974, 1.
126. Den Haan, G. A., Gorter de Vries, H., and Duysens, L. N. M., *Biochim. Biophys. Acta,* 430, 265, 1976.
127. Duysens, L. N. M. and Sweers, H. E., Studies on Microalgae and Photosynthetic Bacteria, Jpn. Soc. Plant Physiol., Ed., Tokyo University Press, Tokyo, 1963, 353.
128. Döring, G., Renger, G., Vater, J., and Witt, H. T., *Z. Naturforsch.,* 24b, 1139, 1969.
129. Gläser, M., Wolff, C., Buchwals, H.-E., and Witt, H. T., *FEBS Lett.,* 42, 81, 1974.
130. Gläser, M., Wolff, C., and Renger, G., *Z. Naturforsch.,* 31c, 712, 1976.
131. van Best, J. A. and Mathis, P., *Biochim. Biophys. Acta,* 503, 178, 1978.
132. van Gorkom, H. J., Pulles, M. P. J., and Wessels, J. S. C., *Biochim. Biophys. Acta,* 408, 331, 1975.

133. Pulles, M. P. J., van Gorkom, H. J., and Verschoor, G. A. M., *Biochim. Biophys. Acta,* 440, 98, 1976.
134. Malkin, R. and Bearden, A. J., *Biochim. Biophys. Acta,* 396, 250, 1975.
135. Visser, J. W. M., Rijgersberg, C. P., and Gast, P., *Biochim. Biophys. Acta,* 460, 36, 1977.
136. Stiehl, H. N. and Witt, H. T., *Z. Naturforsch.,* 24b, 1588, 1969.
137. van Gorkom, H. J., *Biochim. Biophys. Acta,* 347, 439, 1974.
138. Klimov, V. V., Klevanik, A. V., Shuvalov, V. A., and Krasnovskii, A. A., *FEBS Lett.,* 82, 183, 1977.
139. Amesz, J. and van Gorkom, H. J., *Annu. Rev. Plant. Physiol.,* 29, 47, 1978.
140. van Best, J. A. and Duysens, L. N. M., *Biochim. Biophys. Acta,* 459, 187, 1977.
141. Kok, B., *Biochim. Biophys. Acta,* 48, 527, 1961.
142. Döring, G., Bailey, J. L., Kreutz, W., Weikard, J., and Witt, H. T., *Naturwissenschaften,* 55, 219, 1968.
143. Inoue, Y., Ogawa, T., and Shibata, K., *Biochim. Biophys. Acta,* 305, 483, 1973.
144. Mathis, P. and Vermeglio, A., *Biochim. Biophys. Acta,* 369, 371, 1975.
145. Borg, D. C., Fajer, J., Felton, R. H., and Dolphin, D., *Proc. Natl. Acad. Sci. U.S.A.,* 67, 813, 1970.
146. Norris, J. R., Scheer, H., Druyan, M. E., and Katz, J. J., *Proc. Natl. Acad. Sci. U.S.A.,* 71, 4897, 1974.
147. Norris, J. R., Uphaus, R. A., Crespi, H. L., and Katz, J. J., *Proc. Natl. Acad. Sci. U.S.A.,* 68, 625, 1971.
148. Katz, J. J. and Norris, J., *Curr. Top. Bioenerg.,* 5, 41, 1973.
149. Warden, J. T. and Bolton, J. R., *Photochem. Photobiol.,* 20, 251, 1974.
150. Ke, B., Sugahara, K., and Shaw, E. R., *Biochim. Biophys. Acta,* 408, 1975.
151. Knaff, D. B. and Malkin, R., *Arch. Biochem. Biophys.,* 159, 555, 1973.
152. Evans, M. C. W., Sihra, C. K., and Slabas, A. R., *Biochem. J.,* 162, 75, 1977.
153. Shuvalov, V. A., Klimov, V. V., and Krasnovskii, A. A., *Mol. Biol.,* 10, 326, 1976.
154. Haehnel, W., Döring, G., and Witt, H. T., *Z. Naturforsch.,* 26b, 1171, 1971.
155. Bouges-Bocquet, B. and Delosme, R., *FEBS Lett.,* 94, 100, 1978.
156. Sauer, K., Mathis, P., Acker, S., and van Best, J. A., *Biochim. Biophys. Acta,* 503, 120, 1978.
157. Shuvalov, V. A., Dolan, E., and Ke, B., *Proc. Natl. Acad. Sci. U.S.A.,* in press.
158. Shuvalov, V. A., Ke, B., and Dolan, E., *FEBS Lett.,* submitted.
159. Evans, M. C. W., Sihra, C. K., and Cammack, R., *Biochem. J.,* 158, 71, 1976.
160. McIntosh, A. R. and Bolton, J. R., *Biochim. Biophys. Acta,* 430, 555, 1976.
161. McIntosh, A. R., Chu, M., and Bolton, J. R., *Biochim. Biophys. Acta,* 376, 308, 1975.
162. Evans, M. C. W., Sihra, C. K., Bolton, J. R., and Cammack, R., *Nature (London),* 256, 668, 1975.
163. Goldbeck, J. H., Velthuys, B. R., and Kok, B., *Biochim. Biophys. Acta,* 504, 226, 1978.
164. Ke, B., Dolan, E., Sugahara, K., Hawkridge, F. M., Demeter, S., and Shaw, E. R., *Plant Cell Physiol.,* 184, 1977.
165. Shuvalov, V. A., *Biochim. Biophys. Acta,* 430, 113, 1976.
166. Hiyama, T. and Ke, B., *Proc. Natl. Acad. Sci. U.S.A.,* 68, 1010, 1971.
167. Ke, B., *Biochim. Biophys. Acta.,* 301, 1, 1973.
168. Malkin, R. and Bearden, A. J., *Proc. Natl. Acad. Sci. U.S.A.,* 68, 16, 1971.
169. Ke, B., Hansen, R. E., and Beinert, M., *Proc. Natl. Acad. Sci. U.S.A.,* 70, 2941, 1973.
170. Evans, M. C. W., Reeves, S. G., and Cammack, R., *FEBS Lett.,* 49, 111, 1974.
171. Cammack, R. and Evans, M. C. W., *Biochem. Biophys. Res. Commun.,* 67, 544, 1975.
172. Dismukes, G. C., McGuire, A., Blankenship, R. E., and Sauer, K., *Biophys. J.,* 21, in press.
173. Dismukes, G. C., Friesner, R., and Sauer, K., *Biophys. J.,* 17 (Abstr. 228a), 228, 1977.
174. Duysens, L. N. M., van Grondelle, R., and del Valle-Tascón, S., *Proc. 4th Int. Congr. on Photosynthesis,* Hall, D. O., Coombs, J., and Goodwin, T. W., Eds., The Biochemical Society, London, 1977, 173.
175. Ross, R. T. and Calvin, M., *Biophys. J.,* 7, 595, 1967.
176. Knox, R. S., *Biophys. J.,* 9, 1351, 1969.
177. Almgren, M., *Photochem. Photobiol.,* 27, 603, 1978.
178. Moore, W. J., *Physical Chemistry,* Longman Group Ltd., London, 1972.
179. Duysens, L. N. M., *Brookhaven Symp. Biol.,* 11, 10, 1959.

QUANTUM REQUIREMENT OF PHOTOSYNTHETIC BACTERIA AND PLANTS

S. B. Ku and G. E. Edwards

INTRODUCTION

With present concern over energy resources, it is appropriate to summarize data on the efficiency of plants and photosynthetic bacteria in converting solar energy into chemical energy (the food and fiber on which mankind is dependent). This section covers the efficiency of utilization of absorbed energy by photosynthetic organisms. This can be expressed as the quantum requirement (QR), and for photosynthesis *per se*, it is the number of quanta required per molecule of O_2 evolved or CO_2 assimilated. (An alternate expression is the quantum efficiency which is the reciprocal of the QR, i.e., the number of O_2 molecules evolved or CO_2 fixed per quantum of light absorbed). It is logical to consider utilization of solar energy on a quantum basis, since in the chloroplast's photochemical reactions, a single quantum is absorbed by a single chlorophyll molecule exciting a single electron.

Broadly speaking, the QR for photosynthetic processes is dependent on the energy needed for converting inorganic matter (i.e., CO_2, N_2, NO_3^-, $SO_4^=$) into organic matter (i.e., carbohydrates, proteins, and lipids). Generally, the QR for photosynthesis is expressed on the basis of CO_2 assimilation or O_2 evolution because CO_2 fixation usually accounts for the major input of energy, and either CO_2 uptake or O_2 evolution can be readily measured. However, it is obvious that the amount of energy required per unit of CO_2 assimilated will be much higher, for example, if the organism is fixing N_2 or if a large percentage of the photosynthate is protein rather than carbohydrate.

The overall efficiency of solar energy utilization by a photosynthetic organism depends on the percentage of incident light absorbed and on the organism's QR for photosynthesis. The efficiency of absorption of incident light depends on several factors. The chlorophyll of plants absorbs light between 400 to 700 nm (as determined from an absorption spectrum). The action spectrum for photosynthesis (rate of assimilation per unit of incident light at varying wavelengths) also indicates that plants utilize solar energy in this part of the spectrum. Thus, the portion of the solar energy falling outside the 400- to 700-nm part of the spectrum is not available for plant photosynthesis (photosynthetic bacteria have pigments which absorb at higher wavelengths). In addition, some losses of radiant energy will occur by transmittance and reflectance. On the average, leaves of higher plants absorb about 80%, reflect about 10%, and transmit about 10% of the available energy between 400 to 700 nm. In a natural stand, the leaf angle of the plant also plays an important role in determining utilization of incident energy.

THEORY OF ENERGY REQUIREMENT FOR PHOTOSYNTHESIS

The simplest abbreviation for the photosynthetic process in plants is given by the equation: $CO_2 + 2H_2O \rightarrow O_2 + [CH_2O] + H_2O$, requiring about 110 kcal of energy per mole of CO_2 assimilated[2,3] (see Table 1). The amount of energy per quantum of light between 400 and 700 nm ranges from 70 to 40 kcal/mol, respectively (see Figure 1). Theoretically, 2 quanta of light (on average about 55 kcal/mol between 400 to 700 nm) have an energy content equivalent to that gained during CO_2 fixation (2 × 55 kcal = 110 kcal). In photosynthetic bacteria, the energy gained during photosynthesis de-

Table 1
CALCULATED AND OBSERVED QR FOR PHOTOSYNTHESIS (IN THE CONVERSION OF 1 MOL OF CO_2 TO 1 MOL OF CARBOHYDRATE) BY HIGHER PLANTS

Method of calculation	Energy stored/mol[2,3] (kcal)	QR Calculated[a]	QR Observed
Δ in bond energies of reactants and products	∿105	1.91	8—20
Δ in free energy (ΔF)	∿120	2.18	8—20
Oxidation-reduction gradient per 4 electrons transferred	∿110	2.00	8—20

[a] Calculated on the basis of an average 55 kcal/quantum between 400 to 700 nm.

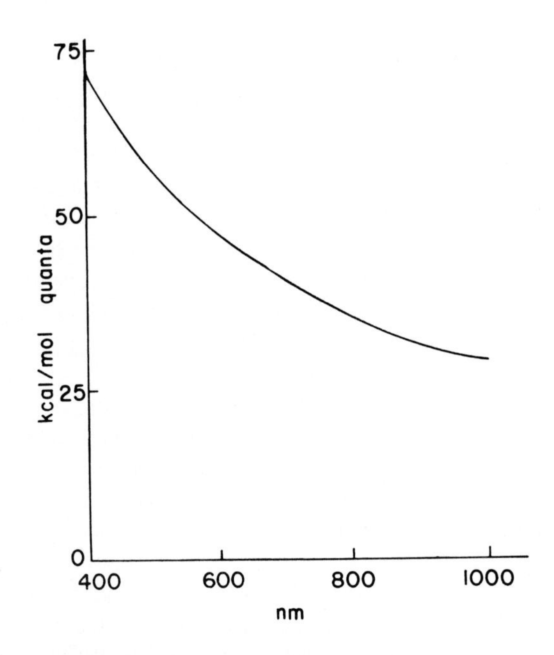

FIGURE 1. Energy content per mole of quanta at varying wavelengths of radiation (400 to 1000 nm).

pends upon the nature of the accessory hydrogen donor[1] (see Table 2). In this case, calculated values for the quantum requirement range from 0.04 to 0.8. In fact, the measured QR for photosynthesis by plants is on the order of four- to fivefold higher than the calculated value and is even greater in photosynthetic bacteria. Thus, the efficiency of conversion of absorbed energy into chemical energy in plants is about 20 to 25%. For example, a QR of 10 for converting 1 mol of CO_2 to 1 mol of carbohydrate would have an efficiency of 20% [10 mol of quanta is on average about 550 kcal (400 to 700 nm), 1 mol carbohydrate = 110 kcal, therefore 110/550 (100) = 20%].

Table 2
CALCULATED AND OBSERVED QR FOR PHOTOSYNTHESIS BY BACTERIA WITH VARIOUS ACCESSORY ELECTRON DONORS

Method of calculation	Electron donor	Calculated energy requirement (kcal)	QR	
			Calculated[a]	Observed
Δ in free energy (ΔF)	H_2	1.7	0.04	8—15
	S_2O_3	29.2	0.75	8—15
	S_4O_6	30.2	0.77	8—15

[a] Calculated on the basis of 39 kcal/quantum in the region of 870 nm which sensitizes the bacteria chlorophyll.[1]

QUANTUM REQUIREMENT FOR GENERATION OF ASSIMILATORY POWER: ATP AND NAD(P)H

Energy in the form of ATP and reduced pyridine nucleotides (NADPH in plants, NADH in photosynthetic bacteria) is generated through photochemical reactions driven by the absorption of light. Their energy is then used in metabolism of inorganic matter to organic matter (an energy requiring process). Thus, the quantum requirement for photosynthesis can be considered in two phases: (1) the quanta needed for generating assimilatory power (QR/ATP and QR/NAD(P)H and (2) the amount of assimilatory power used per CO_2 fixed.

Plants

In algae and higher plants, the series scheme (Z scheme) with two energy-trapping pigment systems (Photosystem I and II) is the commonly accepted means of reducing pyridine nucleotides. As illustrated in Figure 2, 8 quanta are required for the splitting of $2H_2O$ to yield $1O_2$ and 2NADPH. ATP synthesis is powered by a protonmotive force which is generated at certain sites in the electron transport chain. At each of two proton generating sites in the scheme, one electron gives an equivalent of two protons such that four electrons passing through the chain generate eight protons inside the thylakoid compartment of the chloroplasts. As protons are transported through the coupling factor on the thylakoid membrane, ATP is synthesized (see Figure 3). The stoichiometry of H^+/ATP is uncertain with measurements ranging between $4H^+$/ATP to $2H^+$/ATP. Thus, the most efficient coupling ($2H^+$/ATP) would result in 4ATP synthesized as four electrons pass through the Z scheme to reduce 2NADP with a QR of 8. (For sample calculations in other sections, a H^+/ATP ratio of two is used for both plants and photosynthetic bacteria.) This is theoretically enough energy to fix $1CO_2$ in the reductive pentose phosphate (RPP) pathway (3ATP, 2NADPH/CO_2 fixed), although the measured QR is often higher. In some circumstances, additional ATP may be required, and this could come from pseudocyclic electron flow (similar to noncyclic, except O_2 is the electron acceptor rather than pyridine nucleotides) or cyclic electron flow around Photosystem I (see Figures 4 and 5).

In cyclic electron flow in plants, the reaction chain between Fd^- and P_{700}^+ spans about 0.8 V/e^-. This is sufficient energy to synthesize two ATP/e^- or two ATP per quantum where ADP + Pi → ATP has an oxidation-reduction potential, $E'o = +0.3$ V. The highest efficiency suggested for cyclic photophosphorylation is generally 1ATP/quantum which according to the chemiosmotic hypothesis would require two proton transporting sites (see Figure 5). In particular, this has been concluded from

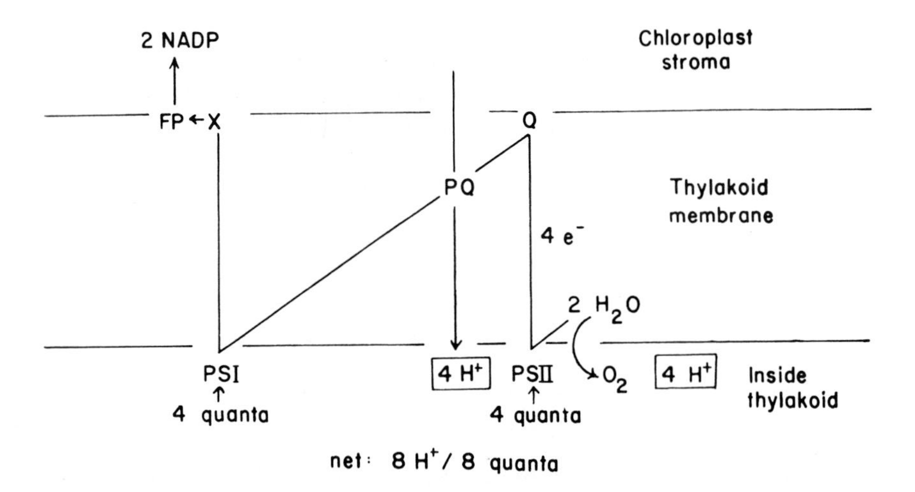

FIGURE 2. Illustration of stoichiometry of quanta absorbed, electrons transported, protons generated, and NADP reduced in the Z scheme for plant photosynthesis.

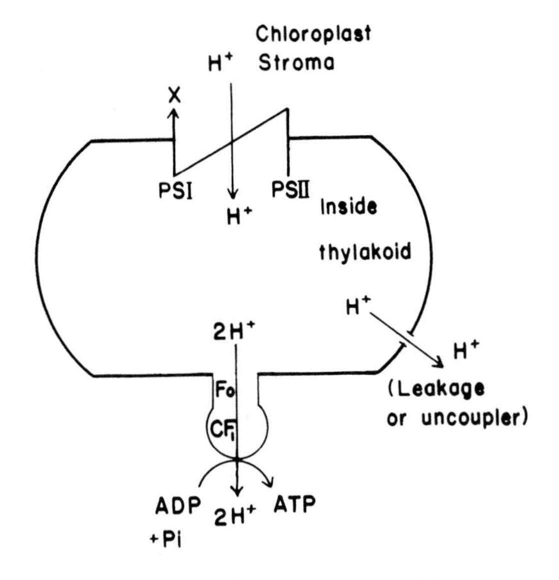

FIGURE 3. Generation and utilization of the proton-motive force in the thylakoid compartment of plants.

experimentation on cyclic photophosphorylation with photosynthetic bacteria.[4] Considering the general lack of conclusive evidence in this area, we have chosen to use the lowest expected QR in vivo for the various types of photophosphorylation (1ATP/quantum in cyclic; 1ATP/2 quanta in pseudocyclic and noncyclic).

As illustrated, in noncyclic and pseudocyclic photophosphorylation, 1 quantum results in the generation of 1 proton inside the thylakoids and with efficient coupling, 1 ATP would be generated per two quanta. Accordingly, the QR for noncyclic or pseudocyclic per ATP synthesized would be twice that for cyclic photophosphorylation. With isolated chloroplasts, Chain and Arnon[5] found the QR for noncyclic photophosphorylation to be twice that for cyclic photophosphorylation. This is consistent with the above models, although their efficiencies with the isolated chloroplasts are lower.

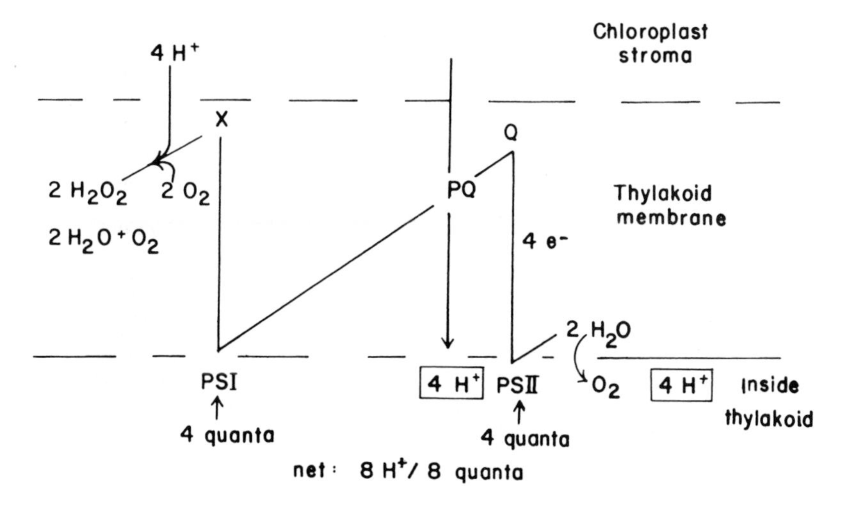

FIGURE 4. Relationship between quantum absorption and proton generation in the thylakoid compartment during pseudocyclic electron transport.

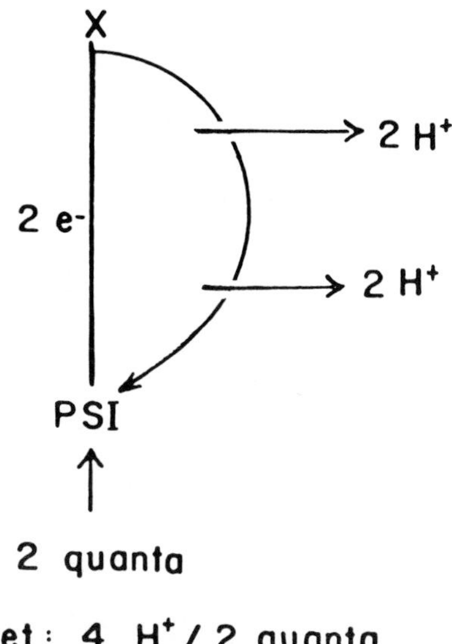

FIGURE 5. Illustration of quantum absorption and cyclic electron flow through Photosystem I in plant chloroplasts with two proton loops or coupling sites.

Photosynthetic Bacteria

In some reports, the QR for generation of assimilatory power in photosynthetic bacteria is less certain than that of higher plants. A number of alternative electron transport schemes have been proposed. One common feature of photosynthetic bacteria is their generation of ATP through cyclic photophosphorylation. With two coupling sites in the cyclic pathway, 1ATP would be synthesized per quantum as illustrated for plants in Figure 5.

Unlike higher plants, photosynthetic bacteria use reductants other than water (e.g.,

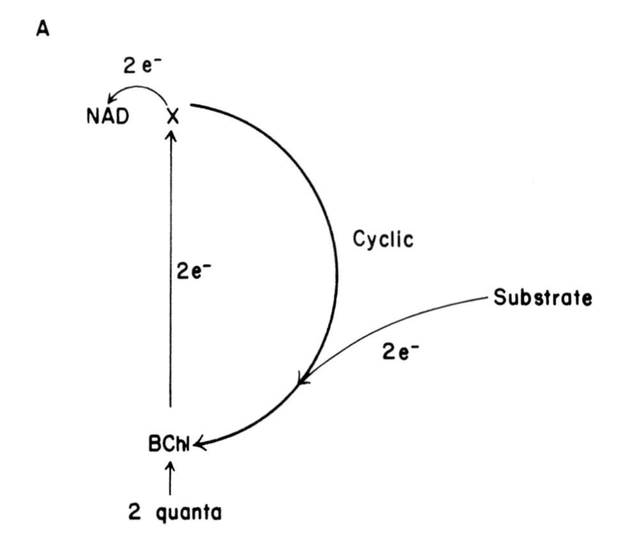

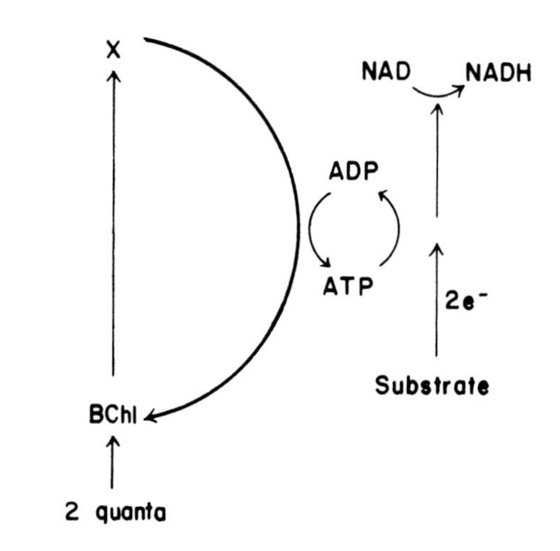

FIGURE 6. Schemes for the photochemical reduction of NAD in photosynthetic bacteria. (A) Substitute is oxidized; electrons are donated into the photosystem and NAD is reduced through noncyclic electron flow. (B) Substrate oxidation and NAD reduction is driven by ATP generated photochemically from cyclic photophosphorylation. Electrons from the substrate are not donated into the photosystem.

H_2S, thiosulfate, tetrathionate, or organic acids). During oxidation of these reductants, NAD is reduced. However, there is a QR during this oxidation-reduction process, and several hypotheses exist for electron transport pathways. For example, donation of electrons into the photosynthetic reaction center (RC) could result in noncyclic electron flow to NAD (see Figure 6A). Alternatively, reduction of NAD by the donor may occur through a series of carriers in an uphill manner, driven by ATP or a high energy intermediate (see Figure 6B). The QR in these various options are somewhat speculative. However, an example can be given for the purple bacterium *Chromatium*.[4] Assume 3ATP and 2NADH are required for CO_2 fixation in the RPP pathway. Cyclic

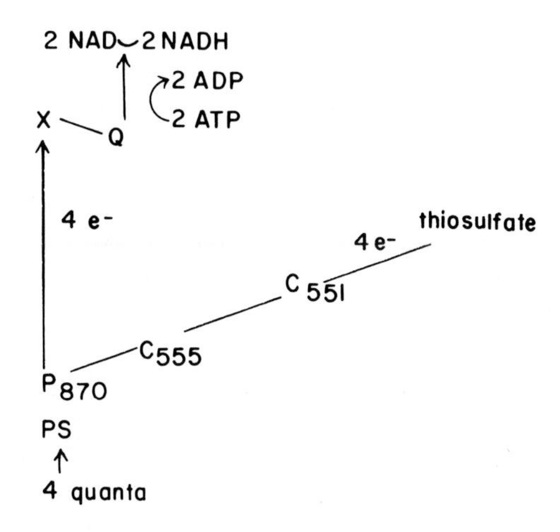

FIGURE 7. Suggested means of NAD reduction in the photosynthetic bacterium *Chromatium* with thiosulfate as the donor.[4]

photophosphorylation is suggested to have two coupling sites, giving 1 quantum per ATP or a requirement of 3 quanta for 3ATP. For NAD reduction, thiosulfate would serve as the electron donor through the pigment system (see Figure 7). The reduction of 2NAD would then require 6 quanta (4 quanta absorbed by the pigment system per 4 electrons transported and 2 quanta required in cyclic photophosphorylation for generating 2ATP). Thus, altogether the QR would be 9 per CO_2 fixed. Variations on this requirement could occur with only one coupling site in the cyclic path ($1H^+/e^-$) or if the pigment system generated a strong enough reductant in the noncyclic path such that additional ATP were not required for reducing the NAD (i.e., QR would then range from 7 to 14).

REQUIREMENT FOR ASSIMILATORY POWER

Plants assimilate CO_2 by the RPP pathway which requires 3ATP and 2NAD(P)H per CO_2 fixed to the level of triose phosphate. Photosynthetic bacteria also assimilate CO_2 by this cycle. However, *Chromatium* and *Chlorobium* have in addition a reductive carboxylic-acid cycle for CO_2 fixation resembling a reversal of the TCA cycle.[6] Carbon can be taken out of the cycle as C_2, C_3, C_4, C_5, or C_6 compounds, with amino acids as the primary products. Per carbon fixed, the energy requirement appears to be lower than in the RPP pathway. For example, fixation of $3CO_2$ in the cycle to phosphoenolpyruvate requires 3ATP and the equivalent of 5NAD(P)H, whereas in the RPP pathway, fixation of $3CO_2$ to triose phosphate requires 9ATP and 6NADPH.

In practice, the energy requirement per CO_2 fixed through the RPP pathway is often higher than the theoretical 3ATP, 2NAD(P)H. Higher plants are divided into three photosynthetic groups based on differences in the mechanism of carbon assimilation which are termed C_3, C_4, and CAM (Crassulacean acid metabolism). The initial product of atmospheric CO_2 fixation in C_3 plants is the 3-carbon compound, 3-phosphoglycerate; while in C_4 plants, it is 4-carbon compounds, oxaloacetate, malate, and aspartate. CAM plants are named after the family Crassulaceae in which this type of photosynthetic metabolism was first identified.

In C_3 plants, atmospheric CO_2 is fixed directly into the RPP pathway via the enzyme riboluse 1,5-bisphophate carboxylase. In these species, oxygen inhibits photosynthesis and increases the QR. In 1966, Bjorkman[7] showed that O_2 inhibited the quantum efficiency under low light to the same extent that it inhibits photosynthesis under high light. It is now known that O_2 reacts with RuBP, a metabolite of the RPP pathway. This occurs in a competitive manner with CO_2 through the enzyme RuBP carboxylase-oxygenase. When O_2 reacts, the carboxylation step is bypassed. The products of the reaction are 3-phosphoglycerate and phosphoglycolate. Phosphoglycolate is metabolized in the glycolate pathway, resulting in respirating loss of CO_2 and synthesis of PGA. In this O_2-dependent metabolism, most of the carbon is returned to the cycle as 3-phosphoglycerate.

Under atmospheric levels of O_2 (21%) and CO_2 at 30°C with C_3 plants, there is about 30% inhibition of photosynthesis by O_2 and an equivalent inhibition of quantum yield (Calculated from measurements at 2 vs. 21% O_2). Calculations based on current knowledge of photosynthetic metabolism under O_2 in C_3 plants indicates there is a similar increase in the requirement for ATP and NADPH/CO_2 fixed. Furthermore, the percentage inhibition of quantum yield can be accounted for by such analysis.[2,8] Thus, when O_2 causes a 30% inhibition of quantum yield, this can be conveniently estimated as follows: 3ATP, 2NADPH/0.7 CO_2 fixed, or 4.3ATP, 2.9NADPH/CO_2 fixed. This is 43% higher than the theoretical requirements of the RPP pathway, changing the QR from the theoretical value of 8 to about 11.

In C_4 plants, atmospheric CO_2 is fixed in the C_4 cycle and donated to the RPP pathway. The C_4 cycle serves to increase CO_2 from a low concentration in the atmosphere to a high concentration in the leaf at the site of its fixation in the RPP pathway. This results in a high rate of photosynthesis as CO_2 is no longer a rate-limiting factor, and O_2 inhibition is repressed (apparently by increasing the CO_2/O_2 ratio). This CO_2 concentrating mechanism is driven at the expense of 2ATP/CO_2, and the total energy requirements for CO_2 fixation through the RPP pathway increases to 5ATP, 2NADPH/CO_2 fixed. This would increase the QR from a theoretical value of 8 to 9 or 10 (depending on whether cyclic or pseudocyclic supplies the additional ATP) for C_4 photosynthesis.

CAM plants fix CO_2 at night into malic acid, and during the day carbon is donated from malate to the RPP pathway. This adaption allows the stomata to be open during the night and closed during the day, thus serving to conserve water. It also occurs at the expense of additional ATP, although actual measurements of the QR in these species are generally not available and interpretation of such may be difficult.[8]

The reasons for the increased requirements for assimilatory power for CO_2 fixation in these three photosynthetic groups are quite different. However, in each case, analysis of requirements for photosynthetic metabolism suggests the QR is similar.

The total requirements for assimilatory power in photosynthetic organisms is naturally higher than calculations based on CO_2 fixation through the RPP pathway. The products of this pathway, triose phosphate, are further metabolized to make up major cellular constituents of various carbohydrates, lipids, and proteins. Energy generated photochemically and from respiration in plants is used in uncertain proportions for synthesizing these macromolecules. Penning de Vries et al.[9] suggest that with energy provided from plant respiration, glucose can be used to synthesize 0.83 g of carbohydrates, 0.4 g of proteins (nitrate as nitrogen source), or 0.33 g of lipid. Obviously, the higher the proportion of carbon flowing into protein or lipid synthesis, the greater will be the QR/CO_2 assimilated.

Per CO_2 fixed in the RPP pathway, 3 additional ATP may be required to synthesize cellular material in the form of carbohydrate.[10] If cyclic or pseudocyclic photophosphorylation generates this additional ATP, this would require 3 (cyclic) to 6 (pseudocyclic) additional quanta. Thus, a QR of 8 to 11 for CO_2 fixation through the RPP pathway (depending on species and environmental conditions) could increase to a value of 11 to 17.

One half of the dry matter of algae can be in the form of protein giving a carbon to nitrogen ratio of about 7:1 (atom to atom).[11] Algae using nitrate as the nitrogen source are estimated to require 2.8 reducing equivalents (NADPH) and 5ATP per CO_2 fixed.[12]

Algal photosynthesis under high CO_2 (a condition generally used in measurements of the QR in algae) would theoretically require 3ATP, 2NADPH per CO_2 fixed in the RPP pathway with a QR of 8. However, when assimilating nitrate into protein, the QR per CO_2 fixed should increase (theoretically $4 \times 2.8 = 11.2$), and values of 12 for algae have been measured. When assimilating nitrate, the photosynthetic quotient (CO_2 fixed per O_2 evolved) is lowered, and values of 0.7 have been measured in algae.[13] This could result in an increased QR per CO_2 fixed without effecting the QR per O_2 evolved. Thus, the basis of measurement is important when the photosynthetic quotient deviates substantially from the value of 1.0. In leaves of higher plants, the nitrogen content is relatively low (roughly C/N = 16/1) and would have less influence on the QR per CO_2 assimilated than in algae. Energy requirements for transport of ions or metabolites in both plants and photosynthetic bacteria are uncertain although this will obviously influence the QR.

MEASUREMENTS OF QUANTUM REQUIREMENT

Photosynthetic Bacteria

The measured values of QR for photosynthesis by bacteria range from 8 to 16, with an average minimum QR of approximately 9 (see Table 3). The metabolic conversions of organic compound during growth and thus the overall energy balance of photosynthetic bacteria is closely related to the nature and oxidation-reduction level of the accessory hydrogen donor. Theoretically, the QR for CO_2 reduction by photosynthetic bacteria would be influenced by the nature of the accessory hydrogen donor. However, Larsen et al. (cited in Reference 1) (see Table 3) found that in *Chlorobium thiosulfertophilum*, the QR for photochemical CO_2 assimilation with different inorganic accessory hydrogen donors were similar. These results have been interpreted that ATP cannot be derived from transfer of electrons from the accessory donor to the oxidized product of photochemical reaction and that the accessory hydrogen donor cannot serve as a source of reducing power for CO_2 reduction (see Reference 1). With inorganic hydrogen donors, the QR can be measured per CO_2 assimilated. Organic hydrogen donors may serve both as an electron donor and source of carbon. Thus, carbon from the donor can compete with CO_2 in utilizing energy during carbon assimilation. In this case, measurement of the QR per carbon assimilated is more complicated. The QR for photophosphorylation and NAD reduction and interaction between the two with various hydrogen donors needs further investigation. The nitrogen sources used for bacterial growth have a profound effect on metabolic conversions and subsequently on the QR for photosynthesis. Under some conditions, photosynthetic bacteria may produce H_2 which must also be accounted for in consideration of energetics.

Higher Plants — Isolated Chloroplasts

A direct means of getting a measure of the QR of photosynthesis is with isolated chloroplasts. With broken chloroplasts (isolated thylakoids), the QR can be measured for ATP synthesis and NADP reduction. With isolated intact chloroplasts, the QR for CO_2 fixation or reduction of other substrates can be measured. This has the distinct advantage of eliminating certain problems encountered in measuring the QR with whole cells. With cells, the measurement of QR will be influenced by the amount of respiratory consumption of O_2. The QR for O_2 evolution or CO_2 consumption by cells is not a precise measure of QR for storage of chemical energy because differences will

Table 3
QR FOR PHOTOSYNTHESIS BY BACTERIA

Species	Hydrogen donor	Experimental environment conditions: Light intensity	Temperature (°C)	CO_2 (%)	O_2 (%)	Method of measurement	QR	Ref.[a]
Streptococcus varians	Hydrogen	Limiting IR light intensities	25	5	0	Manometric[b]	∿12	French
Spirillum rubrum	Butyrate	Limiting IR light intensities	25	5	0	Manometric	∿15	French
Chromatium	Thiosulfate	White light	29	5	0	Manometric	Optimal 10—12	Eymers and Wassink
Chromatium	Hydrogen	High white light intensities	19	5	0	Manometric	12.8	Wassink et al.
			29				8.5—13.8	
	Thiosulfate		22				12.1	
			29				8.5—16.0	
Chlorobium thiosulfatophilum	Hydrogen	Limiting IR light intensities	—	2—5	0	Manometric	8.2—9.3	Larsen et al.
	Thiosulfate						8.9—10.0	
	Tetrathionate						8.9—9.4	

[a] See Gest and Kaman for citation.[14]
[b] Manometric, measure of ΔO_2.

Table 4
MINIMAL QRs OBTAINED WITH ISOLATED SPINACH (C_3) CHLOROPLASTS USING VARIOUS SUBSTRATES

Preparation	Substrates	Energy required or generated	QR	Ref.
Intact chloroplasts	HCO_3^- [a]	3ATP, 2NADPH	12[b]	15
Intact chloroplasts	3 Phosphoglycerate	1ATP, 1NADPH	4	15
Intact chloroplasts	Oxaloacetate	1NADPH	4	15
Broken chloroplasts	ADP, cyclic	ATP	2[c]	5
Broken chloroplasts	ADP, noncyclic	ATP	4[d]	5
Broken chloroplasts	ADP, NADP	3ATP, 2NADPH	12	5

[a] The bicarbonate concentration was 4 m*M*; O_2 concentration was approximately 100 μ*M*.
[b] Average values obtained in the presence of 3 m*M* dithiothreitol. In the absence of dithiothreitol, the QR was higher (16 to 30), with bicarbonate as the substrate.
[c] Measured with 715-nm light.
[d] Measured with 554-nm light.

exist in the cellular composition of organisms and their energy requirements for biosynthesis and the degree to which other inorganic material is reduced (e.g., nitrate, free nitrogen, sulfite, and sulfate).

With isolated C_3 chloroplasts, the primary product of CO_2 fixation is triose phosphate, and under saturating bicarbonate concentrations, O_2 inhibition of the quantum yield will be repressed such that the requirements are 3ATP, 2NADPH per CO_2 fixed. With isolated spinach chloroplasts, Heber[15,16] obtained a QR value of 12 per CO_2 fixed (see Table 4). Chain and Arnon[5] found a requirement of 2 quanta per ATP synthesized in cyclic and 4 quanta per ATP synthesized in noncyclic electron flow. Overall, the QR for generating 3ATP, 2NADPH with isolated chloroplasts was 12, similar to that for CO_2 fixation by whole chloroplasts (see Table 4). If these measurements represent the true in vivo QR for generating assimilatory power, obviously the QR in vivo per CO_2 fixed would be higher (considering other cellular synthetic activities and respiratory losses).

However, the QR for generating a given amount of ATP may be lower in vivo than with isolated chloroplasts. The reason for this is photophosphorylation may not be as tightly coupled in isolated chloroplasts as in vivo. One interpretation is chloroplasts can become loosely coupled during isolation (presumably because of induced leakiness of the thylakoid membrane to protons, see Figure 3) whereby H^+ transport per ATP synthesized increases. This could also cause a relatively high basal rate of photosynthetic O_2 evolution occurring with oxaloacetate in the absence of an uncoupler. Others would suggest this could represent the in vivo state of coupling. In any case, it is possible that 8 quanta could generate 3ATP, 2NADPH in vivo.

Higher Plants — Leaves

QRs for whole leaf photosynthesis obtained from earlier measurements (up to 1949) range from 10.8 to 43.5, with a minimum QR of approximately 11 to 16 (see Table 5). In retrospect, the plant materials employed in these studies were all C_3 species. Since high CO_2 (0.5 to 9%) concentrations were used in these measurements, except that by Gabrielson (1947), the QR obtained must be the values for true photosynthesis (i.e., O_2 inhibition prevented by high CO_2 and respiration generally accounted for in calculations).[8]

Table 5
QR FOR PHOTOSYNTHESIS BY HIGHER PLANTS

Plant	Photosynthesis pathway	Experimental environment conditions: Light intensity	Temperature (°C)	CO_2	O_2 (%)	Method of Measurement[a]	QR	Ref.
Wolffiella lingulata leaf	C_3[b]	Sodium lamps, 45 × 10^{-9} Einstein/ $cm^2 \cdot min$	20	5%	Air	CO_2 exchange	16.6	Emerson and Lewis[c]
Strawberry leaf disc	C_3	Sodium lamps up to 104 $erg/cm^2 \cdot sec$	17 and 25	1—9%	Air	Manometric	12.1—17.8	Wassink,[d]
Asparagus leaf disc							14.6—18.2	
Kohlrabi leaf disc							13.5—15.4	
Chinese cabbage leaf disc							10.8—14.7	
White saccory leaf disc							13.6—16.8	
Tomato leaf disc							16.0—43.5	
Cucumber leaf disc							13.8—24.6	
Endivia leaf disc							21.8	
sinapis alba leaf	C_3	40 $cal/dm^2 \cdot hr$	—	Air	Air	CO_2 exchange	~13.2	Gabrielson,[d]
Corylus maxima leaf							~13.9	
Fraxinus excelsior leaf							~13.2	
Avocado leaf	C_3	Limiting intensities at 660 nm	—	Carbonate buffer	—	Calorimetric	15.6	Arnold,[c]
Mimulus cardinalis	C_3	Limiting intensities	22	280—320 ppm	21	IR CO_2 analysis	~17	Björkman et al.[d]
Radish leaf	C_3	4 × $10^4 erg/cm^2 \cdot sec$	~24	300 ppm	2 and 21	IR CO_2 analysis	For radish, action spectrum had 40% higher	Bulley et al.[d]

							activity in 2% O_2	
Corn leaf	C_4[b]							
Solanum dulcamara	C_3	Limiting intensities at 665 nm	24	300 ppm	21	IR CO_2 analysis		17
Shaded clone							15	
Exposed clone							15	
Bean Leaf	C_3	1.98 μ Einstein/ $cm^2 \cdot sec$ (400—700 nm)	—	Air	Air	IR CO_2 analysis	$\sim$12	Balegh and Biddulph[d]
Atriplex patula leaf	C_3	Limiting intensities	25	310—320 ppm	1.2	IR CO_2 analysis	13.7	Björkman et al.[d]
					21		19.5	
Atriplex rosea leaf	C_4	Limiting intensities	25	310—320 ppm	1.2	IR CO_2 analysis	18.7	
					21		18.4	
A. patula leaf	C_3	Limiting intensities	26	310 ppm	1.5	IR CO_2 analysis	12.5	Björkman[d]
				770 ppm	1.5		12.5	
				320 ppm	21		18.0	
				720 ppm	21		15.6	
Oat leaf	C_3	Limiting intensities (400—700 nm)	11	330—370 ppm	21	IR CO_2 analysis	$\sim$18.5	McCree[d]
			28				$\sim$21.2	
			38				$\sim$23.4	
Sugarbeet leaf	C_3	Limiting intensities (400—700 nm)	28	200 ppm	21	IR CO_2 analysis	$\sim$30.3	
				350 ppm			$\sim$20.5	
				600 ppm			$\sim$15.4	
Corn leaf	C_4	Limiting intensities (400—700 nm)	28	$\sim$350 ppm	21	IR CO_2 analysis		
Young							$\sim$15.4	
Average							$\sim$16.0	
Old							$\sim$19.7	
A. patula	C_3	Limiting intensities at 654 nm	25	320 ppm	21	IR CO_2 analysis		18
Grown at high light							12.4	
Grown at intermediate light							12.4	
Grown at low light							12.4	

Table 5 (continued)
QR FOR PHOTOSYNTHESIS BY HIGHER PLANTS

Plant	Photosynthesis pathway	Experimental environment conditions: Light intensity	Temperature (°C)	CO_2	O_2 (%)	Method of Measurement[a]	QR	Ref.
Ferocactus acanthodes	CAM[b]	1 to 2 μ Einstein/cm^2	—	Air	Air	IR CO_2 analysis	68 For photosynthesis at night	Nobel[d]
Leaves of 7 species	C_3	Limiting intensities (400—700 nm)	30	325 ppm	21	IR CO_2 analysis	~19	Ehleringer and Björkman[d]
			30		2		~13.6	
Leaves of 5 species	C_4	Limiting intensities (400—700 ppm)	30	325 ppm	21	IR CO_2 analysis	~18.7	
			30		2		~18.5	
Encelia california leaf	C_3	Limiting intensities (400—700 nm)	14	325 ppm	21	IR CO_2 analysis	14.5	
			30		21		19.2	
			39		21		23.8	
A. rosea leaf	C_4	Limiting intensities (400—700 nm)	12—39	325 ppm	21	IR CO_2 analysis	Constant ~18.8	
Triticum aestivum leaf	C_3	Limiting intensities (400—700 nm)	15	315 ppm (external)	21	IR CO_2 analysis, polarographic O_2 analysis	16.4	19
			20				17.0	
			25				17.7	
			30				18.7	
			35				21.3	
T. aestivum leaf	C_3	Limiting intensities (400—700 nm)	15	Increasing CO_2 with temperature (constant sol-	21	IR CO_2 analysis, polarographic O_2 analysis	15.9	
			20				16.1	
			25				16.5	
			30				16.7	

			35	ubility ratio of O_2/CO_2 in the leaf)			17.5
T. asetivum leaf	C_3	Limiting intensities (400—700 nm)	15	315 ppm (external)		IR CO_2 analysis, polarographic O_2 analysis	15.6
			20	Decreasing O_2 with temperature (constant solubility ration of O_2/CO_2 in the leaf)			16.3
			25				16.4
			30				17.0
			35				17.3
T. asetivum leaf	C_3	Limiting intensities (400—700 nm)	15—35	315 ppm (external)	1.5	IR CO_2 analysis	Constant, ~12.1
Zea mays leaf	C_4	Limiting intensities	15—35	315 ppm	1.5 and 21	IR CO_2 analysis	Constant, ~16.8

[a] Manometric measure of ΔO_2; calorimetric, measure of Δheat; polargraphic, measure of ΔO_2; IR CO_2 analysis, measure of ΔCO_2.
[b] C_3, plants having the reductive pentose phosphate pathway of photosynthesis; C_4, plants having the C_4 pathway and reductive pentose phosphate pathway of photosynthesis; CAM, plants having the Crassulasean acid metabolism.
[c] See Kok for citation.[1]
[d] See Campbell and Black for citation.[8]

After introduction of IR CO_2 analyzer into plant science research in the 1950s, the study of whole leaf photosynthesis and measurements of the QR could be readily made and with more precision. Under atmospheric conditions (21% O_2, 0.03% CO_2) and at 30°C, the QR for C_3 photosynthesis is around 17. The requirement of quantum energy for photosynthesis by C_3 plants is decreased by increasing the atmospheric level of CO_2, by decreasing the O_2 concentration, and by decreasing the temperature. Under low O_2, the QR for C_3 photosynthesis is reduced to about 12 to 14. Theoretically, the QR of C_3 higher plants under low O_2 (values of 12 to 14) should be comparable to those of algae measured under high CO_2 concentration (average values of 8 to 10). Differences in value could occur, depending on whether CO_2 or O_2 is used for the basis of measurement and the extent of nitrate assimilation. Higher plants might use more energy than algae in transport processes, respiration in heterotrophic tissue of the leaf (epidermal and vascular), and nonspecific absorption of energy by cellular material which could lead to higher QR in the higher plants. Certainly, age of the leaf also exerts a marked effect on QR, usually the QR being lowest with young leaves.

In contrast to C_3 plants, C_4 plants have a QR of about 17 for whole leaf photosynthesis, regardless of O_2, CO_2 concentration, and temperature. The higher QR of C_4 photosynthesis in comparison with that of C_3 photosynthesis under low O_2 is thought to be partly because the extra expense of 2ATP/CO_2 fixed through the operation of the C_4 pathway. Relatively little information is available on the QR for photosynthesis by CAM plants. A value of 68 quanta absorbed the previous day per CO_2 fixed at night by cactus *Ferocactus acanthodes* was reported by Nobel, (cited in Reference 8) 1977 (see Table 5). This value is much higher than that expected for true photosynthesis by CAM plants.

In current practice, the maximum efficiency of converting solar energy into chemical energy in crops is about 3 to 5%. This requires a dense canopy, so that no losses of the photosynthetically active radiation occurs by transmittance. An estimate of the potential daily production in a canopy of maize receiving a total daily solar radiation of 500 cal/cm^2·day is given in Table 6.

With increasing light intensity, the rate of canopy photosynthesis deviates from linearity, particularly in C_3 crops.[18] This reduces the efficiency of utilization of solar energy. As photosynthesis in the upper leaves of the canopy tends to become light saturated, absorbed energy will be lost as heat, resulting in an increased QR. This can occur as CO_2 becomes rate limiting as in C_3 plants and as the level of absorbed energy exceeds the photochemical capacity.

ACKNOWLEDGMENT

The authors appreciate information provided from Drs. W. H. Campbell and C. C. Black prior to publication. This work was supported in part by National Science Foundation Grant PCM77-09384.

Table 6
AN EXAMPLE OF MAXIMUM PHOTOSYNTHETIC EFFICIENCY OF A MAIZE CROP [20,21]

Average daily total solar radiation	500 cal/cm²·day
PAR (400—700 nm)	222 cal/cm²·day
Total quanta available, 400-700 nm (using value of 8.64 μmol of quanta/calorie of total solar radiation	4320 μmol quanta/cm²·day
Loss by reflection (8%)	−360 μmol quanta/cm²·day
Total quanta absorbed	3960 μmol quanta/cm²·day
Assume quantum efficiency = 1CO_2 fixed/17 quanta absorbed = 1 μmol CO_2 fixed/17 μmol quanta. Then μmol CO_2 fixed/3960 μmol quanta =	233 μmol CO_2/cm²·day
Respiratory loss (estimated at about one third of photosynthesis)	78μmol CO_2/cm²·day
Net production of carbohydrate (CO_2 fixed)	155 μmol CO_2/cm²·day
1 mol CO_2 fixed to level of sucrose has 120 kcal energy, or 1 μmol CO_2 converted to sucrose = 0.12 cal. Chemical energy stored/day = 0.12(155) =	18.6 cal/cm²·day
Efficiency of use of total incident radiation =	

$$\frac{18.6\ \text{cal/cm}^2\cdot\text{day}\ (100)}{500\ \text{cal/cm}^2\cdot\text{day}} = 3.7\%$$

REFERENCES

1. **Kok, B.**, Efficiency of photosynthesis, in *Handbook der Pflanzenphysiologie,* Vol. 5 (Part 1), Ruhland, W., Ed., Springer-Verlag, Berlin, 1960, 566.
2. **Edwards, G. E. and Walker, D. A.**, in *C_3, C_4: An Introduction to the Biochemistry of Photosynthesis,* Packard, Sussex, in press.
3. **Govindjee, S. R. and Govindjee, R.**, Introduction to photosynthesis, in *Bioenergetics of Photosynthesis,* Govindjee, S. R., Ed., Academic Press, New York, 1975, 1.
4. **Duysens, L. N. M., Grondelle, R. V., and Valle-Tascón, S. D.**, Electron transport and photophosphorylation associated with primary reactions in purple bacteria., in *Photosynthesis 77,* Proc. 4th Int. Congr. Photosynthesis, Hall, D. O., Coombs, J., and Goodman, T. W., Eds., The Biochemical Society, London, 173, 1978.
5. **Chain, R. K. and Arnon, D. I.**, Quantum efficiency of photosynthetic energy conversion, *Proc. Natl. Acad. Sci. U.S.A.,* 74, 3377, 1977.
6. **Evans, M. C. W., Buchanan, B. B., and Arnon, D. I.**, A new ferredoxin-dependent carbon reduction cycle in a photosynthetic bacterium, *Proc. Natl. Acad. Sci. U.S.A.,* 55, 928, 1966.
7. **Björkman, O.**, The effect of oxygen concentration on photosynthesis in higher plants, *Physiol. Plant.,* 19, 618, 1966.
8. **Campbell, W. H. and Black, C. C.**, The relationship of CO_2 assimilation pathways and photorespiration to the physiological quantum requirement of green plant photosynthesis, *Biosystems,* 10, 253, 1978.
9. **Penning de Vries, F. W. T., Brunsting, A. H. M., and van Laar, H. H.**, Products, requirements, and efficiency of biosynthesis: a quantitative approach, *J. Theor. Biol.,* 45, 339, 1974.
10. **Kok, B.**, Photosynthesis: the path of energy, in *Plant Biochemistry,* 3rd ed., Bonner, J. and Varner, J. E., Eds., Academic Press, New York, 845, 1976.
11. **Samejima, H. and Meyers, J.**, On the heterotrophic growth of *Chlorella pyrenoidosa, J. Gen. Microbiol.,* 18, 107, 1958.
12. **Raven, J. A.**, Division of labour between chloroplast and cytoplasm, in *The Intact Chloroplast,* Barker, J., Ed., Elsevier, Amsterdam, 1976, 403.
13. **Syrett, P. J.**, Nitrogen assimilation, in *Physiology and Biochemistry of Algae,* Lewin, R. A., Ed., Academic Press, New York, 1962, 171.
14. **Gest, H. and Kaman, M. D.**, The photosynthetic bacteria, in *Handbook der Pflanzenphysiologie,* Vol. 5 (Part 2), Ruhland, W., Ed., Springer-Verlag, Berlin, 1960, 568.

15. **Heber, U.**, Stoichiometry of reduction and phosphorylation during illumination of intact chloroplasts, *Biochim. Biophys. Acta,* 140, 305, 1973.
16. **Heber, U.**, Energy coupling in chloroplasts, *J. Bioenerg. Biomembr.,* 8, 157, 1976.
17. **Gauhl, E.**, Differential photosynthetic performance of *Solanum dulcamara* ecotypes from shaded and exposed habitats, *Carnegie Inst. Washington Yearb.,* 67, 482, 1969.
18. **Björkman, O., Boardman, N. K., Anderson, J. M., Thorne, S. W., Goodchild, D. J., and Pyliotis, N. A.**, Effect of light intensity during growth of *Atriplex patula* on the capacity of photosynthetic reactions, chloroplast components and structure, *Carnegie Inst. Washington Yearb.,* 71, 115, 1972.
19. **Ku, S. B. and Edwards, G. E.**, Oxygen inhibition of photosynthesis, III. Temperature dependence of quantum yield and its relation to O_2/CO_2 solubility ratio, *Planta,* 140, 1, 1978.
20. **Loomis, R. S. and Williams, W. A.**, Maximum crop productivity: an estimate, *Crop Sci.,* 3, 67, 1963.
21. **Zelitch, I.**, in *Photosynthesis, Photorespiration and Plant Productivity,* Academic Press, New York, 1971, 279.

EFFICIENCY OF INCIDENT LIGHT UTILIZATION AND QUANTUM REQUIREMENT OF MICROALGAE

H. Senger

INTRODUCTION

The light intensity of the incident light should be the only limiting factor in measurements of quantum yield of photosynthetic reactions. Thus, quantum yield is the best indication of the state of the photochemical part of the photosynthetic apparatus. Since the pioneering work of Warburg and co-workers,[1-3] determinations of the quantum yield and quantum requirement (QR) (the reciprocal value of quantum yield) have been extended from total photosynthesis to partial reactions. In addition, it has been demonstrated that physiological and development states considerably influence the quantum yield.

MEASUREMENTS OF QUANTUM YIELD

Three parameters are necessary for determining the quantum yield: the rate of the photosynthetic reaction, the flux density of incident radiation, and the fraction of radiant energy absorbed. Little has changed in the technique since the first measurements were made by Warburg and Negelein.[1] The most accurate measurement of the flux density of the incident light is still with an actinometer.[4] The introduction of all-glass differential manometers (Gilson Medical Electronics, Madison, Wis.), the specific Eastman white reflectance paint (Distillation Products Industries, Eastman Kodak Company, Rochester, N.Y.) with a constant quantum yield in reflection over the visible spectrum, and the method of moving the light field, together with the shaking reaction vessel, all improved the measurement techniques.[5]

ALL-OVER PHOTOSYNTHESIS

The first measurements of QR of photosynthetic oxygen evolution were made with *Chlorella*[1-3] and yielded very low values. These values apparently are incorrect (see Kok[6] for a critical review). Subsequent measurements with algae resulted in values of 8 quanta per molecule O_2 evolved or higher (see Table 1). Although occasional reports of lower values appeared,[7-10] it is now generally accepted that the theoretical lowest value is 8.

PARTIAL REACTIONS

Knowing that two light reactions work in sequence in photosynthesis, QR measurements of separate photoreactions have been performed. Other than in QR determinations of total photosynthesis, one has to consider that light absorbed by the complementary photosystem might not be involved in the photoreaction specifically measured. Thus, values for the partial reaction are often higher than theoretically expected. Table 2 summarizes the QRs of PS I and PS II reactions, as well as for photophosphorylation.

PHYSIOLOGICAL CHANGES

During the early measurements of quantum yield, it was assumed that physiological

Table 1
QUANTUM REQUIREMENT OF UNICELLULAR ALGAE

QR	Organism	Ref.
10	*Chlorella*	11
9—16	*Chlorella*	12
7—10	*Chlorella*	9
8—14	*Chlorella*	13
12	*Chlorella*	14
8	*Chlorella*	15
10	*Chlorella*	16
10	*Scenedesmus*	14
10	*Scenedesmus*	17
8	*Scenedesmus*	5
10	*Chlamydomonas*	18
10	*Chlorococcus*	11
9	*Stichococcus*	11
12	*Chroococcus*	11
9	*Navicula*	19
11	*Porphyridium*	20
12	*Cyanidium*	21

Table 2
QUANTUM REQUIREMENT OF UNICELLULAR ALGAE (PARTIAL REACTIONS OF PHOTOSYNTHESIS)

Reaction	Organism	QR	Photosystem	Ref.
Photoreduction	*Scenedesmus*	9	PS I	14
Photoreduction	*Scenedesmus*	10	PS I	17
P_{700} photooxidation	*Anacystis*	1	PS I	22
Chinon-Hill reaction	*Chlorella*	9—15	PS II	23
Hill reactions	*Anacystis*	20	PS II, PS I	24
Photophosphorylation	*Anacystis*	2	PS I, PS II	25

changes, such as aging of the algae, influenced the measurements.[11] This assumption has been confirmed for cultures of *Chlorella* and *Scenedesmus*.[26] Extensive studies of the aging of *Scenedesmus* showed that the QR increased more than 20-fold during a range of 30 days.[27]

It was speculated that higher plants growing under sun or shade conditions have different quantum yields. In strong and weak light-grown cultures of *Scenedesmus*, a change in quantum yield could not be found.[28]

During greening of *Scenedesmus* pigment mutants and regreening of glucose-bleached *Chlorella protothecoides*, changes in the QR have been reported.[29,30] Very drastic changes in quantum yield occur during the individual live cycles of the cells. This was most successfully studied in synchronized cultures. Measurements of relative quantum yield were reported for *Chlorella* by Nihei et al.,[31] and by Sorokin and Krauss.[32] Extensive studies on changes of the QR during synchronous growth have been carried out with *Scenedesmus*.[5,33-35] It was shown that only one stage during development yields the theoretical number of 8 quanta required for the photosynthetic process. Cells of different developmental stages of synchronous cultures, from homocontinuous or batch cultures, measured under the same conditions, have a higher QR.[35] Efficiency changes in the photochemical apparatus under different development stages are most probably because of structural changes in the thylakoids causing a different arrangement of the antenna pigments in the photosynthetic unit, as can be deduced from the tripartite model of Butler and Strasser,[36] or causing changes in excitation energy transfer from unit to unit, as described in Strasser's[37] grouping concept. In addition, it has been proposed that the capability of the reaction center to transform excitation energy into a charge separation can vary the efficiency of the primary reaction of photochemistry.[37]

REFERENCES

1. **Warburg, O. and Negelein, E.,** Über den Energieumsatz bei der Kohlensäureassimilation, *Z. Physik. Chem.*, 102, 235, 1922.
2. **Warburg, O. and Negelein, E.,** Über den Einfluß der Wellenlange auf den Energieumsatz bei der Kohlensäuressimilation, *Z. Physik. Chem.*, 106, 191, 1923.
3. **Warburg, O. and Schröder, W.,** Quantenbedarf der Photosynthesese, *Z. Naturforsch.*, 12b, 716, 1957.
4. **Warburg, O. and Schocken, V.,** A manometric actinometer for the visible spectrum, *Arch. Biochem.*, 21, 363, 1949a.
5. **Senger, H.,** Quantum yield of photosynthesis and the Emerson enhancement effect, *Proc. 2nd Int. Congr. Photosynthesis Research*, Stresa, G. Forti, Avron, A., and Melandri, A., Eds., Dr. W. Junk, The Hague, The Netherlands, 1972, 723.
6. **Kok, B.,** Efficiency of photosynthesis, *Handbuch Pflanzenphysiologie*, Ruhland, W., Ed., Springer-Verlag, Berlin, 1960, 566.
7. **Bassham, J. A., Shibata, K., and Calvin, M.,** Quantum requirement in photosynthesis related to respiration, *Biochim. Biophys. Acta*, 17, 332, 1955.
8. **Emerson, R. and Chalmers, R.,** Transient changes in cellular gas exchange and the problem of maximum efficiency of photosynthesis, *Plant Physiol.*, 30, 504, 1955.
9. **Kok, B.,** A critical consideration of the quantum yield of *Chlorella* photosynthesis, *Enzymologia*, 13, 1, 1948.
10. **Yuan, E. L., Evans, R. W., and Daniels, F.,** Energy efficiency of photosynthesis by *Chlorella*, *Biochim. Biophys. Acta*, 17, 185, 1955.
11. **Emerson, R. and Lewis, C. M.,** Carbon dioxide exchange and the measurement of the quantum yield of photosynthesis, *Am. J. Bot.*, 28, 789, 1941.
12. **French, C. S. and Rabideau, G. S.,** The quantum yield of oxygen production by chloroplasts suspended in solution containing ferric oxalate, *J. Gen. Physiol.*, 28, 239, 1945.
13. **Moore, W. E. and Duggar, B. M.,** Quantum Efficiency in photosynthesis in *Chlorella*, in *Photosynthesis in Plants*, Franck, J. and Loomis, W. E., Eds., Iowa State College Press, Ames, 1949, 239.
14. **Riecke, F. F.,** Quantum efficiencies for photosynthesis and photoreduction in green plants, in *Photosynthesis in Plants*, Franck, J. and Loomis, W. E., Eds., Iowa State College Press, Ames, 1949, 251.
15. **Govindjee, R., Rabinowitch, E., and Govindjee, S. R.,** Maximum quantum yield and action spectrum of photosynthesis and fluorescence in *Chlorella*, *Biochim. Biophys. Acta*, 162, 539, 1968.
16. **Senger, H.,** Changes in photosynthetic activities in synchronous cultures of *Scenedesmus*, *Chlorella* and *Chlamydomonas*, in *Colloques Internationaux du C.N.R.S. No. 240 "Les Cycles Cellulaires et leurs Blocages"*, Gif-sur-Yvette, France, 1974, 101.
17. **Bishop, N. I.,** Comparison of the action spectra and quantum requirements for photosynthesis and photoreduction of *Scenedesmus*, *Photochem. Photobiol.*, 6, 621, 1967.
18. **Senger, H.,** Preparation and photosynthetic properties of synchronous cultures of *Chlamydomonas*, in *Methods in Enzymology*, San Pietro, A., Ed, Academic Press, New York, in press.
19. **Tanada, T.,** The photosynthetic efficiency of carotenoid pigments in *Nivicula minima*, *Am. J. Bot.*, 38, 276, 1951.
20. **Brody, M. and Emerson, R.,** The effect of wavelength and intensity of light on the proportion of pigments in *Porphyridium cruentum*, *Am. J. Bot.*, 46, 433, 1959.
21. **Volk, S. L. and Bishop, N. I.,** Photosynthetic efficiency of a phycocyanin-less mutant of *Cyanidium*, *Photochem. Photobiol.*, 68, 312, 1968.
22. **Kok, G. E. and Hoch, G.,** Spectral changes in photosynthesis, in *Light and Life*, McElroy, W. D. and Glass, B., Eds., Johns Hopkins, Baltimore, 1967, 397.

23. **Ehrmantraut, H. and Rabinowitch, E.**, Kinetics of Hill-reaction, *Arch. Biochem.*, 38, 67, 1952.
24. **Susor, W. A., Duane, W. C., and Krogmann, D. W.**, Studies on photosynthesis using cell free preparations of blue-green algae, in *Biochemical Dimensions of Photosynthesis*, Krogmann, D. W. and Powers, W. H., Eds., Wayne State University Press, Detroit, 1965, 36.
25. **Bornefeld, T.**, The quantum requirement of photophosphorylation and its stoichiometric relation to reduction in *Anacystis* in vivo, *Z. Pflanzenphysiol.*, 85, 393, 1977.
26. **Werthmüller, K. and Senger, H.**, Changes in the photosynthetic apparatus during ageing of algal cultures, in *Proc. 2nd Int. Congr. Photosynthesis Research*, Vol. 3, Stresa, G. Forti, Avron, A., and Melandri, A., Eds., Dr. W. Junk, The Hague, The Netherlands, 1972, 2643.
27. **Kulandaivelu, G. and Senger, H.**, Changes in the reactivity of the photosynthetic apparatus in heterotrophic ageing cultures of *Scenedesmus obliquus*. I. Changes in photochemical activities, *Physiol. Plant.*, 36, 157, 1976.
28. **Senger, H. and Fleischhacker, Ph.**, Adaptation of the photosynthetic apparatus of *Scenedesmus obliquus* to strong and weak light conditions. I. Difference in pigments, photosynthetic capacity, quantum yield and dark reactions, *Physiol. Plant.*, 43, 35, 1978.
29. **Bishop, N. I. and Senger, H.**, The development of structure and function in chloroplasts of greening mutants of *Scenedesmus*. II. Development of the photosynthetic apparatus, *Plant Cell Physiol.*, 13, 937, 1972.
30. **Senger, H. and Oh-hama, T.**, Quantum yield and conformational changes during greening and bleaching of *Chlorella protothecoides*, *Plant Cell Physiol.*, 17, 551, 1976.
31. **Nihei, T., Sasa, T., Miyachi, S., Suzuki, K., and Tamiya, H.**, Change of photosynthetic activity of *Chlorella* during the course of their normal life cycle, *Arch. Micro biol.*, 21, 156, 1954.
32. **Sorokin, C. and Krauss, R. W.**, Relative efficiency of photosynthesis in the course of cell development, *Biochim. Biophys. Acta*, 48, 314, 1961.
33. **Senger, H. and Bishop, N. I.**, Changes in the quantum yield and photoreduction during the synchronous life cycle of *Scenedesmus obliquus*, *Nature (London)*, 214, 140, 1967.
34. **Senger, H.**, Quantenausbeute und unterschiedliches Verhalten der beiden Photosysteme des Photosyntheseapparates während des Entwicklungsablaufes von *Scenedesmus obliquus* in Synchronkulturen, *Planta*, 92, 327, 1970b.
35. **Senger, H.**, Quantum yield of photosynthesis in synchronous cultures of algae, in *Proc. Ist Eur. Biophysics Congr.*, Broda, E. and Locker, Al., and Springer-Lederer, Eds., Wiener Medizin. Akad., Wien, 1971, 33.
36. **Butler, W. L. and Strasser, R.** Tripartite model for the photochemical apparatus of green plant photosynthesis, *Proc. Natl. Acad. Sci. U.S.A.*, 74, 3382, 1977.
37. **Strasser, R.**, The grouping model of plant photosynthesis, in *Chloroplast Development*, Akoyunoglou, G., Ed., Elsevier/North-Holland, Amsterdam, 1978, 513.

CHLOROPHYLLS OF PHOTOSYNTHETIC BACTERIA

Norbert Pfennig

All phototrophic bacteria with anoxygenic photosynthesis and only one photosystem (green and purple bacteria) possess bacteriochlorophylls in their membrane-bound photosynthetic apparatus. The major types presently known and their characteristic absorption maxima in living cells are listed in Table 1.

Bacteriochlorophyll *a* (Bchl *a*) is most widely distributed among the purple bacteria, Chromatiaceae and Rhodospirillaceae. With the exception of the following species which contain bacteriochlorophyll *b* (Bchl *b*), all other species of the purple bacteria possess Bchl *a*: *Ectothiorhodospira halochloris, Rhodopseudomonas sulfoviridis, Rhodopseudomonas viridis*, and *Thiocapsa pfennigii*.

The Bchl *a* of all known purple bacteria except *Rhodospirillum rubrum* is Bchl a_p in which the propionic acid side chain is esterified with phytol.[2,7] In *R. rubrum* strains, the Bchl *a* is esterified with geranyl-geraniol (Bchl a_{Gg}).[7]

Bacteriochlorophylls *c*, *d*, and *e* (Bchl *c,d*, and *e*) are the characteristic major bacteriochlorophylls of the green bacteria, Chlorobiaceae and Chloroflexaceae. These bacteriochlorophylls are located in the *Chlorobium* vesicles which are attached to the cytoplasmic membrane and serve a light-harvesting function. All green bacteria contain small amounts of Bchl a_p[2] located in the photosynthetic reaction centers in the cytoplasmic membrane near the *Chlorobium* vesicles.

Bchl *c* or *d* occurs in all known green-colored green sulfur bacteria; Bchl *e* is present in the brown-colored green sulfur bacteria: *Chlorobium phaeobacteroides, Chlorobium phaeovibrioides*, and *Pelodictyon phaeum*.

Chloroflexus aurantiacus possess a Bchl c_s which is esterified with stearyl alcohol[4] instead of the isoprenoid alcohol, farnesol.

The chemical structures of all bacteriochlorophylls are given in Figure 1, together with Table 2. The different substituents R_1 to R_7 indicated in Figure 1 are presented in Table 2 for the different bacteriochlorophylls.

Table 1
BACTERIOCHLOROPHYLLS OF PHOTOTROPHIC GREEN AND PURPLE BACTERIA

Designation[a]	Typical absorption maxima of living cells			
Bch *a* (a_p or a_{Gg})[b]	375	590	800—810	830—890
Bchl *b*	400	605	835—850	1015—1035
Bchl *c* (c_F or c_s)[c]	335	460	745—760	812[d]
Bchl *d*	325	450	725—745	805[d]
Bchl *e*	345	450—460	715—725	805[d]

[a] Bchls *a* to *d* were named by Jensen et al.;[6] Bchl *e* was named by Gloe et al.[3]

[b] Bchl a_p contains phytol; Bchl a_{Gg} contains geranyl-geraniol.[3]

[c] Bchl c_F contains farnesol; Bchl c_s contains stearyl alcohol.[4]

[d] These are very small absorption peaks which are because of Bchl *a* present in small amounts in the cells.

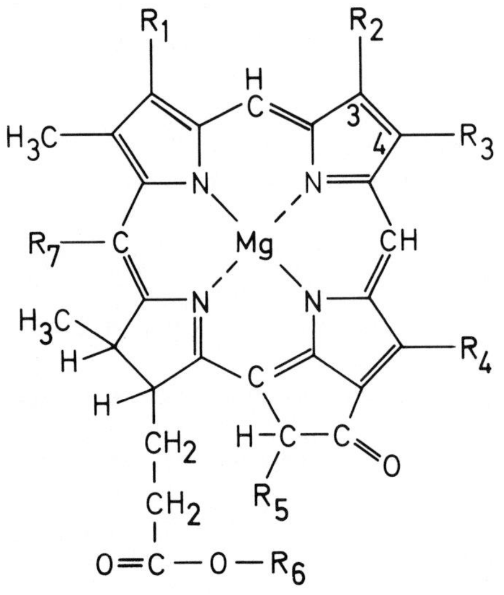

FIGURE 1. Structure of bacteriochlorophylls. The different substituents in the positions R_1 to R_7 are given in Table 2.

Table 2
PRESENTATION OF THE SUBSTITUENTS R_1 TO R_7 OF FIGURE 1 FOR ALL BACTERIOCHLOROPHYLLS

Pigment	R_1	R_2	R_3	R_4	R_5	R_6	R_7	Ref.
Bchl *a*	$-C(=O)-CH_3$	$-CH_3$ [a]	$-CH_2-CH_3$ [a]	$-CH_3$	$-C(=O)-O-CH_3$	Phytol or Geranyl-geraniol	$-H$	1, 8
Bchl *b*	$-C(=O)-CH_3$	$-CH_3$ [b]	$=C(H)-CH_3$ [b]	$-CH_3$	$-C(=O)-O-CH_3$	Phytol	$-H$	10
Bchl *c*	$-CH(OH)-CH_3$	$-CH_3$	$-C_2H_5$ $-C_3H_7$ $i-C_4H_9$	$-C_2H_5$ $-CH_3$ (?)	$-H$	Farnesol or Stearyl alcohol	$-CH_3$	4, 5
Bchl *d*	$-CH(OH)-CH_3$	$-CH_3$	$-C_2H_5$ $-C_3H_7$ $i-C_4H_9$	$-C_2H_5$ $-CH_3$ (?)	$-H$	Farnesol	$-H$	9
Bchl *e*	$-CH(OH)-CH_3$	$-C(=O)-H$	$-C_2H_5$ $-C_3H_7$ $i-C_4H_9$	$-C_2H_5$	$-H$	Farnesol	$-CH_3$	3

[a] No double bond between C-3 and C-4; additional H atoms are in position C-3 and C-4.
[b] No double bond between C-3 and C-4; an additional H atom is in position C-3.[10]

REFERENCES

1. **Brockmann, H., Jr. and Kleber, I.**, Zur absoluten Konfiguration der Chlorophylle. III. Mitteilung. Zur absoluten Konfiguration des Bacteriochlorophylls *a*, *Angew. Chem.*, 81, 626, 1969.
2. **Gloe, A. and Pfennig, N.**, Das Vorkommen von Phytol und Geranylgeraniol in den Bacteriochlorophyllen roter und gruner Schwefelbakterien, *Arch. Microbiol.*, 96, 93, 1974.
3. **Gloe, A., Pfennig, N., Brockmann, H., Jr., and Trowitzsch, W.**, A new bacteriochlorophyll from brown-colored Chlorobiaceae, *Arch. Microbiol.*, 102, 103, 1975.
4. **Gloe, A. and Risch, N.**, Bacteriochlorophyll c_s, a new bacteriochlorophyll from *Chloroflexus aurantiacus*, *Arch. Microbiol.*, 118, 153, 1978.
5. **Holt, A. S., Purdie, J. W., and Wasley, J. W. F.**, Structures of *Chlorobium* Chlorophylls (660), *Can. J. Chem.*, 44, 88, 1966.
6. **Jensen, A., Aasmundrud, O., and Eimhjellen, K. E.**, Chlorophylls of photosynthetic bacteria, *Biochim. Biophys. Acta*, 88, 466, 1964.
7. **Künzler, A. and Pfennig, N.**, Das Vorkommen von Bacteriochlorophyll a_P und a_{Gg} in Stammen aller Arten der Rhodospirillaceae, *Arch. Microbiol.*, 91, 83, 1973.
8. **Mittenzwei, H.**, Uber Bakteriochlorophylle, *Hoppe-Seylers Z. Physiol. Chem.*, 275, 93, 1942.
9. **Purdie, J. W. and Holt, A. S.**, Structures of *Chlorobium* chlorophylls (650), *Can. J. Chem.*, 43, 3347, 1965.
10. **Scheer, H., Svec, W. A., Cope, B. T., Studier, M. H., Scott, R. G., and Katz, J. J.**, Structure of bacteriochlorophyll *b*, *J. Am. Chem. Soc.*, 96, 3714, 1974.

CHLOROPHYLLS OF HIGHER PLANTS AND ALGAE

J. S. Brown

SPECTRAL PROPERTIES IN VITRO, MEASUREMENT AND DISTRIBUTION

Chlorophyll *a*

Chlorophyll *a* (Chl *a*) is the primary photosynthetic pigment in all higher plants and algae. Its molecular structure and empirical formula are shown in Figure 1. The chlorophylls are insoluble in water, but they can be readily extracted from most plant tissues by acetone-water mixtures or by methanol. Following solubilization in organic solvents, the chlorophylls and carotenoids can be separated by use of several chromatographic procedures.[1] However, without physical separation, quantitative determination of the chlorophylls can be made from their absorption in the red spectral region where the carotenoids do not absorb.

Chl *a* is the only chlorophyll in blue-green and red algae, and its concentration can be determined simply from the absorption of an 80% acetone extract at 663 nm. The absorption coefficient at that wavelength is 82 mL mg^{-1} cm^{-1}.[2] Figure 2 shows an absorption spectrum of an 80% acetone extract (solid line) of particles prepared from the blue-green alga, *Nostoc*.[3] The absorption band maxima at 663, 618, 580, 432, and 415 are typical of Chl *a* in aqueous acetone. The shoulder near 480 nm is caused by carotenoid absorption; other carotenoid bands in the blue spectral region are masked by chlorophyll absorption.

Chlorophyll *b*

All higher plants (except certain mutants),[4] green algae, and *Euglena* contain Chlorophyll *b* (Chl *b*) in addition to Chl *a*. Chl *b* has an aldehyde group in place of the methyl group attached to ring II shown in Figure 1. The ratio of Chl *a* to Chl *b* may vary between about 2 and 6 in different species and in the same species grown under different conditions. Apparently many plant species[5] (except *Euglena*)[6] form more Chl *b* when grown under low light intensity in order to increase their light-harvesting capacity.

The concentrations of Chl *a* and Chl *b* in 80% acetone extracts can be calculated by measuring the absorbance at 665 and 649 nm and by using the following equations:[1]

$$\text{Chl } a\ (\mu g/mL) = 11.63\ (A_{665}) - 2.39\ (A_{649}) \quad (1)$$

$$\text{Chl } b\ (\mu g/mL) = 20.11\ (A_{649}) - 5.18\ (A_{665}) \quad (2)$$

$$\text{Total Chl}\ (\mu g/mL) = 6.45\ (A_{665}) + 17.72\ (A_{649}) \quad (3)$$

The equations for 100% methanol extracts are[2]

$$\text{Chl } a\ (\mu g/mL) = 16.5\ (A_{665}) - 8.3\ (A_{650}) \quad (4)$$

$$\text{Chl } b\ (\mu g/mL) = 33.8\ (A_{650}) - 12.5\ (A_{665}) \quad (5)$$

$$\text{Total Chl}\ (\mu g/mL) = 4.0\ (A_{665}) + 25.5\ (A_{650}) \quad (6)$$

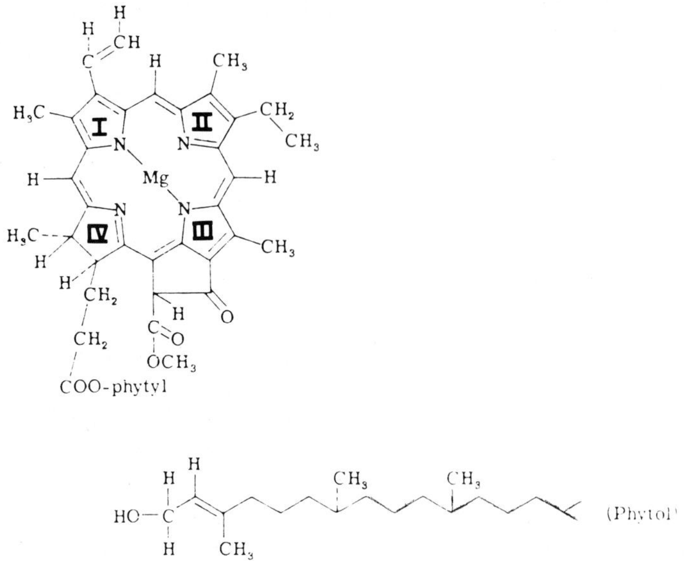

FIGURE 1. The strucutral formula of Chl *a*.

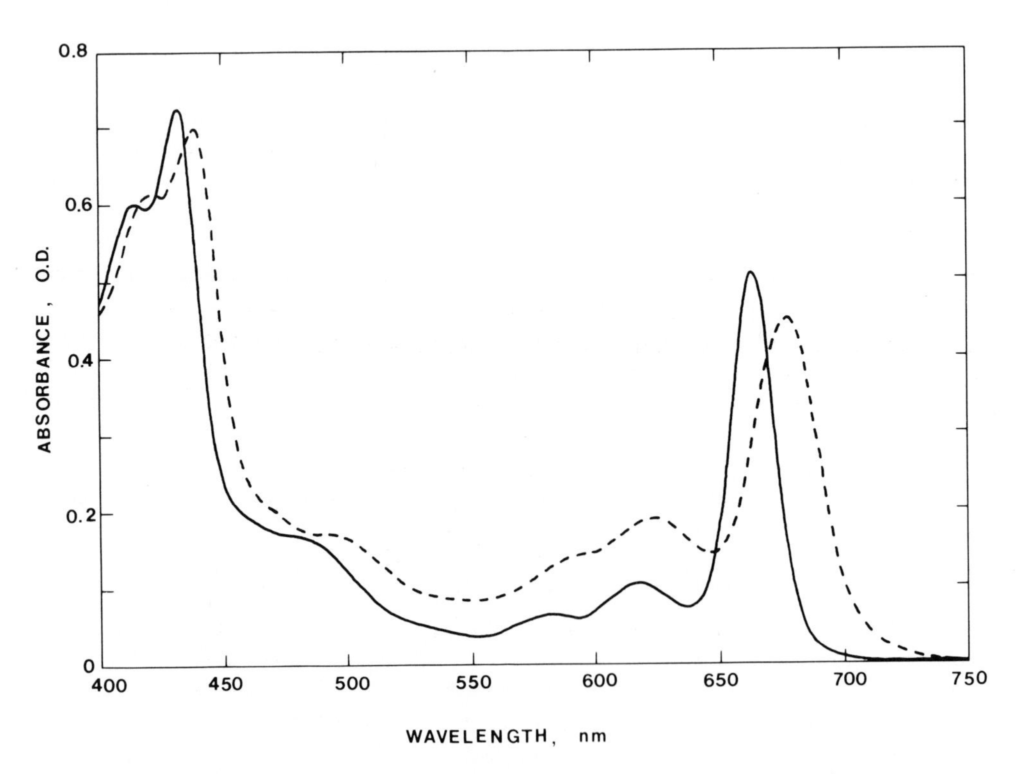

FIGURE 2. Absorption spectra of an aqueous suspension (----) and an 80% acetone-water extract (—) of *Nostoc*[3] particles measured at 20°C.

Precise ratios of Chl *a* and Chl *b* are frequently not obtained because of small wavelength errors in spectrophotometers and the difficulty of measuring on the steep slope of the absorption spectrum near 650 nm. At least two nomograms have been published to facilitate calculation of Chl *a* and Chl *b* concentrations from the absorbance readings.[7,8] When the amount of Chl *b* is very low, the procedures dependent on absorbance are unreliable. A fluorescence method for determining Chl *b* may be used.[9]

Chlorophyll *c*

Chlorophyll *c* (Chl *c*) occurs in large groups of marine algae and freshwater diatoms. Two chemically different forms have been isolated;[10,11] c_1 has a vinyl group on ring IV between the ring and the phytyl chain, and c_2 has an additional vinyl group on ring II in place of the ethyl group shown in Figure 1. All algae that contain Chl *c* have the c_2 form, but some of these are lacking c_1.[12]

Equations 7 and 8 may be used to determine Chl *a* and Chl *c* in 90% acetone-water extracts.[13]

$$\text{Chl } a\ (\mu g/mL) = 11.47\ (A_{664}) - 0.40\ (A_{630}) \qquad (7)$$

$$\text{Chl } c_1 + c_2\ (\mu g/mL) = 24.36\ (A_{630}) - 3.73\ (A_{664}) \qquad (8)$$

Equations 9 to 11 are useful for analyzing 90% acetone-water extracts of phytoplankton in which some algae contain Chl *b* and others contain Chl *c*.

$$\text{Chl } a\ (\mu g/mL) = 11.85\ (A_{664}) - 1.54\ (A_{647}) - 0.08\ (A_{630}) \qquad (9)$$

$$\text{Chl } b\ (\mu g/mL) = -5.43\ (A_{664}) + 21.03\ (A_{647}) - 2.66\ (A_{630}) \qquad (10)$$

$$\text{Chl } c_1 + c_2\ (\mu g/mL) = -1.67\ (A_{664}) - 7.60\ (A_{647}) + 24.52\ (A_{630}) \qquad (11)$$

The occurrence of Chl *b* and Chl *c* in different plant groups is summarized in Table 1.

Chlorophyll *d*

Chlorophyll *d* (Chl *d*) is observed occasionally in methanol extracts of certain red algae and may be recognized by its distinctive absorption near 696 nm. Holt[17] suggested that Chl *d* may be 2-desvinyl-2-formyl-Chl *a*, but the structure remains to be verified.

Chlorophyll Derivatives

The chlorophyllides are chlorophylls without the phytol side chain. Their absorption characteristics are nearly identical to the corresponding chlorophyll, but they can be distinguished by the appropriate chromatography. Plants contain the enzyme chlorophyllase which under certain conditions may convert Chl *a* to chlorophyllide *a* during extraction.[18]

A phaeophytin is chlorophyll without Mg and a phaeophorbide is without both Mg and the phytol. Acid conditions promote the loss of Mg from chlorophyll. Small amounts of phaeophytin *a* can be recognized in absorption spectra of either aqueous or solvent extracts by its characteristic bands near 535 and 410 nm.

ABSORPTION AND FLUORESCENCE IN VIVO

Spectroscopic properties of chlorophyll in vivo differ considerably from those of the pigment dissolved in organic solvents or detergents. Spectra of cells or chloroplasts may be distorted because of light scattering and the sieve effect,[19] but if the membranes

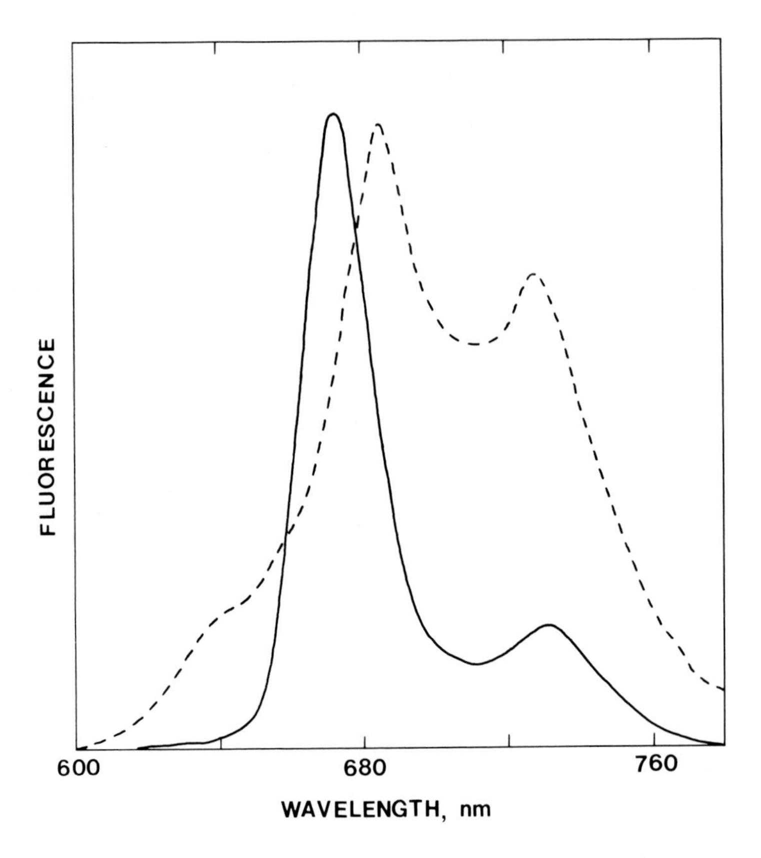

FIGURE 3. Fluorescence emission spectra of an aqueous suspension (---) and an 80% acetone-water extract (—) of *Nostoc*[3] particles measured at 20°C. Excitation at 435 nm, slit width = 10 nm, emission slit width = 5 nm, in relative units.

are separated into sufficiently small particles, spectra representative of the biological state of chlorophyll can be obtained. A spectrum of photochemically active *Nostoc* particles is shown in Figure 2 (dashed line). Typically, the Chl *a* absorption bands are shifted towards longer wavelengths and broadened in vivo. The red absorption maximum is at 678 nm. Part of the absorption near 624 nm is caused by the accessory pigment, phycocyanin, that has not been completely removed from these particles in aqueous suspension, but is lacking in the acetone extract where only the 618-nm band of Chl *a* is visible.

The *Nostoc* particles show two fluorescence emission maxima, the higher near 685 nm and the other near 740 nm (see Figure 3, dashed line). The relatively large emission between 700 and 720 nm probably comes from aggregates or oligomeres of Chl *a*, and the shoulder near 650 nm may be emitted by residual phycocyanin. In vivo, the primary emission of most plants falls near 685 nm; however, the shape of the longer wavelength emission spectrum may vary somewhat in different species and depends strongly upon temperature. By contrast, the primary emission of Chl *a* in 80% acetone is near 672 nm (see Figure 3, solid line), and the long wavelength minor band near 730 nm is from a lower variational level of the molecule. The spectra in Figure 3 have been normalized, but actually the chlorophyll concentration was approximately 50 times lower in the acetone extract. There is always a large increase in fluorescence yield when chlorophyll is solubilized in organic solvents or detergent micelles.

Table 1
DISTRIBUTION OF CHL *b* AND CHL *c*

Plant group	Chl *b*	Chl *c*
Vascular plants	+	−
Bryophyta	+	−
Chlorophyta	+	−
Euglenophyta	+	−
Phaeophyta	−	+
Pyrrophyta	−	+
Chrysophyta	−	±[a]
Rhodophyta	−	−
Cyanophyta	−	−

[a] The Chrysophyta is a varied group, including the diatoms Chrysophyceae, and Xanthophyceae. It was thought that the Xanthophyceae (yellow-green algae) have only Chl *a*, but Guillard and Lorenzen[14] have found Chl *c* in some yellow-green species. Those genera of Xanthophyceae that contain only Chl *a* may be limited to the new class, Eustigmatophyceae.[15] Another genus of this division, *Ochromonas*, also lacks Chl *c*.[16]

In chloroplasts, Chl *b* has absorption maxima near 470 and 650 nm,[20] Chl *c* has absorption maxima near 460, 595, and 640 nm.[21] Thus, they enable a plant to make use of incident light energy falling in the wavelength region between 450 and 670 nm where Chl *a* has low absorption. Because light energy absorbed by these chlorophylls is transferred to Chl *a* with high efficiency, only Chl *a* fluorescence is observed in undamaged, biological systems.

Careful examination of the absorption of Chl *a* in vivo has revealed that the band centered near 678 nm in Figure 2 may actually be composed of several (probably four) overlapping bands at approximately 663, 670, 678, and 684 nm.[22,23] Evidence for this hypothesis has come from both derivative and low-temperature spectroscopy of over 50 different species of plants. Spectra of aqueous, particulate chlorophyll preparations from higher plants are all very similar to each other, but those of different algal groups are distinctly different.[24] However, all the spectra can be resolved into the four major components listed above in varying proportion. None of these components or biological forms of Chl *a* has been separated from the others, but membrane fragments having different proportions of the forms can be obtained from a single plant species.[25] Recent evidence indicates that the molecular environment of the chlorophyll — in particular its attachment to proteins — causes the different absorption bands.[26] Theoretical model calculations[27,28] have shown how the existence of these Chl *a* forms could enhance the efficiency of light energy transfer to the special chlorophylls that function in the reaction centers where charge separation ultimately occurs.

Other chapters in the book cited in Reference 1 provide much information about topics that are dealt with only briefly here, and Meeks[29] has reviewed the algal chlorophylls in more detail.

REFERENCES

1. **Strain, H. H. and Svec, W. A.**, Extraction separation, estimation, and isolation of the chlorophylls, in *The Chlorophylls*, Vernon, L. P. and Seely, G. R., Eds., Academic Press, New York, 1966, chap. 2.
2. **Mackinney, G.**, Absorption of light by chlorophyll solutions, *J. Biol. Chem.*, 140, 315, 1941.
3. **Arnon, D. I., McSwain, B. D., Tsujimoto, H. Y., and Wada, K.**, Photochemical activity and components of membrane preparations from blue-green algae, *Biochim. Biophys. Acta*, 357, 231, 1974.
4. **Highkin, H. R. and Frenkel, A. W.**, Studies of growth and metabolism of a barley mutant lacking chlorophyll *b*, *Plant Physiol.*, 37, 814, 1962.
5. **Thornber, J. P.**, Chlorophyll-proteins: light-harvesting and reaction center components of plants, *Annu. Rev. Plant Physio.*, 26, 127, 1975.
6. **Brown, J. S.**, Factors influencing the proportion of the forms of chlorophyll in algae, *Carnegie Inst. Washington Year.*, 59, 330, 1960.
7. **Kirk, J. T. O.**, Studies on the dependence of chlorophyll synthesis on protein synthesis in *Euglena gracilis*, together with a nomogram for determination of chlorophyll concentration, *Planta*, 78, 200, 1968.
8. **Sestak, Z.**, Determination of chlorophylls *A* and *B*, in *Plant Photosynthetic Production — Manual of Methods*, Sestak, Z., Catsky, J., and Jarvis, P. G., Eds., Dr. W. Junk, The Hague, The Netherlands, 1971, chap. 18.
9. **Boardman, N. K. and Thorne, S. W.**, Sensitive fluorescence method for the determination of chlorophyll *a*/chlorophyll *b* ratios, *Biochim. Biophys. Acta*, 253, 222, 1971.
10. **Jeffrey, S. W.**, Preparation and some properties of crystalline chlorophyll c_1 and c_2 from marine algae, *Biochim. Biophys. Acta*, 279, 15, 1972.
11. **Strain, H. H., Cope, B. T., Jr., McDonald, G. N., Svec, W. A., and Katz, J. J.**, Chlorophylls c_1 and c_2, *Phytochemistry*, 10, 1109, 1971.
12. **Jeffrey, S. W.**, The occurrence of chlorophyll c_1 and c_2 in algae, *J. Phycol.*, 12, 349, 1976.
13. **Jeffrey, S. W. and Humphrey, G. F.**, New spectrophotometric equations for determining chlorophylls *a*, *b*, c_1 and c_2 in higher plants, algae and natural phytoplankton, *Biochem. Physiol. Pflanz.*, 167, 191, 1975.
14. **Guillard, R. L. and Lorezne, C. J.**, Yellow-green algae with chlorophyllide *C*, *J. Phycol.*, 8, 10, 1972.
15. **Hibberd, D. J. and Leedale, G. F.**, Observations on the cytology and ultrasturcture of the new algal class, Eustigmatophyceae, *Ann. Bot.*, 36, 49, 1972.
16. **Allen, M. B., French, C. S., and Brown, J. B.**, Native and extractable forms of chlorophyll in various algal groups, in *Comparative Biochemistry of Photoreactive Systems*, Allen, M. B., Ed., Academic Press, New York, 1960, chap. 3.
17. **Holt, A. S.**, Further evidence of the relation between 2-desvinyl-2-formyl-chlorophyll-*a* and chlorophyll-*d*, *Can. J. Bot.*, 39, 327, 1961.
18. **Terpstra, W.**, A study of properties and activity of chlorophyllase in photosynthetic membranes, *Z. Pflanzenphysiol.*, 85, 139, 1977.
19. **Butler, W. L.**, Absorption spectroscopy *in vivo*: theory and application, *Annu. Rev. Plant Physiol.*, 15, 451, 1964.
20. **Yentsch, C. S. and Guillard, R. R. L.**, The absorption of chlorophyll-*b in vivo*, *Photochem. Photobiol.*, 9, 385, 1969.
21. **Mann, J. E. and Myers, J.**, On pigments, growth, and photosynthesis of *Phaeodactylum tricornutum*, *J. Phycol.*, 4, 349, 1968.
22. **French, C. S., Brown, J. S., and Lawrence M. C.**, Four universal forms of chlorophyll *a*, *Plant Physiol.*, 49, 421, 1972.
23. **Brown, J. S.**, Forms of chlorophyll *in vivo*, *Annu. Rev. Plant Physiol.*, 23, 73, 1972.
24. **Brown, J. S.**, Biological forms of chlorophyll *a*, *Methods Enzymol.*, 23, 477, 1971.
25. **Brown, J. S. and Gasanov, R.**, Photosynthetic activity and chlorophyll absorption spectra of fractions from the alga, *Dunaliella*, *Photochem. Photobiol.*, 19, 139, 1974.
26. **Brown, J. S.**, Spectroscopy of chlorophyll in biological and synthetic systems, *Photochem. Photobiol.*, 26, 319, 1977.
27. **Seely, G. R.**, Effects of spectral variety and molecular orientation on energy trapping in the photosynthetic unit: a model calculation, *J. Theor. Biol.*, 40, 173, 1973.
28. **Seely, G. R.**, Energy transfer in a model of the photosynthetic unit of green plants, *J. Theor. Biol.*, 40, 189, 1973.
29. **Meeks, J. C.**, Chlorophylls, in *Algal Physiology and Biochemistry*, Stewart, W. D. P., Ed., University of California, Berkeley, 1974, chap. 5.

CAROTENOIDS OF PHOTOSYNTHETIC BACTERIA

Karin Schmidt

Photosynthetic bacteria synthesize a large number of different carotenoids. Of the approximately 400 known carotenoids occurring in nature, about 70 are typical of photosynthetic bacteria.[1] These carotenoids differ from those found in nonphotosynthetic bacteria, algae, fungi, and higher plants by the following properties:

1. They are without exception C_{40}-carotenoids.
2. They mostly occur as aliphatic compounds.
3. The typical substituents are tertiary hydroxyl and methoxyl groups.
4. They frequently have double bonds in C-3,4 position.
5. Carbonyl functions are in conjugation with the polyene chain either in C-2 (aerobic conditions) or in C-4 (anaerobic conditions — exception *Chloroflexus*).
6. Aldehyde groups are found only in C-20 position (strict anaerobic reaction).
7. Cyclic carotenoids commonly have aromatic rings: either 1,2,3-trimethylphenyl or 1,2,5-trimethylphenyl end groups.

A selection of the main characteristic carotenoids which have been found in the different species of phototrophic bacteria are represented in Figure 1, including their trivial names and absorption maxima in light petroleum. Most of the given structures are the characteristic end-products of a special biosynthetic pathway or an important intermediate which often is accumulated rather than the real end-product. A complete list of all carotenoids found to occur in photosynthetic bacteria is given by Liaaen-Jensen.[1]

Four main pathways of carotenoid biosynthesis have been suggested[2,3]

1. (a) Spirilloxanthin (via rhodopin) and (b) rhodopinal (± spirilloxanthin via rhodopin).
2. Spheroidene and spheroidenone (± spirilloxanthin via chloroxanthin)
3. (a) Okenone and (b) oxo-carotenoids of *Rhodopseudomonas globiformis* (R.g.-keto-carotenoids),
4. Isorenieratene (including chlorobactene and γ- and β-carotene synthesis).

The spirilloxanthin and rhodopinal series are the most unspecific pathways within the taxa of photosynthetic bacteria (see Table 1). Both occur in many species of the Rhodospirillaceae as well as in Chromatiaceae. Rhodopinal synthesis, however, is restricted to anaerobic growth conditions. It seems that the formation of aromatic carotenoids as okenone, chlorobactene, and isorenieratene is typical of organisms whose growth is dependent on reduced sulfur compounds (see Tables 1 and 2). More detailed data about the distribution and quantitative composition of carotenoids and their biosynthesis have been given by Schmidt.[3]

Name of Carotenoids	Chemical Structures	Absorption Maxima (nm)
Lycopene		446 472 503
Rhodopin	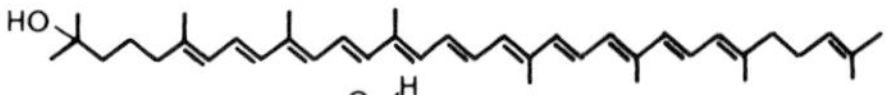 HO	446 472 503
Rhodopinal	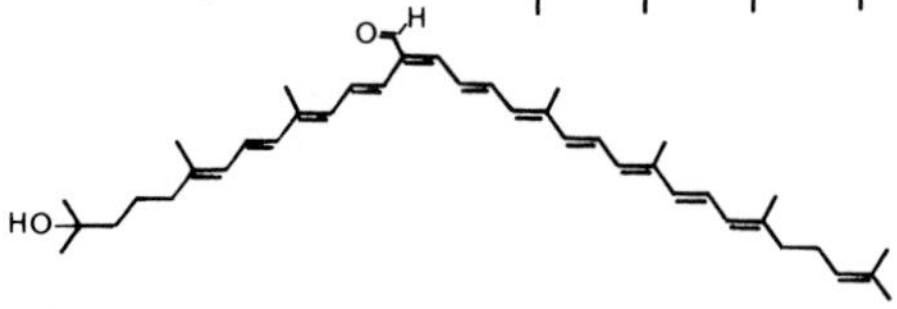 O H, HO	492
Spirilloxanthin	 H_3CO, OCH_3	462 494 528
R.g.-keto III	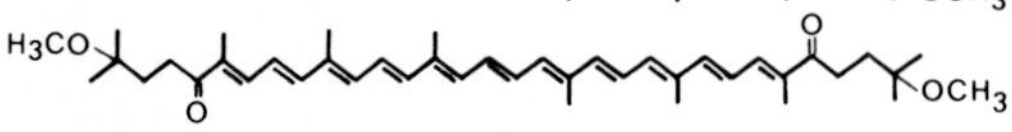 H_3CO, O, O, OCH_3	461 489 522
Chloroxanthin	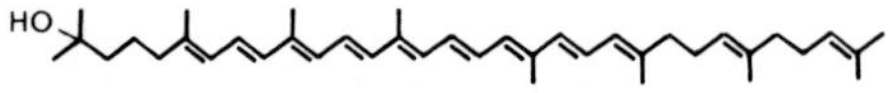 HO	415 439 469
Spheroidene	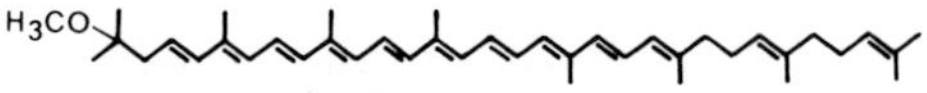 H_3CO	429 455 486
Spheroidenone	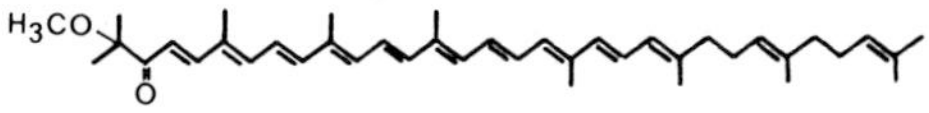 H_3CO, O	(460) 483 515
Okenone	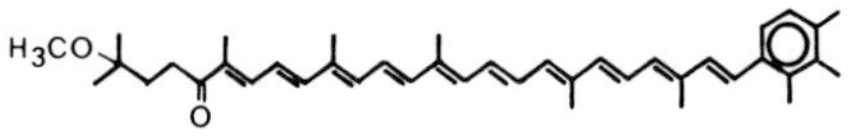 H_3CO, O	(460) 484 516
Chlorobactene	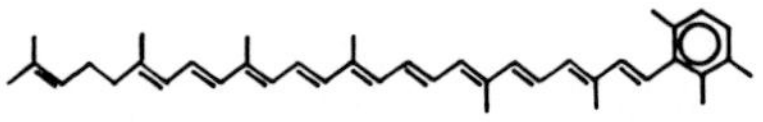	434 460 491
Isorenieratene	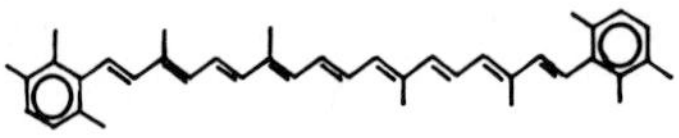	(430) 452 480

FIGURE 1. Chemical structures and spectral properties of some representative carotenoids of photosynthetic bacteria. (The absorption maxima are taken from spectra in light petroleum from different sources).[4-10]

Table 1
DISTRIBUTION OF CAROTENOIDS WITHIN SPECIES OF PURPLE PHOTOSYNTHETIC BACTERIA

Type of carotenoids	Type of bacteria	
	Rhodospirillaceae	**Chromatiaceae**
Spirilloxanthin (via Rhodopin)		
1. Complete pathway end-product spirilloxanthin	*Rhodospirillum rubrum* *Rhodospirillum tenue* *Rhodopseudomonas palustris* *Rhodopseudomonas acidophila* *Rhodomicrobium vanniellii*	*Amoebobacter roseus* *Amoebobacter pendens* *Thiocapsa roseopersicina* *Chromatium vinosum* *Chromatium gracile* *Chromatium minutissimum* *Ectothiorhodospira mobilis* *Ectothiorhodospira shaposhnikovii* *Ectothiorhodospira halophila*
2. Incomplete pathway end-product rhodopin and lycopene	*Rhodospirillum fulvum* *Rhodospirillum molischianum* *Rhodospirillum photometricum* *Rhodopseudomonas viridis*	*Thiospirillum jenense*
Rhodopinal (via Rhodopin)		
1. Complete pathway + spirilloxanthin	*Rhodopseudomonas acidophila*	*Thiocystis* spp.
2. Incomplete pathway − spirilloxanthin	*Rhodospirillum tenue* *Rhodocyclus purpurea*	*Chromatium warmingii* *Chromatium buderi ?* *Thiocystis violacea* *Thiodictyon elegans* *Thiodictyon bacillosum* *Lamprocystis* sp
Spheroidene + Spirilloxanthin (via Chloroxanthin)	*Rhodopseudomonas spheroides* *Rhodopseudomonas gelatinosa* *Rhodopseudomonas capsulata*	
Okenone		*Chromatium okenii* *Chromatium weissii* *Chromatium minus* *Thiocystis gelatinosa* *Thiopedia rosea*
R.g.-keto Carotenoids	*Rhodopseudomonas globiformis*	

Table 2
DISTRIBUTION OF CAROTENOIDS WITHIN SPECIES OF GREEN PHOTOSYNTHETIC BACTERIA

Type of carotenoids	Type of bacteria: Chloroflexaceae	Type of bacteria: Chlorobiaceae
Isorenieratene		
1. Complete pathway end-product isorenieratene (via β-carotene or chlorobactene)		*Chlorobium phaeobacteroides* *Chlorobium phaeovibrioides*
2. Incomplete pathway end-product chlorobactene (via γ-carotene)		*Chlorobium limicola*[a] *Chlorobium vibrioforme*[a] *"Chloropseudomonas ethylica"* *Pelodictyon luteolum* *Pelodictyon chlathratiforme* *Prosthecochloris aestuarii*
3. Incomplete pathway end-product γ- and β-carotene	*Chloroflexus aurantiacus*	

[a] Formathiosulfatophilum.

REFERENCES

1. **Liaaen-Jensen, S.,** Chemistry of carotenoids, in *The Photosynthetic Bacteria,* Clayton, R. K. and Sistrom, W. R., Eds., Plenum Press, New York, 1978, chap. 8.
2. **Schmidt, K., Pfennig, N., and Liaaen-Jensen, S.,** Carotenoids of Thiorhodaceae. IV. The carotenoid composition of 25 pure isolates, *Arch. Mikrobiol.,* 52, 132, 1965.
3. **Schmidt, K.,** Biosynthesis of carotenoids, in *The Photosynthetic Bacteria,* Clayton, R. K. and Sistrom, W. R., Eds., Plenum Press, New York, 1978, chap. 39.
4. **Aasen, A. J. and Liaaen-Jensen, S.,** Bacterial carotenoids. XXIV. The carotenoids of Thiorhodaceae 7. Cross-conjugated carotenals, *Acta Chem. Scand.,* 21, 2185, 1967.
5. **Davies, B. H.,** Analysis of carotenoid pigments, in *Chemistry and Biochemistry of Plant Pigments,* Goodwin, T. W., Ed., Academic Press, New York, 1965, 489.
6. **Liaaen-Jensen, S.,** The constitution of some bacterial carotenoids and their bearing of biosynthetic problems, *Kg. Norske Vit. Selsk. Skr.,* No. 8, 1962.
7. **Liaaen-Jensen, S., Hegge, E., and Jackman, L. M.,** Bacterial carotenoids XVII. The carotenoids of photosynthetic green bacteria, *Acta Chem. Scand.,* 18, 1703, 1964.
8. **Liaaen-Jensen, S.,** Bacterial carotenoids. XVIII. Aryl-carotenes from *Phaeobium, Acta Chem. Scand.,* 19, 1025, 1965.
9. **Liaaen-Jensen, S.,** Bacterial carotenoids. XXII. The carotenoids of Thiorhodaceae 5. Structural elucidation of okenone, *Acta Chem. Scand.,* 21, 961, 1967.
10. **Schmidt, K. and Liaaen-Jensen, S.,** Bacterial carotenoids. XLII. New keto-carotenoids from *Rhodopseudomonas globiformis* (Rhodospirillaceae), *Acta Chem. Scand.,* 27, 3040, 1973.

CAROTENOIDS OF ALGAE AND HIGHER PLANTS

B. H. Davies

INTRODUCTION

The carotenoids are a group of some 500 lipid-soluble isoprenoid pigments, normally yellow to red, which are widely distributed throughout nature. They invariably occur with chlorophylls in photosynthetic organisms, where they function both as photoprotectants[1,2] and, with varying degrees of efficiency, in trapping light energy for photosynthesis.[2,3] Algae and the photosynthetic tissues of higher plants thus contain carotenoids; although each species usually contains only 5 to 10 different main carotenoids (with a few minor components), the total number of carotenoids in all these classes of organisms is about 60. The common names of these are listed alphabetically in Table 1.

STRUCTURES AND NOMENCLATURE

The structures of all the relevant carotenoids are given in Table 1, using the current standard semisystematic carotenoid nomenclature.[4,5] Of seven different natural carotenoid end groups, only three β,ε, and ψ; see Figure 1) occur in the pigments of algae and green tissues of plants. The end groups at either extremity of the (fully unsaturated all-*trans*) polyene chain are used to create the C_{40} carotene (hydrocarbon) stem names (e.g., β,ε-carotene). These are modified by appropriate prefixes and suffixes (with locant designation; see Figure 1) to denote changes in the carbon skeleton (apo-, nor-), hydrogenation level (hydro-, dehydro-), oxygenation (of xanthophylls; e.g., epoxy-, -ol, -one) and stereochemistry (e.g., *cis*).

ABSORPTION SPECTRA AND CAROTENOID IDENTIFICATION

As the most characteristic feature of the carotenoids is their absorption of visible light, their electronic absorption spectra are crucial as a means of identification. Their wavelengths of maximal absorption are defined primarily by the number of conjugated double bonds in the molecule; this influence is modified, however, not only by the nature of the end groups, but also by the presence of some oxygen substituents, such as epoxides (which shorten the chromophore) and carbonyl groups.[6] Carotenoids usually have three main absorption peaks (see Table 1), but the presence of β-rings (especially if two are present, e.g., in β-carotene) tends to attenuate the lowest wavelength peak to a inflexion (shown in brackets in Table 1). The effect of a conjugated carbonyl group is also to decrease spectral persistence (resolution), so that carotenoids with two such groups (e.g., canthaxanthin) have a single symmetrical absorption peak.

The spectral maxima shown in Table 1 are selected from a comprehensive collection of such data,[6] and, where possible, values for the absorption maxima in light petroleum have been used. Where this has not been possible, values for the absorption maxima in other common solvents (e.g., acetone) have been included. It should be noted that a given carotenoid has its maxima at the lowest wavelengths in hexane, light petroleum, or in a primary alcohol (ethanol or methanol). The use of acetone as a solvent increases the wavelengths by about 2 nm and the use of other solvents by more.[6] Peridinin (Dinophyceae) and related carotenoids are exceptional in having a bathochromic shift (to higher wavelengths) of some 20 nm in ethanol. It should also be

End groups (R) =

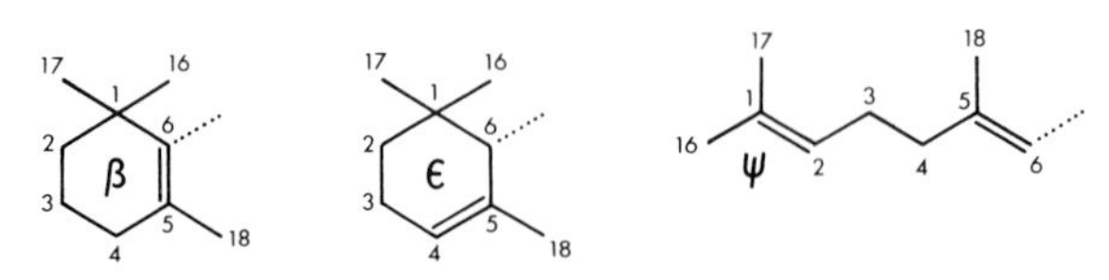

FIGURE 1. End groups occurring in algal and plant (green tissue) carotenoids and the numbering of the carbons of the carotenoid molecule. The order of end groups shown is the order of priority in the stem name of the carotenoid, and the first listed takes unprimed numerals.

noted that the wavelength reproducibility of the average recording spectrophotometer (including those used for obtaining the data in Table 1) is only of the order of ±1 nm; the wavelength calibration of the instrument used should always be checked.

Since the shape (persistence/resolution) of the absorption spectrum is diagnostic, not only of the structure of the carotenoid, but also of its purity, it is imperative that any carotenoid examined should be rigorously purified by chromatography. Chromatographic behavior itself is a guide to the degree of unsaturation and to the nature and number of oxygen substituents of the carotenoid.[6] The differences between the structures of many carotenoids (e.g., alloxanthin, diatoxanthin, and zeaxanthin) cannot be determined by such means alone. Thus, proton magnetic resonance spectroscopy and, particularly, mass spectrometry are required for unambiguous identification, since other techniques cannot differentiate the many features of algal carotenoid structure (see Table 2).

Quantitative determinations of individual or total carotenoids are carried out spectrophotometrically by using the appropriate values of either ε or $A^{1\%}_{1\,cm}$.[6] Other analytical data have been collected for carotenoids,[6-8] and literature references are available for all such data collected up to 1976.[9]

ALGAL CAROTENOIDS

Algal carotenoids have been investigated sufficiently to allow a number of reviewers[10-13] to attempt to correlate algal evolution with the degree of sophistication of carotenoid biosynthesis. Although hypothetical schemes for algal evolution based on such an approach[11,12] still have a few inconsistencies, it is now clear that, as long as their identification is unambiguous, carotenoids have a considerable potential in algal taxonomy.[13]

Table 1 includes information on the distribution of all the individual carotenoids in the various algal classes, and Table 2 is a generalized correlation of structural features with class. Both tables rely on information from Liaaen-Jensen's recent considerations[12,13] of carotenoid structure and algal taxonomy, and, for consistency, the same system of algal classification[14] is used.

HIGHER PLANT CAROTENOIDS

The chloroplasts of actively photosynthetic tissues of higher plants, apparently irre-

Table 1
CAROTENOIDS OF ALGAE AND PLANTS: STRUCTURE, OCCURRENCE[12] IN THE VARIOUS ALGAL CLASSES,[14] AND ABSORPTION MAXIMA[6]

Common names	Structure	Algal classes in which carotenoid occurs[a]	Absorption maxima (nm)[b] in the stated solvent			
Alloxanthin	7,8,7′,8′-Tetradehydro-β,β-carotene-3,3′-diol	3	(427)	450	479	Lt. pet.[c]
Antheraxanthin	5,6-Epoxy-5,6-dihydro-β,β-carotene-3,3′-diol	(2), 5, (6), (11), 14	422	445	472	Lt. pet.
Aphanizophyll	2′-(β-L-Rhamnopyranosyloxy)-3′,4′-didehydro-1′,2′-dihydro-β,ψ-carotene-3,4,1′-triol	1	450	476	507	Acetone
Astaxanthin	3,3′-Dihydroxy-β,β-carotene-4,4′-dione	(4), 12,[c] 14[d]		468		Lt. pet.
Aurochrome	5,8,5′,8′-Diepoxy-5,8,5′,8′-tetrahydro-β,β-carotene	(2)	382	400	426	Lt. pet.
Auroxanthin	5,8,5′,8′-Diepoxy-5,8,5′,8′-tetrahydro-β,β-carotene-3,3′-diol	(2)	382	402	427	Lt. pet.
Caloxanthin	β,β-Carotene-2,3,3′-triol	1	426	449	475	Ethanol
Canthaxanthin	β,β-Carotene-4,4′-dione	1, (7), 12,[c] 14[d]		463		Lt. pet.
α-Carotene	β,ε-Carotene	2, 3, (4), (7), 13, 14	422	444	473	Lt. pet.
β-Carotene	β,β-Carotene	1, 2, 3, 4, 5, 6, 7, 8, 9, 10, 11, 12, 13, 14	(425)	448	475	Lt. pet.
γ-Carotene	β,ψ-Carotene	(1), (13), (14)	437	462	494	Lt. pet.
ε-Carotene	ε,ε-Carotene	3, 8, (11), (14)	416	440	470	Lt. pet.
ζ-Carotene	7,8,7′,8′-Tetrahydro-ψ,ψ-carotene	14[e]	378	400	425	Lt. pet.
β-Carotene diepoxide	5,6,5′,6′-Diepoxy-5,6,5′,6′-tetrahydro-β,β-carotene	(9), (10)	418	443	471	Lt. pet.
β-Carotene epoxide	5,6-Epoxy-5,6-dihydro-β,β-carotene	(7)	420	447	478	Lt. pet.
β,β-Caroten-2-ol	β,β-Caroten-2-ol	(14)	(430)	452	479	Acetone
β,ε-Caroten-2-ol	β,ε-Caroten-2-ol	(14)	(425)	447	475	Acetone
β,β-Carotene-2,2′-diol	β,β-Carotene-2,2′-diol	(14)	425	452	480	Acetone
α-Cryptoxanthin	β,ε-Caroten-3′-ol	(2)	421	446	475	Lt. pet.
β-Cryptoxanthin	β,β-Caroten-3-ol	1, (2), (6), (7), (12), (14)	425	449	476	Lt. pet.
Cryptoxanthin diepoxide	5,6,5′,6′-Diepoxy-5,6,5′,6′-tetrahydro-β,β-caroten-3-ol	(7), (9), (10)	423	442	472	Ethanol
Cryptoxanthin epoxide	5,6-Epoxy-5,6-dihydro-β,β-caroten-3-ol	(6), (9), (10), (12), (14)	424	445	477	Ethanol
Crocoxanthin	7,8-Didehydro-β,ε-caroten-3-ol	3	421	443	472	Ethanol
Diadinoxanthin	5,6-Epoxy-7′,8′-didehydro-5,6-dihydro-β,β-carotene-3,3-diol	4, 7, 8, 9, 10, (11), (12)	(424)	445	474	Ethanol
Diatoxanthin	7,8-Didehydro-β,β-carotene-3,3′-diol	4, 7, 8, 9, 10, (11), 12	(425)	449	475	Ethanol
Dinoxanthin	3′-Acetoxy-5,6-epoxy-6′,7′-didehydro-5,6,5′,6′-tetrahydro-β,β-carotene-3,5′-diol (i.e., neoxanthin-3-acetate)	4	418	442	470	Acetone
Echinenone	β,β-Caroten-4-one	1, (7), 12,[c] 14[d]		456	(482)	Lt. pet.
Fucoxanthin	5,6-Epoxy-3,3′,5′-trihydroxy-6′,7′-didehydro-5,6,7,8,5′,6′-hexahydro-β,β-caroten-8-one 3′-acetate	(2), (4), 6, 7, 8, 11	427	450	476	Hexane
Fucoxanthinol	5,6-Epoxy-3,3′,5′-trihydroxy-6′,7′-didehydro-5,6,7,8,5′,6′-hexahydro-β,β-caroten- 8-one	(7), (11)	425	448	476	Lt. pet.
Heteroxanthin	7′,8′-Didehydro-5,6-dihydro-β,β-carotene-3,5,6,3′-tetrol	9, 10, 12	(423)	448	478	Ethanol

Table 1 (continued)
CAROTENOIDS OF ALGAE AND PLANTS: STRUCTURE, OCCURRENCE[12] IN THE VARIOUS ALGAL CLASSES,[14] AND ABSORPTION MAXIMA[6]

Common names	Structure	Algal classes in which carotenoid occurs[a]	Absorption maxima (nm)[b] in the stated solvent			
19′-Hexanoyloxy-fucoxanthin	5,6-Epoxy-3,3′,5′,19′-tetrahydroxy-6′,7′-didehydro-5,6,7,8,5′,6′-hexahydro-β,β-caroten-8-one 3′-acetate 19′-hexanoate	(7)	423	450	(478)	Acetone
19-Hexanoyloxy-paracentrone 3-acetate	3,5,19-Trihydroxy-6,7-didehydro-5,6,7′,8′-tetrahydro-7′-apo-β-caroten-8′-one 3-acetate 19-hexanoate	(7)	423	450	(478)	Acetone
3-Hydroxycan-thaxanthin	3-Hydroxy-β,β-carotene-4,4′-dione	14[d]		463		Lt. pet.
3-Hydroxyech-inenone	3-Hydroxy-β,β-caroten-4-one	12[c]		457		Lt. pet.
3′-Hydroxyech-inenone	3′-Hydroxy-β,β-caroten-4-one	1		456	(482)	Lt. pet.
4′-Hydroxyech-inenone	4′-Hydroxy-β,β-caroten-4-one	14[d]		460		Ethanol
4-Keto-myxol-2′-O-methyl-methyl pentoside	3,1′-Dihydroxy-2′-(O-methyl-5-C-methylpentosyloxy)-3′,4′-didehydro-1′,2′-dihydro-β,ψ-caroten-4-one	1		483	(510)	Acetone
4-Keto-myxox-anthophyll	3,1′-Dihydroxy-2′-(β-L-rhamno-pyranosyloxy)-3′,4′-didehydro-1′,2′-dihydro-β,ψ-caroten-4-one	1		483	(510)	Acetone
Loroxanthin	β,ε-Carotene-3,19,3′-triol	(14)	(425)	446	473	Ethanol
Lutein	β,ε-Carotene-3,3′-diol	2, 13, 14	421	445	474	Lt. pet.
Lutein epoxide	5,6-Epoxy-5,6-dihydro-β,ε-carotene-3,3′-diol	(2), 5, (14)	420	442	471	Ethanol
Lycopene	ψ,ψ-Carotene	(1), (13)	446	472	505	Lt. pet.
Manixanthin	9,9′-Di-*cis*-7,8,7′,8′-Tetradehydro-β,β-carotene-3,3′-diol (i.e., 9,9′-di-*cis* alloxanthin)	3	423	444	473	Acetone
Monadoxanthin	7,8-Didehydro-β,ε-carotene-3,3′-diol	3	425	448	476	Ethanol
Mutatochrome	5,8-Epoxy-5,8-dihydro-β,β-carotene	1	409	428	452	Lt. pet.
Myxol-2′-O-methyl-methyl pentoside	2′-(O-Methyl-5-C-methylpentosyloxy)-3′,4′-didehydro-1′,2′-dihydro-β,ψ-carotene-3,1′-diol	1	450	476	508	Acetone
Myxoxanthophyll	2′-(β-L-Rhamnopyranosyloxy)-3′,4′-didehydro-1′,2′-dihydro-β,ψ-carotene-3,1′-diol	1	(450)	478	510	Acetone
Neoxanthin	5′,6′-Epoxy-6,7-didehydro-5,6,5′,6′-tetrahydro-β,β-carotene-3,5,3′-triol	(2), (6), 8, 9, (10), (11), 12, 13, 14	418	442	467	Lt. pet.
Nostoxanthin	β,β-Carotene-2,3,2′,3′-tetrol	1	(426)	448	475	Ethanol
Oscillaxanthin	2,2′-Bis(β-L-rhamnopyranosyl-oxy)-3,4,3′,4′-tetradehydro-1,2,1′,2′-tetrahydro-ψ,ψ-carotene-1,1′-diol	1	470	499	534	Acetone

Table 1 (continued)
CAROTENOIDS OF ALGAE AND PLANTS: STRUCTURE, OCCURRENCE[12] IN THE VARIOUS ALGAL CLASSES,[14] AND ABSORPTION MAXIMA[6]

Common names	Structure	Algal classes in which carotenoid occurs[a]	Absorption maxima (nm)[b] in the stated solvent			
Oscillol-2,2′-(o-methyl-methyl pentoside)	2,2′-Bis(O-methyl-5-C-methyl-pentosyloxy)-3,4,3′,4′-tetradehydro-1,2,1′,2′-tetrahydro-ψ,ψ-carotene-1,1′-diol	1	469	496	530	Acetone
Peridinin	3′-Acetoxy-5,6-epoxy-3,5′-dihydroxy-6′,7′-didehydro-5,6,5′,6′-tetrahydro-12′,13′,20′-trinor-β,β-caroten-19,11-olide	4		446		Acetone
Peridininol	5,6-Epoxy-3,3′,5′-trihydroxy-6′,7′-didehydro-5,6,5′,6′-tetrahydro-12′,13′,20′-trinor-β,β-caroten-19,11-olide	(4)		466		Acetone
Phytoene	7,8,11,12,7′,8′,11′,12′-Octahydro-ψ,ψ-carotene	(4), 14[e]	276	286	297	Lt. pet.
Phytofluene	7,8,11,12,7′,8′-Hexahydro-ψ,ψ-carotene	(4), 14[e]	331	348	367	Lt. pet.
Pyrrhoxanthin	3′-Acetoxy-5,6-epoxy-3-hydroxy-7′,8′-didehydro-5,6-dihydro-12′,13′,20′-trinor-β,β-caroten-19,11-olide	(4)		458		Acetone
Pyrrhoxanthinol	5,6-Epoxy-3-hydroxy-7′,8′-didehydro-5,6-dihydro-12′,13′,20′-trinor-β,β-caroten-19,11-olide	(4)		458		Acetone
Siphonaxanthin	3,19,3′-Trihydroxy-7,8-dihydro-β,ε-caroten-8-one	13, (14)		446	(468)	Lt. pet.
Siphonein	3,19,3′-Trihydroxy-7,8-dihydro-β,ε-caroten-8-one 19-acylate	13, (14)		450	(473)	Lt. pet.
Vaucheriaxanthin	5′,6′-Epoxy-6,7-didehydro-5,6,5′,6′-tetrahydro-β,β-carotene-3,5,19,3′-tetrol	9, 10	419	442	471	Ethanol
Violaxanthin	5,6,5′,6′-Diepoxy-5,6,5′,6′-tetrahydro -β,β-carotene-3,3′-diol	(2), (6), 11, 13, 14	418	442	466	Lt. pet.
Zeaxanthin	β,β-Carotene-3,3′-diol	1, 2, (6), (11), 13, 14	(424)	449	476	Lt. pet.

Note: Lt. pet. = Light petroleum.

[a] Brackets indicate that the carotenoid is a minor (or questionable) component in the algae of a particular class. The classes are denoted as follows:

1. Cyanophyceae
2. Rhodophyceae
3. Cryptophyceae
4. Dinophyceae
5. Raphidophyceae
6. Chrysophyceae
7. Haptophyceae
8. Bacillariophyceae
9. Xanthophyceae
10. Eustigmatophyceae
11. Phaeophyceae
12. Euglenophyceae

Table 1 (continued)
CAROTENOIDS OF ALGAE AND PLANTS: STRUCTURES, OCCURRENCE[12] IN THE VARIOUS ALGAL CLASSES,[14] AND ABSORPTION MAXIMA[6]

13. Prasinophyceae/Loxophyceae
14. Chlorophyceae

[b] Brackets indicate an inflexion rather than an absorption peak at the stated wavelength.
[c] Eyespot pigment.
[d] Accumulates under low nitrogen conditions.
[e] Normally detectable only in mutants.

spective of the type and habitat of the plant,[15] contain β-carotene and a number of unesterified xanthophylls, lutein, violaxanthin, and neoxanthin (all of which become esterified in autumnal necrosis[16]). The chloroplast carotenoids occur in two granal chromoproteins[17] corresponding to photosystems I and II.[18] α-Carotene, β-cryptoxanthin, zeaxanthin, and antheraxanthin may also be present, usually in small quantities. Although the carotenoid content is qualitatively constant, quantitative variations (occasionally considerable) may occur in some species.[19] It is thought that the constancy of the distribution pattern reflects not only the common ancestry of higher plants, but also that any departure from this pattern is lethal.[19] Mutations have been observed, both in higher plants[20-22] and in green algae,[23] in which acyclic precursor carotenes more saturated than α- or β-carotene (phytoene, phytofluene, and ζ-carotene) accumulate (of these, only phytoene and phytofluene have been detected as normal trace components of photosynthetic tissue[24]). Such carotenoids, having chromophores of less than the nine conjugated double bonds required,[2] confer no photoprotection on the mutants which are therefore vulnerable to illumination under aerobic conditions.

Table 2
SOME STRUCTURAL FEATURES OF THE MAJOR CAROTENOIDS[a] OF THE VARIOUS ALGAL CLASSES[14] AND OF THE GREEN TISSUES OF HIGHER PLANTS[19]

	Cyanophyceae	Rhodophyceae	Cryptophyceae	Dinophyceae	Raphidophyceae	Chrysophyceae	Haptophyceae	Bacillariophyceae	Xanthophyceae	Eustigmatophyceae	Phaeophyceae	Euglenophyceae	Prasinophyceae/Loxophyceae	Chlorophyceae	Higher plant green tissues
Carbon skeleton															
β-End group	+	+	+	+	+	+	+	+	+	+	+	+	+	+	+
ε-End group	−	+	+	−	+	−	+	+	−	−	−	−	+	+	+
Monocyclic	+[b]	−	−	−	−	−	−	−	−	−	−	−	−	−	−
Acetylenic bond	−	−	+	+	−	−	+	+	+	−	−	+	−	−	−
Allenic linkage	−	−	−	+	−	+	+	+	+	+	+	+	+	+	+
Loss of C_3 (C_{37})	−	−	−	+	−	−	−	−	−	−	−	−	−	−	−
Oxygenation															
3-OH-β	+	+	+	+	+	+	+	+	+	+	+	+	+	+	+
3-OH-ε	−	+	+	−	+	−	−	−	−	−	−	−	+	+	+
2-OH-β	−	−	−	−	−	−	−	−	−	−	−	−	−	+	−
2,3-DiOH-β	+	−	−	−	−	−	−	−	−	−	−	−	−	−	−
3-Acetate	−	−	−	+	−	+	+	+	−	+	+	−	−	−	−
5,6-DiOH-β	−	−	−	−	−	−	−	−	+	−	−	+	−	−	−
5,6-Epoxy-β	−	−	−	+	+	+	+	+	+	+	+	+	+	+	+
17-OH	−	−	−	−	−	−	−	−	+	+	−	−	+	+	−
4-Keto-β	+	−	−	−	−	−	−	−	−	−	−	+[c]	−	+[d]	−
8-Keto	−	−	−	−	−	+	+	+	−	+	+	−	−	+	−
O-Acyl	−	−	−	−	−	−	−	−	+	+	−	−	+	−	−
Glycoside	+[b]	−	−	−	−	−	−	−	−	−	−	−	−	−	−
Butenolide	−	−	−	+	−	−	−	−	−	−	−	−	−	−	−

Table 2 (continued)
SOME STRUCTURAL FEATURES OF THE MAJOR CAROTENOIDS[a] OF THE VARIOUS ALGAL CLASSES[14] AND OF THE GREEN TISSUES OF HIGHER PLANTS[19]

[a] Generalized from Liaaen-Jensen[13]
[b] Monocyclic glycosides (e.g., myxoxanthophyll) are peculiar to Cyanophyceae. Since carotenoid glycosides are typically bacterial, the carotenoid composition of these organisms supports their classification as Cyanobacteria.
[c] Eyespot pigments.
[d] Formed as a response to low nitrogen levels.

REFERENCES

1. **Burnett, J. H.,** Functions of carotenoids other than in photosynthesis, in *Chemistry and Biochemistry of Plant Pigments,* Vol. 1, 2nd ed., Goodwin, T. W., Ed., Academic Press, New York, 1976, 655.
2. **Krinsky, N. I.,** Function, in *Carotenoids,* Isler, O., Ed., Birkhäuser Verlag, Basel, 1971, 669.
3. **Whittingham, C. P.,** Function in photosynthesis, in *Chemistry and Biochemistry of Plant Pigments,* Vol. 1, 2nd ed., Goodwin, T. W., Ed., Academic Press, New York, 1976, 624.
4. Commission on Biochemical Nomenclature: IUPAC Commission on the Nomenclature of Organic Chemistry and IUPAC-IUB Commission on Biochemical Nomenclature, Tentative rules for the nomenclature of carotenoids, *Biochemistry,* 10, 4827, 1971.
5. Commission on Biochemical Nomenclature: IUPAC Commission on the Nomenclature of Organic Chemistry and IUPAC-IUB Commission on Biochemical Nomenclature, Nomenclature of carotenoids (Recommendations 1974), *Biochemistry,* 14, 1803, 1975.
6. **Davies, B. H., Carotenoids,** in *Chemistry and Biochemistry of Plant Pigments,* Vol. 2, 2nd ed., Goodwin, T. W., Ed., Academic Press, New York, 1976, 38.
7. **Moss, G. P. and Weedon, B. C. L.,** Chemistry of carotenoids, in *Chemistry and Biochemistry of Plant Pigments,* Vol. 1, 2nd ed., Goodwin, T. W., Ed., Academic Press, New York, 1976, 149.
8. **Vetter, W., Englert, G., Rigassi, N., and Schwieter, U.,** Spectroscopic methods, in *Carotenoids,* Isler, O., Ed., Birkhäuser Verlag, Basel, 1971, 189.
9. **Straub, O.,** *Key to Carotenoids: Lists of Natural Carotenoids* , Birkhäuser Verlag, Basel, 1976.
10. **Stransky, H. and Hager, A.,** Das Carotenoidmuster und die Verbreitung des lichtinduzierten Xanthophyllcyclus in verschiedenen Algenklassen. VI. Chemosystematische Betrachtung, *Arch. Mikrobiol.,* 73, 315, 1970.
11. **Goodwin, T. W.,** Algal carotenoids, in *Aspects of Terpenoid Chemistry and Biochemistry,* Goodwin, T. W., Ed., Academic Press, New York, 1971, 315.
12. **Liaaen-Jensen, S.,** Algal carotenoids and chemosystematics, in *Marine Natural Products Chemistry,* Faulkner, D. J. and Fenical, W. H., Eds., Plenum Press, New York, 1977, 239.
13. **Liaaen-Jensen, S.,** Carotenoids — a chemosystematic approach, *Pure Appl. Chem.,* 51, 661, 1979.
14. **Christensen, T.,** *Systematisk Botanik, Alger,* University of Copenhagen, 1962.
15. **Strain, H. H.,** Fat-soluble chloroplast pigments; their identification and distribution in various Australian plants, in *Biochemistry of Chloroplasts,* Vol. 1, Goodwin, T. W., Ed., Academic Press, New York, 1966, 387.
16. **Goodwin, T. W.,** Studies in carotenogenesis. 24. The changes in carotenoid and chlorophyll pigments in the leaves of deciduous trees during Autumn necrosis, *Biochem. J.,* 68, 503, 1958.
17. **Thornber, J. P., Stewart, J. C., Hatton, M. W. C., and Bailey, J. L.,** Studies on the nature of chloroplast lamellae. II. Chemical composition and further physical properties of two chlorophyll-protein complexes, *Biochemistry,* 6, 2006, 1966.
18. **Tevini, M. and Lichtenthaler, H. K.,** Untersuchungen uber die Pigment- und Lipochinonaustattung der zwei photosynthetischen Pigmentsysteme, *Z. Pflanzenphysiol.,* 62, 17, 1970.
19. **Goodwin, T. W.,** Distribution of carotenoids, in *Chemistry and Biochemistry of Plant Pigments,* Vol. 1, 2nd ed., Goodwin, T. W., Ed., Academic Press, New York, 1976, 225.
20. **Anderson, I. C. and Robertson, D. S.,** Role of carotenoids in protecting chlorophyll from photodestruction, *Plant. Physiol.,* 35, 531, 1960.
21. **Faludi-Dániel, A., Láng, F., and Fradkin, L. I.,** The state of chlorophyll *a* in leaves of carotenoid mutant maize, in *Biochemistry of Chloroplasts,* Vol. 1, Goodwin, T. W., Ed., Academic Press, New York, 1966, 269.
22. **Treharne, K. J., Mercer, E. I., and Goodwin, T. W.,** Carotenoid biosynthesis in some maize mutants, *Phytochemistry,* 5, 581, 1966.
23. **Powls, R. and Britton, G.,** A series of mutant strains of *Scenedesmus obliquus* with abnormal carotenoid compositions, *Arch. Microbiol.,* 113, 275, 1977.
24. **Mercer, E. I., Davies, B. H., and Goodwin, T. W.,** Studies in carotene carotenogenesis. 29. Attempts to detect lycopersene in higher plants, *Biochem. J.,* 87, 317, 1963.

PHYCOBILINS AND PHYCOBILIPROTEINS OF ALGAE

Jürg R. Gysi and David J. Chapman

The photosynthetic apparatus of red (Rhodophyta), cryptomonad (Cryptophyta), and the blue-green algae (Cyanophyta) or Cyanobacteria contains phycobiliproteins, intensely colored proteinaceous accessory pigments.[1,2] Their general distribution in the algae is listed in Table 1. The phycobiliproteins fall into two basic groups: the phycoerythrins and the phycocyanins. The phycoerythrins are characterized by the presence of a linear tetrapyrrole pigment, phycoerythrobilin (see Figure 1), while the phycocyanins are characterized by a similar pigment, phycocyanobilin.[3,4] There are two exceptions to this generalization: R-phycocyanin and phycoerythrocyanin which possess both chromophores. Within the two groups, the differences depend upon the amino acid compositions and sequences of the α and β subunits of the proteins, the number of chromophores per subunit, and the nature of the protein-bilin interaction. These in turn cause the characteristic absorption spectra differences (see Table 2) between the various biliproteins. These linear tetrapyrroles are covalently bound to the polypeptide. Some workers, consider that a third bilin, phycourobilin, is responsible for the absorption band in the 495- to 500nm region of some of the phycoerythrins,[2] while others consider the absorption band to be the result of a specific phycoerythrobilin-protein interaction.[5] Cleavage from the phycoerythrin and unambiguous identification of the presumptive phycourobilin has not yet been achieved because phycoerythrobilin, during extraction and as the free chromophore, is unstable and produces significant amounts of urobilinoid artifacts. Current research admittedly limited to a few of the biliproteins, indicates that the phycobilins are covalently bound to the peptide by a thio-ether linkage through the ethylidene group of ring A and possibly an ester or amide linkage to the propionic side chain of ring C (see Figure 2).[6,7] Much of the misunderstanding about the structural identity of the free and "native" chromophore can be traced to the various methods used to liberate the free bilin, the chemical reactivity of the ethylidene function resulting from cleavage and the vinyl group. It is not necessary to distinguish between the free and bound chromophore by applying separate names. The relationship between the free structures "phycoerythrobilin" and "phycocyanobilin" and their bound form is obvious, and these terms should be retained for the "chromophore" (the compounds with structures in Figure 1) of the biliprotein irrespective of whether it is free or bound; in this latter case, the bonding to amino acids makes it impossible to distinguish a "separate in vivo bilin".

The basic monomeric unit of all biliproteins thus far reported consists of one α- and one β-polypeptide chain, with the only possible exception being B-phycoerythrin.[8] The molecular weights of these ($\alpha\beta$) units all lie within the range of 26,000 to 38,000. With the exception of the cryptomonads, the proposed in vivo form of the phycobiliproteins is either a trimeric or hexameric aggregate of the (α β) monomer (see Table 3), the majority of which have isoelectric points between 4.2 and 5.1.

A representative number of the numerous amino acid compositions of biliproteins thus far determined is given in Table 4. Wherever possible, a representative of the coccoid and the filamentous blue-green algae as well as the eukaryotic red algae and the cryptomonads was chosen. It is remarkable that the amino acid compositions of proteins from taxonomically unrelated organisms are quite similar. General features are a high ratio of acidic to basic residues and also a high level of aliphatic residues, with alanine being the predominant amino acid in all cases. Histidine is absent in all but one of the investigated allophycocyanins, and its content is low in phycocyanins and phycoerythrins.

Table 1
BILIPROTEIN DISTRIBUTION IN THE ALGAE[a]

Organism	Cyanophyta	Cyanellae[b]	Rhodophyta	Cryptophyta[c]
Phycobiliproteins				
C-phycocyanins (CPC)	+	+	+	
R-phycocyanins (RPC)	−	−	±	
Allophycocyanins (APC)	+	+	+	
Allophycocyanins I or B	±			
Phycoerythrocyanins	±			
B-phycoerythrins (B-PE)			±	
b-phycoerythrins (b-PE)			±	
C-phycoerythrins (CPE)	±			
R-phycoerythrins (RPE)			±	
Cryptomonad phycocyanins (3 types)				±
Cryptomonad phycoerythrins (3 types)				±

[a] +, indicates always present;

−, indicates has never been detected;

±, indicates present only in some species.

[b] Cyanellae are blue-green colored inclusions in various colorless eukaryotic hosts. They are considered variously to be either blue-green algae or chloroplasts.

[c] Cryptophyta always contain only one biliprotein, in contrast to all others which always contain at least two (CPC and APC).

The first complete primary structure of a biliprotein (C-phycocyanin from the cyanobacterium *Mastigocladus laminosus*) was determined by Frank et al.[9] The amino acid sequences of the α- and β-subunits of C-phycocyanin from *M. laminosus* are compared in Table 5 with the N-terminal sequences of the allophycocyanin subunits from the same organism[10] and with additional biliprotein sequence data.[11] As Glazer[11] pointed out, conservation of primary structure is higher within each subunit group than between the two groups, suggesting an early evolutionary divergence between α- and β-subunits. The fact that allophycocyanins show the greatest degree of homology supports the postulate that allophycocyanin was the evolutionary precursor of C-phycocyanin.[12]

The main function of the phycobiliproteins is light-harvesting and energy transfer to chlorophyll *a* (Chl *a*) (see Figure 3). It is now generally accepted that this process is a stepwise migration of energy from phycoerythrin to phycocyanin to allophycocyanin to Chl *a* by the mechanism of inductive resonance of Förster.[13-15] This order of energy transfer can also be anticipated on purely spectroscopic grounds. For effective resonance transfer, the fluorescence spectrum of the donor pigment must overlap with the absorption spectrum of the acceptor pigment. For C-phycocyanin, it has been shown that the energy absorbed by the α-subunit chromophore is reemitted as fluorescence, whereas that absorbed by the β-subunit chromophore sensitizes the fluorescence of the α-chromophore.[16,17] Furthermore, the donor and acceptor molecules must be in close proximity. Such an in vivo system exists in the form of phycobilisomes which are high

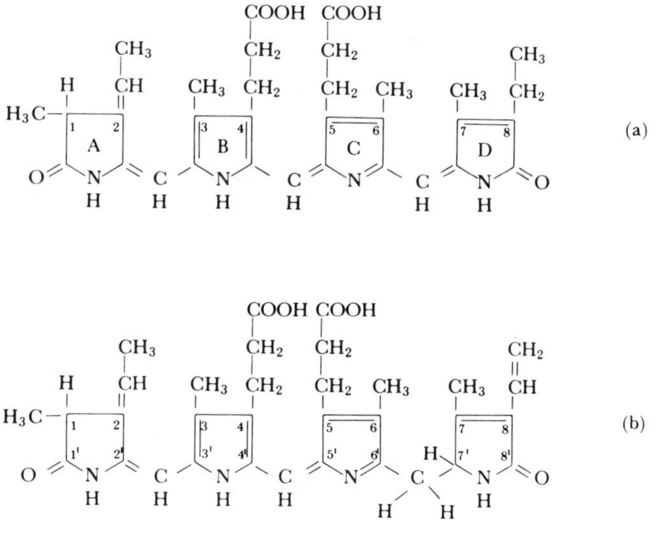

FIGURE 1. Structures of (a) phycocyanobilin and (b) phycoerythrobilin.

molecular weight aggregates of biliproteins located on the outer surfaces of thylakoid membranes,[18-20] except in cryptomonads where they are apparently intrathylakoidal.[21] Gantt et al.[22] recently proposed a model (see Figure 4) for phycobilisomes in which a core of allophycocyanin attached to the phytosynthetic membrane is surrounded by a layer of phycocyanin which in turn is surrounded by a layer of phycoerythrin. Fluorescence emission studies have shown that excitation of intact phycobilisomes with light of 545 nm, absorbed almost exclusively by phycoerythrins, sensitizes predominantly allophycocyanin fluorescence at 675 to 680 nm. Upon dissociation of phycobilisomes, and thus disruption of the energy transfer chain, emission occurs principally as phycoerythrin fluorescence at 575 nm.[23,24] Haxo and Blinks,[25] Blinks,[26] and recently Porter et al.[27] demonstrated that light energy trapped by phycobiliproteins is utilized in photosynthesis with very high efficiency (⩾99%). Table 2 lists the spectroscopic characteristics of the individual biliproteins from various algal species.

The linear tetrapyrrole is biosynthesized[28-30] by the normal porphyrin (or δ-amino levulinic acid) pathway. The bilin moiety is derived from oxidative ring cleavage at the α-bridge of the porphyrin macrocycle, probably protoporphyrin or a metalloporphyrin. The enzyme system involved in plants is probably very similar to the heme-oxygenase system of bile pigment formation in mammalian systems. Little is known about the product-precursor relationships from the point of ring cleavage through to the final bilin and its subsequent attachment to the polypeptide. Regulation of biliprotein synthesis[31] appears to be under photocontrol by one of three methods, each involving a photoreceptive pigment or adaptochrome.

In *Cyanidium caldarium,* the production of biliproteins requires light, although phycocyanobilin (as the free chromophore) can be formed in the dark in the presence of exogenous δ-aminolevulinic acid. In *Tolypothrix tenuis* (CPE, CPC, APC), biliproteins can be formed in the dark. The nature of the immediate preillumination prior to transfer to the dark determines the biliprotein synthesized. Red light induces CPC, while green light induces CPE formation. *Fremyella diplosiphon* forms CPC and APC, but no CPE, when grown in red light (⩾600 nm); it forms principally CPE and little CPC when grown in green light (460 to 550 nm). This phenomenon of light wavelength control is referred to as complementary chromatic adaptation.[31]

Table 2
SPECTROSCOPIC PROPERTIES OF VARIOUS BILIPROTEINS

Biliprotein	Absorption max in visible region (nm)			Fluorescence emission max (nm)			CD max (nm)				Algal source	Ref.
		α	β		α	β	Positive	Negative	α +ve −ve	β +ve −ve		
B-phycoerythrin	(495), 545, 563			575			505, 537, 572	558			*Porphyridium cruentum*	8, 62
b-phycoerythrin	545, (563)			575							*P. cruentum*	8
R-phycoerythrin	498, 540, 568			578			496, 536,572				*Ceramium rubrum*	2, 62
C-phycoerythrin	565			575			548	567			*Fremyella diplosiphon*	33
											Hydrocoleum	63
Cryptomonad phycoerythrins												
Type I	545, (560)	565	531	585 (625)			536	568			*Rhodomonas lens*	34
Type II	556			580							*Hemiselmis rufescens*	35
Type III	566			617 (650)			554	578			*Cryptomonas ovata*	34
Phycoerythrocyanin	568	562	602	610	583	617	578	332			*Anabaena* sp. 6411	36
											Anabaena variabilis	36
R-phycocyanin	555, 618	660	560, 660	634			552, 627	311, 343			*P. cruentum*	37
C-phycocyanin	620	620	608	647			636	340	620, 340	598, 340	*Synechococcus* sp.	17
Cyrptomonad phycocyanins												
Type I	588, 615			637							*Hemiselmis virescens* (Plymouth strain 157)	35
Type II	583, 625—630										*Cryptomonas cyanomagna*	35
Type III	585, (625), 645	644	566	661 (715)			588 (613)	650			*Chromonas* sp.	34
Allphycocyanin	(595), (625), 650	615	615	662	640	630	635	340			*Mastigocladus laminosus*	10, 38
Allophycocyanin-B	617, 671			680							*Anabaena variabilis*	39

Note: The major absorption maximum is underlined; absorption shoulders are in parenthesis. Fluorescence and CD shoulders are in parenthesis.

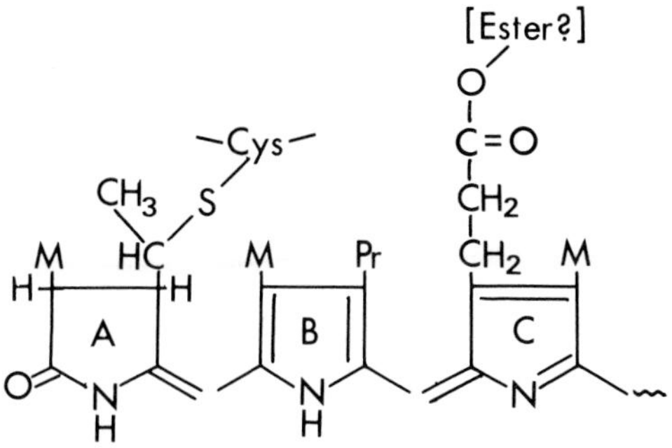

FIGURE 2. Linkage of the bilin chromophore to the popypeptide subunit. Ring A is linked through a thioether to cysteine. Ring C is linked through a presumptive ester or amide bond.

ESTIMATION OF BILIPROTEINS

Biliproteins behave as proteins and can be readily purified by standard protein techniques. Purified biliproteins can be estimated spectrophotometrically from the extinction values given in Table 6. These extinction values are based upon visible (or chromophore) absorption. Slight variations in the values, usually of the order of 1 to 2%, can be expected based upon the source and hence slight variations in the protein structure. Spectrophotometric equations[32] have been developed for some naturally occurring biliprotein mixtures (e.g., from *Fremyella*), isolated without further separation, provided contamination from absorbing species such as chlorophyll-proteins is absent.

$$\text{CPC (mg/mL)} = \frac{\text{E(615 nm)} - 0.474\ \text{E(652 nm)}}{5.34} \quad (1)$$

$$\text{APC (mg/mL)} = \frac{\text{E(652 nm)} - 0.208\ \text{E(615 nm)}}{5.09} \quad (2)$$

$$\text{CPE (mg/mL)} = \frac{\text{E(562 nm)} - 2.41\ \text{(concentration CPC)} - 0.849\ \text{(concentration APC)}}{9.62} \quad (3)$$

These measurements are made in 0.01 *M* sodium phosphate buffer pH 7.0, 0.15 *M* NaCl, for a mixture of APC + CPC + CPE; E = absorbance at appropriate wavelength.

ADDENDUM

Since the completion of this article in early 1978 many papers have appeared. We have provided below in chronological order by thematic section a selective and far from inclusive list of recent articles appropriate to this topic.

Phycobilisomes[66-83]
Phycobiliproteins[84-101]
Bilin chromophores[102-111]
Biosynthesis[112-116]
General reviews[117-123]
Energy transfer[124]

Table 3
CHEMICAL AND PHYSICOCHEMICAL PROPERTIES OF BILIPROTEINS

Organism/Phycobiliprotein	Subunit mol wt		Aggregate size	Sedimentation coefficient $*S_{20,w}$ or $S^0_{20,w}$	Proposed subunit structure	Number and type of chromophores per subunit[a]		pI	Ref.
	α	β				α	β		
Phycoerythrins									
Rhodophyta									
Porphyridium cruentum	17,300								8
(B-PE)	(γ = 30,000)		265,000—280,000						
(b-PE)	17,200		55,000; 110,000						8
Acrochaetium virgatulum									
R-PE I			250,000—299,000	11.6	$(\alpha\beta)_6$				40
R-PE II	19,800	22,500	43,500—44,500	3.8	$\alpha\beta$				40
Cyanophyta									
Phormidium persicinum (CPE)	19,700	22,000							41
Aphanocapsa sp. (6701) (CPE)	20,000	22,000						4.31	42
Fremyella diplosiphon (CPE)	18,300	20,000	180,000—210,000			1 PEB	2 PEB	4.31	33, 43
Cryptophyta									
Cryptomonas sp. (type III)	11,800	19,000	35,000					5.74; 6.35	43
Cryptomonas ovata var. *palustris*	11,000	17,700	30,800					4.9	44
Cryptomonas ovata	9,700	18,200	46,000—53,000	4.4	$\alpha_2\beta_2$				34
Rhodomonas lens (type I)	9,800	17,700	54,300	4.3	$\alpha_2\beta_2$				34
Rhodomonas sp. strain 3-C	11,000	17,700	30,800					4.9	44
Phycoerythrocyanins									
Cyanophyta									
Anabaena sp. 6411	16,000	20,200	101,000			1(?) PXB	2 PCB	5.17	36
Anabaena variabilis	16,000	20,500	104,000			1(?)PXB	2 PCB	5.16	36

Phycocyanins									
Rhodophyta									
Porphyridium cruentum (RPC)	16,400	18,400	127,000; 273,000	*11.4	$(\alpha\beta)_3$ or $[(\alpha\beta)_3]_2$	1 PCB	1 PCB + 1 PEB		8, 45
Porphyridium cruentum (RPC)	18,200	20,500	103,000	*5.98	$(\alpha\beta)_3$	1 PCB	1 PCB + 1 PEB	5.2; 5.3	37
Cyanidium caldarium (CPC)	15,500 17,000	18,300 19,500	228,000		$(\alpha\beta)_6$	1 PCB	2 PCB		46, 47
Cyanidium caldarium (CPC)	12,500	16,500		5.85; 10.75; 17.30				5.11	48
Synechococcus sp. (6301) (CPC)	15,900	19,100	224,000		$(\alpha\beta)_6$	1 PCB	2 PCB	4.20	42, 49
Anabaena sp. (6411) (CPC)	16,500	19,400	105,000	*5.14	$(\alpha\beta)_3$	1 PCB	2 PCB	4.92	36
Synechococcus sp. (6301)	15,200	17,200	96,000; 103,000	5.54	$(\alpha\beta)_3$			4.22; 4.4	42, 64
Synechococcus sp. (6301) allophycocyanin B	16,000	17,000							39
Anabaena sp. (6411)	14,900	14,900	105,000		$(\alpha\beta)_3$	1 PCB	1 PCB	4.91	36
Anabaena variabilis	14,800	14,800	104,000	*5.23	$(\alpha\beta)_3$	1 PCB	1 PCB	4.92	36
Anabaena variabilis allophycocyanin B	15,300	15,300	89,000		$(\alpha\beta)_3$	1 PCB	1 PCB	5.09	39
Phormidium luridum	17,300	19,000	155,000			1 PCB	1 PCB		55
Mastigocladus laminosus									
allophycocyanin II	17,200	17,200	102,500	5.45	$(\alpha\beta)_3$	1 PCB	1 PCB	4.65	10, 38
allophycocyanin I	17,200	17,200	105,000	5.49	$(\alpha\beta)_3$	1 PCB	1 PCB	4.73	10, 56
Cyanophyta									
Aphanocapsa sp. (6701) (CPC)	16,600	20,200						4.17	42
Fremyella diplosiphon (CPC)	16,300	17,600							33
Synechococcus sp. (6301) (CPC)	15,900	19,100	224,000		$\alpha\beta_6$	1 PCB	2 PCB	4.20	42, 49
Anabaena sp. (6411) (CPC)	16,500	19,400	105,000	*5.14	$(\alpha\beta)_3$	1 PCB	2 PCB	4.92	36
Anabaena variabilis (CPC)	16,200	18,800	105,000		$(\alpha\beta)_3$	1 PCB	2 PCB	4.88	36

Table 3 (continued)
CHEMICAL AND PHYSICOCHEMICAL PROPERTIES OF BILIPROTEINS

Organism/Phycobiliprotein	Subunit mol wt		Aggregate size	Sedimentation coefficient*S_{20w} or S^0_{20w}	Proposed subunit structure	Number and type of chromophores per subunit[a]		pI	Ref.
	α	β				α	β		
Anacystis nidulans (CPC)	18,500	20,500	340,000						41
Phormidium luridum (CPC)	11,900	18,500							50
Oscillatoria agardhii (CPC)	12,100	14,100							51
Mastigocladus laminosus (CPC)	16,100	18,300	96,000	5.54	$(\alpha\beta)_3$	1 PCB	2 PCB	4.75	52
Cryptophyta									
Chroomonas sp. (Type III)	10,000	16,000	54,300	4.4	$\alpha_2\beta_2$				34, 53
Hemiselmis virescens (Type I)									
strain Millport 64	10,000	19,000	57,000		$\alpha_2\beta_2$				54
strain Plymouth 157	10,000	19,000	57,000		$\alpha_2\beta_2$			6.5	54
Allophycocyanins									
Rhodophyta									
Porphyridium cruentum	14,600		120,000						8
Cyanidium caldarium	16,300	17,900	196,000		$(\alpha\beta)_6$				47
Cyanophyta									
Aphanocapsa sp. (6701)	16,000	17,000; 17,900						4.26	42
Fremyella diplosiphon	16,000								33
Synechococcus sp. (6301)	15,200	17,200	96,000; 103,000	5.54	$(\alpha\beta)_3$			4.22; 4.4	42, 64
Synechococcus sp. (6301)									
allophycocyanin B	16,000	17,000							39
Anabaena sp. (6411)	14,900	14,900	105,000		$(\alpha\beta)_3$	1 PCB	1 PCB	4.91	36
Anabaena variabilis	14,800	14,800	104,000	*5.23	$(\alpha\beta)_3$	1 PCB	1 PCB	4.92	36
Anabaena variabilis									

allophycocyanin B	15,300	15,300	89,000		$(\alpha\beta)_3$	1 PCB	1 PCB	5.09	39
Phormidium luridum	17,300	19,000	155,000			1 PCB	1 PCB		55
Mastigocladus laminosus									
allophycocyanin II	17,200	17,200	102,500	5.45	$(\alpha\beta)_3$	1 PCB	1 PCB	4.65	10, 38
allophycocyanin I	17,200	17,200	105,000	5.49	$(\alpha\beta)_3$	1 PCB	1 PCB	4.73	10, 56

[a] PEB, phycoerythrobilin; PCB, phycocyanobilin; PXB, bilin chromophore of undetermined structure.

Table 4
AMINO ACID COMPOSITIONS OF DIFFERENT BILIPROTEINS

	Phycoerythrins					Phycoerythrocyanin		
	Porphyra umbilicalis[2] (eukaryote)	*Fremyella diplosiphon*[57] (filamentous prokaryote)			*Rhodomonas lens*[34] (Cryptomonad)	*Anabaena variabilis*[36] (filamentous prokaryote)		
Amino acid	RPE	CPE	α	β	(type I)	PEC	α	β
Lysine	13	11	7	4	16	13	6	7
Histidine	2	2	1	1	1	4	2	1
Arginine	21	16	10	12	9	18	7	11
Aspartic acid	43	33	17	19	27	42	17	28
Threonine	13	24	10	10	11	12	6	5
Serine	30	33	11	23	31	23	13	14
Glutamic acid	24	30	12	14	17	24	12	13
Proline	10	17	8	5	8	10	5	3
Glycine	25	31	18	13	20	27	13	15
Alanine	59	80	34	38	30	50	22	30
Half-cystine[a]	6	2	n.d.[b]	n.d.	9	5	2	2
Valine	30	41	14	23	17	23	10	13
Methionine	7	7	2	6	6	6	3	3
Isoleucine	14	24	9	7	11	16	5	11
Leucine	22	27	11	14	16	28	13	17
Tyrosine	14	14	7	5	6	15	7	6
Phenylalanine	7	7	5	3	6	9	4	5
Tryptophan	0	3	n.d.	n.d.	n.d.	2	n.d.	n.d.
	340	402	176	197	241	327	147	184

Phycocyanins

Porphyridium cruentum[37] (eukaryote)	*Cyanidium caldarium*[46] (eukaryote)	*Synechococcus* sp. (6301)[49] (coccoid; prokaryote)	*Anabaena* sp. (6411)[36] (filamentous prokaryote)

Amino acid	RPC	α	β	CPC	α	β	CPC	α	β	CPC	α	β
Lysine	13	7	5	10	5	6	13	7	5	12	6	6
Histidine	1	1	0	2	1	1	1	1	0	3	3	0
Arginine	17	8	8	17	8	10	20	7	13	18	8	10
Aspartic acid	33	17	16	34	15	20	41	18	21	37	14	24
Threonine	20	13	7	19	9	9	19	10	9	21	12	9
Serine	24	14	11	25	12	14	24	12	11	22	10	11
Glutamic acid	29	17	11	29	15	14	22	11	11	31	16	15
Proline	11	7	4	12	6	5	11	6	4	11	6	5
Glycine	24	15	9	23	12	13	27	13	13	26	13	13
Alanine	48	26	20	44	22	26	59	25	32	48	23	26
Half-cystine[a]	2	2	1	3	1	2	1	n.d.	n.d.	n.d.	n.d.	n.d.
Valine	22	9	12	16	6	10	21	8	13	21	6	16
Methionine	9	4	4	9	4	5	3	1	3	3	0	3
Isoleucine	17	10	6	13	7	6	18	7	10	18	9	9
Leucine	28	15	12	25	12	14	31	16	14	29	13	15
Tyrosine	16	12	4	14	9	5	15	10	5	14	8	5
Phenylalanine	7	4	3	9	3	4	12	6	6	8	4	4
Tryptophan	1	1	0	n.d.	n.d.	n.d.	1	1	0	n.d.	n.d.	n.d.
	322	182	133	304	147	164	339	159	170	322	151	171

	Phycocyanins						Allophycocyanins			
	Mastigocladus laminosus[9] (filamentous prokaryote)			*Chroomonas* sp.[53] (Cryptomonad)			*Cyanidium caldarium*[47] (eukaryote)			*Synechococcus lividus*[58] (coccoid; prokaryote)
Amino acid	CPC	α	β	(type III)	α	β	APC	α	β	APC
Lysine	12	6	6	17	8	8	13	7	8	15
Histidine	2	2	0	1	1	0	1	1	0	1
Arginine	18	8	10	9	4	7	16	8	7	22
Aspartic acid	40	18	22	28	11	17	28	15	14	30
Threonine	19	9	10	10	4	6	16	7	10	17
Serine	21	9	12	24	5	18	22	11	14	23
Glutamic acid	30	15	15	17	9	8	28	17	12	33

Table 4 (continued)
AMINO ACID COMPOSITIONS OF DIFFERENT BILIPROTEINS

	Phycocyanins						Allophycocyanins			
	Mastigocladus laminosus[9] (filamentous prokaryote)			*Chroomonas* sp.[53] (Cryptomonad)			*Cyanidium caldarium*[47] (eukaryote)			*Synechococcus lividus*[58] (coccoid; prokaryote)
Amino acid	CPC	α	β	(type III)	α	β	APC	α	β	APC
Proline	9	5	4	6	3	4	9	5	3	13
Glycine	24	12	12	19	7	13	26	15	14	26
Alanine	56	29	27	33	10	20	31	15	18	44
Half-cystine[a]	4	1	3	11	3	6	3	n.d.	n.d.	6
Valine	17	5	12	16	5	10	16	8	6	24
Methionine	3	0	3	4	2	2	7	1	4	11
Isoleucine	26	13	13	10	4	6	18	9	8	19
Leucine	29	15	14	19	5	13	23	12	13	30
Tyrosine	16	11	5	7	2	5	16	6	10	17
Phenylalanine	7	3	4	5	2	3	6	4	2	8
Tryptophan	1	1	0	n.d.	n.d.	n.d.	n.d.	n.d.	n.d.	n.d.
	334	162	172	236	85	146	279	141	143	339

	Allophycocyanins									
	Synechococcus sp. (6301)[52] (coccoid; prokaryote)			*Mastigocladus laminosus*[38] (filamentous prokaryote)			*Anabaena variabilis*[36] (filamentous prokaryote)			*Anabeana variabilis*[39] (filamentous prokaryote)
Amino acid	APC	α	β	APC	α	β	APC	α	β	Allophycocyanin-B
Lysine	11	5	6	16	7	8	11	5	6	13
Histidine	0	0	0	0	0	0	0	0	0	0
Arginine	20	11	9	18	10	8	15	9	6	14

Aspartic acid	26	12	15	31	16	15	25	12	12	25
Threonine	18	9	9	19	7	12	16	6	9	15
Serine	22	9	15	24	12	12	16	8	10	18
Glutamic acid	28	17	12	30	17	12	26	16	9	27
Proline	7	4	3	12	5	4	7	4	3	9
Glycine	26	15	12	25	15	12	23	14	10	22
Alanine	35	16	21	46	22	24	38	16	22	37
Half-cystine[a]	n.d.	n.d.	n.d.	3	1	2	0	n.d.	n.d.	n.d.
Valine	25	13	11	26	14	12	23	12	9	20
Methionine	5	2	3	7	3	4	4	2	2	5
Isoleucine	20	10	10	24	13	11	17	7	8	23
Leucine	28	12	17	28	13	15	27	13	13	27
Tyrosine	18	7	11	19	8	11	16	7	10	16
Phenylalanine	4	3	1	5	3	2	3	3	1	4
Tryptophan	n.d.	n.d.	n.d.	0	n.d.	n.d.	0	n.d.	n.d.	n.d.
	293	145	155	333	166	164	267	134	130	275

[a] Determined as cysteic acid after performic acid oxidation.

[b] n.d. = not determined

Table 5
THE PRIMARY STRUCTURE OF C-PHYCOCYANIN FROM *MASTIGOCLADUS LAMINOSUS*[9] AND AMINO-TERMINAL SEQUENCES OF BILIPROTEIN SUBUNITS

```
Porphyridium cruentum 11                       1        10
  (eukaryote, Rhodophyta)      b-PE, α:       MKSVIXTVVXAADAAGRFP
                                               1        10
                                   , β:       MLDAFTRVVVNADAKAAYV

Porphyridium cruentum 11                       1        10        20
  (eukaryote, Rhodophyta)      RPC, α:        MKTPITEAIATADNQGRFLXN
                                               1        10
                                   , β:       MLDAFAKVVAQADARGEFL

Cyanidium caldarium 46                         1        10
  (eukaryote, Rhodophyta)      CPC, α:        MKTPITEAIAAABARG
                                               1        10
                                   , β:       MLBAFAKVVAAABARGEFK

Synechococcus sp. 6301 60                      1        10
  (unicellular, Cyanophyta)    CPC, α:        SKTPLXEAVAAABXXG
                                               1        10
                                   , β:       TFDAFTKVVAQADARGEFLS

Mastigocladus laminosus 9                                                                                      PCB
                                               1        10        20        30        40        50        60        70        80   |
  (filamentous, Cyanophyta)    CPC, α:        VKTPITDAIAAADTQGRFLSNTELQAVNGRYQRAAASLEAAR[illegible]LTANAQRLIDGAAQAVYQKFPYLIQTSGPNYAADARGKSKCARDIG
                                               1        10        20        30        40        50        60        70        80
                                   , β:       AYDVFTKVVSQADSRGEFLSNEQLDALANVVKEGNKRLDVVNRITSNASTIVTNAARALFEEQPQLIAPGGS  ATRNGTMAACLRDMEII
                                                                                                                          |
                                                                                                                         PCB
                                              90        100       110       120       130       140       150       160
  (continued)                  CPC, α:        HYLRIITYSLVAGGTGPLDEYLIAGLNEINDAFELSPSWYIEALKYIKANHGLSGQAANEANTYIDYVINALS
                                                90        100       110       120       130       140       150       160       170
                                   , β:       IILRYITYAILAGDASILDDRCLNGLRETYQALGTPGSSVAVGIQKMKEAAINIANDPNGITKGDCSALISEVASYFDRAAAAVA
                                                                                                                                  |
                                                                                                                                 PCB

Porphyridium cruentum 11                         1        10
  (eukaryote, Rhodophyta)      APC, α:          SIVTKXIVNADAEARYL
                                               1        10        20
                                   , β:       MQDAITXVINAADVQGKYLD

Microcystis aeruginosa 11                        1        10        20
  (unicellular, Cyanophyta)    APC, α:          SIVTKXIVNADAEARYLLPGE
                                               1        10        20
                                   , β:       MQDAIXXVINXXDVQXKYLD

Mastigocladus laminosus 10                       1        10        20        30        40        50        60        70
  (filamentous, Cyanophyta)    APC, α:          SIVTKSIVNADAEARYLSPGELDRIKSFVSSGEKRLRIAQILTDNRERIVKQAGDQLFQKRPDVVSPGGNA
                                               1        10        20        30        40
                                   , β:       MQDAITAVINSSDVQGKYLDTAALEKLKSYFSTGELRVRAATTIAANA
```

Note: One-letter abbreviations are used (*Hoppe-Seyler's Z. Physiol. Chem.*, 350, 793, 1969); PCB, phycocyanobilin chromophore covalently bound to a cystein residue.

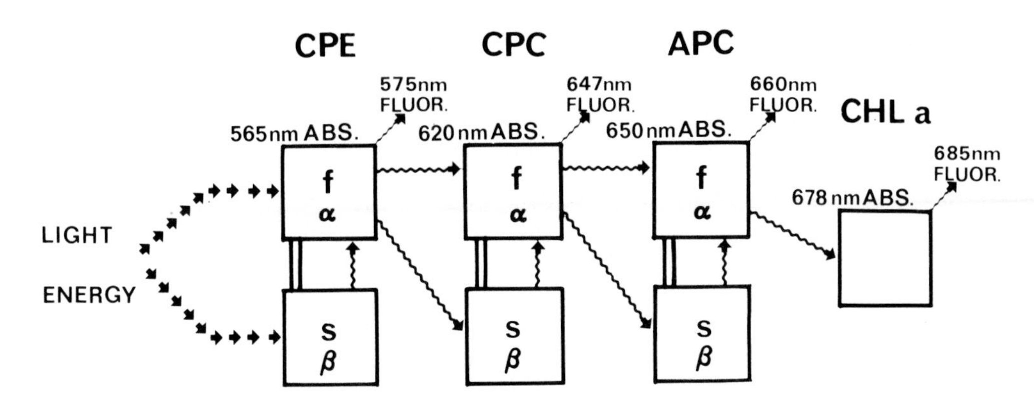

FIGURE 3. Schematic diagram of exciton migration (〰) from C-phycoerythrin (CPE) to C-phycocyanin (CPC) to allophycocyanin (APC) to Chl *a* in a Cyanobacterium. ABS. and FLUOR. refer to the principal absorption and fluorescence emission maximum, respectively, of the intact ($\alpha\beta$) monomer. The designations fα and sβ refer, respectively, to the α subunit with the "fluorescing" chromophore and the β subunit with the "sensitizing" chromophore.

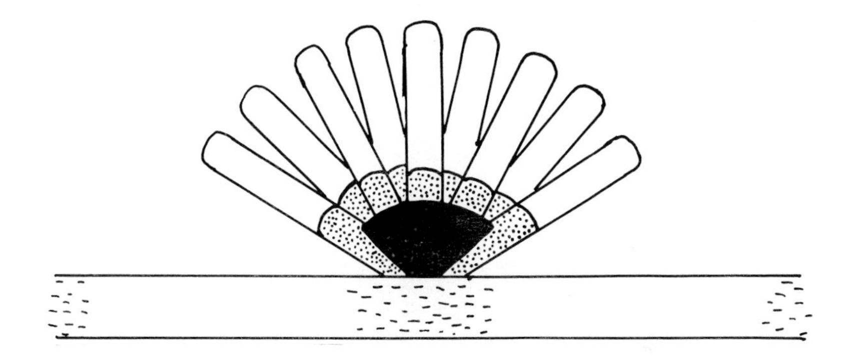

FIGURE 4. Phycobilisome. Model proposed for the red alga *Porphyridium cruentum*, showing an allophycocyanin (black) core surrounded by spoke-like projections consisting of phycocyanin (stippled) and phycoerythrin (clear area). Drawn by Tredwell, C. J., after Gantt et al., *Biochim. Biophys. Acta*, 430, 375-388, 1976.

Table 6
EXTINCTION COEFFICIENTS[a] FOR 1% SOLUTIONS OF BILIPROTEINS AT THEIR VISIBLE ABSORPTION MAXIMUM[2,34,53,61,65]

Biliprotein	$E^{1\%}_{1cm}$
B-phycoerythrin	82
R-phycoerythrin	81
C-phycoerythrin	126
R-phycocyanin	65
C-phycocyanin	65
Allophycocyanin	63.5
Cryptomonad phycoerythrin Type I	126
Cryptomonad phycoerythrin Type II	99
Cryptomonad phycocyanin Type III	114

[a] Extinction values are pH and concentration dependent. Dilute biliprotein solutions give comparatively lower values and therefore cannot simply be multiplied with a corresponding factor.

REFERENCES

1. **O'hEocha, C.**, Biliproteins of algae, *Annu. Rev. Plant Physiol.*, 16, 415, 1965.
2. **O'Carra, P. and O'hEocha, C.**, Algal biliproteins and phycobilins, in *Chemistry and Biochemistry of Plant Pigments*, Vol. 1, 2nd ed., Goodwin, T. W., Ed., Academic Press, New York, 1976, 328.
3. **Chapman, D. J., Cole, W. J., and Siegelman, H. W.**, Chromophores of allophycocyanin and R-phycocyanin, *Biochem. J.*, 105, 903, 1967.
4. **Bennett, A. and Siegelman, H. W.**, Bile pigments in plants, in *The Porphyrins*, Vol. 3, Dolphin, D., Ed., Academic Press, London, 1979.
5. **Chapman, D. J., Cole, W. J., and Siegelman, H. W.**, A comparative study of the phycoerythrin chromophore, *Phytochemistry*, 7, 1831, 1968.
6. **Troxler, R. F., Brown, A. S., and Köst, H.-P.**, Quantitative degradation of radiolabeled phycobiliproteins. Chromic acid degradation of C-phycocyanin, *Eur. J. Biochem.*, 87, 181, 1978.
7. **Köst, H.-P., Rüdiger, W., and Chapman, D. J.**, Über die Bindungen zwischen Chromophor und Protein in Biliproteiden. I. Abbauversuche und Spektraluntersuchungen an Biliproteiden, *Liebigs Ann. Chem.*, 1582, 1975.
8. **Gantt, E. and Lipschultz, C. A.**, Phycobilisomes of *Porphyridium cruentum*: Pigment analysis, *Biochemistry*, 13, 2960, 1974.
9. **Frank, G., Sidler, W., Widmer, H., and Zuber, H.**, The complete amino acid sequence of both subunits of C-phycocyanin from the cyanobacterium *Mastigocladus laminosus*, *Hoppe-Seylers Z. Physiol. Chem.*, 353, 1491, 1978.
10. **Gysi, J.**, Isolation and Characterization of Allophycocyanin I and II and Isolation of C-phycoerythrin from the Thermophilic Blue-Green Alga *Mastigocladus laminosus*, Ph.D. thesis No. 5813, Eidgenössische Technische Hochschule, Zürich, Switzerland, 1976.
11. **Glazer, A. N., Apell, G. S., Hixson, C. S., Bryant, D. A., Rimon, S., and Brown, D. M.**, Biliproteins of cyanobacteria and rhodophyta: homologous family of photosynthetic accessory prigments, *Proc. Natl. Acad. Sci. U.S.A.* 73, 428, 1976.
12. **Glazer, A. N., Phycocyanins: structure and function, in** *Photochemical and Photobiological Reviews*, Vol. 1, Smith, K. C., Ed., Plenum Press, New York, 1976, 71.
13. **French, C. S. and Young, V. K.**, The fluorescence spectra of red algae and the transfer of energy from phycoerythrin to phycocyanin and chlorophyll, *J. Gen. Physiol.*, 35, 873, 1952.

14. **Förster, T.**, Zwischenmolekulare Energiewanderung und Fluoreszenz, *Ann. Physik.*, 2, 55, 1948.
15. **Duysens, L. N. M.**, Transfer of Excitation Energy in Photosynthesis, Ph.D. thesis, University of Utrecht, The Netherlands, 1952.
16. **Teale, F. W. J. and Dale, R. E.**, Isolation and spectral characterization of phycobiliproteins, *Biochem. J.*, 116, 161, 1970.
17. **Glazer, A. N., Fang, S., and Brown, D. M.**, Spectroscopic properties of C-phycocyanin and of its α and β subunits, *J. Biol. Chem.*, 248, 5679, 1973.
18. **Gantt, E. and Conti, S. F.**, Granules associated with the chloroplast lamellae of *Porphyridium cruentum*, *J. Cell Biol.*, 29, 423, 1966.
19. **Gantt, E., Edwards, M. R., and Conti, S. F.**, Ultrastructure of *Porphyridium aerugineum* a blue-green colored Rhodophytan, *J. Phycol.*, 4, 65, 1968.
20. **Gantt, E. and Conti, S. F.**, Ultrastructure of blue-green algae, *J. Bacteriol.*, 97, 1486, 1969.
21. **Gantt, E., Edwards, M. R., and Provasoli, L.**, Chloroplast structure of the Cryptophyceae. Evidence for phycobiliproteins within intrathylakoidal spaces, *J. Cell Biol.*, 48, 280, 1971.
22. **Gantt, E., Lipschultz, C. A., and Zilinskas, B.**, Further evidence for a phycobilisome model from selective dissociation, fluorescence emission, immunoprecipitation and electron microscopy, *Biochim. Biophys. Acta*, 430, 375, 1976.
23. **Gantt, E. and Lipschultz, C. A.**, Energy transfer in phycobilisomes from phycoerythrin to allophycocyanin, *Biochim. Biophys. Acta*, 292, 858, 1973.
24. **Gray, B. H., Lipschultz, C. A., and Gantt, E.**, Phycobilisomes from a blue-green alga *Nostoc* species, *J. Bacteriol.*, 116, 471, 1973.
25. **Haxo, F. T. and Blinks, L. R.**, Photosynthetic action spectra of marine algae, *J. Gen. Physiol.*, 33, 389, 1950.
26. **Blinks, L. R.**, Accessory pigments and photosynthesis, in *Photophysiology*, Giese, A. C., Ed., Academic Press, New York, 1964, 199.
27. **Porter, G., Tredwell, C. J., Searle, G. F. W., and Barber, J.**, Picosecond time-resolved energy transfer in *Porphyridium cruentum*, *Biochim. Biophys. Acta*, 501, 232, 1978.
28. **Troxler, R. F.**, Synthesis of bile pigments in plants. Formation of carbon monoxide and phycocyanobilin in wild type and mutant strains of the alga, *Cyanidium caldarium*, *Biochemistry*, 11, 4235, 1972.
29. **Troxler, R. F., Kelly, P., and Brown, S. B.**, Phycocyanobilin synthesis in the unicellular Rhodophyte *Cyanidium caldarium*, *Biochem. J.*, 172, 569, 1978.
30. **Troxler, R. F.**, Synthesis of bile pigments in plants, in Proc. Int. Symp. Chemistry and Physiology of Bile Pigments, Berk, P. D. and Berlin, N. I., Eds., Department of Health, Education and Welfare Publication No. (NIH) 77-1100, 1977, 431.
31. **Bogorad, L.**, Phycobiliproteins and complementary chromatic adaptation, *Annu. Rev. Plant Physiol.*, 26, 369, 1975.
32. **Bennett, A. and Bogorad, L.**, Complementary chromatic adaptation in a filamentous blue-green alga, *J. Cell Biol.*, 58, 419, 1973.
33. **Bennett, A. and Bogorad, L.**, Properties of subunits and aggregates of blue green algal biliproteins, *Biochemistry*, 10, 3625, 1971.
34. **MacColl, R., Berns, D., and Gibbons, O.**, Characterization of cryptomonad phycoerythrin and phycocyanin, *Arch. Biochem. Biophys.*, 177, 265, 1976.
35. **O'hEocha, C., O'Carra, P., and Mitchell, D.**, Biliproteins of cryptomonad algae, *Proc. R. Ir. Acad. Sect. B*, 63, 191, 1964.
36. **Bryant, D. A., Glazer, A. N., and Eiserling, F. A.**, Characterization and structural properties of the major biliproteins of *Anabaena* sp., *Arch. Mikrobiol.*, 110, 61, 1976.
37. **Glazer, A. N. and Hixson, C. S.**, Characterization of R-phycocyanin; chromophore content of R-phycocyanin and C-phycoerythrin, *J. Biol. Chem.*, 250, 5487, 1975.
38. **Gysi, J. and Zuber, H.**, Isolation and characterization of allophycocyanin II from the thermophilic blue-green alga *Mastigocladus laminosus* Cohn, *FEBS Lett.*, 48, 209, 1974.
39. **Glazer, A. N. and Bryant, D. A.**, Allophycocyanin B (λ max 671, 618 nm). A new cyanobacterial phycobiliprotein, *Arch. Mikrobiol.*, 104, 15, 1975.
40. **Van der Velde, H. H.**, The natural occurrence in red algae of two phycoerythrins with different molecular weights and spectral properties, *Biochim. Biophys. Acta*, 303, 246, 1973.
41. **O'Carra, P. and Killilea, S. D.**, Subunit structures of C-phycocyanin and C-phycoerythrin, *Biochem. Biophys. Res. Commun.*, 45, 1192, 1971.
42. **Glazer, A. N. and Cohen-Bazire, G.**, Subunit structure of the phycobiliproteins of blue-green algae, *Proc. Natl. Acad. Sci. U.S.A.*, 68, 1398, 1971.
43. **Glazer, A. N., Cohen-Bazire, G., and Stanier, R. Y.**, Characterization of phycoerythrin from a *Cryptomonas* sp., *Arch. Mikrobiol.*, 80, 1, 1971.

44. **Brooks, C. and Gantt, E.**, Comparison of phycoerythrins (542, 566 nm) from cryptophycean algae, *Arch. Mikrobiol.*, 88, 193, 1972.
45. **Eriksson-Quensel, I. B.**, The molecular weights of phycoerythrin and phycocyanin, *Biochem. J.*, 32, 585, 1938.
46. **Troxler, R. F., Foster, J. A., Brown, A. S., and Franzblau, C.**, The α and β subunits of *Cyanidium caldarium* phycocyanin: properties and amino acid sequences at the amino terminus, *Biochemistry*, 14, 268, 1975.
47. **Brown, A. S. and Troxler, R. F.**, Properties and N-terminal sequence of allophycocyanin from the unicellular rhodophyte *Cyanidium caldarium*, *Biochem. J.*, 163, 571, 1977.
48. **Kao, O. H. W., Edwards, M. R., and Berns, D. S.**, Physical-chemical properties of C-phycocyanin isolated from an acido-thermophilic eukaryote, *Cyandium caldarium*, *Biochem. J.*, 147, 63, 1975.
49. **Glazer, A. N. and Fang, S.**, Chromophore content of blue green algal phycobiliproteins, *J. Biol. Chem.*, 248, 659, 1973.
50. **Kobayashi, Y., Siegelman, H. W., and Hirs, C. H. W.**, C-phycocyanin from *Phormidium luridum*: isolation of subunits, *Arch. Biochem. Biophys.*, 152, 187, 1972.
51. **Torjesen, P. A. and Sletten, K.**, C-phycocyanin from *Oscillatoria agardhii*. I. Some molecular properties, *Biochim. Biophys. Acta*, 263, 258, 1972.
52. **Binder, A.**, Isolation and Characterization of C-phycocyanin from the Thermophilic Blue-Green Alga *Mastigocladus laminosus*, Ph.D. thesis No. 5051, Eidgenössische Technische Hochschule, Zürich, Switzerland, 1973.
53. **MacColl, R., Habig, W., and Berns, D. S.**, Characterization of phycocyanin from *Chroomonas* species, *J. Biol. Chem.*, 248, 7080, 1973.
54. **Glazer, A. N. and Cohen-Bazire, G.**, A comparison of cryptophytan phycocyanins, *Arch. Mikrobiol.*, 104, 29, 1975.
55. **Brown, A. S., Foster, J. A., Voynow, P. V., Franzblau, C., and Troxler, R. F.**, Allophycocyanin from the filamentous cyanophyte *Phormidium luridum*, *Biochemistry*, 14, 3581, 1975.
56. **Gysi, J. and Zuber, H.**, Allophycocyanin I — a second cyanobacterial allophycocyanin? Isolation, characterization and comparison with allophycocyanin II from the same alga, *FEBS Lett.*, 68, 49, 1976.
57. **Takemoto, J. and Bogorad, L.**, Subunits of phycoerythrin from *Fremyella diplosiphon*: chemical and immunochemical characterization, *Biochemistry*, 14, 1211, 1975.
58. **MacColl, R., Edwards, M. R., and Haaksma, C.**, Some properties of allophycocyanin from a thermophilic blue-green alga, *Biophys. Chem.*, 13, 000, 1978.
59. **Cohen-Bazire, G., Bẽguin, S., Rimon, S., Glazer, A. N., and Brown, D. M,.** Physico-chemical and immunological properties of allophycocyanins, *Arch. Microbiol.*, 111, 225, 1977.
60. **Williams, V. P., Freidenreich, P., and Glazer, A. N.**, Homology of amino-terminal regions of C-phycocyanins from a prokaryote and a eukaryote, *Biochem. Biophys. Res. Commun.*, 59, 462, 1974.
61. **Chapman, D. J.**, Biliproteins and bile pigments, in *The Biology of Blue-Green Algae*, Vol. 9, Carr, N. G. and Whitton, B. A., Eds., Blackwell, Oxford, 1973, 162.
62. **Pecci, J. and Fujimori, E.**, Mercurial-induced circular dichroism changes of phycoerythrin and phycocyanin, *Biochim. Biophys. Acta*, 188, 230, 1963.
63. **Fujimori, E. and Pecci, J.**, Circular dichroism of single- and double-peaked phycoerythrin: mercurial induced changes, *Biochim. Biophys. Acta*, 221, 132, 1970.
64. **MacColl, R., Edwards, M. R., Mulks, M. H., and Berns, D. S.**, Comparison of the biliproteins from two strains of the thermophylic cyanophyte *Synechococcus lividus*, *Biochem. J.*, 141, 419, 1974.
65. **MacColl, R. and Berns, D. S.**, Energy tranfer studies on cryptomonad biliproteins, *Photochem. Photobiol.*, 27, 343, 1978.
66. **Tandeau de Marsac, N. and Cohen-Bazire, G.**, Molecular composition of cyanobacterial phycobilisomes, *Proc. Nat. Acad. Sci. USA*, 74, 1635, 1977.
67. **Gantt, E., Lipschultz, C. A., and Zilinskas, B.**, Probing phycobilisome structure by immuno-electron microscopy, *J. Phycol.*, 13, 185, 1977.
68. **Koller, K. P., Wehrmeyer, W., and Schneider, H.**, Isolation and characterization of disc-shaped phycobilisomes from the red alga *Rhodella violacea*, *Arch. Microbiol.*, 112, 61, 1977.
69. **Mörschel, E., Koller, K. P., Wehrmeyer, W. and Schneider, H.**, Biliprotein assembly in the disc-shaped phycobilisomes of *Rhodella violacea*. I. Electron microscopy of phycobilisomes in situ and analysis of their architecture after isolation and negative staining, *Cytobiologie*, 16, 118, 1977.
70. **Koller, K. P., Wehrmeyer, W., and Mörschel, E.**, Biliprotein assembly in the disc-shaped phycobilisomes of *Rhodella violacea*. On the molecular composition of energy-transfering complexes (tripartite units) forming the periphery of the phycobilisome, *Eur. J. Biochem.*, 91, 57, 1978.
71. **Yamanaka, G., Glazer, A. N., and williams, R. C.**, Cyanobacterial phycobilisomes, *J. Biol. Chem.*, 253, 8303, 1978.

72. **Glazer, A. N., Williams, R. C., Yamanaka, G., and Schachman, H. K.**, Characterization of cyanobacterial phycobilisomes in zwitterionic detergents, *Proc. Natl. Acad. Sci. USA*, 76, 6162, 1979.
73. **Bryant, D. A., Guglielmi, G., Tandeau de Marsac, N., Castets, A. M., and Cohen-Bazire, G.**, Structure of cyanobacterial phycobilisomes, *Arch. Microbiol.*, 123, 113, 1979.
74. **Gantt, E., Lipschultz, C. A., Grabowski, J., and Zimmermann, B. K.**, Phycobilisomes from blue-green and red algae, *Plant Physiol.*, 63, 615, 1979.
75. **Jung, J., Song, P. S., Paxton, R. J., Edelstein, M. S., Swanson, R., and Hazer, E. E., Jr.**, Molecular topography of the phycocyanin photoreceptor from *Chroomonas* sp., *Biochemistry*, 19, 24, 1980.
76. **Yamanaka, G., Glazer, A. N., and Williams, R. C.**, Molecular architecture of a light harvesting antenna, *J. Biol. Chem.*, 255, 11004, 1980.
77. **Wanner, G. and Kost, H. P.**, Investigations on the arrangement and fine structure of *Porphyridium cruentum* phycobilisomes, *Protoplasma*, 102, 97, 1980.
78. **Canaani, O., Lipschultz, C. A., and Gantt, E.**, Reassembly of phycobilisomes from allophycocyanin and a phycocyanin-phycoerythrin complex, *FEBS Lett.*, 115, 225, 1980.
79. **Rigbi, M., Rosinski, J., Siegelman, H. W., and Sutherland, J. C.**, Cyanobacterial phycobilisomes: Selective dissociation monitored by fluorescence and circular dichroism, *Proc. Natl. Acad. Sci. USA*, 77, 1961, 1980.
80. **Yamanaka, G. and Glazer, A. N.**, Dynamic aspect of phycobilisome structure. Phycobilisome turnover during nitrogen starvation in *Synechococcus*, *Arch. Microbiol.*, 124, 39, 1980.
81. **Mörschel, E., Koller, K. P., and Wehrmeyer, W.**, Biliprotein assembly in the disc-shaped phycobilisomes of *Rhodella violacea*. Electron microscopical and biochemical analyses of C-phycocyanin and allophycocyanin aggregates, *Arch. Microbiol.*, 125, 43, 1980.
82. **Mörschel, E., Wehrmeyer, W., and Koller, K. P.**, Biliprotein assembly in the disc-shaped phycobilisomes of *Rhodella violacea*. Electron microscopical and biochemical analysis of B-phycoerythrin and B-phycoerythrin-C-phycocyanin aggregates, *Eur. J. Cell Biol.*, 21, 319, 1980.
83. **Lundell, D. J., Williams, R. C., and Glazer, A. N.**, Molecular architecture of a light harvesting antenna. In vitro assembly of the rod substructures of *Synechococcus* sp. 6301 phycobilisomes, *J. Biol. Chem.*, 256, 3580, 1981.
84. **Mörschel, E. and Wehremeyer, W.**, Multiple forms of phycoerythrin 545 from *Cryptomonas maculata*, *Arch. Microbiol.*, 113, 83, 1977.
85. **Glazer, A. N. and Hixson, C. S.**, Subunit structure and chromophore composition of rhodophytan phycoerythrins, *J. Biol. Chem.*, 252, 32, 1977.
86. **Zilinskas, B. A., Zimmermann, B. K., and Gantt, E.**, Allophycocyanin forms isolated from *Nostoc* sp. phycobilisomes, *Photochem. Photobiol.*, 27, 587, 1978.
87. **MacColl, R., Edwards, M. R., and Haaksma, C.**, Some properties of allophycocyanin from a thermophilic blue-green alga, *Biophys. Chem.*, 8, 369, 1978.
88. **Chen, C. H. and Berns, D. S.**, Comparison of the stability of phycocyanins from thermophilic, mesophilic, psychrophilic and halophilic algae, *Biophys. Chem.*, 8, 203, 1978.
89. **Muckle, G., Otto, J., and Rüdiger, W.**, Amino acid sequence in the chromophore regions of C-phycoerythrin from *Pseudanabaena* W 1173 and *Phormidium persicinum*, *Hoppe-Seyler's Z. Physiol. Chem.*, 359, 345, 1978.
90. **Eder, J., Wagenmann, R., and Rüdiger, W.**, Immunological relationship between phycoerythrins from various blue-green algae, *Immunochemistry*, 15, 315, 1978.
91. **Williams, V. P. and Glazer, A. N.**, Structural studies on phycobiliproteins, *J. Biol. Chem.*, 253, 202, 1978.
92. **Zuber, H.**, Studies on the structure of the light-harvesting pigment-protein-complexes from cyanobacteria and red algae, *Ber. Deutsch. Bot. Ges.*, 91, 459, 1978.
93. **Gysi, J. R. and Zuber, H.**, Properties of allophycocyanin II and its α- and β- subunits from the thermophilic blue-green alga *Mastigocladus laminosus*, *Biochem. J.*, 181, 577, 1979.
94. **Gantt, E.**, Phycobiliproteins of cryptophyceae, in *Biochemistry and Physiology of Protozoa*, 2nd ed., Levandowsky, M. S. and Hutner, S. H., Eds., Academic Press, New York, 1971, 121.
95. **Ohad, I., Clayton, R. K., and Bogorad, L.**, Photoreversible absorbance changes in solutions of allophycocyanin purified from *Fremyella diplosiphon*: Temperature dependence and quantum efficiency, *Proc. Natl. Acad. Sci. USA*, 76, 5655, 1979.
96. **Brown, A. S., Offner, G. D., Ehrhardt, M. M., and Troxler, R. F.**, Phycobilin-apoprotein linkages in the α and β subunits of phycocyanin from the unicellular rhodophyte, *Cyanidium caldarium*, *J. Biol. Chem.*, 254, 7803, 1979.
97. **Troxler, R. F., Greenwald, L. S., and Zilinskas, B. A.**, Allophycocyanin from *Nostoc* sp. phycobilisomes. Properties and amino acid sequences at the NH_2 terminus of the α and β subunits of allophycocyanins I, II, III, *J. Biol. Chem.*, 255, 9380, 1980.
98. **Fisher, R. G., Wood, N. E., Fuchs, H. E., and Sweet, R. M.**, Three dimensional structure of C-phycocyanin and B-phycoerythrin at 5 Å resolution, *J. Biol. Chem.*, 255, 5082, 1980.

99. **Nies, M. and Wehrmeyer, W.**, Isolation and biliprotein characterization of phycobilisomes from the thermophilic cyanobacterium *Mastigocladus laminosus, Planta,* 150, 330, 1980.
100. **Zickendraht-Wendelstadt, B., Friedrich, J., and Rüdiger, W.**, Spectral characterization of monomeric C-phycoerythrin from *Pseudanabaena* W 1173 and its α and β subunits: Energy transfer in isolated subunits and C-phycoerythrin, *Photochem. Photobiol.*, 31, 367, 1980.
101. **Sidler, W., Gysi, J., Isker, E., and Zuber, H.**, The complete amino acid sequence of both subunits of allophycocyanin, a light haresting protein-pigment complex from the cyanobacterium *Mastigocladus laminosus, Hoppe-Seyler's Z. Physiol. Chem.*, in press.
102. **Gossauer, A. and Hirsch, W.**, Totalsynthese des racemischen Phycocyanobilins (Phycobiliverdins) sowie eines "Homophycobiliverdins," *Liebigs Ann. Chem.*, 1496, 1974.
103. **Beuhler, J., Pierce, R. C., Friedman, L., and Siegelman, H. W.**, Cleavage of phycocyanobilin from C-phycocyanin, *J. Biol. Chem.*, 251, 2405, 1976.
104. **Scheer, H. and Kufer, W.**, Conformational studies on C-phycocyanin from *Spirulina platensis, Z. Naturforsch.*, 32c, 513, 1977.
105. **Gossauer, A. and Hinze, R. P.**, An improved chemical synthesis of racemic phycocyanobilin dimethylester, *J. Org. Chem.*, 43, 283, 1978.
106. **Klein, G. and Rüdiger, W.**, Stereochemie von Modell-Imiden, *Liebigs Ann. Chem.*, 267, 1978.
107. **Fu, E., Friedman, L., and Siegelman, H. W.**, Mass spectral identification and purification of phycoerythrobilin and phycocyanobilin, *Biochem. J.*, 179, 1, 1979.
108. **Gossauer, A. and Klahr, E.**, Totalsynthese des racem. Phycoerythrobilin-dimethylesters, *Chem. Ber.*, 112, 2243, 1979.
109. **Lagarias, J. C., Glazer, A. N., and Rapoport, H.**, Chromopeptides from C-phycocyanin. Structure and linkage of a phycocyanobilin bound to the β-subunit, *J. Am. Chem. Soc.*, 101:17, 5030, 1979.
110. **Scheer, H., Formanek, H., and Rüdiger, W.**, The conformation of bilin chromophores in biliproteins: Ramachandran-type calculations, *Z. Naturforsch.*, 34c, 1085, 1979.
111. **MacColl, R., Csatorday, K., Berns, D. S., and Traeger, E.**, Chromophore interaction in allophycocyanin, *Biochemistry,* 19, 2817, 1980.
112. **Schneider, H. A. W. and Bogorad, L.**, Spectral response curves for the formation of phycobiliproteins, chlorophyll and -aminolevulinic acid in *Cyanidium caldarium, Z. Pflanzenphysiol.*, 94, 449, 1979.
113. **Troxler, R. F., Brown, A. S., and Brown, S. B.**, Bile pigment synthesis in plants, *J. Biol. Chem.*, 254, 3411, 1979.
114. **Gendel, S., Ohad, I., and Bogorad, L.**, Control of phycoerythrin synthesis during chromatic adaptation, *Plant Physiol.*, 64, 786, 1979.
115. **Brown, S. B., Holroyd, A. J., and Troxler, R. F.**, Mechanism of bile pigment synthesis in algae. ^{18}O incorporation into phycocyanobilin in the unicellular rhodophyte, *Cyanidium caldarium, Biochem. J.*, 190, 445, 1980.
116. **Brown, S. B., Holroyd, A. J., Troxler, R. F., and Offner, G. D.**, Bile pigment synthesis in plants. Incorporation of heme into phycocyanobilin and phycobiliproteins in *Cyanidium caldarium, Biochem. J.*, 194, 137, 1981.
117. **Gantt, E.**, Recent contributions in phycobiliproteins and phycobilisomes, *Photochem. Photobiol.*, 26, 685, 1977.
118. **Glazer, A. N.**, Structure and molecular organization of the photosynthetic accessory pigments of cyanobacteria and red algae, *Mol. Cell Biochem.*, 18, 125, 1977.
119. **Rüdiger, W.**, Struktur and Spektraleigenschaft von Phycobilinen und Biliproteiden, *Ber. Deutsch. Bot. Ges.*, 92, 413, 1979.
120. **MacColl, R. and Berns, D. S.**, Evolution of the biliproteins, *Trends Biochem. Sci.*, 2, 44, 1979.
121. **Gantt, E.**, Structure and function of phycobilisomes: Light harvesting pigment complexes in red and blue-green algae, *Int. Rev. Cytology,* 66, 45, 1980.
122. **Scheer, H.**, Biliproteins, *Angew. Chem. Int. Ed. Engl.*, 20, 241, 1981.
123. **Gantt, E.**, Phycobilisomes. *Ann. Rev. Plant Physiol.*, 32, 327, 1981.
124. **Kobayashi, T., Degenkolb, E. O., Bersohn, R., Rentzepis, P. M., MacColl, R., and Berns, D. S.**, Energy transfer among chromophores in phycocyanin measured by picosecond kinetics, *Biochemistry,* 18, 5073, 1979.

BASIC ELECTRON TRANSPORT AND ATP SYNTHESIS PATHWAYS IN PHOTOSYNTHETIC BACTERIA

Donald L. Keister

INTRODUCTION

Kamen defined photosynthesis as a series of processes in which electromagnetic energy is converted to chemical free energy which can be used for biosynthesis.[1] Energy transformations in photosynthesis may be divided into several stages including

1. Initial absorption of light quanta
2. Separation of charge
3. Dark electron transport
4. Energy conversion (ion movements and formation of energy-rich phosphate compounds)

INITIAL LIGHT ABSORPTION AND CHARGE SEPARATION

Photosynthesis in bacteria is a cyclic electron transfer process that is coupled to the synthesis of ATP or PP_i. No net oxidant or reductant is consumed or formed. Photosynthesis begins with the adsoprtion of a light quantum by antenna bacteriochlorophyll (Bchl) complexes which act as a funnel to transfer the light energy to the reaction center. Much of our knowledge of primary reactions in photosynthesis stems from studies with photosynthetic bacteria and their reaction center (RC) preparations. Reed and Clayton[2] first succeeded in isolating a RC complex from chromatophore membranes of *Rhodopseudomonas sphaeroides*, and RCs have subsequently been isolated, purified, and characterized from several other bacteria.

Purified RCs that have been studied typically have four molecules of Bchl, two of bacteriopheophytin (BPh), and one carotenoid if prepared from a carotenoid-containing strain. In addition, RCs contain one or more quinones, either ubiquinone or menaquinone and one nonheme iron atom. Of the Bchl molecules in the RC, two appear to interact with each other particularly closely and are often referred to as a "special pair," $Bchl_2$ or P_{870}, the P standing for pigment and 870 for a wavelength at which bleaching of a major adsorption band occurs when the complex loses an electron. The electron transfer events that occur in the RC are depicted in Figure 1. Following the absorption of a quantum of light, $Bchl_2$ is raised to the excited singlet state, $Bchl_2$*, and loses an electron within 10 psec to one of the BPh molecules (the intermediary acceptor, I) to form the radical cation dimer $Bchl_2^{\dot{+}}$ and the radical anion $BPh^{\dot{-}}$. Within 100 to 200 psec, the electron is transferred to the primary acceptor, Q_IFe^-, and then on to a secondary acceptor with a half-time of 10 to 200 μsec depending on the source of the reaction center preparation. The primary acceptor QFe is a quinone closely associated with a nonheme iron molecule. In most bacteria, Q is ubiquinone, but appears to be menaquinone in *Chromatium vinosum* and *Rhodospseudomonas viridis*. The cytochrome composition of reaction center preparations vary markedly. Some bacteria, such as *Rhodopseudomonas gelatinosa* and *Rhodopseudomonas sphaeroides*, have no cytochromes tightly bound to the RC, whereas *C. vinosum* and *R. viridis* have tightly bound cytochromes which donate electrons to the $Bchl_2^{\dot{+}}$ within 3 μsec. A significant property of bacterial photosynthesis which makes these reactions possible is the stabilization of the radical anion, $BPh^{\dot{-}}$. The formation of this species is complete in

$$Cyt^{2+}\,Bchl_2^{*}BPh\,Q_IFeQ_{II} \underset{10\ msec}{\overset{<10\ psec}{\rightleftharpoons}} Cyt^{2+}Bchl_2^{\dot{+}}\,BPh^{\dot{-}}\,Q_IFeQ_{II} \xrightarrow{100\text{–}200\ psec} Cyt^{2+}Bchl_2^{\dot{+}}\,BPh\,Q_I^{\dot{-}}\,FeQ_{II}$$

$$\xrightarrow{10\text{–}200\ \mu sec} Cyt^{2+}Bchl_2^{\dot{+}}\,BPh\,Q_IFeQ_{II}^{\dot{-}} \xrightarrow{2\text{–}1000\ \mu sec} Cyt^{3+}Bchl_2BPh\,Q_IFeQ_{II}^{\dot{-}}$$

FIGURE 1. Sequence of electron transfer events in the bacterial reaction center. Adsorption of a quantum of light excites the $Bchl_2$ dimer to the singlet state and an electron is transferred to bacteriopheophytin within 10 psec, creating the radical cation, $Bchl_2^{\dot{+}}$ and radical anion $BPh^{\dot{-}}$.Within 200 psec, the electron is transferred to the primary acceptor Q_I. The sequence of events from this point depends on the species of bacteria. In some species, the electron is transferred to the second acceptor Q_{II} very rapidly, whereas in others, the transfer of an electron from cytochrome to $(Bchl)_2^{\dot{+}}$ is more rapid.

<10 psec, whereas the back reaction is only half complete in 10 nsec. This property is responsible for the high quantum yield of these primary reactions.

Restrictions on this summary prevent an adequate compilation of credits, and therefore the reader is referred to the recent reviews of Parson,[3] Blankenship and Parson,[4] and Prince et al.[5] for more complete references on primary reactions and RCs (Table 1), to Dutton and Wilson[6] and Prince and Dutton[7] for redox components of the membranes, and to Baltscheffsky et al.,[8] Gromet-Elhanan,[9] and Sistrom and Clayton[10] for electron transport.

ELECTRON TRANSPORT

Electron transport through the cyclic chain of carriers is shown in Figure 2. From the secondary electron acceptor, Q_{II}, the electron is transferred into the pool of quinones and cytochrome b_{560} of the membrane. There is interaction between these components, but the mechanism or sequence of electron transfer is not known. Electrons from the oxidation of NADH and succinate also enter the electron transfer chain through this pool of UQ·cytochrome *b*. The electron acceptor following UQ · cytochrome b has not been definitely established, but the existence of an autooxidizable component(s) between cytochromes *b* and *c* was postulated by Thore et al.[11] and Keister.[12] More recently Cogdell et al.[13] and Prince and Dutton[14] proposed a component, Z, that may be ubiquinone[15,16] for this site. Electrons are transferred from Z to cytochrome *c*, completing the cycle. Artificial electron donors, such as TMPD or exogenous cytochrome *c*, can enter the cycle through cytochrome c_2.

The information on the aerobic electron transfer from NADH and succinate is not as well founded as for photosynthetic electron transport but the minimum number of components is shown in Figure 2. In *Rhodospirillium rubrum*, the terminal oxidase appears to be a cytochrome of the "o" type,[17] while in *R. sphaeroides* and *R. viridis*, cytochromes *a* and a_3 are found.

ENERGY CONVERSION

Photophosphorylation

Chromatophores of various species of bacteria are capable of high rates of photophosphorylation and the coupling site appears to be in the Q·b → cytochrome c_2 segment of the chain. In addition, there is a coupling site between NADH and Q·b in the respiratory portion of the chain. The generation of reducing power by the photoreduction or ATP-driven reduction of NAD^+ occurs through this coupling site.[18] The

Table 1
REDOX COMPONENTS OF THE CHROMATOPHORE MEMBRANE OF *RHODOPSEUDOMONAS SPHAEROIDES* [5,6]

Component	E_m(mV, pH 7.2)	Probable function
Reaction center		
P_{870} or $Bchl_2$	+450	Primary electron donor
I or (BPh)	−550	Intermediary electron carrier
$UQ_I FeQ_{II}$	−180	Primary and secondary electron acceptors
UQ	+20	Cyclic electron transport (and aerobic substrate oxidation)
Cyt b_{560}	+50	Cyclic electron transport (and aerobic substrate oxidation)
Z		Cyclic electron transport
Cyt c_2	+295	Secondary electron donor
FeS g = 1.94	−400	(NADH oxidation/reduction)
FeS g = 1.94	−260	(NADH oxidation/reduction)
FeS g = 1.94	0	Succinate oxidation
FeS g = 1.9	+285	"Rieske" FeS protein
Cyt *a*	+200	Aerobic oxidation
Cyt a_3	+375	Aerobic oxidation
Cyt b_{564}	−90	Unknown
Cyt b_{558}	+155	(Substrate oxidation)
Cyt *c'*	0	(Alternate oxidase)

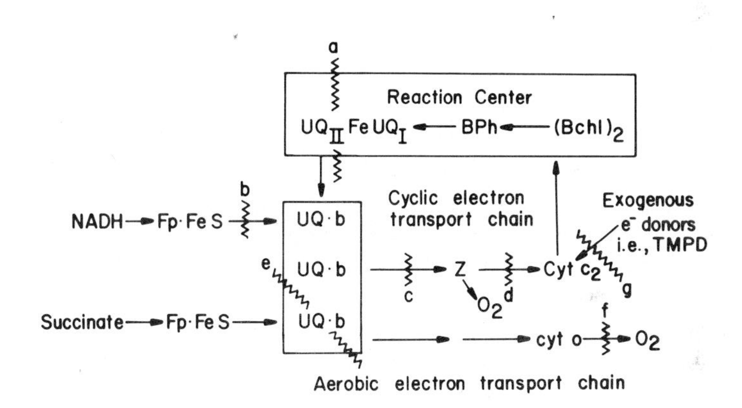

FIGURE 2. The electron transport chain of *R. rubrum*. The cyclic photosynthetic electron transport chain and the pathway of aerobic oxidation are depicted. The two chains are shown to interact at the UQ · cyt *b* pools. Inhibitor sites are as follows: a, *o*-phenanthroline; b, amytal, rotenone; c, 2-hydroxy-3-ω-cyclohexyloctyl)-1,4-naphthoquinone (HHNQ); d, antimycin a, 2-heptyl-4-hydroxyquinoline N-oxide (HOQNO), 2,5-dibromo-3-methyl-6-isopropylbenzoquinone (DBMIB), 4,5,6,7-tetrachloro-2-trifluoro-methylbenzimidazole (TTFB); e, TTFB; f, KCN, NaN_3; g, NaN_3. Other abbreviations are F_p, flavoprotein; FeS, iron-sulfur protein.

mechanism of photophosphorylation is not considered here for the primary mechanism is not yet known. It is certain that energy from electron transfer in coupling membranes or from ATP cleavage can be reversibly transduced to proton and charged-group transfer. However, the mechanism and what is considered to be the primary form of energy conservation is yet to be determined. The reader is referred to a number of recent scholarly reviews by Boyer et al.,[19] Jagendorf,[20] and Williams[21] for general views and to Crofts and Wood[22] for a consideration of proton binding and proton pumps in photosynthetic bacteria.

REFERENCES

1. **Kamen, M. D.,** *Primary Processes in Photosynthesis,* Academic Press, New York, 1963.
2. **Reed, D. W. and Clayton, R. B.,** Isolation of a reaction center fraction from *Rhodopseudomonas spheroides, Biochem. Biophys. Res. Commun.,* 30, 471, 1968.
3. **Parson, W. W.,** Bacterial photosynthesis, *Annu. Rev. Microbiol.,* 28, 41, 1974.
4. **Blankenship, R. E. and Parson, W. W.,** The photochemical electron transfer reactions of photosynthetic bacteria and plants, *Annu. Rev. Biochem.,* 47, 635, 1978.
5. **Prince, R. C., Tiede, D. M., Thornber, J. P., and Dutton, P. L.,** Spectroscopic properties of the intermediary electron carrier in the reaction center of *Rhodopseudomonas viridis, Biochim. Biophys. Acta,* 462, 467, 1977.
6. **Dutton, P. L. and Wilson, D. W.,** Redox potentiometry in mitochondrial and photosynthetic bioenergetics, *Biochim. Biophys. Acta,* 346, 165, 1974.
7. **Prince, R. C. and Dutton, P. L.,** Single and multiple turnover reactions in the ubiquinone cytochrome b-c_2 oxidoreductase of *Rhodopseudomonas sphaeroides, Biochim. Biophys. Acta,* 462, 731, 1977.
8. **Baltscheffsky, H., Baltscheffsky, M., and Thore, A.,** Energy conversion reactions in bacterial photosynthesis, *Curr. Top. Bioenerg.,* 4, 273, 1971.
9. **Gromet-Elhanan, Z.,** Electron transport and photophosphorylation in photosynthetic bacteria, in *Encyclopedia of Plant Physiology,* Vol. 5 (New series), Trebst, A. and Avron, M., Eds., Springer-Verlag, New York, 1977.
10. **Sistrom, W. R. and Clayton, R. K.,** *The Photosynthetic Bacteria,* Plenum Press, New York, 1978.
11. **Thore, A., Keister, D. L., and San Pietro, A.,** Studies on the respiratory system of aerobically (dark) and anaerobically (light) grown *Rhodospirillum rubrum, Arch. Mikrobiol.,* 67, 378, 1969.
12. **Keister, D. L. and Minton, N. J.,** Effect of light on respiration in *Rhodospirillum rubrum* chromatophores, in *Energy Transduction in Respiration and Photosynthesis,* Quagliariello, E., Papa, S., and Rossi, C. S., Eds., Adriatica Editrice, Bari, Italy, 1971, 375.
13. **Cogdell, R. J., Jackson, J. B., and Crofts, A. R.,** The effect of redox potential on the coupling between rapid hydrogen-ion binding and electron transport in chromatophores from *Rhodopseudomonas spheroides, J. Bioenerg.,* 4, 211, 1973.
14. **Prince, R. C. and Dutton, P. L.,** The physical chemistry of the major electron donor to cytochrome c_2, and its coupled reactions, *Biochim. Biophys. Acta,* 462, 731, 1977.
15. **Baccarini-Melandri, A. and Melandri, B. A.,** A role for ubiquinone-10 in the *b*-c_2 segment of the photosynthetic bacterial electron transport chain, *FEBS Lett.,* 80, 459, 1977.
16. **Gromet-Elhanan, Z. and Gest, H.** A comparison of electron transport and photophosphorylation systems of *Rhodopseudomonas capsulata* and *Rhodospirillum rubrum, Arch. Microbiol.,* 116, 29, 1978.
17. **Taniguchi, S. and Kamen, M. D.,** The oxidase system of heterotrophically grown *Rhodospirillum rubrum, Biochim. Biophys. Acta,* 96, 395, 1965.
18. **Keister, D. L. and Minton, N. J.,** Further studies on energy-linked NAD^+ reduction by *Rhodospirillum rubrum* chromatophores, *Biochemistry,* 8, 167, 1969.
19. **Boyer, P. D., Chance, B., Ernster, L., Mitchell, P., Racker, E., and Slater, E. C.,** Oxidative phosphorylation and photophosphorylation, *Annu. Rev. Biochem.,* 46, 955, 1977.
20. **Jagendorf, A. T.,** Mechanisms of photophosphorylation, in *Bioenergetics of Photosynthesis,* Govindjee, S. R., Ed., Academic Press, New York, 1975, 413.
21. **Williams, R. J. P.,** The multifarious couplings of energy transduction, *Biochim. Biophys. Acta,* 505, 1, 1978.
22. **Crofts, A. R. and Wood, P. M.,** Photosynthetic electron-transport chains of plants and bacteria and their role as proton pumps, *Curr. Top. Bioenerg.,* 7, 175, 1978.

BASIC MECHANISMS OF PHOTOSYNTHETIC ELECTRON TRANSPORT AND ATP SYNTHESIS IN OXYGEN-EVOLVING ORGANISMS

S. Izawa

GENERATION OF REDUCING POWER AND ATP*

In plant and algal photosynthesis, the light energy captured by chlorophylls and other pigments is utilized to remove electrons from water (which is oxidized to O_2) and to transfer them to $NADP^+$ through a chain of redox reaction. This electron transport process which takes place in the thylakoids of the chloroplast is coupled to phosphorylation and generates the energy carrier ATP along with the reductant NADPH. These two products are subsequently used for CO_2 fixation in the stroma of the chloroplast (see Figure 1). The development of this modern concept of photosynthesis owes much to three groups of workers: Hill,[1] who demonstrated the splitting of water into O_2 and reducing equivalents in illuminated chloroplasts, Arnon and associates,[2] who discovered photosynthetic phosphorylation in isolated chloroplasts, and Calvin and associates,[3] who elucidated the pathway of CO_2 reduction (the C_3 cycle).

Since water is a very weak reductant (E'_o = +0.81 V for O_2/H_2O system at pH 7) and $NADP^+$ is a weak oxidant (E'_o = −0.32 V), the transfer of electrons from water to $NADP^+$ requires a large energy input ($\Delta E'_o$ = 1.13 V or $\Delta G'_o$ = +52 kcal/mol NADPH). Furthermore, in order to drive the C_3 cycle, this process must produce 1.5 mol of ATP from ADP and orthophosphate ($\Delta G'_o$ = +7.5 kcal/mol ATP) for each mole of NADPH produced. Thus, the overall process of chloroplast electron transport and phosphorylation is a highly endergonic reaction (hence the light energy requirement):

$$2H_2O + 2\ NADP^+ + 3\ ADP + 3\ P_i = 2\ NADPH + 3\ ATP + O_2 + 2\ H^+ \quad (\Delta G'_o = +127\ kcal) \tag{1}$$

However, the NADPH and ATP formed can now drive CO_2 fixation in an exergonic reaction:

$$CO_2 + 2\ NADPH + 3\ ATP + 2\ H^+ = CH_2O + 2\ NADP^+ + 3\ ADP + 3\ P_i + H_2O \quad (\Delta G'_o = -13\ kcal) \tag{2}$$

The net result of Equations 1 and 2 is photosynthesis:

$$H_2O + CO_2 = CH_2O + O_2 \quad (\Delta G'_o = +114\ kcal) \tag{3}$$

To accomplish photosynthetic electron transport and associated phosphorylation (Equation 1), O_2-evolving organisms have invented a mechanism in which two photons cooperate to move one electron from water to $NADP^+$.

THE PATHWAY OF ELECTRON TRANSPORT

Two-Photosystem Model

Figure 2 diagrams the widely accepted two-photosystem model ("Z scheme") of

* Because of the very broad nature of the subject matter dealt with in this article, literature coverage will be limited to reviews and a few selected research papers. For the literature and other information on the individual components of the electron transport chain (pigments, cytochromes, etc.), the reader is referred to those other articles in this volume which are directly concerned with them.

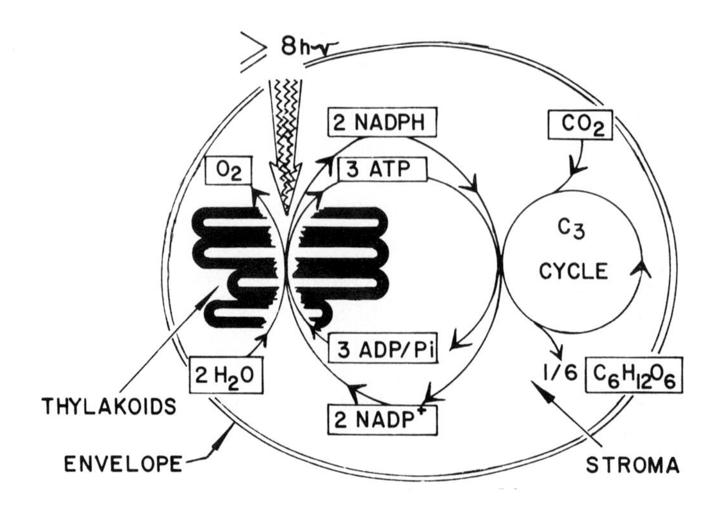

FIGURE 1. A simplified scheme for photosynthetic reactions in the chloroplast of C_3 plants and algae. The stoichiometries shown are theoretically valid, but the reactions are grossly simplified. See text for details of $NADP^+$ photoreduction and associated phosphorylation.

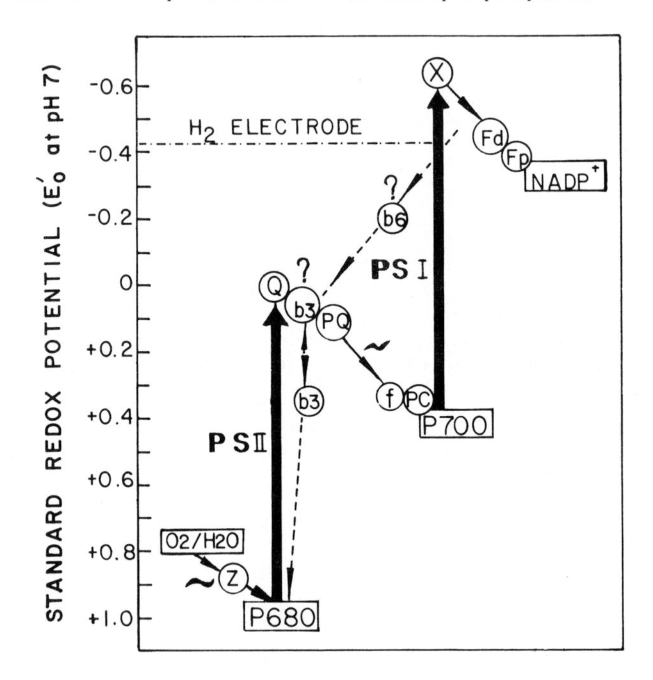

FIGURE 2. The basic pathway of photosynthetic electron transport. Abbreviations used: b_3, cytochrome b_3 (or b-559); upper, low-potential form; lower, high-potential form; b_6, cytochrome b_6 (b-563); f, cytochrome *f*; Fd, ferredoxin; Fp, flavoprotein (ferredoxin-NADP reductase); P680, reaction center chlorophyll of Photosystem II; P700, reaction center chlorophyll of Photosystem I; PC, plastocyanin; PQ, plastiquinone pool; PS I (II), Photosystem I (II); Q, primary electron acceptor of Photosystem II; X, primary electron acceptor of Photosystem I; Z, O_2 evolving center. The squiggles (~) represent "sites" for ATP formation, and the broken lines represent possible cyclic electron transport pathways.

photosynthetic electron transport. The basic framework of this model originates from the theoretical considerations of Hill and Bendall[4] and the experiments of Duysens and associates[5] on the behavior of a cytochrome in algal cells exposed to two different colors of light. In plants and green algae, each photosystem contains light-gathering or antennae pigments consisting of 200 to 300 chlorophylls (majority of chlorophyll *b* (Chl *b*) in Photosystem II) plus carotenoids, a reaction center (RC) chlorophyll (P_{700} in I and P_{680} in II), and a primary electron acceptor (X in I and Z in II). In red and blue-green algae, Chl *b* is replaced by phycobilins. In a photosystem, light energy captured by the antennae pigments is quickly and efficiently transfered to the RC, where photochemical charge separation takes place. The charge separation occurs in the form of a transfer of an electron from the excited RC chlorophyll (= primary electron donor) to the primary acceptor:

$$P700 \cdot X \xrightarrow{h\nu} P700^{*} \cdot X \longrightarrow P700^{+} \cdot X^{-}$$

$$P680 \cdot Q \xrightarrow{h\nu} P680^{*} \cdot Q \longrightarrow P680^{+} \cdot Q^{-}$$

It is these primary charge separation reactions (vertical arrows in Figure 2) that drive photosynthetic electron transport.

Oxidation of Water to O_2 by Photosystem II

Each individual Photosystem II has one complete O_2-evolving center of its own (Z in Figure 2). Four oxidizing equivalents, which are needed to oxidize $2H_2O$ to O_2, are successively accumulated within the center (or its complex with water) through a 4-quantum process. In the simplest version of the currently popular model, this process is expressed as a four-step "state" change of the O_2-evolving center:

$$S_0 \xrightarrow{h\nu} S_1 \xrightarrow{h\nu} S_2 \xrightarrow{h\nu} S_3 \xrightarrow{h\nu} S_4 \longrightarrow O_2 \quad (S_4 \rightarrow S_0)$$

This kinetic model, which is now more elaborated,[6-8] was derived from Jolliot's repetitive flash experiments[9] which clearly demonstrated that the O_2/flash yield oscillates with a periodicity of four. As yet, however, little is known about the chemical mechanism of water oxidation (for a recent model, see Reference 10) or the chemical identity of the O_2-evolving center, except for the fact that the center has at least four atoms of Mn in its catalytic site(s)[11] and that its function requires Cl^- as a cofactor.[12]

Electron Transport Between Photosystems

The primary electron acceptor Q (also named A_1 or X-320) of Photosystem II is believed to be a species of plastoquinone acting as a one-electron acceptor.[13] Its semi-reduced form Q^- donates its electron, probably through an intermediate electron carrier[14] (which may also be a plastoquinone[15]), to the large plastoquinone pool. This step requires HCO_3^-.[16] Electron transfer from the plastoquinone pool to cytochrome *f* is quite slow (10^{-2} sec) and thus forms the primary rate-limiting step of photosynthetic electron transport. Recent evidence suggests the involvement of a nonheme iron center in this step.[17] The reduced cytochrome *f* transfers its electron to the copper protein, plastocyanin, which is the direct electron donor to P_{700}^+. Unsolved questions pertaining to this interphotosystem electron transport include the relation of cytochrome b_3 (b-559) to the main pathway of electrons[18] and the existence of kinetic data which appear to be inconsistent with the obligatory involvement of cytochrome *f*[19] and P700.[20]

Electron Transport from Photosystem I to $NADP^+$

The sequence of electron transfer $X^- \rightarrow$ ferredoxin $\rightarrow$ ferredoxin-NADP reductase $\rightarrow NADP^+$ is well established, except that X has yet to be chemically identified. It probably contains several components,[21] including an iron sulfur protein ("P_{430}").[22] An extensive review on the entire electron transport pathway has appeared,[23] as well as reviews specifically dealing with electron transport (and phosphorylation) in eukaryotic algae[24] and blue-green algae.[25]

PHOTOSYNTHETIC PHOSPHORYLATION

"Sites" of Energy Coupling

As in oxidative phosphorylation, those redox steps in photosynthetic electron transport which are linked to the mechanism of ATP synthesis are often called phosphorylation sites or energy coupling sites. So far two such sites have been recognized: the electron transfer step between the plastoquinone pool and cytochrome *f* (Site I) and the water-oxidizing step (Site II).[26,27] It is very unlikely that a third site exists. In Figure 2, these sites are indicated by the squiggles (~) along the electron transport chain.

Mechanism of Energy Coupling

Mitchell's chemiosmotic coupling hypothesis[28] has now been accepted by majority of workers in the field as the most plausible model of energy coupling in photophosphorylation. According to this hypothesis, the role of electron transport in photophosphorylation is to translocate external protons into the osmotic space of the thylakoid (see Figure 3). The resultant transmembrane electrochemical potential, named proton motive force or pmf, represents the "high energy intermediate" of phosphorylation. The pmf is expressed by the formula:*

$$\text{pmf (V)} = 2.3(RT/F)\Delta pH + \Delta\psi \qquad (4)$$

where ΔpH and $\Delta\psi$ denote the transmembrane pH gradient and the membrane potential (electrical potential), respectively; R, T and F are the gas constant, the absolute temperature, and the Faraday constant, respectively. ATP synthesis is supported by the outflow of protons through the membrane-bound ATPase complex (CF_1 in Figure 3). It is proposed that the protons driven by the pmf create, within the ATPase complex, a local pH gradient which acts to remove H_2O (as H^+ and OH^-) from ADP plus

* Using Equation 4, one may examine the thermodynamic feasibility of chemiosmotic coupling in chloroplasts; Equation 4 can be rewritten as:

$$\Delta G' = 2.3RT\ \Delta\ pH + F\ \Delta\psi$$

where $R \cong 2$ cal/T·mol, $T \cong 300°K$, and $F = 23$ kcal/V. The highest experimental values for ΔpH is in the range 3 to 3.5 pH units. Putting numerical values for R, T, and ΔpH into the equation and ignoring the uncertain and probably minor $F\Delta\psi$ term, one obtains:

$$\Delta G' \cong 4.1-4.8 \text{ kcal/mol } H^+$$

This is the approximate amount of energy which 1 mol of effluxing protons can deliver to CF_1. Kinetic experiments indicate that synthesis of 1 ATP at CF_1 requires cooperation of 2 to 3 protons. Assuming 3 to be the more reliable value, one obtains $\Delta G' \cong (4.1 \text{ to } 4.8 \text{ kcal}) \times 3 \cong 12$ to 14 kcal. This is the amount of energy which 3 mol of protons carry under the pmf. A separate experiment indicates that $\Delta G'$ (not $\Delta G'_o$) for ATP synthesis under actual conditions for photophosphorylation is $\sim +14$ kcal/mol ATP. Thus, chemiosmotic coupling is thermodynamically feasible. (The experimental values used are from References 29 and 36.)

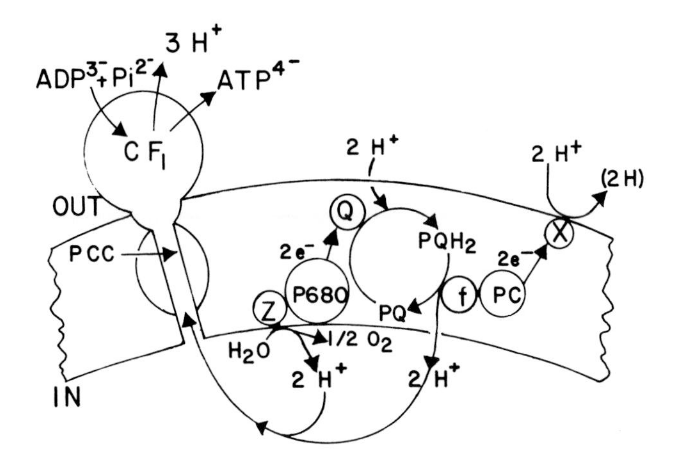

FIGURE 3. A chemiosmotic model of photosynthetic electron transport and phosphorylation in the thylakoid membrane. CF_1, chloroplast coupling factor I (ATPase complex); PCC, proton conducting channel; PQH_2, reduced form of plastoquinone (i.e., plastoquinol). See Figure 2 for other abbreviations. Note that the two sites of energy coupling shown in Figure 2 correspond to redox steps where internal proton discharge occur.

orthophosphate. Aside from the molecular mechanism of ATP formation by CF_1, which is currently under vigorous investigation,[29-31] the main framework of this model has been substantiated by various lines of experiments,[32-35] including the classical demonstration of the light-induced proton uptake and the acid-bath phosphorylation in isolated chloroplasts.[36]

Stoichiometry of Photophosphorylation

If the ATP requirement of the C_3 cycle (ATP:NADPH:CO_2 = 3:2:1) were to be met solely by phosphorylation coupled to NADP photoreduction ("noncyclic photophosphorylation"), then the coupling efficiency (ATP/e_2 ratio) of photosynthetic electron transport would have to be 1.5,* or even higher if additional ATP-requiring processes, e.g., polysaccharide synthesis, were to occur. Indeed, ATP/e_2 ratios approaching two (the presumed limit of two-site phosphorylation) have been reported to be obtained with chloroplasts freshly stripped of their envelopes.[37] Indirect measurements indicated, however, that the intrinsic ATP/e_2 ratio may not exceed 1.4, even in intact chloroplasts with high CO_2-fixing capability.[38] This value is similar to what most workers encounter with their best chloroplast preparations (ATP/e_2 1.3). There is evidence which suggests that chloroplasts in vivo may regulate and, when necessary, elevate the ATP/NADP ratio at the expense of part of the reducing power — i.e., by allowing some of the electrons from Photosystem I to cycle around the photosystem (i.e., through "cyclic photophosphorylation")[39,40] or by diverting them to O_2 (i.e., through "pseudocyclic photophosphorylation").[41,42] The existence of cyclic photophosphorylation in algae is well known.[43]

* In C_4 plants, the ATP/e_2 ratio necessary to sustain the overall process of photosynthesis has been calculated to be 2.5 (Reference 44), a value which seems almost certainly beyond the capability of noncyclic photophosphorylation. Available data on photophosphorylation in chloroplasts isolated from C_4 plants (mesophyll or bundle-sheath), however, are not sufficient to allow one to assess their intrinsic phosphorylation capabilities with any degree of confidence.

REFERENCES

1. **Hill, R.**, Oxygen produced by isolated chloroplasts, *Proc. R. Soc. London Ser. B,* 127, 192, 1939.
2. **Arnon, D. I., Whatley, F. R., and Allen, M. B.**, Photosynthesis by isolated chloroplasts II. Photosynthetic phosphorylation, the conversion of light into phosphate bond energy, *J. Am. Chem. Soc.,* 76, 6324, 1954.
3. **Calvin, M. and Bassham, J. A.**, *The Path of Carbon in Photosynthesis,* Prentice-Hall, New York, 1957.
4. **Hill, R. and Bendall, F.**, Function of the two cytochrome components in chloroplasts: a working hypothesis, *Nature (London),* 186, 136, 1960.
5. **Duysens, L. N. M., Amesz, J., and Kamp, B. M.**, Two photochemical systems in photosynthesis, *Nature (London),* 190, 510, 1961.
6. **Kok, B., Forbush, B., and McGloin, M.**, Cooperation of charges in photosynthetic O_2 evolution. I. A linear four-step model, *Photochem. Photobiol.,* 11, 457, 1970.
7. **Radmer, R. and Kok, B.**, Energy capture in photosynthesis: Photosystem II, *Annu. Rev. Biochem.,* 44, 409, 1975.
8. **Diner, B. A. and Jolliot, P.**, Oxygen evolution and manganese, in *Encyclopedia of Plant Physiol.,* Vol. 5 (New Ser.), Trebst, A. and Avron, M., Eds., Springer, New York, 1977, 187.
9. **Jolliot, P., Barbieri, G., and Chabaud, R.**, Un nouveau modele des centres photochimique du systeme II, *Photochem. Photobiol.,* 10, 309, 1969.
10. **Renger, G.**, A model for the molecular mechanism of photosynthetic oxygen evolution, *FEBS Lett.,* 81, 223, 1977.
11. **Cheniae, G. M.**, Photosystem II and O_2 evolution, *Annu. Rev. Plant Physiol.,* 21, 467, 1970.
12. **Kelley, P. M. and Izawa, S.**, The role of chloride ion in Photosystem II. I. Effects of chloride ion on Photosystem II electron transport and on hydroxylamine inhibition, *Biochim. Biophys. Acta,* 502, 198, 1978.
13. **van Gorkom, H. J.**, Identification of the reduced primary electron acceptor of Photosystem II as a bound semiquinone anion, *Biochim. Biophys. Acta,* 347, 439, 1974.
14. **Borges-Bocquet, B.**, Electron transfer between the two Photosystems in spinach chloroplasts, *Biochim. Biophys. Acta,* 314, 250, 1973.
15. **Pulles, M. P. J., van Gorkom, H. J., and Willemsen, J. G.**, Absorbance changes due to charge-accumulating species in system 2 of photosynthesis, *Biochim. Biophys. Acta,* 449, 536, 1976.
16. **Khanna, R., Govindjee, S. R., and Wydrzynski, T.**, Site of bicarbonate effect in Hill reaction. Evidence from the use of artificial electron acceptors and donors, *Biochim. Biophys. Acta,* 462, 208, 1977.
17. **Malkin, R. and Posner, H. B.**, On the site of function of the Rieske iron-sulfur center in the chloroplast electron transport chain, *Biochim. Biophys. Acta,* 501, 552, 1978.
18. **Cramer, W. A. and Whitmarsh, J.**, Photosynthetic cytochromes, *Annu. Rev. Plant Physiol.,* 28, 133, 1977.
19. **Haehnel, W.**, Electron transport between photosystems and chlorophyll a_I in chloroplasts. II. Reaction kinetics and the function of plastocyanin *in situ, Biochim. Biophys. Acta,* 459, 418, 1977.
20. **Rurainski, H. J., Randles, J., and Hoch, G. E.**, The relationship between P700 and NADP reduction in chloroplasts, *FEBS Lett.,* 13, 98, 1971.
21. **Sauer, K., Mathis, P., Acker, S., and van West, J. A.**, Electron acceptors associated with P700 in Triton-solubilized Photosystem I particles from spinach chloroplasts, *Biochim. Biophys. Acta,* 503, 120, 1978.
22. **Ke. B., Dolan, E., Sugahara, K., Hawkridge, F. M., Demeter, S., and Shaw, E. R.**, Electrochemical and kinetic evidence for a transient electron acceptor in the photochemical charge separation in Photosystem I, *Plant Cell Physiol.,* Special Issue (Photosynthetic Organelles), 187, 1977.
23. **Golbeck, J. H., Lien, S., and San Pietro, A.**, Electron transport in chloroplasts, in *Encyclopedia of Plant Physiol.,* Vol. 5, New Ser., Trebst, A. and Avron, M., Eds., Springer, New York, 1977, 94.
24. **Krogmann, D. W.**, Blue-green algae, in *Encyclopedia of Plant Physiol.,* Vol. 5 (New Ser.), Trebst, A. and Avron, M., Eds., Springer, New York, 1977, 625.
25. **Urbach, W.**, Eukaryotic algae, in *Encyclopedia of Plant Physiol.,* Vol. 5 (New Ser.), Trebst, A. and Avron, M., Eds., Springer, New York, 1977, 603.
26. **Izawa, S., Gould, J. M., Ort, D. R., Felker, P., and Good, N. E.**, Electron transport and photophosphorylation as a function of the electron acceptor III. A dibromothymoquinone-insensitive phosphorylation associated with Photosystem II, *Biochim. Biophys. Acta,* 305, 119, 1973.
27. **Trebst, A.**, Energy conservation in photosynthetic electron transport of chloroplasts, *Annu. Rev. Plant Physiol.,* 25, 423, 1974.

28. **Mitchell, P.**, *Chemiosmotic Coupling and Energy Transduction,* Glynn Research Ltd., Bodmin, Cornwall, 1968.
29. **Jagendorf, A. T.**, Photophosphorylation, in *Encyclopedia of Plant Physiol.,* Vol. 5 (New Ser.), Trebst, A. and Avron, M., Eds., Springer, New York, 1977, 307.
30. **Nelson, N.**, Structure and function of chloroplast ATPase, *Biochim. Biophys. Acta,* 456, 314, 1976.
31. **McCarty, R. E.**, The reaction of coupling factor 1 in chloroplast thylakoid with N-substituted maleimides, *Proc. 4th Int. Congr. Photosynthesis,* Biochemical Society, London, 1978, 571.
32. **Witt, H. T.**, Primary acts of energy conservation in the functional membrane of photosynthesis, in *Bioenergetics of Photosynthesis,* Govindjee, S. R., Ed., Academic Press, New York, 1975, 493.
33. **Junge, W.**, Membrane potentials in photosynthesis, *Annu. Rev. Plant Physiol.,* 28, 503, 1977.
34. **Dilley, R. A. and Giaquinta, R. T.**, Ion transport and energy transduction in chloroplasts, *Curr. Top. Membr. Transp.,* 7, 503, 1977.
35. **Avron, M.**, Energy transduction in chloroplasts, *Annu. Rev. Biochem.,* 46, 143, 1977.
36. **Jagendorf, A. T.**, Mechanism of photophosphorylation, in *Bioenergetics of Photosynthesis,* Govindjee, S. R., Ed., Academic Press, New York, 1975, 413.
37. **Reeves, S. G. and Hall, D. O.**, Photophosphorylation in chloroplasts, *Biochim. Biophys. Acta,* 463, 275, 1978.
38. **Heber, U. and Kirk, M. R.**, Flexibility of coupling and stoichiometry of ATP formation in intact chloroplasts, *Biochim. Biophys. Acta,* 376, 136, 1975.
39. **Arnon, D. I. and Chain, R. K.**, ferredoxin-catalyzed photophosphorylations: concurrence, stoichiometry, regulation, and quantum efficiency, *Plant Cell Physiol.,* Special Issue (Photosynthetic Organelles), 1977, 129.
40. **Hind, G., Mills, J. D., and Slovacek, R. E.**, Cyclic electron transport in photosynthesis, *Proc. 4th Int. Congr. Photosynthesis,* Biochemical Society, London, 1978, 591.
41. **Allen, J. F.**, Oxygen reduction and optimum production of ATP in photosynthesis, *Nature (London),* 256, 599, 1975.
42. **Egneus, H., Heber, U., Matthesen, U., and Kirk, M.**, Reduction of oxygen by the electron transport chain of chloroplasts during assimilation of carbon dioxide, *Biochim. Biophys. Acta,* 408, 252, 1975.
43. **Simonis, W. and Urbach, W.**, Photophosphorylation in vivo, *Annu. Rev. Plant Physiol.,* 24, 89, 1973.
44. **Hatch, M. D. and Slack, C. R.**, Photosynthetic CO_2 fixation pathways, *Annu. Rev. Plant Physiol.,* 21, 141, 1970.

FERREDOXIN-NADP REDUCTASE

Masateru Shin

INTRODUCTION

Ferredoxin-NADP reductase (EC 1.18.1.2, reduced NADP:ferredoxin oxidoreductase) is a FAD-containing flavoprotein which catalyzes a reversible electron transport between ferredoxin and NADP. In photosynthetic organisms, ferredoxin-NADP reductase participates in the reduction of NADP to produce reducing power for CO_2 assimilation. Photosynthetic ferredoxin-NADP reductase is widely distributed in photosynthetic organisms including higher plants, algae, blue-green algae, and photosynthetic bacteria. However, information concerning the molecular properties of the flavoprotein is mostly available on the spinach enzyme.

HISTORICAL VIEW

In 1962, the existence of an enzyme connecting ferredoxin and NADP was proposed by Tagawa and Arnon[1] from their findings on the physiological roles of ferredoxin in photosynthesis. They found that reduced ferredoxin, obtained through an H_2-hydrogenase system did not directly reduce NADP, but it reduced NADP if a flavoprotein fraction of chloroplast extracts was added. During successive investigations, Shin et al.[2] isolated a flavoprotein from the flavoprotein fraction in crystalline form. They demonstrated that photoreduction of NADP by chloroplasts required not only ferredoxin, but also the crystallized flavoprotein. Since the flavoprotein was fairly tightly bound to chloroplasts, it had to be removed before addition of enzyme. Acetone is a useful reagent for a preparative extraction of the enzyme,[3] but it destroys simultaneously the photochemical activity of chloroplasts. Therefore, the chloroplasts used for the experiment were incubated in 0.05 *M* Tris buffer at pH 7.6. The mild extraction released the flavoprotein from chloroplasts without causing a loss in activity of Photosystem I. Then, a reconstituted NADP photoreduction system was constituted by mixing the enzyme-extracted chloroplasts, the crystalline flavoprotein, ferredoxin, NADP, and the ascorbate-DCPIP couple as the electron donor. From its physiological role, the chloroplast flavoprotein was named as "ferredoxin-TPN reductase". Dependency on both ferredoxin and ferredoxin-NADP reductase for NADP photoreduction was independently recognized by Davenport,[4] who used pea chloroplasts from which the enzyme was easier to extract than from spinach chloroplasts.

San Pietro and Lang[5] had found earlier that illuminated chloroplasts reduced not only NADP, but also NAD if high concentration of NAD and chloroplast extracts were added. They considered that the NAD reduction was because of a chloroplast transhydrogenase which might catalyze a hydrogen transfer from free or bound NADP to added NAD.[3,6] However, crystalline ferredoxin-NADP reductase reduced NAD as well as NADP via the same mechanism, but not by a transhydrogenase reaction. The Michealus constants were 3.77 m*M* for NAD and 7.72 μM for NADP.[7,8] As indicated by the K_m for NAD, the NAD reduction required nonphysiologically high concentrations of NAD and also needed a four times greater quantity of the enzyme to obtain the same maximum velocity. The properties of NAD reduction by ferredoxin-NADP reductase clearly explains the early observations of San Pietro and Lang.

Prior to the discovery of ferredoxin-NADP reductase, the chloroplast flavoprotein had been studied as a "TPNH diaphorase" by Avron and Jagendorf,[9] a "Transhydrogenase'' by Keister et al.,[3] and a ''Gelbes Ferment'' by Gewitz and Völker.[10] These enzyme activities are all because of secondary activities of ferredoxin-NADP reductase. The flavoprotein reduces many physiological and nonphysiological substances from NADPH in the absence or in the presence of ferredoxin. In the absence of ferredoxin, it reduces indophenols, viologens, tetrazoliums, methylene blue, menadione, ferricyanide, NAD, and cytochrome b_5. Chloroplast components, cytochrome *f*,[11] and plastocyanin[12] also are directly reduced from NADPH, but the physiological significance of the activities in vivo are not clear. In the presence of ferredoxin, it also reduces metmyoglobin, methomoglobin, peroxidase, and cytochrome *c*.

MOLECULAR PROPERTIES OF THE SPINACH ENZYME

Spinach ferredoxin-NADP reductase has a mol wt of 33,000 by gel filtration[13] or by sedimentation equilibrium using a partial specific volume of 0.717 mL/g.[14] The flavoprotein contains 1 FAD per mole of protein[15] and shows a typical flavoprotein absorption spectrum with absorption maxima at 275, 385, and 456 nm and minima at 321 and 410 nm[2] (see Figure 1). The molar extinction coefficient at 456 nm is 10,740 $M^{-1} \cdot cm^{-1}$ in the oxidized state and 2.160 in the reduced form.[16] The highest value of absorbance ratio, $A_{456\ nm}/A_{275\ nm}$, so far obtained is 0.133.[17] The fully reduced enzyme is obtained by reduction with $Na_2S_2O_4$ or an H_2-hydrogenase-ferredoxin system. The reduction rate by $Na_2S_2O_4$ proceeds slowly, but it goes rapidly if catalytic amounts of ferredoxin are added.[8] With NADPH under anaerobic conditions, the enzyme is reduced partially, showing that a flavin semiquinone is formed.[7,8] An ESR signal of the partially reduced enzyme is observed at $g = 2.0042$.[18] These observations suggest that oxidation-reduction potential of the flavoprotein is quite low, approximately −370 mV. The semiquinone formation is not observed during reduction by $Na_2S_2O_4$ or by the H_2-hydrogenase-ferredoxin system.[8] The circular dichroism spectrum of the enzyme exhibits positive peaks at 270 and 376 nm and exhibits negative peaks at 286 and 475 nm.[19] Flavin fluorescence of the enzyme is extremely quenched.[19,20] The isoelectric point is roughly between pH 6.2 and 6.7,[15] but Gozzer et al. recently isolated five components of the enzyme possessing different isoelectric points of pI 6.0, 5.5, 5.2, 5.0, and 4.8 (by isoelectric focusing).[21] The amino acid compositions reported do not show definite values (see Table 1). This may be because of the existence of multiple forms of the molecule.

Ferredoxin-NADP reductase formed a complex with spinach ferredoxin, bacterial ferredoxin, flavodoxin or rubredoxin,[19,22-25] and with NADP.[19,22] The observation that photoreduction of NADP is sensitive to ionic strength suggests that the complexes served a functional role in photosynthetic electron transport because the complexes are present only under low ionic conditions.

MULTIPLE FORMS

Spinach leaves contain two molecular forms of ferredoxin-NADP reductase[17,26] which have different molecular weights. Small form, FNR II, shows a mol wt of 33,000 and the large form, FNR I, shows a mol wt of about 75,000 by gel filtration. The two forms are also separable by a DEAE-cellulose chromatography. The small form is considered to be the usual ferredoxin-NADP reductase described previously, and the large form is a new component. When SDS-polyacrylamide electrophoresis was used to determine the molecular weight, the two forms migrated as a single band, showing that both forms have a mol wt of 35,000. This result suggests that FNR I consists of two subunits which are likely the small form.

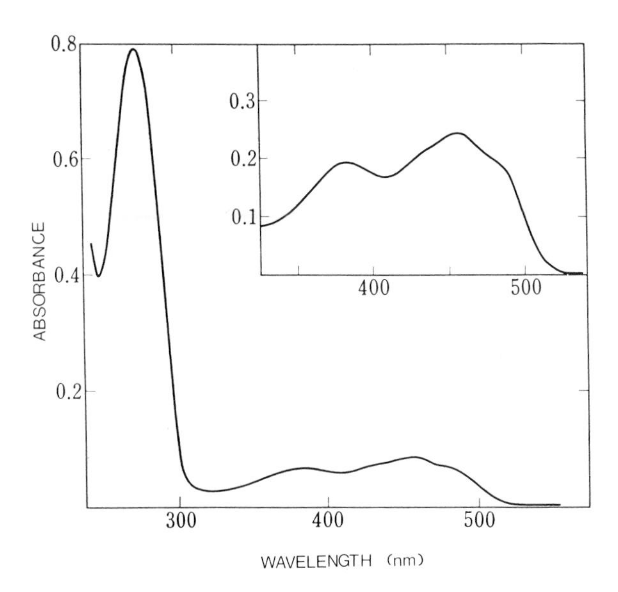

FIGURE 1. Absorption spectrum of spinach ferredoxin-NADP reductase.

Table 1
AMINO ACID COMPOSITIONS OF SPINACH FERREDOXIN-NADP REDUCTASE

Amino acid	Hasumi and Sato[14]	Gozzer et al.[21] Form b	Gozzer et al.[21] Form c	Forti and Sturani[27]
Lysine	32	34	32	35.98
Histidine	5	5	6	5.29
Arginine	9	9	9	9.55
Aspartic acid	27	26	29	29.37
Threonine	14	14	16	16.19
Serine	15	17	20	15.72
Glutamic acid	30	32	35	35.72
Proline	14	13	16	16.38
Glycine	25	25	27	27.17
Alanine	19	17	20	21.77
Cysteine	5	6	6	6.25
Valine	17	17	19	19.53
Methionine	11	11	11	11.27
Isoleucine	13	13	13	14.50
Leucine	22	21	21	25.00
Tyrosine	12	11	11	11.34
Phenylalanine	12	12	13	13.03
Tryptophan	5	6	6	5.72

The small form was found to contain two components, FNR II-a and FNR II-b, which are detectable on polyacrylamide gel electrophoresis.[26] Heterogeneity in the small form of ferredoxin-NADP reductase also was reported by Gozzer et al.[21]

REFERENCES

1. **Tagawa, K. and Arnon, D. I.**, Ferredoxins as electron carriers in photosynthesis and in the biological production and consumption of hydrogen gas, *Nature (London)*, 195, 537, 1962.
2. **Shin, M., Tagawa, K., and Arnon, D. I.**, Crystallization of ferredoxin-TPN reductase and its role in the photosynthetic apparatus of chloroplasts, *Biochem. Z.*, 338, 84, 1963.
3. **Keister, D. L., San Pietro, A., and Stolzenback, F. E.**, Pyridine nucleotide transhydrogenase from spinach, *J. Biol. Chem.*, 235, 2989, 1960.
4. **Davenport, H. E.**, Pathway of reduction of metmyoglobin and nicotinamide adenine dinucleotide phosphate by illuminated chloroplasts, *Nature (London)*, 199, 151, 1963.
5. **San Pietro, A. and Lang, H. M.**, Accumulation of reduced pyridine nucleotide by illuminated grana, *Science*, 124, 118, 1956.
6. **Keister, D. L., San Pietro, A., and Stolzenbach, F. E.**, Pyridine nucleotide transhydrogenase from spinach. II. Requirement of enzyme for photochemical accumulation of reduced pyridine nucleotides, *Arch. Biochem. Biophys.*, 98, 235, 1962.
7. **Shin, M. and Arnon, D. I.**, Enzymic mechanisms of pyridine nucleotide reduction in chloroplasts, *J. Biol. Chem.*, 240, 1405, 1965.
8. **Shin, M.**, Ferredoxin-NADP reductase, in *Flavins and Flavoproteins*, Yagi, K., Ed., University of Tokyo Press, Tokyo and University Park Press, Baltimore, 1968, 1.
9. **Avron, M. and Jagendorf, A. T.**, A TPNH diaphorase from chloroplasts, *Arch. Biochem. Biophys.*, 65, 475, 1956.
10. **Gewitz, H. S. and Völker, W.**, Über die Atmungsfermente der *Chlorella*, *Hoppe-Seylers Z. Physiol. Chem.*, 330, 124, 1962.
11. **Zanetti, G. and Forti, G.**, Studies on the triphosphopyridine nucleotide-cytochrome f reductase of chloroplasts, *J. Biol. Chem.*, 241, 279, 1966.
12. **Katoh, S.**, Flavoprotein in *Chlorolla ellipsoidea* catalyzing TPNH-linked reduction of plastocyanin, *Plant Cell Physiol.*, 2, 165, 1961.
13. **Shin, M. and Oshino, R.**, Isolation of two molecular forms of ferredoxin-NADP reductase from spinach, in *Flavins and Flavoproteins*, Yagi, K. and Yamano, T., Eds., Japan Scientific Societies Press, Tokyo and University Park Press, Baltimore, 1980, 537.
14. **Hasumi, H. and Nakamura, S.**, Studies on ferredoxin-ferredoxin reductase complex: Kinetic and solvent perturbation studies on the location of sulfhydryl and aromatic amino acid residues, *J. Biochem.*, 84, 707, 1978.
15. **Avron, M. and Jagendorf, A. T.**, Some further investigations on chloroplast TPNH diaphorase, *Arch. Biochem. Biophys.*, 72, 17, 1957.
16. **Shin, M.**, Ferredoxin-NADP reductase from spinach, in *Methods in Enzymology*, Vol. 23, San Pietro, A., Ed., Academic Press, New York, 1971, 440.
17. **Shin, M. and Oshino, R.**, Ferredoxin-Sepharose 4B as a tool for the purification of ferredoxin-$NADP^+$ reductase, *J. Biochem.*, 83, 357, 1978.
18. **Huang, K., Tu, S.-I., and Wang, J. H.**, Stabilized flavin radical in chloroplast $NADP^+$ reductase, *Biochem. Biophys. Res. Commun.*, 34, 48, 1969.
19. **Shin, M.**, Complex formation by ferredoxin-$NADP^+$ reductase with ferredoxin or NADP, *Biochim. Biophys. Acta*, 292, 13, 1973.
20. **Forti, G.**, Studies on NADPH-cytochrome f reductase of chloroplasts, in *Energy Conversion by the Photosynthetic Apparatus*, Brookhaven National Laboratory Associated University, Inc., Upton, N.Y., 1966, 195.
21. **Gozzer, C., Zanetti, G., Galliano, M., Sacchi, G. A., Minchiotti, L., and Curti, B.**, Molecular heterogeneity of ferredoxin-$NADP^+$ reductase from spinach leaves, *Biochim. Biophys. Acta*, 485, 278, 1977.
22. **Shin, M. and San Pietro, A.**, Complex formation of ferredoxin-NADP reductase with ferredoxin and with NADP, *Biochem. Biophys. Res. Commun.*, 33, 38, 1968.
23. **Foust, G. P., Mayhew, S. G., and Massey, V.**, Complex formation between ferredoxin triphosphopyridine nucleotide reductase and electron transfer proteins, *J. Biol. Chem.*, 244, 964, 1969.
24. **Nelson, N. and Neumann, J.**, Interaction between ferredoxin and ferredoxin nicotinamide adenine dinucleotide phosphate reductase in pyridine nucleotide photoreduction and some partial reactions I, *J. Biol. Chem.*, 244, 1926, 1969.
25. **Nelson, N. and Neumann, J.**, Interaction between ferredoxin and ferredoxin nicotinamide adenine dinucleotide phosphate reductase in pyridine nucleotide photoreduction and some partial reactions II, *J. Biol. Chem.*, 244, 1932, 1969.
26. **Fredricks, W. W. and Gehl, J. M.**, Multiple forms of ferredoxin-nicotinamide adenine dinucleotide phosphate reductase from spinach, *Arch. Biochem. Biophys.*, 174, 666, 1976.
27. **Forti, G. and Sturani, E.**, On the structure and function of reduced nicotinamide adenine dinucleotide phosphate-cytochrome f reductase of spinach chloroplasts, *Eur. J. Biochem.*, 3, 461, 1968.

FERREDOXIN

Herbert Böhme and Peter Böger

INTRODUCTION

Iron-sulfur proteins (nonheme iron proteins) play a key role in many important electron transfer reactions, including processes like photosynthesis, respiration, nitrogen fixation, carboxylation, and hydroxylation. They are present in all living organisms and are particularly involved in photosynthetic and oxidative energy conservation. Ferredoxins are members of this group of iron-sulfur proteins, usually water-soluble enzymes of low molecular weight (6,000 to 12,000), possessing iron and labile (inorganic) sulfur in equimolar amounts of 2, 4, or 8 per molecule. Their oxidation-reduction potential is sufficiently low to transfer electrons to pyridine nucleotides, generally forming an electron transport chain of the type: ferredoxin $\rightleftharpoons$ flavoprotein $\rightleftharpoons$ NAD $(P)^+$. Ferredoxins may be detected and characterized by their electron paramagnetic resonance (EPR) signal in organelles or cells at low temperatures. This method showed that mitochondria, chloroplasts, and bacterial chromatophores contain, in addition to soluble iron-sulfur proteins, membrane-bound iron-sulfur centers. However, their exact function in primary photochemical processes and mitochondrial and bacterial energy conservation is a subject of active research as is the nature of their iron-sulfur center.[1]

This report is limited to soluble ferredoxins from plants, procaryotic and eucaryotic algae, and photosynthetic bacteria, i.e. biosolar energy converters. A number of reviews and books on iron-sulfur proteins are available and may be consulted for further detail.[2-8]

CHEMISTRY OF FERREDOXINS

Nature of the Active Center

The nature of the [Fe-S]-chromophore has been elucidated by data from various spectroscopic, chemical, and crystallographic studies. So far, two types of iron-sulfur centers containing equal amounts of iron and acid-labile sulfur have been recognized:

1. Binuclear iron clusters, with two bridging inorganic sulfur ligands, bonded to the protein via four cysteine sulfurs: plant-type ferredoxins
2. Tetranuclear iron clusters, with four bridging inorganic sulfur atoms, bonded to the protein via four cysteine residues: bacterial-type ferredoxins (One- and two-cluster ferredoxins can be distinguished. To describe, for example, a two-cluster structure, the designation $[4Fe4S^*]_2$ is used, where S* represents an acid-labile sulfur atom.)

Plant-Type Ferredoxins [2Fe-2S*]

These proteins were first isolated from chloroplasts of higher plants. They appear to be present in all oxygen-evolving organisms as part of the photosynthetic electron transport. Two-iron ferredoxins have also been isolated from nonphotosyntetic sources (putidaredoxin, adrenodoxin) which are involved in hydroxylase reactions and from other nonphotosynthetic bacteria of unknown function.[4-8]

The red-colored ferredoxins are acidic proteins with isoelectric points ≈ pH 4.0 and can be isolated and purified by repeated ion-exchange and gel-filtration chromatography.[3] Recently, a fast high-yield procedure has been described involving precipitation by polyethyleneimine.[9] Many ferredoxins have been crystallized from ammonium sulfate solutions.[3] The molecular weight of these soluble proteins is around 10,000. In the oxidized state, they exhibit typical absorption maxima at 465, 420, 330, and 280 nm, mainly because of ligand-metal charge transfer bands of the chromophore. The molar extinction coefficient of spinach ferredoxin e.g., at 420 nm, is 9700 [$M^{-1} \times cm^{-1}$][3]. Ferredoxin may be reduced by chloroplasts and light or anaerobically by dithionite, leading to a decrease of absorption. The midpoint potentials (E_o') range from −310 to −450 mV.[10] The reduced form is paramagnetic and exhibits characteristic EPR spectra at low temperatures (<77°K). The spectrum is broad, with rhombic symmetry and an average g value of 1.94.[5,11] EPR signals are especially useful for the detection of ferredoxins in multicomponent systems. Ferredoxins show characteristic circular dichroism spectra, with an intense optically active d-d band in the near IR.[6] This together with other physicochemical data leads to the interpretation that the iron atoms most likely are in a tetrahedral all-sulfur environment.[12,13] Both Mössbauer and Endor spectroscopy of the oxidized form indicates two high-spin Fe^{3+} atoms in a similar chemical environment. Upon reduction one atom is high-spin ferrous, and the other is high-spin ferric.[4] To account for the observed diamagnetism in the oxidized state, it was proposed that the two iron atoms are antiferromagnetically coupled.[14] Since X-ray crystallographic data are not yet available, the structure depicted in Figure 1 must be regarded as provisional.[13] Chemical analysis also shows two ferric atoms per molecule. Dilute acids liberate an equal amount of noncysteine sulfur, and the apoprotein thus produced may be reconstituted by adding iron and sulfide. The process of reconstitution has been proven useful for enriching the protein in ^{57}Fe for Mössbauer spectroscopy.[15] Apoproteins have also been successfully used for interprotein cluster transfer which allows the nature of the active center to be determined.[8]

Amino acid analysis reveals 95 to 100 amino acid residues per molecule, 4 to 6 of which are cysteines. Four cysteines are found to be invariant in sequence studies and are obviously necessary to chelate the two iron atoms.[16] They do not react within thiol reagents unless the iron is removed. Comparison of amino acid sequences are used to elucidate the evolutionary and structure-function relationships.[17]

Bacterial-Type Ferredoxins

Two-Cluster Ferredoxins [4Fe-4S]₂*

Ferredoxins containing eight iron atoms have been isolated from both photosynthetic and nonphotosynthetic bacteria. Actually, the name "ferredoxin" was first coined for a low molecular weight "iron-redox-protein" from the obligate anaerobe nitrogen-fixing bacterium *Clostridium pasteurianum*.[18] These brown-colored acidic proteins have a rather featureless optical spectrum, with a single absorption peak around 385 nm and in the UV at 280 nm (shoulder at 300 nm). The molar extinction coefficient at 385 nm is about 30,000. The visible absorption decreases by approximately 50% upon reduction with dithionite and is restored after aeration of the sample. A low redox potential is characteristic, with the lowest value reported for *Chromatium* ferredoxin at −490 mV[5]. In contrast to plant ferredoxin, two electrons may be transferred. Titration of the protein with dithionite shows that the EPR signals appear sequentially during reduction.[4] Recently, two-step high-yield procedures were described for isolation and purification. One is based on tight binding of the protein to DEAE cellulose in 60% ammonium sulfate, the other method involves precipitation by polyethylenediimine.[19,20] The molecular weight is typically around 6000, corresponding to about 55 amino acid residues.

Ferredoxin from *C. pasteurianum* yields a broad complex of EPR signals around g

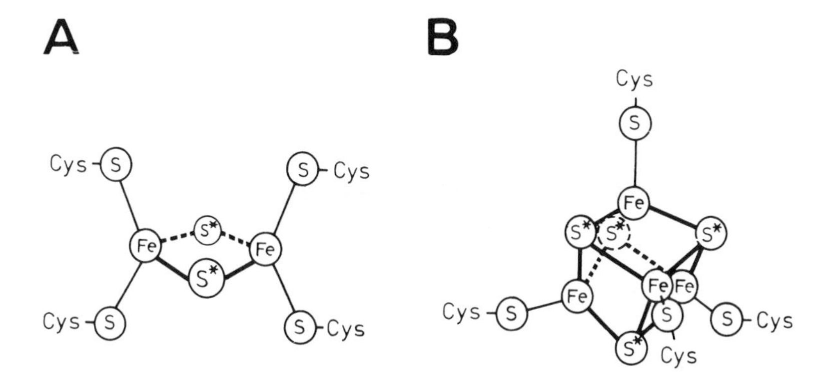

FIGURE 1. Active centers of ferredoxins. Proposed models for (A) [2Fe-2S*] clusters of plant-type ferredoxin,[12] (B) [4Fe-4S*] clusters of bacterial ferredoxin.[26]

= 2 after reduction; a single band at g = 2.01 is typical for the oxidized form. NMR studies confirm that the irons in both clusters are antiferromagnetically exchange coupled.[4]

Iron and labile sulfur can be removed by acidification, and the resulting apoprotein can be reconstituted.[21] Amino acid sequencing has shown the invariance of the cysteine positions. Crystallographic studies on the ferredoxin of *Peptococcus aerogenes* reveals two tetrameric iron-sulfur clusters. They appear to be distorted cubes, composed of two interpenetrating iron and sulfur tetrahedra (see Figure 1).

Single-Cluster Ferredoxins [4Fe-4S]*

The photosynthetic bacterium *Rhodospirillum rubrum* contains in addition to an eight-iron protein (FdI, present in light-grown cells) a single-cluster ferredoxin which is always present (FdII).[24] The optical spectrum of this single-cluster iron protein resembles that with two clusters. The molar extinction coefficient at 400 nm is 17,000, corresponding to about half the value obtained for 8-iron proteins. The mol wt is 14,500, the E_o' = 430 mV, and 1 electron is transferred. Again, the EPR spectrum exhibits a single resonance line at g = 2.01; when photoreduced a rhombic signal around g = 1.94 appears, resembling spinach ferredoxin. Single-cluster Fe-S proteins have also been isolated from nonphotosynthetic bacteria.[24]

There is another group of single-cluster proteins with a surprisingly high midpoint potential of E_o' = +350 mV. Such a high potential iron protein (HiPIP) was first isolated and characterized from *Chromatium vinosum,* a red sulfur bacterium.[25] Its mol wt is about 10,000. The reduced form shows a characteristic absorption band at 388 nm (ε_{388} = 15,000 [$M^{-1} \times cm^{-1}$]) and three maxima after oxidation at 325, 375, and 450 nm. In contrast to other ferredoxins, the *reduced* form is diamagnetic and becomes paramagnetic upon oxidation, with a principal g value at 2.04.[4] The amino acid sequence shows no obvious relationship to other iron sulfur proteins. X-ray crystallography has shown that the tetranuclear cubane-like Fe-S clusters in reduced *Chromatium* (HiPIP) and oxidized eight-iron *Peptococcus* ferredoxin have indistinguishable geometries.[26] Nevertheless, a redox-potential difference of about 750 mV exists between both proteins. This unusual behavior has been explained in terms of a "three-state hypothesis".[27]

It was proposed that the [4Fe-4S*] clusters in oxidized 8-Fe ferredoxin and reduced high-potential iron protein are in an isoelectronic state C and hence not EPR active. In this state, HiPIP is oxidizable to the paramagnetic state C^+ (formal charges 2−/1−), whereas the 8-Fe ferredoxin can be only reduced to C^- (2−/3−), another paramagnetic state.[8] With respect to the high midpoint potential, it is notable that the 4Fe cluster in HiPIP appears to be well preserved from contact with water.

BIOLOGICAL FUNCTIONS

This discussion is limited primarily to the functions of ferredoxin in photosynthetic organisms (compare Figure 2). In plants, procaryotic and eucaryotic algae, the two-iron ferredoxins function as catalysts in photosynthetic electron transport as an essential component of $NADP^+$ photoreduction. The terminal electron transport chain of Photosystem I involves bound iron-sulfur centers, (soluble) ferredoxin, and ferredoxin-$NADP^+$ reductase.[28,29] Reductase, which is membrane-bound, forms an equimolar complex with ferredoxin and $NADP^+$. Soluble ferredoxin was also found to be an effective catalyst of cyclic photophosphorylation under aerobic conditions which involves only Photosystem I. More specifically, it appears to catalyze cytochrome b_{563} photoreduction by Photosystem I.[29,31] Quantitative studies have shown that spinach ferredoxin is in fivefold excess over P700, the reaction center (RC) chlorophyll.[32] From inhibitor experiments, analysis of the mode of inheritance and in vitro biosynthesis it can be concluded that plant ferredoxin is coded for in nuclear DNA.[33,34] Other functions of soluble ferredoxin in chloroplasts involve a control over CO_2 assimilation by regulation of certain key enzymes via the ferredoxin/thioredoxin system.[35] Reduced ferredoxin may serve as electron donor for nitrite and sulfite reduction.[36,37] Recently it was found that ferredoxin forms a 1:1 complex with ferredoxin-nitrite oxidoreductase.[38] In certain algae, ferredoxin is involved in hydrogen metabolism and in nitrogen fixation in blue-green algae.

Bacterial photosynthesis involves only one photoreaction, mainly for generation of ATP. The role of ferredoxins in this process is still unclear. There is evidence that Fe-S proteins are involved in photosynthetic reduction of NAD^+ and formation of ATP by cyclic phosphorylation.[39,40] Reduced ferredoxin is directly involved in CO_2 assimilation through the so-called reductive carboxylic acid cycle which has common features with the citric acid cycle. The strong reducing power of ferredoxin is used to drive energetically difficult carboxylations of acyl-coenzyme A intermediates to their respective α-ketocarboxylic acids (e.g., pyruvate, α-ketoglutarate).[2,41] Bacterial ferredoxins are generally involved in sulfate, nitrate, hydrogen, and nitrogen metabolism.

EVOLUTION

The ubiquitous occurrence and the low molecular weight of ferredoxins provide a powerful tool for the investigation of possible evolutionary relationships between organisms. Mainly, two approaches were used to elucidate the evolution of this group of proteins: amino acid sequencing and immunological cross-reactivity.[16,17,42] Ferredoxins have been isolated and characterized from various bacterial, algal, and higher-plant sources, including photosynthetic green and purple bacteria, halobacteria, and blue-green, red, yellow-green, and green algae.[3,16,43-45] Phylogenetic trees have been constructed on the basis of ferredoxin-sequence data.[16,17] Isolation of two species of ferredoxins from one organism with similar (same active center) but not identical properties was taken as evidence for gene duplication during evolution.[17] Differences were found in the biological activity, amino acid composition, and redox properties, suggesting a physiological differentiation of ferredoxins within one organism. The [4Fe-4S*]-cluster ferredoxins have been claimed to be among the earliest proteins formed under primitive Earth conditions.[5]

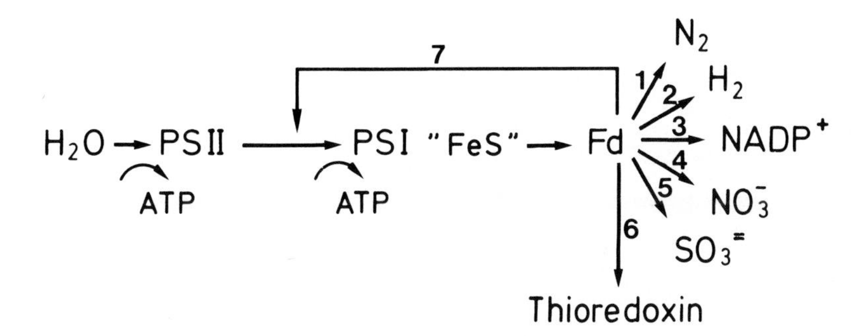

FIGURE 2. Functions of plant-type ferredoxin catalyzed by (1) nitrogenase, (2) hydrogenase, (3) ferredoxin-$NADP^+$ reductase, (4) nitrite reductase, (5) sulfite reductase, (6) thioredoxin reductase, and (7) cytochrome b-563 (cyclic electron transport).

REFERENCES

1. **Malkin, R. and Bearden, A. J.,** Membrane-bound iron-sulfur centers in photosynthetic systems, *Biochim. Biophys. Acta,* 505, 147, 1978.
2. **Buchanan, B. B. and Arnon, D. I.,** Ferredoxins: chemistry and function in photosynthesis, nitrogen fixation, and fermentative metabolism, in *Advances in Enzymology,* Vol. 33, Nord, F. F., Ed., Interscience, New York, 1971, 120.
3. **Buchanan, B. B. and Arnon, D. I.,** Ferredoxins from photosynthetic bacteria, algae and higher plants, in *Methods in Enzymology,* Vol. 23a, San Pietro, A., Ed., Academic Press, New York, London, 1971, 413.
4. **Orme-Johnson, W. H.,** Iron sulfur proteins: structure and function, *Annu. Rev. Biochem.,* 42, 159, 1973.
5. **Hall, D. O., Rao, K. K., and Cammack, R.,** The iron sulfur proteins: structure, function and evolution of a ubiquitous group of proteins, *Sci. Prog.,* 62, 285, 1975.
6. **Hall, D. O. and Rao, K. K.,** Ferredoxin, in *Encyclopedia of Plant Physiology,* New Series, Vol. 5, Trebst, A. and Avron, M., Eds., Springer-Verlag, Heidelberg, 1977, 206.
7. **Lovenberg, W., Ed.,** *Iron-Sulfur Proteins,* Vol. 1 to 3, Academic Press, New York, 1973 and 1977.
8. **Averill, B. A. and Orme-Johnson, W. H.,** Iron sulfur proteins and their synthetic analogs, in *Metal Ions in Biological Systems,* Vol. 7, Sigel, H., Ed., Marcel Dekker, New York, 1978, 127.
9. **Thauer, R. K., Schirrmacher, H., Schymanski, W., and Schönheit, P.,** A rapid procedure for the purification of ferredoxin from spinach using polyethyleneimine, *Z. Naturforsch. Teil.,* 33, 495, 1978.
10. **Cammack, R., Rao, K. K., Bargeron, C. P., Hutson, K. G., Andrew, P. W., and Rogers, L. J.,** Midpoint redox potentials of plant and algal ferredoxins, *Biochem. J.,* 168, 205, 1977.
11. **Orme-Johnson, W. H. and Sands, R. H.,** Probing iron sulfur proteins with EPR and ENDOR spectroscopy, in *Iron-Sulfur Proteins,* Vol. 2, Lovenberg, W., Ed., Academic Press, New York, 1973, 195.
12. **Brintzinger, H., Palmer, G., and Sands, R. H.,** On the ligand field of iron in ferredoxin from spinach chloroplasts and related non heme iron enzymes, *Proc. Natl. Acad. Sci. U.S.A.,* 56, 987, 1966.
13. **Palmer, G.,** Current insights into the active center of spinach ferredoxin and other iron sulfur proteins, in *Iron-Sulfur Proteins,* Vol. 2, Lovenberg, W., Ed., Academic Press, New York, 1973, 285.
14. **Gibson, J. G., Hall, D. O., Thornley, J. H. M., and Whatley, F. R.,** The iron complex in spinach ferredoxin, *Proc. Natl. Acad. Sci. U.S.A.,* 56, 987, 1966.
15. **Bearden, A. J. and Dunham, W. R.,** Mössbauer spectroscopy of iron sulfur proteins, in *Iron-Sulfur Proteins,* Vol. 2, Lovenberg, W., Ed., Academic Press, New York, 1973, 239.
16. **Yasunobu, K. T. and Tanaka, M.,** The types, distribution in nature, structure-function, and evolutionary data of the iron sulfur proteins, in *Iron-Sulfur Proteins,* Vol. 2, Lovenberg, W., Ed., Academic Press, New York, 197, 29.
17. **Matsubara, H., Hase, T., Wakabayashi, S., and Wada, K.,** Gene duplications during evolution of chloroplast-type ferredoxins, in *Evolution of Protein Molecules,* Matsubara, H. and Yamanaka, T., Eds., Scientific Societies Press, Tokyo, 1978, 209.
18. **Mortenson, L. E., Valentine, R. C., and Carnahan, J. E.,** An electron transport factor from *Clostridium pasteurianum, Biochem. Biophys. Res. Commun.,* 7, 448, 1962.

19. **Mayhew, S. G.**, Nondenaturing procedure for rapid preparation of ferredoxin from *Clostridium pasteurianum*, *Anal. Biochem.*, 42, 191, 1971.
20. **Schönheit, P., Wäscher, C., and Thauer, R. K.**, A rapid procedure for the purification of ferredoxin from clostridia using polyethyleneimine, *FEBS Lett.*, 89, 219, 1978.
21. **Hong, J. and Rabinowitz, J. C.**, Preparation and properties of clostridial apoferredoxins, *Biochem. Biophys. Res. Commun.*, 29, 246, 1967.
22. **Adman, E. T., Sieker, L. C., and Jensen, L. H.**, The structure of a bacterial ferredoxin, *J. Biol. Chem.*, 248, 3987, 1973.
23. **Yoch, D. C., Arnon, D. I., and Sweeney, W. V.**, Characterization of two soluble ferredoxins as distinct from bound iron-sulfur proteins in the photosynthetic bacteria *Rhodospirillum rubrum*, *J. Biol. Chem.*, 250, 8330, 1975.
24. **Shetna, Y. I., Stombaugh, N. A., and Burris, R. H.**, Ferredoxin from *Bacillus polymyxa*, *Biochem. Biophys. Res. Commun.*, 42, 1108, 1971.
25. **Bartsch, R. G.**, Non-heme iron proteins and *Chromatium* iron protein, in *Bacterial Photosynthesis*, Gest, H., San Pietro, A., and Vernon, L. P., Eds., Antioch Press, Yellow Springs, Ohio, 1963, 315.
26. **Carter, C. W.**, New stereochemical analogies between iron-sulfur electron transport proteins, *J. Biol. Chem.*, 252, 7802, 1977.
27. **Carter, C. W., Kraut, J., Freer, S. T., Alden, R. A., Sieker, L. C., Adman, E., and Jensen, L. H.**, A comparison of Fe_4S_4* clusters in high potential iron protein and in ferredoxin, *Proc. Natl. Acad. Sci. U.S.A.*, 69, 3526, 1972.
28. **Davenport, H. E., Hill, F. R. S., and Whatley, F. R.**, A natural factor catalyzing reduction of methaemoglobin by isolated chloroplasts, *Proc. R. Soc. London Ser. B*, 139, 346, 1952.
29. **Arnon, D. I.**, Photosynthesis 1950—75: changing concepts and perspectives, in *Encyclopedia of Plant Physiology*, New Series, Vol. 5, Trebst, A. and Avron, M., Eds., Springer-Verlag, Heidelberg, 1977, 7.
30. **Bookjans, G. and Böger, P.**, Complex-forming properties of butanedione-modified ferredoxin-$NADP^+$ reductase with $NADP^+$ and ferredoxin, *Arch. Biochem. Biophys.*, 194, 327, 1979.
31. **Böhme, H.**, On the role of ferredoxin and ferredoxin-$NADP^+$ reductase in cyclic electron transport of spinach chloroplasts, *Eur. J. Biochem.*, 72, 283, 1977.
32. **Böhme, H.**, Quantitative determination of ferredoxin, ferredoxin-$NADP^+$ reductase and plastocyanin in spinach chloroplasts, *Eur. J. Biochem.*, 83, 137, 1978.
33. **Haslett, B. G., Cammack, R., and Whatley, F. R.**, Quantitative studies on ferredoxin in greening bean leaves, *Biochem. J.*, 136, 697, 1973.
34. **Huisman, J. G., Moorman, A. F. M., and Verkley, F. N.**, In vitro synthesis of chloroplast ferredoxin as a high molecular weight precursor in a cell-free protein synthesizing system from wheat germs, *Biochem. Biophys. Res. Commun.*, 82, 1121, 1978.
35. **Wolosiuk, R. A. and Buchanan, B. B.**, Thioredoxin and glutathione regulate photosynthesis in chloroplasts, *Nature (London)*, 266, 565, 1977.
36. **Schmidt, A. and Trebst, A.**, The mechanism of photosynthetic sulfate reduction by isolated chloroplasts, *Biochim. Biophys. Acta*, 180, 529, 1969.
37. **Losada, M., Paneque, A., Ramirez, J. M., and Del Campo, F. F.**, Mechanism of nitrite reduction in chloroplasts, *Biochem. Biophys. Res. Commun.*, 10, 298, 1963.
38. **Knaff, D. B., Smith, J. M., and Malkin, R.**, Complex formation between ferredoxin and nitrite reductase, *FEBS Lett.*, 90, 195, 1978.
39. **Buchanan, B. B. and Evans, M. C. W.**, Photoreduction of ferredoxin and its use in $NAD(P)^+$ reduction by a subcellular preparation from the photosynthetic bacteriaum *Chlorobium thiosulfatophilum*, *Biochim. Biophys. Acta*, 180, 123, 1969.
40. **Shanmugam, K. T. and Arnon, D. I.**, Effect of ferredoxin on bacterial photophosphorylation, *Biochim. Biophys. Acta*, 256, 487, 1972.
41. **Kerscher, L. and Oesterhelt, D.**, Ferredoxin is the co-enzyme of α-ketoacid oxidoreductases in *Halobacterium halobium*, *FEBS Lett.*, 83, 197, 1977.
42. **Tel-Or, E., Cammack, R., Rao, K. K., Rogers, L. J., Stewart, W. D. P., and Hall, D. O.**, Comparative immunochemistry of bacterial, algal and plant ferredoxins, *Biochim. Biophys. Acta*, 490, 120, 1977.
43. **Böger, P.**, Ferredoxin aus *Bumilleriopsis filiformis*, *Planta*, 92, 105, 1970.
44. **Andrew, P. W., Rogers, L. J., Boulter, D., and Haslett, B. G.**, Ferredoxin from a red alga, *Porphyra umbilicalis*, *Eur. J. Biochem.*, 69, 243, 1976.
45. **Kerscher, L., Oesterhelt, D., Cammack, R., and Hall, D. O.**, A new plant-type ferredoxin from *Halobacteria*, *Eur. J. Biochem.*, 72, 101, 1976.

PRIMARY ELECTRON DONORS*

Bacon Ke

INTRODUCTION

The primary electron donors of the bacterial reaction center (RC), Photosystem I (PS I), and Photosystem II (PS II) are designated as P_{870} (or P_{890}), P_{700}, and P_{690}, respectively (the letter P stands for "pigment", and the numeral refers to the wavelength location of its major absorption band). The primary electron donors are formed by a small fraction (e.g., ∼1/400 in the case of PS I) of the total light-absorbing chlorophyll molecules in each photosystem. These chlorophyll molecules are present in a special environment different from that of the bulk pigment molecules, and they serve the purpose of accepting excitation energy transferred from the bulk pigment molecules.

P_{700} of PS I and P_{870} of the bacterial RC are known to consist of a pair of chlorophyll molecules (dimer). Energy absorbed by the bulk chlorophyll molecules and transferred to P (P_{700}, P_{690}, or P_{870}) raises the latter to an excited singlet state P* which releases an electron and becomes oxidized. The released electron goes first to reduce the so-called "primary" electron acceptor and is subsequently transferred to secondary acceptors along the electron-transport chain. Oxidation of P is accompanied by a change in its absorption spectrum. The resultant oxidized dimer is usually in the form of a free-radical cation which can be detected by EPR spectroscopy. At present, much less is known about P_{690} of PS II. Since the three species have many common features, their properties are compared in tables with references, each followed by brief comments.

SPECTRAL (OPTICAL) PROPERTIES

Table 1
SPECTRAL (OPTICAL) PROPERTIES: LIGHT-INDUCED ABSORPTION CHANGES

	P700 [1-8]	P690 [4,9-12]	P870 [13-16]
Major absorption changes (nm)	(−) 700 (−) 430 (−) 682 (+) 810	(−) 687 (−) 435 (+) 825	(−) 865 (−)/(+) 803/797 (−) 600 (−) 365 (+) 430 (+) 1250
Extinction coefficient, $mM^{-1} \cdot cm^{-1}$ (nm)	64 (700)[5] 70 (701)[5] 67 (703)[17]	—	112 (865)[18]

Note: Increase, (+); decrease, (−).

The absolute absorption spectra of the primary electron donors cannot be measured directly. However, their spectra can be constructed from the light-induced absorption

* Contribution No. 637 from the Charles F. Kettering Research Laboratory.

changes.[1-16] Note that the major absorption bands of each primary electron donor are located on the longer wavelength side of the main band of the bulk chlorophyll molecules which is consistent with the notion that the excitation energy of the bulk chlorophyll travels toward the RC chlorophyll with a lower-energy excited state. Thus, the RC chlorophyll has often been termed the energy-trapping center. The light-induced absorption decreases (bleaching) of the major absorption band correspond to a loss of an electron (oxidation) by the primary electron donor. Photooxidation of the primary donors occurs with a high quantum efficiency[19] and at cryogenic temperatures[10,20-24] which is consistent with their role in a primary photochemical event.

REFERENCES

1. **Kok, B.**, Absorption changes induced by the photochemical reaction of photosynthesis, *Nature (London)*, 179, 583, 1957.
2. **Ke, B.**, Light-induced rapid absorption changes during photosynthesis. V. Digitonin-treated chloroplasts, *Biochim. Biophys. Acta*, 88, 297, 1964.
3. **Rumberg, B. and Witt, H. T.**, Die Photooxydation von Chlorophyll-a_I-430-703, *Z. Naturforsch. Teil B*, 19, 693, 1964.
4. **Döring, G., Stiehl, H. H., and Witt, H. T.**, A second chlorophyll reaction in the electron chain of photosynthesis-registration by the repetitive excitation technique, *Z. Naturforsch. Teil B*, 22, 639, 1967.
5. **Hiyama, T. and Ke, B.**, Difference spectra and extinction coefficients of P700, *Biochim. Biophys. Acta*, 267, 160, 1972.
6. **Ke, B.**, The primary electron acceptor of photosystem I, *Biochim. Biophys. Acta*, 301, 1, 1973.
7. **Inoue, Y., Ogawa, T., and Shibata, K.**, Light-induced spectral changes of P700 in the 800-nm region in *Anacystis* and spinach lamellae, *Biochim. Biophys. Acta*, 369, 371, 1973.
8. **Mathis, P. and Vermeglio, A.**, Chlorophyll radical cation in photosystem II of chloroplasts. Millisecond decay at low temperature, *Biochim. Biophys. Acta*, 369, 371, 1975.
9. **Doring, G., Renger, G., Vater, J., and Witt, H. T.**, Properties of the photoactive chlorophyll-a_{II} in photosynthesis, *Z. Naturforsch. Teil B*, 24, 1139, 1969.
10. **Ke, B., Sahu, S., Shaw, E. R., and Beinert, H.**, Further characterization of a photosystem-II particle isolated from spinach chloroplasts by Triton treatment: the reaction-center components, *Biochim. Biophys. Acta*, 347, 36, 1974.
11. **Van Gorkom, H. J., Tamminga, J. J., and Haveman, J.**, Primary reactions, plastoquinone and fluorescence yield in subchloroplast fragments prepared with deoxycholate, *Biochim. Biophys. Acta*, 347, 417, 1974.
12. **Gläser, M., Wolff, C., Buchwald, H., and Witt, H. T.**, On the photoactive chlorophyll reaction in system II of photosynthesis. Detection of a fast and large component, *FEBS Lett.*, 42, 81, 1974.
13. **Duysens, L. N. M.**, Transfer of Excitation Energy in Photosynthesis, Thesis, University of Utrecht, 1952.
14. **Clayton, R. K.**, Primary reactions in bacterial photosystem II. The nature of light-induced absorbance changes in chromatophores: evidence for a special bacteriochlorophyll component, *Photochem. Photobiol.*, 1, 201, 1962.
15. **Reed, D. W.**, Isolation and composition of a photosynthetic reaction-center complex from *Rps. spheroides*, *J. Biol. Chem.*, 244, 4936, 1969.
16. **Feher, G.**, Some chemical and physical properties of a bacterial reaction-center particle and its primary photochemical reactants, *Photochem. Photobiol.*, 14, 373, 1971.
17. **Haehnel, W.**, The ratio of the two light reactions and their coupling in chloroplasts, *Biochim. Biophys. Acta*, 423, 499, 1976.
18. **Straley, S. C., Parson, W. W., Mauzerall, D., and Clayton, R. K.**, Pigment content and molar extinction coefficients of photochemical reaction centers from *Rps. spheroides*, *Biochim. Biophys. Acta*, 305, 597, 1973.
19. **Hiyama, T. and Ke, B.**, A further study of P430: a possible primary electron acceptor of Photosystem I, *Arch. Biochem. Biophys.*, 147, 99, 1971.
20. **Ke, B., Sugahara, K., and Sahu, S.**, Light-induced absorption changes in Photosystem I at low temperatures, *Biochim. Biophys. Acta*, 449, 84, 1976.
21. **Floyd, R., Chance, B., and DeVault, D.**, Low-temperature photo-induced reactions in green leaves and chloroplasts, *Biochim. Biophys. Acta*, 226, 103, 1971.
22. **Lozier, R. and Butler, W. L.**, Light-induced absorbance changes in chloroplasts mediated by Photosystem I and Photosystem II at low temperature, *Biochim. Biophys. Acta*, 333, 465, 1974.

23. **Mathis, P. and Vermeglio, A.,** Chlorophyll radical cation in Photosystem II of chloroplasts. Millisecond decay at low temperature, *Biochim. Biophys. Acta,* 396, 371, 1975.
24. **W. Arnold and Clayton, R. K.,** The first step in photosynthesis: evidence for its electronic nature, *Proc. Natl. Acad. Sci. U.S.A.,* 46, 769, 1960.

SPECTRAL (EPR) PROPERTIES

Table 2
SPECTRAL (EPR) PROPERTIES

	P700 [1-5]	**P690** [5-7]	**P870** [8-10]
g-value	2.0025	2.0026	2.0025
ΔH(G)	7.2	8	9.6

The *g*-values of the primary-donor EPR signals are typical of a paramagnetic free radical. The linewidths of the EPR signals of P_{700}^{+} and P_{870}^{+} observed in vivo are narrower than that of the respective chlorophyll cataions in vitro by a factor of $\sqrt{2}$ which can be accounted for by a spin delocalization over a special pair (dimer) of chlorophyll molecules.[3] Further evidence for the dimer configuration comes from the electron-nuclear double resonance (ENDOR) study which confirmed the prediction that the electron-nuclear hyperfine splitting constant for the in vivo chlorophyll dimer is smaller than that found for the chlorophyll monomer cation in solution by a factor of two.[11,12]

The light-induced free-radical EPR signal and the absorption-change signal, which supposedly represent a given primary electron donor, have been correlated in a number of ways, notably the kinetic correlation[9,13,14] for P_{870}^{+} and P_{700}^{+} and a correlation of spins[15,10,16] for P_{700}^{+} and P_{870}^{+}.

REFERENCES

1. **Beinert, H., Kok, B., and Hoch, G.,** The light-induced electron paramagnetic resonance signal of photocatalyst P700, *Biochem. Biophys. Res. Commun.,* 7, 209, 1962.
2. **Malkin, R. and Bearden, A. J.,** Detection of a free radical in the primary reaction of chloroplast photosystem II, *Proc. Natl. Acad. Sci. U.S.A.,* 70, 294, 1973.
3. **Norris, J. R., Uphaus, R. A., Crespi, H. C., and Katz, J. J.,** Electron spin resonance of chlorophyll and the origin of signal I in photosynthesis, *Proc. Natl. Acad. Sci. U.S.A.,* 68, 625, 1971.
4. **Weaver, E. C. and Weaver, H. E.,** Electron resonance studies of photosynthetic systems, in *Photophysiology,* Vol. 7, Giese, A. C., Ed., Academic Press, New York, 1972, 1.
5. **Van Gorkom, H. J., Tamminga, J. J., Haveman, J., and Van der Linden, I. K.,** Primary reactions, plastoquinone and fluorescence yield in subchloroplast fragments prepared with deoxycholate, *Biochim. Biophys. Acta,* 347, 417, 1974.
6. **Bearden, A. J. and Malkin, R.,** Oxidation-reduction potential dependence of low-temperature photoreactions of chloroplast Photosystem II, *Biochim. Biophys. Acta,* 325, 266, 1973.
7. **Ke, B., Sahu, S., Shaw, E. R., and Beinert, H.,** Further characterization of a Photosystem-II particle isolated from spinach chloroplasts by Triton treatment. The reaction-center components, *Biochim. Biophys. Acta,* 347, 36, 1974.
8. **Schleyer, H.,** Electron paramagnetic resonance studies on photosynthetic bacteria. I. Properties of photoinduced EPR signals of Chromatium D, *Biochim. Biophys. Acta,* 153, 427, 1968.
9. **McElroy, M. D., Feher, G., and Mauzerall, D. C.,** On the nature of the free radical formed during the primary process of bacterial photosynthesis, *Biochim. Biophys. Acta,* 172, 180, 1969.
10. **Bolton, J. R., Clayton, R. K., and Reed, D. W.,** An identification of the radical giving rise to the light-induced electron spin resonance signal in photosynthetic bacteria, *Photochem. Photobiol.,* 9, 209, 1969.

11. **Norris, J. R., Scheer, H., Druyan, M. E., and Katz, J. J.,** An electron-nuclear double resonance (ENDOR) study of the special pair model for photoreactive chlorophyll in photosynthesis, *Proc. Natl. Acad. Sci. U.S.A.*, 71, 4987, 1974.
12. **Feher, G., Hoff, A. J., Issacson, R. A., and Ackerson, L. C.,** ENDOR experiments on chlorophyll and bacteriochlorophyll in vitro and in the photosynthetic unit, *Ann. N.Y. Acad. Sci.*, 244, 239, 1974.
13. **Warden, J. T. and Bolton, J. R.,** Simultaneous optical and EPR detection of the primary photoproduct P700 in green-plant photosynthesis, *J. Am. Chem. Soc.*, 94, 4352, 1972.
14. **Ke, B., Sugahara, K., Shaw, E. R., Hansen, R. E., Hamilton, W. D., and Beinert, H.,** Kinetics of appearance and disappearance of light-induced EPR signal of $P700^+$ and iron-sulfur protein(s) at low temperature, *Biochim. Biophys. Acta*, 368, 401, 1974.
15. **Baker, R. A. and Weaver, E. C.,** A correlation of EPR spins with P700 in spinach chloroplast particles, *Photochem. Photobiol.*, 18, 237, 1973.
16. **Loach, P. and Walsh, K.,** Quantum yield for the photoproduced ERP signal in chromatophores from *R. rubrum*, *Biochemistry*, 8, 1908, 1969.

REDOX PROPERTIES

Table 3
REDOX PROPERTIES

	P_{700} [1-7]	P_{690} [8]	P_{870} [9-16] (purple bacteria)
E_m, mV	+430	>+850	+443
	+460		+450
	+480		+490
	+520		+500
	+468		+330 (*Chlorobium*)[17]
	+493		+250 (*Chlorobium*)[18]
	+375		+390—+450 (*R. viridis*)[19]
Electron change (n)	1	—	1

All primary electron donors appear to be relatively strong oxidizing one-electron carriers, and their redox potentials are independent of pH. The redox-potential values reported by different workers for P_{700} vary over a wide range; this discrepancy cannot yet be reconciled. The redox potential of P_{690} has not yet been measured directly. It was only estimated from the observation that light-induced free-radical EPR signal in Photosystem II can still be measured at a potential of +850 mV (pH 7.8).[8]

REFERENCES

1. **Kok, B.,** Partial purification and determination of oxidation reduction potential of the photosynthetic chlorophyll complex absorbing at 700 nm, *Biochim. Biophys. Acta*, 48, 527, 1961.
2. **Rumberg, B.,** Die Eigenschaften des Reaktionscyclus von Chlorophyll-a_I-430-703, *Z. Naturforsch. Teil B*, 19, 707, 1964.
3. **Yamamoto, H. and Vernon, L. P.,** Characterization of a partially purified photosynthetic reaction center from spinach chloroplasts, *Biochemistry*, 10, 4131, 1969.
4. **Knaff, D. B. and Malkin, R.,** The oxidation-reduction potential of electron carriers in chloroplast Photosystem I fragments, *Arch. Biochem. Biophys.*, 159, 555, 1973.
5. **Ke, B., Sugahara, K., and Shaw, E. R.,** Further purification of "Triton subchloroplast fraction I" (TSF-I particles). Isolation of a cytochrome-free high-P700 particle and a complex containing cytochromes *f* and b_6, plastocyanin and iron-sulfur protein(s), *Biochim. Biophys. Acta*, 408, 12, 1975.

6. **Shuvalov, V. A., Klimov, V. V., and Krasnovsky, A. A.,** Primary photo-processes in light fragments of chloroplasts, *Mol. Biol.*, 10, 326, 1976.
7. **Evans, M. C. W., Sihra, C. K., and Slabas, A. R.,** The oxidation-reduction potential of the reaction-centre chlorophyll (P700) in photosystem I, *Biochem. J.*, 162, 75, 1977.
8. **Bearden, A. J. and Malkin, R.,** Oxidation-reduction potential dependence of low-temperature photoreactions of chloroplast photosystem II, *Biochim. Biophys. Acta*, 325, 266, 1973.
9. **Loach, P. A., Androes, G. M., Maksim, A. F., and Calvin, M.,** Variation in electron paramagnetic resonance signals of photo synthetic systems with redox level of their environment, *Photochem. Photobiol.*, 2, 443, 1963.
10. **Dutton, P. L. and Jackson, J. B.,** Thermodynamic and kinetic characterization of electron-transfer components *in situ* in *Rhodopseudomonas spheroides* and *Rhodospirillum rubrum*, *Eur. J. Biochem.*, 30, 495, 1972.
11. **Cusanovich, M. A. and Kamen, M. D.,** Light-induced electron transfer in Chromatium strain D. I. Isolation and characterization of *Chromatium* chromatophores, *Biochim. Biophys. Acta*, 153, 376, 1968.
12. **Case, G. D. and Parson, W. W.,** Thermodynamics of the primary and secondary photochemical reactions in *Chromatium*, *Biochim. Biophys. Acta*, 253, 187, 1971.
13. **Dutton, P. L.,** Oxidation-reduction potential dependence of the interaction of cytochromes, bacteriochlorophyll and carotenoids at 77°K in chromatophores of *Chromatium* D and *Rhodopseudomonas gelatinosa*, *Biochim. Biophys. Acta*, 226, 63, 1971.
14. **Thronber, J. P. and Olson, J. M.,** Chlorophyll-proteins from reaction-center preparations from photosynthetic bacteria, algae and higher plants, *Photochem. Photobiol.*, 14, 329, 1971.
15. **Carithers, R. P. and Parson, W. W.,** Delayed fluorescence from *Rhodopseudomonas viridis* following single flashes, *Biochem. Biophys. Acta*, 387, 194, 1975.
16. **Prince, R. C., Leigh, J. S., and Dutton, P. L.,** Thermodynamic properties of the reaction center of *Rhodopseudomonas viridis*; in vivo measurement of the reaction enter bacteriochlorophyll-primary acceptor intermediary electron carrier, *Biochim. Biophys. Acta.*, 440, 622, 1976.
17. **Prince, R. C. and Olson, J. M.,** Some thermodynamic and kinetic properties of the primary photochemical reactants in a complex from a green photosynthetic bacterium, *Biochim. Biophys. Acta*, 423, 357, 1976.
18. **Knaff, D. B., Buchanan, B. B., and Malkin, R.,** Effect of oxidation-reduction potential on light-induced cytochrome and bacteriochlorophyll reactions in chromatophores from the photosynthetic green bacterium *Chlorobium*, *Biochim. Biophys. Acta*, 325, 934, 1973.
19. **Thornber, J. P. and Olson, J. M.,** Chlorophyll-protein and reaction-center preparations from photosynthetic bacteria, algae and higher plants, *Photochem. Photobiol.*, 14, 329, 1971.

KINETIC BEHAVIOR

Table 4
KINETIC BEHAVIOR

	P700	P690	P870
Photooxidation time	<100 nsec[1] <20 nsec[2] ≤50 psec[3] ≤14 psec[4]	<20 nsec[5]	≤10 psec[6-10]
Reduction time, $t_{1/2}$ or k (electron donor)	~20 μsec (cytochrome *f*)[11] ~200 μsec (plastocyanin)[12] ~20 msec (plastoquinone)[12] 1.5×10^7 $M^{-1} \cdot s^{-1}$ (PMSH·)[13] 1.7×10^4 $M^{-1} \cdot s^{-1}$ (TMPD)[14] 3.0×10^9 $M^{-1} \cdot s^{-1}$ ($P430^-$)[14]	100—200 μsec (20°C) 3 msec (77°K) (reduced primary acceptor)[15-17] ~35 μsec (an unknown donor)[17-19] ~30 nsec (the first physiological donor)[20,21]	2 μs (high-potential *c*-type cytochrome in *Chromatium*; 20°C)[22,23] 20 msec (Fe · Q·; 20°C)[24-27]

Determination of the photooxidation time of P_{870} in photosynthetic bacteria was one of the first applications of picosecond spectroscopy in biology.[6-10] Picosecond measurements for P_{700} photooxidation have been performed only very recently.[3,4] As can be seen from Table 4, rapid (nsec and μsec) spectroscopic techniques are very important tools for examining the relatively slower subsequent electron-transfer reactions involving either the oxidized primary donor or the reduced primary acceptor or for studying their interactions with artificial electron carriers.

REFERENCES

1. **Ke, B.**, The risetime of photoreduction, difference spectrum, and oxidation-reduction potential of P430, *Arch. Biochem. Biophys.*, 152, 70, 1972.
2. **Witt, K. and Wolff, C.**, Risetime of the absorption changes of chlorophyll-a_I and carotenoids in photosynthesis, *Z. Naturforsch. Teil B*, 25, 387, 1970.
3. **Shuvalov, V. A., Ke, B., and Dolan, E.**, Kinetic and spectral properties of the intermediary electron acceptor A_I in Photosystem I: subnanosecond spectroscopy, *FEBS Lett.*, 100, 5, 1979.
4. **Netzel, T. L.**, personal communication.
5. **Witt, H. T.**, Coupling of quanta, electrons, fields, ions and phosphorylation in the functional membrane of photosynthesis, *Q. Rev. Biophys.*, 4, 365, 1971.
6. **Netzel, T. L., Rentzepis, P. M., and Leigh, J. S.**, Picosecond kinetics of reaction-center bacteriochlorophyll, *Science*, 182, 238, 1973.
7. **Kaufmann, K. J., Dutton, P. L., Netzel, T. L., Leigh, J. S., and Rentzepis, P. M.**, Picosecond kinetics of events leading to reaction-center bacteriochlorophyll oxidation, *Science*, 188, 1301, 1975.
8. **Rockley, M. G., Windsor, M. W., Cogdell, R. J., and Parson, W. W.**, Picosecond detection of an intermediate in the photochemical reaction of bacterial photosynthesis, *Proc. Natl. Acad. Sci. U.S.A.*, 72, 2251, 1975.
9. **Dutton, P. L., Kaufmann, K. J., Chance, B., and Rentzepis, P. M.**, Picosecond kinetics of the 1250 nm band of the *Rps. spheroides* reaction center: the nature of the primary photochemical intermediate state, *FEBS Lett.*, 60, 275, 1975.
10. **Peters, K., Avouris, P., and Rentzepis, P. M.**, Picosecond dynamics of primary electron-transfer processes in bacterial photosynthesis, *Biophys. J.*, 23, 207, 1978.
11. **Haehnel, W. and Witt, H. T.**, The reaction between chlorophyll-a_I and its primary electron donors, in *Proc. 2nd Int. Congr. on Photosynthesis*, Forti, G., Eds., Dr. Junk, The Hague, The Netherlands, 469, 1972.
12. **Haehnel, W.**, Electron transport between plastoquinone and chlorophyll-a_I and the function of plastocyanin *in situ*, *Biochim. Biophys. Acta*, 459, 418, 1977.
13. **Rumberg, B. and Witt, H. T.**, Die Photooxydation von Chlorophyll-a_I-430-703, *Z. Naturforsch. Teil B*, 19, 693, 1964.
14. **Hiyama, T. and Ke, B.**, A further study of P430: a possible primary electron acceptor of Photosystem I, *Arch. Biochem. Biophys.*, 147, 99, 1971.
15. **Döring, G., Renger, G., Vater, J., and Witt, H. T.**, Properties of the photoactive chlorophyll-a_{II} in photosynthesis, *Z. Naturforsch. Teil B*, 24, 1139, 1969.
16. **Gläser, M., Wolff, C., Buchwald, H. E., and Witt, H. T.**, On the photoactive chlorophyll reaction in photosystem II of photosynthesis. Detection of a fast and large component, *FEBS Lett.*, 42, 81, 1974.
17. **Mathis, P. and Vermeglio, A.**, Chlorophyll radical cation in photosystem II of chloroplasts. Millisecond decay at low temperature, *Biochim. Biophys. Acta*, 396, 371, 1975.
18. **Zankel, K. L.**, Rapid delayed luminescence from chloroplasts: kinetic analysis of components; the relationship to the O_2 evolving system, *Biochim. Biophys. Acta*, 245, 373, 1971.
19. **Lavorel, J.**, Kinetics of the luminescence in the 10^{-6}—10^{-4} s range in *Chlorella*, *Biochim. Biophys. Acta*, 325, 213, 1973.
20. **Van Best, J. A. and Mathis, P.**, Kinetics of reduction of the oxidized primary electron donor of Photosystem II in spinach chloroplasts and in *Chlorella* cells in the microsecond and nanosecond time ranges following a flash excitation, *Biochim. Biophys. Acta*, 503, 178, 1978.
21. **Mauzerall, D.**, Light-induced fluorescence changes in *Chlorella* and the primary photoreduction for the production of oxygen, *Proc. Natl. Acad. Sci. U.S.A.*, 69, 1358, 1972.
22. **Parson, W. W.**, The role of P870 in bacterial photosynthesis, *Biochim. Biophys. Acta*, 153, 248, 1968.
23. **DeVault, D. and Chance, B.**, Temperature dependence of cytochrome oxidation rate in *Chromatium*. Evidence for tunneling, *Biophys. J.*, 6, 825, 1966.
24. **Ke, B., Garcia, A. F., and Vernon, P. L.**, Light-induced absorption changes in *Chromatium* subchromatophore particles exhaustively extracted with nonpolar solvents, *Biochim. Biophys. Acta*, 292, 226, 1973.

25. **Dutton, L. P. and Leigh, J. S.,** Electron spin resonance characterization of *Chromatium* D hemes, nonheme irons, and the components involved in primary photochemistry, *Biochim. Biophys. Acta,* 314, 178, 1973.
26. **McElroy, J. D., Mauzerall, D. C., and Feher, G.,** Kinetic studies of the light-induced EPR signal ($g = 2.0026$) and the optical absorption changes at cyrogenic temperatures, *Biochim. Biophys. Acta,* 333, 261, 1974.
27. **Hsi, E. S. P. and Bolton, J. R.,** Flash photolysis-EPR studies of the decay time of the ESR signal B1 in reaction-center preparations and chromatophores of mutant and wild strains of *Rps. spheroids* and *R. rubrum, Biochim. Biophys. Acta,* 347, 126, 1973.

PRIMARY ELECTRON ACCEPTORS

Bacon Ke

INTRODUCTION

The "primary" electron acceptor in a photosynthetic reaction center (RC), by definition, should be the first component receiving an electron released by the primary electron donor during the photochemical charge-separation process. Because of the photochemical nature of this electron-transfer process from the donor to the acceptor molecule, its rate is expected to be extremely rapid. Until very recently, because of a lack of time resolution of the detection (mostly spectroscopic) methods, the primary electron acceptor in all three photosystems has invariably been identified with the component that has a relatively long lifetime (in the order of milliseconds). These are, respectively, P_{430} or membrane-bound iron-sulfur protein(s) in Photosystem I (PS I), bound plastoquinone in Photosystem II (PS II), and bound ubiquinone or an iron-ubiquinone complex in purple photosynthetic bacteria. As more rapid kinetic-spectroscopic techniques, particularly picosecond spectroscopy, have become available, earlier, intermediary electron acceptors have been found, first in the bacterial RC and, very recently, in PS I. An intermediary electron acceptor (pheophytin *a*) has also been suggested for PS II, largely based on spectral evidence.

At the moment, the subject of early electron acceptors is rapidly undergoing changes as new results are unveiled. This section can only serve as a survey of the present status on the subject of early electron acceptors in the various reaction centers. For the moment, the components originally identified as the "primary" electron acceptors have retained the name of "stable primary" acceptors, as they are the first readily identifiable reduced species which are stable for milliseconds, and the earlier acceptor(s) are termed transient "intermediary" acceptors because of their extremely short lifetimes. In the following sections, tables are included with all available information on early acceptors.

REFERENCES

1. **Parson, W. W. and Cogdell, R. J.,** The primary photochemical reactions of bacterial photosynthesis, *Biochim. Biophys. Acta,* 416, 105, 1975.
2. **Dutton, P. L., Prince, R. C., Tiede, D. M., Petty, K. M., Kaufmann, K. J., Netzel, T. L., and Rentzepis, P. M.,** Electron transfer in the photosynthetic reaction center, *Brookhaven Symp. Biol.,* 28, 213, 1977.
3. **Ke, B.,** The primary electron acceptors in green-plant photosystem I and photosynthetic bacteria, *Curr. Top. Bioenerg.,* 8, 75, 1978.
4. **Amesz, J. and Duysens, L. N. M.,** Primary and associated reactions of Photosystem II, in *Primary Processes of Photosynthesis,* Vol. 3, Barber, J., Ed., Elsevier/North-Holland, Amsterdam, 1978, 149.

LIGHT-INDUCED ABSORPTION CHANGES

In retrospect, the light-induced spectral changes measured under conditions to reflect the reduction of the electron-acceptor species together with spectral information obtained with model compounds in solution had often helped decisively in identifying the chemical nature of the acceptor molecule (see Table 1). The first cases for such

Table 1
SPECTRAL (OPTICAL) PROPERTIES: LIGHT-INDUCED ABSORPTION CHANGES

	PS I			PS II		Bacterial RC	
	P_{430}	A_2[6]	A_1[6]	Q(or X320)	Pheo[9]	X(Fe·UQ)10,11	I (Bpheo) (*Chromatium*[12-14] (*virdis*)[15-17])
Major absorption changes (nm)	(−) 430[1-4]	(−) 400—500		(−)/(+)[7]	(−) 422	(−) 270	(−) 370
	(−) 460	(4°C)	(−) 425	265/325	(+) 450	(+) 320	(−) 390
	(−) 380		(−) 450		(−) 515	(+) 450	(+) 425
	(−) 325		(+) 480	(−)/(+)[8]	(−) 545		(−) 543
			(+) 670	270/330	(+) 660		(−) 595
	(−) 420[5]		(−) 700		(−) 667		(+) 665 (680)
	(−) 445		(5°K)		(−) 685		(−) 757 (790)
							(+)/(−) 785/802
Extinction coefficient, $mM^{-1} \cdot cm^{-1}$ (nm)	12 (430 nm)[2]	—	—	17 (275 nm)[8] 12 (330 nm)[8]	—	3.5 (450 nm)[10]	25 (543 nm)[14]

Note: Increase, (+); decrease, (−).

spectral correlations involve the stable primary electron acceptors in Photosystem II[7,8] and photosynthetic bacteria.[10,11] The identification of plastoquinone and ubiquinone, respectively, as the primary acceptor for the two photosystems gained a firm support from the difference spectra for plastosemiquinone anion and ubisemiquinone anion vs. their respective neutral species, induced by pulse radiolysis in methanolic solution.[18]

In the case of photosynthetic bacteria, absorption changes obtained by picosecond spectroscopy[19,20] quickly led to the suggestion of an involvement of bacteriopheophytin, again based on spectral and electrochemical information on bacteriopheophytin in organic solvents.[21,22] A similar situation exists for the recent observation of pheophytin absorption changes in Photosystem II.[9]

The recent assignment of a chlorophyll *a* (Chl *a*) dimer to the transient intermediary electron acceptor in PS I follows a similar route, namely, by correlating the light-minus-dark difference spectrum for the acceptor in photosystem I RCs,[6,23] with those obtained with Chl-*a* anion radical[24] and Chl-*a* dimer[25] in solution.

Last but not least is the correlation of light-induced absorption changes of certain electron acceptors with the difference spectrum of iron-sulfur proteins. For the moment, this appears better established for P_{430},[1,2] while more information is needed for the A_2 species.[6,23]

It should also be noted that in all cases cited above, invariably a shift in the position was observed for the in vivo spectrum, compared with that obtained with model compounds in organic solvents. Such a shift is usually attributable to the special environment for the pigment molecules in the RC.

Another spectral species, C_{550}[26,27] has also been identified with the primary acceptor Q of PS II. Although extensive work has been done on C_{550}, its exact nature remains unclear.[28]

REFERENCES

1. **Hiyama, T. and Ke, B.,** A new photosynthetic pigment, "P430": its possible role as the primary electron acceptor of Photosystem I, *Proc. Natl. Acad. Sci. U.S.A.*, 68, 1010, 1971.
2. **Hiyama, T. and Ke, B.,** A further study of P430: a possible primary electron acceptor of Photosystem I, *Arch. Biochem. Biophys.*, 147, 99, 1971.
3. **Hiyama, T. and Ke, B.,** Difference spectra and extinction coefficients of P700, *Biochim. Biophys. Acta*, 267, 160, 1972.
4. **Ke, B.,** The risetime of photooxidation, difference spectrum, and oxidation-reduction potential of P430, *Arch. Biochem, Biophys.*, 152, 70, 1972.
5. **Shuvalov, V. A.,** The study of the primary photoprocesses in Photosystem I of chloroplasts. Recombination luminescence, chlorophyll triplet state and triplet-triplet annihilation, *Biochim. Biophys. Acta*, 430, 113, 1976.
6. **Shuvalov, V. A., Dolan, E., and Ke, B.,** Spectral and kinetic evidence for two early electron acceptors in Photosystem I, *Proc. Natl. Acad. Sci. U.S.*, 76, 770, 1979.
7. **Stiehl, H. H. and Witt, H. T.,** Quantitative treatment of the function of plastoquinone in photosynthesis, *Z. Naturforsch. Teil B.*, 24, 1588, 1969.
8. **Van Gorkom, H. J.,** Identification of the reduced primary electron acceptor of Photosystem II as a bound semiquinone anion, *Biochim. Biophys. Acta*, 347, 439, 1974.
9. **Klimov, V. V., Klevanik, A. V., Shuvalov, V. A., and Krasnovsky, A. A.,** Reduction of phoephytin in the primary light reaction of Photosystem II, *FEBS Lett.*, 82, 183, 1977.
10. **Clayton, R. K. and Straley,** Photochemical electron transport in photosynthetic reaction centers. IV. Observations related to the reduced photoproducts, *Biophys. J.*, 12, 1221, 1972.
11. **Slooten, L.,** Electron acceptors in reaction center preparations from photosynthetic bacteria., *Biochem. Biophys. Acta*, 275, 208, 1972.
12. **Shuvalov, V. A. and Klimov, V. V.,** The primary photoreactions in the complex cytochrome-P890-P760 (bacteriopheophytin-760) of *Chromatium minutissimum* at low redox potentials, *Biochim. Biophys. Acta*, 440, 587, 1976.
13. **Tiede, D. M., Prince, R. C., and Dutton, P. L.,** EPR and optical spectroscopic properties of the electron carrier intermediate between the reaction-center bacteriochlorophylls and the primary acceptor in *Chromatium vinosum*, *Biochim. Biophys. Acta*, 449, 447, 1976.

14. **Van Grondelle, R., Romijn, J. C., and Holmes, N. G.,** Photoreduction of the long wavelength bacteriopheophytin in reaction centers and chromatophores of the photosynthetic bacterium *Chromatium vinosum, FEBS Lett.,* 72, 187, 1976.
15. **Shuvalov, V. A., Krakhmaleva, I. N., and Klimov, V. V.,** Photooxidation of P960 and photoreduction of P800 (bacteriopheophytin *b*-800) in reaction centers from *Rhodopseudomonas viridis, Biochim. Biophys. Acta,* 449, 597, 1976.
16. **Prince, R. C., Tiede, D. M., Thornber, J. P., and Dutton, P. L.,** Spectroscopic properties of the intermediary electron carrier in the reaction center of *Rhodopseudmonas viridis, Biochim. Biophys.,* 426, 467, 1977.
17. **Holten, D., Windsor, M. W., Parson, W. W., and Thornber, J. P.,** Primary photochemical processes in isolated reaction centers of *Rhodopseudomonas viridis, Biochim. Biophys. Acta,* 501, 112, 1978.
18. **Bensasson, B. and Land, E. J.,** Optical and kinetic properties of semireduced plastoquinone and ubiquinone: electron acceptors in photosynthesis, *Biochim. Biophys. Acta,* 325, 175, 1973.
19. **Rockley, M. G., Windsor, M. W., Cogdell, R. J., and Parson, W. W.,** Picosecond detection of an intermediate in the photochemical reaction of bacterial photosynthesis, *Proc. Natl. Acad. Sci. U.S.A.,* 72, 2251, 1975.
20. **Dutton, P. L., Kaufmann, K. J., Chance, B., and Rentzepis, P. M.,** Picosecond kinetics of the 1250 nm band of the *Rps. spheroides* reaction center: the nature of the primary photochemical intermediary state, *FEBS Lett.,* 60, 275, 1975.
21. **Fajer, J., Brune, D. C, Davis, M. S., Forman, A., and Spaulding, L. D.,** Primary charge separation in bacterial photosynthesis: oxidized bacteriochlorophylls and reduced pheophytin, *Proc. Natl. Acad. Sci. U.S.A.,* 72, 4956, 1975.
22. **Krasnovsky, A. A. and Vinovskaya, K. K.,** Reversible photochemical reduction and oxidation of bacteriochlorophyll and bacteriopheophytin, *Dokl. Acad. Nauk,* 81, 879, 1951.
23. **Shuvalov, V. A., Ke, B., and Dolan, E.,** Kinetic and spectral properties of the intermediary electron acceptor A_1 in Photosystem I. Subnanosecond spectroscopy, *FEBS Lett.,* 100, 5, 1979.
24. **Fujita, I., Davis, M. S., and Fajer, J.,** Anion radicals of pheophytin and chlorophyll *a*: their role in the primary charge separations in plant photosynthesis, *J. Am. Chem. Soc.,* 100, 6280, 1978.
25. **Fong, F. K., Koester, V. J., and Polles, J. S.,** Optical spectroscopic study of $(Chl\ a \cdot H_2O)_2$ according to the proposed C_2 symmetrical molecular structure for the P700 photoactive aggregate in photosynthesis, *J. Am. Chem. Soc.,* 98, 6406, 1976.
26. **Knaff, D. B. and Arnon, D. I.,** Spectral evidence for a new photoactive component of the oxygen-evolving system in photosynthesis, *Proc. Natl. Acad. Sci. U.S.A.,* 63, 963, 1969.
27. **Butler, W. L.,** Primary photochemistry of Photosystem II of photosynthesis, *Acc. Chem. Res.,* 6, 177, 1973.
28. **Radmer, R. and Kok, B.,** Energy capture in photosynthesis: Photosystem II, *Annu. Rev. Biochem.,* 44, 409, 1975.

EPR SPECTRAL PROPERTIES

Table 2
SPECTRAL (EPR) PROPERTIES

	PSI			PSII		Bacterial RC	
	Membrane-bound iron-sulfur proteins[1-3]	$X(A_2?)$[4-7]	A_1[6]	Q	Pheo	X(Fe·UQ)	I (BPheo)
g values	2.05	2.08	2.004	2.0044	2.0035	~1.8[9,10]	2.034[14]
	1.94	1.88				2.005[11,12]	2.003
	1.86	1.78					1.976 (*Chromatium*; 8°K)
	2.05						2.06[15,16]
	1.92						2.003
	1.89						1.96 *R* (*for viridis*; 7.5°K)
ΔH (G)			11	9.2	12.5	600 (for *g* = 1.8 at 2°K)[13] 8.1 (for *g* = 2.0046 at 1.3°K)[12]	13 for *g* = 2.003 signal[15-17]

The past decade has been a renaissance period for EPR spectroscopy in photosynthetic research of primary acceptors, particularly the iron-sulfur centers and pigment radicals (see Table 2). The reason for the absence of an EPR signal for the plastoquinone radical, the primary acceptor in PS II has only recently been resolved.[18]

The line width ΔH of the broad acceptor signal at g = 1.8 apparently reflects the environment of the acceptor. Evidently the presence of the paramagnetic iron perturbs the magnetic environment of the ubiquinone radical ion, leading to the very broad EPR spectrum.[11-13] Removal of the iron from the ubiquinone-iron complex causes its spectrum to become identical with that of the ubisemiquinone in vitro.[11,12]

The line width of the EPR spectrum of I^- (13 G) appears consistent with the view that the radical consists of a bacteriopheophytin monomer. The 13-G wide signal at g = 2.003 is accompanied by an additional doublet with a splitting of about 60 G in *Chromatium*[14] and 120 G in *Rhodopseudomonas viridis*[15]. The broad split signal appears to arise from an interaction, via exchange coupling and dipolar effects, between $I^{\bar{\cdot}}$ and $Fe{\cdot}UQ^{\bar{\cdot}}$. In RCs prepared with SDS, which lacks the $Fe{\cdot}UQ^{\bar{\cdot}}$ signal at g = 1.8, the broad split signal is also lacking. Instead, the $I^{\bar{\cdot}}$ signal appears at g = 2.003 with ΔH = 13 G.[15]

REFERENCES

1. **Malkin, R. and Bearden, A. J.,** Primary reactions of photosynthesis: photoreduction of a bound chloroplast ferredoxin at low temperature as detected by EPR spectroscopy, *Proc. Natl. Acad. Sci. U.S.A.,* 68, 16, 1971.
2. **Evans, M. C. W., Telfer, A., and Lord, V. A.,** Evidence for the role of a bound ferredoxin as the primary electron acceptor of Photosystem I in spinach chloroplasts, *Biochim. Biophys. Acta,* 267, 530, 1972.
3. **Ke, B., Hansen, R. E., and Beinert, H.,** Oxidation-reduction potentials of bound iron-sulfur proteins of Photosystem I, *Proc. Natl. Acad. Sci. U.S.A.,* 70, 2941, 1973.
4. **McIntosh, A. R., Chu, M., and Bolton, J. R.,** Flash photolysis-EPR studies of the electron acceptor species at low temperatures of Photosystem I in spinach subchloroplast particles, *Biochim. Biophys. Acta,* 376, 308, 1975.
5. **Evans, M. C. W., Sihra, C. K., Bolton, J. R., and Cammack, R.,** Primary electron acceptor complex of Photosystem I in spinach chloroplasts, *Nature (London),* 256, 668, 1975.
6. **Shuvalov, V. A., Dolan, E., and Ke, B.,** Spectral and kinetic evidence for two early electron acceptors in Photosystem I, *Proc. Natl. Acad. Sci. U.S.A.,* 76, 770, 1979.
7. **Dismukes, G. C. and Sauer, K.,** The orientation of membrane-bound radicals: an EPR investigation of magnetically ordered spinach chloroplasts, *Biochim. Biophys. Acta,* 504, 1978.
8. **Van Gorkom, H. J.,** Identification of the reduced primary electron acceptor of Photosystem II as a bound semiquinone anion, *Biochim. Biophys. Acta,* 347, 439, 1974.
9. **Leigh, J. S. and Dutton, P. L.,** The primary electron acceptor in photosynthesis, *Biochim. Biophys. Res. Commun.,* 46, 414, 1972.
10. **Feher, G., Issacson, R. A., McElroy, J. D., Ackerson, L. C., and Okamura, M. Y.,** On the question of the primary acceptor in bacterial photosynthesis: maganese substitution for iron in reaction centers of *Rhodopseudomonas spheroides* R-26, *Biochim. Biophys. Acta,* 368, 135, 1974.
11. **Locah, P. A. and Hall, R. L.,** The question of the primary electron acceptor in bacterial photosynthesis, *Proc. Natl. Acad. Sci. U.S.A.,* 69, 786, 1972.
12. **Feher, G., Okamura, M. Y., and M. C. McElroy, J. D.,** Identification of an electron acceptor in reaction centers of *Rhodopseudomonas spheroides* by EPR spectroscopy, *Biochim. Biophys. Acta,* 267, 222, 1972.
13. **Okamura, M. Y., Issacson, R. A., and Feher, G.,** Primary acceptor in bacterial photosynthesis: obligatory role of ubiquinone in photoactive reaction centers of *Rhodopseudomonas spheroides, Proc. Natl. Acad. Sci. U.S.A.,* 72, 4391, 1975.
14. **Tiede, D. M., Prince, R. C., and Dutton, P. L.,** EPR and optical spectral properties of the electron carrier intermediate between the reaction center bacteriopchlorophylls and the primary acceptor in *Chromatium vinosum, Biochim. Biophys. Acta,* 449, 447, 1976.
15. **Prince, R. C., Tiede, D. M., Thornber, J. P., and Dutton, P. L.,** Spectroscopic properties of the intermediary electron carrier in the reaction center of *Rhodopseudomonas viridis, Biochim. Biophys. Acta,* 462, 467, 1977.

16. Thronber, J. P., Dutton, P. L., Fajer, J., Forman, A., Holten, D., Olson, J. M., Parson, W. W., Prince, R. C., Tiede, D. M., and Windsor, M. W., Isolated photochemical reaction centers from bacteriochlorophyll *b*-containing organisms, *Proc. 4th Int. Conf. on Photosynthesis,* The Chemical Society, London, 55.
17. Shuvalov, V. A. and Klimov, V. V., The primary photoreaction in the complex cytochrome-P890·P760 (bacteriopheophytin-760) of *Chromatium minutissimum* at low redox potentials, *Biochim. Biophys. Acta,* 440, 587, 1976.
18. Klimov, V. V., Dolan, E., Shaw, E. R., and Ke, B., Interaction between the intermediary electron acceptor (pheophytin) and a possible plastoquinone-iron complex in photosystem II reaction centers, *Proc. Natl. Acad. Sci.,* 77, 7227, 1980.

REDOX PROPERTIES

Table 3
REDOX PROPERTIES

	Membrane-bound iron-sulfur proteins[1,2]	PS I		PS II		Bacterial reaction center	
		X (A_2?)[3]	A_1[3]	Q (X320)	Pheo	X[8,9] (Fe·UQ)	I (BPheo)
E_m, mV	−530−	∼−730(?)	⩽730(?)	−35/−270[4]	⩽−490[7]	−180 (pK = 10) (*R. spheroides*)	⩽−430[13]
	−550			−325[5]	∼−610[17]	−160 (pK = 8) (*Chromatium*)	−550[14]
	<−580			−45/−247[6]		−200 (pK = 9) (*R. rubrum*)	
						−150 (pK = 7.8) (*R. viridis*)	−400[9,10]
							−620[11,12]

The redox potentials of PS-I acceptors are probably the most negative values known for a biological species (see Table 3).[1-3] The exact potential values for the earlier acceptors A_1 and A_2 remain to be determined.

The redox-potential values of the PS-II acceptors have been controversial. Nevertheless, an additional quencher with a redox potential much lower than ∼−40 mV appears well documented, but its function remains to be elucidated.[4-6] The more recent observation on the effect of redox potentials on the light-induced pheophytin absorption changes again suggests a rather negative potential for the PS-II primary acceptor.

The redox-potential values of the primary acceptor, the Fe·UQ complex, of bacterial chromatophores is pH dependent (−dE/dpH = 60 mV), becoming constant (∼−180 mV) at pH 9. Thus, at physiological pH the equilibrium reaction of the primary acceptor involves both an electron and a proton; at higher pH, only an electron is involved in the reduction. The effective, operating midpoint potential during light-induced electron flow is probably near −180 mV because protonation of Fe·UQ apparently is not fast enough to compete with electron transfer to the secondary quinone.[8] The discrepancy between the redox-potential values reported for the intermediary electron acceptor I in bacterial reaction centers remains to be resolved.[9-14]

The primary electron acceptor in green bacteria (Chlorobiceae) appears similar to those in PS I,[15] and their midpoint potentials have been estimated to be about −550 mV.[15,16]

REFERENCES

1. Ke, B., Hansen, R. E., and Beinert, H., Oxidation-reduction potentials of bound iron-sulfur proteins of Photosystem I, *Proc. Natl. Acad. Sci. U.S.A.,* 70, 2941, 1973.
2. Evans, M. C. W., Reeves, S. G., and Cammack, R., Determination of the oxidation-reduction potential of the bound iron-sulfur proteins of the primary electron acceptor complex of Photosystem I in spinach chloroplasts, *FEBS Lett.,* 49, 111, 1974.

3. **Ke, B., Dolan, E., Sugahara, K., Hawkridge, F., Demeter, S., and Shaw, E. R.,** Electrochemical and kinetic evidence for a transient electron acceptor in the photochemical charge separation in Photosystem I, *Plant and Cell Physiol.*, (special issue on Photosynthetic Organelles), 1977, 189.
4. **Cramer, W. A. and Butler, W. L.,** Potentiometric titrations of the fluorescence yield of spinach chloroplasts, *Biochim. Biophys. Acta*, 172, 503, 1969.
5. **Ke, B., Hawkridge, F. M., and Sahu, S.,** Redox titration of fluorescence yield of Photosystem II, *Proc. Natl. Acad. Sci. U.S.A.*, 73, 2211, 1976.
6. **Holten, P. and Croze, E.,** Characterization of two quenchers of chlorophyll fluorescence with different midpoint oxidation-reduction potentials in chloroplasts, *Biochim. Biophys. Acta*, in press.
7. **Klimov, V. V., Klevanik, A. V., Shuvalov, V. A., and Krasnovsky, A. A.,** Reduction of pheophytin in the primary light reaction of Photosystem II, *FEBS Lett.*, 82, 183, 1977.
8. **Prince, R. C. and Dutton, R. C.,** The primary acceptor of bacterial photosynthesis: its operating midpoint potential, *Arch. Biochem. Biophys.*, 172, 329, 1976.
9. **Prince, R. C., Leigh, J. S., and Dutton, P. L.,** Thermodynamic properties of the reaction center of *Rhodopseudomonas viridis.* In vivo measurement of the reaction center bacteriochlorophyll-primary acceptor intermediary electron carrier, *Biochim. Biophys. Acta*, 440, 622, 1976.
10. **Prince, R. C., Tiede, D. M., Thornber, J. P., and Dutton, P. L.,** Spectroscopic properties of the intermediary electron carrier in the reaction center of *Rhodopseudomonas viridis.* Evidence for its interaction with the primary acceptor, *Biochim. Biophys. Acta*, 462, 467, 1977.
11. **Klimov, V. V., Shuvalov, V. A., Krakhmaleva, I. N., Klevanik, A. V., and Krasnovsky, A. A.,** Photoreduction of bacteriopheophytin *b* in the primary light reaction of the chromatophores of *Rhodopseudomonas viridis*, *Biokhimiya*, 42, 519, 1977.
12. **Shuvalov, V. A., Krakhmaleva, I. N., and Klimov, V. V.,** Photooxidation of P960 and photoreduction of P800 (bacteriopheophytin *b*-800) in reaction centers from *Rhodopseudomonas viridis*, *Biochim. Biophys. Acta*, 449, 597, 1976.
13. **Dutton, P. L., Kaufmann, K. J., Chance, B., and Rentzepis, P. M.,** Picosecond kinetics of the 1250 nm band of the *Rps. spheroides* reaction center: the nature of the primary photochemical intermediate state, *FEBS Lett.*, 60, 275, 1975.
14. **Fajer, J., Brune, D. C., Davis, M. S., Forman, A., and Spaulding, L. D.,** Primary charge separation in bacterial photosynthesis: oxidized bacteriochlorophylls and reduced pheophytin, *Prod. Natl. Acad. Sci. U.S.A.*, 72, 4956, 1975.
15. **Knaff, D. B. and Malkin, R.,** Iron-sulfur proteins in the green photosynthetic bacterium *Chlorobium*, *Biochim. Biophys. Acta*, 430, 244, 1976.
16. **Prince, R. C. and Olson, J. M.,** Some thermodynamic and kinetic properties of the primary photochemical reactants in a complex from a green photosynthetic bacterium, *Biochim. Biophys. Acta*, 423, 357, 1976.
17. **Klimov, V. V., Allakherdiev, S. I., Demeter, S., and Krasnovsky, A. A.,** Pheophytin reduction in photosystem II of chloroplasts in relation to oxidation-reduction potential of medium, *Dakl. Akad. Nauk. SSSR*, 249, 227.

KINETIC BEHAVIOR

The electron-transfer time (<10 psec) from the excited primary donor to the intermediary acceptor in bacterial photosynthesis is one of the fastest biological processes known (see Table 4).[14,15] Apparently, prior reduction of X does not affect this transfer time.[21] When I is reduced, P* is stable for about 20 psec.[22] A similar electron-transfer process in PS I has recently been measured to be $\leqslant$50 psec.[2] An interesting resemblance exists between PS I and the bacterial RC in that the intermediary acceptor is reoxidized in ∿200 psec and that, at low redox potentials, when the secondary acceptors are reduced, a recombination between the oxidized primary donor and reduced intermediary acceptor takes place, via triplet formation (called P^R state in bacterial RC) and return to the ground state in 10 nsec and 5 to 50 μsec, respectively.[20,22]

In bacterial RCs, the reoxidation of the primary acceptor X by secondary ubiquinone displays an oscillatory behavior with a periodicity of two. This is interpreted by the presence of a two-electron gate between the one-electron primary quinone X and the two-electron secondary ubiquinone pool, i.e., each odd-numbered flash leads to the formation of an ubisemiquinone of X, and after each even-numbered flash, the

Table 4
KINETIC BEHAVIOR

	PS I			PS II		Bacterial RC	
	P430	A_2(X)	A_1	Q(X320)	Pheo	X(Fe · UQ)	I(BPheo)
Photoreduction time	≤100 nsec[1]	200 psec(?)[2]	≤50 psec[2]	<30 μsec[8] <1 μsec[9]	<1 ns	∼200 psec[14,15]	≤10 psec[14,15]
Re-oxidation time, $t_{1/2}$ or k (acceptor)	3.0×10^9 $M^{-1} \cdot s^{-1}$ (P700+)[3] 1.0×10^8 $M^{-1} \cdot S^{-1}$ (safranine T)[3] 9.6×10^6 $M^{-1} \cdot s^{-1}$ (methyl viologen)[3] 1.8×10^7 $M^{-1} \cdot s^{-1}$ (benzyl viologen)[3] 1.5×10^8 $M^{-1} \cdot s^{-1}$ (methylene blue)[3] 4.0×10^8 $M^{-1} \cdot s^{-1}$ (TMPD)[3]	250 μsec (20°C)[4] 130 ms (5°K)[5-7] (P700+)	200 psec (A_2?; 20°C)[2] 3 μsec (20°C)[4-7] 1.3 msec (50°K)[5,6] (P700+)	0.6 msec (20°C)[10] (secondary plasto-quinone) 100—200 μsec (20°C)[11] 3 msec (100°K)[12,13] (P690+)	∼4 ns (P680+)	60 μsec (secondary quinone)[16] 20 msec (20°C) (P870+)[17-19]	200 psec (Fe · UQ)[14,15] 10 nsec (P870+)[20]

semiquinone disappears and a molecule of the fully reduced quinone appears which is followed by a rapid transfer of a pair of electrons to a molecule in the quinone pool.[23,24] A similar situation exists on the reducing side of PS II.[25,26]

Compared with knowledge available on PS II and the bacterial system, the kinetic behavior on the reducing side of PS I is relatively unknown.

REFERENCES

1. **Ke, B.**, The risetime of photoreduction, difference spectrum, and oxidation-reduction potential of P430, *Arch. Biochem. Biophys.*, 152, 70, 1972.
2. **Shuvalov, V. A., Ke, B., and Dolan, E.**, Kinetic and spectral properties of the intermediary electron acceptor A_1 in Photosystem I: subnanosecond spectroscopy, *FEBS Lett.*, 100, 5, 1979.
3. **Hiyama, T. and Ke, B.**, A further study of P430: a possible primary electron acceptor of Photosystem I, *Arch. Biochem. Biophys.*, 147, 99, 1971.
4. **Sauer, K., Mathis, P., Acker, S., and Van Best, J.**, Electron acceptors associated with P700 in Triton solubilized photosystem-I particles from spinach chloroplasts, *Biochim. Biophys. Acta*, 503, 120, 1978.
5. **Shuvalov, V. A., Dolan, E., and Ke, B.**, Spectral and kinetic evidence for two early electron acceptors in Photosystem I, *Proc. Natl. Acad. Sci. U.S.A.*, 76, 770, 1979.
6. **Ke, B., Shuvalov, V. A., and Dolan, E.**, Recent developments on the "primary" electron acceptors in Photosystem I, in *Proc. Frontiers of Biological Energetics: Electrons to Tissues*, (Johnson Research Foundation), Academic Press, New York, 1979.
7. **Mathis, P., Sauer, K., and Remy, R.**, Rapidly reversible flash-induced electron transfer in a P700 chlorophyll-protein complex isolated with SDS, *FEBS Lett.*, 88, 275, 1978.
8. **Witt, K.**, Further evidence for X320 as a primary electron acceptor of Photosystem II in photosynthesis, *FEBS Lett.*, 38, 115, 1973.
9. **Renger, G. and Wolff, Ch.**, The existence of a high photochemical turnover rate at the reaction centers of system II in Tris-washed chloroplasts, *Biochim. Biophys. Acta*, 423, 610, 1976.
10. **Stiehl, H. H. and Witt, H. T.**, Quantitative treatment of the function of plastoquinone in photosynthesis, *Z. Naturforsch. Teil B*, 24, 1588, 1969.
11. **Gläser, M., Wolff, Ch., Buchwald, H. E., and Witt, H. T.**, On the photoactive chlorophyll reaction in Photosystem II of photosynthesis. Detection of a fast and large component, *FEBS Lett.*, 42, 81, 1974.
12. **Mathis, P. and Vermeglio, A.**, Chlorophyll radical cation in Photosystem II of chloroplasts. Millisecond decay at low temperature, *Biochim. Biophys. Acta*, 396, 371, 1975.
13. **Haveman, J., Mathis, P., and Vermeglio, A.**, Light-induced absorption changes in the near ultraviolet of the primary electron acceptor of Photosystem II at liquid nitrogen temperature, *FEBS Lett.*, 58, 259, 1975.
14. **Rockley, M. G., Windsor, M. W., Cogdell, R. J., and Parson, W. W.**, Picosecond detection of an intermediate in the photochemical reaction of bacterial photosynthesis, *Proc. Natl. Acad. Sci., U.S.A.*, 72, 2251, 1975.
15. **Dutton, P. L., Kaufmann, K. J., Chance, B., and Rentzepis, P. M.**, Picosecond kinetics of the 1250 nm band of the *Rps. spheroides* reaction center: the nature of the primary photochemical intermediary state, *FEBS Lett.*, 60, 275, 1975.
16. **Parson, W. W.**, The reaction between primary and secondary electron acceptors in bacterial photosynthesis, *Biochim. Biophys. Acta*, 189, 384, 1969.
17. **Ke, B., Garcia, A. F., and Vernon, L. P.**, Light-induced absorption changes in *Chromatium subchromatophore* particles exhaustively extracted with nonpolar solvents, *Biochim. Biophys. Acta*, 292, 226, 1973.
18. **Dutton, L. P. and Leigh, J. S.**, Electron spin resonance characterization of Chromatium D hemes and components involved in primary photochemistry, *Biochim. Biophys. Acta*, 314, 178, 1973.
19. **McElroy, J. D., Mauzerall, D. C., and Feher, G.**, Kinetic studies of the light-induced EPR signals ($g = 2.0026$) and the optical absorption changes at cryogenic temperatures, *Biochim. Biophys. Acta*, 333, 261, 1974.
20. **Parson, W. W., Clayton, R. K., and Cogdell, R. J.**, Excited states of photosynthetic reaction centers at low redox potentials, *Biochim. Biophys. Acta*, 387, 265, 1975.
21. **Kaufmann, K. J., Petty, K. M., Dutton, P. L., and Rentzepis, P. M.**, Picosecond kinetics in reaction centers of *Rps. spheroides* and the effect of ubiquinone extraction and reconstitution, *Biochem. Biophys. Res. Commun.*, 70, 839, 1976.
22. **Holten, D., Windsor, M. W., Parson, W. W., and Thornber, J. P.**, Primary photochemical processes in isolated reaction centers of *Rps. viridis*, *Biochim. Biophys. Acta*, 501, 112, 1978.

23. **Vermeglio, A.,** Secondary electron transfer in reaction centers of *Rps. spheroides.* Out-of-phase periodicity of two for the formation of ubisemiquinone and fully reduced ubiquinone, *Biochim. Biophys. Acta,* 459, 516, 1977.
24. **Wraight, C. A.,** Electron acceptors of photosynthetic bacterial reaction centers. Direct observation of oscillatory behavior suggesting two closely equivalent ubiquinones, *Biochim. Biophys. Acta,* 459, 525, 1977.
25. **Bouges-Bocquet, B.,** Electron transfer between the two photosystems in spinach chloroplasts, *Biochim. Biophys. Acta,* 314, 250, 1973.
26. **Velthuys, B. R. and Amesz, J.,** Charge accumulation at the reducing side of system 2 of photosynthesis, *Biochim. Bipophys. Acta,* 333, 85, 1974.
27. **Shuvalov, V. A., Klimov, V. V., Dolan, E., Parson, W. W., and Ke, B.,** Nanosecond fluorescence and absorbance changes in photosystem II at low redox potential, *FEBS Lett,* 118, 279, 1980.

CYTOCHROMES OF PHOTOSYNTHETIC BACTERIA

Tomisaburo Kakuno, Jinpei Yamashita, and Takekazu Horio

In general, photosynthetic bacteria contain *c*-type and *b*-type cytochromes, and any of the *c*-type cytochromes in the cytosol have been purified and characterized. The fundamental properites of the *c*-type cytochromes, such as redox potential ($E_{m,7}$), vary remarkably from one bacterial species to another. In contrast, mitochondrial *c*-type cytochromes are extremely similar among species.

The *c*-type cytochromes purified from photosynthetic bacteria have $E_{m,7}$ ranging from +380 to −254 mV (see Table 1). In addition to $E_{m,7}$, their molecular weights, isoelectric points, and wavelengths of absorbance peaks are remarkably different from one another.

Although cytochrome c_2 of *Rhodospirillum rubrum* resembles horse heart mitochondrial cytochrome *c* in its three-dimensional structure,[1] the former is 48 mV more positive in $E_{m,7}$ than the latter[2] and hardly reacts with mitochondrial cytochrome *c* oxidase in the same manner as other cytochrome c_2.[3] Cytochrome c' shows characteristic absorbance spectra of the high-spin type, and at the reduced form, it readily binds CO.[4] Flavocytochromes *c* are composed of the heme-protein subunit and the flavoprotein subunit.[5] Two are found to exhibit the activity for sulfide dehydrogenase.[6] The *b*-type cytochromes purified from photosynthetic bacteria are few, but similar to cytochrome b_1 of *Escherichia coli* with respect to absorbance peaks and $E_{m,7}$. The cytochrome b_{558} purified from *R. rubrum* does not bind CO.[7]

In *R. rubrum*, cytochrome c' shows different properties between the free and membrane-bound states.[2] At the bound state, the cytochrome shows an absorption spectrum of the low-spin type which is similar to that of typical *b*-type cytochromes and is involved in the succinate oxidation system. On the other hand, the bound cytochrome *b* is involved in the NADH oxidation system.[2,8] Both oxidation systems can supply electrons to the photosynthetic cyclic electron transport system. The action spectrum for the inhibition of oxygen respiration by CO with dark-grown cells is similar to the absorption spectrum of the CO-binding cytochrome c' and cytochrome *b*.[9-11] However, only a small part of the membrane-bound cytochrome c' and *b* can bind CO. The absorption spectrum of the free cytochrome c' changes from the high-spin type to the low-spin type in the presence of organic solvents as well as at alkaline pH.[12] Cytochrome c' exists as a dimer either at the free state or at the bound state in *R. rubrum*,[2] whereas it exists as a monomer at the free state in *Rhodopseudomonas palustris*.[13] Also, cytochrome c_2 is bound to the membrane. The cytochrome bound to chromatophore membrane functions as the primary electron donor to the reaction center (RC).[14] *R. rubrum* cells, if grown aerobically in the dark, contain cytochrome *b*, c' and c_2, half the amounts of which are bound with the cytoplasmic membrane. When dark-grown cells are illuminated, cytochrome c' and c_2 increase significantly in the cytosol, but not in the membrane.[15] The function of these cytosol cytochromes is not known, although they are the same as the bound ones.[8]

Table 1
PROPERTIES OF CYTOCHROMES PURIFIED FROM PHOTOSYNTHETIC BACTERIA

Cytochrome	Organism	$E_{m,7}$ (mV)	Molecular weight	α-peak[a] (nm)	γ-peak[a] (nm)	pI[b]	Ref.
		High-potential cytochromes c ($E_{m,7}$ > + 200 mV)					
Cytochrome c_2	*Rhodomicrobium vannielii*	+356	11,800	551	416	7.9	16,17
	Rhodopseudomonas capsulata	+340	12,800	550	415		17,18
	Rhodopseudomonas palustris (st. 6)	+346	12,830	552	418	7.7	17—19
	R. palustris (st. 37)	+370	13,500	552	418	9.7	17—19
	Rhodopseudomonas sphaeroides	+352	14,090	550	417	5.5	16,17
	Rhodopseudomonas viridis	+296	12,300	551	416		17,18
	Rhodospirillum fulvum	+375		550	415		17,18
		+290		550	415		17,18
	Rhodospirillum molischianum	+380	10,740	549	415	9.8	18,20
		+305	10,560	549	415	9.4	18,20
	Rhodospirillum photometricum	+340	12,000	551	416		17,18
Cytochrome c_2	*Rhodospirillum rubrum*	+308	12,800	550	421	6.2	2,21,22
Cytochrome c_{553}	*Chromatium vinosum*	+320	19,000	553	418	4.4	17,23
Cytochrome c_{556}	*R. palustris* (st. 37)	+230	12,000	556	420		17
		Low-potential cytochromes c ($E_{m,7}$ < +200 mV)					
Cytochrome c'	*C. vionsum*	−5	28,000	547	426	4.6	17,24
	R. capsulata	0	28,000	550	425	4.7	17
	R. palustris (st. 37)	+102	14,820	552	426	9.4	13,17,25
	R. sphaeroides	+30	25,000	546	425	4.9	4,17,26
	R. rubrum	+9	28,000	550	423	5.6	2,21,22,27
Cytochrome c_3	*R. palustris* (st. 37)	−150	13,000	552	419	6.1	17,26
	R. sphaeroides	−254	21,000	552	419	4.1	17,26
Cytochrome c_{551}	*Chlorobium thiosulfatophilum*	+135	43,600	551	416	6.0	26,28
Cytochrome c_{554}	*R. palustris* (st. 37)	−6	58,000	554	419		17,26
	R. sphaeroides	+120	44,000	554	419	4.1	26,29
Cytochrome c_{555}	*Chlorobium limicola*	+103	11,150	555	418	4.7	17,26,30,31
	C. thiosulfatophilum	+145	9,470	555	419	10.5	17,26,31,32
Flavocytochrome c	*C. thiosulfatophilum*	+98	58,000	554	417	7.1	6,26,28
	C. vinosum	+10	72,000	552	417	4.5	5,23
		Cytochromes b					
Cytochrome b_{558}	*R. palustris*	−200		558	426		17
	R. rubrum	−204	23,000	558	425	4.6	17,21

[a] Absorbance peaks of reduced form.

[b] pI, isoelectric point.

REFERENCES

1. **Salemme, F. R., Kraut, J., and Kamen, M. D.**, Structural bases for function in cytochromes *c*, an interpretation of comparative X-ray and biochemical data, *J. Biol. Chem.*, 248, 7701, 1973.
2. **Kakuno, T., Hosoi, K., Higuti, T., and Horio, T.**, Electron and proton transport in *Rhodospirillum rubrum* chromatophores, *J. Biochem. (Tokyo)*, 74, 1193, 1973.
3. **Yamanaka, T.**, Evolution of cytochrome *c* molecule, *Adv. Biophys.*, 3, 227, 1972.
4. **Vernon, L. P. and Kamen, M. D.**, Hematin compounds in photosynthetic bacteria, *J. Biol. Chem.*, 211, 643, 1954.
5. **Bartsch, R. G., Meyer, T. E., and Robinson, A. B.**, Complex *c*-type cytochromes with bound flavin, in *Structure and Function of Cytochromes*, Okunuki, K., Kamen, M. D., and Sekuzu, I., Eds., University of Tokyo Press, Tokyo, 1968, 443.
6. **Kusai, A. and Yamanaka, T.**, The oxidation mechanisms of thiosulphate and sulphide in *Chlorobium thiosulphatophilum*: role of cytochrome *c*-551 and cytochrome *c*-553, *Biochim. Biophys. Acta*, 325, 304, 1973.
7. **Bartsch, R. G., Kakuno, T., Horio, T., and Kamen, M. D.**, Preparation and properties of *Rhodospirillum rubrum* cytochrome c_2, cc', and $b_{557.5}'$, and flavin mononucleotide protein, *J. Biol. Chem.*, 246, 4489, 1971.
8. **Kakuno, T., Bartsch, R. G., Nishikawa, K., and Horio, T.**, Redox components associated with chromatophores from *Rhodospirillum rubrum*, *J. Biochem.*, 70, 79, 1971.
9. **Horio, T. and Taylor, C. P. S.**, The photochemical determination of an oxide of the photoheterotroph *Rhodospirillum rubrum*, and the action spectrum of the inhibition of respiration by light, *J. Biol. Chem.*, 240, 1772, 1965.
10. **Taniguchi, S. and Kamen, M. D.**, The oxidase system of heterotrophically-grown *Rhodospirillum rubrum*, *Biochim. Biophys. Acta*, 96, 395, 1965.
11. **Nisimoto, Y., Kakuno, T., Yamashita, J., and Horio, T.**, Two different NADH dehydrogenases in respiration of *Rhodospirillum rubrum* chromatophores, *J. Biochem.*, 74, 1205, 1973.
12. **Imai, Y., Imai, K., Sato, R., and Horio, T.**, Three spectrally different states of cytochrome cc' and c' and their interconversion, *J. Biochem.*, 65, 225, 1969.
13. **Dus, K., de Klerk, H., Bartsch, R. G., Horio, T., and Kamen, M. D.**, On the monoheme nature of cytochrome c' (*Rhodopseudomonas palustris*), *Proc. Natl. Acad. Sci. U.S.A.*, 57, 367, 1967.
14. **Horio, T. and Yamashita, J.**, Electron transport system in facultative photoheterotroph: *Rhodospirillum rubrum*, in *Bacterial Photosynthesis*, Gest, H., San Pietro, A., and Vernon, L. P., Eds., Antioch Press, Yellow Springs, 1963, 275.
15. **Ishimoto, M. and Yamashita, J.**, Development of chromatophore in photosynthetic bacteria, *Protein Nucleic Acid Enzyme*, 16, 822, 1971.
16. **Pettigrew, G. W., Meyer, T. E., Bartsch, R. G., and Kamen, M. D.**, pH dependence of the oxidation-reduction potential of cytochrome *c*, *Biochim. Biophys. Acta*, 430, 197, 1975.
17. **Bartsch, R. G.**, Cytochromes, in *The Photosynthetic Bacteria*, Clayton, R. K. and Sistrom, W. R., Eds., Plenum Press, New York, 1979.
18. **Pettigrew, G. W., Bartsch, R. G., Meyer, T. E., and Kamen, M. D.**, Redox potentials of the photosynthetic bacterial cytochromes c_2 and the structural bases for variability, *Biochim. Biophys. Acta*, 503, 509, 1978.
19. **Henderson, R. W. and Nankiville, D. D.**, Electrophoretic and other studies on haem pigments from *Rhodopseudomonas palustris:* cytochrome 552 and cytochromoid *c*, *Biochem. J.*, 98 587, 1966.
20. **Dus, K., Flatmark, T., de Klerk, H., and Kamen, M. D.**, Isolation and chemical properties of two *c*-type cytochromes of *Rhodospirillum molischianum*, *Biochemistry*, 9, 1984, 1970.
21. **Horio, T. and Kamen, M. D.**, Preparation and properties of three pure crystalline bacterial haem proteins, *Biochim. Biophys. Acta*, 48, 266, 1961.
22. **Dus, K., Sletten, K., and Kamen, M. D.**, Cytochrome c_2 of *Rhodospirillum rubrum*: complete amino acid sequence and phylogenetic relationships, *J. Biol. Chem.*, 243, 5507, 1968.
23. **Cusanovich, M. A. and Bartsch, R. G.**, A high potential cytochrome *c* from *Chromatium* chromatophores, *Biochim. Biophys. Acta*, 189, 245, 1969.
24. **Bartsch, R. G. and Kamen, M. D.**, Isolation and properties of two soluble heme proteins in extracts of the photoanaerobe *Chromatium*, *J. Biol. Chem.*, 235, 825, 1960.
25. **Cusanovich, M. A.**, Molecular weights of some cytochrome cc', *Biochim. Biophys. Acta*, 236, 238, 1971.
26. **Meyer, T. E.**, Comparative Studies on Soluble Iron-Containing Proteins in Photosynthetic Bacteria and Some Algae, Ph.D. thesis, University of California, San Diego, 1970.
27. **Meyer, T. E., Ambler, R. P., Bartsch, R. G., and Kamen, M. D.**, The amino acid sequence of cytochrome c' from the purple photosynthetic bacterium *Rhodospirillum rubrum* S1, *J. Biol. Chem.*, 250, 8416, 1975.

28. **Meyer, T. E., Bartsch, R. G., Cusanovich, M. A., and Mathewson, J. H.,** The cytochromes of *Chlorobium thiosulfatophilum, Biochim. Biophys. Acta,* 153, 854, 1968.
29. **Orlando, J. A.,** *Rhodopseudomonas sphaeroides* cytochrome 553 *Biochim. Biophys. Acta,* 57, 373, 1962.
30. **Shioi, Y., Takamiya, K., and Nishimura, M.,** Studies on energy and electron transfer systems in the green photosynthetic bacterium *Chloropseudomonas ethylica* strain 2-2K: composition of pigments and electron transfer systems, *J. Biochem.,* 76, 241, 1974.
31. **Van Beeuman, J. and Ambler, R. F.,** Homologies in the sequence of cytochrome c-555 from the green photosynthetic bacteria *Chloropseudomonas ethylica* and *Chlorobium thiosulfatophilum, Antonio van Leeuwenhoek, J. Microbiol. Serol.,* 39, 355, 1973.
32. **Gibson, J.,** Cytochrome pigments from the green photosynthetic bacterium *Chlorobium thiosulfatophilum, Biochem. J.,* 79, 151, 1961.
33. **Kennel, S. J.,** Cytochromes and Their Relation to Electron Transport in *Chromatium vinosum,* Ph.D. thesis, University of California, San Diego, 1971.

CYTOCHROMES OF ALGAE

Peter Böger and Herbert Böhme

INTRODUCTION

Cytochromes (Cyt) possess heme as the prosthetic group, with the iron in the Fe^{III} state (oxidized) and Fe^{II} state (reduced). The latter always has four dominant absorption bands ($\alpha,\beta,\gamma,\delta$) in addition to the protein band. All Cyt function in redox reactions; four main groups (a,b,c,d) can be distinguished by the heme groups.[1]* Two of them, namely the *b*- and *c*-type Cyt, are of particular interest because they are the only ones that have been described in algae. A *b*-type Cyt has a protoheme IX not attached by covalent bonding to the protein (with two free vinyl groups in Positions 2 and 4), whereas a *c*-type Cyt has its heme (mesoheme) covalently linked to the protein via thioether bridges at the vinyl groups.

A standard nomenclature is not established. It has been suggested that the position of the α-band of the (alkaline) pyridine hemochromogenes be used to denote different Cyt types.[1] In this paper, following general use, the position of the α-band of the (reduced) native protein designates a particular *b*- or *c*-type Cyt. The positions of the α peaks of these two types overlap (see Table 1 of Reference 2). However, by adding the source (e.g., organelle) or a typical property (e.g., soluble or bound) to the name, this nomenclature should be appropriate. Membrane-bound plastidic *c*-type Cyt are denoted "Cyt *f*".

In procaryotic and eucaryotic algae, Cyt *b*-559, *b*-563, and many different *c*-type Cyt have been described and some of them isolated and characterized in more detail. Undoubtedly, most of them belong either to the respiratory or photosynthetic electron transport chain. This section gives a short survey of biochemical properties. As far as their function is concerned, only their role in algal photosynthesis has been investigated (for recent reviews on this topic see Cramer and Whitmarsh,[3] Crofts and Wood,[4] Knaff,[5] and Cramer[6]).

In algal thylakoids, *b*- and *c*-type Cyt are either tightly membrane bound and can be extracted only by comparatively harsh treatment of the membranes (e.g., extraction with acetone, detergents, or sonication) or they occur in soluble form which is the predominant feature of most thylakoid-located algal *c*-type Cyt described. They can be easily removed from the membranes by freezing/thawing or by treating the membranes with glass beads and osmotic shock. Accordingly, little is known about the membrane-bound *b*- and *c*-type Cyt, whereas much is known about the soluble *c*-type Cyt.

b-TYPE CYTOCHROMES

Cytochrome *b*-559 (bound)

Cyt *b*-559 is tightly bound to the photosynthetic membrane and present in at least two different states. In fresh chloroplasts, it occurs mainly as a "high-potential" (HP) form with a midpoint potential (E_o') of +350 to +400 mV (reducible by hydroquinone) and (in a lesser amount) in a "low-potential" form (LP) with an E_o' of about +80 mV (reducible by ascorbate). Aging or chaotropic reagents convert the HP form into the LP one. There are indications that forms with a potential between HP and LP may exist. Typically, the HP form is photooxidized *in situ* at 77°K by 650-nm light

* Only recent papers and reviews are cited. See those for older or detailed references.

indicating close association with photosystem II. In chloroplasts from Bumilleriopsis Cyt *b*-559 HP is reduced by PSII through plastoquinone.[7a] Some Cyt *b*-559 (presumably the LP form) may contribute electrons to Photosystem I. Most of the functional data stem from higher-plant chloroplasts, but are generally in accordance with findings in algae (*Anabaena,*[7] *Euglena,*[8] *Nostoc,*[9] *Chlamydomonas,*[10,11] and Bumilleriopsis[7a]). The conclusions are conflicting, leaving unclear the role of this Cyt and also its possible involvement in DCMU inhibition.[3]

In *Bumilleriopsis filiformis* (Xanthophyceae), the molar ratio of chlorophyll to P700 to Cyt *b*-559 to Cyt *b*-563 was found to be 1000:1.5:3.2:2.7. For spinach, this ratio was 1000:1:2.6:2.3[13] which is comparable to other findings.[3]

Cyt *b*-559 was isolated from *Bumilleriopsis* by urea and Triton treatment in a homogeneous LP form.[12] In the reduced form, the α-band is at 559, β is at 530, and the γ-band is at 429nm; the oxidized γ-band at 415 nm. The ratio of the γ/α absorbance (= A) is approximately 6. The spectral data are similar to the spinach cytochrome.[14] Mol wt (determined by SDS gel electrophoresis) is 17,000. Under the same conditions, spinach Cyt *b*-559 exhibits a mol wt of 37,000. The lipid content (with little chlorophyll) was variable between 5 to 25%. The protein could be kept in solution only with 2% Triton® X-100 present. The isolated protein[12] was photooxidized at room temperature by Photosystem I and II particles (from spinach). Further, it could be enzymatically reduced by NADPH via plastidic ferredoxin-$NADP^+$ reductase.[15] Other preparations from algae have not been reported.

Cytochromes *b*-563, *b*-562 (Bound and Soluble)

Cyt *b*-563 (= *b*6) is of plastidic origin, membrane bound, and functions as a redox carrier of cyclic electron transport donating electrons to plastoquinone.[3,5,16] It has been isolated from spinach and contains lipids, the protein part having a mol wt of 40,000. The E_o' is variable and found to be between −80 to −120 mV. An E_o' of 0 mV was determined *in situ.*[16] It is reduced by dithionite: α,563; β,536, and γ,434 nm; (A (γ/α) ≈5.5.[17] From the alga *Bumilleriopsis*, the preparation of a plastidic Cyt *b*-563/Cyt *f*-553 particle with a variable cytochrome content has been reported.[18] Absorption bands for Cyt *b*-563 in the visible are: α,563; β,533, and γ,432 nm; A(γ/α) is about 7.[19]

A bound *b*-type Cyt (reducible by dithionite) has been isolated from the marine diatom *Phaeodactylum* (reduced: α,560; β,529, γ,427 nm; A(γ/α) ≈4.9).[20] Some authors reported isolation of a soluble algal Cyt *b*-562[21] which is ascorbate reducible, with an E_o' near +175 mV. Absorption bands (reduced) are α,562; β,531; and γ,431 nm. Furthermore, a Cyt *b*-555 (reduced by dithionite) was detected. Whether these cytochromes are artifacts is unclear, and so is their possible role or organelle localization.

c-TYPE CYTOCHROMES

Cytochrome *c*-550/555 (Soluble)

These are low molecular weight proteins which have been isolated and partially characterized from 60 algal species, ranging from blue-greens through all classes (except minor ones such as the Cryptophyceae), including the marine groups (References up to 1978 on these preparations are found in References 4, 21-23, 27). Until recently, they were often denoted as (algal) Cyt *f*, since their functional role was thought to be analogous to that of the (membrane-bound) Cyt *f* of higher plant thylakoids. It is now clear that algal thylakoids have a bound *c*-type Cyt (herein also called Cyt *f*).[28] Soluble (low molecular weight) *c*-types fulfill the role of plastocyanin.[28-30] Accordingly, there are algae having no plastocyanin under different applied physiological conditions, e.g., the xanthophycean *B. filiformis,*[30] *Chlamydomonas mundana,*[31] and, presumably, *Eu-*

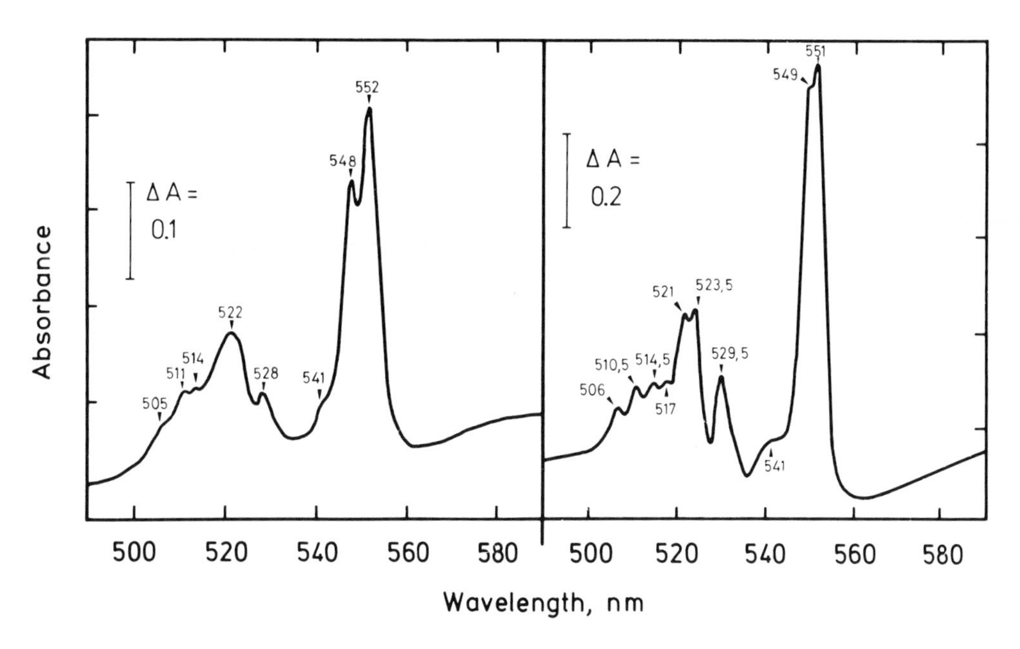

FIGURE 1. Comparison of low-temperature difference spectra of purified thylakoid *c*-type Cyt from *Scenedesmus acutus*; α- and β-bands of the reduced forms are shown at 77°K (bandwidth 1 nm; instrument: Aminco® DW-2). Left: Cyt c_{553} (soluble). At room temperature, the band absorbance ratios are (γ,416/α, 553) = 6.4; (416/β, 523) = 9; (416/δ, 319) = 4; (416/protein, 274) = 4.5. γ-Band (oxidized) peaks at 411 nm (compare, for example, these details with soluble Cyt $c_{552/3}$ from *Euglena* and *Bumilleriopsis*[35]). Right: Cyt f_{553} (bound). Note the high asymmetry of the β-band splitting into 7 maxima. At room temperature, the band absorbance ratios are (γ,422/α, 553) = 8.1; (422/β, 524) = 15.6; (422/;δ, 331) = 7; (422/protein, 275) = 7.5.[45]

glena.[32] Instead, *Bumilleriopsis* has only a soluble Cyt *c*-553[35] and bound Cyt *f*-554, respectively. Further, in some algal species, plastocyanin and plastidic Cyt *c* form a mutually interchangeable pair. Copper deficiency in *Scenedesmus acutus* (Chlorophyceae) leads to cells with high amounts of soluble Cyt *c*-553, with no detectable plastocyanin. A copper content of >0.1 μ*M* in the culture medium allows for plastocyanin formation, but "represses" synthesis of soluble Cyt *c*. The amount of (bound) Cyt *f*-553 remains constant whether or not copper is present.[33] Similar findings were reported for *Chlamydomonas reinhardtii.*[31]

The maximum amount of *c*-Cyt in the thylakoids was determined in *Bumilleriopsis* at the following molar ratios: Chl *a* to P700 to Cyt *f*-554 to Cyt *c*-553 = 1000:1.5:1.5:5.[34] The concentration of the soluble cytochrome could be varied by the iron content of the medium, but not that of the bound Cyt.[29] In *S. acutus* molar ratios were determined with the following maximum figures: chlorophyll (a + b) to P700 to Cyt *f*-553 to Cyt *c*-553 = 1000:1.1:1.1:3.5.[34,44] When the copper protein replaces Cyt *c*-553 (see above), a maximum of plastocyanin to P700 = 4:1.1 is reached. A ratio of Cyt *f*-554.5 to Cyt *c*-553 of about 1:1 was reported for the marine alga *Bryopsis.*[36]

General properties of soluble Cyt *c* are

1. Acidic proteins of low molecular weight, ranging roughly between 10,000 to 14,000, having 1 heme per molecule and 86 to 92 amino acids.[37]
2. An E_o' of between +340 and +390 mV in vitro and *in situ* (that is some 100 mV more positive than that of mitochondrial Cyt *c*-550) pH dependency of −60 mV/pH unit above pH 8).
3. α-, β-, γ-bands at 550 to 555, 521 to 523, and at 415 to 417 nm (The A(γ/α) generally is between 5.4 and 7.7; the α-band is often asymmetrical (see Figure 1). The extinction coefficient of the α-peak is about 25, Δεα is 17 [mM^{-1} × cm^{-1}].

4. No combination with CO or CN^-.
5. A relatively low reactivity with cow Cyt oxidase (Most react with bacterial nitrite reductase, except for *Anacystis*.)[22]

Six Cyt have been sequenced.[37] The only conserved residue is methionine in Position 62; its sulfur is the sixth ligand of the heme, causing a small absorption band at 695 nm which is typical of all soluble *c*-type Cyt.[38] Further, the heme attachment site is conserved, i.e., the two cysteine residues in Positions 14 and 17.

Cytochrome *c*-549/550 (Soluble)

During isolation of the plastidic HP *c*-Cyt, several authors have found an acidic autoxidizable Cyt (E_o' −200 to −260 mV). The absorption bands (reduced) are α,549 to 550; β,520 to 522; and γ,417 to 418 nm. $A(\gamma/\alpha)$ is between 7 to 8.1. It also combines with CO. The mol wt is 17,000 to 26,000.

This protein was isolated in high amounts from *Anacystis* (Cyt *c*-549)[39] and *Phaeodactylum* (Cyt *c*-550).[20,25] Apparently the same kind of Cyt was also found in *Bryopsis* (Cyt *c*-549LP),[21] *Navicula* (Cyt *c*-550),[40] and *Bumilleriopsis* (Cyt *c*-549) (this laboratory, unpublished). Its function is not known.

Cytochrome *c*-558 (Soluble)

This is a basic protein (isoelectric point at pH 8 to 9.6) having an E_o' of +244 to +307 mV and most probably being of mitochondrial origin. Reduced bands are α,558; β,525, and γ,421 to 422 nm. $A(\gamma/\alpha)$ is about 6. It has been isolated (and sequenced) from *Euglena*.[41,42] The mol wt is about 11,000 (102 to 103 amino acids). The sequence pattern rules out that it might belong to the plastidic *c*-types (e.g., residue Number 14 is an alanine). The covalent link is made through one vinyl group.

Other Soluble *c*-Type Cytochromes

A basic Cyt *c*-549 was found in *Scenedesmus*[43] and is apparently the same as Cyt *c*-549 found in *Bryopsis*.[21] Both have a low $A(\gamma/\alpha)$ ratio of 4.3 to 4.6 and are not yet characterized.

Cytochrome *f*-553/554.5 (Bound)

As already mentioned, algal chloroplast thylakoids also contain a bound *c*-type Cyt *f*. This had been claimed before, but was first proven by Wood through isolation from *Chlamydomonas*, *Euglena*, and *Anacystis*.[28] It was confirmed for *Bumilleriopsis*,[29] *Scenedesmus*,[44,45] *Bryopsis*,[36] and the blue-green alga *Spirulina*.[48]

Properties are similar to those of Cyt *f* from higher plants.[46,47] Reduced *Chlamydomonas* Cyt *f*-554 has an α-band at 554, β at 523, and γ at 422 nm. $A(\gamma/\alpha)$ = approximately 8.[28] The α-band has little asymmetry which, however, is very large in the β-band. The same is found for *Bryopsis* Cyt *f*-554.5[36] and Cyt *f*-553 from *Scenedesmus* (see Figure 1).[48] The reduced form of the latter cytochrome has peaks at α 554; β 531.3/524, γ 421.5; δ 331 nm. A (γ/α) is 7.5. Ferricytochrome *f* has a maximum at 410.5 nm. The extinction coefficient (ε) at 553 (α) nm is 25.3, the differential $\Delta\varepsilon_\alpha$: 17.3 ($mM^{-1} \times cm^{-1}$).

The midpoint potential (at pH 7.0, E_o' = +380 mV) shows a pH dependency above pH 8.0 of −60 mV/pH unit. The isoelectric point is at 5.1 pH. Cytochrome *f* from the blue-green alga *Spirulina platensis*[48] can be distinguished from those isolated from eucaryotic organisms by (1) an asymmetrical, red-shifted α band at 556.5 nm (shoulder at 550 nm); (2) a more negative redox potential (E_o' = +318 mV) which is pH-independent between pH 6 to 10; (3) a more acidic IEP = 3.6 pH; (4) a different amino acid composition and a slightly higher molecular weight (34,000).

An antiserum against plastidic Cyt *c*-552 from *Euglena* did not react with the corresponding Cyt *f*-554. The reaction rate of Cyt *f*-552 with plastocyanin was found to be 100 times faster than that of soluble plastidic Cyt *c*-552 from the same alga.[28]

REFERENCES

1. **Lemberg, R. and Barrett, J.,** *Cytochromes,* Academic Press, London, 1973.
2. **Dickerson, R. E. and Timkovich, R.,** Cytochromes c, in *The Enzymes,* Vol. 11, 3rd ed., Boyer, P. D., Ed., Academic Press, New York, 1975, chap. 7.
3. **Cramer, W. A. and Whitmarsh, J.,** Photosynthetic cytochromes, *Annu. Rev. Plant Physiol.,* 28, 133, 1977.
4. **Crofts, A. R. and Wood, P. M.,** Photosynthetic electron transport chains of plants and bacteria and their role as proton pumps, in *Current Topics in Bioenergetics,* Vol. 7, Sanadi, D. R. and Vernon, L. P., Eds., Academic Press, 1978, 175.
5. **Knaff, D. B.,** The cytochromes of higher plants and algae, *Coord. Chem. Rev.,* 26, 47, 1978.
6. **Cramer, W. A.,** Cytochromes, in *Encyclopedia of Plant Physiology,* Vol. 5 (New Series), Trebst, A. and Avron, M., Eds., Springer, Berlin, 1977, 227.
7. **Fujita, Y.,** The light-induced oxidation-reduction reaction of cytochrome b-559 in membrane fragments of the blue-green alga *Anabaena variabilis, Plant Cell Physiol.,* 15, 861, 1974.

7a. **Böhme, H. and Kumert, K.-J.,** Photoreactions of cytochromes in algal chloroplasts, *Eur. J. Biochem.,* 106, 329, 1980.

8. **Ikegami, I., Katoh, S., and Takamiya, A.,** Light-induced changes of b-type cytochromes in the electron transport chain of *Euglena* chloroplasts, *Plant Cell Physiol.,* 11, 777, 1970.
9. **Aparicio, P. J., Ando, K., and Arnon, D. I.,** Photochemical activity and components of membrane preparations from the blue-green algae, *Biochim. Biophys. Acta,* 357, 246, 1974.
10. **Maroc, J. and Garnier, J.,** La photooxidation du cytochrome b-559, en presence de CCCP et de DBMIB ou de *p*-benzoquinone, chez trois mutants non-photosynthetiques de *Chlamydomonas reinhardti, Biochim. Biophys. Acta,* 387, 52, 1975.
11. **Epel, B. L., Butler, W. L., and Levine, R. P.,** A spectroscopic analysis of low-fluorescent mutants of *Chlamydomonas reinhardti* blocked in their water-splitting oxygen-evolving apparatus, *Biochim. Biophys. Acta,* 275, 395, 1972.
12. **Lach, H. J. and Böger, P.,** Isolation and some molecular properties of plastidic algal cytochrome b-559, *Z. Naturforsch. Teil C,* 32, 75, 1977.
13. **Böhme, H.,** Photoreactions of cytochrome b_6 and cytochrome f in chloroplast photosystem I fragments, *Z. Naturforsch. Teil C,* 31, 68, 1976.
14. **Garewal, H. S. and Wasserman, A. R.,** Triton X-100-4 *M* urea as an extraction medium for membrane proteins. I. Purification of chloroplast cytochrome b_{559}, *Biochemistry,* 13, 4063, 1974.
15. **Lach, H. J., Böhme, H., and Böger, P.,** Some photoreactions of isolated cytochrome b-559, *Biochim. Biophys. Acta,* 462, 12, 1977.
16. **Böhme, H. and Cramer, W. A.,** Uncoupler-dependent decrease in midpoint potential of the chloroplast cytochrome b_6, *Biochim. Biophys. Acta,* 325, 275, 1973.
17. **Stuart, A. L. and Wasserman, A. R.,** Purification of cytochrome b_6, a tightly bound protein in chloroplast membranes, *Biochim. Biophys. Acta,* 314, 284, 1973.
18. **Lach, H. J. and Böger, P.,** Variable composition of cytochrome b_6-f particles, *Z. Naturforsch. Teil C,* 31, 606, 1976.
19. **Lach, H. J. and Böger, P.,** Some properties of plastidic cytochrome b-563, *Z. Naturforsch. Teil C,* 32, 877, 1977.
20. **Shimazaki, K., Takamiya, K., and Nishimura, M.,** Studies on electron transfer systems in the marine diatom, *Phaeodactylum tricornutum.* II. Identification and determination of quinones, cytochromes, and flavins, *J. Biochem. (Tokyo),* 83, 1639, 1978.
21. **Kamimura, Y., Yamasaki, T., and Matsuzaki, E.,** Cytochrome components of green alga, *Bryopsis maxima, Plant Cell Physiol.,* 18, 317, 1977.
22. **Yamanaka, T. and Okunuki, K.,** Cytochromes, in *Microbial Iron Metabolism,* Neilands, J. B., Ed., Academic Press, New York, 1974, chap. 14.
23. **Sugimura, Y., Toda, F., Murata, T., and Yakushiji, E.,** Studies on algal cytochromes, in *Structure and Function of Cytochromes,* Okunuki, K., Kamen, M. D., and Sekuzu, I., Eds., University of Tokyo Press, 1968, 452.
24. **Mehard, C. W., Prézelin, B. L., and Haxo, F. T.,** Isolation and characterization of dinoflagellate and chrysophyte cytochrome-f (553-4), *Phytochemistry,* 14, 2379, 1975.
25. **Shimazaki, K., Takamiya, K., and Nishimura, M.,** Studies on electron transfer systems in the marine diatom, *Phaeodactylum tricornutum.* I. Isolation and characterization of cytochromes, *J. Biochem. (Tokyo),* 83, 1631, 1978.

26. **Yamanaka, T., Fukumori, Y., and Wada, K.**, Cytochrome c-554 derived from the blue-green alga *Spirulina platensis*, *Plant Cell Physiol.*, 19, 117, 1978.
27. **Laycock, M. V.**, The amino acid sequence of cytochrome f from the brown alga *Alaria esculenta* (L.) Grev., *Biochem. J.*, 149, 271, 1975.
28. **Wood, P. M.**, The roles of c-type cytochromes in algal photosynthesis, *Eur. J. Biochem.*, 72, 605, 1977.
29. **Böhme, H., Kunert, K. J., and Böger, P.**, The role of plastidic cytochrome c in algal electron transport and photophosphorylation, *Biochim. Biophys. Acta*, 501, 275, 1978.
30. **Kunert, K. J. and Böger, P.**, Absence of plastocyanin in the alga *Bumilleriopsis* and its replacement by cytochrome 553, *Z. Naturforsch. Teil C*, 30, 190, 1975.
31. **Wood, P. M.**, Interchangeable copper and iron proteins in algal photosynthesis, *Eur. J. Biochem.*, 87, 9, 1978.
32. **Wildner, G. F. and Hauska, G.**, Localization of the reaction site of cytochrome 552 in chloroplasts from *Euglena gracilis*, *Arch. Biochem. Biophys.*, 164, 127, 1974.
33. **Bohner, H. and Böger, P.**, Reciprocal formation of cytochrome c-553 and plastocyanin in *Scenedesmus*, *FEBS Lett.*, 85, 337, 1978.
34. **Böger, P.**, Some properties of plastocyanin and its function in algal chloroplasts, in *Proc. 4th Int. Congr. Photosynthesis*, Hall, D. O., Coombs, J., and Goodwin, T. W., Eds., The Biochemical Society, London, 1978, 755.
35. **Lach, H. J., Ruppel, H. G., and Böger, P.**, Cytochrom 553 aus der Alge *Bumilleriopsis filiformis*, *Z. Pflanzenphysiol.*, 70, 432, 1973.
36. **Kamimura, Y. and Matsuzaki, E.**, Cytochrome components of green alga, *Bryopsis maxima*: purification and properties of cytochrome f from membrane fragments, *Plant Cell Physiol.*, 19, 1175, 1978.
37. **Aitken, A.**, Protein evolution in cyanobacteria, *Nature* (London), 263, 793, 1976.
38. **Ben-Hayyim, G. and Schejter, A.**, Heme-linked properties of *Euglena* cytochrome f, *Eur. J. Biochem.*, 46, 569, 1974.
39. **Holton, R. W. and Myers, J.**, Water-soluble cytochromes from a blue-green alga. I. Extraction, purification, and spectral properties of cytochromes c ($_{549,552}$, *Anacystis nidulans*); II. Physicochemical properties of cytochromes c ($_{549,552}$, and $_{554}$, *Anacystis nidulans*), *Biochim. Biophys. Acta*, 131, 362 and 375, 1967.
40. **Yamanaka, T., De Klerk, H., and Kamen, M. D.**, Highly purified cytochromes c derived from the diatom, *Navicula pelliculosa*, *Biochim. Biophys. Acta*, 143, 416, 1967.
41. **Pettigrew, G. W.**, The amino-acid sequence of cytochrome c from *Euglena gracilis*, *Nature* (London), 241, 531, 1973.
42. **Lin, D. K., Niece, R. L., and Fitch, W. M.**, The properties and amino-acid sequence of cytochrome c from *Euglena gracilis*, *Nature* (London), 241, 533, 1973.
43. **Powls, R., Wong, J., and Bishop, N. I.**, Electron transfer components of wild-type and photosynthetic mutant strains of *Scenedesmus obliquus* D_3, *Biochim. Biophys. Acta*, 180, 490, (1969).
44. **Bohner, H.**, Wechselseitiger Austausch von Plastocyanin und Cytochrom c-553 in Grünalgen, Thesis, University Konstanz, West Germany, 1979.
45. **Böhme, H., Brütsch, S., Weithmann, G., and Böger, P.**, Isolation and characterization of soluble cytochrome *c*-553 and membrane-bound cytochrome *f*-553 from thylakoids of the green alga *Scenedesmus acutus*, *Biochim. Biophys. Acta*, 590, 248, 1980.
46. **Bendall, D. S., Davenport, H. E., and Hill, R.**, Cytochrome components in chloroplasts of higher plants, *Methods Enzymol.*, 23, 327, 1971.
47. **Gray, J. C.**, Purification and properties of monomeric cytochrome f from charlock, *Sinapis arvensis* L., *Eur. J. Biochem.*, 82, 133, 1978.
48. **Böhme, H., Pelzer, B., and Böger, P.**, Purification and characterization of cytochrome *f*-556.5 from the blue-green alga *Spirulina platensis*, *Biochim. Biophys. Acta*, 592, 528, 1980.

CHLOROPLAST CYTOCHROMES OF HIGHER PLANTS

D. S. Bendall

Four cytochrome components can be readily distinguished in chloroplast preparations from higher plants: cytochrome *f* (belonging to group *c* and sometimes referred to as c_6) and cytochromes *b*-559_{HP} (high-potential component), *b*-559_{LP} (low-potential component), and *b*-563. In early work, only one *b*-component was recognized and termed b_6 by Hill,[1] but this designation should be regarded as obsolete, as it is now evident that it refers to a mixture of components, although the literature can usually be understood by reading *b*-563 for b_6. All four components are integral membrane proteins. Cytochrome *f* can be solubilized by extraction with organic solvents and then purified by conventional methods; its molecular properties are now being studied in detail. The cytochromes *b* are exceptionally difficult to bring into solution and require extraction with a mixture of Triton® X-100 and urea; there is still considerable uncertainty about their properties.

The individual components are mainly distinguished by their absorption spectra (see Table 1), especially the positions of the α-peaks of the reduced form, and by their oxidation-reduction potentials (see Table 2). Neither criterion is adequate by itself. The four components can be differentiated by redox potential into two groups separated by 300 mV or more. Cytochromes *f* and *b*-559_{HP} both have midpoint potentials close to +370 mV, but may be distinguished by their α-peaks at 554 and 559 nm, respectively, and by the fact that various treatments, such as exposure to detergents, selectively lower the potential of cytochrome *b*-559_{HP}.

In the middle of the potential range, around 0 mV, are found cytochromes *b*-559_{LP} and *b*-563. Some of the values recorded in Table 2 would make the two components difficult to distinguish by potential alone, although recent determinations favor a potential for cytochrome *b*-563, about 100 mV lower than that of cytochrome *b*-559_{LP}, in which case they could be distinguished by redox poising at suitable potentials. Their spectra can also be recorded separately by making use of the fact that reduction of cytochrome *b*-563 by dithionite is much slower than reduction of *b*-559_{LP}.[3] A further useful technique is to record spectra at the temperature of liquid nitrogen (77 K) when the bands sharpen and, in the case of the α-bands, shift 2 to 3 nm towards the violet. Under these conditions, the α-band of cytochrome *f* shows a characteristic splitting into a main peak at 552 nm and a minor peak at 548 nm. The cytochromes *b* show simple bands at 77 K; recent work[14,15] has confirmed that this is true for cytochrome *b*-563, despite suggestions that it splits into a double α-band at 561 and 557 nm.

Some confusion may be caused by the fact that a variety of treatments can lower the potential of the HP component, sometimes by at least 300 mV, and so apparently increase the complement of cytochrome *b*-559_{LP}. Nevertheless, it is clear that a LP 559 component is always present in fresh chloroplast preparations and can be distinguished in principle from a modified form of the HP component, not only by small differences in midpoint redox potentials, but also by the fact that, on fractionation of thylakoid membranes with detergents, cytochrome *b*-559_{HP} remains closely associated with Photosystem II, whereas the low-potential component fractionates with Photosystem I (D-144 particles)[34] or with cytochrome *f*.[14,35] On general grounds, the two cytochromes *b*-559 of normal chloroplasts are likely to involve distinct polypeptides, but so far only one 559 component has been purified.

Relative molecular masses determined on purified components are recorded in Table 3. A notable feature in the case of cytochrome *f* is the wide range of values reported for different species; this diversity remains when preparations from different species

Table 1
ABSORPTION SPECTRA AND EXTINCTION COEFFICIENTS

Component	Absorption peaks (nm)					Isosbestic points (nm)	Extinction coefficients of α-peak		Temperature	Plant source[a]	Ref.
	Reduced				Oxidized		ε_{mM}	ε_{mM}(reduced - oxidized)			
	α	β	γ	δ							
f	554.5	526	423	331			26.1	19.7	RT[b]	*Petroselinum crispum*	2
	554.5	524	422	333	526, 410	560, 543.5	26.0	17.7[c]	RT	*P. crispum*	3
	552	529	421						LT[d]	*P. crispum*	4,5
	548	524									
	553.5	522	421	328	532, 410		29.0	20.8	RT	*Spinacea oleracea*	6
	551.5	531							LT	*S. oleracea*	6
	548	522.5									
	553.5	522	420						RT	*Vicia faba*	7
	553.5	524	420.5	328					RT	*Raphanus sativus*	8
	554.5	524	422	328		560, 542, 534, 506, 433, 415			RT	*Brassica komatsuna*	9
b-559	559	530	429		415		21	14.7[e]	RT	*S. oleracea*	10
	556	529	429						LT	*S. oleracea*	11

b-563	563[f]	536	434	413	21	14.7[e]	RT	*S. oleracea*	12
	563[g]	533	432	414			RT	*S. oleracea*	13
	561[h]						LT	*S. oleracea* D-144 particles	14
	561[h]						LT	*S. oleracea* Digitonin cytochrome *b/f* preparation	15

[a] Data obtained with purified component unless otherwise stated.
[b] Room temperature.
[c] This figure should be increased to 19.3 to allow for the presence of a trace of reduced cytochrome.
[d] Temperature of liquid nitrogen (77 K).
[e] Calculated from spectra reported.
[f] This preparation showed a double α-band at 77 K.
[g] The spectrum at 77 K was not reported.
[h] Only a single α-peak was detected.

Table 2
OXIDATION-REDUCTION POTENTIALS[a]

Component	E_m (mV)	pH	Plant source	Ref.
f	+365	6—8	*Petroselinum crispum* purified protein	16
	$\Delta E_m/\Delta pH = -60$	>pH 8.4		
	+350	7.5	*Pisum sativum* chloroplasts	17
	+350	6.5	*Hordeum vulgare* etioplasts	17
	+330	7	*Lactuca sativa* cytochrome *b/f* particle	18
	+340	8	*Spinacea oleracea* chloroplasts	19
	+390	8.2	*S. oleracea* chloroplasts	20
	+385	7.8	*S. oleracea* chloroplasts	21
	+403—431[b]	7	*S. oleracea* chloroplasts	22
	+380	7.8	Unspecified D-144 particles	23
	+370	7	*Brassica komatsuna* purified protein	24
	+365	7	*Sinapis arvensis* purified protein	25
b-559$_{HP}$	+370	6.5, 7.5	*Pisum sativum* chloroplasts	17
	+383	7.8	*P. sativum* chloroplasts	26
	+350	7.2	*S. oleracea* chloroplasts	27
	+325	8.2	*S. oleracea* chloroplasts	20
	+395	8	*S. oleracea* chloroplasts	28
	+335[c]	5		
	+450[b]	7	*S. oleracea* chloroplasts	22
	+376	6—8	Unspecified chloroplasts	29
	+400	5		
b-559$_{LP}$	+65	7	*Hordeum vulgare* etioplasts	17
	+100	8.2	*S. oleracea* D-144 particles	23
b-559$_{LP}$	+77	7.8	*P. sativum* chloroplasts	25
	+20	7	*L. sativa* chloroplasts	30
	+85	6.2, 7.8	*L. sativa* cytochrome *b/f* particle	30
b-563	−16[d]	8	*S. oleracea* chloroplasts	31
	−120	8.2	*S. oleracea* D-144 particles	23
	−5	7.8	*S. oleracea* D-144 particles	32
	−80	7.8	*S. oleracea* purified protein + Triton	33
	−120	7.8	*S. oleracea* purified protein − Triton	33
	−110	7	*L. sativa* chloroplasts	30
	−90	6.2, 7.8	*L. sativa* Digitonin cytochrome *b/f* particle	30
	−100	7	*L. sativa* Triton cytochrome *b/f* particle	18

[a] Determined at room temperature unless otherwise specified.
[b] Determined at 0°C.
[c] Hysteresis between oxidative and reductive titrations.
[d] A contribution from cytochrome *b*-559 was ignored. A corrected value for cytochrome *b*-563 would be approximately −30 mV.

are examined by an identical procedure. All the components show a marked tendency to aggregate, but the values given in Table 3 refer to a monomer containing one heme group. In the case of cytochrome *f*, the tendency to aggregate is strongly species dependent. These variations are derived, at least in part, from variations in the properties of the proteins themselves, but are not correlated with relative molecular mass. Members of the family Cruciferae tend to yield essentially monomeric preparations. The common cereals *Avena sativa*, *Hordeum vulgare*, and *Triticum aestivum* and also to a lesser extent parsley (*Petroselinum crispum*) give fairly high proportions of monomers, but the cytochrome from spinach (*Spinacea oleracea*), beet (*Beta vulgaris*), pea (*Pisum sativum*), tobacco (*Nicotiana tabacum*), and maize (*Zea mays*) is normally ex-

Table 3
RELATIVE MOLECULAR MASSES[a]

Component	Relative molecular mass	Method	Plant source	Ref.
f	33,000	Gel filtration	*Raphanus sativus*	8
	32,000	Gel filtration	*Brassica komatsuna*	9
	27,700	Sedimentation equilibrium	*Sinapis arvensis*	25
	26,900	SDS	*S. arvensis*	25
	27,000	Amino acid analysis	*S. arvensis*	25
	33,900	Gel filtration	*S. arvensis*	25
	28,500	SDS	*S. arvensis*	36
	28,500	SDS	*Brassica napus*	36
	28,500	SDS	*Brassica rapa*	36
	32,700	SDS	*Nicotiana tabacum*	36
	32,700	SDS	*Petroselinum crispum*	36
	34,800	SDS	*Hordeum vulgare*	36
	34,800	SDS	*Triticum aestivum*	36
	37,300	SDS	*Pisumsativum*	36
	38,000	SDS	*P. sativum*	37
	31,000	SDS	*Vicia faba*	7
	32,500—34,000	SDS	*Spinacea oleracea*	6, 38—40
	32,700	SDS	*S. oleracea*	36
	26,200	Sedimentation equilibrium	*S. oleracea*	6
	36,500	Amino acid analysis	*S. oleracea*	38
b-559	45,900	Protein (g)/mol heme	*S. oleracea*	41
	37,000	SDS	*S. oleracea*	42
	38,000	Amino acid analysis	*S. oleracea*	41
b-563	40,000	Protein (g)/mol heme	*S. oleracea*	12, 33
	20,000[b]	SDS	*S. oleracea*	33
	18,000	SDS	*S. oleracea*	13

[a] The values given refer to the relative molecular mass of the monomer, excluding lipid.
[b] Polypeptides of 20,000, 9,600, and 6,600 were reported to occur in the molar ratio 1:1:2.

tracted entirely in aggregated form.[36] Table 4 provides amino acid compositions of purified cytochromes.

Table 4
AMINO ACID COMPOSITION

	Cytochrome *f*						Cytochrome b-559
	Nicotiana[36]				*Spinacea oleracea*		
	Tabacum (mol/ 32,700 g protein)	*Glutinosa* (mol/ 32,700 g protein)	*Petroselinum crispum*[43] (mol/ 10^5 g protein)	*Sinapis arvensis*[25] (mol/mol heme)	(mol/mol heme)[38]	(mol/10^5 g protein)[43]	S. oleracea[10] *(mol/ mol cys)*[12]
Asp	29	30	100.1	27	35	103.2	9
Thr	11	12	37.7	8	12	32.9	10
Ser	15	16	54.1	11	18	51.2	15
Glu	32	32	132.3	30	37	125.9	7
Pro	26	26	63.9	24	20	63.2	9
Gly	30	28	86.3	24	28	85.4	12
Ala	21	20	74.7	22	24	67.0	9
Val	22	21	59.1	16	25	56.6	7
Met	3	3	7.6	1	5	14.3	2
Ile	20	19	53.5	17	20	46.0	9
Leu	26	24	79.5	19	30	80.7	11
Tyr	10	10	28.1	8	11	27.1	2
Phe	13	12	32.9	7	13	35.5	10
His	3	3	6.4	4	3	6.6	0
Lys	20	22	72.1	19	24	76.2	2
Arg	13	15	31.9	7	11	33.9	4
Cys	2[b]	2[b]	8.6	2	2	8.4	1
Try	—	—	—	2	6	—	—

[a] 3 mol cys/heme.
[b] Assumed as the minimum requirement for heme attachment.

REFERENCES

1. **Hill, R.**, The cytochrome *b* component of chloroplasts, *Nature (London)*, 174, 501, 1954.
2. **Forti, G., Bertolè, M. L., and Zanetti, G.**, Purification and properties of cytochrome *f* from parsley leaves, *Biochim. Biophys. Acta*, 109, 33, 1965.
3. **Bendall, D. S., Davenport, H. E., and Hill, R.**, Cytochrome components in chloroplasts of the higher plants, *Methods Enzymol.*, 23, 327, 1971.
4. **Hill, R. and Bonner, W. D.**, The nature and possible function of chloroplast cytochromes, in *Light and Life*, McElroy, W. D. and Glass, B., Eds., Johns Hopkins, Baltimore, 1961, 424.
5. **Bonner, W. D.**, The cytochromes of plant tissues, in *Haematin Enzymes*, Falk, J. E., Lemberg, M. R., and Morton, R. K., Eds., Pergamon, Oxford, 1961, 479.
6. **Singh, J. and Wasserman, A. R.**, The use of disc gel electrophoresis with nonionic detergents in the purification of cytochrome *f* from spinach grana membranes, *J. Biol. Chem.*, 246, 3532, 1971.
7. **Süss, K.-H.**, Identification of chloroplast thylakoid membrane polypeptides: coupling factor of photophosphorylation (CF_1) and cytochrome *f*, *FEBS Lett.*, 70, 191, 1976.
8. **Takahashi, M. and Asada, K.**, Purification of cytochrome *f* from Japanese-radish leaves, *Plant Cell Physiol.*, 16, 191, 1975.
9. **Matsuzaki, E., Kamimura, Y., Yamasaki, T., and Yakushiji, E.**, Purification and properties of cytochrome *f* from *Brassica komatsuna* leaves, *Plant Cell Physiol.*, 16, 237, 1975.
10. **Garewal, H. W. and Wasserman, A. R.**, Triton X-100 - 4*M* urea as an extraction medium for membrane proteins. I. Purification of chloroplast cytochrome b_{559}, *Biochemistry*, 13, 4063, 1974.
11. **Garewal, H. S., Singh, J., and Wasserman, A. R.**, Purification of chloroplast cytochrome b_{559}, *Biochem. Biophys. Res. Commun.*, 44, 1300, 1971.
12. **Stuart, A. L. and Wasserman, A. R.**, Purification of cytochrome b_6. A tightly bound protein in chloroplast membranes, *Biochim. Biophys. Acta*, 314, 284, 1973.
13. **Lach, H. J. and Böger, P.**, Some properties of plastidic cytochrome *b*-563, *Z. Naturforsch. Teil C*, 32, 877, 1977.
14. **Anderson, J. M. and Boardman, N. K.**, Localization of low potential cytochrome *b*-559 in photosystem I, *FEBS Lett.*, 32, 157, 1973.
15. **Cox, R. P.**, Some properties of chloroplast cytochrome *b*-563 in an enriched preparation obtained by digitonin treatment, in *Abstracts 4th Int. Congress on Photosynthesis*, Coombs, J., Ed., UK ISES, London, 1977, 76.
16. **Davenport, H. E. and Hill, R.**, The preparation and some properties of cytochrome *f*, *Proc. R. Soc. London Ser. B*, 139, 327, 1952.
17. **Bendall, D. S.**, Oxidation-reduction potentials of cytochromes in chloroplasts from higher plants, *Biochem. J.*, 109, 46P, 1968.
18. **Nelson, N. and Neumann, J.**, Isolation of a cytochrome b_6-*f* particle from chloroplasts, *J. Biol. Chem.*, 247, 1817, 1972.
19. **Fan, H. N. and Cramer, W. A.**, The redox potential of cytochromes *b*-559 and *b*-563 in spinach chloroplasts, *Biochim. Biophys. Acta*, 216, 200, 1970.
20. **Knaff, D. B. and Arnon, D. I.**, On two photoreactions in system II of plant photosynthesis, *Biochim. Biophys. Acta*, 226, 400, 1971.
21. **Malkin, R., Knaff, D. B., and Bearden, A. J.**, The oxidation-reduction potential of membrane-bound chloroplast plastocyanin and cytochrome *f*, *Biochim. Biophys. Acta*, 305, 675, 1973.
22. **Erixon, K., Lozier, R., and Butler, W. L.**, The redox state of cytochrome b_{559} in spinach chloroplasts, *Biochim. Biophys. Acta*, 267, 375, 1972.
23. **Knaff, D. B. and Malkin, R.**, The oxidation-reduction potentials of electron carriers in chloroplast photosystem I fragments, *Arch. Biochem. Biophys.*, 159, 555, 1973.
24. **Matsuzaki, E., Kamimura, Y., Yamasaki, T., and Yakushiji, E.**, Purification and properties of cytochrome *f* from *Brassica komatsuna* leaves, *Plant Cell Physiol.*, 16, 237, 1975.
25. **Gray, J. C.**, Purification and properties of monomeric cytochrome *f* from charlock, *Sinapis arvensis* L, *Eur. J. Biochem.*, 82, 133, 1978.
26. **Horton, P. and Croze, E.**, The relationship between the activity of chloroplast photosystem II and the midpoint oxidation-reduction potential of cytochrome *b*-559, *Biochim. Biophys. Acta*, 462, 86, 1977.
27. **Boardman, N. K., Anderson, J. M., and Hiller, R. G.**, Photooxidation of cytochromes in leaves and chloroplasts at liquid nitrogen temperature, *Biochim. Biophys. Acta*, 234, 126, 1971.
28. **Horton, P., Whitmarsh, J., and Cramer, W. A.**, On the specific site of action of 3-(3,4-dichlorophenyl)-1,1-dimethylurea in chloroplasts: inhibition of a dark acid-induced decrease in midpoint potential of cytochrome *b*-559, *Arch. Biochem. Biophys.*, 176, 519, 1976.
29. **Knaff, D. B.**, The effect of pH on the midpoint oxidation-reduction potentials of components associated with plant photosystem II, *FEBS Lett.*, 60, 331, 1975.

30. **Rich, P. R. and Bendall, D. S.,** The redox potentials of the *B*- type of cytochromes of higher plant chloroplasts, *Biochim. Biophys. Acts,* 591, 153, 1980.
31. **Böhme, H. and Cramer, W. A.,** Uncoupler-dependent decrease in midpoint potential of the chloroplast cytochrome b_6, *Biochim. Biophys. Acta,* 325, 275, 1973.
32. **Böhme, H.,** Photoreactions of cytochrome b_6. and cytochrome *f* in chloroplast photosystem I fragments, *Z. Naturforsch. Teil C,* 31, 68, 1976.
33. **Stuart, A. L. and Wasserman, A. R.,** Chloroplast cytochrome b_6 Molecular composition as a lipoprotein, *Biochim. Biophys. Acta,* 376, 561, 1975.
34. **Boardman, N. K. and Anderson, J. M.,** Fractionation of the photochemical systems of photosynthesis. II. Cytochrome and carotenoid contents of particles isolated from spinach chloroplasts, *Biochim. Biophys. Acta,* 143, 187, 1967.
35. **Wood, P. M. and Bendall, D. S.,** The reduction of plastocyanin by plastoquinol-1 in the presence of chloroplasts. A dark electron transfer reaction involving components between the two photosystems, *Eur. J. Biochem.,* 61, 337, 1976.
36. **Gray, J. C.,** personal communication, 1978.
37. **Eaglesham, A. R. J. and Ellis, R. J.,** Protein synthesis in chloroplasts. II. Light-driven synthesis of membrane proteins by isolated pea chloroplasts, *Biochim. Biophys. Acta,* 335, 396, 1974.
38. **Nelson, N. and Racker, E.,** Partial resolution of the enzymes catalysing photophosphorylation. X. Purification of spinach cytochrome *f* and its photooxidation by resolved photosystem I particles, *J. Biol. Chem.,* 247, 3848, 1972.
39. **Klein, S. M. and Vernon, L. P.,** Protein composition of spinach chloroplasts and their photosystem I and photosystem II subfragments, *Photochem. Photobiol.,* 19, 43, 1974.
40. **Wessels, J. S. C. and Borchert, M. T.,** Polypeptide profiles of chlorophyll-protein complexes and thylakoid membranes of spinach chloroplasts, *Biochim. Biophys. Acta,* 503, 78, 1978.
41. **Garewal, H. S. and Wasserman, A. R.,** Triton X-100 - 4 *M* urea as an extraction medium for membrane proteins. II. Molecular properties of pure cytochrome b_{559}: a lipoprotein containing small polypeptide chains and a limited lipid composition, *Biochemistry,* 13, 4072, 1974.
42. **Lach, H.-J. and Böger, P.,** Isolation and some molecular properties of plastidic algal cytochrome *b*-559, *Z. Naturforsch. Teil C,* 32, 75, 1977.
43. **Davenport, H. E.,** Some observations on cytochrome *f*, in *Proc. 2nd Int. Congress on Photosynthesis Research,* Forti, G., Avron, M., and Melandri, A., Eds., Dr. W. Junk, The Hague, 1972, 1593.

PLASTOCYANIN

Sakae Katoh

DISTRIBUTION

Plastocyanin is a blue copper protein functioning as an electron carrier in photosynthetic electron transport in higher plants and algae.

Since 1960 when plastocyanin was first discovered in *Chlorella ellipsoidea*,[1] the protein has been shown to occur in a wide variety of higher plants and green algae.[2-6] Plastocyanin is also present in red[7] and blue-green algae,[8] but not in photosynthetic bacteria. In higher plants, plastocyanin is found in green chlorophyll-containing tissues, and the protein is localized in chloroplasts.[3] Chloroplasts contain 3 to 5 atoms of copper of plastocyanin per 1000 molecules of chlorophyll, which are often twice as much as the content of P_{700}, the reaction center (RC) chlorophyll of Photosystem I (see Table 1).

EXTRACTION, PURIFICATION AND ASSAY

The association of plastocyanin with the thylakoid membrane is rather loose, so that a small but significant part of the copper protein is solubilized by a hypotonic treatment of chloroplasts.[3,13] Plastocyanin is usually extracted from chloroplasts by acetone treatment, detergent treatment, or sonication.[3,9,10] Plastocyanin is highly purified by ammonium sulfate fractionation and column chromatography with diethylaminoethyl cellulose.[18,19] Other techniques, such as gel filtration or electrophoresis, can be also used for the purification of plastocyanin.

The concentration of plastocyanin in solution is determined by measuring the absorbance of oxidized plastocyanin at 597 nm (see Figure 1).[18] The extinction coefficient, $\varepsilon_{mg}^{c\,m}$ atom copper, of oxidized plastocyanin isolated from spinach is 4.9 at this wavelength. The reduced protein has no absorption in the visible region. With crude extracts which might contain colored substances, the protein content is estimated from the difference between the absorbance determined after the addition of a few crystals of potassium ferricyanide (to oxidize plastocyanin completely) and that determined after the subsequent addition of a few milligrams of sodium ascorbate (to fully reduce the protein).

Three additional sensitive procedures for the assay of plastocyanin have been reported:

1. Enzymatic assays, based on a linear dependency of rate of a Photosystem I-mediated electron transport on the concentration of plastocyanin, are at least ten times as sensitive as the direct spectrophotometric method. Photooxidation of reduced mammalian cytochrome *c* and ascorbate or photoreduction of $NADP^+$ by a Photosystem I preparation was employed for the assay.[5,9,10,20] Note, however, that these reactions are also stimulated by a soluble algal cytochrome of *c*-type.
2. More specific is the technique of quantitative immunoelectrophoresis in an agarose gel containing an antibody against plastocyanin. Crude chloroplast extract has been directly analyzed by this method.[13]
3. Plastocyanin in the cells or chloroplasts can be estimated by low-temperature electron paramagnetic resonance (EPR) spectroscopy by comparing the EPR signal of the soluble plastocyanin.[11] The oxidized protein gives a characteristic signal with g = 2.05 and g = 2.23.

Table 1
PLASTOCYANIN CONTENTS IN PLANT CHLOROPLASTS AND ALGAL CELLS

Plant	Assay method	Atom copper of plastocyanin per 1000 molecules of chlorophyll	Atom copper of plastocyanin per one molecule of P_{700}	Ref.
Spinach	Spectroscopy	3.3		3
Spinach	Enzymatic	5.1		5
Spinach	Enzymatic	4.9		9
Spinach	Enzymatic	4.5		10
Spinach	EPR	5.0		11
Spinach	EPR	5.0	2	7
Spinach	Kinetics *in situ*	—	2	14
Spinach	Immunoelectrophoresis	4.0	4	13
Spinach	Binding of antibody	3.3		12
Antirrhinum	Binding of antibody	2.0		12
Pea	Enzymatic	5.6		5
Barley	Enzymatic	4.7		5
Orache	Enzymatic	8.0		5
Tobacco (green)	Enzymatic	2.0		5
Tobacco (pale)	Enzymatic	5.1		5
Chlamydomonas	Spectroscopy	2.5		4
Chlamydomonas	Spectroscopy	6.5		16
Scenedesmus	Spectroscopy	2.7	3	15
Scenedesmus	EPR	5.0	1.6	7
Chlorella	EPR	5.0	1.8	7
Bumillieriopsis	Spectroscopy, EPR	0.0		17
Porphyridium	EPR	6.0	2.4	7
Anacystis	EPR	4.5	1.4	7

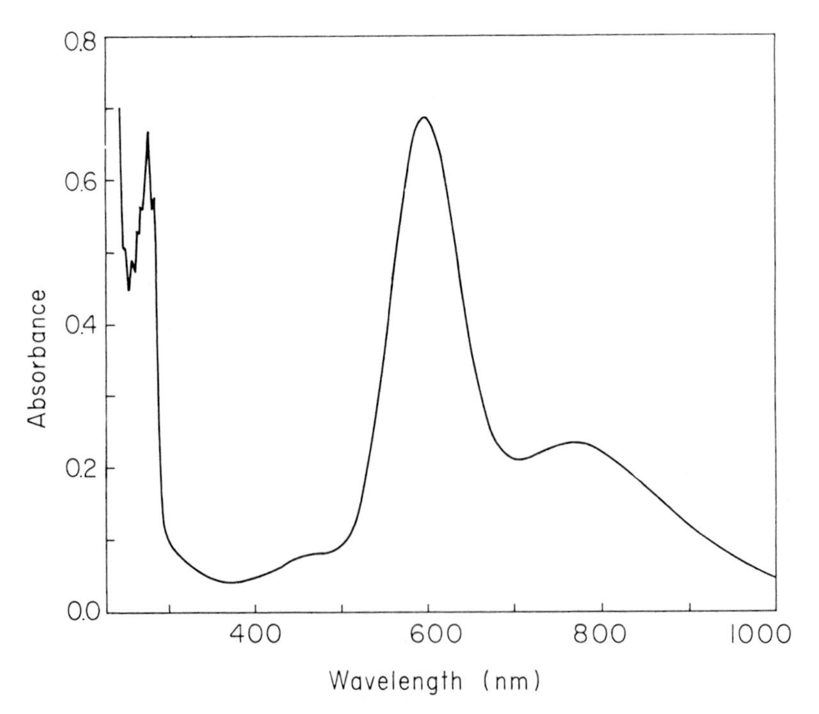

FIGURE 1. Absorption spectrum of oxidized spinach plastocyanin.

MOLECULAR PROPERTIES

Plastocyanin is identified by its absorption spectrum (see Figure 1).[18] The oxidized protein is blue in color and shows a main peak at 597 nm and two minor peaks at 770 and 460 nm. The protein peak in the UV region shows a maximum at 278 nm. In addition, there are four vibrational fine structure bands at 253, 259, 265, and 269 nm because of phenylalanine and a band at 284 nm because of tyrosine. The blue color of the oxidized protein is due to of Cu^{2+}.[21] Spinach plastocyanin contains 0.58% copper.[18] The copper of plastocyanin does not react with molecular oxygen, but undergoes reversible oxidation-reduction on addition of various redox reagents. The oxidation reduction potential of spinach plastocyanin is 370 mV between pH 5.4 and 9.9.[18]

A single peptide of plastocyanin has a mol wt of about 10,000 and consists of about 100 amino acid residues and 1 copper atom.[6-18,19,21-24] In addition, the protein may contain a small amount of carbohydrate.[18,23,25] There are indications that the protein occurs in dimeric or even in tetrameric form.[12,18,26] The amino acid composition of plastocyanin from higher plants and algae is similar in that the protein contains two methionine residues and one cysteine residue but no tryptophan nor arginine residues, except *Chlorella* plastocyanin which has one methionine and one tryptophan residue (Table 2). Because of relative abundance of acidic amino acids, plastocyanin is an acidic protein with an isoelectric point of about 4. However, plastocyanin isolated from a blue-green alga is a basic protein.[8] The amino acid sequence of plastocyanin from various plants and algae has been determined.[25,27-34]

Three-dimensional structure of plastocyanin from poplar leaves has been determined at a resolution of 2.7Å.[35] The shape of the molecule resembles a slightly flattened cylinder with dimensions 40 × 32 × 28 Å. The copper atom is embedded in one end of the cylinder and is coordinated by sulfur atoms of Cys 84 and Met 92 and by nitrogen atoms of the imidazole groups of His 37 and His 87. There is a hydrophobic patch surrounding an opening beneath which the copper site is located which might be of interest with respect to the association with the membrane or aggregation of the protein.

Spinach plastocyanin is a stable protein and can be stored at −20°C for a year or two without any significant change in the absorption spectrum or solubility in water. No change in color was detected after incubation of the protein at temperature up to 60°C for 5 min.[18]

FUNCTION

Plastocyanin is an electron carrier functioning between Photosystems I and II in chloroplasts.[11,12,14,19,26-44] In higher plants, it is well established that plastocyanin serves as the electron donor for the RC of Photosystem I. Photooxidized P_{700} is directly reduced by plastocyanin. Several lines of evidence indicate that oxidized plastocyanin is, in turn, reduced by cytochrome *f*, although spectrophotometric studies on light-induced redox changes of electron carriers *in situ* suggest that a significant part of electrons from Photosystem II to plastocyanin does not pass cytochrome *f*.[14,45]

The situation appears to be more complicated in algae which have, in addition to plastocyanin and cytochrome *f*, a soluble *c*-type cytochrome showing the maximum of α-band at 552 to 553 nm.[46,47] The soluble *c*-type cytochrome is similar to plastocyanin in size and charge of the molecule and also in redox potential. It has been suggested that in some algae, in which plastocyanin has not been detected, the soluble algal cytochrome *c* serves in place of plastocyanin as electron donor for P_{700}.[17] The plastocyanin and soluble cytochrome *c* contents in *Scenedesmus*[15] and *Chlamydomonas* cells[16] varied approximately reciprocally with changing concentrations of copper in the cul-

Table 2
AMINO ACID COMPOSITION OF PLASTOCYANIN

	Chlorella fusca[25]	*Scenedesmus acutus*[26]	Broad bean (*Vicia faba*)[27]	French bean (*Phaseolus vulgaris*)[28]	Dock (*Rumex obtusifolius*)[32]	Potato (*Solanum tuberosum*)[30]	*Solanum crispum*[33]	Marrow (*Cucurbita pepo*)[29]	Shepherd's purse (*Capsella bursa-pastoris*)[34]	Dog's mercury (*Mercurialis prennis*)[34]	Spinach (*Spinacia oleracea*)[31]
Aspartic acid	6	10	13	5	10	13	12	15	6	8	11
Asparagine	5			5					7	7	
Threonine	8	5	4	5	5	6	5	2	4	4	5
Serine	7	5	7	8	6	7	9	7	6	8	7
Glutamic acid	8	8	9	9	10	8	9	9	9	6	10
Glutamine	2			1					1	1	
Proline	5	5	6	7	5	4	6	5	4	5	5
Glycine	11	10	11	11	12	12	12	12	12	10	13
Alanine	11	8	9	5	11	10	7	7	10	9	7
Cysteine	1	1	1	1	1	1	1	1	1	1	1
Valine	9	9	12	14	10	9	9	11	10	10	11
Methionine	1	2	2	2	2	2	2	2	2	2	2
Isoleucine	3	3	3	3	5	4	5	5	5	4	3
Leucine	4	4	6	7	5	7	6	6	6	7	7
Tyrosine	4	4	3	3	3	3	3	3	3	3	3
Phenylalanine	5	5	6	6	6	6	6	6	6	6	6
Lysine	4	5	5	5	6	5	5	6	5	5	6
Histidine	3	4	2	2	2	2	2	2	2	2	2
Arginine	0	0	0	0	0	0	0	0	0	0	0
Tryptophan	1		0	0	0	0	0	0	0	0	0
Total	98		99	99	99	99	99	99	99	99	99

ture media. It is likely therefore that plastocyanin and soluble algal cytochrome *c* are interchangeable in function in algal photosynthesis.

The participation of plastocyanin in electron transport in chloroplasts can be tested with $HgCl_2$,[41] KCN,[42] or an antibody against plastocyanin[12] which affect, more or less specifically, the copper protein *in situ.* In treated chloroplasts, electron transport through Photosystem I is blocked and photooxidation of cytochrome *f* is suppressed, whereas photooxidation of P_{700} and photoreduction of cytochrome *f* remain active.

REFERENCES

1. **Katoh, S.,** A new copper protein from *Chlorella ellipsoidea, Nature (London),* 186, 533, 1960.
2. **Katoh, S.,** A new leaf copper protein 'plastocyanin', a natural Hill oxidant, *Nature (London),* 189, 665, 1961.
3. **Katoh, S., Suga, I., Shiratori, I., and Takamiya, A.,** Distribution of plastocyanin in plants, with special reference to its localization in chloroplasts, *Arch. Biochem. Biophys.,* 94, 136, 1961.
4. **Gorman, D. S. and Levine, R. P.,** Photosynthetic electron transport chain of *Chlamydomonas reinhardi.* IV. Purification and Properties of plastocyanin, *Plant Physiol.,* 41, 1637, 1966.
5. **Plensničar, M. and Bendall, D. S.,** The plastocyanin content of chloroplasts from some higher plants estimated by a sensitive enzymatic assay, *Biochim. Biophys. Acta,* 216, 192, 1970.
6. **Ramshaw, J. A. M., Brown, R. H., Scawen, M. D., and Boulter, D.,** Higher plant plastocyanin, *Biochim. Biophys. Acta,* 303, 269, 1973.
7. **Visser, J. W., Amesz, J., and Van Gelder, B. F.,** EPR signals of oxidized plastocyanin in intact algae, *Biochim. Biophys. Acta,* 333, 279, 1974.
8. **Lightbody, J. J. and Krogmann, D. W.,** Isolation and properties of plastocyanin from *Anabaena variabilis, Biochim. Biophys. Acta,* 131, 508, 1967.
9. **Baszynski, T., Brand, J., Krogmann, D. W., and Crane, F. L.,** Plastocyanin participation in chloroplast photosystem I, *Biochim. Biophys. Acta,* 234, 537, 1971.
10. **Sane, P. V. and Hauska, G. A.,** The distribution of photosynthetic reactions in the chloroplast lamellar system. I. Plastocyanin content and reactivity, *Z. Naturforsch. Teil B,* 27, 932, 1972.
11. **Malkin, R. and Bearden, A. J.,** Light-induced changes of bound chloroplast plastocyanin as studied by EPR spectroscopy: the role of plastocyanin in noncyclic photosynthetic electron transport, *Biochim. Biophys. Acta,* 292, 169, 1973.
12. **Schmid, G. H., Radunz, A., and Menke, W.,** The effect of an antiserum to plastocyanin on various chloroplast preparations, *Z. Naturforsch. Teil C,* 30, 201, 1975.
13. **Böhme, H.,** Structural and quantitative analysis of membrane-bound components of the photosystem I complex of spinach chloroplasts by immunological methods, in *Bioenergetics of Membranes,* Packer, L., Papageorgiou, G. C., and Trebst, A., Eds., Elsevier/North-Holland, Amsterdam, 1977, 329.
14. **Haehnel, W.,** Electron transport between plastoquinone and chlorophyll a_I in chloroplasts. II. Reaction kinetics and the function of plastocyanin *in situ, Biochim. Biophys. Acta,* 459, 418, 1977.
15. **Bohner, H. and Böger, P.,** Reciprocal formation of cytochrome *c*-553 and plastocyanin in *Scenedesmus, FEBS Lett.,* 85, 337, 1978.
16. **Wood, P. M.,** Interchangeable copper and iron proteins in algal photosynthesis. Studies on plastocyanin and cytochrome *c*-552 in *Chlamydomonas, Eur. J. Biochem.,* 87, 9, 1978.
17. **Kunert, K-J. and Böger, P.,** Absence of plastocyanin in the alga *Bumilleriopsis* and its replacement by cytochrome 553, *Z. Naturforsch. Teil C,* 30, 190, 1975.
18. **Katoh, S., Shiratori, I., and Takamiya, A.,** Purification and some properties of spinach plastocyanin, *J. Biochem. (Tokyo),* 51, 32, 1962.
19. **Katoh, S.,** Plastocyanin, in *Methods in Enzymology,* Vol. 23, San Pietro, A., Ed., Academic Press, New York, 1971, 408.
20. **Hauska, G. A., McCarty, R. E., Berzborn, R. J., and Racker, E.,** Partial resolution of the enzymes catalyzing photophosphorylation. VII. The function of plastocyanin and its interaction with a specific antibody, *J. Biol. Chem.,* 240, 3524, 1971.
21. **Katoh, S. and Takamiya, A.,** Nature of copper-protein binding in spinach plastocyanin, *J. Biochem. (Tokyo),* 55, 378, 1964.
22. **Milne, P. R. and Wells, J. R. E.,** Structure and molecular weight studies on the small copper protein, plastocyanin, *J. Biol. Chem.,* 245, 1566, 1970.
23. **Mutuskin, A. A., Pshenova, K. V., Alekhina, S. K., and Kolesnikov, P. A.,** Characteristics of plastocyanin from wheat leaves, *Biokimia,* 36, 236, 1971.

24. **Scawen, M. D. and Hawitt, E. J.**, Plastocyanin from *Cucurbita pepo* L., *Biochem. J.*, 124, 32p, 1971.
25. **Kelly, J. and Ambler, R. P.**, The amino acid sequence of plastocyanin from *Chlorella fusca*, *Biochem. J.*, 143, 681, 1974.
26. **Siegelman, H. H., Rasched, I. R., Kunert, K. J., Kroneck, P., and Böger, P.**, Plastocyanin: possible significance of quaternary structure, *Eur. J. Biochem.*, 64, 131, 1978.
27. **Ramshaw, J. A. H., Scawen, M. D., and Boulter, D.**, The amino acid sequence of plastocyanin from *Vicia faba* L. (Broad bean), *Biochem. J.*, 141, 835, 1974.
28. **Milne, P. R., Wells, J. R. E., and Ambler, R. P.**, The amino acid sequence of plastocyanin from French bean (*Phaseolus vulgaris*), *Biochem. J.*, 143, 691, 1974.
29. **Scawen, M. D. and Boulter, D.**, The amino acid sequence of plastocyanin from *Cucurbita pepo* L. (Vegetable Marrow), *Biochem. J.*, 143, 257, 1974.
30. **Ramshaw, J. A. M., Scawen, M. D., Bailey, C., and Boulter, D.**, The amino acid sequence of plastocyanin from *Solonum tuberosum* L. (Potato), *Biochem. J.*, 139, 583, 1974.
31. **Scawen, M. D., Ramshaw, J. A. M., and Boulter, D.**, The amino acid sequence of plastocyanin from spinach (*Spinacia oleracea* L.) , *Biochem. J.*, 147, 343, 1975.
32. **Haslett, B. G., Bailey, C. J., Ramshaw, J. A. M., Scawen, M. D., and Boulter, D.**, The amino acid sequence of plastocyanin from *Rumex obtusifolius*, *Phytochemistry*, 17, 615, 1978.
33. **Haslett, B. G., Evans, M., and Boulter, D.**, Amino acid sequences of plastocyanin from *Solanum crispum* using automatic methods, *Phytochemistry*, 17, 735, 1978.
34. **Scawen, M. D., Ramshaw, J. A. M., Brown, R. H., and Boulter, D.**, The amino acid sequences of plastocyanin from *Mercurialis perennis* and *Capsella bursapastoris*, *Phytochemistry*, 17, 901, 1978.
35. **Colman, P. M., Freeman, H. C., Guss, J. M., Murata, M., Norris, V. A., Ramshaw, J. A. M., and Venkatappa, M. P.**, X-ray crystal structure analysis of plastocyanin at 2.7 Å resolution, *Nature (London)*, 272, 319, 1978.
36. **Katoh, S. and Takamiya, A.**, Light-induced reduction and oxidation of plastocyanin by chloroplast preparations, *Plant Cell Physiol.*, 4, 335, 1963.
37. **Katoh, S. and Takamiya, A.**, Photochemical reactions of plastocyanin in chloroplasts, in *Photosynthetic Mechanisms of Green Plants*, Jagendorf, A. T. and Kok, B., Eds., National Academy of Science, - National Research Council, Washington, D.C., 1964, 262.
38. **Katoh, S. and Takamiya, A.**, Restoration of NADP photoreducing activity of sonicated chloroplasts by plastocyanin, *Biochim. Biophys. Acta*, 99, 156, 1965.
39. **Gorman, D. S. and Levine, R. P.**, Photosynthetic electron transport chain of *Chlamydomonas reinhardi*. VI. Electron transport in mutant strains lacking either cytochrome 553 or plastocyanin, *Plant Physiol.*, 41, 1648, 1966.
40. **Avron, M. and Shneyour, A.**, On the site of action of plastocyanin in isolated chloroplasts, *Biochim. Biophys. Acta*, 226, 498, 1971.
41. **Kimimura, M. and Katoh, S.**, Studies on electron transport associated with photosystem I. I. Functional site of plastocyanin; inhibitory effects of $HgCl_2$ on electron transport and plastocyanin in chloroplasts, *Biochim. Biophys. Acta*, 283, 279, 1972.
42. **Izawa, S., Kraayenhof, R., Ruuge, E. K., and Devault, D.**, The site of KCN inhibition in the photosynthetic electron transport pathway, *Biochim. Biophys. Acta*, 314, 328, 1973.
43. **Brand, J. and San Pietro, A.**, Site of polylysine inhibition of photosystem I in spinach chloroplasts, *Arch. Biochem. Biophys.*, 152, 426, 1972.
44. **Bouqes-Bocquet, B.**, Cytochrome *f* and plastocyanin kinetics in *Chlorella pyrenoidosa*, *Biochim. Biophys. Acta*, 462, 371, 1977.
45. **Katoh, S.**, Studies on the algal cytochrome of *c*-type, *J. Biochem. (Tokyo)*, 46, 629, 1959.
46. **Katoh, S.**, Studies on algal cytochrome. II. Physico-chemical properties of crystalline *Porphyra tenera* cytochrome 553, *Plant Cell Physiol.*, 1, 91, 1960.

QUINONES

Richard A. Dilley

Quinones are involved as electron and probably proton carriers in all photosynthetic systems so far investigated. The ease of oxidation-reduction derives from the molecular structure (see Figure 1), and the general redox equation is

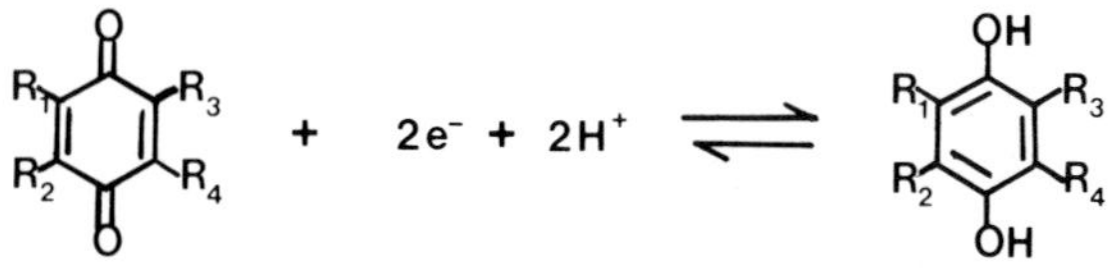

$$\text{Quinone}(R_1, R_2, R_3, R_4) + 2e^- + 2H^+ \rightleftharpoons \text{Hydroquinone}(R_1, R_2, R_3, R_4)$$

Figure 1 shows the structures of some naturally occurring quinones known, or believed, to be involved in redox reactions of the photosynthetic process. Higher plants and algae chloroplasts have plastoquinone[1] as a redox carrier, while photosynthetic bacteria have ubiquinone[2] and possibly menaquinone.[3] It is not certain whether tocopherol quinone or any naphtoquinones are involved in redox reactions of higher plants, although the latter are likely redox agents in certain photosynthetic bacteria.[3] Figure 2 shows the oxidized and reduced spectra of four quinones.

Table 1 lists various properties of the quinones, including molecular weight, spectral absorption peaks and molar absorption coefficients, and redox potentials. Nomenclature approved by the International Union of Biochemistry is used in the table.

Plastoquinone A-9 is shown in Figure 1. PQ-C has a hydroxyl group in one of the isoprene side chains.[9] PQ-B is PQ-C with a fatty acid esterified to the sidechain hydroxyl. PQ-D is an isomer of PQ-C.[10]

Vitamin K, (phylloquinone) one of the nathoquinones, has a phytol sidechain (20 carbon) similar to the tocopherols, while Vitamin K_2 (menaquione) has an isoprene side chain.[11] Chlorobiumquinone is a naphtoquinone derivative having a carbonyl on the first carbon of the side chain.[5]

Rhodospirillum rubrum contains an amino quinone given the trivial name rhodoquinone.[13] It is similar to ubiquinone,[10] except that one of the ring methoxyl groups is replaced by an amino group.

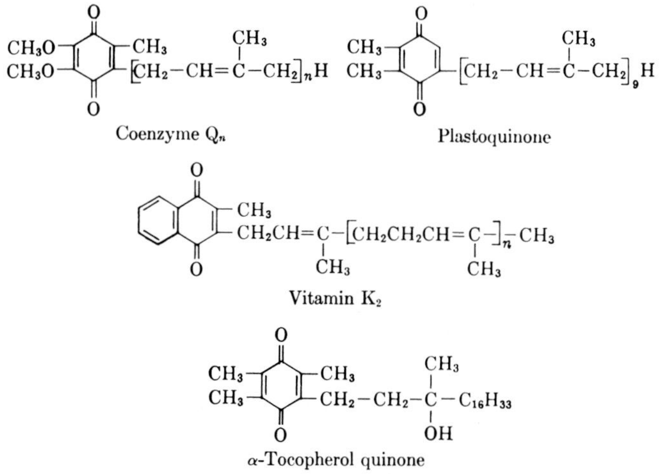

Coenzyme Q_n — Plastoquinone — Vitamin K_2 — α-Tocopherol quinone

FIGURE 1. Structures of some naturally occurring quinones.

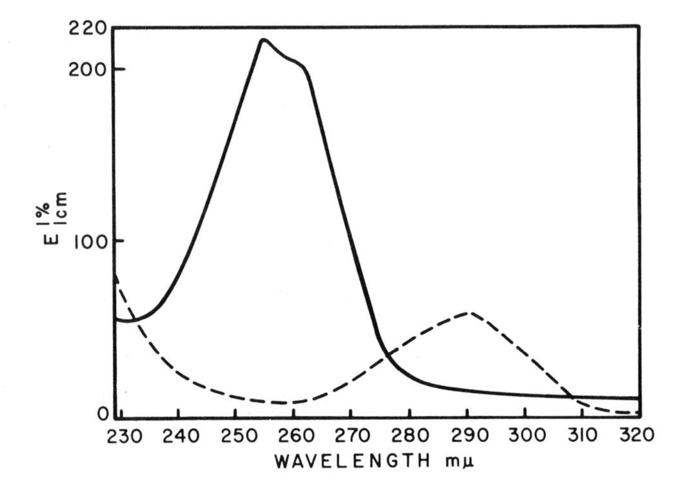

2A

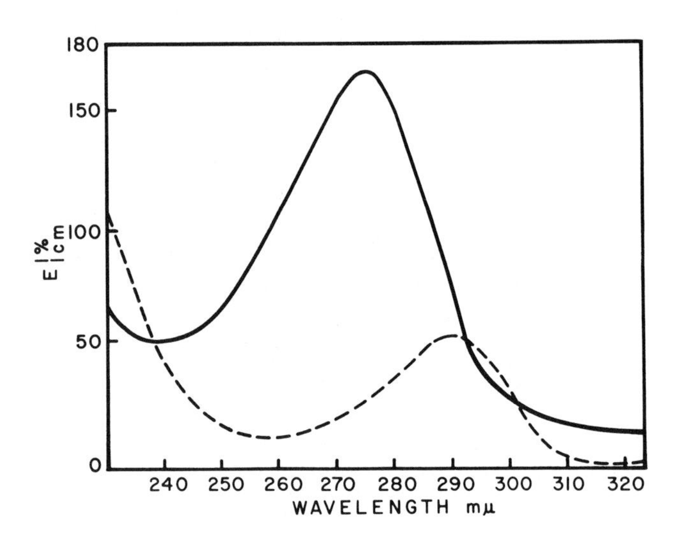

2B

FIGURE 2. UV absorption spectra of four naturally occurring quinones. (A) plastoquinone, (B) ubiquinone, (C) α-Tocopherol quinone, (D) Vitamin K_2 (menaquinone). The solid line is the oxidized form; the dashed line is the reduced form.

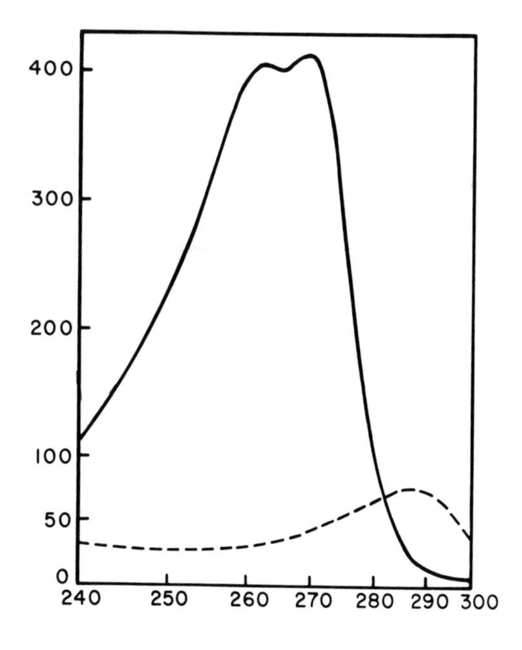

FIGURE 2C

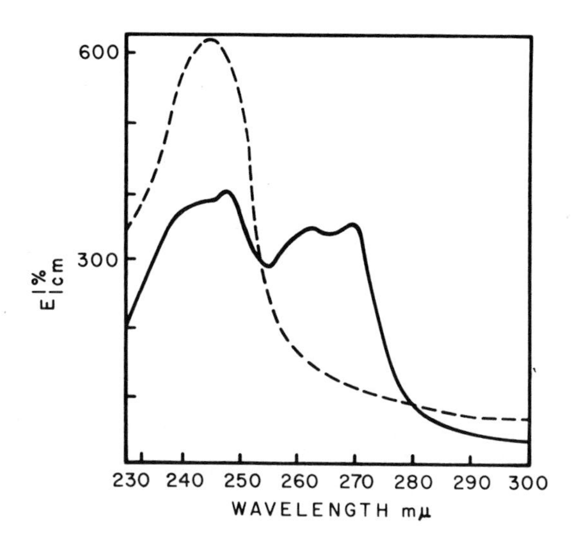

FIGURE 2D

Table 1
PHYSICAL PROPERTIES OF VARIOUS QUINONES

Quinone	Mol wt	Spectral absorption peak, λ max, in units of nm	Molar absorption coefficient ε, abs. per m/L	Melting point (°C)	Redox potential E_{m7} (V)
Higher Plants and Algae					
Plastoquinone A-9	748	255 262 shoulder (ethanol)	15,200	42—43	+0.113
Plastoquinone B-9[a]	1003.57	Same as PQ-A	Same as PQ-A	—	—
Plastoquinone C-9[b]	764	Same as PQ-A	Same as PQ-A	—	+0.055
Plastoquinone D-9[b]	764	Same as PQ-A	Same as PQ-A	−20	−0.06
Vitamin K_1 phylloquinone[b]	450.7	245,248,264,271,329	19,144(245 nm)	—	—
α-Tocopherol	430.69	292 (Ethanol)	3,185	2.5—3.5	+0.187(8)
α-Tocopherol quinone	447.70	262,269 (Ethanol)	15,100	—	—
Photosynthetic Bacteria					
Ubiquinone-10 (Rhodospirillum rubrium, Chromatium D)	849.3	275 (Ethanol)	14,000	48—49	+0.050
Menaquinone-7 (*Chloropseudomonas ethylicum*)[6]	—	243,248,261,270 (Cyclohexane)	15,900 ΔE (270-290 n*M*) in cyclohexane	50—51	−0.081
Chlorobiumquinone (*C. ethylicum*)[6]	663.0	251,245,257,267 (Cyclohexane) 254,265 (Ethanol)	16,300(251 nm) in cyclohexane	42—43[c] 47—48	+0.39

[a] For the palmitate ester.
[b] See text for structural details.
[c] Chlorobiumquinone-10.

REFERENCES

1. **Amez, J.**, plastoquinones in photosynthesis, *Biochim. Biophys. Acta*, 301, 35, 1973.
2. **Halsey, Y. D. and Parson, W. W.**, Identification of ubiquinone as the secondary electron acceptor in the photosynthetic apparatus of *Chromatium Vinosum*, *Biochim. Biophys. Acta*, 347, 404, 1974.
3. **Takamiya, K.**, Light induced oxidation-reduction reactions of menaquinone in intact cells of a green photosynthetic bacterium, *Biochim. Biophys. Acta*, 234, 390, 1971.
4. **Feher, G., Isaacson, R. A., McElroy, J. D., Ackerson, L. C., and Okamura, M. Y.**, On the question of the primary acceptor in bacterial photosynthesis, *Biochim. Biophys. Acta*, 368, 135, 1974.
5. **Powls, R., Redfearn, E., and Trippett, S.**, The structure of chlorobiumquinone, *Biochem. Biophys. Res. Commun.*, 33, 408, 1968.
6. **Powls, R. and Redfearn, E. R.**, Quinones of the Chlorobacteriaceae, *Biochim. Biophys. Acta*, 172, 429, 1969.
7. **Amez, J.**, Plastoquinones in photosynthesis, *Biochim. Biophys. Acta*, 301, 35, 1973.
8. **Smith, L. I., Spillane, L. J., and Kolthoff, I. M.**, The chemistry of vitamin E XXV, *J. Am. Chem. Soc.*, 64, 447, 1942 .
9. **Das, B. D., Lounasmaa, M., Tendille, C., and Lederer, E.**, Mass spectrometry of plastoquinones, *Biochem. Biophys. Res. Commun.*, 21, 318, 1965.
10. **Das, B. D., Lounasmaa, M., Tendille, C., and Lederer, E.**, The structure of PQ B and C, *Biochem. Biophys. Res. Commun.*, 26, 211, 1965.
11. **Morton, R. A.**, Ubiquinones, plastoquinones and vitamins K, *Biol. Rev.*, 46, 47, 1971 .
12. **Clark, W. M.**, *Oxidation-Reduction Potentials of Organic Systems*, Williams & Wilkins, Baltimore, 1960, 376.
13. **Okayama, S., Yamamoto, N., Nishikawa, K., and Horio, T.**, Roles of UQ-10 and rhodoquinone in photosynthetic formation of ATP in *R. rubrum* chromatophores, *J. Biol. Chem.*, 243, 2995, 1968.

Section 2
Assimilation and Dissimilation of Major Elements

CARBON DIOXIDE ASSIMILATION IN PHOTOSYNTHETIC BACTERIA

Bob B. Buchanan

INTRODUCTION

Photosynthetic bacteria appear to assimilate CO_2 via two basically different pathways: (1) the reductive pentose phosphate cycle (C_3 cycle) and its associated reactions, as in green plants, and (2) ferredoxin-linked reactions of the reductive carboxylic acid cycle that are unique to prokaryotic cells. Each of these routes is discussed below.

REDUCTIVE PENTOSE PHOSPHATE CYCLE AND ASSOCIATED REACTIONS

Aside from differences in the properties of specific enzymes of the pathway, the reductive pentose phosphate cycle of bacterial photosynthesis appears to be identical with that of plant photosynthesis. Because evidence for the cycle in plants is summarized elsewhere in this volume, the cycle will not be described here.

The reductive pentose phosphate cycle appears to function as the main path of CO_2 assimilation in two of the three major groups of photosynthetic bacteria, viz., the purple sulfur bacteria (*Chromatium vinosum*) and the purple nonsulfur bacteria (*Rhodospirillum rubrum*),[1-3] but, as described below, not in the third major group of these organisms, viz., the green sulfur bacteria (*Chlorobium thiosulfatophilum*). The reductive pentose phosphate cycle appears to be the main cyclic pathway of CO_2 assimilation in the cyanobacteria (blue-green algae).[1]

In photosynthetic bacteria in which the reductive pentose phosphate cycle is predominant, the main accessory carboxylase is phosphoenolpyruvate carboxylase. In these organisms, phosphoenolpyruvate carboxylase appears to function in the formation of C_4 acids analogous to its role in C_3 plants.

FERREDOXIN-LINKED CO_2 ASSIMILATION

A role for ferredoxin in the assimilation of CO_2 emerged from studies in our laboratory in 1964.[4] Since that time, five different carboxylation reactions have been described. In the discussion below, the ferredoxin-linked carboxylases are collectively referred to as "synthases" and are identified for individual reactions by the name of the α-keto acid product formed. In the equations, reversibility of the reactions is not indicated.

Synthesis of Pyruvate

$$\text{Acetyl-CoA} + CO_2 + \text{ferredoxin}_{\text{reduced}} \xrightarrow[\text{synthase}]{\text{pyruvate}} \text{pyruvate} + \text{CoA-SH} + \text{ferredoxin}_{\text{oxidized}}$$

Pyruvate synthase[4] (pyruvate:ferredoxin oxidoreductase) is widely distributed in anaerobic organisms. It has been found in each of the three main types of photosynthetic bacteria, in various types of fermentative bacteria, and in cyanobacteria.[2,3] In most of these organisms, pyruvate synthase appears to be important in the assimilation of exogenous acetate and CO_2. There is evidence that CO_2 rather than bicarbonate is the active species fixed by pyruvate synthase.[5] Based on inhibition studies with glyoxylate, pyruvate synthase appears to occupy the key position of CO_2 assimilation in the green sulfur bacteria.[6]

Pyruvate synthase has been purified from two photosynthetic bacteria, *C. thiosulfatophilum* and *C. vinosum*, and, following specific treatment to release cofactor from the enzyme, it was shown to require thiamine pyrophosphate.[7] A similar enzyme has been highly purified from the fermentative bacterium *Clostridium acidi-urici* and shown to contain an iron-sulfur chromophore in addition to thiamine pyrophosphate.[8,9] This chromophore appears to couple directly to ferredoxin in both the synthesis and breakdown of pyruvate. It is noteworthy that a corresponding enzyme purified from mixed cultures of rumen microorganisms showed only synthase activity.[10,11] A second enzyme, also partially purified, was required for pyruvate breakdown. These findings raise the possibility that at least some organisms may use one enzyme for pyruvate synthesis and another enzyme for pyruvate breakdown.

Synthesis of α-Ketoglutarate

$$\text{Succinyl-CoA} + CO_2 + \text{ferredoxin}_{\text{reduced}} \xrightarrow[\text{synthase}]{\alpha\text{-ketoglutarate}} \alpha\text{-ketoglutarate} + \text{CoA-SH} + \text{ferredoxin}_{\text{oxidized}}$$

α-Ketoglutarate synthase[12] (α-ketoglutarate:ferredoxin oxidoreductase) occurs in *C. thiosulfatophilum, R. rubrum*, and in a mixed culture of *Prostheocochloris aesturii.*[2,3] α-Ketoglutarate synthase has also been demonstrated in fermentative bacteria, namely those of the rumen,[13,14] but so far there is no evidence for this enzyme in photosynthetic purple sulfur bacteria such as *C. vinosum.*

Apart from its role in the reductive carboxylic acid cycle described below, α-ketoglutarate synthase appears to function in the assimilation of exogenous succinate and CO_2. The principal products formed from succinate and CO_2 by this reaction are amino acids—especially glutamate which is derived directly from α-ketoglutarate by transamination.[2,3]

α-Ketoglutarate synthase has been purified from *C. thiosulfatophilum.*[15] The enzyme at its highest state of purity was free of pyruvate synthase activity and catalyzed the breakdown as well as the synthesis of α-ketoglutarate. Like pyruvate synthase, α-ketoglutarate synthase shows a requirement for thiamine pyrophosphate.

Synthesis of α-Ketobutyrate

$$\text{Propionyl-CoA} + CO_2 + \text{ferredoxin}_{\text{reduced}} \xrightarrow[\text{synthase}]{\alpha\text{-ketobutyrate}} \alpha\text{-ketobutyrate} + \text{CoA-SH} + \text{Ferredoxin}_{\text{oxidized}}$$

α-Ketobutyrate synthase[16] (α-ketobutyrate:ferredoxin oxidoreductase) occurs in both photosynthetic and fermentative bacteria.[10] In these organisms, α-ketobutyrate synthase appears to function in a novel pathway for the biosynthesis of isoleucine and α-aminobutyrate. Several lines of evidence indicate that α-ketobutyrate synthase is a separate enzyme and is not associated with pyruvate synthase.

Synthesis of Phenylpyruvate

$$\text{Phenylacetyl-CoA} + CO_2 + \text{ferredoxin}_{\text{reduced}} \xrightarrow[\text{synthase}]{\text{phenylpyruvate}} \text{phenylpyruvate} + \text{CoA-SH} + \text{ferredoxin}_{\text{oxidized}}$$

Phenylpyruvate synthase,[17] which has been described in green photosynthetic sulfur bacteria, appears to function in the synthesis of aromatic amino acids via a pathway that is independent of the shikimate pathway established for aerobic cells.[2] Phenylpyruvate synthase has not been purified, but evidence suggests that this activity is due to a specific enzyme.

Synthesis of α-Ketoisovalerate

$$\text{Isobutyryl-CoA} + CO_2 + \text{ferredoxin}_{\text{reduced}} \xrightarrow[\text{synthase}]{\alpha\text{-ketoisovalerate}} \alpha\text{-ketoisovalerate} + \text{CoA-SH} + \text{ferredoxin}_{\text{oxidized}}$$

α-Ketoisovalerate synthase[18] was found in cell-free extracts from two different fermentative bacteria. The α-ketoisovalerate formed in this reaction, which is dependent on thiamine pyrophosphate, is converted to valine by transamination. As with the other ferredoxin-linked carboxylation reactions that lead to amino acids, the α-ketoisovalerate synthase mechanism for valine biosynthesis does not involve steps of the pathway previously established for aerobic cells.[2] The presence of α-ketoisovalerate synthase in photosynthetic cells has not been reported.

Synthesis of Formate

$$CO_2 + \text{ferredoxin}_{\text{reduced}} \xrightarrow[\text{reductase}]{\text{carbon dioxide}} \text{formate} + \text{ferredoxin}_{\text{oxidized}}$$

Carbon dioxide reductase[19] (reduced ferredoxin:CO_2 oxidoreductase) was discovered in cell-free extracts of the fermentative bacterium *Clostridium pasteurianum* and has so far not been reported to occur in photosynthetic bacteria. Like pyruvate synthase, the active species fixed by CO_2 reductase is CO_2 rather than bicarbonate.[5] Growth and inhibitor studies suggest that, despite the reversibility of the reaction, CO_2 reductase functions in the synthesis of formate rather than in its degradation[20] and that molybdenum is an essential component of the enzyme.[21] CO_2 reductase is the only known case in which reduced ferredoxin specifically promotes the fixation of CO_2 via a reaction that does not involve an acyl coenzyme derivative.

The Reductive Carboxylic Acid Cycle

The reductive carboxylic acid cycle was proposed in 1966 as a cyclic pathway for the assimilation of CO_2 by the photosynthetic bacteria *C. thiosulfatophilum*[22] and *R. rubrum*.[23] On the basis of the influence of different colors of light on photosynthetic products, it has been suggested that the reductive carboxylic acid cycle functions also in higher plants.[24,25] The confirmation of this proposal awaits a demonstration in leaves of the enzymes associated with the cycle.

The reductive carboxylic acid cycle is in effect a reversal of the oxidative citric acid cycle of Krebs and in one turn yields one molecule of acetyl coenzyme A from two molecules of CO_2 (see Figure 1). Reduced ferredoxin is needed to form (via α-ketoglutarate synthase) α-ketoglutarate, a key intermediate of the cycle. The formation of pyruvate from acetyl coenzyme A and CO_2 by pyruvate synthase is also driven by reduced ferredoxin. The pyruvate formed in this manner is used for a variety of biosynthetic reactions, including the synthesis of amino acids and carbohydrates.[26,27] In both cases, *C. thiosulfatophilum* would use the enzyme pyruvate, P_i dikinase, for the synthesis of phosphoenolpyruvate prior to the formation of sugars by a reversal of glycolysis[28] or of amino acids by carboxylation/transamination reactions.[26] Pyruvate, P_i dikinase, is also found in *C. vinosum* and *R. rubrum*.[28]

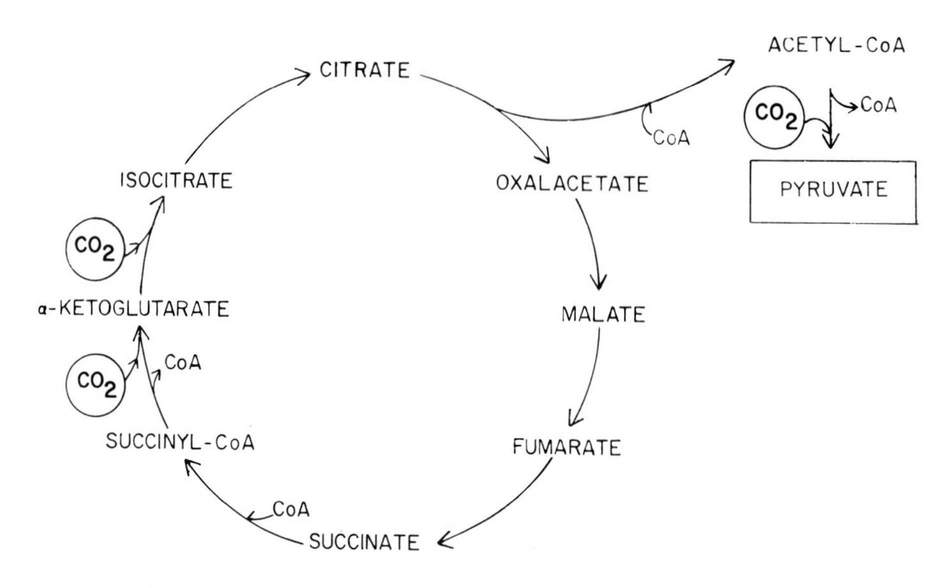

FIGURE 1. Reductive carboxylic acid cycle of bacterial photosynthesis.

Aside from demonstration of the formation of the intermediates of the reductive carboxylic acid cycle in $^{14}CO_2$ short-exposure experiments with whole cells, evidence was presented for the occurrence of the enzymes of, and associated with, the cycle in cell-free extracts of *C. thiosulfatophilum*[22] and *R. rubrum*[23] (acetyl coenzyme A synthetase; pyruvate synthase; pyruvate, P_i dikinase; phosphoenolpyruvate carboxylase; malate dehydrogenase; fumarate hydratase; succinate dehydrogenase; succinyl coenzyme A synthetase; α-ketoglutarate synthase; isocitrate dehydrogenase; aconitate hydratase; and citrate lyase). The presence of citrate lyase in these two organisms has, however, been questioned.[29] As a consequence, we are currently examining evidence for the existence of this enzyme. It is noteworthy in this connection that a nonphotosynthetic methanogenic bacterium was recently reported to contain all of the enzymes of the reductive carboxylic acid cycle except citrate lyase.[30] In this case, it is believed that the acetate feeds into an incomplete reductive carboxylic acid cycle (i.e., ending with α-ketoglutarate formation) for biosynthesis[31] and that this acetate is formed by direct reduction of CO_2.[30]

PATH OF CO_2 ASSIMILATION IN PHOTOSYNTHETIC GREEN BACTERIA

In light of the widely held view that the reductive pentose phosphate cycle (C_3 cycle) is present universally in photosynthetic cells, a comment on the existence of this pathway in the photosynthetic green bacteria seems appropriate. Although there is no unanimity on the issue,[1] there is a growing body of evidence that the green bacteria lack this carbon reduction mechanism that is otherwise considered to be present universally in autotrophic cells. Evidence for this conclusion rests on the absence in these organisms of the two enzymes peculiar to the reductive pentose phosphate cycle (viz., ribulose 1,5-biphosphate carboxylase and phosphoribulokinase)[26,27,32-35] and on $^{12}C/^{13}C$ isotope discrimination studies that indicate that CO_2 assimilation via the cycle is minimal at best.[33,35,36] Thus, in line with recent evidence,[6] it now appears that the ferredoxin-linked carboxylation reactions, in particular pyruvate synthase, constitute the major routes of CO_2 assimilation in this group of photosynthetic organisms.

REFERENCES

1. **McFadden, B.**, Autotrophic CO_2 assimilation and the evolution of ribulose 1,5-diphosphate carboxylase, *Bacteriol. Rev.*, 37, 289, 1973.
2. **Buchanan, B. B.**, Ferredoxin-linked carboxylation reactions, in *The Enzymes*, Vol. 6, 3rd ed., Boyer, P. D., Ed., Academic Press, New York, 1972, 193.
3. **Buchanan, B. B.**, Ferredoxin and carbon assimilation, in *Iron-Sulfur Proteins*, Vol. 1, Lovenberg, W., Ed., Academic Press, New York, 1973, 129.
4. **Bachofen, R., Buchanan, B. B., and Arnon, D. I.**, Ferredoxin as a reductant in pyruvate synthesis by a bacterial extract, *Proc. Natl. Acad. Sci. U.S.A.*, 51, 690, 1964.
5. **Thauer, R. K., Käufer, B., and Fuchs, G.**, The active species of 'CO_2' utilized by reduced ferredoxin:CO_2 oxidoreductase from *Clostridium pasteurianum*, *Eur. J. Biochem.*, 55, 111, 1975.
6. **Quandt, L., Pfennig, N., and Gottschalk, G.**, Evidence for the key position of pyruvate synthase in the assimilation of CO_2 by *Chlorobium*, *FEMS Microbiol. Lett.*, 3, 227, 1978.
7. **Buchanan, B. B., Evans, M. C. W., and Arnon, D. I.**, Ferredoxin-dependent pyruvate synthesis by enzymes of photosynthetic bacteria, in *Non-Heme Iron Proteins: Role in Energy Conversion*, San Pietro, A., Ed., Antioch Press, Yellow Springs, Ohio, 1965, 175.
8. **Uyeda, K. and Rabinowitz, J. C.**, Pyruvate-ferredoxin oxidoreductase. III. Purification and properties of the enzyme, *J. Biol. Chem.*, 246, 3111, 1971a.
9. **Uyeda, K. and Rabinowitz, J. C.**, Pyruvate-ferredoxin oxidoreductase. IV. Studies on the reaction mechanism, *J. Biol. Chem.*, 246, 3120, 1971b.
10. **Bush, R. S. and Sauer, F. D.**, Enzymes of 2-oxo acid degradation and biosynthesis in cell-free extracts of mixed rumen microorganisms, *Biochem. J.*, 157, 325, 1976.
11. **Sauer, F. D., Bush, R. S., and Stevenson, L. L.**, The separation of pyruvate-pyruvate-ferredoxin oxidoreductase from *Clostridium pasteurianum* into two enzymes catalyzing different reactions, *Biochim. Biophys. Acta*, 445, 518, 1976.
12. **Buchanan, B. B. and Evans, M. C. W.**, The synthesis of α-ketoglutarate from succinate and carbon dioxide by a subcellular preparation of a photosynthetic bacterium, *Proc. Natl. Acad. Sci. U.S.A.*, 54, 1212, 1965.
13. **Allison, M. J. and Robinson, I. M.**, Biosynthesis of α-ketoglutarate by the reductive carboxylation of succinate in *Bacteroides ruminicola*, *J. Bacteriol.*, 104, 50, 1970.
14. **Milligan, L. P.**, Carbon dioxide fixing pathways of glutamic acid synthesis in the rumen, *Can. J. Biochem.*, 48, 463, 1970.
15. **Gehring, U. and Arnon, D. I.**, Purification and properties of α-ketoglutarate synthase from a photosynthetic bacterium, *J. Biol. Chem.*, 247, 6963, 1972.
16. **Buchanan, B. B.**, Role of ferredoxin in the synthesis of α-ketogluturate from propionyl coenzyme A and carbon dioxide by enzymes from photosynthetic and nonphotosynthetic bacteria, *J. Biol. Chem.*, 244, 4218, 1969.
17. **Gehring, U. and Arnon, D. I.**, Ferredoxin-dependent phenylpyruvate synthesis by cell-free preparations of photosynthetic bacteria, *J. Biol. Chem.*, 246, 4518, 1971.
18. **Allison, M. J. and Peel, J. L.**, The biosynthesis of valine from isobutyrate by *Peptostreptococcus elsdenii* and *Bacteroides ruminicola*, *Biochem. J.*, 121, 431, 1971.
19. **Jungermann, K., Kirshniawy, H., and Thauer, R. K.**, Ferredoxin dependent CO_2 reduction to formate in *Clostridium pasteurianum*, *Biochem. Biophys. Res. Commun.*, 41, 682, 1970.
20. **Thauer, R. K., Fuchs, G., and Jungermann, K.**, Reduced ferredoxin:CO_2 oxidoreductase from *Clostridium pasteurianum*: its role in formate metabolism, *J. Bacteriol.*, 118, 758, 1974.
21. **Thauer, R. K., Fuchs, G., Schnitker, U., and Jungermann, K.**, CO_2 reductase from *Clostridium pasteurianum*: molybdenum dependence of synthesis and inactivation by cyanide, *FEBS Lett.*, 38, 45, 1973.
22. **Evans, M. C. W., Buchanan, B. B., and Arnon, D. I.**, A new ferredoxin-dependent carbon reduction cycle in a photosynthetic bacterium, *Proc. Natl. Acad. Sci. U.S.A.*, 55, 928, 1966.
23. **Buchanan, B. B., Evans, M. C. W., and Arnon, D. I.**, Ferredoxin-dependent carbon assimilation in *Rhodospirillum rubrum*, *Arch. Mikrobiol.*, 59, 23, 1967.
24. **Punnett, T.**, A complete pathway for C_4 photosynthetic CO_2 fixation occurring in vascular plants, *Fed. Proc. Fed. Am. Soc. Exp. Biol.*, 35, 1597, 1976a.
25. **Punnett, T. and Kelly, J. H.**, Environmental control over C_3 and C_4 photosynthesis in vascular plants, *Plant Physiol.* (Abstr.), 59, 1976.
26. **Buchanan, B. B., Schurmann, P., and Shanmugam, K. T.**, Role of the reductive carboxylic acid cycle in a photosynthetic bacterium lacking ribulose 1,5-diphosphate carboxylase, *Biochim. Biophys. Acta*, 283, 136, 1972.
27. **Sirevåg, R.**, Further studies on carbon dioxide fixation in *Chlorobium*, *Arch. Microbiol.*, 98, 3, 1974.
28. **Buchanan, B. B.**, Orthophosphate requirement for the formation of phosphoenolpyruvate from pyruvate by enzyme preparations from photosynthetic bacteria, *J. Bacteriol.*, 119, 1066, 1974.

29. **Beuscher, N. and Gottschalk, G.,** Lack of citrate lyase — the key enzyme of the reductive carboxylic acid cycle — in *Chlorobium, Z. Naturforsch. Teil B,* 27, 967, 1972.
30. **Fuchs, G. and Stupperich, E.,** Evidence for an incomplete reductive carboxylic acid cycle in *Methanobacterium thermoautotroplicum, Arch. Microbiol.,* 118, 121, 1978.
31. **Fuchs, G., Stupperich, E., and Thauer, R. K.,** Acetate assimilation and the synthesis of alanine, aspartate, and glutamate in *Methanobacterium thermoautotrophicum, Arch. Microbiol.,* 117, 61, 1978.
32. **Buchanan, B. B. and Sirevåg, R.,** Ribulose 1,5-diphosphate carboxylase and *Chlorobium thiosulfatophilum, Arch. Microbiol.,* 109, 15, 1976.
33. **Quandt, L., Gottschalk, G., Ziegler, H., and Stichler, W.,** Isotope discrimination by photosynthetic bacteria, *FEMS Microbiol Lett.,* 1, 125, 1977.
34. **Takabe, T. and Akazawa, T.,** A comparative study of the effect of O_2 on photosynthetic carbon metabolism by *Chlorobium thiosulfatophilum* and *Chromatium vinosum, Plant Cell Physiol.,* 18, 753, 1977.
35. **Bondar, V. A., Gogotova, G. I., and Ziakum, A. M.,** Fractionation of carbon isotopes by photoautotrophic microorganisms having different pathways of carbon dioxide assimilation, *Dokl. Acad. Nauk. SSSR,* 228, 223, 1976.
36. **Sirevåg, R., Buchanan, B. B., Berry, J. A., and Troughton, J. H.,** Mechanisms of CO_2 fixation in bacterial photosynthesis studied by the carbon dioxide isotope fractionation technique, *Arch. Microbiol.,* 112, 35, 1977.

PHOTOSYNTHETIC CARBON DIOXIDE ASSIMILATION VIA THE REDUCTIVE PENTOSE PHOSPHATE CYCLE (C_3 CYCLE)

James A. Bassham

All photosynthetic green plants and cyanobacteria (blue-green algae) capable of using sunlight to convert CO_2 and H_2O to O_2 and carbohydrates employ the reductive pentose phosphate cycle (also called the C_3 cycle or the Calvin cycle).[1-3] Most available evidence indicates that the C_3 cycle is the principal pathway for initial fixation of CO_2 in all plants except those which have either the C_4 pathway[4,5] or the Crassulacean acid metabolism (CAM)[6] pathway or both. Even in those species in which most of the CO_2 is initially incorporated by C_4 metabolism or by CAM, most of the CO_2 is released inside chloroplasts where it is reincorporated via the C_3 cycle.[7] This is necessary, since neither the C_4 cycle nor CAM can bring about a net conversion of CO_2 to sugar phosphate or carbohydrates.

The C_3 cycle can be considered as consisting of three phases (see Figure 1). The first phase is the conversion of a C_5 sugar, ribulose 5-phosphate, with ATP to ribulose 1,5-bisphosphate (Reaction M), followed by carboxylation to give two molecules of the C_3 acid, 3-phosphoglycerate (3-PGA), in Reaction A. The second phase is the conversion of 3-PGA to triose phosphate which occurs in two steps: formation of the acyl phosphate with ATP (Reaction B) and reduction of the acyl phosphate to aldehyde with release of inorganic phosphate (Reaction C). The third phase of the RPP cycle consists of all the remaining reactions (D through L) which convert five molecules of C_3 sugar phosphate to three molecules of pentose phosphate.

One complete cycle (in which each reaction occurs at least once) requires that three ribulose 5-phosphate molecules are phosphorylated, carboxylated, and split to give a total of six molecules of 3-PGA which are in turn converted to six molecules of the triose phosphate, glyceraldehyde-phosphate. Since only five triose phosphate molecules are required to regenerate the three pentose phosphates, one triose phosphate is left over and represents the product of the fixation and reduction of three molecules of CO_2. The overall reaction of one complete cycle may therefore be represented as

$$3CO_2 + 9ATP^{4-} + 6NADPH \longrightarrow 9ADP^{3-} + \underset{\text{(glyceraldehyde-3-phosphate)}}{OCH{-}CHOH{-}CH_2OPO_3H^-}$$

$$+ 8\,HPO_3^{2-} + 2H^+$$

ATP is regenerated from ADP and P_i by photophosphorylation in the chloroplast thylakoids. NADPH is regenerated from $NADP^+$ via reduced ferredoxin which is in turn reduced by electrons from water generated in the photochemical and electron transport reactions in thylakoids.

The triose phosphate product of the C_3 cycle is to a large extent translocated out of the chloroplast (in exchange for inorganic phosphate)[8] to be used for various biosynthetic needs of the plant, including synthesis of sucrose for export to other parts of the plant in the case of multicellular plants. Under conditions where photosynthesis exceeds the needs of the plant, some of the triose phosphate is converted to fructose-6-phosphate (via reactions D, E, and F), then to glucose phosphates, and finally to the storage product, starch, which accumulates in the chloroplasts. Some biosynthetic reactions starting with intermediate compounds of the C_3 cycle also occur in the chloroplasts. For example, triose phosphate may be reduced to glycerol phosphate and used in fat synthesis, and ribose 5-phosphate can be used in ribonucleotide and deoxyribonucleotide biosynthesis.

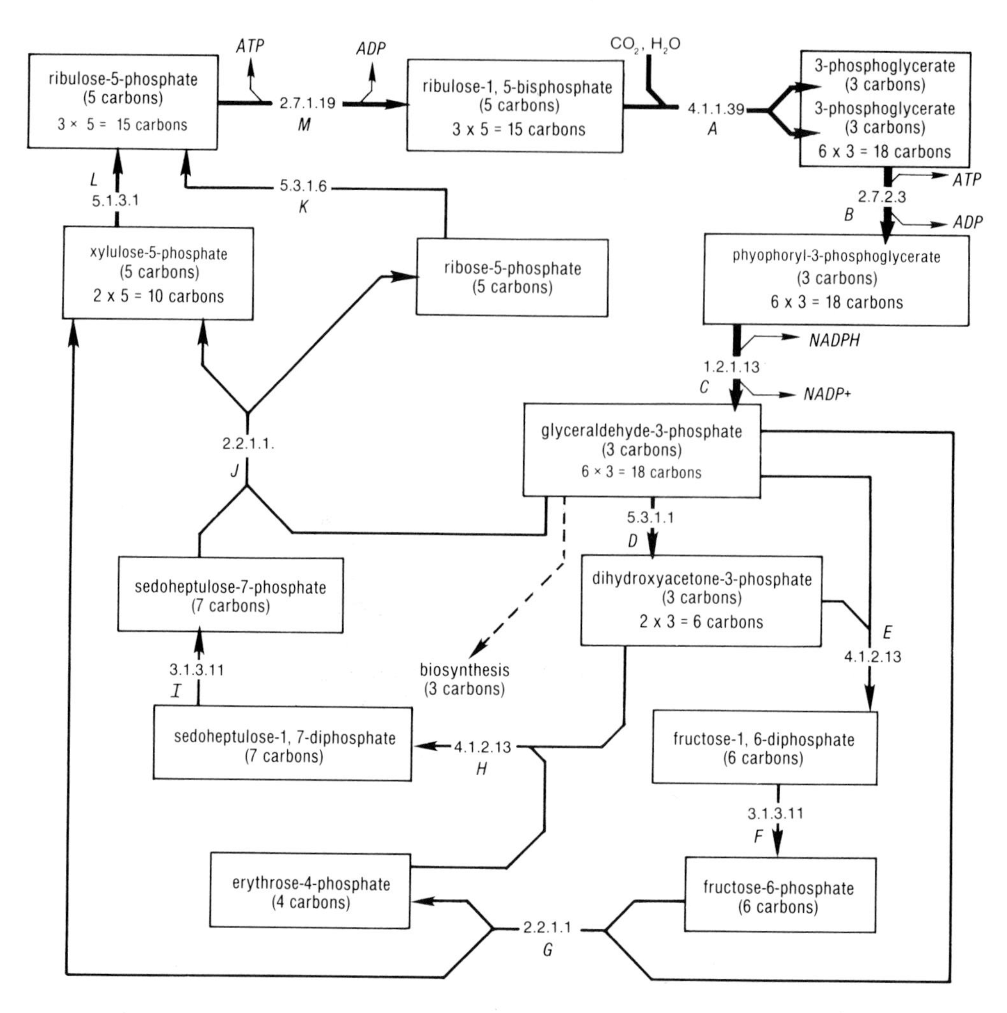

FIGURE 1. Reductive pentose phosphate cycle of photosynthesis.

Reactions

A, Ribulose 1,5-bisphosphate adds CO_2 at C-2 and splits hydrolytically to give 2 molecules of 3-phosphoglycerate; B, The carboxyl group of 3-phosphoglycerate is converted to an acyl phosphate in a reaction utilizing the terminal phosphate of ATP; C, The acyl phosphate group is reduced in the presence of NADPH to give P_i and an aldehyde group, thus converting the 3-C-atom compound to glyceraldehyde 3-phosphate; D, The aldo-sugar is converted into a keto-sugar by the transfer of 2 H atoms from C-2 to C-1; E, The aldol condensation between aldotriose and ketotriose gives fructose 1,6-bisphosphate; F, The phosphate group on C-1 is removed by hydrolysis; G, 2 H atoms plus the glycolyl group (C-1 and C-2) of fructose 6-phosphate are transferred to glyceraldehyde 3-phosphate to form xylulose 5-phosphate, leaving erythrose 4-phosphate; H, An aldol condensation between erythrose 4-phosphate and ketotriose gives sedoheptulose 1,7-bisphosphate; I, Hydrolysis of the phosphate on C-1 gives sedoheptulose 7-phosphate; J, Transfer of the glycolyl group (C-1 and C-2) plus 2 H atoms from sedoheptulose 7-phosphate to C-1 of glyceraldehyde 3-phosphate gives xylulose 5-phosphate and ribose 5-phosphate; K, Isomerization of ribose 5-phosphate gives ribulose 5-phosphate; L, Epimerization of C-3 of xylulose 5-phosphate gives ribulose 5-phosphate; M, Phosphorylation of C-1 of ribulose 5-phosphate in the presence of ATP gives ribulose 1,5-bisphosphate, thus completing the reductive pentose phosphate cycle.

Enzyme (synonym) key

Reductive pentose phosphate cycle (photosynthesis)—2.7.1.19 = Phosphoribulokinase; 3.1.3.11 = Hexosebisphosphatase (heptosebisphosphatase); 4.1.1.39 = Ribulosebisphosphate carboxylase; 4.1.2.13 = Fructosebisphosphate aldolase; 2.2.1.1 = Transketolase; 2.7.2.3 = Phosphoglycerate kinase; 5.1.3.1 = Ribulosephosphate 3-epimerase; 5.3.1.1 = Triosephosphate isomerase; 5.3.1.6 = Ribosephosphate isomerase; 1.2.1.13 = Triosephosphate dehydrogenase (NADP specific).

The most distinctive reaction of the C_3 cycle is the carboxylation of ribulose 1,5 bisphosphate to give two molecules of 3-PGA.[9] The enzyme, ribulose 1,5-bisphosphate carboxylase (RuBPCase) constitutes a major fraction of the soluble protein in green cells and may be the most abundant protein in the biosphere. It is generally accepted that the enzyme mediates the addition of CO_2 to the carbon atom 2 of the RuBP molecule,[10] forming an enzyme-bound 6-carbon intermediate. This undergoes an internal oxidation-reduction reaction (the enzyme was once called "carboxydismutase"),[11] so that hydrolytic splitting of the molecule results in the formation of two identical molecules of 3-PGA. One 3-PGA molecule is composed of carbon atoms 1 and 2 of the RuBP plus the newly incorporated CO_2 which becomes the carboxyl group. The other 3-PGA product is made from carbons 3, 4, and 5 of RuBP, with carbon 3 becoming the carboxyl group.

An interesting and important property of the RuBPCase is its ability to function as an oxygenase.[12,13] At low levels of CO_2 and atmospheric levels of O_2 (21%), O_2 can bind competitively at the active site of the enzyme and react with the RuBP at the carbon 2 position, oxidatively splitting it to one molecule of 3-PGA (from carbon atoms 3, 4, and 5) and one molecule of 2-phosphoglycolate. The latter compound is converted to glycolate in the chloroplast[14] from which it is mainly exported. Glycolate is widely believed to be the principal if not the exclusive substrate for photorespiration.[15] This alternate oxygenase activity of RuBPCase thus competes with carbon assimilation and at the same time leads to conversion of some of the reduced carbon back to CO_2, via photorespiration. This limits the maximum rate of photosynthetic carbon assimilation under conditions of high light intensity, high temperature, low CO_2, and atmospheric O_2 level in plants lacking C_4 metabolism.

REFERENCES

1. **Bassham, J. A., Benson, A. A., Kay, L. D., Harris, A. Z., Wilson, A. T., and Calvin, M.,** The path of carbon in photosynthesis. XXI. The cyclic regeneration of carbon dioxide acceptor, *J. Am. Chem. Soc.,* 76, 1760, 1954.
2. **Bassham, J. A. and Calvin, M.,** *The Path of Carbon in Photosynthesis,* Prentice-Hall, Inc., Englewood Cliffs, N.J., 1957.
3. **Norris, L., Norris, R. E., and Calvin, M.,** A survey of the rates and products of short-term photosynthesis in plants of nine phyla, *J. Exp. Bot.,* 6, 64, 1955.
4. **Kortschak, H. P., Hartt, C. E., and Burr, G. O.,** Carbon dioxide fixation in sugarcane leaves, *Plant Physiol.,* 40, 209, 1965.
5. **Hatch, M. D. and Slack, C. R.,** Photosynthesis by sugar-cane leaves, *Biochem. J.,* 101, 103, 1966.
6. **Osmond, C. B.,** CO_2 assimilation and dissimilation in the light and dark in CAM plants, in *CO_2 Metabolism and Plant Productivity,* Burris, R. H. and Black, C. C., Eds., University Park Press, Baltimore, 1975, 217.
7. **Black, C. C., Jr.,** Photosynthetic carbon fixation in relation to net CO_2 uptake, *Annu. Rev. Plant Physiol.,* 24, 253, 1973.
8. **Heldt, H. W. and Rapley, L.,** Unspecific permeation and specific uptake of substances in spinach chloroplasts, *FEBS Lett.,* 10, 143, 1970.
9. **Weissback, A., Horecker, B. L., and Hurwitz, J.,** The enzymatic formation of phosphoglyceric acid from ribulose diphosphate and carbon dioxide, *J. Biol. Chem.,* 218, 795, 1956.
10. **Mulhoffer, G. and Rose, I. A.,** The position of carbon-carbon bond cleavage in the ribulose diphosphate carboxydismutase reaction, *J. Biol. Chem.,* 240, 1341, 1965.
11. **Quale, R. R., Fuller, R. C., Benson, A. A., and Calvin, M.,** Enzymatic carboxylation of ribulose diphosphate, *J. Am. Chem. Soc.,* 76, 3610, 1954.
12. **Bowes, G., Ogren, W. L., and Hageman, R. H.,** Phosphoglycolate production catalyzed by ribulose diphosphate carboxylase, *Biochem. Biophys. Res. Commun.,* 45, 716, 1971.
13. **Lorimer, G. H., Andrews, T. J., and Tolbert, N. E.,** Ribulose diphosphate oxygenase. II. Further proof of reaction products and mechanism of action, *Biochemistry,* 12, 18, 1973.
14. **Richardson, K. E. and Tolbert, N. E.,** Phosphoglycolic acid phosphatase, *J. Biol. Chem.,* 236, 1285, 1961.
15. **Tolbert, N. E.,** Microbodies—peroxisomes and glyoxysomes, *Annu. Rev. Plant Physiol.,* 22, 45, 1971.

PHOTOSYNTHETIC CARBON DIOXIDE ASSIMILATION VIA THE C_4 PATHWAY

M. D. Hatch

Most photosynthetic organisms assimilate atmospheric CO_2 directly into 3-phosphoglyceric acid via the enzyme ribulose 1,5-bisphosphate (RuBP) carboxylase and reduce this carbon through a sequence of reactions termed the Photosynthetic carbon reduction cycle (C_3 or PCR cycle). Relatively recently in evolutionary terms there has evolved among certain families of flowering plant species (angiosperms) a highly modified form of photosynthesis known as the C_4 pathway. The C_3 cycle still operates in these plants, but instead of fixing CO_2 directly from the atmosphere, the C_3 cycle is provided with metabolically generated CO_2 by a complex additional metabolic cycle. During the operation of this cycle, CO_2 is assimilated into C_4 acids in mesophyll cells through the carboxylation of phosphoenolpyruvate (PEP). These acids then move to bundle sheath cells where their decarboxylation provides CO_2 at high concentrations for assimilation by the C_3 cycle which is confined exclusively to these cells.

Figure 1 demonstrates the essential metabolic features of the C_4 pathway and their relationship to the specialized mesophyll and bundle sheath cell types that are unique to C_4 plants. The main points to note are the carboxylation reaction in mesophyll cells, the decarboxylation process in bundle sheath cells, and return of the residual C_3 compound from bundle sheath to mesophyll cells where conversion to phosphoenolpyruvate completes the cycle. This sequence of events is common to all C_4 plants, but the exact mechanisms of C_4 acid decarboxylation, in particular, have proved to be surprisingly variable in different species.

Three decarboxylating enzymes have been shown to operate in different species: a chloroplast NADP malic enzyme, a cytoplasmic PEP carboxykinase, and a mitochondrial NAD malic enzyme. Detailed reaction sequences for these alternative processes are described in Figure 2, where the terms NADP-ME-type, PCK-type, and NAD-ME-type refer, respectively, to the reaction sequences involving the three enzymes referred to above.

The legend to Figure 2 lists the enzymes responsible for particular reactions. The sequence of reactions are as follows: CO_2 is assimilated in mesophyll cells by carboxylation of PEP, giving oxaloacetate which is either reduced to malate in chloroplasts or aminated to aspartate. Where the NADP-ME-type mechanism operates, malate is decarboxylated in bundle sheath chloroplasts to yield CO_2 (which is fixed by the C_3 cycle), NADPH (which is reoxidized by coupling to the reductive step of the C_3 cycle), and pyruvate (which is returned to mesophyll cells). For the PCK-type mechanism, aspartate is the major C_4 acid metabolized; it is transformed to oxaloacetate which is then decarboxylated in bundle sheath cytoplasm in a PEP-carboxykinase-mediated reaction to give PEP and ADP. The fate of PEP is uncertain. There is evidence favoring the view that PEP is initially transformed to alanine, but some may be returned directly to mesophyll cells. In the case of the NAD-ME-type mechanism, aspartate derived from mesophyll cells is transformed in mitochondria to oxaloacetate. The oxaloacetate is reduced to malate which then undergoes oxidative decarboxylation to yield CO_2, NADH (which serves as continuing reducing source for malate formation), and pyruvate (which is transformed to alanine in a reaction coupled to the aspartate/oxaloacetate conversion through cycling of glutamate/oxaloacetate). Where alanine is returned to mesophyll cells, it is initially converted to pyruvate which is then transformed to the primary CO_2-acceptor PEP via pyruvate, $P_i$dikinase. This complex reaction also produces AMP and PP_i which are further processed via adenylate kinase and pyrophosphatase, respectively.

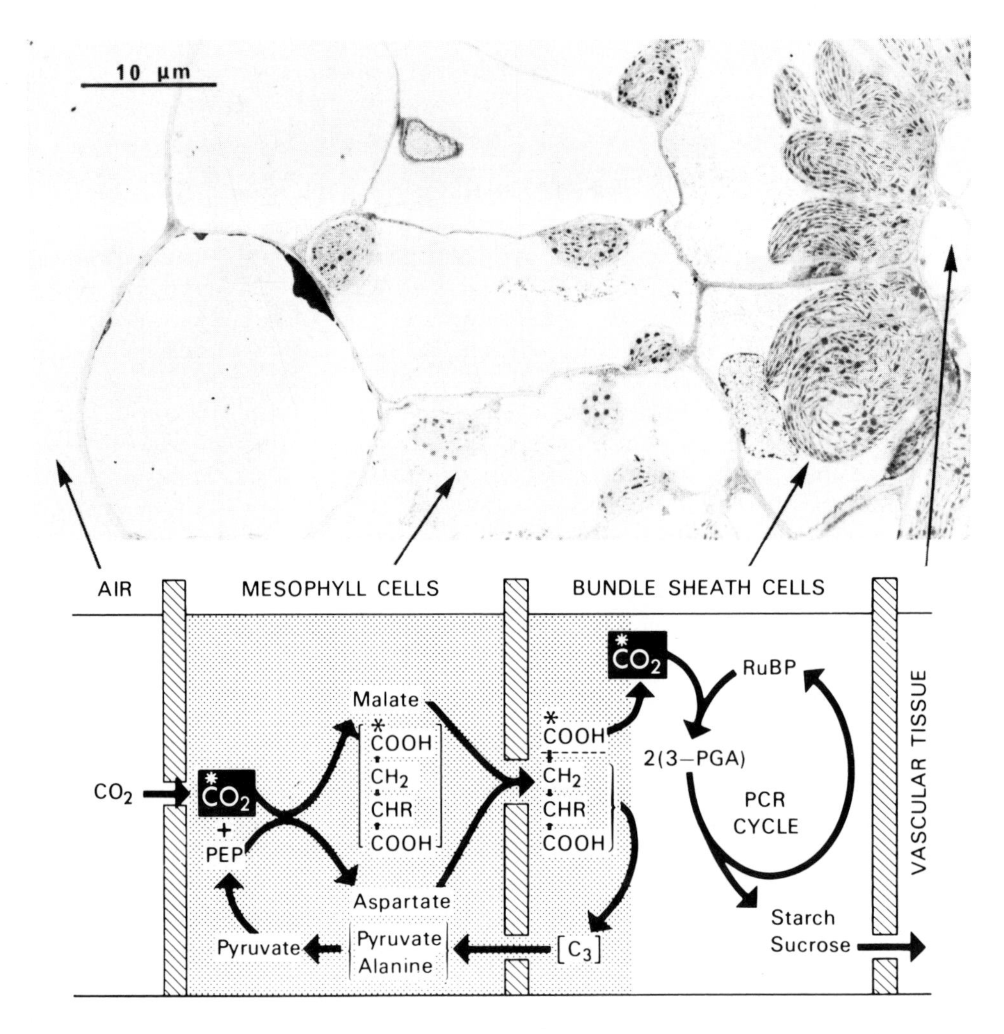

FIGURE 1. Basic steps of C_4 photosynthesis and the intercellular location of these reactions shown in relation to the arrangement of mesophyll and bundle sheath cells in an electron micrograph of *Panicum miliaceum*. PCR cycle = C_3 cycle. (From Hatch, M.D., *Current Topics in Cellular Regulation*, Vol. 14, Horecker, B. L. and Stadtman, E. R., Eds., Academic Press, New York, in press.

As shown in summary form in Table 1, most C_4 species can be clearly classified on the basis of their C_4 acid decarboxylase content.

However, some other species apparently contain significant levels of more than one decarboxylase. Furthermore, even the complex schemes illustrated in Figure 2 for metabolism of C_4 acids in bundle sheath cells may be an oversimplification of the processes operating in some species. For instance, isolated bundle sheath cells from some NADP-ME-type species decarboxylate aspartate at substantial rates, and the bundle sheath cells from some PCK-type species rapidly decarboxylate malate through PEP carboxykinase.

Many hundreds of C_4 species are now known, and these come from 14 dicotyledon and 2 monocotyledon families. These species appeared quite recently even in terms of angiosperm evolution, and, surprisingly, the process has apparently developed quite separately on many occasions; it may have even evolved separately within different genera of some families.

The C_4 pathway modification appears to function as a mechanism for concentrating

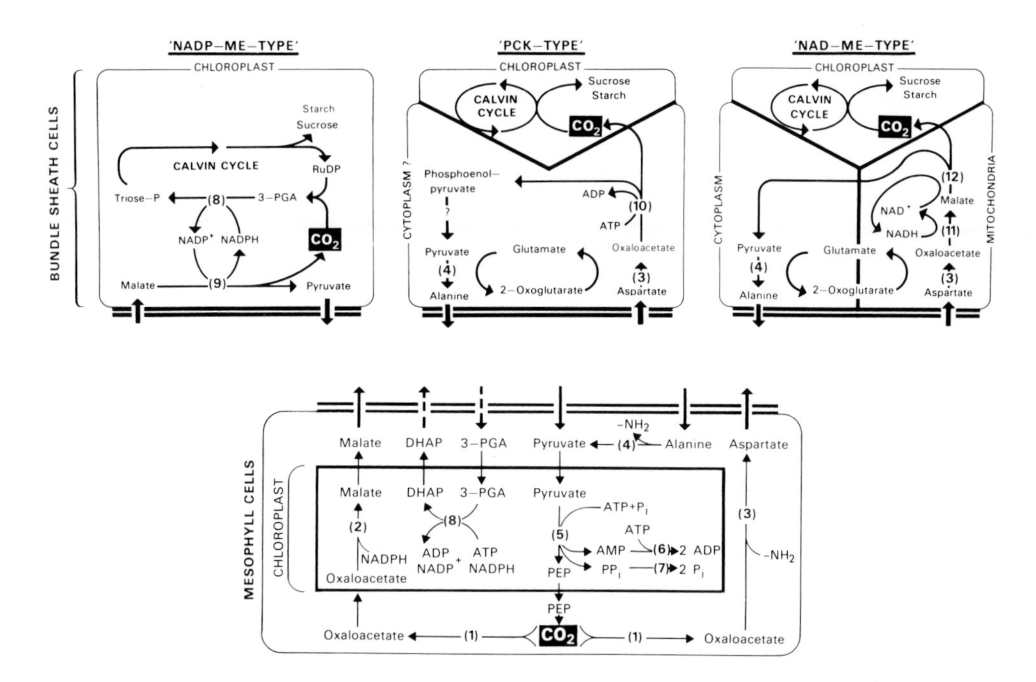

FIGURE 2. Reactions of the C_4 pathway and their inter- and intracellular location. The basis for defining three separate schemes for bundle sheath cell metabolism is explained in the text. The enzymes involved are (1) PEP carboxylase (EC.4.1.1.31); (2) NADP malate dehydrogenase (EC.1.1.1.82); (3) aspartate aminotransferase (EC.2.6.1.2); (5) pyruvate, P_i dikinase (EC.2.7.9.1); (6) adenylate kinase (EC.4.6.1.1); (7) pyrophosphatase (EC.3.6.1.1); (8) 3-PGA kinase (EC.2.7.2.3) and NADP glyceraldehyde-3-P dehydrogenase (EC.1.2.1.13); (9) NADP malic enzyme (EC.1.1.1.40); (10) PEP carboxykinase (EC.4.1.1.49); (11) NAD malate dehydrogenase (EC.1.1.1.37); and (12) NAD malic enzyme (EC.1.1.1.39). Calvin cycle = C_3 cycle. (From Hatch, M.D., *Current Topics in Cellular Regulation,* Vol. 14, Horecker, B. L. and Stadtman, E. R., Eds., Academic Press, New York, in press. With permission.)

CO_2 in bundle sheath cells where the C_3 cycle operates. The combination of biochemical and anatomical features characteristic of C_4 species explains the unique physiological features of these plants, including their higher potential for photosynthesis, higher water-use efficiency, and reduced photorespiration.

Recent detailed accounts of the C_4 pathway can be found in the references cited.

Table 1
ACTIVITIES OF C_4 ACID DECARBOXYLASES AND OTHER C_4 PATHWAY ENZYMES IN SUBGROUPS OF C_4 PLANTS RELATIVE TO ACTIVITIES IN C_3 CYCLE PLANTS[3,6]

Subgroup of C_4 plants	Enzymes with similar activities in each subgroup	Average ratio of activity in C_4 plants/C_3 cycle plants					
		NADP malic enzyme	NADP malate dehydrogenase	PEP carboxykinase	NAD malic enzyme	Aspartate aminotransferase	Alanine aminotransferase
NADP-ME-type (6 species)	PEP carboxylase (30×); pyruvate, P_i dikinase (α); AMP kinase and PPase (>20×) PCR cycle enzymes (similar to activities in C_3 cycle plants)	35	18	1	1[a]	3	1.5
PCK-type (8 species)		1	2	>60	5[b]	15	14
NAD-ME-type (15 species)		1	1	1	40	12	16

[a] Corrected for NAD-dependent activity not activated by CoA or fructose 1,6-bisphosphate and mostly because of secondary activity of NADP malic enzyme.[4]

[b] Higher NAD malic enzyme activity has been reported for some other PCK-type species.[5]

REFERENCES

1. **Burris, R. H. and Black, C. C.,** *Carbon Dioxide Metabolism and Plant Productivity,* University Park Press, Baltimore, 1976.
2. **Hatch, M. D.,** Photosynthesis: the path of carbon, in *Plant Biochemistry,* Bonner, J. and Varner, J., Eds., Academic Press, New York, 1976, 797.
3. **Hatch, M. D. and Osmond, C. B.,** Compartmentation and transport in C_4 photosynthesis, in *Encyclopedia of Plant Physiology, New Series,* Vol. 3, Heber, U. and Stocking, C. R., Eds., Springer-Verlag, Heidelberg, 1976, 144.
4. **Hatch, M. D.,** Regulation of enzymes in C_4 photosynthesis, in *Current Topics in Cellular Regulation,* Vol. 14, Horecker, B. L. and Stadtman, E. R., Eds., Academic Press, New York, 1978, 1—28.
5. **Edwards, G. E. and Walker, D. A.,** *C_3 and C_4: An Introduction to the Biochemistry of Photosynthesis,* Packard, Sussex, in press.
6. **Oliver, I. T. and Hatch, M. D.,** unpublished data.

CRASSULACEAN ACID METABOLISM

Clanton C. Black, Jr.

Crassulacean acid metabolism (CAM) plants are distinguished from other higher plants by their ability to fix large quantities of CO_2 at night via PEP carboxylase, resulting in the night accumulation of malic acid. The next day, malic acid furnishes CO_2 for the light-dependent operation of the reductive pentose phosphate cycle (the C_3 cycle). In addition to these distinctive activities, CAM plants are characterized by the features listed in Table 1.

The adaptative biochemical and anatomical features listed in Table 1 result in CAM plants being very efficient in utilizing water. Thus, CAM plants are an adaptation to arid or dry environments. For example, the cactus in deserts or spanish moss hanging in trees in the southeastern U.S. are CAM plants. This contribution will center on CO_2 fixation in CAM plants, but discussions and extensive literature citations for each feature in Table 1 are available elsewhere.[1-5]

A general pattern of daily CO_2 fixation measured in CAM plants is shown in Figure 1. Clearly, most net CO_2 fixation occurs at night, and concurrently the green tissue accumulates acid. The small CO_2 fixation early and late in a day may occur under well-watered growth conditions, but under dry conditions no day CO_2 fixation will occur. Acid accumulation may be measured by titration with base over 24 hr, and the malic acid (titratable acidity) pattern shown in Figure 1 for the whole green tissue will be obtained. Similarly, one can measure the starch or glucan content and a reciprocal pattern relative to titratable acid or malic acid content will be obtained.

The participation of starch as a major component of CAM can be calculated from Figure 1 by knowing that the tissue is near 90% H_2O.[6] The daily change of 17 g of starch (see Figure 1) is nearly 17% of the total composition of the plant, excluding H_2O. Thus, each day, green CAM tissues devote a major portion of their organic constituents to the synthesis of malic acid from starch and to the synthesis of starch from pyruvic acid on PEP. This diurnal cycle is a dominant feature of CAM, and it furnishes most of the CO_2 each day for photosynthesis.

On a cellular level, the biochemical details of CAM are outlined in Figure 2. CO_2 is fixed at night by PEP carboxylase to form oxalacetate which is reduced to malate. Malic acid is stored in the large vacuole at night, removed the next day, and decarboxylated to furnish CO_2 for C_3 photosynthesis in the chloroplast. Otherwise, as far as is known today, the C_3 cycle operates in CAM plants quite similarly to that given by Bassam in this volume.

The depletion of the malic acid via a decarboxylation each day occurs through the action of either PEP carboxykinase or $NADP^+$ malic enzyme in specific CAM plants.[2-5] An NAD^+ malic enzyme is known to be present in some CAM plants which also contain high activities of either PEP carboxykinase or $NADP^+$ malic enzyme; however, the role of NAD^+ malic enzyme is unknown. These decarboxylases are similar to those described by Hatch in this volume.

CAM plants exhibit one other unique feature when compared to other higher plants. CAM plants can change the amount of CO_2 fixed from the air via PEP or RuBP carboxylase (see Table 1, Item 7). This response to environmental changes such as water status is not exhibited by other plants. The most direct measurement of this feature is the variable $\delta^{13}C$ values found with a given CAM plant grown in various environments.[4-5,7]

CAM is known to be widely distributed taxonomically, since it has been found in some 18 families of flowering plants and in *Welwitschia mirabilis*, a gymnosperm, and in two ferns.[3]

Table 1
ADAPTATIVE FEATURES OF CAM PLANTS

1. Major quantities of CO_2 fixed at night in green cells
2. Major utilization of starch or glucan supply at night in green cells to form the CO_2 acceptor, phosphoenolpyruvate
3. Formation of a large storage pool of malic acid at night in green cells which is depleted the next day to furnish CO_2 for C_3 photosynthesis inside the green tissue
4. The features above combined in succulent green cells (Succulence refers to green cells with large vacuoles.)
5. Ability to reduce stomatal apertures as a standard function in daylight
6. Plants with the lowest known H_2O requirement for dry matter production ranging from 18 to 155 g of H_2O transpired per g of dry matter produced (Therefore CAM is an adaptation to arid or otherwise dry habitats.)
7. Ability to shift the relative amounts of atmospheric CO_2 assimilated either via PEP carboxylase or RuBP carboxylase in response to environmental changes

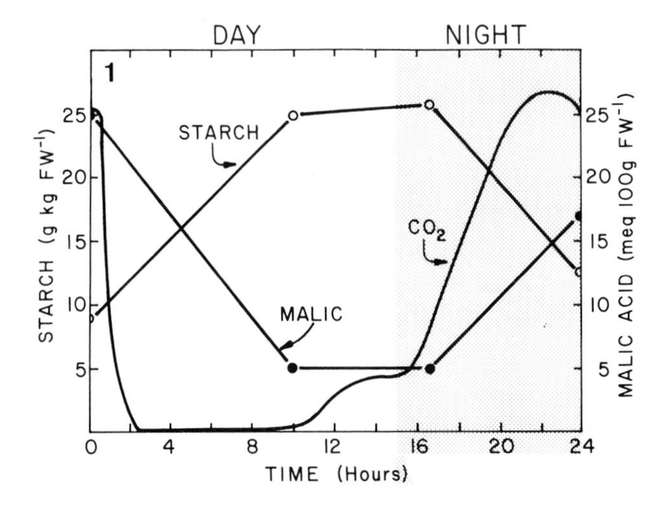

FIGURE 1. Diurnal changes in malic acid and starch content of *Bryophyllum* leaves[6] and a general curve of net CO_2 uptake over a day for a well-watered CAM plant.

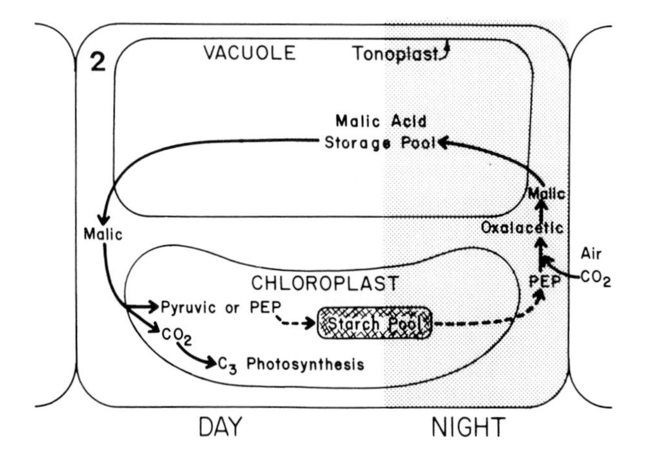

FIGURE 2. Intercellular localization of various processes involved in CAM within a green cell over 24 hr.

In summary, most CO_2 fixation in CAM plants proceeds by the night fixation of CO_2 and its storage as malic acid in the vacuole of green cells. The next day, the malic acid furnishes CO_2 for photosynthetic CO_2 fixation by the C_3 cycle. This temporal separation of CO_2 metabolism within an individual green cell distinguishes CAM CO_2 fixation from C_4 cycle CO_2 fixation which occurs only during the day and which involves the spatial separation of β-carboxylation in green mesophyll cells from decarboxylation followed by the carboxylation of RuBP in green bundle sheath cells.[2]

REFERENCES

1. **Ranson, S. L. and Thomas, M.,** Crassulacean acid metabolism, *Annu. Rev. Plant Physiol.,* 11, 81, 1960.
2. **Black, C. C.,** Photosynthetic carbon fixation in relationship to net CO_2 uptake, *Annu. Rev. Plant Physiol.,* 24, 253, 1973.
3. **Burris, R. H. and Black, C. C.,** *Carbon Dioxide Metabolism and Plant Productivity,* University Park Press, Baltimore, 1976.
4. **Osmond, C. B.,** Crassulacean acid metabolism: a curiosity in context, *Annu. Rev. Plant Physiol.,* 29, 379, 1978.
5. **Kluge, M. and Ting, I. P.,** *Crassulacean Acid Metabolism (Analysis of an Ecological Adaptation),* Springer-Verlag, Berlin, 1978.
6. **Pucher, G. W., Vickery, H. B., Abrahams, M. D., and Leavenworth, C. S.,** Studies in the metabolism of crassulacean plants: diurnal variation in organic acids and starch in excised plants *Bryophyllum calycinum, Plant Physiol.,* 24, 610, 1949.
7. **Bender, M. M., Rouhani, I., Vines, H. M., and Black, C. C.,** $^{13}C/^{12}C$ ratio changes in Crassulacean acid metabolism plants, *Plant Physiol.,* 52, 427, 1973.

RIBULOSE 1,5-BISPHOSPHATE CARBOXYLASE FROM PLANTS AND PHOTOSYNTHETIC BACTERIA

Louise E. Anderson

The enzyme ribulose 1,5-bisphosphate (RuBP) carboxylase catalyzes the conversion of one molecule of ribulose 1,5-bisphosphate and one molecule of CO_2 to two molecules of 3-phosphoglyceric acid (see Equation 1).

$$\begin{array}{l} CH_2O(P) \\ | \\ C{=}O \\ | \\ CHOH \\ | \\ CHOH \\ | \\ CH_2O(P) \end{array} + CO_2 \rightarrow 2 \begin{array}{l} COOH \\ | \\ CHOH \\ | \\ CH_2O(P) \end{array} \qquad (1)$$

In eucaryotic green plants, this enzyme is localized in the chloroplast.[1,2] In higher plants, eucaryotic algae, and some procaryots, it consists of eight large and eight small subunits. Smaller carboxylases lacking small subunits are found in some other photo- and chemosynthetic procaryots (see Table 1).

Carbon dioxide is the active substrate for the enzyme[3] and also acts, with Mg^{++}, as an activator.[4] The pH dependence curve for the enzyme catalyzed reaction is very broad, with a maximum at 8.2.[5] The K_m for CO_2 for the spinach enzyme is about 20 μM.[5,6] Most published pH dependency curves and K_m values for CO_2 reflect both activation and catalysis.[4]

The enzyme requires a divalent cation such as Mg^{++} for activity (see Tables 2 and 3). The one exception is the *Rhodospirillum rubrum* enzyme which, when isolated by the method of Anderson and Fuller[7] with $MnCl_2$ precipitation of nucleic acids, is active in the absence of added divalent cations.

K_m values for RuBP range from 15 to 250 μM (see Table 2).

Oxygen is a potent inhibitor and an alternative substrate for the enzyme. In this case, one molecule of 2-phosphoglycolate and one of 3-phosphoglycerate is formed from one molecule of RuBP and one molecule of O_2 (Equation 2).

$$\begin{array}{l} CH_2O(P) \\ | \\ C{=}O \\ | \\ CHOH \\ | \\ CHOH \\ | \\ CH_2O(P) \end{array} + O_2 \rightarrow \begin{array}{l} COOH \\ | \\ CH_2O(P) \end{array} + \begin{array}{l} COOH \\ | \\ CHOH \\ | \\ CH_2O(P) \end{array} \qquad (2)$$

The pH dependency curve for this reaction is very broad.[5] K_m for O_2 is 0.22 mM for the spinach enzyme.[5] Carbon dioxide is an activator for the enzyme when it acts as an oxygenase[8,9] and a competitive inhibitor with respect to O_2.[10,11]

Oxygenase activity has been demonstrated for the enzyme from *Spinacea oleracea,*[12] *Glycine max,*[10,13] *Hordeum vulgare,*[14] *Nicotiana tabacum,*[15] *Zea mays,*[13,16] *Panicum milioides,*[14] *Panicum maximum,*[14] *Atriplex spongiosa,*[17] *Kalanchoe daigremontiana,*[17] *Chlorella fusca,*[18,19] *Chlamydomonas reinhardi,*[20] *Euglena gracilis,*[21] *Chromatium,*[22] *R. rubrum,*[23,24] and *Hydrogenomas eutropha.*[25,26] Carbon dioxide to oxygen fixation ratios seem to be the same for the enzyme from organisms ranging from chemolitho-

Table 1
COMPOSITION OF RIBULOSE 1,5-BISPHOSPHATE CARBOXYLASE IN PHOTO- AND CHEMOSYNTHETIC SPECIES

	Mol			Number of subunits		Ref.
Source of enzyme	Native enzyme	Large subunit	Small subunit	Large	Small	
Higher plants						
Spinacea oleracea	475,000—559,000					44
	557,000					45
		55,800	12,000	8	8—10	46
Beta vulgaris		55,400	12,100			47
Phaseolus vulgaris		55,000	15,300			48
Rumex crispus		55,000—57,000				49
Eucaryotic algae						
Chlorella ellipsoidea	588,000	58,200	15,300	8	6—8	50
Chlorella fusca	530,000	53,000	14,000	8	8	18
Chlamydomonas reinhardtii		55,000	16,500			51
Euglena gracilis	525,000	50,000	15,000			21
Blue-green algae						
Agmenellum quadruplicatum	456,000	56,000		8	0	52
Microcystis aeruginosa	518,000	50,000	14,000	8	8	53
Aphanocapsa 6308	525,000	51,000	15,000	8	8	54
Anabaena cylindrica	449,000	54,000			0	55
Anabaena variabilis		52,000	13,000			56
Plectonema boryanum		54,000	13,000			56
Scenedesmus quadricauda	520,000	51,000	14,000			57
Photosynthetic bacteria						
Chlorobium thiosulfatophilum	361,000	53,000		6	0	58
Chromatium strain D	550,000	57,000	12,000	8	6—8	59
Ectothiorhodospira halophila	601,000	56,000	18,000			60
Rhodopseudomonas palustris	360,000					61
Rhodopseudomonas spheroides	360,000					61
Photosynthetic Bacteria						
R. spheroides	360,000	52,000—55,000			0	62
	550,000	52,000	11,000	8	8	62
Rhodospirillum rubrum	120,000					7
	114,000	56,000		2	0	63
Chemosynthetic bacteria						
Hydrogenomonas eutropha	515,000	40,700			0	64
		54,000	13,500	8	8	65
		54,000	13,000		0	66
		56,000 and 52,000	15,000			67
Hydrogenomonas facilis	551,000	38,000			0	64
Paracoccus denitrificans	513,000	52,000	11,500			68
Micrococcus dentificans		57,000			0	69
Thiobacillus novellus	496,000	60,000	12,250			70,71
Thiobacillus nepolitanus	500,000	55,000	11,000			72
Thiobacillus intermedius	462,000	54,500		8	0	73

Table 2
K_m(RuBP) AND K_m(M_g^{++}) OF RIBULOSE 1,5-BISPHOSPHATE CARBOXYLASE FROM PHOTO- AND CHEMOSYNTHETIC SPECIES

	K_m		
Source of enzyme	RuBP μ*M*	M_g^{++} m*M*	Ref.
Higher plants			
Spinacea oleracea	250	—	74,75
	120	1.1	76[a]
	70	—	77
Brassica petsai	100	—	78
Ricinis communis (endosperm)	17	0.5	79
Zea mays	17	—	80
Eucaryotic algae			
Chlorella ellipsoidea	119	—	77
Euglena gracilis	15.1	2.2	21
Photosynthetic bacteria			
Rhodopseudomonas spheroides	—	0.23	103
Rhodospirillum rubrum	83	—	7
	53	0.21	63
	—	0.37	103
Chemosynthetic bacteria			
Paracoccus denitrificans	210	0.06	68
Hydrogenomonas eutropha	125	—	81
Hydrogenomonas facilis	235	1.39	81,64
Thiobacillus novellus	14.8	0.61	70,71
Chemosynthetic Bacteria			
Thiobacillus A2	122	0.02	82
Thiobacillus denitrificans	120[a]	—	83
Thiobacillus intermedius	76	3.6	73
Thiobacillus neopolitanus	90	—	72
Blue-green algae			
Aphanocapsa 6308	—	0.35	54

[a] $K_m Mn^{++}$ of 0.039 m*M*.

trophs through angiosperms.[27] Photorespiration may be partly or almost totally the result of the oxygenase activity of the enzyme.

A number of metabolic intermediates have been shown to affect the enzyme (see Table 4). In addition, compounds reactive toward cysteine, lysine, and arginine are inhibitory (see Table 5) which suggests that these amino acids are part of, or are located near, the active site of the enzyme.

In eucaryotic species, the large subunit of the enzyme is coded by chloroplastic DNA[28,29] and is translated in the chloroplast.[30-32] The 20,000-dalton precursor of the small subunit is coded by nuclear DNA[33] and is translated in the cytoplasm.[34,35] During, or after, transport into the chloroplast, a piece of the precursor peptide is cleaved from the N-terminal end, giving rise to the small subunit.[36-38]

The enzyme has been the subject of several reviews[39-42] and one recent symposium.[43]

Table 3
METALLIC CATIONS EFFECTIVE AS COFACTORS

Source of enzyme	Mg^{++}	Mn^{++}	Ni^{++}	Co^{++}	Fe^{++}	Ref.
Higher plants						
Spinacea oleracea	+	+	+	+	+	74
	+	+				76
Tetragonia expansa	+	+	+	+		100
Ricinis communis (endosperm)	+					80
Eucaryotic algae						
Chlamydonomas reinhardi	+					101
Euglena gracilis	+	+	+	+	+	21
Chlorella ellipsoidea	+					77
Chlorella fusca	+					18
Blue-green algae						
Aphanocapsa 6308	+					54
Photosynthetic bacteria						
Chromatium strain D	+					102
Rhodopseudomonas spheroides	+	−	−	−		103
Rhodospirillum rubrum	−					7
R. rubrum	+	−	−	−		104
Chemosynthetic bacteria						
Hydrogenomonas facilis	+					105
	+	+	−	+		64
Hydrogenomonas eutropha	+	−	−	+		64
Hydrogenomonas pantotropha	+					106
Paracoccus denitrificans	+					68
Thiobacillus novellus	+	−	−	−		71
Thiobacillus A2	+	+				82
Thiobacillus intermedius	+	−		+		73

Note: Activation denoted by +; no effect denoted by −.[a]

[a] Ca^{++},[64,71,103] Cd^{++},[64,73] Cu^{++},[71] and Zn^{++}[71,73] had no effect.

Table 4
METABOLIC EFFECTORS OF RIBULOSE 1,5-BISPHOSPHATE CARBOXYLASE

Effector	Source of Enzyme	Effect	Ref.
Sugar phosphates			
Sedoheptulose 1,7-bisphosphate	*Thiobacillus neopolitanus*	Inhibition	72
Sedoheptulose 7-phosphate	*Spinacea oleracea*	Activation	84
Fructose 1,6-bisphosphate	*S. oleracea*	Inhibition, competitive with RuBP	85
		Inhibition, also reverses fructose 6-phosphate effect	84
	Glycine max	Inhibition	10
Fructose 1,6-bisphosphate	*Nicotiana tabacum*	Activation at low pH	15
	Zea mays	Reverses fructose 6-phosphate activation	84
	Chlorella pyrenoidosa	Reverses fructose 6-phosphate activation	84
	Nostoc muscorum	Reverses fructose 6-phosphate activation	84
	Thiobacillus novellus	Inhibition	71

Table 4 (continued) METABOLIC EFFECTORS OF RIBULOSE 1,5-BISPHOSPHATE CARBOXYLASE

Effector	Source of Enzyme	Effect	Ref.
Fructose 6-phosphate	*S. oleracea*	Activation	84
	G. max	Inhibition	10
	N. tabacum	No effect	15
	Z. mays	Activation	84
	C. pyrenoidosa	Activation	84
	N. muscorum	Activation	84
Fructose 1-phosphate	*S. oleracea*	Inhibition	84
	G. max	Inhibition	10
Glucose 1,6-bisphosphate	*S. oleracea*	Inhibition	84
Glucose 1-phosphate	*S. oleracea*	Inhibition	84
	G. max	Inhibition	10
Glucose 6-phosphate	*S. oleracea*	Activation	84
ADP-glucose	*S. oleracea*	Activation	84
Xylulose 1,5-bisphosphate	*S. oleracea*	Inhibition	86
Xylitol 1,5-bisphosphate	*S. oleracea*	Inhibition	87
Xylulose 5-phosphate	*S. oleracea*	Activation	84
Ribulose 5-phosphate	*S. oleracea*	Activation	84
	N. tabacum	Activation at low pH	15
	G. max	Inhibition	10
	T. novellus	Inhibition, competitive with RuBP	71
Ribose 5-phosphate	*S. oleracea*	Activation	84
	N. tabacum	Inhibition	15
	T. novellus	Inhibition, competitive with RuBP	71
	T. neopolitanus	Activation	72
Threose-phosphate	*Tetragonia expansa*	Inhibition	88
Erythrose 4-phosphate	*S. oleracea*	Activation	84
Dihydroxyacetone-phosphate	*S. oleracea*	Activation	84
Glyceraldehyde 3-phosphate	*S. oleracea*	Inhibition, competitive with HCO_3^-	89
6 phosphate-gluconate	*S. oleracea*	Inhibition, competitive with RuBp	90
		Inhibition	91
		Activation at low HCO_3^-, Mg^{++}	84
		Inhibition at saturation HCO_3^-, Mg^{++}	84
		Activation at low levels under certain conditions; inhibition always at high levels	92
	N. tabacum	Activation, effective at low pH	15
	Euglena gracilis	Inhibition	21
	Aphanocapsa 6308	Inhibition	54
Chromatium		Inhibition	90
Ectothiorhodospira halophila		Inhibition	90
	Rhodopseudomonas spheroides		
	550,000 mol wt enzyme	Inhibition	62
	360,000 mol wt enzyme	No effect	62
	Hydrogenomonas eutropha	Inhibition	90
	Paracoccus denitrificans	Inhibition	68
Other intermediary metabolites			
6-Phosphate-gluconate	*T. novellus*	Inhibition, noncompetitive with RuBP, mixed inhibition with respect to HCO_3^-	71

Table 4 (continued)
METABOLIC EFFECTORS OF RIBULOSE 1,5-BISPHOSPHATE CARBOXYLASE

Effector	Source of Enzyme	Effect	Ref.
	Thiobacillus A2	Inhibition, noncompetitive with RuBP, HCO_3^-, Mg^{++}	93
	T. neopolitanus	Inhibition, competitive with RuBP	72
	Thiobacillus denitrificans	No effect	90
	Thiobacillus intermedius	Inhibition, competitive with RuBP	73
	Chlorella fusca	Inhibition, when HCO_3^- high; Activation, when HCO_3^- low	18
3 Phosphate-glycerate	*S. oleracea*	Inhibition, competitive with HCO_3^-	76
	G. max	Inhibition	10
	Rhodospirillum rubrum	Inhibition, competitive with HCO_3^-	7
	T. neopolitanus	Activation	72
	Hydrogenomonas facilus	Inhibition, competitive with HCO_3^-	64
Glycerate	*G. max*	Inhibition	10
2-phosphate-glycolate	*G. max*	Inhibition	10
Other intermediary metabolites			
Phospho-enolpyruvate	*S. oleracea*	Inhibition	94
	T. novellus	Inhibition	71
Citrate	*R. rubrum*	Competitive with RuBP	7
Nucleotides			
NADPH	*N. tabacum*	Activation at low pH	15
	G. max	Inhibition	10
	S. oleracea	Activation	85
ATP	*S. oleracea*	Activation	84
	G. max	Inhibition	10
	T. novellus	Inhibition	70
ADP	*S. oleracea*	Activation	94
	G. max	Inhibition	10
	T. novellus	Inhibition	70
	Thiobacillus A2	Inhibition	82
	T. neopolitanus	Activation	72
AMP	*S. oleracea*	Activation	94
	T. novellus	Inhibition	70
	Thiobacillus X	Inhibition	95
ATP/ADP/AMP	*Thiobacillus* A2	U-type dependence on energy charge	82
GDP	*T. novellus*	Inhibition	70
	Thiobacillus A2	Inhibition, competitive with RuBP	82
IDP	*T. novellus*	Inhibition	70
	Thiobacillus A2	Inhibition, competitive with RuBP	82
Inorganic anions			
PO_4^{3-}	*S. oleracea*	Inhibition, competitive with RuBP	76
	S. oleracea	Inhibition	74
R. rubrum		Inhibition, noncompetitive with RuBP, CO_2	7
	H. facilis	Inhibition	64

Table 4 (continued)
METABOLIC EFFECTORS OF RIBULOSE 1,5-BISPHOSPHATE CARBOXYLASE

Effector	Source of Enzyme	Effect	Ref.
H. eutropha		Stimulation	64
AsO_4^{3-}	*S. oleracea*	Inhibition	74
$SO_4^{=}$	*S. oleracea*	Inhibition (as ammonium salt)	76
	H. facilis	Inhibition, competitive with RuBP	64
NO_3^-	*Z. mays*	Activation	96
$SO_3^{=}$	*S. oleracea*	Inhibition competitive with CO_2	98
Chlorella		Inhibition, competitive with CO_2	99
	Pseudevernia furfuracea	Inhibition, competitive with CO_2	99
Ferredoxin	*Scenedesmus obliquus*	Activation	97

Table 5
INHIBITORS OF RIBULOSE 1,5-BISPHOSPHATE CARBOXYLASE FROM PHOTO- AND CHEMOSYNTHETIC SPECIES

Thiol Group Inhibitors

Source of enzyme	*p*-Chloromercuri benzoate	Hg^{++}	Iodoacetamide	N-ethylmaleimide	CN^-	Ref.
Spinacea oleracea	+	+				74
	+					107
Glycine max	+					10
Tetragonia expansa	+		+			108
			+			109—111
Avena sativa	+					112
Rhodopseudomonas spheroides	+					103
Thiobacillus novellus	+	+				71
Thiobacillus A2	+	+		+	+	82

Active Site Inhibitors

	Lysine or cysteine			Arginine		Not directed	
Source of enzyme	Lysine Pyridox-al-5′-P	3-Bromo-1,4-dihydroxy-2-butanone-1, 4-bisphosphate	N-bromacetyle-thanol-P	2,3-butanedione	Phenylgly-oxal	Ketoribitol-1,5-bisphosphate	Ref.
S. oleracea	+						113
		+					114
			+				115—117
						+	118—119
					+		120
Nicotiana tabacum				+			127
Hordeum vulgare				+			125
Chlamydomonas reinhardi	+				+		120,121
Rhodospirillum rubrum	+						78,121—123

Hydrogenomonas eutropha	+			121
Pseudomonas oxalaticus			+	125

Anions

		Cl^-		
S. oleracea		+		126
C. reinhardi		+		20
Thiobacillus A2		+		82

REFERENCES

1. Fuller, R. C. and Gibbs, M., *Plant Physiol.*, 34, 324, 1959.
2. Heber, U., Pon, N. G., and Heber, M., *Plant Physiol.*, 38, 355, 1963.
3. Cooper, T. G., Filmer, D., Wishnick, M., and Lane, M. D., *J. Biol. Chem.*, 244, 1081, 1969.
4. Lorimer, G. H., Badger, M., and Andrews, T. J., *Biochemistry*, 15, 529, 1976.
5. Andrews, T. J., Badger, M. R., and Lorimer, G. H., *Arch. Biochem. Biophys.*, 171, 93, 1975.
6. Bahr, J. T. and Jensen, R. G., *Plant Physiol.*, 53, 39, 1974.
7. Anderson, L. E. and Fuller, R. C., *J. Biol. Chem.*, 244, 3105, 1969.
8. Badger, M. R. and Lorimer, G. H., *Arch. Biochem. Biophys.*, 175, 723, 1976.
9. Lorimer, G. H., Badger, M. R., and Andrews, T. J., *Anal. Biochem.*, 78, 66, 1977.
10. Bowes, G. and Ogren, W. L., *J. Biol. Chem.*, 247, 2171, 1972.
11. Badger, M. R. and Andrews, T. J., *Biochim. Biophys. Acta*, 60, 204, 1974.
12. Andrews, T. J., Lorimer, G. H., and Tolbert, N. E., *Biochemistry*, 12, 11, 1973.
13. Bowes, G., Ogren, W. L., and Hageman, R. H., *Biochem. Biophys. Res. Commun.*, 45, 716, 1971.
14. Kestler, D. P., Mayne, B. C., Ray, T. B., Goldstein, L. D., Brown, R. H., and Black, C. C., *Biochem. Biophys. Res. Commun.*, 66, 1439, 1975.
15. Chollet, R. and Anderson, L. L., *Arch. Biochem. Biophys.*, 176, 344, 1976.
16. Bahr, J. T. and Jensen, R. G., *Biochem. Biophys. Res. Commun.*, 57, 1180, 1974.
17. Badger, M. R., Andrews, T. J., and Osmond, C. B., Detection in C_3, C_4 and CAM plant leaves of a low-$K_m(CO_2)$ form of RuDP carboxylase having high RuDP oxygenase activity at physiological pH, in *Proc. 3rd Int. Congr. Photosynthesis*, Avron, M., Ed., Elsevier, Amsterdam, 1974, 1421.
18. Lord, J. M. and Brown, R. H., *Plant Physiol.*, 55, 360, 1975.
19. Lorimer, G. H., Osmond, C. B., Akazawa, T., and Asami, S., *Arch. Biochem. Biophys.*, 185, 49, 1978.
20. Nelson, P. E. and Surzycki, S. J., *Eur. J. Biochem.*, 61, 475, 1976.
21. McFadden, B. A., Lord, J. M., Rowe, A., and Dilks, S., *Eur. J. Biochem.*, 54, 195, 1975.
22. Takabe, T. and Akazawa, T., *Biochem. Biophys. Res. Commun.*, 53, 1173, 1973.
23. Ryan, F. J., Jolly, S. O., and Tolbert, N. E., *Biochem. Biophys. Res. Commun.*, 59, 1233, 1974.
24. McFadden, B. A., *Biochem. Biophys. Res. Commun.*, 60, 312, 1974.
25. Purohit, K. and McFadden, B. A., *J. Bacteriol.*, 129, 415, 1977.
26. Bowien, B., Mayer, F., Codd, G. A., and Schlegel, H. G., *Arch. Microbiol.*, 110, 157, 1976.
27. Andrews, T. J. and Lorimer, G. H., *FEBS Lett.*, 90, 1, 1978.
28. Chan, P-H. and Wildman, S. G., *Biochim. Biophys. Acta*, 277, 677, 1972.
29. Gelvin, S., Heizmann, P., and Howell, S. H., *Proc. Natl. Acad. Sci., U.S.A.*, 74, 3193, 1977.
30. Blair, G. E. and Ellis, R. J., *Biochim. Biophys. Acta*, 319, 223, 1973.
31. Gooding, L. R., Roy, H., and Jagendorf, A. T., *Arch. Biochem. Biophys.*, 159, 324, 1973.
32. Alscher, R., Smith, M. A., Petersen, L. W., Huffaker, R. C., and Criddle, R. S., *Arch. Biochem. Biophys.*, 174, 216, 1976.
33. Kawashima, N. and Wildman, S. G., *Biochim. Biophys. Acta*, 262, 42, 1972.
34. Roy, H., Patterson, R., and Jagendorf, A. T., *Arch. Biochem. Biophys.*, 172, 64, 1976.
35. Dobberstein, B., Blobel, G., and Chua, N-H., *Proc. Natl. Acad., Sci. U.S.A.*, 74, 1082, 1977.
36. Highfield, P. E. and Ellis, R. J., *Nature (London)*, 271, 420, 1978.
37. Cashmore, A. R., Broadhurst, M. K., and Gray, R. E., *Proc. Natl. Acad. Sci. U.S.A.*, 75, 655, 1978.
38. Chua, N-H. and Schmidt, G. W., *Brookhaven Symp. Biol.* 30, in press.
39. Kawashima, H. and Wildman, S. G., *Annu. Rev. Plant Physiol.*, 21, 325, 1970.
40. Siegel, M. I., Wishnick, M., and Lane, M. D., Ribulose-1,5-diphosphate carboxylase, in *The Enzymes*, Vol. 6, 3rd ed., Boyer, P. D., Ed., Academic Press, New York, 1972, chap. 5.
41. McFadden, B. A., *Bacteriol. Rev.*, 37, 289, 1973.
42. Jensen, R. G. and Bahr, J. T., *Annu. Rev. Plant Physiol.*, 28, 379, 1977.
43. Photosynthetic carbon assimilation {(ribulose 1,5-bisphosphate carboxylase/oxygenase (EC. 4.1.1.39)}, *Brookhaven Symp. Biol.*, 30, 1978.
44. Pon, N. G., *Arch. Biochem. Biophys.*, 119, 179, 1967.
45. Rutner, A. C. and Lane, M. D., *Biochem. Biophys. Res. Commun.*, 28, 531, 1967.
46. Rutner, A. C., *Biochem. Biophys. Res. Commun.*, 39, 923, 1970.
47. Moon, K. E. and Thompson, E. O. P., *Aust. J. Biol. Sci.*, 24, 755, 1971.
48. Gray, J. C. and Kekwick, R. G. O., *Eur. J. Biochem.*, 44, 481, 1974.
49. Goldthwaite, J. J. and Bogorad, L., *Anal. Biochem.*, 41, 57, 1971.
50. Sugiyama, T., Ito, T., and Akazawa, T., *Biochemistry*, 10, 3406, 1971.
51. Iwanij, V., Chua, N-H, and Siekevitz, P., *Biochim. Biophys. Acta*, 358, 329, 1974.
52. Tabita, F. R., Stevens, S. E., Jr., and Quijano, R., *Biochem. Biophys. Res. Commun.*, 61, 45, 1974.

53. Stewart, R., Auchterlonie, C. C., and Codd, G. A., *Planta,* 136, 61, 1977.
54. Codd, G. A. and Stewart, W. D. P., *Arch. Microbiol.,* 113, 105, 1977.
55. Tabita, F. R., Stevens, S. E., Jr., and Gibson, J. L., *J. Bacteriol.,* 125, 531, 1976.
56. Takabe, T., Nishimura, M., and Akazawa, T., *Biochem. Biophys. Res. Commun.,* 68, 537, 1976.
57. Stewart, R. and Codd, G. A., *Plant Physiol.,* 57 (Suppl.), 6, 1976.
58. Tabita, F. R., McFadden, B. A., and Pfennig, N., *Biochim. Biophys. Acta,* 341, 187, 1974.
59. Akazawa, T., Kondo, H., Shimazue, T., Nishimura, M., and Sugiyama, T., *Biochemistry,* 11, 1298, 1972.
60. Tabita, F. R. and McFadden, B. A., *J. Bacteriol.,* 126, 1271, 1976.
61. Anderson, L. E., Price, G. B., and Fuller, R. C., *Science,* 161, 482, 1968.
62. Gibson, J. L. and Tabita, F. R., *J. Biol. Chem.,* 252, 943, 1977.
63. Tabita, F. R. and McFadden, B. A., *J. Biol. Chem.,* 249, 3459, 1974.
64. Kuehn, G. D. and McFadden, B. A., *Biochemistry,* 8, 2403, 1969.
65. McFadden, B. A. and Tabita, F. R., *BioSystems,* 6, 93, 1974.
66. Bowien, B. and Mayer, F., *Eur. J. Biochem.,* 88, 97, 1978.
67. Purohit, K. and McFadden, B. A., *Biochem. Biophys. Res. Commun.,* 71, 1220, 1976.
68. Bowien, B., *FEMS Microbiol. Lett.,* 2, 263, 1977.
69. Shively, J. M., Saluja, A. K., and McFadden, B. A., *Abstr. Ann. Meeting Am. Soc. Microbiol.,* 19, 156, 1977.
70. McCarthy, J. T. and Charles, A. M., *FEBS Lett.,* 37, 329, 1973.
71. McCarthy, J. T. and Charles, A. M., *Arch. Microbiol.,* 105, 51, 1975.
72. Snead, R. M. and Shively, J. M., *Abstr. Ann. Meeting Am. Soc. Microbiol.,* 19, 156, 1977.
73. Purohit, K., McFadden, B. A., and Cohen, A. L., *J. Bacteriol.,* 127, 505, 1976.
74. Weissbach, A., Horecker, B. L., and Hurwitz, J., *J. Biol. Chem.,* 218, 795, 1956.
75. Bassham, J. A., Sharp, P., and Morris, I., *Biochim. Biophys. Acta,* 153, 898, 1968.
76. Paulsen, J. M. and Lane, M. D., *Biochemistry,* 5, 2350, 1966.
77. Sugiyama, T., Matsumoto, C., and Akazawa, T., *Arch. Biochem. Biophys.,* 129, 597, 1969.
78. Kieras, F. J. and Haselkorn, R., *Plant Physiol.,* 43, 1264, 1968.
79. Osmond, C. B., Akazawa, T., and Beevers, H., *Plant Physiol.,* 55, 226, 1975.
80. Andrews, T. J. and Hatch, M. D., *Phytochemistry,* 10, 9, 1971.
81. Kuehn, G. D. and McFadden, B. A., *Biochemistry,* 8, 2394, 1969.
82. Charles, A. M. and White, B., *Arch. Microbiol.,* 108, 195, 1976.
83. McFadden, B. A. and Denend, A. R., *J. Bacteriol.,* 110, 633, 1972.
84. Buchanan, B. B. and Shürmann, P., *J. Biol. Chem.,* 248, 4956, 1973.
85. Chu, D. K. and Bassham, J. A., *Plant Physiol.,* 55, 720, 1975.
86. McCurry, S. D. and Tolbert, N. E., *J. Biol. Chem.,* 252, 8344, 1977.
87. Ryan, F. J., Barker, R., and Tolbert, N. E., *Biochem. Biophys. Res. Commun.,* 65, 39, 1975.
88. Park, R. B., Pon, N. G., Louwrier, K. P., and Calvin, M., *Biochim. Biophys. Acta,* 42, 27, 1960.
89. Wishnick, M. and Lane, M. D., *Methods Enzymol.,* 23, 570, 1971.
90. Tabita, F. R. and McFadden, B. A., *Biochem. Biophys. Res. Commun.,* 48, 1153, 1972.
91. Chu, D. K. and Bassham, J. A., *Plant Physiol.,* 50, 224, 1971.
92. Chu, D. K. and Bassham, J. A., *Plant Physiol.,* 52, 373, 1973.
93. Charles, A. M. and White, B., *Arch. Microbiol.,* 108, 203, 1976.
94. Laing, W. A. and Christeller, J. T., *Biochem. J.,* 159, 563, 1976.
95. Aleem, M. I. H. and Huang, E., *Biochem. Biophys. Res. Commun.,* 20, 515, 1965.
96. Vaklinova, S. G., Popova, L., and Gushchina, L., *C. R. Acad. Bulg. Sci.,* 24, 111, 1971.
97. Vaklinova, S. G. and Popova, L. P., *C. R. Acad. Bulg. Sci.,* 25, 263, 1972.
98. Ziegler, I., *Planta,* 103, 155, 1972.
99. Ziegler, I., *Oecologia,* 29, 63, 1977.
100. Pon, N. G., Rabin, B. R., and Calvin, M., *Biochem. Z.,* 338, 7, 1963.
101. Nelson, P. E. and Surzycki, S. J., *Eur. J. Biochem.,* 61, 465, 1976.
102. Akazawa, T., Sato, K., and Sugiyama, T., *Plant Cell Physiol.,* 11, 39, 1970.
103. Akazawa, T., Sugiyama, T., and Kataoka, H., *Plant Cell Physiol.,* 11, 541, 1970.
104. Akazawa, T., Sato, K., and Sugiyama, T., *Arch. Biochem. Biophys.,* 132, 255, 1969.
105. McFadden, B. A. and Tu, C. L., *Biochem. Biophys. Res. Commun.,* 19, 728, 1965.
106. Romanova, A. K. and Mikelsaar, P. C., *Biokimya,* 37, 1030, 1972.
107. Sugiyama, T., Akazawa, T., and Nakayana, N., *Arch. Biochem. Biophys.,* 121, 522, 1967.
108. Mayadoun, J., Benson, A. A., and Calvin, M., *Biochim. Biophys. Acta,* 23, 342, 1957.
109. Rabin, B. R. and Trown, P. W., *Proc. Natl. Acad. Sci. U.S.A.,* 51, 497, 1964.
110. Argyroudi-Akoyunoglou, J. H. and Akoyunoglou, G., *Nature (London),* 213, 287, 1967.
111. Argyroudi-Akoyunoglou, J. and Akoyunoglou, G., *Biochem. Biophys. Res. Commun.,* 32, 15, 1968.
112. Sugiyama, T. and Akazawa, T., *J. Biochemistry, (Tokyo),* 62, 474, 1967.

113. Paech, C., Ryan, F. J., and Tolbert, N. E., *Arch. Biochem. Biophys.*, 179, 279, 1977.
114. Hartman, F. C., Welch, M. H., and Norton, I. L., *Proc. Natl. Acad. Sci. U.S.A.*, 70, 3721, 1973.
115. Norton, I. L., Welch, M. H., and Hartman, F. C., *J. Biol. Chem.*, 250, 8062, 1975.
116. Schloss, J. V. and Hartman, F. C., *Biochem. Biophys. Res. Commun.*, 77, 230, 1977.
117. Schloss, J. V., Stringer, C. D., and Hartman, F. C., *J. Biol. Chem.*, 253, 5707, 1978.
118. Wishnick, M., Lane, M. D., and Scrutton, M. C., *J. Biol. Chem.*, 245, 4939, 1970.
119. Siegel, M. I. and Lane, M. D., *Biochem. Biophys. Res. Commun.*, 48, 508, 1972.
120. Schloss, J. V., Norton, I. L., Stringer, C. D., and Hartman, F. C., *Fed. Proc. Fed. Am. Soc. Exp. Biol.*, 37, 1310, 1978.
121. Whitman, W. and Tabita, F. R., *Biochem. Biophys. Res. Commun.*, 71, 1034, 1976.
122. Whitman, W. B. and Tabita, F. R., *Biochemistry*, 17, 1282, 1978.
123. Whitman, W. B. and Tabita, F. R., *Biochemistry*, 17, 1288, 1978.
124. Schloss, J. V. and Hartman, F. C., *Biochem. Biophys. Res. Commun.*, 75, 320, 1977.
125. Lawlis, V. B. and McFadden, B. A., *Biochem. Biophys. Res Commun.*, 80, 580, 1978.
126. Trown, P. W., *Biochemistry*, 4, 908, 1965.
127. Chollet, R., *Biochem. Biophys. Res. Commun.*, 83, 1267, 1978.

PEP CARBOXYLASE FROM PLANTS AND PHOTOSYNTHETIC BACTERIA

Tatsuo Sugiyama

Phosphoenolpyruvate carboxylase {orthophosphate:oxalacetate carboxylase (phosphorylating) EC. 4.1.1.31} catalyzes the divalent cation-dependent and practically irreversible reaction:

$$\underset{\text{(PEP)}}{\text{phosphoenolpyruvate}} + CO_2 + H_2O \xrightarrow{Me^{2+}} \underset{\text{(OAA)}}{\text{oxalacetate}} + \underset{\text{(Pi)}}{\text{orthophosphate}}$$

Bicarbonate rather than CO_2 is the reactive "CO_2" species in this reaction.[1-4] In almost all cases, the maximal activity of the plant enzyme is observed when Mg^{2+} is used as the divalent cation. Divalent Mn is less effective, but still active, and Co^{2+} has occasionally been found to be substantially active.[5] The optimal pH for most of the different species of the enzyme lie on the alkaline side of neutrality and can be as high as 8 or 8.5 in the case of the plant-derived enzyme.[5]

PEP carboxylase was first found in extracts of spinach leaves in 1953[6] and is known to be widely distributed in photosynthetic organisms. Different metabolic functions of the enzymes in these organisms are summarized in Table 1 with observed PEP K_m values. Multiple forms of the plant PEP carboxylase have been separated from various sources. These include *Chlamydomonas reinhardii,*[15] *Euglena,*[16] and different plant species.[7,17-20] It seems conceivable that a form of the enzyme from various portions of a plant may have a different function. The K_m values for HCO_3^- in most cases are similar for the enzymes from a variety of plant sources, ranging between 0.1 and 0.4 m*M* at near optimal pH. However, significant variations have been recorded on the K_m for PEP. In the case of the enzyme from maize leaves,[10] the K_m for PEP is decreased by glucose 6-phosphate, an activator for plant PEP carboxylases,[21,22] and by raising the pH. Kinetic studies of the spinach leaf enzyme using Mg^{2+} or Mn^{2+} as an activator indicate that the K_m for PEP is greatly dependent on the concentration of the divalent cation.[11] In addition to glucose 6-P, glycine also activates the enzyme from leaves of several monocotyledoneous C_4 species, but it is without effect on the enzyme from dicotyledoneous C_4 species or C_3 species.[23]

The cooperative binding of PEP and a Hill coefficient of approximately two have been reported for the enzyme from leaves of some C_4 plant species,[7,10,24] whereas simple hyperbolic responses have been observed in other investigations.[19,23,25,26] For the enzyme from maize leaves, sigmoidicity depends on pH, as well as the concentration of glucose 6-P.[10] This could reconcile those cases where different responses have been observed with the enzyme from the same species. No reports exist on cooperativity using the enzyme from photosynthetic organisms other than C_4 plant species.

Oxalacetate, a product of the carboxylase-catalyzed reaction, inhibits the enzymes from C_3 and C_4 plant species at concentration of less than 1 m*M*;[13,24,26] the inhibitory effects of other organic acids are not equivocal. Conflicting observations using these enzymes on the effects of organic acids, such as malate and aspartate, could be mainly because of variations in assay conditions, such as pH, concentrations of Mg^{2+}, and PEP.[27]

The PEP carboxylases from *Atriplex spongiosa*, *Atriplex hastata*, and *Sedum praeltum* are inhibited by several bisulfite compounds.[28] Analogs of PEP, D- or L-phospholactate, and phosphoglycolate are competitive inhibitors of the spinach leaf enzyme.[11] The enzyme from C_4 plant leaves has been shown to be sensitive to high concentration of NaCl, while other sources of the plant enzyme are tolerant to salts.[29]

Table 1
DIFFERENT FUNCTIONS AND K_m VALUES FOR PEP CARBOXYLASES IN PHOTOSYNTHETIC ORGANISMS

Type of enzyme[7-9]	Metabolic function[7-9]	K_m for PEP[a] (*M*)	
C_4 photosynthesis	Primary carboxylation in photosynthesis	2.60×10^{-3}, pH 7.0 1.16×10^{-3}, pH 8.0 0.60×10^{-3}, pH 8.0 with G-6-P	maize leaf[10]
C_3 photosynthesis	Generation of malate as a photosynthetic product	1.0×10^{-3} without Mg^{2+} 0.11×10^{-3} with Mg^{2+} (pH 7.5)	spinach leaf[11]
Crassulacean acid metabolism	Generation of "CO_2" donor at night for subsequent photosynthesis	1.94×10^{-4}, pH 7.4	*Kalanchoe* leaf[12]
Nonphotosynthetic	Generation of malate for ionic balance and NADPH production	0.08×10^{-3}, pH 7.4	maize root[13]
		0.077×10^{-3} with Mn^{2+} 0.177×10^{-3} with Mg^{2+} (pH 8.5)	potato tuber[14]
Photosynthetic bacteria	Photosynthetic carboxylation in reductive carboxylic acid cycle		

[a] Obtained from data using a highly purified enzyme preparation with each type of enzyme.

Table 2
STRUCTURAL PROPERTIES OF PEP CARBOXYLASES FROM VARIOUS PHOTOSYNTHETIC ORGANISMS

Source	$S_{20,w}$	Mol wt	Estimated number of subunits	Ref.
Euglena gracillis strain Z	—	183,000[a,d]	—	32
Peanut cotyledon	13.9	350,000[a]	6[e]	1
Potato tuber	10	265,000[b]	—	14
Spinach leaf	—	560,000[a]	4[f]	11
Maize leaf	12.3	400,000[c]	4[f]	10
	9.7(I)	225,650[b]	—	20
	11.6(II)	270,800[b]	—	20

[a] Estimated by gel filtration.
[b] Estimated by sucrose density gradient.
[c] Estimated by sedimentation equilibrium.
[d] Forming a multienzyme complex with acetyl-CoA carboxylase and malate dehydrogenase.
[e] Based on Mn^{2+} binding sites as determined by water proton relaxation rates.
[f] Based on monomer as estimated by SDS-gel electrophoresis.

PEP carboxylase from C_4 plant leaves has been shown to slightly fractionate the stable carbon isotopes, ^{12}C and ^{13}C. [4,30] The slight ($\sim -3^0/_{00}$) fractionation of carbon almost completely accounts for the $\delta^{13}C$ values of C_4 plants.[4]

The structural data available from studies using highly purified PEP carboxylases from various plant sources are summarized in Table 2. The data are too limited to permit confident comments about the structures of various enzymes, but there is a possibility that the plant carboxylase may exist in several forms with different molecular weights.

REFERENCES

1. **Maruyama, H., Easterday, R. L., Chang, H., and Lane, M. D.,** The enzymatic carboxylation of phosphoenolpyruvate I. Purification and properties of phosphoenolpyruvate carboxylase, *J. Biol. Chem.,* 241, 2405, 1966.
2. **Waygood, E. R., Mache, R., and Tan, C. K.,** Carbon dioxide, the substrate for phosphoenolpyruvate carboxylase from leaves of maize, *Can. J. Bot.,* 47, 14 and 551, 1969.
3. **Cooper, T. G. and Wood, H.,** The carboxylation of phosphoenolpyruvate and pyruvate, *J. Biol. Chem.,* 246, 5488, 1971.
4. **Reibach, P. H. and Benedict, C. R.,** Fractionation of stable carbon isotopes by phosphoenolpyruvate carboxylase from C_4 plants, *Plant Physiol.,* 59, 564, 1977.
5. **Utter, M. F. and Kolenbrander, H. M.,** Formation of oxalacetate by CO_2 fixation on phosphoenolpyruvate, in *The Enzymes,* Vol. 6, Boyer, P. D., Ed., Academic Press, New York, 1972, chap. 4.
6. **Bandurski, R. S. and Greiner, C. M.,** The enzymic synthesis of oxalacetate from phosphoenolpyruvate and carbon dioxide, *J. Biol. Chem.,* 204, 781, 1953.
7. **Ting, I. P. and Osmond, C. B.,** Multiple forms of plant phosphoenolpyruvate carboxylase associated with different metabolic pathways, *Plant Physiol.,* 51, 448, 1973.
8. **Black, C. C.,** Photosynthetic carbon fixation in relation to net CO_2 uptake, *Annu. Rev. Plant Physiol.,* 24, 253, 1973.
9. **Buchanan, B. B.,** Ferredoxin-linked carboxylation reactions, in *The Enzymes,* Vol. 6, Boyer, P. D., Ed., Academic Press, New York, 1972, chap. 6.
10. **Uedan, K. and Sugiyama, T.,** Purification and characterization of phosphoenolpyruvate carboxylase from maize leaves, *Plant Physiol.,* 57, 906, 1976.
11. **Miziorko, H. M., Nowak, T., and Mildvan, A. S.,** Spinach leaf phosphoenolpyruvate carboxylase urification, properties, and kinetic studies, *Arch. Biochem. Biophys.,* 163, 378, 1974.
12. **Walker, D. A.,** Physiological studies on acid metabolism. IV. Phosphoenolpyruvic carboxylase activity in extracts of Crassulacean plants, *Biochem. J.,* 67, 73, 1957.

13. **Ting, I. P.**, CO_2 metabolism in corn roots. III. Inhibition of P-enolpyruvate carboxylase by L-malate, *Plant Physiol.*, 43, 1919, 1968.
14. **Smith, T. E.**, Partial purification and characteristics of potato phosphoenolpyruvate carboxylase, *Arch. Biochem. Biophys.*, 125, 178, 1968.
15. **Chen, J. H. and Jones, R. F.**, Multiple forms of phosphoenolpyruvate carboxylase from *Chlamydomonas reinhardti*, *Biochem. Biophys. Acta*, 214, 318, 1970.
16. **Perl, M.**, Phosphoenolpyruvate carboxylating enzyme in dark-grown and in green *Euglena* cells, *J. Biochem.*, 76, 1095, 1974.
17. **Mukerji, S. K. and Ting, I. P.**, Phosphoenolpyruvate carboxylase isoenzymes:separation and properties of three forms from cotton leaf tissue, *Arch. Biochem. Biophys.*, 143, 297, 1971.
18. **Hatch, M. D., Osmond, C. B., Troughton, J. H., and Bjorkman, O.**, Physiological and biochemical characteristics of C_3 and C_4 *Atriplex* species and hybrids in relation to the evolution of the C_4 pathway, *Carnegie Inst. Washington Yearb.*, 71, 135, 1972.
19. **Goatly, M. B. and Smith, H.**, Differential properties of phosphoenolpyruvate carboxylase from etiolated and green sugar cane, *Planta*, 117, 67, 1974.
20. **Mukerji, S. K.**, Corn leaf phosphoenolpyruvate carboxylase
urification and properties of two isoenzymes, *Arch. Biochem. Biophys.*, 182, 343, 1977.
21. **Coombs, J. and Baldry, C. W.**, C-4 pathway in *Pennisetum purpureum*, *Nature (London) New Biol.*, 238, 268, 1972.
22. **Ting, I. P. and Osmond, C. B.**, Activation of plant P-enolpyruvate carboxylases by glucose-6-phosphate: a particular role in Crassulacean acid metabolism, *Plant Sci. Lett.*, 1, 123, 1973.
23. **Nishikido, T. and Takanashi, H.**, Glycine activation of PEP carboxylase from monocotyledoneous C_4 plants, *Biochem. Biophys. Res. Commun.*, 53, 126, 1973.
24. **Coombs, J., Baldry, C. W., and Bucke, C.**, The C-4 pathway of *Pennisetum purpureum*. I. The allosteric nature of PEP carboxylase, *Planta*, 110, 95, 1973.
25. **Bhagwat, A. S. and Sane, P. V.**, Studies on enzymes of C_4 pathway: partial purification and kinetic properties of maize phosphoenolpyruvate carboxylase, *Indian J. Exp. Biol.*, 14, 155, 1976.
26. **Lowe, J. and Slack, C. R.**, Inhibition of maize leaf phosphoenolpyruvate carboxylase by oxalacetate, *Biochim. Biophys. Acta*, 235, 207, 1971.
27. **Huber, S. C. and Edwards, G. E.**, Inhibition of phosphoenolpyruvate carboxylase from C_4 plants by malate and aspartate, *Can. J. Bot.*, 53, 1925, 1975.
28. **Osmond, C. B. and Avadhani, P. N.**, Inhibition of the β-carboxylation of CO_2 fixation by bisulfite compounds, *Plant Physiol.*, 45, 228, 1970.
29. **Osmond, C. B. and Greenway, H.**, Salt responses of carboxylation enzymes from species differing in salt tolerance, *Plant Physiol.*, 49, 260, 1972.
30. **Whelan, T., Sackett, W. M., and Benedict, C. R.**, Enzymatic fractionation of carbon isotopes by phosphoenolpyruvate carboxylase from C_4 plants, *Plant Physiol.*, 51, 1051, 1973.
31. **Wolpert, J. S. and Ernst-Fonberg, M. L.**, A multienzyme complex for CO_2 fixation, *Biochemistry*, 14, 1095, 1975.
32. **Wolpert, J. S. and Ernst-Fonberg, M. L.**, Dissociation and characterization of enzymes from a multienzyme complex involved in CO_2 fixation, *Biochemistry*, 14, 1103, 1975.

C_4 ACID DECARBOXYLASES

M. D. Hatch

INTRODUCTION

Both the C_4 pathway and Crassulacean acid metabolism (CAM) modifications of photosynthetic CO_2 assimilation depend on C_4 acid decarboxylation as a key component process. In a remarkable turn of events, it appears that not only C_4 pathway species but also CAM species employ one of three alternative enzymes to achieve this step. These are NADP malic enzyme (EC.1.1.1.40) (Equation 1), PEP carboxykinase (EC.4.1.1.49) (Equation 2), and NAD malic enzyme (EC.1.1.1.39) (Equation 3).

$$\text{Malate} + \text{NADP}^+ \xrightarrow{Mg^{2+}} \text{pyruvate} + \text{NADPH} + CO_2 \qquad (1)$$

$$\text{Oxaloacetate} + \text{ATP} \xrightarrow{Mn^{2+}} \text{PEP} + \text{ATP} + CO_2 \qquad (2)$$

$$\text{Malate} + \text{NAD}^+ \xrightarrow{Mn^{2+}} \text{pyruvate} + \text{NADH} + CO_2 \qquad (3)$$

As shown in Table 1, one or another of these enzymes generally predominates in a particular C_4 species, while the activity of the other decarboxylases is comparable to the low activities present in species utilizing conventional photosynthetic carbon reduction cycle (C_3 cycle) photosynthesis.[1-6] This feature has been used to divide C_4 plants into three subgroups.[3] CAM plants can also be clearly subdivided according to whether they contain PEP carboxykinase or not. In most species containing PEP carboxykinase, there is substantially reduced NADP malic enzyme activity and very low NAD malic enzyme activity. NAD malic enzyme is particularly active in members of the family Crassulaceae where it occurs in combination with high NAD malic enzyme.

NADP MALIC ENZYME

In one group of C_4 species, NADP malic enzyme activity averages about 30 times the activity found in other C_4 species or C_3 cycle species (see Table 1). In this group (NADP-ME-type), the enzyme is specifically located in the chloroplasts of bundle sheath cells. Divalent metal ions (Mg^{2+} or Mn^{2+}) are required for activity, and the enzyme is essentially specific for NADP at its pH optimum of 8.3; at lower pH, the C_4 NADP malic enzyme shows substantial activity with NAD^+. This activity with NAD^+ is readily distinguished from the NAD-specific mitochondrial malic enzyme (see below) by its lack of response to activation by CoA or fructose 1,6-bisphosphate and its sensitivity to NADP (completely inhibited by less than 1 μM). The *Zea mays* enzyme shows simple hyperbolic responses to varying malate (K_m, 0.15 mM) and NADP (K_m, 0.025 mM) concentrations. However, a complex interaction between malate and pH, with the pH optimum shifting from 7.4 to 8.5 as malate increases from 0.1 to 10 mM, could have regulatory significance. Inhibition by fructose 1,6-bisphosphate and oxaloacetate may also provide fine control of activity in vivo.[7] The predominant control of this enzyme in the light is probably through its dependency on the reducing step of the C_3 cycle (NADP glyceraldehyde-3-P dehydrogenase) with which it is coupled by a NADPH/NADP cycle.

NADP malic enzyme of CAM species is probably located in chloroplasts and shows allosteric characteristics with respect to malate binding and inhibition by fructose 1,6-

Table 1
RANGE OF C_4 ACID DECARBOXYLASE ACTIVITIES IN SPECIES WITH DIFFERING PHOTOSYNTHETIC PATHWAYS

Photosynthetic pathway and subgroup[a]	Activity range [μmol min^{-1} (mg chlorophyll)$^{-1}$] NADP malic enzyme	PEP carboxykinase	NAD malic enzyme
C_3 cycle plants			
10 spp.	0.1—0.8	<0.2	0.1—0.5
C_4 pathway plants			
NADP-ME-type (6 spp.)	10—23	<0.2	<0.3[b]
PCK-type (8 spp.)	0.2—1.0	10—24	0.2—2.0
NAD-ME-type (15 spp.)	0.2—0.8	<0.2	4—9
CAM plants			
NADP-ME-type			
Crassulaceae	1.4—3.5	0	1.9—3.1
Others	1.5—6.0	0	0.2—0.8
PCK-type	0.3—3.5	4.5—12	0.1—0.5

[a] C_3 cycle and C_4 pathway species data summarized from Hatch and Osmond[3] and unpublished data of I. R. Oliver and M. D. Hatch. Data for CAM species from surveys of a large number of species referred to and summarized by Osmond.[5] In a few instances, unusually low values for single species were ignored. Abbreviations: NADP-ME-type, NADP malic enzyme type; PCK-type, PEP carboxykinase type; NAD-ME-type, NAD malic enzyme type.

[b] After subtracting NAD-dependent activity associated with NADP malic enzyme (not activated by CoA or fructose 1,6-bisphosphate and inhibited by NADP)[4].

bisphosphate which may be significant for regulation in vivo. As with C_4 plants, coupling with the reducing step of the C_3 cycle could be the major factor influencing malate decarboxylation.

PEP CARBOXYKINASE

The very high activities of PEP carboxykinase in certain groups of C_4 and CAM species contrast with the negligible levels of this enzyme in leaves of other plants, including those of the C_4 and CAM type (see Table 1). In C_4 plants, PEP carboxykinase is located in bundle sheath cytoplasm where its operation is light dependent, presumably through a requirement for ATP. The enzyme is Mn^{2+} specific, and ATP is the preferred nucleotide. Oxaloacetate and ATP show simple hyperbolic response curves, but the curve for increasing Mn^{2+} is sigmoidal. Inhibition of the C_4 enzyme by the C_3 cycle intermediates fructose 1,6-bisphosphate, 3-phosphoglycerate, and dihydroxyacetone phosphate may serve to integrate decarboxylation with CO_2 assimilation via the C_3 cycle.

PEP carboxykinase from a CAM species has recently been characterized.[8] This enzyme is essentially inactive at pH values higher than 7.3, suggesting a cytoplasmic location. Generally, its other properties are similar to those of the C_4 enzyme. Other characteristics indicate that the primary role of the CAM enzyme in vivo is as a decarboxylase rather than as a carboxylase.

NAD MALIC ENZYME

NAD malic enzyme is the predominant C_4 acid decarboxylating enzyme in one group of C_4 species (NAD-ME-type) and at least a substantial contributor to decarboxylation in some CAM species (see Table 1). The enzyme is confined to mitochondria in both C_4 and CAM species. In NAD-ME-type C_4 plants, this activity is located specifically in bundle sheath cells which are distinguished by unusually large numbers of mitochondria.

The NAD malic enzyme from different C_4 species are specific for Mn^{2+}, and most are also specific for NAD^+; others show some activity with $NADP^+$. Individual enzymes are activated to an equal extent by CoA, acetyl-CoA, and fructose 1,6-bisphosphate$_2$. However, activation varies from about 4- to 15-fold with enzymes from different sources. Surprisingly, activation is because of an effect on malate affinity (K-type) with some enzymes and an effect on V_{max} with others. The former group of enzymes, but not the latter, show pronounced positive cooperativity for malate binding. They also show a marked inhibition by HCO_3^- (or CO_2) which is competitive with respect to malate and activators. For NAD malic enzyme extracted from C_4 plants, both Mn^{2+} and thiol compounds are essential for maintaining activity.

NAD malic enzyme from the CAM plant *Kalanchoe diagremontiana* is also specific for Mn^{2+} and NAD^+ and is activated by CoA and acetyl-CoA. This activation is of the K-type through an effect on the affinity of the enzyme for malate. Responses to increasing concentrations of malate, CoA, and Mn^{2+} are all highly sigmoidal.

REFERENCES

1. **Burris, R. H. and Black, C. C.,** *Carbon Dioxide Metabolism and Plant Productivity,* University Park Press, Baltimore, 1976.
2. **Hatch, M. D.,** Photosynthesis: the path of carbon, in *Plant Biochemistry,* Bonner, J. and Varner, J., Eds., Academic Press, New York, 1976, 797.
3. **Hatch, M. D. and Osmond, C. B.,** Compartmentation and transport in C_4 photosynthesis, in *Encyclopedia of Plant Physiology, New Series,* Vol. 3, Heber, U. and Stocking, C. R. Eds., Springer-Verlag, Heidelberg, 1976, 144.
4. **Hatch, M. D.,** Regulation of enzymes in C_4 photosynthesis, in *Current Topics in Cellular Regulation,* Vol. 14, Horecker, B. L. and Stadtman, E. R., Eds., Academic Press, New York, 1978, 1—28.
5. **Osmond, C. B.,** Crassulacean acid metabolism: a curiosity in context, *Annu. Rev. Plant Physiol.,* 29, 379, 1978.
6. **Edwards, G. E. and Walker, D. A.,** *C_3 and C_4: An Introduction to the Biochemistry of Photosynthesis,* Packard, Sussex, in press.
7. **Bhagwat, A. S., Mitra, J., and Sane, P. V.,** Studies on enzymes of C_4 pathway. III. Regulation of malic enzyme of *Zea mays* by fructose-1,6-diphosphate and other metabolites, *Ind. J. Exp. Biol.,* 15, 1008, 1977.
8. **Daley, L. S., Ray, T. B., Vines, H. M., and Black, C. C.,** Characterization of phosphoenolpyruvate carboxykinase from pineapple leaves *Ananas comosus* (L.) Merr, *Plant Physiol.,* 59, 618, 1977.

CARBONIC ANHYDRASES (CARBONATE DEHYDRATASES) FROM PLANTS

D. Graham

ENZYMIC REACTION AND ASSAY

Carbonic anhydrase (carbo-anhydrase, carbonate dehydratase, or dehydrolyase) (E.C. 4.2.1.1) catalyzes the reversible hydration of carbon dioxide above about pH 6:

$$CO_2 + H_2O \rightleftharpoons H^+ + HCO_3^- \qquad (1)$$

Below this pH, carbonic acid is formed to a significant extent:

$$CO_2 + H_2O + \rightleftharpoons H_2CO_3 \qquad (2)$$

The nonenzymic (uncatalyzed) rate of the hydration reaction (Equation 1) is relatively slow in the pH range 6.5 to 10 compared to the very rapid dehydration reaction.[1]

The enzyme can be assayed in several ways, including manometric,[2] electrometric,[3] and colorimetric[3-5] techniques. Earlier methods have been reviewed.[6] Enzyme units (E.U.) are defined for:

1. The manometric method as

$$\frac{R - R_o}{R_o}$$

where R and R_o are the catalyzed and uncatalyzed rates, a unit being the amount of enzyme required to double the rate of the uncatalyzed reaction at a given temperature and dilution

2. The colorimetric method as arbitrary units given by

$$10\left[\frac{t_b}{t_e} - 1\right]$$

where t_b and t_e are the times taken for the change from pH 8.2 to pH 6.3 at 0°C for boiled enzyme and unboiled enzyme, respectively

3. The electrometric method follows the change in pH (log H^+ concentration) which may be used to calculate E.U. as for the colorimetric method

Enzymic activities may also be calculated as micromoles substrate (usually H^+) produced (or consumed).

PROPERTIES

Characterization and Inhibition

The enzyme has been isolated and characterized from several species of higher plants (angiosperms), algae, and a fern (Pteridophyta) (see Table 1). In the angiosperms, all dicotyledons so far examined have two high molecular weight isozymes separable by gel electrophoresis.[8,18] Monocotyledons usually have two low molecular weight forms.[8,18] The dicotyledon enzyme is usually a hexamer with a mol wt of about 180,000

Table 1
Properties of plant carbonic anhydrases

Species	Purification (fold)	Mol wt	Subunit mol wt	No. of Sub-units	Zn content g-atom per molecule	Inhibition				Comment	Ref.
						Acetazolamide	Azide	Ethoxyzolamide	NO_3^-		
Algae											
Serraticardia maxima (Rhodophyta)		150,000	?30,000			I_{50} 5×10^{-8} *M*	5×10^{-3} *M*		2×10^{-2} *M*		7
Ulva pertusa (Chlorophyta)		150,000	?30,000			I_{50} 7×10^{-9} *M*	3×10^{-3} *M*		9×10^{-3} *M*		7
Chara coronata		90,000	?30,000			I_{50} 10^{-4} *M*	2×10^{-4} *M*		8×10^{-4} *M*		7
Pteridophyta (ferns)											
Matteuccia struthiopteris		180,000				I_{50} 6×10^{-8} *M*	5×10^{-3} *M*		6×10^{-3} *M*		7
Monocotyledons											
Tradescantia albiflora (Wandering Jew)	24	42,000 GF	27,500 SDS			I_{50} 2.7×10^{-4}	13×10^{-6} *M*	18×10^{-6} *M*	1.75×10^{-4} *M*		8
Hordeum vulgare (Barley, leaf)		45,000 GF				I_{50} 2×10^{-6} *M*	2.2×10^{-4} *M*				9
Dicotyledons											
Spinacea oleracea (Spinach, leaf)	(1)330	180,000 PAG[a]	30,000 SDS[e]	6	6	K_i 2×10^{-4} *M*	K_i 3.3×10^{-5} *M*			(pH 6.82, 25°C)	10,11
	(2)	160,000	26,500 AA[f]	6							12
	(3)139	140,000 SV[b] 145,000 SE[c] 148,000 GF[d]	n.d[g]	 n.d n.d	 none none						13
	(4)>59	212,000 GF	26,000 SDS	8	8	K_i 5.56×10^{-5} *M*	K_i 3.1×10^{-5} *M*			(pH 7.9, 22°C)	14
Petroselinum crispum var. (*latifolium* (Parsley, leaf)	133	180,000 SE,GF	29,000 SDS	6	6						15
Pisum sativum L. (Garden pea, leaf)	(1) 56	188,000 GF	28,000 SDS	6	6	I_{50} 4.5×10^{-4}	9×10^{-6} *M*	5×10^{-6} *M*	3.8×10^{-5} *M*	lg-atom Zn per 32,500 dalton subunit.	8
	(2) 200	193,000 GF	30,000 SDS	6		I_{50} 2.8×10^{-5} *M*					16
Lactuca sativa (Lettuce, leaf)	900	(a)195,000 PAG, GF (b)250,000 PAG	34,000 SDS	6	5—6					(Fraction a) located in chloroplast.	17

Phaseolus vulgaris (French or kidney bean, leaf)	205,000 GF	n.d.	4.6×10^{-6} *M*	Enzyme not located in bacteriods.	9
Phaseolus vulgaris L. (root nodule)	45,000 GF	n.d.	1.6×10^{-5} *M*		

[a] Mol wt determined by polyacrylamide gel electrophoresis.
[b] By sedimentation velocity.
[c] By sedimentation equilibrium.
[d] By gel filtration (Sephadex® G-200).
[e] Mol wt determined by sodium dodecyl sulphate (SDS) and PAG electrophoresis.
[f] Mol wt based on amino acid analysis.
[g] Not determined.

and monomer subunits of about 30,000, with 1 atom of Zn per subunit molecule. However, recently Pocker and Miksch[12] and Kandel et al.[14] have shown subunits of 26,500 and 26,000, respectively, in spinach, giving the former group a hexamer close to 160,000 and the latter group an octamer of 212,000. The only monocotyledon carbonic anhydrases so far characterized are from *Tradescantia*, where the enzyme has a mol wt of about 42,000, with a subunit size of 27,500,[8] and from barley leaf, where the mol wt of the enzyme is about 45,000.[9] The enzyme from *Phaseolus vulgaris* root nodule is also about 45,000.[9] With 1 Zn atom per 34,000, it is not possible to conclude whether the *Tradescantia* enzyme has 1 Zn atom per native enzyme molecule or per polypeptide, although no smaller molecule of around 14,000 was found on SDS gel electrophoresis. The only fern (pteridophyte) carbonic anhydrase so far characterized has a mol wt around 180,000.[7]

The algae show an array of molecular weights for the enzyme with *Serraticardia*, a red alga, and *Ulva*, a green alga, having values of 150,000, and *Chara* (Charophyta), having a mol wt of about 90,000.[7] These appear to be multiples of a 30,000 subunit. A number of microalgae have carbonic anhydrases with mol wt of about 30,000 and 60,000. These include the green alga *Dunaliella* sp. and the diatoms *Nitschia* and *Cylindrotheca*, with mol wt about 30,000, and the green alga *Scenedesmus*, with mol wt of about 60,000.[19]

The plant enzymes therefore provide an interesting contrast with the mammalian enzyme, all of which have mol wt of about 30,000.[20]

The plant enzyme is inhibited by sulphonamides and related compounds (acetazolamide, sulfanilamide, and ethoxyzolamide) and inorganic ions, such as azide, nitrate, and chloride, which also inhibit the animal carbonic anhydrases[20] (see Table 1). However, the plant enzyme is less sensitive to many of these inhibitors, e.g., acetazolamide (Diamox), than the animal enzyme which suggests that the active site is different in the two types of enzymes.[11] Care is necessary in interpreting the effects of acetazolamide action on photosynthesis,[21,22] since the inhibitor is now known to affect photosynthetic electron transport[23,24] as well as inhibit carbonic anhydrase.

Kinetic Properties of the Enzyme

These have been investigated for enzymes from three higher plants and a cyanophyte. The K_mCO_2 and $K_mHCO_3^-$ are shown in Table 2. The values obtained are usually dependent on the pH and buffer used. The values for K_mCO_2 and $K_mHCO_3^-$ are about 30 to 40 m*M* at around pH 7.0 for the higher plants. The cytoplasmic and chloroplastic forms of the enzyme have very similar properties, with K_mCO_2 about 32 to 34 m*M*.

DISTRIBUTION AMONG SPECIES

The enzyme is very widely distributed through the plant kingdom, including the cyanobacteria (blue-green algae) micro-and macroalgae, ferns, and higher plants, including monocotyledons and dicotyledons and plants with C_3 or C_4 photosynthetic pathways (see Table 3). There are no reports of carbonic anhydrase in photosynthetic bacteria, although the enzyme is present in the chemosynthetic bacterium *Thiobacillus thioxidans*[41] and in *Neisseria sicca*.[42] The wide variation in enzymic activities may reflect the difficulties of extracting the enzyme rather than inherently low activities. The values for many marine macroscopic algae on a chlorophyll basis give an apparently falsely high value, since the chlorophyll concentration per unit fresh weight is usually extremely low.[30] Calcareous algae, in which calcium carbonate is deposited, do not seem to have particularly high activities of this enzyme which could be involved in creating the high pH necessary for deposition.

Table 2
KINETICS OF PLANT CARBONIC ANHYDRASES

Species	K_mCO_2 m*M*	$K_mHCO_3^-$ m*M*	Experimental Conditions pH	Buffer	Temperature (°C)	Comments	Ref.
Spinacea oleracea L. (Spinach leaf)	1.5		6.0—9.0	Dimethyl imidazole	25	pH independent 6.0 to 9.0	
	179		6.25	15 m*M* phosphate		Extrapolated to zero buffer conc.	10
	41		7.00		25	Not corrected for proton uptake by HCO_3^- solution	
	18		7.43				
		34	6.35 —	20 m*M* *N*-methyl imidazole	25	pH independent in this range	25
		58.9	7.37			Corrected for proton uptake by HCO_3^- solution	
		46.7	6.23	20 m*M* phosphate	25		
		29.2	7.03				
			7.40				
Petroselinum crispum var. (Parsley leaf)	42.4	138	7.15	50 m*M* Imidazole sulfate Imidazole sulfate	25		15
	1.5		6.51				
	6.4		6.80			Extrapolated to zero buffer conc.	
	8.3		7.07				
	10.5		7.32				
Pisum sativum (Garden pea-leaf) (1)		30					16
(2 a)	31 ± 8		7.0	40 m*M* phosphate	30		
(2 b)	34 ± 6		7.4	24 m*M* sodium dimethyl barbiturate	0	(a)Cytoplasmic isoenzyme; pH optimum CO_2 hydration 7.0; dehydration 7.5	26
					0	(b)Chloroplast isoenzyme; pH optimum CO_2 hydration 7.0; dehydration 7.5	
Spirulina platensis (Cyanophyta)	4.9		7.2		0		27

Table 3
DISTRIBUTION OF CARBONIC ANHYDRASE AMONG PLANT SPECIES

Division	Species	Common name	Enzyme activity E.U.[a] mg chl^{-1}	Enzyme activity E.U.[a] mg protein^{-1}	Enzyme activity E.U.[a] g fresh wt	Assay[b] method	Ref.
Cyanophyta (Blue-green algae)	*Anacystis nidulans* (Göttingen Strain L 1402-1)			211 (Air)[c]		2	28
				66.7 (3% CO_2)[c]		2	
	A. nidulans Richt			13.6 (Air)		2	29
				<1 (5% CO_2)		2	
	Anabaena flos-aquae (Lyngbye) Breb.			16.1 (Air)		2	29
				<1 (5% CO_2)		2	
	Coccochloris peniocystis Kütz			33.0 (Air)		2	29
				<1 (5% CO_2)		2	
	Microcoleus lyngbyaceus (Kuetzing) Crouan			256		2	30
	Oscillatoria sp.			8.1 (Air)		2	29
				<1 (5% CO_2)		2	
	Spirulina platensis			2.5 μmol H^+ min^{-1}		1	27
Micro algae Chlorophyta	*Chlorella pyrenoidosa*		188			2	31
			(Air) <4 (5% CO_2)			2	
	Chlamydomonas reinhardi		2041			2	32
			(Air) 9 (5% CO_2)			2	
	Scenedesmus obliquus		200			2	33
			(Air) 0 (2% CO_2)			2	
Cryptophyta	Phytoplankton (mixed marine) *Gymnodinium microadriaticum*		770			2	30
			1332				
	Zooanthellae from Coral—*Pocillopora damicornis*					2	30
			340			2	30
	from Clam—*Tridacna maxima*						
Euglenophyta	*Euglena gracilis*		56			2	32
			(Air) <4 (5% CO_2)			2	

Macroalgae Chlorophyta	*Boergesenia forbesii* (Harvey) Feldmann	762	2	30
	Boodlea composita (Harvey) Brand	610	2	30
	Enteromorpha sp.	1063	2	30
	Caulerpa lentillifera J. Agardh	100	2	30
	C. racemosa var. *clavifera* (Turner) Weber-van Bosse	1540	2	30
	C. verticillata J. Agardh	316	2	30
	C. sertularioides (Gmlin) Howe	2100	2	30
	Avrainvillea erecta (Berkeley) Gepp and Gepp	1320	2	30
	Chlorodesmis fastigiata (C. Agardh) Ducker	1440	2	30
	Halimeda cylindracea[d] Decaisne	207	2	30
	H. opuntia[d] *(L.) Lamouroux*	1190	2	30
	H. macroloba[d] *Decaisne*	3400	2	30
	Udotea argentea[d] *Zanardini*	2820	2	30
Phaeophyta	*Dictyota dichotoma* (Hudson) Lamouroux	1110	2	30
	Padina tenuis Bory	368	2	30
	Sargassum sp.	442	2	30
Rhodophyta	*Liagora pinnata*[d] (Farmosa) Lamouroux	4800	2	30
	Peyssonnelia capensis[d] Montagne	7.8	2	30
	Amphiroa crassa[d] Lamouroux	1070	2	30
	Jania adhaerens[d] Lamouroux	1760	2	30
	Halymenia durvillaei Bory	201	2	30
	Acanthophora spicata (spicifera?)	510	2	30
	Laurencia papillosa (C. Agardh) Greville	290	2	30

Table 3 (continued)
DISTRIBUTION OF CARBONIC ANHYDRASE AMONG PLANT SPECIES

Angiosperms (Higher Plants)

Monocotyledons

Family	Species	Common name	Enzyme Activity E.U.[a]			Assay[b] method	Ref.
			mg chl^{-1}	mg $protein^{-1}$	g fresh wt		
Alismataceae	*Siggitaria graminea* Michx		6420		6320	2	18
Amaryllidaceae	*Amaryllis belladona* L.		330		800	2	18
Andropogoneae	*Sorghum halepense* (L.) Pers.	Johnson grass	618			1	34
	S. bicolor	sorghum	741			2	32
Araceae	*Zantedeschia aethiopica* L. Spreng		3710		3040	2	18
Aveneae	*Avena sativa* L.	oat	2200			2	5
Cannaceae	*Canna indica* L.		3600		3200	2	18
Chlorideae	*Cynodon dactylon* (L.) Pers.	Bermuda grass	760			1	34
	Eleusine indica (L.) Gaertn.	goose grass	1020			1	34
	E. coracana Gaertn.		4570			2	35
Commelinaceae	*Tradescantia albiflora* Kunth.	wandering jew	6630		3300	2	18
Cyperaceae	*Cyperus rotundus* L.	purple nut-sedge	1230			1	34
Festucaceae	*Festuca arundinacea* Schreb.	fescue	1420			1	34
	Lolium multiflorum Lam.	Italian ryegrass			$13\text{-}17 \times 10^{3}$[e]	1	36
Hordeae	*Triticum sativum*	wheat	1880			2	5
	T. vulgare L.	wheat	980		1400	2	18
Tridaceae	*Dietes iridioides* weet		2740		2790	2	18
Liliaceae	*Chlorophytum comosum* L.		3060		2840	2	18
Orchidaceae	*Cymbidium* sp.	Cymbidium orchid	4820		4000	2	18
Oryzaceae	*Oryza sativa* L.	rice			79	3	37
Palmae	*Chamaedorea erumpeus*, H. E. Moore		3790		7120	2	18
Paniceae	*Echinocloa colonum* L.	jungle rice	625			1	34
	E. crusgalli (L.) Beauv.	barnyard grass	1020			1	34

Panicoideae	*Saccharum officinarum* L.	sugar cane	380			2	5
	Zea mays	corn (maize)	350			2	5
			1200			2	32
			1135			2	22
			4780			2	38
Phalarideae	*Phalaris arundinaceae* L.	reed canary grass	3840			1	34
Typhaceae	*Typha* sp.		9040		4740	2	18
		Dicotyledons					
Amaranthaceae	*Amaranthus viridis* L.		2675			2	22
	A. retroflexus L.	redroot pig-weed	1480			1	34
	A. hybridus L. Sens. lat.		3220		2290	2	18
	A. palmeri		387			2	5
	Gomphrena globosa		520			2	5
Araliaceae	*Hedera canariensis* Willd.	variegated ivy	3700			2	39
				184.8		2	22
Chenopodiaceae	*Spinacea oleraceae* L.	spinach	6950	5340		3	37
Compositae	*Beta vulgaris* L.	spinach beet	5930			2	18
	Lactuca sativa L.	garden lettuce	1700			2	5
	Helianthus annuus L.	sunflower	4000	2300		2	18
Convolvulaceae	*Convolvulus mauritanicus* Boiss		10030	8630		2	18
Cruciferae	*Raphanus sativus* c. L.	radish	4080	2000		2	18
	Brassica chinensis c. L.			180		3	37
	B. chinensis	Chinese cab-bage (pak-choi)		108		3	37
	B. chinensis	Chinese spin-ach	1800			2	37
	R. sativus I. var. longipinnatus	Japanese rad-ish		86		2	39
Cucurbitaceae	*Cucumis sativus* L.	cucumber			48	3	37
	Cucurbita pepo L.	pumpkin			96	3	37

Table 3(continued)
DISTRIBUTION OF CARBONIC ANHYDRASE AMONG PLANT SPECIES

Family	Species	Common name	Enzyme Activity E.U.[a]			Assay[b] method	Ref.
			mg chl^{-1}	mg protein^{-1}	g fresh wt		
Leguminoseae	*Pisum sativum* L.	garden pea	7880		7630	2	18
				148(125)[f]	7900(3820)[f]	2	9
	Vicia faba L.	broad bean		258(107)	5420(1740)	2	9
					222	3	37
	V. sativa L.	common vetch		397(96)	16150(3060)	2	9
	Glycine max L.	soya bean		116(55)	3250(2490)	2	9
	Trifolium repens L.	white clover		365(190)	7670(7690)	2	9
	Medicago sativa L.	lucerne (alfalfa)		500(255)	20200(6600)	2	9
	Phaseolus vulgaris L.	French or kidney bean		471(49)	14000(1525)	2	9
	Lupinus L. spp.	lupin		430(77)	10200(1730)	2	9
	Melilotus Hill spp.	melilot		132(114)	2640(3450)	2	9
Passifloraceae	*Passiflora edulis* Sims	purple passion-fruit	6480		6280	2	18
Phytolaccaceae	*Phytolacca americana* L.	pokeweed	5960			1	34
Polygonaceae	*Rheum rhaponticum* L.	rhubarb	2120		1700	2	18
Proteaceae	*Grevillea rosmarinifolia* A. Cunn. in Field	spider flower	813		900	2	18
Rutaceae	*Fortunella japonica* L.	kumquat	3430		3650	2	18
Saxifragaceae	*Hydrangea macrophylla* (Thunb.) Ser.	hydrangea	3550		2160	2	18
Solanaceae	*Capsicum annum* L.	capsicum	6240		9800	2	18
Verbenaceae	*Lantana camara* L.	lantana	13850		12000	2	18
Marine, submerged							
Hydrocharitaceae	*Halophila ovata*, Gaud		540			2	30
	Thalassia hemprichii (Ehrenb.)Aschers.		2160			2	30
Potamogetonaceae	*Cymodocea rotundata* Ehrenb. and Hempr. ex Aschers. and Schweinf.		500			2	30
Aquatic freshwater submerged							

Haloragaceae	*Myriophyllum spicatum* L.	spiked water	207	2	40
Hydrocharitaceae	*Hydrilla verticillata* (L. f.) Royle	milfoil	253	2	40
Nymphaeaceae	*Ceratophyllum demersum* L.	hornwort	292	2	40

[a] Enzyme Units, for definition see text.

[b] Method 1, electrometric assay of Wilbur and Anderson.[3]
Method 2, colorimetric assay after Rickli et al.[4]
Method 3, manometric assay of Meldrum and Roughton.[2] (Methods 1 and 2 give equivalent values in this table except where noted.)

[c] Culture grown in air (0.03% CO_2) or in elevated concentration of CO_2.

[d] Calcareous alga.

[e] Units g^{-1} (dry wt.)s^{-1}.

[f] Values in brackets for root nodules.

LOCALIZATION OF ENZYME

Cells

The two high molecular weight forms of carbonic anhydrase, separable by electrophoresis, in dicotyledons[18,26] are chloroplastic and cytoplasmic in location.[24,39] A major portion of the enzymic activity is found in the chloroplast.[5,21,35,43-46]

Most workers have concluded that the enzyme is highly soluble and present mainly in the chloroplast stroma.[43] Attempts to locate the enzyme in membrane fractions have usually been unsuccessful.[24,44] However, there is one report[35] that the enzyme is principally located with the chloroplast envelope membrane fraction of both mesophyll and bundle sheath cells of a C_4 plant. This finding may accord with the observations[32] that some of the carbonic anhydrase from C_4 plants remains at the origin in polyacrylamide gel electrophoresis.

Tissues

Carbonic anhydrase is present in highest concentration in green leaves of higher plants (see Table 3), where it may comprise 1% of the total protein.[15] The enzyme is also present in etiolated tissues in the dicotyledons *Spinacea oleracea* L. and *Brassica chinensis* L. and in albino tissues in *Hedera canariensis* Willd. var. *marginata* and the monocotyledon *Tradescantia albiflora* Kunth.[39] Only the cytoplasmic form is present in these tissues. Root tissues have not been extensively investigated, but the enzyme appears to be absent, or present in only very low amounts, in roots of legumes[9] and a range of monocotyledons and dicotyledons,[47] although it is present in some calcifuge species.[48]

Initial reports[5,32,34] indicated relatively low activities of the enzyme in plants with the C_4 photosynthetic pathway. Most or all of the enzyme was present in mesophyll cells. The use of N_2 during extraction resulted in an increase in the level of activity extracted,[32] and progressive grinding techniques have shown activities in maize as high as for other higher plants,[38] with most of the activity in the mesophyll cell fractions. However, it has been shown in another C_4 plant, *Eleusine corcarana*, that approximately half the activity is in the bundle sheath fraction.[35]

FUNCTION IN PLANTS

The exact function of this enzyme in plants remains unknown. Since the enzyme is found in highest concentration in the chloroplast, it is generally assumed to have a function related to photosynthesis. However, the evidence to support such a function is rather meager.

It has been shown for algae, such as *Chlorella*, *Chlamydomonas*, and *Scendesmus*, and cyanophytes grown in high (>1%) CO_2 that the subsequent slow induction of photosynthesis in low CO_2 concentrations is correlated with the development of carbonic anhydrase activity.[28,29,31,33,49-51] Raven and Glidewell[52] suggest a "CO_2 concentrating mechanism" associated with carbonic anhydrase is involved in these chlorophyte algae and cyanophytes. This may represent a special adaptation compared with other C_3 plants. The effect is equivalent to the CO_2-concentrating system in bundle sheath cells of C_4 photosynthetic plants which may account for the low levels of carbonic anhydrase usually found in bundle sheath cells.

In higher plants there is, as yet, little direct evidence to support a role for carbonic anhydrase in photosynthesis. Although Zn-deficient plants have only 1% of the normal carbonic anhydrase activity without affecting the rate of photosynthesis,[53] this low level of enzyme may not be rate limiting, since there is usually a 100-fold excess of enzyme above the level necessary to sustain the photosynthetic rate.[43] Furthermore, the evidence adduced from reduced rates of photosynthesis in the presence of sulphonamide inhibitors which block the carbonic anhydrase catalyzed reaction has not been substantiated, since these compounds are also inhibitory to photosynthetic electron transport.[23,24]

The following summarizes potential roles for the enzyme in plant cells.

1. The enzyme may facilitate CO_2 transport through cell membranes.[5,22,31] Such enhanced transport through artificial membranes has been shown,[54,55] but such a role in the chloroplast has been rejected on theoretical grounds.[44]
2. When the K_m HCO_3^- for ribulose bisphosphate carboxylase was believed to be very high (up to 22 m*M*), it was suggested that carbonic anhydrase may be associated with the carboxylase in some way to facilitate CO_2 fixation.[5,31,43,56] Since CO_2 is the species fixed by the carboxylase and the K_mCO_2 is very much lower than previously thought (for discussion see Black[57] and Walker[58]), it is unlikely that involvement of carbonic anhydrase is necessary at the carboxylase site. However, the light-induced increase in stroma pH results in an increase in HCO_3^- at the expense of CO_2, so it has been suggested that carbonic anhydrase may facilitate the production of CO_2 in the acid microenvironment of the carboxylase active site during photosynthesis.[56] An alternative concept is that carbonic anhydrase maintains a high HCO_3^- concentration which stabilizes the low K_mCO_2 form of the carboxylase.[33]
3. Carbonic anhydrase may effect a rapid buffering action by virtue of its interconversion of charged and uncharged species.[22,31,44] This could have an important protective role, especially in the chloroplast where rapid fluctuations in H^+ could be expected because of transient changes in light intensity in natural situations. The suggestion[31] that the enzyme has a role in the maintenance of the pH gradient according to the Mitchell scheme has not been substantiated.[24]

A special role for the enzyme in aquatic plants is not evident. The activities of the enzyme in marine algae and submerged angiosperms which live in an environment high in HCO_3^- ($\sim$1 m*M*) at high pH ($\sim$8.2) are generally lower than terrestrial plants on a fresh weight basis.[30,59,60] Calcareous algae do not appear to have substantially higher activities than other algae.[30] Freshwater aquatic angiosperms also have relatively low activities of the enzyme.[61]

REFERENCES

1. **Kern, D. M.,** The hydration of carbon dioxide, *J. Chem. Educ.,* 37, 14, 1960.
2. **Meldrum, N. U. and Roughton, F. J. W.,** Carbonic anhydrase. Its preparation and properties, *J. Physiol. (London),* 80, 113, 1933.
3. **Wilbur, K. M. and Anderson, N. G.,** Electrometric and colorimetric determination of carbonic anhydrase, *J. Biol. Chem.,* 176, 147, 1948.
4. **Rickli, E. E., Ghazanfar, S. A. S., Gibbons, B. H., and Edsall, J. T.,** Carbonic anhydrases from human erythrocytes. Preparation and properties of two enzymes, *J. Biol. Chem.,* 239, 1065, 1964.
5. **Everson, R. G. and Slack, C. R.,** Distribution of carbonic anhydrase in relation to the C_4 pathway of photosynthesis, *Phytochemistry,* 7, 581, 1968.
6. **Waygood, E. R.,** Carbonic anhydrase (plant and animal), in *Methods in Enzymology,* Vol. 2, Colowick, S. P. and Kaplan, N. O., Eds., Academic Press, New York, 1955, 836.
7. **Okazaki, M.,** Properties of carbonic anhydrases from various sources with special reference to their molecular weights and behaviours towards inhibitors, *Bull. Tokyo Gakugei Univ.,* (Ser. IV), 26, 209, 1974.
8. **Atkins, C. A., Patterson, B. D., and Graham, D.,** Plant carbonic anhydrases. II. Preparation and some properties of monocotyledon and dicotyledon enzyme types, *Plant Physiol.,* 50, 218, 1972.
9. **Atkins, C. A.,** Occurrence and some properties of carbonic anhydrases from legume root nodules, *Phytochemistry,* 13, 93, 1974.

10. **Pocker, Y. and Ng, J. S. Y.**, Plant carbonic anhydrase. Properties and carbon dioxide hydration kinetics, *Biochemistry,* 12, 5127, 1973.
11. **Pocker, Y. and Ng, J. S. Y.**, Plant carbonic anhydrase. Hydrase activity and its reversible inhibition, *Biochemistry,* 13, 5116, 1974.
12. **Pocker, Y. and Miksch, R. R.**, Plant carbonic anhydrase. Properties and bicarbonate dehydration kinetics, *Biochemistry,* 17, 1119, 1978.
13. **Rossi, C., Chersi, A., and Cortivo, M.**, Studies on carbonic anhydrase from spinach leaves: isolation and properties, in CO_2: Chemical, Biochemical and Physiological Aspects, Forster, R. E., Edsall, J. T., Otis, A. B., and Roughton, F. J. W., Eds., National Aeronautics and Space Administration, Washington, D. C., 1969, 131.
14. **Kandel, M., Gornall, A. G., Cybulsky, D. L., and Kandel, S. I.**, Carbonic anhydrase from spinach leaves. Isolation and some chemical properties, *J. Biol. Chem.,* 253, 679, 1978.
15. **Tobin, A. J.**, Carbonic anhydrase from parsley leaves, *J. Biol. Chem.,* 245, 2656, 1970.
16. **Kisiel, W. and Graf, G.**, Purification and characterisation of carbonic anhydrase from *Pisum sativum, Phytochemistry,* 11, 113, 1972.
17. **Walk, R-A. and Metzner, H.**, Reinigung und charakterisierung von chloroplasten-carbonat-dehydratase (Isoenzym I) aus blattern von *Lactuca sativa, Hoppe-Seylers Z. Physiol. Chem.,* 356, 1733, 1975.
18. **Atkins, C. A., Patterson, B. D., and Graham, D.**, Plant carbonic anhydrases. I. Distribution of types among species, *Plant Physiol.,* 50, 214, 1972.
19. **Dwyer, M. R., Hockley, D. G., and Graham, D.**, unpublished data, 1979.
20. **Pocker, Y. and Sarkanen, S.**, Carbonic anhydrase: structure, catalytic versatility and inhibition, *Adv. Enzymol. Relat. Subj. Biochem.,* 47, 149, 1978.
21. **Everson, R. G.**, Carbonic anhydrase and CO_2 fixation in isolated chloroplasts, *Phytochemistry,* 9, 25, 1970.
22. **Everson, R. G.**, Carbonic anhydrase in photosynthesis, in *Photosynthesis and Photorespiration,* Hatch, M. D., Osmond, C. B., and Slatyer, R. O., Eds., Wiley-Interscience, New York, 1971, 275.
23. **Swader, J. A. and Jacobson, B. S.**, Acetazolamide inhibition of photosystem II in isolated spinach chloroplasts. *Phytochemistry,* 11, 65, 1972.
24. **Graham, D., Perry, G. L., and Atkins, C. A.**, In search of a role for carbonic anhydrase in photosynthesis, In Mechanisms of Regulation of Plant Growth, Bieleski, R. L., Ferguson, A. R., and Cresswell, M. M., Eds., *R. Soc. N.Z. Bull.,* 12, 251, 1974.
25. **Pocker, Y. and Miksch, R. R.**, Plant carbonic anhydrase. Properties and bicarbonate dehydration kinetics, *Biochemistry,* 17, 1119, 1978.
26. **Kachru, R. B. and Anderson, L. E.**, Chloroplast and cytoplasmic enzymes. V. Pea-leaf carbonic anhydrases, *Planta,* 118, 235, 1974.
27. **Komarova, Y. M., Terekhova, I.V., Doman, N. G., and Al'bitskaya, O. N.**, Carboanhydrase of blue-green algae *Spirulina platensis, Biochemistry,* 41, 150, 1976.
28. **Döhler, G.**, Carbonic anhydrase levels and enzymes of the glycolate pathway in the blue-green alga *Anacystis nidulans, Planta,* 117, 97, 1974.
29. **Ingle, R. K. and Colman, B.**, Carbonic anhydrase levels in blue-green algae, *Can. J. Bot.,* 53, 2385, 1975.
30. **Graham, D. and Smillie, R. M.**, Carbonate dehydratase in marine organisms of the Great Barrier Reef, *Aust. J. Plant Physiol.,* 3, 113, 1976.
31. **Graham, D. and Reed, M. L.**, Carbonic anhydrase and the regulation of photosynthesis, *Nature (London) New Biol.,* 231, 81, 1971.
32. **Graham, D., Atkins, C. A., Reed, M. L., Patterson, B. D., and Smillie, R. M.**, Carbonic anhydrase, photosynthesis and light-induced pH changes, in *Photosynthesis and Photorespiration,* Hatch, M. D., Osmond, C. B., and Slatyer, R. O., Eds., John Wiley & Sons, New York, 1971, 267.
33. **Findanegg, G. R.**, Correlations between accessibility of carbonic anhydrase for external substrate and regulation of photosynthetic use of CO_2 and HCO_3^- by *Scenedesmus obliquus, Z. Pflanzenphysiol.,* 79, 428, 1976.
34. **Chen, T. C., Brown, R. H., and Black, C. C.**, CO_2 compensation concentration, rate of photosynthesis, and carbonic anhydrase activity of plants, *Weed Sci.,* 18, 399, 1970.
35. **Rathnam, C. K. M. and Das, V. S. R.**, Inter- and intracellular distribution of carbonic anhydrase, PEP carboxylase and RuDP carboxylase in leaves of *Eleusine coracana,* a C-4 plant, *Z. Pflanzenphysiol.,* 75, 360, 1975.
36. **Reyss, A. and Prioul, J. L.**, Carbonic anhydrase and carboxylase activities from plants (*Lolium multiflorum*) adapted to different light regimes, *Plant Sci. Lett.,* 5, 189, 1975.
37. **Kondo, K., Yonezawa, D., and Chiba, H.**, Studies on plant carbonic anhydrase. I. Confirmation of plant carbonic anhydrase and preliminary isolation, Bulletin of the Research Institute for Food Science, Kyoto University, Kyoto, Japan, No. 8, 1, 1952.
38. **Poincelot, R. P.**, The distribution of carbonic anhydrase and ribulose diphosphate carboxylase in maize leaves, *Plant Physiol.,* 50, 336, 1972.

39. **Reed, M. L.**, The intracellular location of carbonate dehydratase (carbonic anhydrase) in leaf tissue, *Plant Physiol.*, 63, 216, 1979.
40. **Van, T. K., Haller, W. T., and Bowes, G.**, Comparison of the photosynthetic characteristics of three submerged aquatic plants, *Plant Physiol.*, 58, 761, 1976.
41. **Romanova, A. K., Rusinova, N. G., and Kornitskaya, V. M.**, Participation of carbonic anhydrase in chemosynthetic assimilation of carbonic anhydrase in *Thiobacillus thioxidans* 58R, *Dokl. Biochem.*, 203, 125, 1972.
42. **Brundell, J., Falkbring, S. O., and Nyman, P. O.**, Carbonic anhydrase from *Neisseria sicca*, Strain 6021. II. Properties of the purified enzyme, *Biochim. Biophys. Acta*, 284, 311, 1972.
43. **Poincelot, R. P.**, Intracellular distribution of carbonic anhydrase in spinach leaves, *Biochim. Biophys. Acta*, 258, 637, 1972.
44. **Jacobson, B. S., Fong, F., and Heath, R. L.**, Carbonic anhydrase of spinach. Studies on its location, inhibition, and physiological function, *Plant Physiol.*, 55, 468, 1975.
45. **Chang, C. C.**, Carbonic anhydrase of the cotton plant, *Phytochemistry*, 14, 119, 1975.
46. **Nishimura, M., Graham, D., and Akazawa, T.**, Isolation of chloroplasts and other cell organelles from spinach leaf protoplasts, *Plant Physiol.*, 58, 309, 1976.
47. **Atkins, C. A., Hockley, D. G., and Graham, D.**, unpublished data, 1972.
48. **Champagnol, F.**, Mise en évidence d'une activité anhydrase carbonique dans les broyats de racines de quelque plantes calcifuges, *C. R. Acad. Sci. Ser. D*, 282, 1273, 1976.
49. **Reed, M. L. and Graham, D.**, Control of photosynthetic carbon dioxide fixation during an induction phase in *Chlorella*, *Plant Physiol.*, 43, S29, 1968.
50. **Reed, M. L. and Graham, D.**, Carbon dioxide and the regulation of photosynthesis: activities of photosynthetic enzymes and carbonate dehydratase (carbonic anhydrase) in Chlorella after growth or adaptation in different carbon dioxide concentrations, *Aust. J. Plant Physiol.*, 4, 87, 1977.
51. **Nelson, E. B., Cenedella, A., and Tolbert, N. E.**, Carbonic anhydrase levels in *Chlamydomonas*, *Phytochemistry*, 8, 2305, 1969.
52. **Raven, J. A. and Glidewell, S. M.**, C_4 characteristics of photosynthesis in the C_3 alga *Hydrodictyon africanum*, *Plant, Cell Environ.*, 1, 185, 1978.
53. **Randall, P. J. and Bouma, D.**, Zinc deficiency, carbonic anhydrase and photosynthesis in leaves of spinach, *Plant Physiol.*, 52, 229, 1973.
54. **Enns, T.**, Facilitation by carbonic anhydrase of carbon dioxide transport, *Science*, 155, 44, 1967.
55. **Broun, G., Selegny, E., Tran Minh, C., and Thomas, D.**, Facilitated transport of CO_2 across a membrane bearing carbonic anhydrase, *FEBS Lett.*, 7, 223, 1970.
56. **Werdan, K. and Heldt, H. W.**, Accumulation of bicarbonate in intact chloroplasts following a pH gradient, *Biochim. Biophys. Acta*, 283, 430, 1972.
57. **Black, C. C.**, Photosynthetic carbon fixation in relation to net CO_2 uptake, *Annu. Rev. Plant Physiol.*, 24, 253, 1973.
58. **Walker, D. A.**, Interaction between cytoplasm and plastids, in *Encyclopedia of Plant Physiology*, Vol. 3, Heber, U., Ed., Springer, Heidelberg, 1976, 85.
59. **Bowes, G.**, Carbonic anhydrase in marine algae, *Plant Physiol.*, 44, 726, 1969.
60. **Litchfield, C. D. and Hood, D. W.**, Evidence of carbonic anhydrase in marine and freshwater algae, *Int. Ver. Theor. Angew. Limnol. Verh.*, 15, 817, 1964.
61. **Van, T. K., Haller, W. T., and Bowes, G.**, Comparison of photosynthetic characteristics of three submersed aquatic plants, *Plant Physiol.*, 58, 761, 1976.

ASSIMILATION OF NITRATE, NITRITE, AND AMMONIA IN ALGAE

Masayuki Ohmori

Nitrate assimilation in algae proceeds in two separate steps.[1] The first step is the reduction of nitrate to nitrite catalyzed by nitrate reductase(NADH: nitrate oxidoreductase, EC 1.6.6.1) with NADH as an electron donor. The second step is a further reduction of nitrite to ammonia catalyzed by nitrite reductase (ferredoxin: nitrite oxidoreductase, EC 1.6.6.4) with ferredoxin(Fd) as the electron donor.

Nitrate reductase is a complex of two enzymes participating sequentially in the reduction of nitrate by NADH: (1) a NADH-specific diaphorase and (2) nitrate reductase proper. The former contains FAD and the latter contains molybdenum. A *b*-type cytochrome is associated with this enzyme. In green algae, nitrate reductase appears in the soluble fraction of cell-free extract and is unable to accept electrons directly from reduced ferredoxin. On the other hand, nitrate reductase in blue-green algae appears in a particulate fraction and can accept electrons from reduced ferredoxin. Some biochemical characteristics of nitrate reductases are shown in Tables 1 and 2.

Ammonia is assimilated by algal cells mainly by two different pathways.[2] One involves glutamate dehydrogenase (L-glutamate: NAD(P) oxidoreductase, EC 1.4.1.3) (see Equation 1).

$$NH_3 + \text{2-oxoglutarate} + NADPH + H^+ \rightarrow \text{glutamate} + NADP^+ + H_2O \quad (1)$$

The other is mediated by glutamine synthetase (L-glutamate: ammonia ligase (ADP), EC 6.3.1.2) (see Equation 2) and glutamate synthase (L-glutamine: ferredoxin oxidoreductase (transaminating), EC 1.4.7.1.) (see Equation 3).

$$NH_3 + \text{glutamate} + ATP \rightarrow \text{glutamine} + ADP + Pi \quad (2)$$

$$\text{Glutamine} + \text{2-oxoglutarate} + Fd(red) \rightarrow 2\ \text{glutamate} + Fd(ox) \quad (3)$$

Ammonia is preferentially assimilated by algae when nitrate or nitrite are present simultaneously. Ammonia inhibits the activity of nitrate reductase in vivo and also represses the synthesis of the enzyme (see Table 3). Assimilation of nitrate, nitrite, and ammonia is enhanced by light (see Table 4).

Table 1
CHARACTERISTICS OF NITRATE REDUCTASE OF ALGAE

Source	Mol Wt	Km(μM)	Inhibitors and Ki (μM)	Other properties	Ref.
Green algae and diatom					
Chlorella vulgaris	356,000	NO^-_3(84)	NO^-_2(280) CNS^-(0.3) CNO^-(0.2) N^-_3(0.05)	Subunit: 100,000 × 3 FAD:Heme:Mo = 1:1:0.8 $S_{20,w}$ = 9.7S Activated by NO^-_3 or ferri-cyanide.	3—5
Chlorella fusca (= pyrenoidosa)	500,000	NADH(10) FAD(2.6)	*p*-CMB KCN	Inactivated at 45°C, 5 min Inactivated by exchanging molybdenum for tungstate	6—8
Ankistrodesmus braunii	475,000	NO^-_3(190) NADH(11) $FMNH_2$(750)	NO^-_2(400) N^-_3(0.8) CNS^-(12)	Subunit: 60,000	9
Chlamydomonas reinhardii	500,000	NO^-_3(125) NADH(7) $FMNH_2$(45) FAD(80)	N^-_3(0.55) CNO^-(4.4) ClO^-(256) *p*-HMB, KCN	Optimum pH: 7.3 Inactivated at 40°C, 6 min FAD protects the heat inactivation	10, 11
Thalassiosira pseudonana	330,000	NO^-_3(62) NADH(15) FAD(420)	CN^- N^-_3 *p*-HMB	Inactivated by 40°C, 5 min Optimum pH: 7—8 S_{20w} = 10.3—10.5	12
Dunaliella tertiolecta	500,000	NO^-_3(260) NADH(180) FAD(60)	CN^- *p*-CMB	Inactivated by 55°C, 5 min Optimum pH: 7.5—7.7	13
Blue-green algae					
Anabaena cylindrica	—	NO^-_3(50) NADH(100) FAD(300)		Ferredoxin dependent Solubilized by Triton® X-100 treatment	14, 15
Anacystis nidulans	—	NO^-_3(60)		Ferredoxin dependent Optimum pH: 10.0—10.5	16

Table 2
CHARACTERISTICS OF NITRITE REDUCTASE OF ALGAE

Source	Mol wt	K_m for NO^-_2 (μM)	Electron donor	Specific activity (μmol/min/mg protein)	Optimum pH	Inhibitors	Ref.
Chlorella fusca	63,000	—	Reduced Fd, flavodoxin	51.7	—	99% inhibition by 1 mM KCN	17,18
Dunaliella tertiolecta	70,000	110	Reduced Fd	2	7—8	—	19
Anabaena cylindrica	68,000	50	Reduced Fd (K_m = 5 μM)	0.2	7.4—7.6	100% inhibition by 3.8 mM 2,4-DNP; 85% inhibition by 1.2 mM KCN	20,21

Table 3
EFFECT OF AMMONIA ON NITRATE REDUCTASE ACTIVITY OF ALGAE

Source	Substrate (m*M*)	Incubation periods (hr)	Relative activity			Ref.
			NADH-diaphorase	NADH-NO^-_3 reductase	$FMNH_2$-NO^-_3 reductase	
Chlorella fusca	NO_3^-(8)	1.5	100	100	100	22
	NO_3^- + NH_4^+(8)	1.5	181	11	7	
Chlamydomonas reinhardii	NO_3^-(8)	1	100	100	100	23
	NO_3^- + NH_4^+(16)	1	104	5	4	
Chlamydomonas reinhardii	NO_3^-(16)	24	100	100	100	24
	NO_3^- + NH_4^+(16)	24	10	5	3	

Table 4
EFFECT OF LIGHT ON NITRATE, NITRITE, AND AMMONIA ASSIMILATION IN ALGAE

Source	Light	Dark	Comments	Ref.
Dunaliella tertiolecta				
NO_3^- assimilation	2.7	0.4	μmol/hr/mg Chl, light intensity: 27,000—32,000; 1x	25
NO_2^- assimilation	4.5	1.8		
Ditylum brightwellii				
NH_4^+ assimilation	8.5	3.1	μmol/hr/L culture; light intensity: 2.1mW/cm²	26
Chlamydomonas reinhardii				
NO_3^- assimilation	0.117	0.013	μmol/hr/mg dry weight; light intensity: 1.4 mW/cm²	
NH_4^+ assimilation	0.127	0.000		27

REFERENCES

1. **Hewitt, E. J.,** Assimilatory nitrate-nitrite reduction, *Annu. Rev. Plant Physiol.,* 26, 73, 1975.
2. **Miflin, B. J. and Lea, P. J.,** The pathway of nitrogen assimilation in plants, *Phytochemistry,* 15, 873, 1976.
3. **Solomonson, L. P. and Vennesland, B.,** Properties of a nitrate reductase of *Chlorella, Biochim. Biophys. Acta,* 267, 544, 1972.
4. **Solomonson, L. P., Jetschmann, K., and Vennesland, B.,** Reversible inactivation of the nitrate reductase of *Chlorella vulgaris* Beijerinck, *Biochim. Biophys. Acta,* 309, 32, 1973.
5. **Solomonson, L. P., Lorimer, G. H., Hall, R. L., Borchers, R., and Bailey, J. L.,** Reduced nicotinamide adenine dinucleotide-nitrate reductase of *Chlorella vulgaris*: purification, prosthetic groups, and molecular properties, *J. Biol. Chem.,* 250, 4120, 1975.
6. **Zumft, W. G., Paneque, A., Aparicio, P. J., and Losada, M.,** Mechanism of nitrate reduction in *Chlorella, Biochem. Biophys. Res. Commun.,* 36, 980, 1969.
7. **Paneque, A., Vega, J. M., Cárdenas, J., Herrera, J., Aparicio, P. J., and Losada, M.,** ^{185}W-labelled nitrate reductase from *Chlorella, Plant Cell Physiol.* 13, 175, 1972.
8. **Schloemer, R. H. and Garrett, R. H.,** Partial purification of the NADH-nitrate reductase complex from *Chlorella pyrenoidosa, Plant Physiol.,* 51, 591, 1973.
9. **Ahmed, J. and Spiller, H.,** Purification and some properties of the nitrate reductase from *Ankistrodesmus braunii, Plant. Cell Physiol.,* 17, 1, 1976.
10. **Barea, J. L. and Cárdenas, J.,** The nitrate-reducing enzyme system of *Chlamydomonas reinhardii, Arch. Microbiol.,* 105, 21, 1975.
11. **Barea, J. L., Maldonado, J. M., and Cárdenas J.,** Further characterization of nitrate and nitrite reductases from *Chlamydomonas reinhardii, Physiol. Plant.,* 36, 325, 1976.
12. **Amy, N. K. and Garrett, R. H.,** Purification and characterization of the nitrate reductase from the diatom *Thalassiosira pseudonana, Plant Physiol.,* 54, 629, 1974.
13. **LeClaire, J. A. and Grant, B. R.,** Nitrate reductase from *Dunaliella tertiolecta,* purification and properties, *Plant Cell Physiol.,* 13, 899, 1972.
14. **Hattori, A. and Myers, J.,** Reduction of nitrate and nitrite by subcellular preparations of *Anabaena cylindrica,* II. Reduction of nitrate to nitrite, *Plant Cell Physiol.,* 8, 327, 1967.
15. **Hattori, A.,** Solubilization of nitrate reductase from the blue-green alga *Anabaena cylindrica, Plant Cell Physiol.,* 11, 975, 1970.
16. **Manzano, C., Candau, P., Gomez-Moreno, C., Relimpio, A. M., and Losada, M.,** Ferredoxin-dependent photosynthetic reduction of nitrate and nitrite by particles of *Anacystis nidulans, Mol. Cell. Biochem.,* 10, 161, 1976.
17. **Zumft, W. G. and Spiller, H.,** Characterization of a flavodoxin from the green alga *Chlorella, Biochem. Biophys. Res. Commun.,* 45, 112, 1971.
18. **Zumft, W. G.,** Ferredoxin:nitrite oxidoreductase from *Chlorella,* purification and properties, *Biochim. Biophys. Acta,* 276, 363, 1972.
19. **Grant, B. R.,** Nitrite reductase in *Dunaliella tertiolecta*: isolation and properties, *Plant Cell Physiol.,* 11, 55, 1970.

20. **Hattori, A. and Myers, J.,** Reduction of nitrate and nitrite by subcellular preparations of *Anabaena cylindrica.* I. Reduction of nitrite to ammonia, *Plant Physiol.,* 41, 1031, 1966.
21. **Hattori, A. and Uesugi, I.,** Purification and properties of nitrite reductase from the blue-green alga *Anabaena cylindrica, Plant Cell Physiol.,* 9, 689, 1968.
22. **Losada, M., Paneque, A., Aparicio, P. J., Vega, J. M., Cárdenas, J., and Herrera, J.,** Inactivation and repression by ammonium of the nitrate reducing system in *Chlorella, Biochem. Biophys. Res. Commun.,* 38, 1009, 1970.
23. **Losada, M., Herrera, J., Maldonado, J. M., and Paneque, A.,** Mechanism of nitrate reductase reversible inactivation by ammonia in *Chlamydomonas, Plant Sci. Lett.,* 1, 31, 1973.
24. **Herrera, J., Paneque, A., Maldonado, J. M., Barea, J. L., and Losada, M.,** Regulation by ammonia of nitrate reductase synthesis and activity in *Chlamydomonas reinhardi, Biochem. Biophys. Res. Commun.,* 48, 996, 1972.
25. **Grant, B. R.,** The action of light on nitrate and nitrite assimilation by the marine chlorophyte, *Dunaliella tertiolecta* (Butcher), *J. Gen. Microbiol.,* 48, 379, 1967.
26. **Eppley, R. W. and Rogers, J. N.,** Inorganic nitrogen assimilation of *Ditylum brightwellii,* a marine plankton diatom, *J. Phycol.,* 6, 344, 1970.
27. **Thacker, A. and Syrett, P. J.,** The assimilation of nitrate and ammonium by *Chlamydomonas reinhardi, New Phytol.,* 71, 423, 1972.

ASSIMILATION OF NITRATE, NITRITE, AND AMMONIA IN VASCULAR PLANTS

K.W. Joy

INTRODUCTION

Plants and some microorganisms utilize inorganic forms of nitrogen, whereas animals require an organic nitrogen supply. This ability further stresses the role of plants as primary producers in the biosphere, since nitrogen as well as carbon must cycle through primary assimilatory processes in plants. The energy required for inorganic nitrogen assimilation is derived, in some cases directly, from energy trapped in photosynthetic reactions.

The main source of inorganic nitrogen available to terrestrial plants in most soils is nitrate; occasionally ammonia may also be available.* Utilization of nitrate involves reduction first to nitrite, then further reduction to ammonia, in which form it is then incorporated into an organic compound.

The term assimilation, in relation to plant nitrogen metabolism, has been used in several ways. It is used in a restricted sense to refer to a single step in which inorganic nitrogen (ammonia) enters a primary organic compound which can then be utilized in the synthesis of the other nitrogenous components required by the plant. In a broader sense, it refers to a sequence of related reactions, including nitrate reduction, which result in the incorporation of inorganic nitrogen into organic form. (The term is also used occasionally, incorrectly, to denote the uptake of inorganic nitrogen into a cell or organism.) Over the last few years, several reviews have covered nitrate reduction,[1,2] assimilation of ammonia,[3,4] and the specific role of chloroplasts in these processes.[5,6]

NITRATE REDUCTION

Two enzyme systems operate in sequence in the reduction of nitrate to ammonia.
Nitrate Reductase (NR)

$$NO_3^- + NADH + H^+ \rightarrow NO_2^- + H_2O + NAD^+$$

Various molecular weights have been reported for this rather unstable enzyme, ranging from 160,000 to 600,000. The enzyme contains FAD and molybdenum which are involved in electron transport within the protein. Plants supplied with tungstate synthesize an inactive protein in which tungsten replaces molybdenum. Experimentally a number of electron donors (such as methyl and benzyl viologens, $FADH_2$) can be utilized by NR, but the most effective natural donor is usually found to be NADH. Additional activity with NADPH is found in some plants and has been attributed to the presence of nucleotide phosphatase, although recent work describes the presence of a distinct NADPH dependent NR which is resistant to an inhibitor of NADH-NR.[7]

Nitrate reductase has been found in both root and leaf tissue. Within the cell, NR is usually found in the soluble cytoplasmic fraction,[8,9] although in leaf tissue, association with chloroplasts and the chloroplast envelope[10] has been reported. An association with the chloroplast envelope is attractive, in view of the probable role of triosephosphate from photosynthesis in the provision of reducing power[11] and the need to transfer nitrite to the nitrite reductase within the chloroplast (see Figure 1). However, the number of negative observations and reports of nonspecific binding to membranes[8] urge caution in the interpretation of localization data.

* For simplicity, "NH_3" is used throughout to denote ammonia or the ammonium ion.

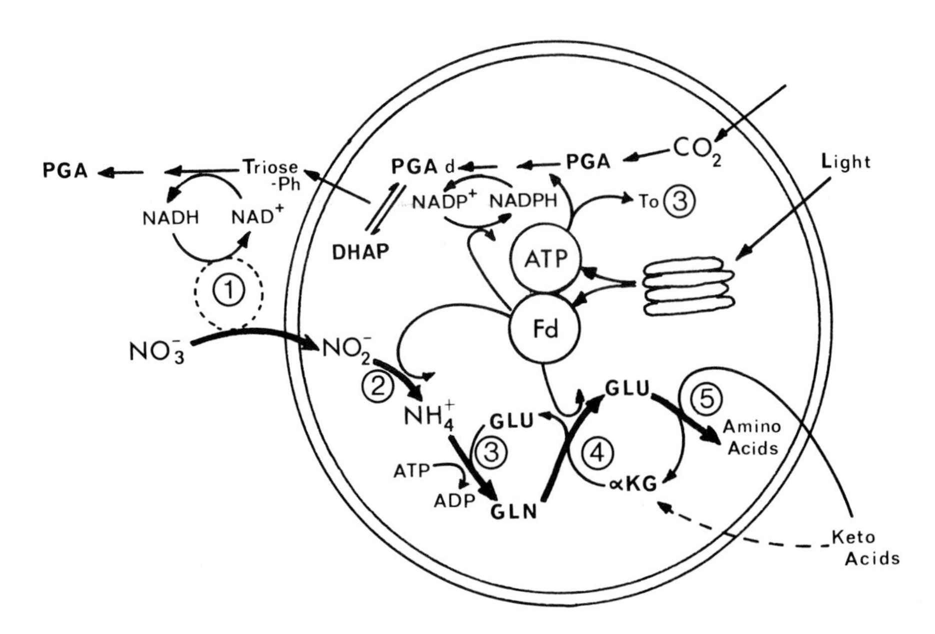

FIGURE 1. Pathways of nitrate assimilation in the chloroplast. 1, Nitrate reductase; 2, nitrite reductase; 3, glutamine synthetase; 4, glutamate synthase; 5, aminotransferases. The heavy arrows show the net flow of assimilate.

Regulation — In leaves, presence of NR is related to illumination, and levels fall in darkness. The enzyme is substrate inducible, and one effect of light may be in mediating transfer of nitrate to a site where it is effective as inducer; it has been shown that tissue may contain a nitrate pool that is ineffective as inducer.[12] Light may also influence the enzyme through an effect on protein synthesis.[13] The enzyme turns over rapidly, and in cultured pith cells, a half-life of about 4 hr has been measured;[14] inactivation of NR by a specific protein with proteolytic properties has been reported.[15] The decrease in leaf NR activity in darkness has generally been assumed to result from irreversible turnover or inactivation, but recently a reversible, dark activated inhibitor for NADH dependent NR has been found in soybean leaves.[7] Although the nitrate reductase from microorganisms is subject to repression by ammonia, end-product effects in plants are less well established. Some effects of ammonia, supplied in the presence of nitrate, may be because of restriction of nitrate uptake rather than a direct regulatory influence on NR. In green tissues, the availability of photosynthetically generated triose phosphate is likely to play a major role in regulation of nitrate reduction; in roots, end-product effects from ammonia or amino acids may be expected to be more important.

Nitrite Reductase (NIR)

$$NO_2^- + 6e^- + 7H^+ \rightarrow NH_3 + 2H_2O$$

This complex reduction takes place without the appearance of any free intermediates. The early suggestion that hydroxylamine was an intermediate is now thought unlikely, although hydroxylamine reductase activity is associated with some NIR preparations. The enzyme is smaller and more stable than NR and contains iron. NIR requires an electron donor of quite low redox potential. In leaf tissue, ferredoxin is the electron donor, and this is in accord with the now well-accepted localization of this enzyme in the chloroplast.[8,16] NADPH is ineffective as electron donor unless diaphorase and an intermediate carrier are added.[17] In roots, NIR is undoubtedly present and active and is located in plastids,[9,16] but no natural electron donor has been identified for this system.

Regulation — NIR is substrate induced; induction in the presence of nitrate may be because of direct nitrate induction or the rapid production of nitrite by nitrate reduction. No regulatory process seems to moderate the activity of the enzyme, and this is not surprising in view of the high toxicity of nitrite which must be metabolized continuously to prevent build up in the tissue.

AMMONIA ASSIMILATION

Until 1973, the main route for assimilation of ammonia into amino acids was thought to be through reductive amination of 2-oxoglutarate to glutamate by the enzyme glutamate dehydrogenase (GDH). However, GDH is located mainly in mitochondria[18] and has a low affinity for ammonia.[18,19] Although GDH has been detected in chloroplasts,[19] the principal location and low affinity continuously raised doubts about the function of the enzyme in assimilation of ammonia arising from nitrite reduction in chloroplasts; this is particularly important in view of the high toxicity of ammonia for photosynthetic processes. The discovery that chloroplasts contained considerable amounts of glutamine synthetase[20,21] suggested the possibility of a more efficient route for ammonia assimilation. Convincing evidence for the operation of this route was presented by the demonstration that a ferredoxin-dependent glutamate synthase was also present in chloroplasts,[22] thus providing a convenient mechanism for the conversion of the amide group (newly assimilated into glutamine) to the α-amino group (in glutamate), where it is more available for further amino acid synthesis.

Glutamine Synthetase (GS)

$$\text{Glutamate} + \text{ATP} + \text{NH}_3 \rightarrow \text{glutamine} + \text{ADP} + \text{H}_2\text{O} + \text{P}_i$$

GS from a number of plant sources has been investigated. It has a molecular weight around 360,000, requires Mg^{++} (or Mn^{++}), and is rather unstable.[23] The K_m for ammonia[24] is around $2 \times 10^{-5} M$; thus the affinity for this compound is over a hundred-fold greater than that of glutamate dehydrogenase which would therefore be unable to compete with GS if both enzymes were present together at a site of ammonia production. ADP and AMP inhibit GS from leaves, and the enzyme is regulated by energy charge;[25] measurements on roots show a correlation between in vivo energy charge and GS activity.[26] Carbamyl phosphate and some amino acids also modulate the activity of GS.[25] In leaves GS is found predominantly in chloroplasts,[20,21] although recent work suggests that the enzyme is also found in cytoplasm;[27] in roots, the enzyme is also found to be associated with plastids.[16]

Glutamate Synthase (GOGAT: *glutamine amide: 2-oxoglutarate amino transferase oxido-reductase*)

$$\text{Glutamine} + \alpha\text{-(2-)oxoglutarate} + 2[\text{H}]\text{-carrier} \rightarrow 2\ \text{glutamate} + \text{oxidized carrier}$$

Although GOGAT is not directly involved in assimilation, it is closely associated with the process as it provides a path for the flow of nitrogen from glutamine, forming glutamate which is readily used in the distribution of nitrogen to secondary amino acids.

In leaves, the enzyme is found in chloroplasts and is dependent on ferredoxin as the electron donor, but is inactive with NAD(P)H.[22] In contrast, the root enzyme utilizes NAD(P)H (and is also active with ferredoxin which is, however, not found in roots).[28] NAD(P)H-dependent GOGAT activity has also been found in tissue cultures and in developing pea seeds, where it may function in the utilization of glutamine transported to the growing pods.[29]

Other Enzymes with Assimilatory Potential

Glutamate Dehydrogenase (GDH)

$$NH_3 + \alpha\text{-oxoglutarate} + NAD(P)H + H^+ \rightleftharpoons \text{glutamate} + NAD(P)^+ + H_2O$$

GDH is found in roots[18] and leaves[19] of higher plants. As discussed above, the affinity for ammonia of this enzyme makes it unlikely to be the main route for primary assimilation of ammonia following nitrate reduction. The ubiquitous occurrence of GDH suggests that the enzyme does have an important function. It is possible that it may have an assimilatory role in conditions when high ammonia concentrations arise in cytoplasm — for example, during photorespiration (see below) or when ammonia is available as nitrogen source for the tissue. Labeling studies with ^{15}N do not give clear cut results in such cases, since glutamate is also formed rapidly via glutamine synthetase and GOGAT.

The enzyme reaction is reversible, and GDH is likely to be involved in catabolic processes (for example, during mobilization of nitrogen in tissue senescence and in germination).

Asparagine Synthetase

The synthesis of asparagine was originally thought to be similar to the glutamine synthetase reaction. However, careful investigation[30-32] showed that glutamine was the most effective amide donor; the enzyme does have activity with ammonia, but has a considerably greater affinity for glutamine. AMP is produced in the reaction.[31]

$$ATP + \text{aspartate} + \text{glutamine} \rightarrow \text{asparagine} + \text{glutamate} + AMP + PP_i$$

Labeling experiments with ^{15}N in pea leaves confirmed that glutamine acts as the amide donor for asparagine in vivo,[33] and the results suggest that asparagine is effective in distribution and storage of the nitrogen assimilated through glutamine. It is possible, however, that this enzyme could have a role in primary assimilation in some circumstances.

Carbamoyl Phosphate Synthetase

$$NH_3 + CO_2 + 2ATP + H_2O \rightarrow \text{carbamoyl phosphate} + 2ADP + P_i$$

Although it is capable of using ammonia, this enzyme also is found to be more effective with glutamine as the nitrogen donor. The affinity for glutamine is about 100-fold greater than for ammonia,[34] and glutamine is thought to be the donor under normal physiological conditions.

Other Amino Acid Dehydrogenases

Reductive amination reactions, similar to that of glutamate dehydrogenase, have been reported for aspartate, alanine, and a number of other amino acids. However, there is little evidence to suggest that these activities are significant in ammonia assimilation.

OVERVIEW OF THE ASSIMILATORY PROCESS

In the Leaf

Many studies suggest that assimilatory processes in the leaf are largely light dependent, and this is in accord with the relationships of the enzymic steps as outlined in Figure 1. It is likely that nitrate reductase, the enzyme which appears to be most strongly modulated by environmental effects, regulates flow of nitrogen into the assimilatory pathway. Once nitrate is reduced to nitrite, there should be no need for control in the pathway, since the intermediates nitrite and ammonia are both toxic; however, rapid and reversible changes in GS activity have been found (in *Lemna*) in response to changes in availability of ammonia and light.[35] In rice leaves, there are two forms of GS, one of which is low in etiolated plants and increases to become the major form on illumination.[36]

The chloroplast must be regarded as the main site for assimilation in the leaf. In C4 plants, two types of photosynthetic cells and chloroplasts exist, with an unequal distribution of the enzymes of nitrate assimilation. NR and NIR are found predominantly in mesophyll cells;[37,38] GS and GOGAT are also found in the mesophyll, but there is some controversy about the presence of the latter two enzymes in the bundle sheath.[37,38]

In darkness, some assimilation of nitrate[39] and ammonia[40] can occur, although this will be limited by availability of energy and also possibly by the nitrate supply, since delivery in the transpiration stream will be curtailed at night.

In addition to providing a net increase of organic nitrogen, assimilatory processes must have a further important function in leaves of C3 plants. A key reaction in photorespiration generates ammonia:

$$2\text{ Glycine} + H_2O \rightarrow \text{serine} + CO_2 + NH_3 + 2[H]$$

and flux through this pathway may considerably exceed the rate of net assimilation of nitrogen. The breakdown of glycine occurs in mitochondria, and the essential reassimilation of ammonia may provide a role for glutamate dehydrogenase, however, recent work suggests that glutamine synthetase is the principal enzyme involved.[41]

The relationship of nitrogen assimilation to carbon metabolism should be mentioned. Although nitrogen assimilation occurs in the chloroplast, recently photoassimilated carbon does not readily enter into primary amino acids. Evidence suggests that keto acids for the synthesis of glutamate and other amino acids are supplied from elsewhere in the cell.[6,42,43]

In the Root

The enzymes involved in assimilation in the leaf are also present in roots, although the electron donor preferences (and availability) are different. Although roots may assimilate a major part of incoming nitrogen, unfortunately recent investigations have given much more attention to the leaf than to the root, and a number of questions remain. In particular, it is not yet clear(1) whether NIR and GOGAT utilize NAD(P)H as a source of reductant or whether an unknown more strongly reduced compound is required; (2) whether GDH may take part in assimilation in the root — particularly when ammonia is available at relatively high concentrations.

Site of Assimilation and Transport of Primary Assimilates

Inorganic nitrogen must be transported to the sites of assimilation, and then further transport of primary assimilates will be necessary to the main sites of nitrogen utilization (e.g., growing leaves, developing seeds). In vascular plants, movement from root to shoot occurs predominantly in the transpiration stream within the xylem, while redistribution in the shoot involves phloem transport. The proportion of total assimilation carried out by roots varies greatly between plants. This is reflected in the composition of nitrogenous compounds in xylem sap moving to the shoot, where the proportion of nitrate may vary from less than 10 to more than 90% of the nitrogen leaving the root; the organic nitrogen is transported in a range of forms, predominantly as amides.[44] Ammonia is not found as a significant component of xylem sap, and all evidence suggests that nitrate reduction and ammonia assimilation occur together at the same location.

Although glutamine is the main product of primary assimilation, glutamic acid and asparagine are clearly of fundamental importance in transport and in the reactions of redistribution of nitrogen to secondary compounds; they must be regarded as being closely related, as accessory compounds, with the assimilatory process.

REFERENCES

1. **Beevers, L. and Hageman, R. H.,** The role of light in nitrate metabolism of higher plants, *Photophysiology*, 7, 85, 1972.
2. **Hewitt, E. J.,** Assimilatory nitrate-nitrite reduction, *Annu. Rev. Plant Physiol.*, 26, 73, 1975.
3. **Miflin, B. J. and Lea, P. J.,** The pathway of nitrogen assimilation in plants, *Phytochemistry*, 15, 873, 1976.
4. **Miflin, B. J. and Lea, P. J.,** Amino acid metabolism, *Annu. Rev. Plant Physiol.*, 28, 299, 1977.
5. **Leech, R. M. and Murphy, D. J.,** The cooperative function of chloroplasts in the biosynthesis of small molecules, in *The Intact Chloroplast*, Barber, J., Ed., Elsevier, Amsterdam, 1976.
6. **Givan, C. V. and Harwood, J. L.,** Biosynthesis of small molecules in chloroplasts of higher plants, *Biol. Rev.*, 51, 365, 1976.
7. **Jolly, S. O. and Tolbert, N. E.,** NADH-nitrate reductase inhibitor from soybean leaves, *Plant Physiol.*, 62, 197, 1978.
8. **Dalling, M. J., Tolbert, N. E., and Hageman, R. H.,** Intracellular location of nitrate reductase and nitrite reductase. I. Spinach and tobacco leaves, *Biochim. Biophys. Acta*, 283, 505, 1972.
9. **Dalling, M. J., Tolbert, N. E., and Hageman, R. H.,** Intracellular location of nitrate reductase and nitrite reductase. II. Wheat roots, *Biochim. Biophys. Acta*, 283, 513, 1972.
10. **Rathnam, C. K. M. and Das, V. S. R.,** Nitrate metabolism in relation to the aspartate-type C-4 pathway, *Can. J. Bot.*, 52, 2599, 1974.
11. **Klepper, L., Flesher, D., and Hageman, R. H.,** Generation of reduced nicotinamide adenine dinucleotide for nitrate reduction in green leaves, *Plant Physiol.*, 48, 580, 1971.
12. **Heimer, Y. M. and Filner, P.,** Regulation of the nitrate assimilation pathway in cultured tobacco cells, *Biochim. Biophys. Acta*, 230, 362, 1971.
13. **Travis, R. L. and Key, J. L.,** Correlation between polyribosome level and the ability to induce nitrate reductase in dark-grown corn seedlings, *Plant Physiol.*, 48, 617, 1971.
14. **Zielke, H. R. and Filner, P.,** Synthesis and turnover of nitrate reductase induced by nitrate in cultured tobacco cells, *J. Biol. Chem.*, 246, 1772, 1971.
15. **Wallace, W.,** Purification and properties of a nitrate reductase-inactivating enzyme, *Biochim. Biophys. Acta*, 341, 265, 1974.
16. **Miflin, B. J.,** The location of nitrite reductase and other enzymes related to amino acid biosynthesis in plastids of root and leaves, *Plant Physiol.*, 54, 550, 1974.
17. **Joy, K. W. and Hageman, R. H.,** Purification and properties of nitrite reductase from higher plants, and its dependence on ferredoxin, *Biochem. J.*, 100, 263, 1966.
18. **Pahlich, E. and Joy, K. W.,** Glutamate dehydrogenase from pea roots, *Can. J. Biochem.*, 49, 127, 1971.
19. **Lea, P. J. and Thurman, D. A.,** Intracellular location and properties of plant L-glutamate dehydrogenases, *J. Exp. Bot.*, 23, 440, 1972.
20. **O'Neal, D. and Joy, K. W.,** Localisation of glutamine synthetase in chloroplasts, *Nature (London) New Biol.*, 246, 61, 1973.
21. **Haystead, A.,** Glutamine synthetase in the chloroplasts of *Vicia faba*, *Planta*, 111, 271, 1973.
22. **Lea, P. J. and Miflin, B. J.,** Alternative route for nitrogen assimilation in plants, *Nature (London)*, 251, 614, 1974.
23. **O'Neal, D. and Joy, K. W.,** Glutamine synthetase of pea leaves, *Arch. Biochem. Biophys.*, 159, 113, 1973.

24. **O'Neal, D. and Joy, K. W.,** Glutamine synthetase of pea leaves: divalent cation effects, substrate specificity and other properties, *Plant Physiol.*, 54, 773, 1974.
25. **O'Neal, D. and Joy, K. W.,** Pea leaf glutamine synthetase: regulatory properties, *Plant Physiol.*, 55, 968, 1975.
26. **Weissman, G. S.,** Glutamine synthetase regulation by energy charge in sunflower roots, *Plant Physiol.*, 57, 339, 1976.
27. **Wallsgrove, R. M., Lea, P. J., and Miflin, B. J.,** Distribution of the enzymes of nitrogen assimilation within the pea leaf cell, *Plant Physiol.*, 63, 232, 1979.
28. **Miflin, B. J. and Lee, P. J.,** Glutamine and asparagine as nitrogen donors for reductant-dependent glutamate synthesis in pea roots, *Biochem. J.*, 149, 403, 1975.
29. **Beevers, L. and Storey, R.,** Glutamate synthetase in developing cotyledons of *Pisum sativum, Plant Physiol.*, 57, 862, 1976.
30. **Streeter, J. G.,** *In vivo* and *in vitro* studies on asparagine biosynthesis in soybean seedlings, *Arch. Biochem. Biophys.*, 157, 613, 1973.
31. **Rognes, S. E.,** Glutamine-dependent asparagine synthetase from *Lupinus luteus, Phytochemistry,* 14, 1975, 1975.
32. **Lea, P. J. and Fowden, L.,** The purification and properties of glutamine-dependent asparagine synthetase isolated from *Lupinus albus, Proc. R. Soc. London Ser. B.* 192, 13, 1975.
33. **Bauer, A., Joy, K. W. and Urquhart, A. A.,** Amino acid metabolism of pea leaves: labeling studies on utilisation of amides, *Plant Physiol.*, 59, 920, 1977.
34. **O'Neal, T. D. and Naylor, A. W.,** Some regulatory properties of pea leaf carbamoyl phosphate synthetase, *Plant Physiol.*, 57, 23, 1976.
35. **Stewart, G. R. and Rhodes, D.,** in *Regulation of Enzyme Synthesis and Activity,* Smith, H., Ed., Academic Press, London, 1977.
36. **Hirel, B. and Gadal, P.,** Glutamine synthetase in rice. *Plant Physiol.*, 66, 619, 1980.
37. **Rathnam, C. K. M. and Edwards, G. E.,** Distribution of nitrate-assimilating enzymes between mesophyll protoplasts and bundle sheath cells in leaves of three groups of C4 plants, *Plant Physiol.*, 57, 881, 1976.
38. **Harel, E., Lea, P. J., and Miflin, B. J.,** The localisation of enzymes of nitrogen assimilation in maize leaves and their activities during greening, *Planta,* 134, 195, 1977.
39. **Bauer, A., Urquhart, A. A., and Joy, K. W.,** Amino acid metabolism of pea leaves: diurnal changes and amino acid synthesis from ^{15}N-nitrate, *Plant Physiol.*, 59, 915, 1977.
40. **Canvin, D. T. and Atkins, C. A.,** Nitrate, nitrite and ammonia assimilation by leaves: effect of light, carbon dioxide and oxygen, *Planta,* 116, 207, 1974.
41. **Keys, A. J., Bird, I. F., Cornelius, M. J., Lea, P. J., Wallsgrove, R. M., and Miflin, B. J.,** The photorespiratory nitrogen cycle, *Nature (London),* 275, 741, 1978.
42. **Givan, C. V., Givan, A. L., and Leech, R. M.,** Photoreduction of α-ketoglutarate to glutamate by *Vicia faba* chloroplasts, *Plant Physiol.*, 45, 624, 1970.
43. **Kirk, P. R. and Leech, R. M.,** Amino acid biosynthesis by isolated chloroplasts during photosynthesis, *Plant Physiol.*, 50, 228, 1972.
44. **Pate, J. S.,** Uptake, assimilation and transport of nitrogen compounds by plants, *Soil Biol. Biochem.*, 5, 109, 1973.

NITROGEN FIXATION IN PHOTOSYNTHETIC BACTERIA

Michael Madigan and Howard Gest

DISCOVERY OF N_2 FIXATION IN *RHODOSPIRILLUM RUBRUM*

The capacity of *R. rubrum* to fix dinitrogen was discovered in 1949 during the course of an investigation of light-dependent H_2 production.[1,2] Cells growing photosynthetically (anaerobically) on malate with glutamate as the nitrogen source were observed to produce large quantities of H_2, whereas the latter was not produced when glutamate was replaced by an ammonium salt. It was also demonstrated that resting cell suspensions prepared from malate/glutamate cultures catalyzed "photoproduction" of H_2 from malate and related compounds. This process was strongly inhibited by either N_2 or NH_4^+, a major clue indicating that *R. rubrum* is capable of assimilating N_2. Shortly after the initial findings with *R. rubrum* were reported, other studies[3-6] showed that the ability to fix N_2 was widespread among different species of photosynthetic bacteria. It has been established[7] that the photoproduction of H_2, observed in the absence of N_2, is catalyzed by nitrogenase.

GENERAL PROPERTIES OF PHOTOSYNTHETIC BACTERIA

Photosynthetic bacteria are ubiquitously distributed;[8] they can be isolated from natural waters, soil, sewage, and related environments. Taxonomically, the organisms are currently subdivided into four families:[9] Chlorobiaceae and Chloroflexaceae, so called "green bacteria", and Rhodospirillaceae and Chromatiaceae, the purple bacteria. All bacteria belonging to these families can use light energy to produce the ATP required for biosynthesis; reducing power is derived from the oxidation of organic substances or inorganic compounds other than water, and consequently, O_2 is not produced (hence, the term "anoxygenic photosynthesis"). The photoactive pigments of photosynthetic bacteria are produced maximally under anaerobic, low light intensity conditions and are localized in structures consisting of unit membranes (chromatophores or lamellae)[9] or of tube-shaped vesicles that do not have a unit membrane ("*Chlorobium* vesicles"[9] or "chlorosomes"[10]). There is a wide spectrum of metabolic types among photosynthetic bacteria in respect to ability to use energy sources other than light and also in relation to patterns of carbon assimilation. Representatives of the Rhodospirillaceae (also known as the "nonsulfur purple" bacteria) show the greatest versatility in this connection. For this and other reasons (e.g., their relatively fast growth rates), these organisms have been more intensively studied than other types in regard to N_2 fixation and related processes.

ECOLOGICAL ASPECTS OF N_2 FIXATION BY PHOTOSYNTHETIC BACTERIA

Although few data are available describing N_2 fixation activity of photosynthetic bacteria in the field (e.g., acetylene reduction rates by natural populations), they have been implicated as significant contributors to the total fixed nitrogen of rice paddies in certain areas of the world.[11] Presumably, photosynthetic bacteria growing in such environments utilize organic substrates leached from the rice plants as electron donors for reduction of N_2 to ammonia (anaerobically with light as the energy source). Amino acids and other organic nitrogenous compounds excreted by the bacteria can be assimilated by the rice plant.[11] In view of recent findings[12,13] on dark N_2 fixation by photosynthetic bacteria, it may be that the range of their natural habitats and their contributions to the nitrogen cycle of the earth are considerably greater than has been suspected in the past.[14,15]

The ability of photosynthetic bacteria to fix N_2 can be used to advantage in their enrichment and isolation. Enriched cultures of such bacteria can be readily obtained as follows. A liquid mineral salts medium (inorganic nitrogen salts and potential electron acceptors such as nitrate and sulfate are omitted) supplemented with a suitable nonfermentable organic carbon source is inoculated with a soil or water sample, and the suspension is incubated anaerobically under a gas phase of 95% N_2 + 5% CO_2 with illumination from incandescent lamps. These enrichment conditions strongly favor photosynthetic bacteria (especially Rhodospirillaceae) since these organisms can use light to drive synthesis of the ATP (photophosphorylation) required for N_2 fixation and biosynthesis; a number of nonfermentable organic compounds, unavailable to anaerobic heterotrophs as an energy source, can serve admirably as sources of cell carbon for photosynthetic bacteria (e.g., succinate).

N_2 FIXATION BY RHODOSPIRILLACEAE IN DIFFERENT GROWTH MODES

Many Rhodospirillaceae possess mechanisms for energy generation in the absence of light and consequently can grow in alternative modes.[8,16,17,20] Dinitrogen fixation has now been demonstrated for a number of these alternatives. A striking example is given by *Rhodopseudomonas capsulata* which is capable of growing in five different types of nutritional situations. Under anaerobic photoautotrophic or photoheterotrophic growth conditions (with CO_2 or organic substrates serving as sources of cell carbon, respectively), dinitrogen fixation occurs at high rates.[13,18,19] When *R. capsulata* is grown aerobically in darkness as a chemoheterotroph, N_2 reduction can occur, provided that the O_2 concentration is kept low (so called "microaerophilic" conditions[12,13]). In addition, under anaerobic dark conditions, *R. capsulata* can fix N_2 during growth in a medium containing fructose as the source of carbon and energy;[13] the sugar can be fermented when an "accessory" oxidant such as dimethyl sulfoxide is added. Aerobic chemoautotrophic growth in darkness with H_2 serving as sole energy source is the fifth growth mode available to this remarkable bacterium,[20] but studies on N_2 fixation under these conditions have not yet been undertaken.

SUBCELLULAR SYSTEMS

Active cell-free preparations of nitrogenase have been obtained from *R. rubrum*[21] and *Chromatium vinosum*.[22] Purified preparations from these organisms closely resemble those from other bacteria[23] in that the active enzyme complex consists of two proteins:

1. A nonheme iron protein, currently referred to as "nitrogenase reductase"
2. An Fe-Mo protein, referred to as "nitrogenase"

In addition, a low molecular weight membrane-bound "activating factor"has recently been found to be required for all cell-free nitrogenase activities of *R. rubrum*.[24] As in other systems, subcellular nitrogenase activity is also dependent on Mg-ATP[23] and a suitable reductant such as a reduced viologen dye or ferredoxin.[22,23] Although crude preparations of nitrogenase are only moderately sensitive to oxygen inactivation, the highly purified protein components are rapidly and irreversibly inactivated by even very low concentrations of O_2.[23] Since many of the facultatively anaerobic purple bacteria can fix N_2 under microaerophilic conditions in the dark,[12,13] respiratory removal of low levels of O_2 must suffice to maintain the nitrogenases of these bacteria in their native conformations in vivo.

REGULATION OF NITROGENASE AND RELATIONSHIPS WITH H_2 METABOLISM

Many photosynthetic bacteria possess a hydrogenase capable of activating H_2 for use as a biosynthetic reductant.[25] The "uptake" (sometimes called "conventional") hydrogenase is thought to work primarily, if not entirely, in the direction of H_2 oxidation. Studies[7,19] on wild-type and appropriate mutants of *R. capsulata* have conclusively demonstrated independence of the uptake hydrogenase and the nitrogenase-catalyzed H_2-evolving activity in this organism.

During photosynthetic growth of Rhodospirillaceae on malate, lactate, etc. with certain amino acids as the nitrogen source, nitrogenase is derepressed, resulting in production of large amounts of H_2.[18] Addition of ammonium salts prevents H_2 formation through repression of nitrogenase synthesis; NH_4^+ also inhibits activity of the preformed nitrogenase system.[19,26] Thus, nitrogenase synthesis in photosynthetic bacteria is regulated by repression/derepression (as in other N_2 fixers).[15] Ammonia also exerts an additional "fine tuning" control of activity of nitrogenase; thus far, this has not been reported for other types of nitrogenase systems.

Evidence indicates the participation of glutamine synthetase (GS) in regulation of nitrogenase in heterotrophic bacteria.[27] The demonstration[28] of adenylylation/deadenylylation control of the GS of *R. capsulata* in response to changes in exogenous ammonia concentrations strongly suggests operation of a similar control system in this bacterium. A regulatory role of GS is also indicated by the observation that certain glutamine-requiring mutants of *R. capsulata* produce H_2 during photosynthetic growth (with glutamine as N source), even when NH_4^+ is present in excess.[29] It is noteworthy that attempts to demonstrate adenylylation/deadenylylation control of the GS of cyanobacteria have thus far yielded negative results.

ASSIMILATION OF AMMONIA

Ammonia, the product of N_2 fixation, is incorporated into organic compounds by two major routes. In most species of *Rhodopseudomonas* and *Rhodospirillum* (also in *Rhodomicrobium*), NH_3 is assimilated through the coupled activities of GS (glutamate + NH_3 + ATP $\rightarrow$ glutamine) and glutamine-oxoglutarate aminotransferase (GOGAT; glutamine + oxoglutarate $\rightarrow$ 2 glutamate)[30] or through the direct reductive amination of oxoglutarate to glutamate by glutamate dehydrogenase (GDH).[30] In *R. capsulata*[31] and *Rhodopseudomonas acidophila*[32] GDH is absent, and all ammonia assimilation occurs via the GS-GOGAT sequence. Ammonia assimilation systems of the green and purple sulfur bacteria resemble those of the Rhodospirillaceae. The GS-GOGAT couple as well as GDH have been detected in all species of Chromatiaceae and Chlorobiaceae examined to date.[33]

BIOCHEMICAL GENETICS

Biochemical genetic analysis of N_2 fixation in photosynthetic bacteria has been initiated, mainly with *R. capsulata.* Mutants of this organism unable to fix N_2 (Nif^-) are readily isolated,[7] and those studied thus far fall into several categories:

a. Strains with defects in nitrogenase proteins, as indicated by inability to produce H_2 or reduce acetylene (The "gene transfer agent" produced by certain wild-type strains[34,35] can be accepted by Nif^- strains, resulting in "transferants" which have regained the capacity to fix N_2).[7]
b. Mutants blocked in generation of α-ketoglutarate, required for assimilation of the ammonia generated by nitrogenase action[36]
c. A pleiotropic mutant that can use NH^+_4 as the N source when it is supplied at relatively high concentration. (It has been suggested that this mutant may have a defective glutamine amidotransferase.)[37]
d. Glutamine auxotrophs which probably have defective GS proteins and show the property of producing H_2 during photosynthetic growth even when NH^+_4 is present in excess[29] (a glutamate auxotroph of *R. rubrum* with similar phenotype has been reported)[38]

Since facile techniques are now available for genetic mapping in *R. capsulata*,[39,40] a genetic analysis of the nitrogenase system in this organism can be anticipated in the near future.

ADDENDUM

Since completion of this manuscript in November, 1978, the following pertinent papers have been published:

Carithers, R. P. et al.[41]
Hillmer, P. and Fahlbusch, K.[42]
Jouanneau, Y. et al.[43]
Kampf, C. and Pfennig, N.[44]
Kelley, B. C. et al.[45]
Wall, J. D. and Gest, H.,[46]
Yoch, D. C.[47]
Yoch, D. C.[48]
Yoch, D. C. and Cantu, M.[49]

REFERENCES

1. **Gest, H., and Kamen, M. D.**, Photoproduction of molecular hydrogen by *Rhodospirillum rubrum*, *Science*, 109, 558, 1949.
2. **Kamen, M. D. and Gest, H.**, Evidence for a nitrogenase system in the photosynthetic bacterium *Rhodospirillum rubrum*, *Science*, 109, 560, 1949.
3. **Lindstrom, E. S., Burris, R. H., and Wilson, P. W.**, Nitrogen fixation by photosynthetic bacteria, *J. Bacteriol.*, 58, 313, 1949.
4. **Lindstrom, E. S., Tove, S. R., and Wilson, P. W.**, Nitrogen fixation by the green and purple sulfur bacteria, *Science*, 112, 197, 1950.
5. **Lindstrom, E. S., Lewis, S. M., and Pinsky, M. J.**, Nitrogen fixation and hydrogenase in various bacterial species, *J. Bacteriol.*, 61, 481, 1951.
6. **Newton, J. W. and Wilson, P. W.**, Nitrogen fixation and photoproduction of molecular hydrogen by Thiorhodaceae, *Antonie van Leeuwenhoek J. Microbiol. Serol.*, 19, 71, 1953.
7. **Wall, J. D., Weaver, P. F., and Gest, H.**, Genetic transfer of nitrogenase-hydrogenase activity in *Rhodopseudomonas capsulata*, *Nature (London)*, 258, 630, 1975.
8. **Pfennig, N.**, Photosynthetic bacteria, *Annu. Rev. Microbiol.*, 21, 285, 1967.
9. **Pfennig, N. and Trüper, H. G.**, The Rhodospirillaceae (phototrophic or photosynthetic bacteria), in *CRC Handbook of Microbiology*, Vol. 1, 2nd ed., Laskin, A. I. and Lechevalier, H. A., Eds., CRC Press, Boca Raton, Fla., 1977, 119.
10. **Fuller, R. C.**, Further Studies on the Photosynthetic Apparatus of the Green Bacteria - Chlorosome-Membrane Interaction, presented at the 5th Annu. Conf. Molecular Biology of Photosynthetic Procaryotes, Bloomington, Ind., November 19 to 21, 1978, 41.

11. **Kobayashi, M. and Haque, M. Z.**, Contribution to nitrogen fixation and soil fertility by photosynthetic bacteria, *Plant Soil*, Special Volume, 443, 1971.
12. **Siefert, E.**, Nitrogen fixation in facultative aerobic Rhodospirillaceae with photosynthetic or respiratory energy generation, in *Proc. 2nd Int. Symp. Photosynthetic Prokaryotes*, Codd, G. A. and Stewart, W. D. P., Eds., Dundee, Scotland, 1976, 149.
13. **Madigan, M. T., Wall, J. D., and Gest, H.**, Dark anaerobic dinitrogen fixation by a photosynthetic microorganism, *Science*, 204, 1429-1430, 1979.
14. **Stewart, W. D. P.**, Nitrogen fixation by photosynthetic microorganisms, *Annu. Rev. Microbiol.*, 27, 283, 1973.
15. **Smith, B. E.**, Nitrogen fixation by free-living micro-organisms, *Process Biochem.*, 12, 21, 1977.
16. **Uffen, R. L. and Wolfe, R. S.**, Anaerobic growth of purple nonsulfur bacteria under dark conditions, *J. Bacteriol.*, 104, 462, 1970.
17. **Madigan, M. T. and Gest, H.**, Growth of a photosynthetic bacterium anaerobically in darkness, supported by "oxidant-dependent" sugar fermentation, *Arch. Microbiol.*, 117, 119, 1978.
18. **Hillmer, P. and Gest, H.**, H_2 metabolism in the photosynthetic bacterium *Rhodopseudomonas capsulata*: H_2 production by growing cultures, *J. Bacteriol.*, 129, 724, 1977.
19. **Hillmer, P. and Gest, H.**, H_2 metabolism in the photosynthetic bacterium *Rhodopseudomonas capsulata*: production and utilization of H_2 by resting cells, *J. Bacteriol.*, 129, 732, 1977.
20. **Madigan, M. T. and Gest, H.**, Growth of the photosynthetic bacterium *Rhodopseudomonas capsulata* chemoautotrophically in darkness with H_2 as energy source, *J. Bacteriol.*, 137, 524-530, 1979.
21. **Munson, T. O. and Burris, R. H.**, Nitrogen fixation by *Rhodospirillum rubrum* grown in nitrogen-limited continuous culture, *J. Bacteriol.*, 97, 1093, 1969.
22. **Winter, H. C. and Arnon, D. I.**, The nitrogen fixation system of photosynthetic bacteria. I. Preparation and properties of a cell-free extract from *Chromatium*, *Biochim. Biophys. Acta*, 197, 170, 1970.
23. **Winter, H. C. and Burris, R. H.**, Nitrogenase, *Annu. Rev. Biochem.*, 45, 409, 1976.
24. **Ludden, P. W. and Burris, R. H.**, Purification and properties of nitrogenase from *Rhodospirillum rubrum*: evidence for the presence of phosphate, ribose, and an adenine-like unit covalently bound to the Fe protein, *Biochem. J.*, 175, 251, 1978.
25. **Ormerod, J. G. and Gest, H.**, Hydrogen photosynthesis and alternative metabolic pathways in photosynthetic bacteria, *Bacteriol. Rev.*, 26, 51, 1962.
26. **Schick, H. J.**, Substrate and light dependent fixation of molecular nitrogen in *Rhodospirillum rubrum*, *Arch. Microbiol.*, 75, 89, 1971.
27. **Streicher, S. L., Shanmugan, K. T., Ausubel, F., Morandi, C., and Goldberg, R. B.**, Regulation of nitrogen fixation in *Klebsiella pneumoniae*: evidence for a role of glutamine synthetase as a regulator of nitrogenase synthesis, *J. Bacteriol.*, 120, 815, 1974.
28. **Johansson, B. C. and Gest, H.**, Adenylylation/deadenylylation control of the glutamine synthetase of *Rhodopseudomonas capsulata*, *Eur. J. Biochem.*, 81, 365, 1977.
29. **Wall, J. D. and Gest, H.**, Glutamine auxotrophy and control of nitrogenase activities in *Rhodopseudomonas capsulata*, *Abstr. Ann. Meeting Am. Soc. for Microbiol.*, 1978, K22, 130.
30. **Brown, C. M. and Herbert, R. A.**, Ammonia assimilation in members of the Rhodospirillaceae, *FEMS Microbiol. Lett.*, 1, 43, 1977.
31. **Johansson, B. C. and Gest, H.**, Inorganic nitrogen assimilation by the photosynthetic bacterium *Rhodopseudomonas capsulata*, *J. Bacteriol.*, 128, 683, 1976.
32. **Herbert, R. A., Siefert, E., and Pfennig, N.**, Nitrogen assimilation in *Rhodopseudomonas acidophila*, *Soc. Gen. Microbiol. Proc.*, Volume 5 (4), 103, 1978.
33. **Brown, C. M. and Herbert, R. A.**, Ammonia assimilation in purple and green sulfur bacteria, *FEMS Microbiol. Lett.*, 1, 39, 1977.
34. **Marrs, B.**, Genetic recombination in *Rhodopseudomonas capsulata*, *Proc. Natl. Acad. Sci. U.S.A.*, 71, 971, 1974.
35. **Wall, J. D., Weaver, P. F., and Gest, H.**, Gene transfer agents, bacteriophages, and bacteriocins of *Rhodopseudomonas capsulata*, *Arch. Microbiol.*, 105, 217, 1975.
36. **Beatty, J. T., Johansson, B. C., Wall, J. D., and Gest, H.**, Nitrogen assimilation defects in a mutant of *Rhodopseudomonas capsulata* blocked in α-ketoglutarate generation, *FEMS Microbiol. Lett.*, 2, 267, 1977.
37. **Wall, J. D., Johansson, B. C., and Gest, H.**, A pleiotropic mutant of *Rhodopseudomonas capsulata* defective in nitrogen metabolism, *Arch. Microbiol.*, 115, 259, 1977.
38. **Weare, N. M.**, The photoproduction of H_2 and NH^+_4 fixed from N_2 by a derepressed mutant of *Rhodospirillum rubrum*, *Biochim. Biophys. Acta.*, 502, 486, 1978.
39. **Yen, H-C. and Marrs, B.**, Map of genes for carotenoid and bacteriochlorophyll biosynthesis in *Rhodopseudomonas capsulata*, *J. Bacteriol.*, 126, 619, 1976.
40. **Marrs, B.**, Conjugational Analysis of the Photosynthesis Region of the *Rhodopseudomonas capsulata* Chromosome, presented at the 5th Annu. Conf. Molecular Biology of Photosynthetic Procaryotes, Bloomington, Ind., November 19 to 21, 1978, 1.

41. **Carithers, R. P., Yoch, D. C., and Arnon, D. I.,** Two forms of nitrogenase from the photosynthetic bacterium *Rhodospirillum rubrum, J. Bacteriol.,* 137, 779, 1978.
42. **Hillmer, P. and Fahlbusch, K.,** Evidence for an involvement of glutamine synthetase in regulation of nitrogenase activity in *Rhodopseudomonas capsulata. Arch. Microbiol.,* 122, 213, 1979.
43. **Jouanneau, Y., Siefert, E., and Pfennig, N.,** Microaerobic nitrogenase activity in *Thiocapsa* sp. strain 5811. *FEMS Microbiol. Letts.,* 9, 89, 1980.
44. **Kampf, C. and Pfennig, N.,** Capacity of *Chromatiaceae* for chemotrophic growth. Specific respiration rates of *Thiocystis violacea* and *Chromatium vinosum. Arch. Microbiol.,* 127, 125, 1980.
45. **Kelley, B. C., Jouanneau, Y., and Vignais, P. M.,** Nitrogenase activity in *Rhodopseudomonas sulfidophila. Arch. Microbiol.,* 122, 145, 1979.
46. **Wall, J. D. and Gest, H.,** Derepression of nitrogenase activity in glutamine auxotrophs of *Rhodopseudomonas capsulata, J. Bacteriol.,* 137, 1459, 1979.
47. **Yoch, D. C.,** Manganese, an essential trace element for N_2 fixation by *Rhodospirillum rubrum* and *Rhodopseudomonas capsulata:* Role in nitrogenase regulation, *J. Bacteriol.,* 140, 987, 1980.
48. **Yoch, D. C.,** Regulation of nitrogenase A and R concentrations in *Rhodopseudomonas capsulata* by glutamine synthetase. *Biochem. J.,* 187, 273, 1980.
49. **Yoch, D. C., and Cantu, M.,** Changes in the regulatory form of *Rhodospirillum rubrum* nitrogenase as influenced by nutritional and environmental factors. *J. Bacteriol.,* 142, 899, 1980.

NITROGEN FIXATION IN FREE-LIVING ALGAE

E. O. Duerr, S. Kumazawa, and A. Mitsui

INTRODUCTION

Blue-green algae (Cyanobacteria) carry out oxygen-evolving photosynthesis, and certain species also are capable of fixing atmospheric nitrogen. This combination of autotrophy and N_2 fixation capacities in one organism is unique to blue-green algae.

A wide distribution of blue-green algae is observed in the diverse environments where life can be supported.[1] These environments range from boreal[2] to tropical,[3] to terrestrial and aquatic habitats, including soils,[4] forests,[5] deserts,[6] hot springs,[7] lakes,[8] estuaries, and oceans.[9] Nitrogen fixing blue-green algae have similar distributions and appear to contribute to the fertility in these ecosystems through an input of combined nitrogen nutrients.[10,11]

While the capacity to fix nitrogen extends the range of habitable areas to those that are nutrient (nitrogen) poor, certain constraints still exist. Aside from the growth requirements generally ascribed to blue-green algae, two further conditions exist:

1. The nitrogenase enzyme must be produced
2. The enzyme must be protected from O_2 inactivation

It is the oxygen-evolving steps of photosynthesis (Photosystem II) that constitute the apparent paradox of its coexistence with nitrogenase.

GENERA OF BLUE-GREEN ALGAE CAPABLE OF FIXING NITROGEN

Since the first pure cultures of *Nostoc* and *Anabaena* were shown to fix nitrogen,[12] at least 36 additional genera have been reported with this capacity. These genera have been listed in Table 1. Four divisions have been incorporated into the table. These divisions reflect the general categories of nitrogenase protection from O_2 inactivation. They have been termed

1. Heterocystous forms
2. Mutualistic or protocooperative forms
3. Internally protected forms
4. Anoxic forms

Heterocystous Forms

Certain filamentous blue-green algae of the Nostocaceae, Rivulariaceae, and Stigonemataceae are capable of forming specialized cells called heterocysts.[13] With respect to O_2 exclusion, these cells differ from the vegetative cells in two major respects: the cell wall composition and the absence of normal Photosystem II activity. The cell walls contain three additional layers, the innermost incorporating unique glycolipids which are speculated to assist in excluding oxygen.[14,15] Other structural differences from vegetative cells include highly evolved intercellular membranes,[16] polar granules,[17] and the concentration of thylakoid membranes near the poles.[18,19]

While the oxygenic steps of photosynthesis are absent,[20,21] cyclic photophosphorylation of Photosystem I may furnish the ATP demands of the nitrogenase system.[22] Photosynthate transported from vegetative cells to heterocysts is suggested as the source of reducing potential.[23,24]

Table 1
NITROGEN FIXING GENERA OF BLUE-GREEN ALGAE

Group	Genera	Number of strains	Ref.
Heterocystous forms			
	Anabaena	>20	37[a],38
	Anabaenopsis	2	38
	Aphanizomenon	1	38
	Aulosira	1	38
	Calothrix	4	38
	Chlorogloea	1	38
	Cylindrospermum	4	38
	Fischerella	4	38,97
	Gloeotrichia	2	98
	Hapalosiphon	1	38
	Mastigocladus	1	38
	Nodularia	1	99
	Nostoc	12	38,100-102
	Rivularia	2	98
	Scytonema	2	38
	Stigonema	1	38
	Tolypothrix	1	38
	Westiella	1	97
	Westiellopsis	1	38
Mutualistic and proto-cooperative forms[b]			
	Anabaena		6,10
	Calothrix		6,10,40
	Chroococcus		6
	Dichothrix		6,40
	Gloeocapsa		6
	Hapalosiphon		103
	Nostoc		6,10,40
	Richelia		104
	Scytonema		6,40
	Stigonema		6,40
Internally protected forms			
	Aphanothece	1	105
	Gloeothece (Gloeocapsa)	5	48,50,53
	Osciallatoria (Trichodesmium)		47,51,104
	Schizothrix	1	46
Anoxic forms			
	Chroococcidiopsis	8	53
	Dermocarpa	2	53
	Lyngbya [c]		53
	Myxosarcina	1	53
	Oscillatoria	5	53
	Plectonema [c]		53
	Pleurocapsa	7	53
	Phormidium [c]		53
	Pseudoanabaena	4	53
	Synechococcus	3	53
	Xenococcus	1	53

[a] Classified as *Anabaena* type.
[b] For lists of associations, refer to references given.
[c] *Lyngbya, Plectonema, Phormidium* group has 16 strains with nitrogenase activity.

Induction of heterocyst formation occurs under combined nitrogen-limited culture.[25-27] The presence of light, O_2, and organic substrates in the cell (high C/N ratio[26]) also may be required for mature heterocyst formation.[28,29] The fraction of cells which initially differentiate into heterocysts is thought to be maintained through the release of an inhibitor to adjacent cells, so that no further differentiation occurs.[18,30,31] Once heterocysts are formed, regression back to vegetative cells is not thought to occur, even in the presence of fixed nitrogen.[18,32-34] Although limited survey work is at hand, virtually all heterocyst-forming blue-green algae can fix nitrogen aerobically.[35-38]

Mutualistic and Protocooperative Forms

Besides those found in the free-living state, nitrogen-fixing blue-green algae may also be found associated with other organisms.[6,39-41] In such associated forms, both organisms benefit each other. Protocooperative forms are capable of independent growth, while mutualistic forms maintain an obligatory relationship.[42] These organisms appear to be composed of heterocystous blue-green algae, most commonly of the *Nostoc* and *Anabaena* genera. It was felt that a separate category was justified in spite of the overlap in genera, because of the indicated specificity of some of the associations and because of the independent distribution and physiology of these forms from the free-living blue-green algae.[43-45]

Internally Protected Forms

For a number of years, the nonheterocystous blue-green algae *Schizothrix* and *Trichodesmium* had been reported to fix nitrogen under aerobic conditions.[46,47] In neither case were the cultures free of bacteria. *Gloeothece* (also called *Gloeocapsa*) was the first axenic culture shown to fix nitrogen aerobically.[48,49] Broken cell preparations of nitrogenase derived from this last organism were shown to be sensitive to O_2 inactivation.[50] While this group is the least studied and least understood, both filamentous and unicellular forms fit into this category. The in vivo protection mechanism(s) of the nitrogenase complex has not been clarified, and perhaps several different mechanisms of protection may exist.[51,52]

Anoxic Forms

The last group of nitrogen-fixing blue-green algae may be defined as those which possess very limited or no protective mechanisms against O_2 inhibition. Research into this group of organisms in several laboratories has revealed a considerable number of strains of such blue-green algae.[37,53] With the discovery of facultative anoxygenic photosynthesis in a number of blue-green algae,[54] the ecological significance of this group is possibly underestimated.

INDUCTION OF NITROGENASE

The primary product of the nitrogenase reaction is ammonia.[55] In blue-green algae, the major ammonia assimilatory route under nitrogen-fixing conditions is mediated by glutamine synthetase (GS) and glutamate synthase (GOGAT).[56,57] The intracellular pool of ammonia is shown to relate inversely with nitrogenase activity (C_2H_2 reduction).[58] This inhibitory action of ammonia to nitrogenase activity is suggested to occur at the level of nitrogenase synthesis, but does not directly affect nitrogenase itself.[59,60] On the other hand, nitrate seems to have no inhibitory effect like that of ammonia.[58,59]

L-Methionine-D,L-sulphoximine (MSX), a structural analog of L-glutamate, is shown to derepress nitrogenase synthesis in the presence of ammonia in the medium.[61] This derepression is interpreted to occur by an impaired assimilation of ammonia by the

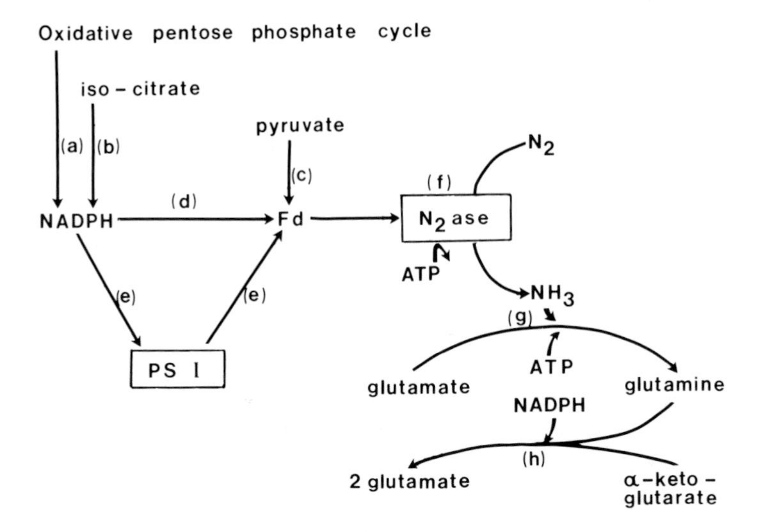

FIGURE 1. Major biochemical pathways in the nitrogenase system. Reactions: (a) glucose-6-phosphate dehydrogenase and 6-phosphogluconate dehydrogenase;[24,78] (b) iso-citrate dehydrogenase;[73,82] (c) pyruvate ferredoxin oxidoreductase;[80,81,125,126] (d) ferredoxin NADP oxidoreductase;[73] (e) Photosystem I-dependent reduction of ferredoxin;[73] (f) nitrogenase enzyme; (g) glutamine synthetase;[55-57] (h) glutamate synthase.[55-57]

inhibition of GS. Thus, ammonia itself does not seem to function as the repressor of nitrogenase synthesis.[62,63]

A relationship between the cellular carbon to nitrogen (C/N) ratio and nitrogen fixation activity has been shown in blue-green algae.[64,65] Nitrogen content within the cells may be reflected by the levels of phycobilisome proteins and cyanophycin granules.[66,67] Low levels of these components were noted when nitrogen fixation was initiated.[29,68,69] For example, heterocyst differentiation and nitrogenase activity were found when cellular C/N ratios increased from 4.5 to 8 in *Anabaena cylindrica.*[26]

ELECTRON DONORS AND BIOCHEMICAL PATHWAYS

The nitrogenase enzyme system requires ATP and electrons to reduce dinitrogen gas to ammonia.[36] In vitro nitrogen fixation was demonstrated using dithionite or ferredoxin as the electron donor compound to the enzyme.[70-74] In addition to ferredoxin, phytoflavin has been shown to mediate electron transfer to nitrogenase, but this is thought to be significant only under iron-deficient conditions.[75] Figure 1 provides a schematic diagram of the major pathways of reductant, energy, and substrate flow in the nitrogenase system.

A number of sources of reductant exist in blue-green algae. For nitrogenase, the principal source is thought to be the oxidative pentose phosphate cycle.[24,76-78] The evidence that many nonheterocystous blue-green algae are capable of fixing nitrogen under anaerobic conditions and when Photosystem II is impaired by DCMU {3-(3,4-dichlorophenyl)-1, 1-dimethylurea},[53,79] suggests that reduced ferredoxin is unlikely to be derived photochemically from water.[70] Other sources of reducing power that have been implicated include pyruvate[80,81] and iso-citrate.[73,82] Glucose-6-phosphate dehydrogenase, 6-phosphogluconate dehydrogenase,[24,78] and isocitrate dehydrogenase coupled to NADP ferredoxin oxidoreductase[73] and pyruvate ferredoxin oxidoreductase reactions[80,81] have been shown to stimulate nitrogen fixation. Photosystem I-dependent reduction of ferredoxin may also contribute as an electron donor for the nitrogenase catalysis.[73]

During nitrogen fixation, hydrogen gas is also formed.[83,84] The loss of electrons as hydrogen gas is shown to be recaptured by hydrogenase, improving the efficiency of the process.[85-87]

ATP required for the nitrogenase catalysis can be formed by cyclic phosphorylation and/or by oxidative phosphorylation.[88-91] Hydrogen gas, trapped by active uptake hydrogenase, is shown to enhance O_2 consumption in several blue-green algae.[85,86] This oxyhydrogen reaction has been demonstrated to support ATP synthesis via oxidative phosphorylation by cytochrome-cytochrome oxidase system.[87]

FACTORS AFFECTING NITROGENASE ACTIVITY

A number of factors have been identified that influence the rate of nitrogen fixation in free-living blue-green algae. External, internal, and experimental variables exist which must be considered in the study of nitrogen fixation. A brief review is given for several of the external and internal variables in Table 2. In addition, several studies on rates of nitrogenase activity have been compiled to provide a perspective on this subject (see Table 3).

Nitrogenase catalyzes the reduction of several compounds, such as dinitrogen, proton, azide, nitrous oxide, acetylene, cyanides, and isocyanides, and their analogous compounds.[83,84] Nitrogenase activity can be measured by the determination of total nitrogen increase in the cell, measurement of nitrogen gas uptake in the flask, measurement of labeled $^{15}N_2$ fixation, determination of ammonia accumulation (cell-free preparation), measurement of hydrogenase activity of nitrogenase, reduction of alternative substrates (C_2H_2, N_2O, N_3^-, CN^-), or measurement of oxidation of dithionite (by purified nitrogenase). These methodologies are described by Burris.[92] Of these methods, acetylene reduction by the nitrogenase enzyme provides probably the most convenient measure of its activity and is widely used for routine measurements.[93] Although the acetylene reduction method is a convenient tool for estimating nitrogen fixation activity, it is generally recommended that each system be measured directly to derive a conversion factor.[94] Theoretically, a stoichiometric requirement of electrons for 1 mol acetylene reduction to 1 mol ethylene would be 2 electrons, while that of 1 mol of dinitrogen fixation to 2 mol of ammonia would be 6 electrons. In vivo measurements, however, have shown that the stoichiometry may not be constant.[94]

Numerous procedural points exist which will influence the value derived for acetylene reduction. These include the pC_2H_2 in the test chamber, the composition of the rest of the gas phase, diffusion rates of C_2H_2, and the length of exposure to C_2H_2. Physiological change in the cellular carbon to nitrogen ratio occurs during long term incubation under acetylene.[95]

Table 2
MAJOR EXTERNAL AND INTERNAL FACTORS AFFECTING NITROGENASE ACTIVITY

Variables	Expressed via	Comments	Ref.
Light quantity	Photosynthetic production of 1. Hexose 2. ATP	Light contributes to general metabolic potential through photosynthesis Dark N_2 fixation occurs where reducing potential and ATP are available	91,106—109
Light quality	Photochemical reaction (Photosystem I and II)	PS I dependent	20,89,90
Combined nitrogen	Metabolic requirement for nitrogenous compounds	Availability of free ammonia for the biosynthesis inversely correlates with in vivo nitrogenase synthesis Nitrite or nitrate may compete for electrons with nitrogenase	58,110
Oxygen	Direct inactivation of N_2ase Photorespiration Oxidative phosphorylation	N_2ase is susceptible to O_2 inactivation. Competition for electrons with photorespiration Oxidative phosphorylation as an alternative ATP source (especially under dark)	106,110—112
Temperature	Combination of intrinsic metabolic potential and thermodynamic reactivity	Species dependent	64,110,113—115
pH		Nitrogenase optima $7 \sim 8$.	44,64,74
Salinity	Degree of energy consumption	ATP and reductant requirement for the maintenance of intracellular ionic potential	96,116
Growth phase	Metabolic requirement for nitrogenous compounds	Change in phycobilisome protein (attenuation of PS II activity) may be observed.	16,65,66,82
Inorganic nutrients	General requirement for nitrogenase synthesis	Fe, Mo required for nitrogenase synthesis	70,117,118

Table 3
EXAMPLES OF NITROGEN FIXATION (ACETYLENE REDUCTION) RATES OF BLUE-GREEN ALGAE

	Maximum rate[a]			Reaction vessel			
Genus/species	Light	Dark	Units	Partial pressure[b]	Reaction time (hr)	Comments	Ref.
Anabaena azollae	ana = 93	ana = 0.35	nmol C_2H_4/mg Chl/ min	0.1	5.0	0.1 atm pC_2H_2 determined to be optimal for acetylene reduction	43
Anabaena cylindrica	ana = 16.4		μl C_2H_4/mg dry wt/hr	0.1	6.0	Intermittent light (5 sec on/5 sec off) increases C_2H_4 reduction 1.35 × the amount reduced under an equal amount of light provided continuously	119
A. cylindrica	ana = 6.7	ana = 0.1	nmol C_2H_4/mg dry wt/min	0.1	0.5	Increasing light (up to 6.4 klx) increased N_2 fixation (through ATP generation)	120
A. cylindrica	ana = 225	ana = 38	μmol C_2H_4/mg dry wt/hr	0.1	0.5		121
A. cylindrica	aer = 1—3.8		μmol C_2H_4/mg N/hr	0.1	0.5	3-hr preincubation with 30-min periods of an alternating atmosphere of air and 10% C_2H_2 in air gives highest rates; control samples under air resulted in a rate of 1.0	95
A. cylindrica	ana = 7.9		nmol C_2H_4/mg protein/min	0.1	0.5	{Mo}: temperature interaction in nitrogenase activity	117
A. cylindrica	aer = 1.9		nmol C_2H_4/mg protein/min	0.1	0.5	pAr: pO_2: pC_2H_2: pCO_2 = 0.7: 0.2: 0.1: 0.04	122
Anabaena flos-aquae	aer = 4.7		nmol C_2H_4/mg protein/min	0.1	0.5	pAr: pO_2: pC_2H_2 CO_2 = 0.7: 0.2: 0.1: 0.04	122
A. flos-aquae	ana = 12		nmol C_2H_4/mg dry wt/min	0.1	0.5		123
A. flos-aquae	ana = 12		nmol C_2H_4/mg protein/min	0.1	0.5	Dark heterotrophic cells can fix N_2 at 10 nmol/mg protein/min	45

Table 3 (continued)
EXAMPLES OF NITROGEN FIXATION (ACETYLENE REDUCTION) RATES OF BLUE-GREEN ALGAE

	Maximum rate[a]			Reaction vessel			
Genus species	Light	Dark	Units	Partial pressure[b]	Reaction time (hr)	Comments	Ref.
Anabaena variabilis	aer = 2.5± 1.2 ana = 4.6± 1.3 mic = 6.2± 1.8		nmol C_2H_4/mg protein/min		0.5		15
A. variabilis	ana = 1040 aer = 930		nmol C_2H_4/mg Chl/min	0.07	0.33	Anaerobic atmosphere was 93% H_2/7% C_2H_2; heterocysts isolated subsequently had an activity of 63% of whole filaments	124
Anabaena sp. CA	ana = 50—100		nmol C_2H_4/mg protein/min	0.2	1.0		96
Anabaena sp. CA	ana = 16		nmol C_2H_4/mg dry wt/min	0.13		Presence of Aza-T[c] in medium increases rate to 20	27
Anabaena sp. (isolated from *Azolla*)	ana = 13		nmol C_2H_4/mg protein/min	0.1	0.5	Dark heterotrophic growth accompanied by fixation was 2—3 times given autotrophic rate	45
Anabaena sp.	ana = 13.2—29.4		nmol C_2H_4/mg protein/min	0.1	1.0		68
Anabaena 7120	ana = 77		nmol C_2H_4/mg N/min	0.17	0.33	pH_2: pC_2H_2 = 0.83:0.17	78
Anabaenopsis circularis	aer = 39	aer = 26	nmol C_2H_4/mg dry wt/hr	0.1	0.5	Rate with addition of glucose to medium in light = 957; dark = 620	91
Cylindrospermum sp.	aer = 0.58		nmol C_2H_4/mg dry wt/min	0.1	0.5		48
Gloeocapsa sp. LB795	aer = 0.11	aer = 0.0	nmol C_2H_2/mg dry wt/min	0.1	0.5		48
Gloeocapsa sp.	ana = 1.0		nmol C_2H_4/mg protein/min	0.1	1.0		49
Gloeocapsa sp. LB795	ana = 3.77	ana = 1.85	nmol C_2H_4/mg protein/min	0.01			82

Mastigocladus laminosus	aer = 1.0		nmol C_2H_4/mg protein/min	0.1	0.5	pAr: pO_2: PC_2H_2: pCO_2 = 0.7: 0.2: 0.1: 0.04	122
Nostoc commune	aer = 0.04		nmol C_2H_4/mg dry wt/min	0.1	0.5		48
Nostoc entophytum	aer = 0.9		nmol C_2H_4/mg protein/min	0.1	0.5	pAr: pO_2: PC_2H_2: PCO_2 = 0.7: 0.2: 0.1: 0.04	122
Nostoc linckia	aer = 17.1		nmol C_2H_4/μg Chl *a*/hr	0.1	0.5		101
Nostoc muscorum	aer = 20.0		nmoles C_2H_4/μg Chl *a*/hr	0.1	0.5		101
N. muscorum	aer = 0.25		nmol C_2H_4/mg dry wt/min	0.1	0.5		48
N. muscorum	aer = 3.7		nmol C_2H_4/mg protein/min	0.1	0.5	pAr: pO_2: pC_2H_2: pCO_2 = 0.7: 0.2: 0.1: 0.04	122
Nostoc sp.	aer = 54 ana = 112		nmol C_2H_2/mg Chl *a*/min (aer = 20% O_2 80% Ar. 04% CO_2) prob = .04%	0.1	1—3	Free living strain derived from Liverwort *Blasia pusilla*; the pO_2 comparison between free living and endophytic gave similar results; pH, light intensity, temperature and dehydration effects differ between the two	44
Plectonema boryanum	ana = 16 aer = 0		nmol C_2H_4/mg protein/min	0.1	7	Aerobic growth required combined nitrogen; DCMU was added; N_2ase synthesis only under complete absence of O_2	53
P. boryanum	ana = 50	ana = 1	nmol C_2H_4/mg Chl/min	0.07	0.5	5 m*M* $Na_2S_2O_4$ in algal suspension enhanced activity 2—7 times; dark activity is inhibited by dithionite; N_2ase induction was observed after 5 hr of preincubation under sparging argon	118
Tolypothrix distorta	aer = 0.18		nmol C_2H_4/mg dry wt/min	0.1	0.5		48
Tolypothrix tenuis	aer = 1.8		nmol C_2H_4/mg protein/min	0.1	0.5	pAr: pO_2: pC_2H_2: pCO_2 = 0.7: 0.2: 0.1: 0.04	122

[a] aer = aerobic; ana = anaerobic; mic = microaerobic.
[b] Partial pressure of substrate; atmospheric pressure = 1.0.
[c] D,L-7-azatryptophan at 20 μmol.

REFERENCES

1. **Holm-Hansen, O.,** Ecology, physiology, and biochemistry of blue-green algae, *Annu. Rev. Microbiol.,* 22, 47, 1968.
2. **Kallio, P. and Kallio, S.,** Adaptation of nitrogen fixation to temperature in the *Peltigera aphthosa* group, in *Environmental Role of Nitrogen-fixing Blue-Green Algae and Asymbiotic Bacteria,* Ecol. Bull., Granhall, U., Ed., Swedish Natural Science Research Center, Stockholm, 26, 225, 1978.
3. **Watanabe, A.,** Nitrogen fixation by algae, in *Advance of Phycology in Japan,* Tokida, J. and Hirose, H., Eds., Dr. W. Junk, The Hague, 1975, 255.
4. **Dooley, F. and Houghton, J. A.,** The nitrogen-fixing capacities and the occurrence of blue-green algae in peat soils, *Br. Phycol. J.,* 8, 289, 1973.
5. **Watanabe, A.,** Distribution of nitrogen-fixing blue-green algae in the forests, in *Biological Solar Energy Conversion,* Mitsui, A., Miyachi, S., San Pietro, A., and Tamura, S., Eds., Academic Press, London, 1977, 323.
6. **Fogg, G. E., Stewart, W. D. P., Fay, P., and Walsby, A. E.,** *The Blue-green Algae,* Academic Press, London, 1973.
7. **Castenholz, R. W.,** Thermophilic blue-green algae and the thermal environment, *Bacteriol. Rev.,* 33, 476, 1969.
8. **Burris, R. H. and Peterson, R. B.,** Nitrogen-fixing blue-green algae: their H_2 metabolism and their activity in freshwater lakes, in *Environmental Role of Nitrogen-fixing Blue-green Algae and Asymbiotic Bacteria,* Ecol. Bull., Granhall, U., Ed., Swedish Natural Science Research Council, Stockholm, 26, 28, 1978.
9. **Fogg, G. E.,** Nitrogen fixation in the oceans, in *Environmental Role of Nitrogen-fixing Blue-green Algae and Asymbiotic Bacteria, Ecol. Bull.,* Granhall, U., Ed., Swedish Natural Science Research Council, Stockholm, 26, 11, 1978.
10. **Burns, R . C. and Hardy, R. W. F.,** *Nitrogen Fixation in Bacteria and Higher Plants,* Springer-Verlag, New York, 1975.
11. **Paul, E. A.,** Contribution of nitrogen fixation to ecosystem functioning and nitrogen fluxes on a global basis, in *Environmental Role of Nitrogen-fixing Blue-green Algae and Asymbiotic Bacteria,* Ecol. Bull., Granhall, U., Ed., Swedish Natural Science Research Council, Stockholm, 26, 282, 1978.
12. **Drewes, K.,** Über die assimilation des luftstickstoffs durch blaualgen, *Zentralbl. Bakteriol. Parasitenk.,* 76 (Abstr. 2), 88, 1928.
13. **Stainer, R. Y. and Cohen-Bazire, G.,** Phototrophic prokaryotes: the cyanobacteria, *Annu. Rev. Microbiol.,* 31, 225, 1977.
14. **Abreu-Grobois, F. A., Billyard, T. C., and Walton, T. J.,** Biosynthesis of heterocyst glycolipids of *Anabaena cylindrica, Phytochemistry,* 16, 351, 1977.
15. **Haury, J. F. and Wolk, C. P.,** Classes of *Anabaena variabilis* mutants with oxygen-sensitive nitrogenase activity, *J. Bacteriol.,* 136, 688, 1978.
16. **Lang, N. J. and Fay, P.,** The heterocysts of blue-green algae. II. Details of ultrastructure, *Proc. R. Soc. London Ser. B,* 178, 193, 1971.
17. **Simon, R. D.,** Measurement of the cyanophycin granule polypeptide contained in the blue-green alga *Anabaena cylindrica, J. Bacteriol.,* 114, 1213, 1973.
18. **Wilcox, M., Mitchison, G. J., and Smith, R. J.,** Pattern formation in the blue-green alga *Anabaena.* I. Basic mechanisms, *J. Cell Sci.,* 12, 707, 1973.
19. **Haselkorn, R.,** Heterocysts, *Annu. Rev. Plant Physiol.,* 29, 319, 1978.
20. **Donze, M., Haveman, J., and Schiereck, P.,** Absence of photosystem 2 in heterocysts of the blue-green alga *Anabaena, Biochim. Biophys. Acta,* 256, 157, 1972.
21. **Tel-Or, E. and Stewart, W. D. P.,** Photosynthetic components and activities of nitrogen-fixing isolated heterocysts of *Anabaena cylindrica, Proc. R. Soc. London Ser. B,* 198, 61, 1977.
22. **Bishop, N. I. and Jones, L. W.,** Alternate fates of the photochemical reducing power generated in photosynthesis: hydrogen production and nitrogen fixation, in *Current Topics in Bioenergetics,* Vol. 8, Sanadi, D. R. and Vernon, L. P., Eds., Academic Press, New York, 1978, 3.
23. **Wolk, C. P.,** Movement of carbon from vegetative cells to heterocysts in *Anabaena cylindrica, J. Bacteriol.,* 96, 2138, 1968.
24. **Winkenbach, F. and Wolk, C. P.,** Activities of enzymes of the oxidative and the reductive pentose phosphate pathways in heterocysts of blue-green alga, *Plant Physiol.,* 52, 480, 1973.
25. **Fogg, G. E.,** Growth and heterocyst production in *Anabaena cylindrica* Lemm. II. In relation to carbon and nitrogen metabolism, *Ann. Bot. (London),* 13, 241, 1949.
26. **Kulasooriya, S. A., Lang, N. J., and Fay, P.,** The heterocysts of blue-green algae. III. Differentiation and nitrogenase activity, *Proc. R. Soc. London Ser. B,* 181, 199, 1972.
27. **Stacey, G., Bottomley, P. J., Van Baalen, C., and Tabita, F. R.,** Control of heterocyst and nitrogenase synthesis in cyanobacteria, *J. Bacteriol.,* 137, 321, 1979.

28. **Bradley, S. and Carr, N. G.,** Heterocyst development in *Anabaena cylindrica*: the necessity for light as an initial trigger and sequential stages of commitment, *J. Gen. Microbiol.,* 101, 291, 1977.
29. **Rippka, R. and Stanier, R. Y.,** The effects of anaerobiosis on nitrogenase synthesis and heterocyst development by Nostocacean cyanobacteria, *J. Gen. Microbiol.,* 105, 83, 1978.
30. **Thomas, J. and David, K. A. V.,** Studies on the physiology of heterocyst production in the nitrogen-fixing blue-green alga *Anabaena* sp. L-31 in continuous culture, *J. Gen. Microbiol.,* 66, 127, 1971.
31. **Wolk, C. P. and Quine, M. P.,** Formation of one-dimensional patterns by stochastic processes and by filamentous blue-green algae, *Dev. Biol.,* 46, 370, 1975.
32. **Kulasooriya, S. A. and Fay, P.,** On the reversibility of heterocyst differentiation, *Br. Phycol. J.,* 9, 97, 1974.
33. **Wilcox, M., Mitchison, G. J., and Smith, R. J.,** Pattern formation in the blue-green alga *Anabaena.* II. Controlled proheterocyst regression, *J. Cell Sci.,* 13, 637, 1973.
34. **Wilcox, M., Mitchison, G. J., and Smith, R. J.,** Mutants of *Anabaena cylindrica* altered in heterocyst spacing, *Arch. Mikrobiol.,* 103, 219, 1975.
35. **Fay, P., Stewart, W. D. P., Walsby, A. E., and Fogg, G. E.,** Is the heterocyst the site of nitrogen fixation in blue-green algae?, *Nature (London),* 220, 810,1968.
36. **Stewart, W. D. P., Haystead, A., and Pearson, H. W.,** Nitrogenase activity in heterocysts of blue-green algae, *Nature (London),* 224, 226, 1969.
37. **Kenyon, C. N., Rippka, R., and Stanier, R. Y.,** Fatty acid composition and physiological properties of some filamentous blue-green algae, *Arch. Mikrobiol.,* 83, 216, 1972.
38. **Stewart, W. D. P.,** Blue-green algae, in *A Treatise on Dinitrogen Fixation, Sect. III, Biology,* Hardy, R. W. F. and Silver, W. S., Eds., John Wiley & Sons, New York, 1977, 63.
39. **Peters, G. A.,** The *Azolla-Anabaena azollae* symbiosis, in *Genetic Engineering for Nitrogen Fixation,* Hollaender, A., Ed., Plenum Press, New York, 1977, 231.
40. **Millbank, J. W.,** Lower plant associations, in *A Treatise on Dinitrogen Fixation, Sect. III, Biology,* Hardy, R. W. F. and Silver, W. S., Eds., John Wiley & Sons, New York, 1977, 125.
41. **Paerl, H. W.,** Role of heterotrophic bacteria in promoting N_2 fixation by *Anabaena* in aquatic habitats, *Microb. Ecol.,* 4, 215, 1978.
42. **Odum, E. P.,** *Fundamentals of Ecology,* W. B. Saunders, Philadelphia, 1971.
43. **Peters, G. A., Toia, R. E., and Lough, S. M.,** *Azolla-Anabaena azollae* relationship. V. $^{15}N_2$ fixation, acetylene reduction, and H_2 production, *Plant Physiol.,* 59, 1021, 1977.
44. **Rogers, G. A.,** The effect of some external factors on nitrogenase activity in the free-living and endophytic *Nostoc* of the liverwort *Blasia pusilla*, *Physiol. Plant.,* 44, 407, 1978.
45. **Newton, J. W. and Herman, A. I.,** Isolation of cyanobacteria from the aquatic fern, *Azolla*, *Arch. Microbiol.,* 120, 161, 1979.
46. **Watanabe, A.,** Production in cultural solution of some amino acids by the atmospheric nitrogen-fixing blue-green algae, *Arch. Biochem. Biophys.,* 34, 50, 1951.
47. **Dugdale, R. C., Goering, J. J., and Ryther, J. H.,** High nitrogen fixation rates in the Sargasso Sea and the Arabian Sea, *Limnol. Oceanogr.,* 9, 507, 1964.
48. **Wyatt, J. T. and Silvey, J. K. G.,** Nitrogen fixation by *Gloeocapsa*, *Science,* 165, 908, 1969.
49. **Rippka, R., Neilson, A., Kunisawa, R., and Cohen-Bazire, G.,** Nitrogen fixation by unicellular blue-green algae, *Arch. Mikrobiol.,* 76, 341, 1971.
50. **Gallon, J. R., LaRue, T. A., and Kurz, W. G. W.,** Characteristic of nitrogenase activity in broken cell preparations of the blue-green alga *Gloeocapsa* sp. LB795, *Can. J. Microbiol.,* 18, 327, 1972.
51. **Carpenter, E. J. and Price, C. C.,** Marine *Oscillatoria (Trichodesmium).* Explanation for aerobic nitrogen fixation without heterocysts, *Science,* 191, 1278, 1976.
52. **Gallon, J. R., Ul-Haque, M. I., and Chaplin, A. E.,** Fluoroacetate metabolism in *Gloeocapsa* sp. LB795 and its relationship to acetylene reduction (nitrogen fixation), *J. Gen. Microbiol.,* 106, 329, 1978.
53. **Rippka, R. and Waterbury, J. B.,** The synthesis of nitrogenase by non-heterocystous cyanobacteria, *FEMS Microbiol. Lett.,* 2, 83, 1977.
54. **Garlick, S., Oren, A., and Padan, E.,** Occurrence of facultative anoxygenic photosynthesis among filamentous and unicellular cyanobacteria, *J. Bacteriol.,* 129, 623, 1977.
55. **Wolk, C. P., Thomas, J., Shaffer, P. W., Austin, S. M., and Galonsky, A.,** Pathway of nitrogen metabolism after fixation of ^{13}N-labeled nitrogen gas by the cyanobacterium, *Anabarna cylindrica*, *J. Biol. Chem.,* 251, 5027, 1976.
56. **Thomas, J., Meeks, J. C., Wolk, C. P., Shaffer, P. W., Austin, S. M., and Chien, W.-S.,** Formation of Glutamine from [^{13}N] ammonia, [^{13}N] dinitrogen, and [^{4}C] glutamate by heterocysts isolated from *Anabaena cylindrica*, *J. Bacteriol.,* 129, 1545, 1977.
57. **Meeks, J. C., Wolk, C. P., Lockau, W., Schilling, N., Shaffer, P. W., and Chien, W.-S.,** Pathways of assimilation of {^{13}N} N_2 and $^{13}NH_4^+$ by cyanobacteria with and without heterocysts, *J. Bacteriol.,* 134, 125, 1978.

58. **Stewart, W. D. P., Haystead, A., and Dharmawardene, M. W. N.,** Nitrogen assimilation and metabolism in blue-green algae, in *Nitrogen Fixation by Free-living Micro-organisms*, IBP 6, Stewart, W. D. P., Ed., Cambridge University Press, Cambridge, 1975, 129.
59. **Ohmori, M. and Hattori, A.,** Effect of nitrate on nitrogen-fixation by the blue-green alga *Anabaena cylindrica, Plant Cell Physiol.,* 13, 589, 1972.
60. **Ohmori, M. and Hattori, A.,** Effect of ammonia on nitrogen fixation by the blue-green alga *Anabaena cylindrica, Plant Cell Physiol.,* 15, 131, 1974.
61. **Stewart, W. D. P. and Rowell, P.,** Effects of L-methionine-DL-sulphoximine on the assimilation of newly fixed NH_3, acetylene reduction and heterocyst production in *Anabaena cylindrica, Biochem. Biophys. Res. Commun.,* 65, 846, 1975.
62. **Rowell, P., Enticott, S., and Stewart, W. D. P.,** Glutamine synthetase and nitrogenase activity in the blue-green alga *Anabaena cylindrica, New Phytol.,* 79, 41, 1977.
63. **Shanmugam, K. T., O'Gara, F., Andersen, K., and Valentine, R. C.,** Biological nitrogen fixation, *Annu. Rev. Plant Physiol.,* 29, 263, 1978.
64. **Cobb, H. D. and Myers, J.,** Comparative studies of nitrogen fixation and photosynthesis in *Anabaena cylindrica, Am. J. Bot.,* 51, 753, 1964.
65. **Ownby, J. D., Shannahan, M., and Hood, E.,** Protein synthesis and degradation in *Anabaena* during nitrogen starvation, *J. Gen. Microbiol.,* 110, 255, 1979.
66. **Allen, M. M. and Smith, A. J.,** Nitrogen chlorosis in blue-green algae, *Arch. Mikrobiol.,* 69, 114, 1969.
67. **Simon, R. D.,** Cyanophycin granules from the blue-green alga *Anabaena cylindrica*: a reserve material consisting of copolymers of aspartic acid and arginine, *Proc. Natl. Acad. Sci. U.S.A.,* 68, 265, 1971.
68. **Neilson, A., Rippka, R., and Kunisawa, R.,** Heterocyst formation and nitrogenase synthesis in *Anabaena* sp. A kinetic study, *Arch. Mikrobiol.,* 76, 139, 1971.
69. **DeVasconcelos, L. and Fay, P.,** Nitrogen metabolism and ultrastructure in Anabaena cylindrica, I. The effect of nitrogen starvation, *Arch. Microbiol.,* 96, 271, 1974.
70. **Bothe, H.,** Photosynthetic nitrogen fixation by a cell-free extract of the blue alga, *Anabaena cylindrica, Ber. Dtsch. Bot. Ges.,* 83, 421, 1970.
71. **Smith, R. V. and Evans, M. C. W.,** Soluble nitrogenase from vegetative cells of the blue-green alga *Anabaena cylindrica, Nature (London),* 225, 1253, 1970.
72. **Smith, R. V. and Evans, M. C. W.,** Nitrogenase activity in cell-free extracts of the blue-green alga, *Anabaena cylindrica, J. Bacteriol.,* 105, 913, 1971.
73. **Smith, R. V., Noy, R. J., and Evans, M. C. W.,** Physiological electron donor systems to the nitrogenase of the blue-green alga *Anabaena cylindrica, Biochim. Biophys. Acta,* 253, 104, 1971.
74. **Haystead, A. and Stewart, W. D. P.,** Characteristics of the nitrogenase system from the blue-green algae *Anabaena cylindrica, Arch. Mikrobiol.,* 82, 325, 1972.
75. **Bothe, H.,** Role of phytoflavin in photosynthetic reactions, in *Progress Photosyn. Research, Proc. Int. Congr.,* Vol. 3, Metzner, H., Ed., Verlag C. Lichtenstern, Munich, 1968, 1483.
76. **Batt, T. and Brown, D. H.,** The influence of inorganic nitrogen supply on carbohydrate and related metabolism in the blue-green alga, *Anabaena cylindrica* Lemm., *Planta,* 116, 197, 1974.
77. **Lex, M. and Carr, N. G.,** The metabolism of glucose by heterocysts and vegetative cells of *Anabaena cylindrica, Arch. Microbiol.,* 101, 161, 1974.
78. **Peterson, R. B. and Burris, R. H.,** Properties of heterocysts isolated with colloidal silica, *Arch. Microbiol.,* 108, 35, 1976.
79. **Stewart, W. D. P. and Lex, M.,** Nitrogenase activity in the blue-green alga *Plectonema boryanum* strain 594, *Arch. Mikrobiol.,* 73, 250, 1970.
80. **Leach, C. K. and Carr, N. G.,** Pyruvaterredoxin oxidoreductase and its activiation by ATP in the blue-green alga *Anabaena variabilis, Biochim. Biophys. Acta,* 245, 165, 1971.
81. **Bothe, H., Falkenberg, B. and Nolteernsting, U.,** Properties and function of the pyruvaterredoxin oxidoreductase from the blue-green alga *Anabaena cylindrica, Arch. Microbiol.,* 96, 291, 1974.
82. **Gallon, J. R., Kurz, W. G. W., and LaRue, T. A.,** Isocitrate supported nitrogenase activity in *Gloeocapsa* sp. LB795, *Can. J. Microbiol.,* 19, 461, 1973.
83. **Hwang, J. C. and Burris, R. H.,** Nitrogenase-catalyzed reactions, *Biochim. Biophys. Acta,* 283, 339, 1972.
84. **Rivera-Ortiz, J. M. and Burris, R . H.,** Interactions among substrates and inhibitors of nitrogenase, *J. Bacteriol.,* 123, 537, 1975.
85. **Bothe, H., Tennigkeit, J., and Eisbrenner, G.,** The utilization of molecular hydrogen by the blue-green alga *Anabaena cylindrica, Arch. Microbiol.,* 114, 43, 1977.
86. **Tel-Or, E., Luijk, L. W., and Packer, L.,** An inducible hydrogenase in cyanobacteria enhances N_2 fixation, *FEBS Lett.,* 78, 49, 1977.
87. **Peterson, R. B. and Burris, R. H.,** Hydrogen metabolism in isolated heterocysts of *Anabaena* 7120, *Arch. Microbiol.,* 116, 125, 1978.

88. **Bottomley, P. J. and Stewart, W. P. D.**, ATP pools and transients in the blue-green alga, *Anabaena cylindrica, Arch. Microbiol.*, 108, 249, 1976.
89. **Fay, P.**, Photostimulation of nitrogen fixation in *Anabaena cylindrica, Biochim. Biophys. Acta,* 216, 353, 1970.
90. **Bothe, H. and Loos, E.**, Effect of far red light and inhibitors on nitrogen fixation and photosynthesis in the blue-green alga *Anabaena cylindrica, Arch. Mikrobiol.*, 86, 241, 1972.
91. **Fay, P.**, Factors influencing dark nitrogen fixation in a blue-green alga, *Appl. Environ. Microbiol.*, 31, 376, 1976.
92. **Burris, R. H.**, Nitrogen fixation — assay methods and technique, in *Methods in Enzymology,* Vol. 24, San Pietro, A., Ed., Academic Press, New York, 1972, 415.
93. **Stewart, W. D. P., Fitzgerald, G. P., and Burris, R. H.**, *In situ* studies on N_2 fixation using the acetylene reduction technique, *Proc. Natl. Acad. Sci. U.S.A.*, 58, 2071, 1967.
94. **Hardy, R. W. F., Burns, R. C., and Holsten, R. D.**, Applications of the acetylene-ethylene assay for measurement of nitrogen fixation, *Soil Biol. Biochem.*, 5, 47, 1973.
95. **David, K. A. V. and Fay, P.**, Effects of long-term treatment with acetylene on nitrogen-fixing microorganisms, *Appl. Environ. Microbiol.*, 34, 640, 1977.
96. **Stacey, G., Van Baalen, C., and Tabita, F. R.**, Isolation and characterization of a marine *Anabaena* sp. capable of rapid growth on molecular nitrogen, *Arch. Microbiol.*, 114, 197, 1977.
97. **Martin, T. C. and Wyatt, J. T.**, Comparative physiology and morphology of six strains of Stigonematacean blue-green algae, *J. Phycol.*, 10, 57, 1974.
98. **Sinclair, C. and Whitton, B. A.**, Influence of nitrogen source on morphology of Rivulariaceae (Cyanophyta), *J. Phycol.*, 13, 335, 1977.
99. **Öström, B.**, Fertilization of the Baltic by nitrogen fixation in the blue-green alga *Nodularia spumigena, Remote Sensing Environ.*, 4, 305, 1976.
100. **Jurgensen, M. F. and Davey, C. B.**, Nitrogen-fixing blue-green algae in acid forest and nursery soils, *Can. J. Microbiol.*, 14, 1179, 1968.
101. **Singh, H. N., Ladha, J. K., and Kumar, H. D.**, Genetic control of heterocyst formation in the blue-green algae *Nostoc muscorum* and *Nostoc linckia, Arch. Microbiol.*, 114, 155, 1977.
102. **Henrikkson, L. E. and DaSilva, E. J.**, Effects of some inorganic elements on nitrogen-fixation in blue-green algae and some ecological aspects of pollution, *Z. Allg. Mikrobiol.*, 18, 487, 1978.
103. **Stewart, W. D. P.**, *Nitrogen Fixation in Plants,* Athlone Press, London, 1966, 68.
104. **Mague, T. H., Weare, N. M., and Holm-Hansen, O.**, Nitrogen fixation in the North Pacific Ocean, *Mar. Biol.*, 24, 109, 1974.
105. **Singh, P. K.**, Nitrogen fixation by the unicellular blue-green alga *Aphanothece, Arch. Mikrobiol.* 92, 59, 1973.
106. **Weare, N. M. and Benemann, J. R.**, Nitrogen fixation by *Anabaena cylindrica.* I. Localization of nitrogen fixation in the heterocysts, *Arch. Mikrobiol.*, 90, 323, 1973.
107. **Stanier, R. Y.**, The utilization of organic substrates by cyanobacteria, *Biochem. Soc. Trans.*, 3, 352, 1975.
108. **Peterson, R. B., Friberg, E. E., and Burris, R. H.**, Diurnal variation in N_2 fixation and photosynthesis by aquatic blue-green algae, *Plant Physiol.*, 59, 74, 1977.
109. **Paerl, H. W.**, Optimization of carbon dioxide and nitrogen fixation by the blue-green alga *Anabaena* in freshwater blooms, *Oecologia,* 38, 275, 1979.
110. **Haystead, A., Robinson, R., and Stewart, W. D. P.**, Nitrogenase activity in extracts of heterocystous and non-heterocystous blue-green algae, *Arch. Mikrobiol.*, 74, 235, 1970.
111. **Lex, M., Silvester, W. B., and Stewart, W. D. P.**, Photorespiration and nitrogenase activity in the blue-green alga, *Anabaena cylindrica, Proc. R. Soc. London Ser. B,* 180, 87, 1972.
112. **Paerl, H. W.**, Light-mediated recovery of N_2-fixation in the blue-green algae *Anabaena* spp. in O_2 supersaturated waters, *Oecologia,* 32, 135, 1978.
113. **Davis, E. B., Tischer, R. G., and Brown, L. R.**, Nitrogen fixation by the blue-green alga, *Anabaena flos-aquae* A-37, *Physiol. Plant.*, 19, 823, 1966.
114. **Fogg, G. E. and Than-Tun**, Interrelations of photosynthesis and assimilation of elementary nitrogen in a blue-green alga, *Proc. R. Soc. London Ser. B,* 153, 111, 1960.
115. **Stewart, W. D. P.**, Nitrogen fixation by blue-green algae in Yellowstone thermal areas, *Phycologia,* 9, 261, 1960.
116. **Raven, J. A.**, Transport in algal cells, in *Transport in Plants II, Part A, Cells,* Lüttge, U. and Pitman, M. G., Eds., Springer-Verlag, Berlin, 1976, 129.
117. **Jacobs, R. and Lind, O.**, The combined relationship of temperature and molybdenum concentration to nitrogen fixation by *Anabaena cylindrica, Microbial Ecol.*, 3, 205, 1977.
118. **Nagatani, H. H. and Haselkorn, R.**, Molybdenum independence of nitrogenase component synthesis in the non-heterocystous cyanobacterium *Plectonema, J. Bacteriol.*, 134, 597, 1978.
119. **Jeffries, T. W. and Leach, K. L.**, Intermittent illumination increases biophotolytic hydrogen yield by *Anabaena cylindrica, Appl. Environ. Microbiol.*, 35, 1228, 1978.

120. **Cox, R. M. and Fay, P.**, Special aspects of nitrogen fixation by blue-green algae, *Proc. R. Soc. London Ser. B,* 172, 357, 1969.
121. **Bradley, S. and Carr, N. G.**, Heterocyst and nitrogenase development in *Ananbaena cyilindrica, J. Gen. Microbiol.,* 96, 175, 1976.
122. **Stewart, W. D. P., Fitzgerald, G. P., and Burris, R. H.**, Acetylene reduction by nitrogen-fixing blue-green algae, *Arch. Mikrobiol.,* 62, 336, 1968.
123. **Bone, D. H.**, Nitrogenase activity and nitrogen assimilation in *Anabaena flos-aquae* growing in continuous culture, *Arch. Mikrobiol.,* 80, 234, 1971.
124. **Peterson, R. B. and Wolk, C. P.**, High recovery of nitrogenase activity and of ^{55}Fe-labeled nitrogenase in heterocysts isolated from *Anabaena variabilis, Proc. Natl. Acad. Sci. U.S.A.,* 75, 6271, 1978.
125. **Codd, G. A., Rowell, P., and Stewart, W. D. P.**, Pyruvate and nitrogenase activity in cell-free extracts of the blue-green alga *Anabaena cylindrica, Biochem. Biophys. Res. Commun.,* 61, 424, 1974.
126. **Bennett, K. J., Silvester, W. B., and Brown, J. M. A.**, The effect of pyruvate on nitrogenase activity in the blue-green alga *Anabaena cylindrica, Arch. Microbiol.,* 105, 61, 1975.

SYMBIOTIC NITROGEN FIXATION IN NODULATED LEGUMINOUS AND NONLEGUMINOUS ANGIOSPERMS*

Karel R. Schubert

INTRODUCTION

The availability of nitrogen in a utilizable form is a prerequisite for plant growth.[1] Although molecular nitrogen is the most abundant element in the atmosphere, plants are unable to use nitrogen in this form and therefore must rely upon sources of oxidized or reduced nitrogen commonly referred to as "fixed" nitrogen. The availability of an adequate supply of this "fixed" nitrogen is the major factor, aside from water, limiting plant productivity.[1,2]

The largest input of fixed nitrogen is derived from biological sources which account for approximately 70% of the nitrogen fixed worldwide.[3] Over half of this is fixed symbiotically.[3] These symbiotic N_2-fixing systems include obligatory associations, such as the *Azolla/Anaebena* or the legume/*Rhizobium* symbiosis, as well as loose associations such as that formed between the tropical grass *Paspalum notatum* and the procaryote *Azotobacter paspali*[3]. In terms of biosolar productivity, the obligatory symbiosis is the most efficient being conducive to maximal and sustained rates of N_2 fixation. The focus of this article is on obligatory associations involving the nodulation of the roots of leguminous and nonleguminous angiosperms.

DISTRIBUTION OF OBLIGATORY N_2-FIXING SYMBIOSES IN HIGHER PLANTS

With the exception of certain angiosperms and gymnosperms capable of forming a symbiotic partnership with specific N_2-fixing microorganisms, vascular plants are unable to reduce dinitrogen. In the case of the angiosperms, the association is apparently obligatory in that both partners are essential for effective N_2 fixation. The host plant provides the energy as well as the proper environment to support fixation which occurs within the endophyte. The ultimate source of energy for this process is sunlight which is captured and converted photosynthetically into chemical energy and stored in the form of carbohydrates. A portion of these carbohydrates is transported to the nodule and used to support the growth and metabolic functions of the endophyte. In return, the microsymbiont provides fixed nitrogen for the growth of the host.

Agriculturally, the legume/*Rhizobium* symbiosis is probably the most significant, fixing an estimated 80×10^6 metric tonnes of N per year.[2] The incidence of nodulation in the family Leguminoseae, which comprises 12,000 to 14,000 species in 700 genera and 4 subfamilies of tropical origin, is widespread. As indicated in Table 1, more than 90% of the species examined from the subfamilies Mimosoideae and Papilionatae and about one third of those from the Caesalpinioideae form root nodules, indicative of the ability to fix N_2.[3,4] The legumes of agricultural importance, such as the grain legumes, soybeans, peanuts, lentils, beans, and peas and the forage legumes, alfalfa and clover, are all members of the Papilionatae.[5]

Nitrogen-fixing symbioses with nonleguminous angiosperms occur in many diverse species from eight families and seven orders (Table 2).[3,5,7] At this time, more than one third of the species within the 14 genera involved are nodulated, and many species have not been examined.[3,5,6] Because of their competitive advantage, the nonlegumes

* This work was supported in part by the National Science Foundation (Grant No. PCM 77-24683) and the Michigan Agriculture Experiment Station (Publication No. 8810).

Table 1
SYMBIOTIC N_2 FIXATION IN THE FAMILY LEGUMINOSEAE

Subfamily	Nodulation: Genera[a]	Nodulation: Species[b]	Distribution
Caesalpinioideae	15/31/91	26/115/1,300	Tropical
Mimosoideae	18/19/30	127/146/1,500	Tropical to temperate
Papilionatae	149/154/308	952/1,024/10,000	Tropical to temperate
Swartzoideae	—	0/0/80	Tropical

[a] Number genera nodulated per number genera examined per total number of genera in subfamily.[3]
[b] Number species nodulated per number species examined per total number of species in subfamily.[4]

are particularly well suited to marginal or disturbed areas of low fertility. The distribution is worldwide for certain genera, such as *Alnus, Coriaria, Elaeagnus*, and *Hippophaë*, while others, such as *Ceanothus* and *Shepherdia*, are more limited in distribution.[5-7] Most of the plants involved are of little agricultural importance, but are valuable in forest and range ecosystems, both as feed for animals and to improve the N status of the soil.[7]

FORMATION OF AN EFFECTIVE N_2-FIXING SYMBIOSIS

During the formation of the symbiosis, the bacterium invades the cells of its specific plant host and induces the formation of a nodule, the actual site of N_2 fixation. The nodule, a proliferation of host cell tissue containing bacteria, is a highly organized structure which provides the stable environment conducive for maximum fixation.

In the case of the legume symbiosis, the invasive N_2-fixing microorganisms are Gram-negative rods (0.5 to 0.9 × 1.2 × 3.0 μm) of the genus *Rhizobium*. Rhizobia are classified into 16 cross-inoculation groups on the basis of the specificity of the host plant/*Rhizobium* interaction.[4,8,9] The classification and characteristics of agronomically important host/endophyte combinations are given in Table 3. Originally each cross-inoculation group was thought to be mutually exclusive. Since that time, infection by heterologous strains has been well documented.[4,9] Moreover, infection by *Rhizobium* is not restricted to the Leguminoseae, as noted in the case of infection of the nonleguminous angiosperm, *Parasporia Trema*, by a promiscuous "cowpea" isolate.[11] For these reasons, a more general classification into fast-growing, acid-producing and slow-growing, base-producing species has been suggested (see Table 3).[4,10]

Details on the process of infection and nodule formation are summarized in a number of articles and reviews.[12-17] These reviews discuss requirements for nodulation,[12-14] the effect of environmental factors,[13,14] and the morphological[12,15,16] and biochemical[16,17] changes in the host plant and the microsymbiont that occur during nodule formation.

The process of nodule formation in the nonleguminous angiosperms is apparently quite similar to that in the legumes.[18] In the case of the nonlegumes, however, the reinoculation of the host plant with pure cultures of the "actinomycetes-like" organisms isolated from nodules has not been successful until recently when Torrey and coworkers[19] used a pure culture of an isolate from *Comptonia gale* to reinfect the host plant. The taxonomic identification and classification of the actinomycete, *Frankia*, which infect the nonlegumes is under investigation.

Table 2
SYMBIOTIC N_2 FIXATION IN NONLEGUMINOUS ANGIOSPERMS

Order	Family	Genus	Nodulation[a]	Distribution[b]	Symbiotic capacity (kg N/ha · year)[c]	Biomass[d]
Casuarinales	Casuarinaceae	*Casuarina*	24/45	Australia, tropical Asia, Pacific Islands	58 (218)	29
Coriales	Coriariaceae	*Coriaria*	13/15	Mediterranean to Japan , New Zealand, Chile to Mexico	129—192	5(30)
Ericales	Ericaceae	*Arctostaphyles*	1/70	Northwest and central America, Europe, Asia	—	—
Fagales	Betulaceae	*Alnus*	32/35	Europe, Siberia, North America, Japan, Andes	40—362	1-20
Myricales	Myricaceae	*Myrica*	26/35	Many tropical, subtropical, and temperate regions	9	—
Rhamnales	Elaeagnaceae	*Elaeagnus*	16/45	Asia, Europe, North America,	(183)	14
		Hippophaë	1/3	Asia, Europe, from Himalayas to Arctic Circle,	2—179	—
		Shepherdia	2/3	North America	—	—
	Rhamnaceae	*Ceanothus*	31/55	North America	60 (91)	6
		Discaria	2/10	Andes, Brazil, New Zealand, Australia	—	—
Rosales	Rosaceae	*Cercocarpus*	4/20	North America	—	—
		Dryas	3/4	Arctic and mountains of north Temperate Zone	12	—
		Purshia	2/2	North America	5	—

[a] Number of species nodulated per total number of species.[6]

[b] Based on data by Bond.[5]

[c] Based on estimated accrual of nitrogen (kg N/ha · year).[5,7] Values within parentheses reflect measurements of total N in litterfall and as such overestimate the amount of N fixed.[7]

[d] Values reported for the production of biomass as litterfall in nodulated nonlegumes (ton dry matter/ha · year).

Table 3
CLASSIFICATION AND CHARACTERISTICS OF *RHIZOBIUM* AND THEIR HOSTS

Host group	Species[a]	Characteristics of *Rhizobium* Growth Rate[b]	Reaction on YMA[c]	Carbohydrate utilization[d] Suc	Glc	Lac	Mal	Rha	Ara	pH tolerance[e] 4.5	9.5	Flagellation[f]	Agronomically important host plants	Symbiotic capacity[g] (Kg N fixed/ha · year)
Pea and vetch	*Rhizobium leguminosarum*	Fast	Acid	+	+	+	+	+	+	0.54	0	Peritrichous	*Pisum arvense* (field pea)	72—85
													Pisum sativum (garden pea)	—
													Vicia faba (broad bean)	—
													Vicia sativa (common vetch)	80
													Lathyrus odoratus (sweet pea)	—
													Lens culinaris (lentil)	85—103
Bean	*R. phaseoli*	Fast	Acid	+	+	+	+	+	+	0.80	0	Peritrichous	*Phaseolus vulgaris* (garden bean)	40
Clover	*R. trifolii*	Fast	Acid	+	+	+	+	+	+	0.68	0	Peritrichous	*Trifolium* spp. (clover)	104—220
Alfalfa	*R. meliloti*	Fast	Acid	+	+	±	±	+	±	0.09	0.90	Peritrichous	*Medicago sativa* (alfalfa)	40—350
													Melilotus alba (sweet clover)	119
Lupin	*R. lupini*	Slow	Alkaline	−	±	−	−	−	+	0.84	0	Subpolar	*Lupinus* spp. (lupinus)	150—169
													Ornithopus sativus (serradella)	—
Soybean	*R. japonicum*	Slow	Alkaline	−	±	−	−	−	+	0.84	0	Subpolar	*Glycine max* (soybean)	59—94
Cowpea	"Cowpea isolates"	Slow	Alkaline	−	±	−	−	−	+	0.84	0	Subpolar	*Vigna unguiculata* (cowpea)	84
													Cajanus cajan (pigeon pea)	—
													Vigna radiata (mung bean)	—
													Phaseolus lunatus (lima bean)	—
													Arachis hypogaea (peanut)	47
Lotus and Leucaena	*Lotus* and *Leucaena* isolates	Fast	Acid	+	+	+	+	+	+	—	—	—	*Lotus corniculatus* (birdsfoot trefoil)	—

Slow	Alkaline	–	±	–	–	–	+	—	—	—	*Leucaena* spp. (leucaena)	—

[a] Based on cross-inoculation classification.[8]

[b] Fast growers have a mean generation time of 2 to 4 hr and give little detectable growth on yeast mannitol agar (YMA) within 24 hr, but generally form relatively large (2 to 4 mm) colonies in 3 to 5 days. Slow growers (generation time 6 to 8 hr) form colonies 1 mm or less, even after 7 to 10 days.

[c] Production of acid or base on yeast mannitol agar (YMA) indicator plates with bromthymol blue.[4,11]

[d] Utilization of carbohydrates vary with species; Suc-sucrose; Glc-glucose; Lac-lactose, Mal-maltose; Rha-rhamnose; and Ara-arabinose.[4,11]

[e] Fraction of strains tested that were tolerant of pH 4.5 or 9.5.[4]

[f] Based on data of Vincent.[4]

[g] Values from Burns and Hardy[3] and Vincent.[4]

THE N_2-FIXING SYSTEM

The first step of biological nitrogen fixation is the reduction of N_2 to form ammonia. The reaction is catalyzed by an enzyme complex composed of two O_2-labile proteins, nitrogenase reductase (Fe protein) and nitrogenase (MoFe protein).[20] In addition to the two proteins, a source of low-potential electrons (about −400 mV) and ATP are required for catalysis to occur.[21-24]

Nitrogenase reductase (mol wt 50,000 to 60,000) and nitrogenase (mol wt 200,000 to 220,000) have been purified and characterized from a variety of bacterial sources, including the bacteroids of soybean and other legume nodules.[21-24] Up to now, the enzyme from nonlegumes has not been purified. The properties of the enzyme from nodules are quite similar to those of the complex isolated from *Azotobacter* or other free-living microorganisms.[21-24]

Composed of two identical subunits (mol wt 30,000) with a single (Fe_4S_4) iron-sulfur cluster, nitrogenase reductase transfers electrons in one-electron steps from reductant to nitrogenase.[20,25] Prior to electron transfer, the reductase binds 2 mol of ATP which are hydrolyzed to ADP and Pi during the process of electron transfer.[21,22,24] The reduced form of nitrogenase then catalyzes substrate reduction.[20,21] Like the reductase, nitrogenase is a metalloprotein containing both Fe and Mo along with acid-labile sulfur.[21-24] The Mo and part of the Fe are associated with a low molecular weight cofactor (MoFe cofactor, mol wt. 2000 to 3000) which may participate directly in the catalytic process.[26]

In addition to N_2, the nitrogenase complex reduces a variety of other substrates, including H_3O^+, C_2H_2, N_3^-, CN^-, and CH_3CN.[3,21] These alternate substrates compete with N_2 for "activated" electrons and function as inhibitors of N_2 reduction.[3,27]

One of the alternate reactions, the ATP and reductant-dependent evolution of H_2, is characteristic of the purified nitrogenase complex from all sources examined.[3,27,28] During catalysis, protons and N_2 compete for electrons.[3,27,28] In the presence of an atmosphere of N_2, H_2 production occurs concomitantly with NH_4^+ formation, with 25 to 35% of the total equivalents of electrons diverted for proton reduction.[28] Based on kinetic studies, Rivera-Ortiz and Burris reported that even at infinite concentrations of N_2, H_2 evolution would still account for 13 to 23% of the total nitrogenase activity.[29] In the absence of N_2 or other substrates, the entire electron flux is expended for proton reduction.[3,24,27,28] The allocation of electrons between the two substrates, H^+ and N_2, is influenced by physiological parameters, such as the ATP/ADP ratio and the ratio of nitrogenase reductase to nitrogenase.[3,24,27,28]

Unlike other nitrogenase-catalyzed substrate reductions, proton reduction is not inhibited by CO.[3,21,22] Acetylene is the only effective inhibitor of nitrogenase reactions such that a saturating concentration of C_2H_2 essentially eliminates both H_2 evolution and N_2 reduction.[3,27,28] For this reason and others, the reduction of C_2H_2 has been assumed to reflect the total rate of N_2 fixation.[3] The key reactions and properties of the nitrogenase complex are summarized in Figures 1 and 2.

Substantial rates of ATP-dependent H_2 evolution in air, equivalent to 30 to 60% of the total nitrogenase activity, have been observed for excised nodules as well as intact nodulated legumes.[27,28,30] Because of the magnitude of H_2 evolution from these symbionts, Schubert and Evans recently postulated that the nitrogenase-catalyzed ATP-dependent evolution of H_2 was a significant pathway whereby energy may be lost during the N_2-fixing process in vivo.[27,28] Most of the nonleguminous and a few of the leguminous symbionts examined, however, exhibit low rates of H_2 evolution in air.[27] These symbionts possess an active hydrogenase system which reoxidizes the H_2 produced during N_2 fixation.[27,28,31,32] The oxidation of H_2 catalyzed by the hydrogenase of bacterial origin may be coupled to ATP or reductant generation and thus may in-

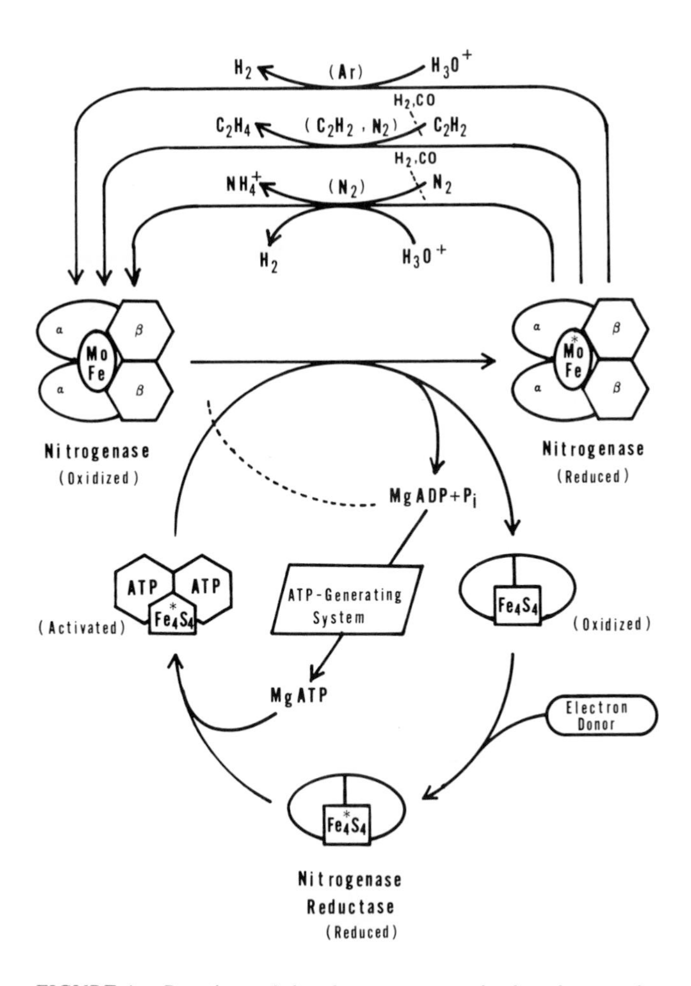

FIGURE 1. Reactions of the nitrogenase complex based on results with purified nitrogenase and nitrogenase reductase under the atmospheres indicated in brackets. Inhibition by CO, H_2, and ADP is indicated by a dashed line. The molybdenum-iron cofactor of nitrogenase and the iron-sulfur cluster of nitrogenase reductase are designated MoFe and Fe_4S_4, respectively.

crease the efficiency of the process by recapturing a portion of the energy otherwise lost.[27,28,31-34]

The pathway of electron transfer to the nitrogenase complex in nodulated symbionts has not been elucidated in its entirety. Electron carrier proteins, ferredoxin and flavodoxin, isolated from soybean nodules function in vitro to transport electrons to nitrogenase reductase using exogenous electron donor systems, such as Photosystem I of chloroplasts.[35] Based on investigations with *Azotobacter*, Veeger and associates[36] have demonstrated that the ratio of reduced to oxidized pyridine nucleotides was not adequate to maintain the oxidation-reduction potential necessary for N_2 reduction. They have suggested that the generation of reductant in vivo in bacteroids requires an intact energized cytoplasmic membrane.[37] To date the reconstruction in vitro of a complete sequence of electron carriers and physiological electron donors from nodules has not been successful.[35]

In relation to other biosynthetic reactions, the ATP requirement for N_2 reduction is unusually high.[38] Although the stoichiometry of ATP hydrolyzed per electron transferred varies with temperature, pH, and the ratio of nitrogenase reductase to nitrogen-

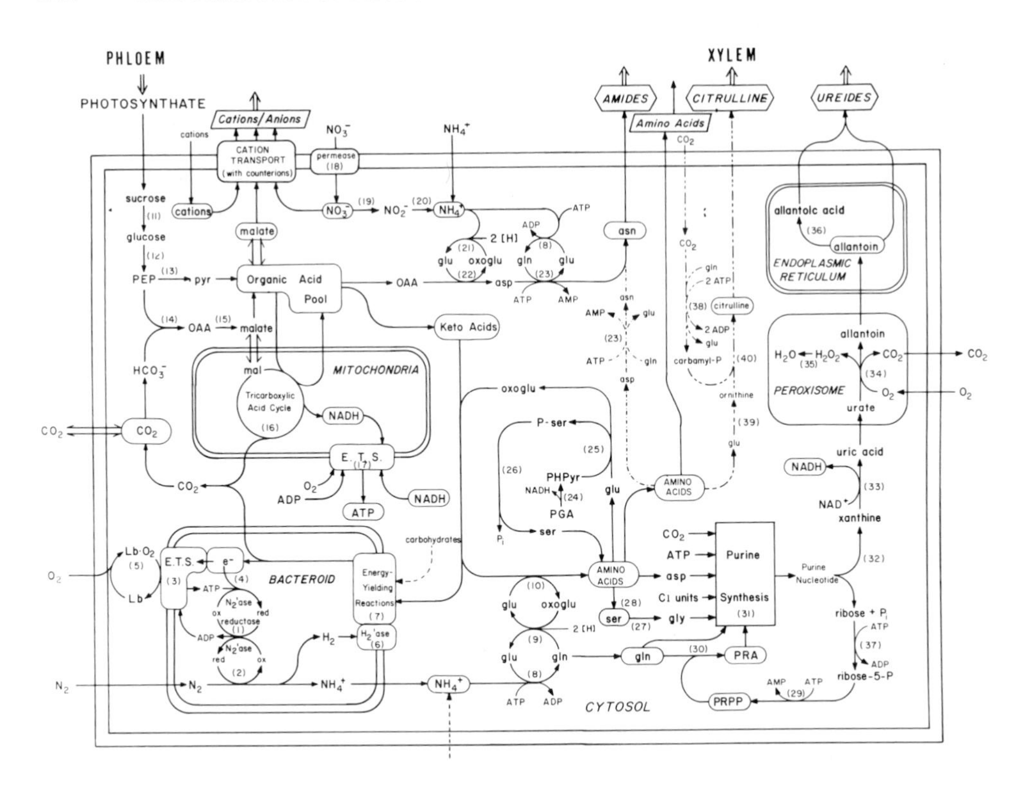

FIGURE 2. Schematic diagram summarizing the proposed pathways of nitrogen and carbon assimilation in the nodules and roots of N_2-fixing plants including the intracellular location of certain pathways. The diagram is based primarily on studies with soybean, a ureide-exporting legume, with specific notations for the synthesis of asparagine in lupin (↑·↑·↑) and citrulline in the actinorhizal plant, *Alnus glutinosa* (— — —). The numbers in parentheses corresponding to the proteins involved in specific reactions are given below: (1) nitrogenase reductase, (2) nitrogenase, (3) bacterial electron transport system and oxidative phosphorylation, (4) formation of membrane potential and low potential reductant, (5) leghemoglobin/O_2 transport, (6) hydrogenase [H_2-oxidizing], (7) reactions of energy-generating metabolism, (8) glutamine synthetase, (9) glutamate synthase [GOGAT], (10) amino transferases, (11) invertase, (12) glycolytic enzymes, (13) pyruvate kinase, (14) phosphoenolpyruvate [PEP] carboxylase, (15) malate dehydrogenase, (16) enzymes of tricarboxylic acid cycle, (17) mitochondrial electron transport system and oxidative phosphorylation, (18) nitrate uptake permease, (19) nitrate reductase, (20) nitrite reductase, (21) glutamate dehydrogenase, (22) aspartate amino transferase, (23) asparagine synthetase, (24) phosphoglycerate dehydrogenase, (25) phosphohydroxypyruvate (PHPyr) amino transferase, (26) phosphatase, (27) serine transhydroxymethyl transferase, (28) NADP-N^5,N^{10}-methylene FH_4 dehydrogenase and cyclohydrolase, (29) phosphoribosylpyrophosphate [PRPP] synthetase, (30) glutamine-PRPP amido transferase [PRA = phosphoribosylamine], (31) enzymes of purine biosynthesis, (32) phosphatase and nucleosidase, (33) xanthine dehydrogenase, (34) uricase, (35) catalase, (36) allantoinase, (37) ribokinase, (38) carbamyl phosphate synthetase, (39) enzymes of ornithine synthesis, (40) ornithine transcarbamylase. Pools of metabolites are encircled and solules for transport in the xylem are enclosed in boxes.

ase, the ratio is independent of substrate.[3,24,28,39] Values range from 1 to 10 mol ATP hydrolyzed per electron transferred to substrate with a minimum value of 2 ATP/e^- under optimum conditions.[3,24,38,39]

ADP functions in vitro as a potent inhibitor of nitrogenase activity.[3,24] For this reason, an ATP-regenerating system such as creatine kinase/creatine phosphate must be included in reaction mixtures. Similarly, there is a strong correlation between nitrogenase activity and the levels of ATP and ADP in intact nodulated plants.[3,40] These observations are consistent with the idea that the energy state of the cell is a primary factor regulating N_2 fixation in vivo.[36,40]

The generation of ATP within bacteroids, the symbiotic form of *Rhizobium*, is the result of oxidative metabolism.[17,41,42] Leghemoglobin which constitutes about 20 to 30% of the soluble protein of the nodule cytosol facilitates the diffusion of O_2 into the bacteroids in order to sustain the O_2 flux essential for the aerobic metabolism of the endophyte while maintaining the low partial pressure of free O_2 necessary to protect the nitrogenase complex from inactivation.[17,41] This protein, similar in structure to myoglobin, is synthesized coordinately by the macro- and microsymbionts.[43,44] There is no substantial evidence for the presence of leghemoglobin in the root nodules of nonleguminous plants, although the existence of other O_2-binding components has not been excluded.[45] In order to function efficiently at the reduced O_2 tensions (0.5 mmHg) within the legume nodule, bacteroids utilize a terminal oxidase system with high affinity for O_2.[17,41] Induced during nodule formation, this system contains cytochrome *c*, but not the a-a_3 complex characteristic of aerobically cultured cells.[17,41]

The nature of the physiological substrate for nodule respiration is as yet unclear.[16,17,41,42,46] Recently-fixed carbon in the form of sucrose, and to a lesser extent glucose, other carbohydrates, and organic acids are transported to the nodule via the phloem.[16,17,41,42,46-48] Within the nodule cytosol, sucrose may be hydrolyzed by invertase.[49] The direct utilization of the glucose formed to sustain bacteroid respiration, however, has not been demonstrated convincingly. A number of reports indicate that glucose does not support respiration in isolated bacteroids, even though free-living *Rhizobium* metabolize glucose.[17,41,50] This apparent anomaly may reflect the loss of a functional glucose transport system during bacteroid isolation.[17] Moreover, there is no consensus of opinion with respect to the metabolic pathways involved in glucose catabolism.[17,41,42,46] On the other hand, succinate and to a lesser extent pyruvate appear to be the best substrates to support respiration of isolated bacteroids. Although, no consistent picture has developed with respect to endogenous energy-yielding substrates and reactions in nodules, several recent lines of evidence, indicate that organic acids, especially dicarboxylic acids, play an essential role in the development and effectiveness of the legume-*Rhizobium* symbiosis.

THE ENERGETICS OF SYMBIOTIC N_2 FIXATION

Because of the high energy requirement for N_2 reduction, symbiotic N_2 fixation in many cases is limited by the availability of photosynthate, as reflected by the intimate relationship between symbiotic N_2 fixation and photosynthesis.[51] Photosynthate is used for

1. Nodule growth
2. Respiration associated with nodule growth and maintenance, N_2 fixation, and ammonia assimilation
3. Carbon skeletons for ammonia assimilation

For these reasons, it has been suggested that the intense demand for photosynthate during periods of active N_2 fixation, which may range from 10 to 30% of the total net carbon assimilated by the plant, may directly reduce the yield potential of legumes.[52]

Estimates of the energy requirements for N_2 fixation in nodulated plants have been based on studies with detached nodules or nodulated root systems.[53] In short-term studies, the rate of O_2 uptake and/or CO_2 efflux by detached nodules has been related to the rate of C_2H_2 reduction. Other more long-term assessments have related the respiratory CO_2 efflux of intact nodulated root systems or detached nodules measured at intervals over a period of days or weeks to the accumulation of organic nitrogen in the plant (in the complete absence of combined nitrogen). Results of a number of these analyses for a variety of symbionts are summarized in Table 4.

Table 4
ESTIMATES OF THE ENERGY REQUIREMENTS FOR SYMBIOTIC N_2 FIXATION

Species	Energy requirement (gC/gN)[a]		Ref.
	Detached nodules	Nodulated root system	
Pisum sativum	1.6	5.9	54
	6.7	—	55
Lupinus albus	1.7	14.5	56
Vigna unguiculata	1.1	5.9	57
	3.3[b]	—	58
	—	6.8	59
	10.1—12.6[c]	—	60
Glycine max	7.7	—	61
	4.1—7.9	6.3	59
	11.8[c]	—	60
Trifolium repens	—	6.6	59
Alnus glutinosa	5.1	—	62
Myrica gale	8.7	—	63

[a] Unless otherwise noted, estimates are based on measurements of CO_2 efflux from detached nodules or nodulated roots.

[b] This value includes components for respiration, nodule growth, and export of fixed nitrogen.

[c] Estimates based on O_2 consumption for excised nodules.

Alternatively, expressing the cost as a percentage of the net photosynthate of the plant, nodules utilize from 9 to 37% of the photosynthate (see Table 5). About one third of this amount is respired, representing only a fraction of the amount respired by the roots alone. Moreover, approximately half is returned to the host in the form of nitrogenous compounds and therefore should not be included in the cost of N_2 fixation. On this basis, the actual cost of fixation for some legumes may be only 5 or 6% of the total photosynthate of the plants.

THE ASSIMILATION AND TRANSPORT OF NITROGEN

Ammonia, the first stable product of N_2 fixation, is exported to the plant in the form of amino acids, amides, and ureides [64-70] (see Table 6). Initially, ammonia is excreted from the bacteroids into the plant cytosol where assimilation occurs via the glutamine synthetase (GS)/glutamate synthase (GOGAT) pathway as shown in Figure 2.[64,71-73] Recent findings indicate that the enzymes essential for NH_4^+ assimilation in bacteriods are repressed, whereas assimilatory enzymes in the plant cytosol are derepressed so that bacteroids are unable to utilize ammonia effectively.[73-78]

The initial products of NH_4^+ assimilation, glutamate and glutamine, are required for the synthesis of other nitrogenous compounds needed for nodule metabolism or for export through the xylem to other plant tissues. Effectively nodulated legumes can be classified into amide-exporting or ureide-exporting species based on the predominant form of organic nitrogen being transported from the roots (Table 6).[69,76-82,84,85-89] A similar classification may be appropriate for actinorhizal species.[76,90,91] The composition of the xylem sap which varies during plant development and with changes in environmental conditions, may serve as the basis for the selective portioning and utilization of products of nitrogen assimilation within various plant tissues.[57,77,79,92-96] In ureide-exporters, the source of nitrogen utilized by the plant may also affect the composition of xylem fluid in that NH_4^+ derived from N_2 is used preferentially for the

Table 5
ESTIMATES OF PHOTOSYNTHATE UTILIZATION IN NODULES AND NODULATED ROOTS OF SEVERAL LEGUMES

Species	Net photosynthate (%)			
	Pisum sativum[54]	*Lupinus albus*[56]	*Vigna unguiculata*[57]	*Phaseolus aureus*[64]
Nodule	32.0	11.3	9.0	5.7
Respiration	12.0	4.5	3.5	2.1
Dry matter	5.0	0.8	1.1	1.1
Export of fixed N	15.0	5.9	4.4	2.4
Roots	42.0	39.9	24.1	—
Respiration	35.0	33.2	15.4	—
Dry matter	7.0	6.7	8.7	—
Nodulated root	74.0	51.2	33.1	—
Respiration	47.0	37.7	18.9	—
Dry matter	12.0	7.5	9.8	—
Export of fixed N	15.0	5.9	4.4	—
Net photosynthate (gC)	0.35	41.45	24.72	3.47
Net N fixed (gN)	0.03	2.08	0.79	0.10

Table 6
THE COMPOSITION OF XYLEM EXUDATE OF SEVERAL N2-FIXING LEGUMINOUS AND ACTINORHIZAL PLANTS

	Composition of xylem exudate (% of total N)[e]						
	Amino acids			Amides[a]		Ureides[b,c,d]	
Species	Glu	Asp	Other	Asn	Gln		Ref.
Pisum arvense	—	15	—	50	21	14	77
Pisum sativum	3	5	12	65	15	+	78
Glycine max	—	—	—	—	—	98	79
Vicia faba	3	23	3	64	7	—	80
Lupinus albus	3	8	22	57	10	—	81
Vigna unquiculata	—	—	6	11	16	67	81
Vigna radiata	—	—	3	4	10	83	81
Phaseolus vulgaris	+	+	+	+	+	80	82
Alnus glutinosa	6	5	20	13	15	41[c]	83

[a] Other legumes in which the amides, asparagine and glutamine, are the major nitrogenous constituents of the xylem include the other temperate legumes, *Lathyrus sativa, Lens esculenta, Trifolium repens* and *Vicia sativa.*[84,85]
[b] Composed of the ureides, allantoin and allantoic acid unless noted otherwise.
[c] Composed of the ureide, citrulline.
[d] Other legumes in which allantoin and allantoic acid are the major nitrogenous constituents in the xylem sap include the other temperate legumes, *Psophocarpus tetragonolobus, Vigna mungo* and *Cyamopsis tetragonoloba.*[84,86]
[e] Less than 1% (-), present but not quantified (+).

synthesis of allantoin and allantoic acid, whereas nitrogen from NH_4^+ or NO_3^- is selectively incorporated into asparagine.[69,79,86,97-99] The molecular mechanism for the selective incorporation of nitrogen from N_2 or NO_3^- (NH_4^+) into ureides and amides, respectively, is not yet known; although compartmentation or differences in the pathways of assimilation represent possible mechanisms.[69,79,99] The export of nitrogen in the form of ureides may be energetically more efficient than the transport of amides.[63,79]

The assimilation of NH_4^+ and the biosynthesis and export of nitrogenous metabolites requires a continuous supply of reduced carbon compounds to be used for carbon skeletons and as energy sources to support biosynthetic reactions. Much of this reduced carbon is derived from recent photosynthesis occurring in the leaves and is transported to nodulated roots in the form of sucrose.[100] In addition, many N_2-fixing plants possess active systems for assimilating CO_2.[82,83,96,99,101-103] In legume nodules, the carboxylation of phosphoenolpyruvate (PEP) to form oxalacetate (OAA) catalyzed by PEP carboxylase appears to be the primary reaction of CO_2 fixation.[82,89,96,99,101,102] The amount of carbon fixed may represent 30% of the carbon respired by nodule.[96] The organic acids formed may be used as carbon skeletons in the synthesis of amino acids, amides, and ureides, as respiratory substrates for the host and/or endophyte, and as counterions for cation transport.[96,99]

Ultimately, the products of symbiotic fixation contribute to the overall composition and quality of legumes and nonlegumes as forage and feed. As a way of comparison, some representative values are given in Table 7 for the composition and yields of various agronomically important crops. Although the yields of legumes are lower than those of certain cereal crops, the legumes through their ability to fix N_2 symbiotically function to efficiently couple solar energy to the production of protein and biomass.

Table 7
SEED YIELDS, COMPOSITION, AND TOTAL PROTEIN OF GRAIN LEGUMES AND CEREAL CROPS

Crop	Seed yields[a] (kg/ha · year)	Composition[b] (% dry weight)				Total protein (kg/ha · year)
		Carbohydrate	Protein	Lipid	Ash	
Grain legumes						
Soybean (*Glycine max*)	1471	38	38	20	4	459
Vetches (*Vicia* spp., *Lathyrus* spp.)	997	59	27	1	3	269
Pea (*Pisum sativum*)	999	68	27	2	3	270
Broad bean (*Vicia faba*)	1166	59	26	1	3	303
Peanut (*Arachis hypogaea*)	976	25	27	45	3	264
Lupin (*Lupinus* spp.)	674	26	40	4	4	270
Pigeon pea (*Cajanus cajan*)	699	69	25	2	4	175
Chick pea (*Cicer arietinum*)	590	68	23	5	4	136
Lentil (*Lens* spp.)	640	67	28	1	4	179
Beans (*Phaseolus* spp.)	535	70	24	2	4	128
Cowpea (*Vigna unguiculata*)	212	69	26	2	3	55
Cereal crops						
Corn (*Zea mays*)	2816	84	10	5	1	282
Oat (*Avena sativa*)	1549	77	13	5	5	201
Rice (*Oryza sativa*)	2441	88	8	2	2	195
Wheat (*Triticum esculentum*)	1557	82	14	2	2	218
Rye (*Secale cereale*)	1601	82	14	2	2	224
Barley (*Hordeum vulgare*)	1695	80	9	1	4	152

[a] Values from FAO Production Yearbook[104] are based on worldwide figures for 1975.

[b] Composition of seed as percent carbohydrate, protein, lipid, and ash.[105,106]

ADDENDUM

There is a growing body of evidence that the ureides are synthesized via the oxidation and hydrolysis of purines[79,98,107-110] and that the purines are synthesized *de novo*.[79,98,110] Recent findings indicate that xanthine may be formed directly as a result of purine synthesis and that hypoxanthine may not be an intermediate in the catabolism of purines leading to the formation of ureides.[110] The reactions of purine synthesis and catabolism are located within different cellular compartments[111] as presented in Figure 2.

REFERENCES

1. **Date, R. A.,** Nitrogen, a major limitation in the productivity of natural communities, crops and pastures in the pacific area, *Soil Biol. Biochem.,* 5, 5, 1973.
2. **Hardy, R. W. F. and Havelka, U. D.,** Nitrogen fixation research: a key to world food, *Science,* 188, 633, 1975.
3. **Burns, R. C. and Hardy, R. W. F.,** *Nitrogen Fixation in Bacteria and Higher Plants,* Springer-Verlag, New York, 1975.
4. **Vincent, J. M.,** Root-nodule symbioses with *Rhizobium,* in *The Biology of Nitrogen Fixation,* Quispel, A., Ed., North-Holland, Amsterdam, 1974, 265.
5. **Bond, G.,** Fixation of nitrogen by higher plants other than legumes, *Annu. Rev. Plant Physiol.,* 18, 107, 1967.
6. **Bond, G.,** The result of the IBP survey of root-nodule formation in non-leguminous angiosperms, in *Symbiotic Nitrogen Fixation in Plants,* Vol. 7, Nutman, P. S., Ed., International Biological Programme, Cambridge University Press, Cambridge, 1976, 443.
7. **Silvester, W. B.,** Dinitrogen fixation by plant associations excluding legumes, in *A Treatise on Dinitrogen Fixation,* Section 4, Hardy, R. W. F. and Gibson, A. H., Eds., John Wiley & Sons, New York, 1977, 141.
8. **Fred, E. B., Baldwin, I. L., and McCoy, E.,** *Root Nodule Bacteria and Leguminous Plants,* University of Wisconsin Press, Madison, 1932.
9. **Dixon, R. O. D., Rhizobia,** *Annu. Rev. Microbiol.,* 23, 137, 1969.
10. **Graham, P. H.,** Identifcation and classification of root nodule bacteria, in *Symbiotic Nitrogen Fixation in Plants,* Vol. 7, Nutman, P. S., Ed., International Biological Programme, Cambridge, University Press, Cambridge, 1976, 99.
11. **Trinick, M. J.,** Symbiosis between *Rhizobium* and the non-legume, *Trema aspera, Nature (London),* 244, 459, 1973.
12. **Nutman, P. S.,** The physiology of nodule formation, in *Nutrition of the Legumes,* Hallsworth, E. G., Ed., Buttersworths, London, 1958, 87.
13. **Dart, P. J.,** The infection process, in *The Biology of Nitrogen Fixation,* Quispel, A., Ed., North-Holland, Amsterdam, 1974, 381.
14. **Lie, T. A.,** Environmental effects on nodulation and symbiotic nitrogen fixation, in *The Biology of Nitrogen Fixation,* Quispel, A., Ed., North-Holland, Amsterdam, 1974, 555.
15. **Libbenga, K. R. and Bogers, R. J.,** Root-nodule morphogenesis, in *The Biology of Nitrogen Fixation,* Quispel, A., Ed., North-Holland, Amsterdam, 1974, 430.
16. **Quispel, A.,** Prerequisites for biological nitrogen fixation in root nodule symbioses, in *The Biology of Nitrogen Fixation,* Quispel, A., Ed., North-Holland, Amsterdam, 1974, 719.
17. **Bergersen, F. J.,** Formation and function of bacteroids, in *The Biology of Nitrogen Fixation,* Quispel, A., Ed., North-Holland, Amsterdam, 1974, 473.
18. **Quispel, A.,** The endophytes of the root nodules in non-leguminous plants, in *The Biology of Nitrogen Fixation,* Quispel, A., Ed., North-Holland, Amsterdam, 1974, 499.
19. **Callaham, D., Tredici, P. D., and Torrey, J. G.,** Isolation and cultivation in vitro of the actinomycete causing root nodulation in *Comptonia, Science,* 199, 899, 1978.
20. **Hageman, R. V. and Burris, R. H.,** Nitrogenase and nitrogenase reductase associate and dissociate with each catalytic cycle, *Proc. Natl. Acad. Sci. U.S.A.,* 75, 2699, 1978.
21. **Winter, H. C. and Burris, R. H.,** Nitrogenase, *Annu. Rev. Biochem.,* 45, 409, 1976.
22. **Ljones, T.,** The enzyme system, in *The Biology of Nitrogen Fixation,* Quispel, A., Ed., North-Holland, Amsterdam, 1974, 617.
23. **Israel, D . W., Howard, R. L., Evans, H. J., and Russell, S. A.,** Purification and characterization of the molybdenum-iron protein component of nitrogenase from soybean root nodules, *J. Biol. Chem.,* 249, 500, 1974.
24. **Orme-Johnson, W. H. and Davis, L. D.,** Current topics and problems in the enzymology of nitrogenase, in *Iron-Sulfur Proteins,* Vol. 3, Lovenberg, W., Ed., Academic Press, New York, 1977, 15.

25. **Ljones, T. and Burris, R. H.,** Evidence for one-electron transfer by the Fe protein of nitrogenase, *Biochem. Biophys. Res. Commun.,* 80, 22, 1978.
26. **Shah, V. K. and Brill, W. J.,** Isolation of an iron-molybdenum cofactor from nitrogenase, *Proc. Natl. Acad. Sci. U.S.A.,* 74, 3249, 1977.
27. **Schubert, K. R. and Evans, H. J.,** Hydrogen evolution: a major factor affecting the efficiency of nitrogen fixation in nodulated symbionts, *Proc. Natl. Acad. Sci. U.S.A.,* 73, 1207, 1976.
28. **Schubert, K. R. and Evans, H. J.,** The relations of hydrogen reactions to nitrogen fixation in nodulated symbionts, in *Recent Developments in Nitrogen Fixation,* Newton, W. R., Postgate, J. R., and Rodriquez-Barrueco, C., Eds., Academic Press, New York, 1977, 469.
29. **Rivera-Ortiz, J. M. and Burris, R. H.,** Interactions among substrates and inhibitors of nitrogenase, *J. Bacteriol.,* 123, 537, 1975.
30. **Bethenfalvay, G. J. and Phillips, D. A.,** Effect of light intensity on efficiency of carbon dioxide and nitrogen reduction in *Pisum sativum* L., *Plant Physiol.,* 60, 868, 1977.
31. **Dixon, R. O. D.,** Hydrogenase and efficiency of nitrogen fixation in aerobes, *Nature (London),* 262, 173, 1976.
32. **Schubert, K. R., Jennings, N. T., and Evans, H. J.,** Hydrogen reactions of nodulated leguminous plants. II. Effects on dry matter accumulation and nitrogen fixation, *Plant Physiol.,* 61, 398, 1978.
33. **McCrae, R. E., Hanus, J., and Evans, H. J.,** Properties of the hydrogenase system in *Rhizobium japonicum* bacteroids, *Biochem. Biophys. Res. Commun.,* 80, 384, 1978.
34. **Maier, R. J., Campbell, N. E. R., Hanus, F. J., Simpson, F. B., Russell, S. A., and Evans, H. J.,** Expression of hydrogenase activity in free-living *Rhizobium japnicum, Proc. Natl. Acad. Sci. U.S.A.,* 75, 3258, 1978.
35. **Evans, H. J. and Phillips, D . A.,** Reductants for nitrogenase and relationships to cellular electron transport, in *Nitrogen Fixation by Free-Living Micro-organisms,* Vol. 6, Stewart, W. D. P., Ed., International Biological Programme, Cambridge University Press, Cambridge, 1975, 389.
36. **Haaker, H., DeKok, A., and Veeger, C.,** Regulation of dinitrogen fixation in intact *Azotobacter vinelandii, Biochim. Biophys. Acta,* 357, 344, 1974.
37. **Laane, C., Haaker, H., and Veeger, C.,** Involvement of the cytoplasmic membrane in nitrogen fixation by *Rhizobium leguminosarum* bacteroids, *Eur. J. Biochem.,* 87, 147, 1978.
38. **Orme-Johnson, W. H.,** Biochemistry of nitrogenase, in *Genetic Engineering for Nitrogen Fixation,* Hollaender, A., Ed., Plenum Press, New York, 1977, 317.
39. **Ljones, T. and Burris, R. H.,** ATP hydrolysis and electron transfer in the nitrogenase reaction with different combinations of the iron protein and the molybdenum-iron protein, *Biochim. Biophys. Acta,* 275, 93, 1972.
40. **Ching, T. M.,** Regulation of nitrogenase activity in soybean nodules by ATP and energy charge, *Life Sci.,* 18, 1071, 1976.
41. **Bergersen, F. J.,** Biochemistry of symbiotic nitrogen fixation in legumes, *Annu. Rev. Plant Physiol.,* 22, 121, 1971.
42. **Dixon, R. O. D.,** Relationship between nitrogenase systems and ATP-yielding processes, in *Nitrogen Fixation by Free-living Micro-organisms,* Vol. 6, Stewart, W. D. P., Ed., International Biological Programme, Cambridge University Press, Cambridge, 1975, 421.
43. **Godfrey, C. A., Coventry, D. R., and Dilworth, M. J.,** Some aspects of leghaemoglobin biosynthesis, in *Nitrogen Fixation by Free-Living Micro-organisms,* Vol. 6, Stewart, W. D. P., Ed., International Biological Programme, Cambridge University Press, Cambridge, 1975, 311.
44. **Verma, D. P. S. and Bal, A. K.,** Intracellular site of synthesis and localization of leghemoglobin in root nodules, *Proc. Natl. Acad. Sci. U.S.A.,* 73, 3843, 1976.
45. **Becking, J. H.,** Plant-endophyte symbiosis in non-leguminous plants, *Plant Soil,* 32, 611, 1970.
46. **Jordan, D. C.,** The bacteroids of the genus *Rhizobium, Bacteriol. Rev.,* 26, 119, 1962.
47. **Bach, M. K., Magee, W. E., and Burris, R. H.,** Translocation of photosynthetic products to soybean nodules and their role in nitrogen fixation, *Plant Physiol.,* 33, 118, 1957.
48. **Wheeler, C. T.,** The causation of the diurnal changes in nitrogen fixation in the nodules of *Alnus glutinosa, New Phytol.,* 70, 487, 1971.
49. **Kidby, D. K.,** Activation of a plant invertase by inorganic phosphate, *Plant Physiol.,* 41, 1139, 1966.
50. **Keele, B. B., Jr., Hamilton, P. B., and Elkan, G. H.,** Glucose catabolism in *Rhizobium japonicum, J. Bacteriol.,* 97, 1184, 1969.
51. **Hardy, R. W. F. and Havelka, U. D.,** Photosynthate as a major factor limiting nitrogen fixation by field-grown legumes with emphasis on soybeans, in *Symbiotic Nitrogen Fixation in Plants,* Vol. 7, Nutman, P. S., Ed., International Biological Programme, Cambridge University Press, Cambridge, 1976, 421.
52. **Ryle, G. J. A., Powell, C. E., and Gordon, A. J.,** Effect of source of nitrogen on the growth of Fiskeby soyabean: the carbon economy of whole plants, *Ann. Bot.,* 42, 637, 1978.

53. **Schubert, K. R. and Ryle, G. J. A.,** The energy requirements for N_2 fixation in nodulated legumes, in *Advances in Legume Science,* Summerfield, R. J. and Bunting, A. H., Eds., Kew Botanical Gardens, London, 1980, 85.
54. **Minchin, F. R. and Pate, J. S.,** The carbon balance of a legume and the functional economy of its root nodules, *J. Exp. Bot.,* 24, 259, 1973.
55. **Mahon, J. D.,** Respiration and the energy requirement for nitrogen fixation in nodulated pea roots, *Plant Physiol.,* 60, 817, 1977.
56. **Pate, J. S. and Herridge, D. F.,** Partitioning and utilization of net photosynthate in a nodulated annual legume, *J. Exp. Bot.,* 29, 401, 1978.
57. **Herridge, D. F. and Pate, J. S.,** Utilization of net photosynthate for nitrogen fixation and protein production in an annual legume, *Plant Physiol.,* 60, 759, 1977.
58. **Halliday, J.,** Ph.D. thesis, University of Western Australia, 1976.
59. **Ryle, G. J. A., Powell, C. E., and Gordon, A. J.,** The respiratory costs of nitrogen fixation in soyabean, cowpea, and white clover, *J. Exp. Bot.,* 30, 145, 1979.
60. **Schubert, K. R., Jennings, N., Evans, H. J., and Tjepkema, J.,** unpublished results.
61. **Bond, G.,** Symbiosis of leguminous plants and nodule bacteria. I. Observations on respiration and on the extent of utilization of host carbohydrates by the nodule bacteria, *Ann. Bot.,* 5, 313, 1941.
62. **Tjepkema, J. D.,** Oxygen regulation and energy consumption in the root nodules of legumes, *Parasponia,* and actinorhizal plants, in *Current Perspectives in Nitrogen Fixation,* Gibson, A. H. and Newton, W. E., Eds., Australian Academy of Sciences, Canberra, 1981, 368.
63. **Schwintzer, C.,** Ecological physiology of nitrogen fixation in *Myrica gale,* in *Current Perspectives in Nitrogen Fixation,* Gibson, A. H. and Newton, W. E., Eds., Australian Academy of Sciences, Canberra, 1981, 254.
64. **Pate, J. S. and Minchin, F. R.,** Comparative studies of carbon and nitrogen nutrition of selected grain legumes, in *Advances in Legume Science,* Summerfield, R. J. and Bunting, A. H., Eds., Kew Botanical Gardens, London, 1980, 105.
65. **Bergersen, F. J. and Turner, G. L.,** Nitrogen fixation by the bacteroid fraction of breis of soybean root nodules, *Biochim. Biophys. Acta,* 141, 507, 1967.
66. **Meeks, J. C., Wolk, C. P., Schilling, N., Shaffer, P. W., Avissar, Y., and Chien, W.-S.,** Initial organic products of fixation of [^{13}N] dinitrogen by root nodule of soybean *(Glycine max), Plant Physiol.,* 61, 980, 1978.
67. **Scott, D. B., Farnden, K. J. F., and Robertson, J. G.,** Ammonia assimilation in lupin nodules, *Nature (London),* 263, 703, 1976.
68. **Miflin, B. J. and Lea, P. J.,** The pathway of nitrogen assimilation in plants, *Phytochemistry,* 15, 873, 1976.
69. **Schubert, K. R. and Coker, G. T.,** III. Studies of nitrogen and carbon assimilation in N_2-fixing plants: Short term studies using [^{13}N] and [^{11}C], in Recent Developments in Biological and Chemical Research with Short-lived Radioisotopes, *Advances in Chemistry Series,* in press.
70. **Robertson, J. G., Farnden, K. J. F., Warburton, M., and Banks, J. M.,** Induction of glutamine synthease during nodule development in lupin, *Aust, J. Plant Physiol.,* 2, 265, 1975.
71. **Robertson, J. G., Warburton, M. P., and Farnden, K. J. F.,** Induction of glutamate synthase during nodule development in lupin, *FEBS Lett.,* 55, 33, 1975.
72. **O'Gara, F. and Shanmugam, K. T.,** Regulation of nitrogen fixation by rhizobia-export of fixed N_2 as NH_4^+, *Biochim. Pbiphys, Acta,* 437, 313, 1975.
73. **O'Gara, F. and Shanmugam, K. T.,** Control of symbiotic nitrogen fixation in rhizobia-regulation of NH_4^+ assimilation, *Biochim, Biophys, Acta,* 451, 342, 1976,
74. **Brown, C. M. and Dilworth, M. J.,** Ammonia assimilation by *Rhizobium* cultures and bacteroids, *J. Gen. Microbiol.,* 86, 39, 1975.
75. **Boland, M. J., Fodyce, A. M., and Greenwood, R. M.,** Enzymes of nitrogen metabolism in legume nodules: A comparative study, *Aust. J. Plant Physiol.,* 5, 553, 1978.
76. **Pate, J. S.,** Uptake, assimilation and transport of nitrogen compounds by plants, *Soil Biol. Biochem.,* 5, 109, 1973.
77. **Pate, J. S. and Wallace, W.,** Movement of assimilated nitrogen rom the root system of the field pea *(Pisum arvense L.), Ann. Bot.,* 28, 80, 1964.
78. **Gunning, B. E. S., Pate, J. S., Minchin, R. F., and Marks, I.,** Quantitative aspects of transfer cell structure in relation to vein loading in leaves and solute transport in legume nodules. Transport at the cellular level, *Symposia Soc. Exp. Biol.* 28, 87, 1974.
79. **Schubert, K. R.,** Enzymes of purine biosynthesis and catabolism in *Glycine max:* Comparison of activities with N_2 fixation and composition of xylem exudate during nodule development, *Plant Physiol.* 68, in press.
80. **Pate, J. S., Gunning, B. E. S., and Briarty, L. G.,** Ultrastructure and functioning of the transport system of the leguminous root nodule, *Planta,* 85, 11, 1969.

81. **Pate, J. S., Atkins, C. A., and Rainbird, R. M.,** Theoretical and experimental costing of nitrogen fixation and related processes in nodules of legumes, in *Current Perspectives in Nitrogen Fixation,* Gibson, A. H. and Newton, W. E., Eds., Australian Academy of Sciences, Canberra, 1981, 105.
82. **Cookson, C., Hughes, H., and Coombs, J.,** Effects of combined nitrogen on anapleurotic carbon assimilation and bleeding sap composition in *Ph aseolus vulgaris* L., *Planta,* 148, 338, 1980.
83. **Schubert, K. R., Coker, G. T III, and Firestone, R. B.,** Ammonia assimilation in *Alnus glutinosa* and *Glycine Max:* Short-term studies using $^{13}NH_4^+$, *Plant Physiol.,* 67, 662, 1981.
84. **Pate, J. S. and Atkins, C. A.,** Nitrogen uptake transport and utilization, in *Ecology of Nitrogen Fixation,* Broughton, W. J., Ed., in press.
85. **Copeland, R. and Pate, J. S.,** Nitrogen metabolism of nodulated white clover in the presence and absence of nitrate nitrogen, in *White Clover Research,* British Grassland Society, 6, 71, 1970.
86. **Pate J. S., Atkins, C. A., White, S. T., Rainbird, R. M. and Woo, K. C.,** Nitrogen nutrition and xylem transport of nitrogen in ureide-producing grain legumes, *Plant Physiol.,* 65, 961, 1980.
87. **McClure, P. R. and Israel, D. W.,** Transport of nitrogen in the xylem of soybean plants, *Plant Physiol.,* 64, 411, 1979.
88. **Streeter, J. G.,** Allantoin and allantoic acid in tissues and stem exudate from field-grown soybean plants, *Plant Physiol.,* 63, 478, 1979.
89. **Rawsthorne, S., Minchin, R. F., Summerfield, R. J., Cookson, C., and Coombs, J.,** Carbon and nitrogen metabolism in legume root nodules, *Phytochemistry,* 19, 341, 1980.
90. **Leaf, G., Gardner, I. C., and Bond, G.,** Observations on the composition and metabolism of the nitrogen-fixing root nodules of *Myrica, Biochem. J.* 72, 662, 1959.
91. **Leaf, G., Gardner, I. C., and Bond, G.,** Observations on the composition and metabolism of the nitrogen-fixing root nodules of *Alnus, J. Exp. Bot.,* 9, 320, 1958.
92. **Herridge, D. F., Atkins, C. A., Pate, J. S., and Rainbird, R. M.,** Allantoin and allantoic acid in the nitrogen economy of the cowpea *(Vigna unguiculata* [L] Walp.), *Plant Physiol.,* 62, 495, 1978.
93. **Pate, J. S., Sharkey, P. J., and Atkins, C. A.,** Nutrition of a developing legume fruit, *Plant Physiol.,* 59, 506, 1977.
94. **McNeil, D. L., Atkins, C. A., Pate, J. S.,** Uptake and utilization of xylem-borne amino compounds by shoot organs of a legume, *Plant Physiol.,* 63, 1076, 1979.
95. **Pate, J. S., Atkins, C. A., Hamel, K., McNeil, D. L., and Layzell, D. B.,** Transport of organic solutes in phloem and xylem of a nodulated legume, *Plant Physiol.,* 63, 1082, 1979.
96. **Coker, G. T. III, and Schubert, K. R.,** Carbon dioxide fixation in soybean roots and nodules. I. Characterization and comparison with N_2-fixation and composition of xylem exudate during early nodule development, *Plant Physiol.,* 67, 691, 1981.
97. **Atkins, C. A., Pate, J. S., and Layzell, D. B.,** Assimilation and transport of nitrogen in non-nodulated (NO_3-grown) *Lupinus albus* L., *Plant Physiol.,* 64, 1078, 1979.
98. **Schubert, K. R., DeShone, G. M., Polayes, D. P. and Hanks, J. F.,** Nitrogen assimilation and partitioning in soybean: The biosynthesis of ureides and amides, in *Current Perspectives in Nitrogen Fixation* , Gibson, A. H. and Newton , W. E. Eds., Australian Academy of Science, Canberra, 1981, 391.
99. **Schubert, K., Tierney, M., Christensen, A., McClure, P. and Coker, G.,** Carbon and nitrogen metabolism in soybean and alder nodules and roots, in *Current Perspectives in Nitrogen Fixation,* Gibson, A. H., and Newton, W. E., Eds., Australian Academy of Sciences, Canberra, 1981, 384.
100. **Lawrie, A. C. and Wheeler, C. T.,** Nitrogen fixation in the root nodules of *Vicia faba* L. in relation to the assimilation of carbon. I. Plant growth and metabolism of photosynthetic assimilates, *New Phytol.,* 74, 429, 1975.
101. **Lawrie, A. C. and Wheeler, C. T.,** Nitrogen fixation in the root nodules of *Vicia faba* L. in relation to the assimilation of carbon II. The dark fixation of carbon dioxide, *New Phytol.,* 74, 437, 1975.
102. **Christeller, J. T., Laing, W. A., and Sutton, W. D.,** Carbon dioxide fixation by lupin root nodules. I. Characterization, association with phosphoenolpyruvate carboxylase, and correlation with nitrogen fixation during nodule development, *Plant Physiol.,* 60, 47, 1977.
103. **Gardner, I. C. and Leaf, G.,** Translocation of Citrulline in *Alnus glutinosa, Plant Physiol.,* 35, 948, 1960.
104. Food and Agricultural Organization of the United Nations, Production Yearbook, Vol. 29, Rome, Italy, 1975.
105. **Sinclair, T. R. and DeWit, C. T.,** Photosynthate and nitrogen requirements for seed production by various crops, *Science,* 189, 565, 1975.
106. **Sprague, H. B.,** Characteristics of Economically Important Food and Forage Legumes and Forage Grasses for the Tropics and Subtropics, Technical Series Bull. No. 14, Agency for International Development, Washington, D. C., 1975.
107. **Reinbothe, H. and Mothes, K.,** Urea, ureides and guanidines in plants, *Ann. Rev. Plant Physiol.,* 13, 129-150, 1962.

108. **Atkins, C. A ., Rainbird, R. M. and Pate, J. S.** Evidence for a purine pathway of ureide synthesis in N_2-fixing nodules of cowpea *(Vigna unguiculata* L. *Walp.) Z. Pflanzenphysiol.*, 97, 249, 1980.
109. **Triplett, E. W., Blevins, D. G. and Randall, D. D.**, Allantoic acid synthesis in soybean root nodule cytosol via xanthine dehydrogenase, *Plant Physiol.*, 65, 1203, 1980.
110. **Boland, M. J. and Schubert, K. R.**, unpublished results
111. **Hanks, J. F., Tolbert, N. E., and Schubert, K. R.**, Localization of enzymes of ureide biosynthesis in peroxisomes and microsomes of nodules, *Plant Physiol.*, 68, in press.

SULFUR ASSIMILATION IN PHOTOSYNTHETIC BACTERIA

Hans G. Trüper

INTRODUCTION

Chlorobiaceae, Chromaticeae, Chloroflexaceae, and several species of the Rhodospirillaceae use reduced sulfur compounds as electron donors for photosynthesis. Like other microorganisms, photosynthetic bacteria further require inorganic sulfur for the biosynthesis of the sulfur-containing amino acids cysteine and methionine and the derivatives thereof, such as biotin, thiamine, lipoate, and coenzyme A. This article deals with the assimilatory pathways of sulfur and biosynthesis of the two S-amino acids. Dissimilatory sulfur metabolism has been considered elsewhere.[1]

WHOLE CELL STUDIES

The Chlorobiaceae and the Chromatiaceae are usually found in anoxic, sulfide-containing habitats. Energy would be wasted if they had to assimilate and reduce sulfate for the purpose of synthesizing S-amino acids. It is therefore not surprising that so far no member of the Chlorobiaceae has been found that is able to assimilate sulfate,[2] and of the Chromatiaceae, *Chromatium okenii,*[3] *Chromatium weissei, Chromatium warmingii,*[4] *Chromatium buderi,*[5] *Chromatium minus, Amoebobacter pendens* (syn. *Rhodothece pendens*), and *Amoebobacter roseus* (syn. *Rhodothece rosea*)[4] are not able to utilize sulfate as a sulfur source.

On the other hand, the Rhodospirillaceae, being predominantly photoorganotrophic, depend on assimilatory sulfate reduction. *Rhodopseudomonas globiformis* requires thiosulfate as a sulfur source for growth, i.e., for the biosynthesis of S-amino acids.[6] *Rhodopseudomonas sulfoviridis* cannot utilize sulfate.

A number of Chromatiaceae and Rhodospirillaceae species hold an intermediary position by being capable of assimilatory sulfate reduction and utilization of H_2S as well. These are *Chromatium vinosum, Chromatium violascens, Thiocapsa roseopersicina, Thiocystis violacea, Rhodopseudomonas palustris, Rhodopseudomonas sulfidophila, Rhodopseudomonas capsulata,* as well as some of the *Ectothiorhodospira* species. Hurlbert and Lascelles[7] were the first to suggest that members of the Chromatiaceae utilize reduced sulfur, sulfide, and thiosulfate as a course of cell sulfur. The utilization of sulfide by Rhodospirillaceae was first found by Hansen and van Gemerden and later carefully studied by Hansen.[9] Kobayashi[10] reported the removal of badly smelling mercaptans during sewage treatment by phototrophic bacteria. This indicated that even sulfur compounds of the composition of CH_3SH, CH_3-S-CH_3 and $(CH_3)_2S_2$ are utilized by *Rhodopseudomonas* species (predominantly *R. palustris*).

A large number of *Chlorobium* strains were studied by Lippert,[2] who found that during growth with molecular H_2 as the electron donor, the strains assimilated cysteine, but not methionine, cystine, cysteic acid, thioglycolate, thioacetamide, or sulfite. Cysteine sulfur was transformed into methionine sulfur.[11] Thiele[4,12] found similar results with *Chromatium vinosum* and *Thiocapsa roseopersicina*, growing with H_2, acetate, or fructose, as the photosynthetic electron donor. Lippert[2] further found that in *Chlorobium* growing phototrophically with H_2 and bicarbonate and with cysteine as the sulfur source, growth ceased after several generations. He concluded that cysteine is an insufficient sulfur source for *Chlorobium* and that these bacteria — being unable to assimilate sulfate — need some sulfur compound at the oxidation level of sulfate that can be synthesized only during growth on reduced sulfur compounds. It is possible that under the conditions tested, certain sulfolipids, for example, cannot be synthesized. Sulfolipids have been shown to occur in phototrophic bacteria[13,14], e.g., 1-*o*- (*β*-6′-deoxy-aldohexopyranosyl-6′-sulfonic acid) -3-*O*-oleoyl-) glycerol.

ENZYMOLOGY

The pathway of assimilatory sulfate reduction in the Rhodospirillaceae and in members of the Chromatiaceae is only partly resolved. At least for the former family, there is no reason to assume a pathway principally different from that in other bacteria, fungi, or plants, although certain regulatory phenomena might be different.

Phototrophic bacteria, growing with sulfide as the photosynthetic electron donor, need only one enzyme to transfer sulfur to the carbon skeleton of a precurser of cysteine. In *Chlorobium limicola, Rhodopseudomonas sphaeroides,*[15] *Rhodospirillum rubrum, Rhodospirillum fulvum, Rhodospirillum tenue, Rhodospirillum photometricum, Rhodopseudomonas capsulata, Rhodopseudomonas sulfidophila, Rhodopseudomonas palustris, Rhodopseudomonas gelatinosa, Rhodopseudomonas globiformis, Chromatium vinosum, Thiocapsa roseopersicina, Thiocapsa pfennigii, Ectothiorhodospira mobilis,* and *Chlorobium vibrioforme,*[16] *O*-acetylserine sulfhydrylase (*O*-acetyl-L-serine acetate lyase, EC 4.2.99.8) has been found to be responsible for cysteine biosynthesis. Of these organisms, only *Rhodospirillium tenue, Rhodopseudomonas gelatinosa, Chromatium vinosum, Thiocapsa roseopersicina, T. pfennigii,* and *E. mobilis* possess the ability to form S-sulfocysteine from *O*-acetylserine and thiosulfate.[16] Because S-sulfocysteine is an easily hydrolyzable unstable compound,[15] these bacteria can synthesize cysteine from thiosulfate directly. This ability is not shared by the thiosulfate-utilizing subspecies of *Chlorobium limicola* and *C. vibrioforme.* For these species, as well as for the thiosulfate-utilizing *Rhodopseudomonas,* a thiosulfate-splitting enzyme — perhaps rhodanese or thiosulfate dehydrogenase — has to be postulated which supplies *O*-acetylserine sulfhydrylase with sulfide. Thiosulfate-splitting enzymes have been shown to occur in many phototrophic bacteria, although mostly in connection with studies of dissimilatory sulfur metabolism.[1,17] Hensel[18] has recently shown for *Chromatium vinosum* and *Rhodospirillium tenue* that each contains two different OAS-sulfhydrylases, one reacts with sulfide only; and the other reacts with both sulfide or thiosulfate. In the latter case, S-sulfocysteine is formed. Both enzymes are repressed differently by sulfide and thiosulfate, a fact that might point towards an in vivo importance of S-sulfocysteine biosynthesis.

The biosynthesis of methionine in phototrophic bacteria obviously proceeds via homocysteine, since a vitamin B_{12}-dependent methionine synthase (N^5-methyl-tetrahydrofolate: homocysteine B_{12}-methyltransferase, EC 2.1.1.13) has been found and studied in *Rhodopseudomonas sphaeroides,*[19] *Rhodospirillium rubrum,* and *Chromatium vinosum.*[20]

Growth with sulfate as single sulfur source, i.e., photoorganotrophic growth, requires the reduction of sulfate to the level of sulfide which then will react with *O*-acetylserine to form cysteine. Therefore, one must postulate ATP sulfurylase (and perhaps adenylylsulfate kinase) as the activating step(s) and two reducing steps, one reducing adenylylsulfate (APS) or 3′-phosphoadenylylsulfate (PAPS) to sulfite and the other reducing sulfite to sulfide (sulfite reductase). Ibanez and Lindstrom[20,21] showed that chromatophores of *Rhodospirillium rubrum* synthesized PAPS from sulfate in the light or upon addition of ATP in the dark. They also demonstrated that light was required for the reduction of sulfate to a volatile form, probably sulfite, and that light enhanced the incorporation of radioactive sulfate sulfur into cysteine of the chromatophores. Also, in *Chromatium vinosum* and *Thiocapsa roseopersicina,* radioactive label was found in cysteine and methionine after growth with ^{35}S-sulfate.[4,12] These observations strongly point towards the metabolic pathway outlined above. The sulfate-activating enzymes ATP-sulfurylase and APS-kinase have been demonstrated in *Rhodopseudomonas capsulata, Rhodopseudomonas sulfidophila, Rhodopseudomonas viridis, Rhodopseudomonas gelatinosa*, and *Rhodospirillium rub.um.*[23]

The first reductive step, i.e., the reduction of APS or PAPS to the level of sulfite has not been resolved definitely. Cells of *Chromatium vinosum* and *Thiocapsa roseopersicina*, after several passages of photoorganotrophic growth with sulfate as sulfur source, possess considerable activities of APS reductase,[24,25] the enzyme responsible for sulfite oxidation during dissimilatory sulfur metabolism in Chromatiaceae[24,26] and Chlorobiaceae.[25,27] It has been proposed therefore that this enzyme participates in the assimilatory sulfate reduction pathway[24,25] in the two families of the phototrophic sulfur bacteria. APS reductase does not occur in the Rhodospirillaceae.[25] The studies so far undertaken with species of the latter family point towards the presence of systems as they have been described from green plants[28,29] and fungi,[30] where the sulfur moiety of APS or PAPS is transferred upon a sulfhydryl carrier protein by an APS or PAPS sulfotransferase and reduced to sulfide level at the carrier. In a preliminary study, Schmidt and Trüper[31] tested extracts of 14 species on the presence of APS or PAPS sulfotransferases detectable by "volatile sulfur" formation from APS or PAPS. It seems that *Rhodospirillum* species react with both sulfur nucleotides, while *Rhodopseudomonas* species apparently prefer PAPS. APS sulfotransferase was purified from *Rhodospirillium rubrum*[32] and also found in *Rhodospirillium tenue,*[33] whereas in *Rhodopseudomonas sulfidophila* and *Rhodopseudomonas viridis,* PAPS is reduced in a reaction requiring a heat-resistant protein of low molecular weight. APS is not reduced.[33]

Indications of the presence of a sulfite reductase in phototrophic bacteria were found first in a series of tests by Peck et al.[34] In cell-free extracts of *Rhodopseudomonas viridis, Rhodopseudomonas gelatinosa, Rhodospirillium rubrum, Rhodomicrobium vannielii, Chromatium vinosum*, and *Chlorobium limicola*, a methylviolegen-dependent reduction of sulfite to sulfide was demonstrated. *Chromatium vinosum* was grown photolithoautotrophically as well as photoorganotrophically (with sulfate). This latter finding must be taken as an indication of a function for sulfite reductase in dissimilatory sulfur metabolism, i.e., in the oxidation of sulfide and/or elemental sulfur to sulfite. Recently, a siroheme containing sulfite reductase has been partially purified from *Chromatium vinosum.*[35,36] The enzyme shows similarities to sulfite reductases from the anaerobic *Thiobacillus denitrificans*[35,37] and *Desulfovibrio vulgaris,*[38] and it is assumed that it is part of the dissimilatory pathway rather than playing a role in assimilatory sulfate reduction.[36,37]

The question remains, however, whether sulfite reductase, i.e., the enzyme-reducing free sulfite to free sulfide, is identical with the so-called "thiosulfonate reductase"[28-30] responsible for the reduction of carrier-bound sulfite to carrier-bound sulfide.

REFERENCES

1. **Trüper, H. G.,** Photolithotrophic sulfur oxidation, in *Biology of Inorganic Nitrogen and Sulfur,* Bothe, H. and Trebst, A., Eds., Springer-Verlag, N.Y., 1981.
2. **Lippert, K. D.,** Die Verwertung von molekularem Wasserstoff durch *Chlorobium thiosulfatophilum,* Doctoral thesis, University of Göttingen, Germany, 1967.
3. **Trüper, H. G. and Schlegel, H. G.,** Sulphur metabolism in Thiorhodaceae. I. Quantitative measurements on growing cells of *Chromatium okenii, Antonie van Leeuwenhoek J. Microbiol. Serol.,* 30, 225, 1964.
4. **Thiele, H. H.,** Wachstumsphysiologische Untersuchungen an Thiorhodaceae: Wasserstoff-Donatoren und Sulfatreduktion, Doctoral thesis, University of Göttingen, Germany, 1966.
5. **Trüper, H. G. and Jannasch, H. W.,** *Chromatium buderi* nov. spec., eine neue Art der "großen" Thiorhodaceae, *Arch. Mikrobiol.,* 61, 363, 1968.

6. **Pfennig, N.**, *Rhodopseudomonas globiformis*, sp. n., a new species of the Rhodospirillaceae, *Arch. Microbiol.*, 100, 197, 1974.
7. **Hurlbert, R. E. and Lascelles, J.**, Ribulose diphosphate carboxylase in Thiorhodaceae, *J. Gen. Microbiol.*, 33, 445, 1963.
8. **Hansen, T. A. and van Gemerden, H.**, Sulfide utilization by purple nonsulfur bacteria, *Arch. Microbiol.*, 86, 49, 1972.
9. **Hansen, T. A.**, Sulfide als electronendonor voor Rhodospirillaceae, Doctoral thesis, University of Groningen, The Netherlands, 1974.
10. **Kobayashi, M.**, Utilization and disposal of wastes by photosynthetic bacteria, in *Microbial Energy Conversion*, Schlegel, H. G. and Barnea, J., Eds., Unitar Symp., Göttingen, 1977.
11. **Lippert, K. D. and Pfennig, N.**, Die Verwertung von molekularem Wasserstoff durch *Chlorobium thiosulfatophilum*: Wachstum und CO_2-Fixierung, *Arch. Mikrobiol.*, 65, 29, 1969.
12. **Thiele, H. H.**, Sulfur metabolism in Thiorhodaceae. IV. Assimilatory reduction of sulfate in *Thiocapsa floridana* and *Chromatium* species, *Antonie van Leeuwenhoek J. Microbiol. Serol.*, 34, 341, 1968.
13. **Benson, A. A., Daniel, H., and Wiser, R.**, A sulfolipid in plants, *Proc. Natl. Acad. Sci. U.S.A.*, 45, 1582, 1959.
14. **Wood, B. J. B., Nichols, B. W., and James, A. Z.**, The lipids and fatty acid metabolism of photosynthetic bacteria, *Biochim. Biophys. Acta*, 106, 261, 1965.
15. **Chambers, L. A. and Trudinger, P. A.**, Cysteine and S-sulphocysteine biosynthesis in bacteria, *Arch. Mikrobio!.*, 77, 165, 1971.
16. **Hensel, G. and Trüper, H. G.**, Cysteine and S-sulfocysteine biosynthesis in phototrophic bacteria, *Arch. Microbiol.*, 109, 101, 1976.
17. **Hashwa, F.**, Die enzymatische Thiosulfatspaltung bei phototrophen Bakterien, Doctoral thesis, University of Göttingen, Germany, 1972.
18. **Hensel, G.**, O-Acetylserin - Sulfhydrylase and S-Sulfocystein synthase bei *Rhodospirillum tenue* und *Chromatium vinosum*, Doctoral thesis, University of Bonn, Germany, 1979.
19. **Cauthen, S. E., Pattison, J. R., and Lascelles, J.**, Vitamin B_{12} in photosynthetic bacteria and methionine synthesis by *Rhodopseudomonas spheroides*, *Biochem. J.*, 102, 774, 1967.
20. **Ohmori, H. and Fukui, S.**, vitamin B_{12}-dependent methionine synthetase in photosynthetic bacteria: partial purification and properties, *Agric. Biol. Chem.*, 38, 1317, 1974.
21. **Ibanez, M. L. and Lindstrom, E. S.**, Photochemical sulfate reduction by *Rhodospirillum rubrum*, *Biochem. Biophys. Res. Commun.*, 1, 224, 1959.
22. **Ibanez, M. L. and Lindstrom, E. S.**, Metabolism of sulfate by the chromatophore of *Rhodospirillum*, *J. Bacteriol.*, 84, 451, 1962.
23. **Cooper, B. P.**, Der enzymatische Mechanismus der Sulfataktivierung bei *Rhodopseudomonas sulfidophila*, Doctoral thesis, University of Bonn, Germany, 1979.
24. **Thiele, H. H.**, Sulfur metabolism in Thiorhodaceae. V. Enzymes of sulfur metabolism in *Thiocapsa floridana* and *Chromatium* species, *Antonie van Leeuwenhoek J. Microbiol. Serol.*, 34, 350, 1968.
25. **Trüper, H. G. and Peck, H. D., Jr.**, Formation of adenylyl sulfate in phototrophic bacteria, *Arch. Microbiol.*, 73, 125, 1970.
26. **Trüper, H. G. and Rogers, L. S.**, Purification and properties of adenylyl sulfate reductase from the phototrophic sulfur bacterium *Thiocapsa roseopersicina*, *J. Bacteriol.*, 108, 1112, 1971.
27. **Kirchhoff, J. and Trüper, H. G.**, Adenylylsulfate reductase of *Chlorbium limicola*, *Arch. Microbiol.*, 100, 115, 1974.
28. **Schmidt, A.**, Sulfate reduction in a cell-free system of *Chlorella*. The ferredoxin dependent reduction of a protein-bound intermediate by a thiosulfate reductase, *Arch. Microbiol.*, 93, 29, 1973.
29. **Abrams, W. R. and Schiff, J. A.**, Studies of sulfate utilization by algae. XI. An enzyme-bound intermediate in the reduction of adenosine-5′-phosphosulfate (APS) by cell-free extracts of wild-type *Chlorella* mutants blocked for sulfate reduction, *Arch. Mikrobiol.*, 94, 1, 1973.
30. **Wilson, L. G. and Bierer, D.**, The formation of exchangeable sulphite from adenosine 3′-phosphate 5′-sulphatophosphate in yeast, *Biochem. J.*, 158, 255, 1976.
31. **Schmidt, A. and Trüper, H. G.**, Reduction of adenylylsulfate and 3′-phosphoadenylsulfate in phototrophic bacteria, *Experientia*, 33, 1008, 1977.
32. **Schmidt, A.**, Adenosine-5′-phosphosulfate (ADS) as sulfate donor for assimilatory sulfate reduction in *Rhodospirillum rubrum*, *Arch. Microbiol.*, 112, 263, 1977.
33. **Imhoff, J. F. and Trüper, H. G.**, unpublished results, 1979.
34. **Peck, H. D., Jr., Tedro, S., and Kamen, M. D.**, Sulfite reductase activity in extracts of various photosynthetic bacteria, *Proc. Natl. Acad. Sci. U.S.A.*, 71, 2404, 1974.
35. **Schedel, M.**, Untersuchungen zur anaeroben Oxidation reduzierter Schwefelverbindungen durch *Thiobacillus denitrificans*, *Chromatium vinosum* und *Chlorobium limicola*, Doctoral thesis, University of Bonn, Germany, 1977.

36. **Schedel, M., Vanselow, M., and Trüper, H. G.,** Siroheme sulfite reductase isolated from *Chromatium vinosum*. Purification and investigation of some of its molecular and catalytic properties, *Arch. Microbiol.*, 121, 29, 1979.
37. **Schedel, M. and Trüper, H. G.,** Purification of *Thiobacillus denitrificans* siroheme sulfite reductase and investigation of some molecular and catalytic properties, *Biochim. Biophys. Acta*, 568, 454, 1979.
38. **Lee, J. P., LeGall, J., and Peck, H. D., Jr.,** Isolation of assimilatory and dissimilatory-type sulfite reductase from *Desulfovibrio vulgaris*, *J. Bacteriol.*, 115, 519, 1973.

SULFUR ASSIMILATION IN HIGHER PLANTS AND ALGAE

A. Trebst

INTRODUCTION

Plants contain a variety of sulfur-containing compounds at several valency states.[1] Most prominent are the sulfur-containing amino acids, cysteine and methionine, and the intermediates in their biosynthesis, homocysteine and cystathionine. In these, sulfur has the valency state of −2, i.e., of sulfide. This also is the valency state of the sulfur in the coenzymes thiamine, biotin, coenzyme A, and lipoic acid, for all of which cysteine is the biosynthetic precursor. Sulfur in the valency state +6, i.e., of sulfate, in nature is in sulfate esters of polysaccharides and in plants predominantly occurring in marine algae. The plant-specific sulfolipid contains sulfoquinovose with an unusual 6-sulfonic acid of a 6-deoxyglucose.[2] Sulfur in the valency state +4, i.e., of sulfite, like in the secondary plant constituent alliin, is not widespread in nature. The valency of sulfite is an intermediate state in the reduction of sulfate to sulfide. The sulfur-containing amino acids and coenzymes are, of course, ubiquitious in all plant parts and organelles, whereas the sulfolipid is a component specific for the inner thylakoid membrane of chloroplasts.

Plants are supplied with sulfur in the form of sulfate, taken up by roots from the soil. The biosynthesis of sulfur-containing compounds, particularly of sulfolipid and of cysteine, on the other hand, seems to be localized in leaves. Therefore, sulfate is transported up from the root and some mechanism for transport of reduced sulfur is needed.[3]

The biosynthesis of cysteine requires a reduction of sulfate to the valency state of sulfide, a process termed assimilatory sulfate reduction. It is assumed now that sulfate reduction is coupled to the photosynthetic apparatus like CO_2 and nitrate reduction. If sulfate is photosynthetically reduced to sulfide, the biosynthesis of cysteine should be light dependent and localized in the chloroplasts. On the other hand, nonphotosynthetic parts of plants are able to reduce sulfate as well.[3,4]

This is particularly obvious in heterotrophically grown algae or plant cell tissue cultures which grow on sulfate in the dark.[4] Hence, assimilatory sulfate reduction depends on products of photosynthetic energy conservation in the leaf being made available to other parts of the plant. Several heterotrophic organisms, like yeast or *Enterobacteria*, but not higher animals, have the potential of "assimilatory sulfate reduction": the reduction of sulfate for biosynthetic purposes.[5] "Dissimilatory sulfate reduction", on the other hand, is a process of energy metabolism in which sulfate replaces oxygen as the terminal acceptor in respiratory electron flow, e.g., in *Desulfovibrio*.[5,6] Several reviews have described the present knowledge about the biochemistry of sulfate reduction in plants.[5,7-10]

PHOTOSYNTHETIC (ASSIMILATORY) SULFATE REDUCTION IN CHLOROPLASTS AND ALGAE*

The reduction of sulfate to provide sulfide for the biosynthesis of cysteine and other sulfur compounds derived from them is termed assimilatory sulfate reduction. The

* Abbreviations: APS, adenosine 5′-phosphosulfate; GSH, glutathione; GSSH, glutathione persulfide; $GSSO_3H$, sulfoglutathione; PAP, adenosine 3′5′-diphosphate; PAPS, 3′-phosphoadenosine-5′-phosphosulfate; XSH, XSSX, low molecular weight carrier (reduced and oxidized form); $XSSO_3H$, carrier bound sulfite = a thiosulfonate; and XSSH, carrier bound sulfide = a persulfide.

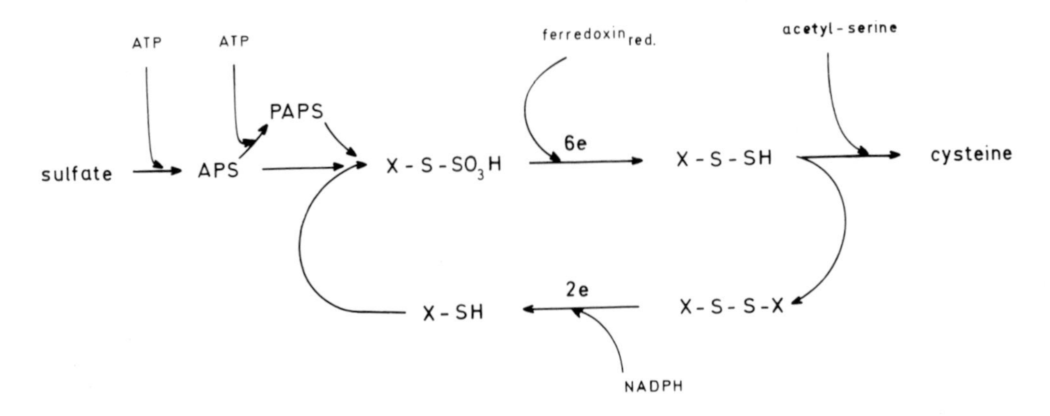

FIGURE 1. Proposed scheme for photosynthetic assimilatory sulfate reduction in chloroplasts. Sulfur is activated to APS and PAPS, one of which will transfer its sulfate group onto a carrier XSH to form a thiosulfonate = bound sulfite. Six electrons, provided by reduced ferredoxin, reduce bound sulfite $XSSO_3H$ to bound sulfide XSSH. This is reductively cleaved to displace the acetate group from serine to form cysteine. The oxidized carrier is reduced by NADPH (or thioredoxin?) to regenerate XSH.

coupling of sulfate reduction to the chloroplast photosynthetic apparatus gradually became accepted from physiological evidence.[11-14] Direct proof for the ability of chloroplasts to reduce sulfate in the light came[15-17] with the preparation of intact chloroplasts with an intact envelope membrane and high CO_2 assimilation rates.[18] Photosynthesis in chloroplasts drives CO_2, $SO_4^{=}$, and NO_2^- assimilation.

The sulfate-reducing system of an intact chloroplast in vitro and the formation of cysteine is strictly light dependent.[16,17,19] Radioactive ^{35}S-sulfate is used to characterize

The sulfate-reducing system of an intact chloroplast in vitro and the formation of cysteine is strictly light dependent.[16,17,19] Radioactive^{35} S-sulfate is used to characterize the intermediates between sulfate and cysteine. The kinetics of the appearance of these intermediates after a pulse of radioactive sulfate indicates the consecutive steps in the reduction pathway.[20] We will discuss the reaction mechanism, cofactor requirements, enzymes, and the coupling mechanism in photosynthesis. Knowledge about photosynthetic coupling comes from studies with broken and reconstituted chloroplasts in which the thylakoid membrane fraction contains the light-driven processes of photosynthesis (formation of reducing equivalents and of ATP), and the supernatant fraction contains soluble and peripheral bound enzymes of the sulfate reduction pathway. Eventually, the still incomplete characterization of the enzymes involved should provide the final details and proof for the scheme proposed in Figure 1. Evidence from the chloroplast studies were strongly supported by work on sulfate reduction by the green algae *Chlorella.*[7,21-24] In particular, mutants in the reduction pathway supported the notion of bound intermediates of sulfite and sulfide,[21,23,24] though a small percentage of sulfur assimilation occurs via free sulfite.[24]

The present knowledge about the mechanism of the eight-electron reduction of sulfate to the valency state of sulfide in cysteine is summarized in Figure 1: sulfate is activated to APS and PAPS. These activated sulfate nucleotides transfer to an acceptor XSH in a (P)APS sulfotransferase/thiosulfonate enzyme complex. The intermediate thiosulfonate $XSSO_3H$, a bound sulfite, is reduced by ferredoxin (via a prosthetic siroheme and nonheme iron group in the active center)[5] to a persulfide XSSH — a bound sulfide. This in turn is reductively cleaved and exchanges its SH group with an *O*-acetyl of serine to form cysteine. The oxidized (disulfide) form of the acceptor X is reduced by a NADPH-dependent disulfide reductase. The photosynthetic reactions in the thylakoid membrane supply the sulfate reduction system with ATP, NADPH, and reduced ferredoxin.

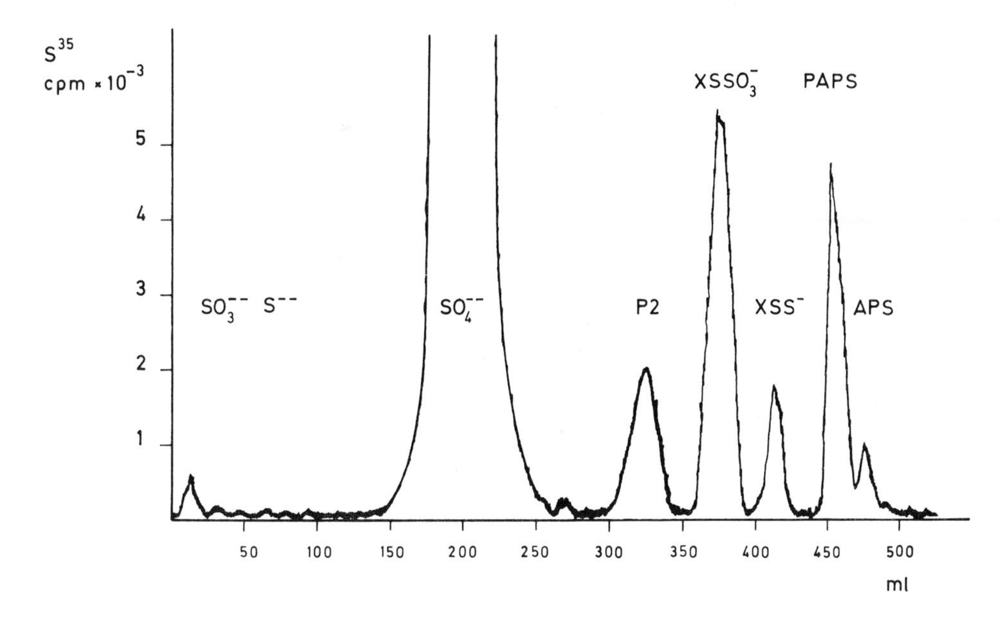

FIGURE 2. ^{35}S radioactive labeled compounds eluted from an ion exchange column. Intact chloroplasts were incubated 30 min in the light with ^{35}S sulfate [16,20] and terminated with ethanol.

The scheme provides for the total eight electrons necessary for the reduction of sulfate (valency state +6) to sulfide (−2) by the thiosulfonate reductase ($6e^-$ transfer) and the disulfide reductase ($2e^-$ transfer).

A mechanism for sulfate assimilation[25] was proposed earlier from studies with *Escherichia coli* [5] and particularly with yeast.[5,26] A sulfite, bound to an enzyme via a low molecular weight carrier, was characterized by Hilz et al.[27] and Torii and Bandurski.[28] In this system, the bound sulfite is proposed to be reductively cleaved ($2e^-$ transfer) to free sulfite which is then reduced by the well-known sulfite reductase ($6e^-$ transfer). The difference in this mechanism and the mechanism proposed for the chloroplast system is the bound sulfide. A bound sulfite in chloroplasts is reduced directly. A thiosulfonate reductase is postulated as replacing the sulfite reductase. The $2e^-$ reductive cleavage occurs at the sulfide level. It should be noted that the postulated thiosulfonate has not been convincingly demonstrated, whereas a sulfite reductase in chloroplasts exists without a doubt.

Recently a cell-free sulfate-reducing system was obtained from plant cell tissue cultures[4] of *Catharanthus.*[29] Such systems produce cysteine with better rates than chloroplasts. Light, ferredoxin, ADP, and NADP were required for cysteine production. It appears that in these systems, which do not necessarily contain thylakoid membranes when grown heterotrophically, the (P)APS reducing system is particle bound.[29]

SULFATE REDUCTION IN INTACT CHLOROPLASTS

Radioactive ^{35}S-labeled sulfate is incorporated into a number of ^{35}S-labeled products, when an intact chloroplast system is illuminated (but not in the dark).[16,20] The compounds are separated from each other by ion-exchange chromatography (a Dowex®-2 nitrate column, eluted by increasing salt concentrations according to Iguchi[30]) after stopping the reaction with ethanol (no addition of carrier sulfite or sulfide). Figure 2 gives the elution profile for such an experiment. After the large sulfate peak, an unidentified compound, P_2, then bound sulfite, denoted $XSSO_3H$, and bound sulfide, denoted XSSH, and finally the activated sulfate nucleotides APS and PAPS ap-

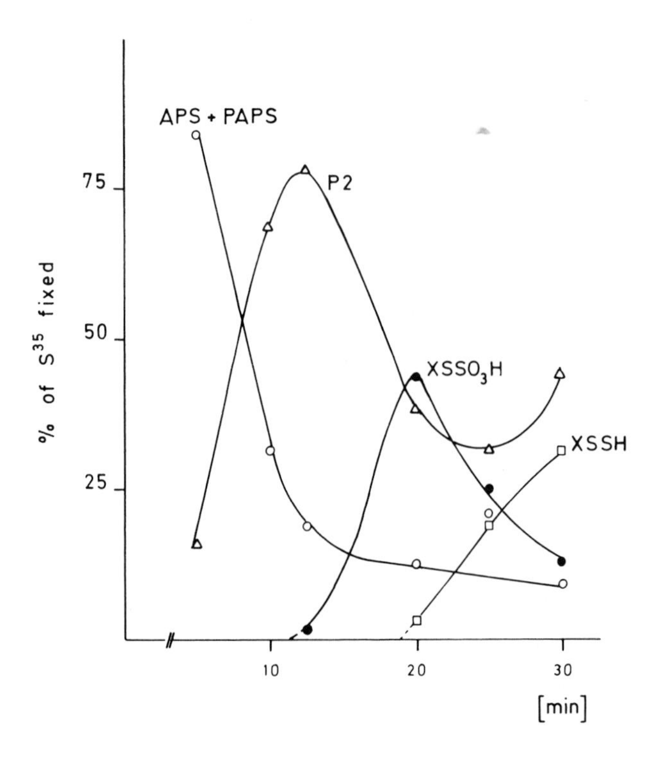

FIGURE 3. The kinetics of ^{35}S appearance in intermediates of sulfate reduction in intact chloroplasts.[20,31]

pear. Significantly, no labeled free sulfite or sulfide are observed.[20] Bound sulfite and bound sulfide are characterized by their ability to exchange their radioactivity with free (unlabeled) sulfite or free sulfide, respectively. If the separation column is run in the presence of carrier sulfite and sulfide, these exchangeable compounds will disappear and labeled free sulfite and sulfide will appear instead. From the exchangeability of $XSSO_3H$ and XSSH, it is suggested that sulfite and sulfide are bound to a carrier via a thiol bridge which is known to readily exchange its substituent, like the sulfite/thiosulfate exchange. This appearance of bound sulfite and sulfide instead of free sulfite and sulfide is the basis for the proposal for a thiosulfonate reductase instead of a sulfite reductase in the sulfate reeducation pathway[16,17,20,30-34] in the chloroplast system which has been adopted for the *Chlorella* system as well.[7,9,23,24] APS and PAPS are well-known substances from early work on sulfate activation. The substance in the peak P_2 is not yet identified, but it seems to contain sulfur at the valency state of the sulfate and may be a chemical artifact.

The reaction kinetics of the appearance of the labeled intermediates of a ^{35}S sulfate reduction in intact chloroplasts is given in Figure 3.[20,31] APS/PAPS are the compounds which contain 100% of the radioactivity fixed when extrapolated to zero incubation time. Therefore, they are the first products of sulfate incorporation into organic compounds. The next compound to appear in time is P_2, followed by bound sulfite, and then finally bound sulfide. From this, a sequence of intermediates is suggested.[20,31]

$$\text{Sulfate} \rightarrow \text{APS} \rightarrow \text{PAPS} \rightarrow P_2 \rightarrow XSSO_3H \rightarrow \text{XSSH} \rightarrow \text{cysteine}$$

Cysteine is not eluted from the Dowex®-2 column described in these experiments.

STUDIES WITH BROKEN CHLOROPLASTS

The first attempts to demonstrate photosynthetic sulfate reduction used broken chloroplasts (thylakoid membrane preparations) supplemented with a supernatant fraction and — at the beginning — even a sulfate-activating system from yeast.[15] Asahi produced the first evidence for a coupling of sulfate reduction to photosynthesis[15] by this method. Furthermore, the presence of sulfite-reducing activity, a soluble sulfite reductase, in chloroplasts also was demonstrated.[36,37] The reductase reduces free sulfite to free sulfide.

The origin of the reducing equivalent for the six-electron step in sulfite or thiosulfonate reduction was clarified with broken chloroplasts. The coupling of partially purified sulfite reductase,[14,37] then of the complete sulfate reduction system,[20,31,34] contained in the chloroplast supernatant, and finally of a thiosulfonate reductase system[22,24,38] onto the light reactions is possible if ferredoxin is present. This shows that ferredoxin, reduced in the light, provides the reducing equivalents ($6e^-$) for sulfite or bound sulfite reduction to sulfide or bound sulfide, respectively. Bound sulfite for the thiosulfonate reductase is generated by sulfate + ATP or by APS and the activating and transferring enzymes in the chloroplast supernatant or by using the model substrate sulfoglutathione.[32,33]

The requirement of a thiol for the production of sulfide in the latter system, but not for sulfite reductase, was taken as indication that an S-S group of the carrier X had to be first reduced to the XSH stage before it would react with activated sulfate.[9,10] The thiol requirement is not needed if NADP, reduced in the light to NADPH by the chloroplasts or by an NADPH generating system, is added.[29] This is taken as evidence that the thiol reduction of the carrier is accomplished by an NADPH-dependent thiol reductase. The carrier XSH with a molecular weight of about 5000 in chloroplasts,[31] but smaller in *Chlorella*,[23] is not yet available in "substrate" amounts. Schmidt therefore introduced glutathione as an artificial carrier for bound sulfite.[33] In particular, the transfer of sulfate from APS onto GSH (to form $GSSO_3H$) was used to check for an APS sulfotransferase.[24,39,40] Sulfoglutathione is then reduced to glutathione persulfide, GSSH, by the broken chloroplasts when ferredoxin is added to the system in light.[32,33] However, the use of glutathione instead of the natural substrate XSH could induce an artifact in that the formation as well as the reduction of sulfoglutathione is not brought about by the enzymes of the sulfate reduction cycle.[34] On the other hand, Tsang and Schiff recently obtained evidence that glutathione is the native acceptor in *Chlorella*.[41] From this observation, a controversy has developed: which activated sulfate nucleotide is transferring its sulfate moiety onto the thiol carrier? From their studies with *Chlorella*, Schiff and Schmidt and colleagues[10,23,24,42] produced evidence that it is APS which is sufficiently activated to provide the sulfate for the formation of sulfoglutathione. Later on, Schmidt found an APS sulfotransferase activity (APS loading glutathione) in all green organisms[39,42] except some blue-green algae.[43] Schwenn and Trebst[9] suggest a PAPS transferase from kinetic studies[20] in an intact chloroplast system[31] and from the regulation of enzyme levels in autotrophic vs. heterotrophic cell tissue cultures with a "natural" enzymic system,[29] where no artificial carrier such as glutathione is added and in which NADPH provides for the reduction of the acceptor molecule. PAPS is not questioned to be the sulfate donor for assimilating sulfate reduction in yeast,[5,26] bacteria,[5,6,25] and some blue-green algae.[43]

The sulfate activation to APS and PAPS was studied in detail in broken chloroplasts. Sulfurylase activity[15,16] is well documented, and also the presence of APS-kinase,[44] though often only the formation of PAPS is taken as evidence for this enzyme.

Several enzymes of the sulfate reduction pathway are peripheral enzymes, bound loosely to the thylakoid membrane.[29,38,44] It is not clear though whether the membrane attachment of these enzymes changes their properties. It might be speculated that the sulfite reductase is a degradation product of a membrane-bound thiosulfonate reductase.

ENZYMES INVOLVED IN ASSIMILATORY SULFATE REDUCTION

Sulfurylase

$$\text{Sulfate} + \text{ATP} \rightarrow \text{APS} + \text{pyrophosphate}$$

Sulfurylase[35] activates sulfate to an "active sulfate anhydride" with AMP. Indeed, APS is so activated that it forms ATP, i.e., the equilibrium is toward the left ($K \approx 10^{-8}$). The reaction is pulled to the right by the pyrophosphatase reaction, by further activation of APS to PAPS, or by the energy drop in consecutive reactions such as sulfate reduction.

Sulfurylase is a widely found enzyme in practically all organisms.[35] Chloroplasts of plants contain a high amount of the enzyme, more than sufficient to account for the biosynthesis of sulfolipid and photosynthetic sulfate reduction to cysteine.[45] The sulfurylase reaction is conveniently measured by the appearance of inorganic phosphate in a reaction in which molybdate substitutes for sulfate and the equilibrium is on the right site because of the instability of the intermediate AMP-molybdate complex.

APS-Kinase

$$\text{APS} + \text{ATP} \rightarrow \text{PAPS} + \text{ADP}$$

The enzyme takes APS to the 3′-phospho-derivative. PAPS is well established as the substrate "activated sulfate" in the biosynthesis of a variety of sulfate esterified carbohydrates. Though implicated in sulfate reduction, APS-kinase escaped measurement in chloroplasts for some time. Burnell and Anderson[44] then showed that chloroplasts do contain the enzyme, where it is possibly peripherally bound to the thylakoid membrane.[44] An APS-kinase fraction from *Chlorella* may be used for the preparation of PAPS.[46]

APS-kinase is essential for the biosynthesis of sulfolipid if PAPS is assumed to be the sulfate donor. However, the enzyme might not be required for sulfate reduction if APS is the sulfate donor. As discussed above and in "APS-sulfotransferase" below, it is not yet settled whether PAPS or APS is the actual substrate species for the formation of bound sulfite in sulfate reduction.

APS Sulfotransferase

$$\text{APS} + \text{acceptor} \rightarrow \text{acceptor-sulfate} + \text{AMP}$$

$$\text{APS} + \text{XSH} \rightarrow \text{XSSO}_3\text{H} = \text{bound sulfite} + \text{AMP}$$

$$\text{APS} + \text{GSH} \rightarrow \text{GSSO}_3\text{H} = \text{sulfoglutathione} + \text{AMP}$$

Schiff and Schmidt and colleagues[23,24,42] have put forward evidence for an APS sulfotransferase activity in sulfate reduction, first identified in the *Chlorella* system. The enzyme is assumed to act in vivo either with itself, the reductase, or a carrier of the sulfate reducing cycle. Very likely the actual functional group is an SH group of an enzyme or a small molecular weight carrier (XSH) of this enzyme. It is checked with glutathione (GSH) as the substrate acceptor. The compound formed, $XSSO_3H$, is a thiosulfonate or thioester of sulfate. Such a compound (S-ester as against an *O*-ester) readily exchanges the sulfonic group with free sulfite. Therefore, the name bound sulfite was coined. It should be clear that the exchange does not indicate that the sulfate group has already been reduced. The reaction could be visualized as an intramolecular reduction of sulfate by the thiol.

The enzyme activity is measured with glutathione as acceptor by the formation of labile bound sulfite (sulfoglutathione), exchangeable with free sulfite. Other dithiols, like dithioerythritol, also may be used as acceptor.[23,24,42] Schmidt has followed enzymic activity in a large variety of higher and lower plants as well as chloroplasts, and, except for some blue-green algae, he finds that APS and not PAPS is the sulfate donor.[39,42,43] An occational activity with PAPS is thought to be because of a phosphatase reaction which splits PAPS to APS.[47] APS as substrate in assimilatory sulfate reduction would like dissimilatory sulfate reduction where also APS rather than PAPS participates.[5,6,25] Schwenn, on the other hand, concludes from kinetic studies as well as results on the regulation of reductive enzymes in assimilatory sulfate reduction in cell tissue cultures,[29] that PAPS transfers the sulfate.

Sulfite and Thiosulfonate Reductase

$$SO_3^{--} + n \cdot fd_{red} \rightarrow S^{--} + n \cdot fd_{ox}$$

$$XSSO_3H + n \cdot fd_{red} \rightarrow XSSH + n \cdot fd_{ox}$$

In the reduction of sulfite or bound sulfite ($XSSO_3H$ = thiosulfonate), six electrons are provided by ferredoxin to form free or bound sulfite (XSSH), respectively. The formation of labeled sulfide — formed also from bound sulfide on exchange with free sulfide — from a labeled sulfite may be followed. The reduction via ferredoxin may couple the system onto the photosynthetic ferredoxin reducing system of chloroplasts.[22,31,38] NADPH may be used as a reductant, when a ferredoxin-NADP reductase also is present. Ferredoxin can be replaced by methylviologen and the reaction followed by the disappearance of the blue radical form.[37] Methylviologen may be reduced by a hydrogenase reaction or electrochemically, but not by the usual method via dithionite because then other sulfite derivatives are introduced into the system. Sulfoglutathione ($GSSO_3H$) has been used as artificial substrate for a thiosulfonate reductase activity.[32,33]

Sulfite reductase is well characterized from bacteria, yeast, and plants.[5] The postulated prosthetic group is a siroheme plus several nonheme iron supplemented in bacterial (but not in the plant) system by a flavin component.[5] Sulfite reductase has been isolated and characterized from chloroplasts.[36,37] The enzyme is likely to participate in assimilatory sulfite reduction; the exact role, however, and the relation to the postulated thiosulfonate reductase remains unclear. This is because free sulfite and free sulfide are not intermediates in sulfate reduction, and therefore a bound sulfite thiosulfonate reductase has been proposed to replace a sulfite reductase. To date, attempts to purify and establish the identity of a thiosulfonate reductase have failed. An early test system[22] via a "dithionite" reductase activity proved to be unreliable and unspecific.[22,34] There is some evidence for a membrane aggregation of the thiosulfonate reductase.[38] It has yet to be established whether a membrane-bound thiosulfonate reductase is degraded to a soluble sulfite reductase.

O-Acetyl-Serine-Sulfhydrylase (Cystein Systhetase)

$$\text{Acetyl-serine} + H_2S \rightarrow \text{cysteine} + \text{acetate}$$

The enzyme performs the final step in assimilatory sulfate reduction. It is a well-characterized enzyme from yeast and bacteria,[1,48] as well as from plants[1,49] and chloroplast sources.[50] It has not been shown whether this enzyme also reacts with bound sulfide, as is postulated for the chloroplast sulfate reduction system. In this system, a reductive cleavage of the XSSH would be required.[22] During purification from *Chlorella*, Schmidt observed an isotopic exchange between cysteine and H_2S,[57] another possibility for erroneous incorporation of labeled ^{35}S.

Disulfide Reductase

X—S—S—X

or $\quad$ + NADPH → XSH

X<(S—S) (cyclic disulfide)

The enzyme is required to regenerate the carrier assumed to be needed in the reduced form for the APS sulfotransferase and thiosulfonate reductase and which is liberated when cysteine is formed from bound sulfide. To account for the NADPH requirement of the sulfate reduction cycle, a NADP-specific enzyme is proposed. Because the carrier X has not been obtained in sufficient amounts, there is only indirect evidence for such an enzyme activity. A NADP-dependent glutathione reductase of appropriate properties has been found in chloroplasts and perhaps might be identical with the postulated enzyme,[52] particularly if glutathione is the carrier (as suggested by Tsang and Schiff).[41] A thioredoxin reductase (see below) also could be the enzyme needed for the disulfide bridge cleavage as proposed for the NADPH-PAPS reductase system of yeast.[53]

Thioredoxin

The disulfide-containing low molecular weight coenzyme of reductive processes, like the biosynthesis of deoxynucleotides, has recently been found in chloroplasts.[54,55] There it participates in the regulation of redox state-controlled enzymes of the Calvin cycle such as fructose 1,6-bisphosphatase.[54] Evidence for a ferredoxin thioredoxin reductase has been found.[54,55] Reduced thioredoxin, in turn, could reduce other disulfide bridges such as oxidized glutathione.[54,55]

The thioredoxin PAPS reductase is implicated in the yeast assimilatory sulfate reduction system, where it splits bound sulfite.[53] It has also been proposed for the *E. coli* system.[56] In the chloroplast system, thioredoxin might participate in the reductive cleavage of bound sulfide. A thioredoxin-dependent PAPS-sulfotransferase has been found in some blue-green algae.[57]

PAP Phosphatase (Nucleosidase)

$$\text{PAP} \rightarrow \text{AMP} + P_i$$

$$\text{PAPS} \rightarrow \text{APS} + P_i$$

Such a 3′-nucleosidase is required to recycle the 3′-phospho-AMP (= PAP) liberated when PAPS is forming *O*-sulfoesters (or when PAPS would be the sulfate donor in the sulfotransferase reaction). Presumably, the enzyme could also remove the 3′-phosphate from PAPS to form APS. Enzymic activity is found in *Chlorella*.[47] In spite of the relevance of the enzyme for sulfate metabolism, it has not been detected in chloroplasts.[44]

REFERENCES

1. **Thompson, J. F.**, Sulfur metabolism in plants, *Annu. Rev. Plant Physiol.*, 18, 59, 1967.
2. **Benson, A. A.**, On the orientation of lipids in chloroplast and all membranes, *J. Am. Oil Chem. Soc.*, 43, 265, 1966.
3. **Raven, J. A.**, Transport in algal cells, in *Encyclopedia of Plant Physiology*, New Series, Vol. 2A, Lüttge, U. and Pitman, M. G., Eds., Springer-Verlag, Heidelberg, 1976, 129.
4. **Hart, J. W. and Filner, P.**, Regulation of sulfate uptake by aminoacids in cultured tobacco cells, *Plant Physiol.*, 44, 1253, 1969.
5. **Siegel, L. M.**, Biochemistry of the sulfur cycle, in *Metabolic Pathway*, Greenberg, H., Ed., Academic Press, New York, 1975, chap. 7.
6. **Thauer, R. K., Jungermann, K., and Decker, K.**, Energy conservation in chemotrophic anaerobic bacteria, *Bacteriol. Rev.*, 41, 100, 1977.
7. **Schiff, J. A. and Hodson, R. C.**, The metabolism of sulfate, *Annu. Rev. Plant Physiol.*, 24, 381, 1973.
8. **Greenberg, D. M.**, Metabolism of sulfur compounds, in *Metabolic Pathway*, Vol. 7, Academic Press, New York, 1975.
9. **Schwenn, J. D. and Trebst, A.**, Photosynthetic sulfate reduction by chloroplasts, in *The Intact Chloroplast*, Barber, J., Ed., Elsevier, Amsterdam, 1976, chap. 9.
10. **Schmidt, A.**, Photosynthetic assimilation of sulfur compounds in *Encyclopedia of Plant Physiology*, New Series, Vol. 6, Gibbs, M. and Latzko, E., Eds., Springer-Verlag, Heidelberg, 1979.
11. **Kylin, A.**, The uptake and metabolism of sulfate by deseeded wheat plants, *Physiol. Plant*, 6, 775, 1953.
12. **Willenbrink, J.**, Lichtabhängiger ^{35}S-Einbau in organische Bindung in Tomatenpflanzen, *Z. Naturforsch. Teil B*, 19, 356, 1964.
13. **Fromageot, P. and Perez-Milan, H.**, Reduction du sulfate en sulfite par la feuille de tabac, *Biochem. Biophys. Acta*, 32, 457, 1959.
14. **Kawashima, N. and Asahi, T.**, Sulfur metabolism in higher plants, *J. Biochem. (Tokyo)*, 49, 52, 1961.
15. **Asahi, T.**, Sulfur metabolism in higher plants, *Biochim. Biophys. Acta*, 82, 58, 1964.
16. **Schmidt, A.**, Untersuchungen zum Mechanismus der photosynthetischen Sulfatreduktion isolierter Chloroplasten, Thesis, University of Göttingen, West Germany, 1968.
17. **Trebst, A. and Schmidt, A.**, Photosynthetic sulfate and sulfite reduction by chloroplasts, in *Progress in Photosynthesis Research*, Vol. 3, Metzner, H., Ed., W. Junk, The Hague, The Netherlands, 1969.
18. **Walker, D. A.**, CO_2 fixation by intact chloroplasts: photosynthetic induction and its relation to transport phenomena and control mechanisms, in *The Intact Chloroplast*, Barber, J., Ed., Elsevier, Amsterdam, 1976, chap. 7.
19. **Schmidt, A. and Trebst, A.**, The mechanism of photosynthetic sulfate reduction by isolated chloroplasts, *Biochim. Biophys. Acta*, 180, 529, 1969.
20. **Schwenn, J. D.**, Untersuchungen zur Kinetik der photosynthetischen Sulfatreduktion isolierter Chloroplasten, Thesis, University of Bochum, West Germany, 1970.
21. **Hodson, R. C., Schiff, J. A., and Mather, J. P.**, Studies of sulfate utilization by algae. X. Nutritional and enzymatic characterization of *Chlorella* mutants impaired for sulfate utilization, *Plant Physiol.*, 47, 306, 1971.
22. **Schmidt, A.**, Sulfate reduction in a cell-free system of *Chlorella*. The ferredoxin dependent reduction of a protein bound intermediate by a thiosulfonate reductase, *Arch. Mikrobiol.*, 93, 29, 1973.
23. **Abrams, W. R. and Schiff, J. A.**, Studies of sulfate utilization by algae, *Arch. Mikrobiol.*, 94, 1, 1973.
24. **Schmidt, A., Abrams, W. R., and Schiff, J. A.**, Reduction of adenosine 5′-phosphosulfate to cysteine in extracts from *Chlorella* and mutants blocked for sulfate reduction, *Eur. J. Biochem.*, 47, 423, 1974.
25. **Roy, A. B. and Trudinger, P. A.**, The biochemistry of inorganic compounds of sulphur, in *The Biochemistry of Inorganic Compounds of Sulphur*, Cambridge University Press, Cambridge, 1970.
26. **Bandurski, R. S.**, Biological reduction of sulfate and nitrate, in *Plant Biochemistry*, Bonner, J. and Varner, J. E., Eds., Academic Press, New York, 1965, 467.
27. **Hilz, H., Kittler, M., and Knape, G.**, Die Reduktion von Sulfat in der Hefe, *Biochem. Z.*, 332, 151, 1959.
28. **Torii, K. and Bandurski, R. S.**, Yeast sulfate-reducing system III. An intermediate in the reduction of 3′-phosphoryl-5′-adenosinephosphosulfate to sulfite, *Biochim. Biophys. Acta*, 136, 286, 1967.
29. **Schwenn, J. D., El-Shagi, H., Kemena, A., and Petrak, E.**, On the role of S: sulfotransferases and assimilatory sulfate reduction by plant cell suspension cultures, *Planta*, 144, 419, 1979.

30. **Iguchi, A.**, The separation of sulfate, sulfite, thiosulfate and sulfite ions with anion exchange resins, *Bull. Chem. Soc. Jpn.*, 31, 600, 1958.
31. **Schmidt, A. and Schwenn, J. D.**, On the mechanism of photosynthetic sulfate reduction, in *Proc. 2nd Int. Congr. Photosynthesis*, Forti, G., Avron, M., and Melandri, A., Eds., W. Junk, The Hague, 1971, 507.
32. **Schmidt, A.**, Uber Teilreaktionen der photosynthetischen Sulfatreduktion in zellfreien Systemen aus Spinatchloroplasten und *Chlorella, Z. Naturforsch. Teil*, 27, 183, 1972.
33. **Schmidt, A.**, On the mechanism of photosynthetic sulfate reduction — an APS sulfotransferase, *Arch. Mikrobiol.*, 84, 77, 1972.
34. **Schwenn, J. D. and Hennies, H.-H.**, Enzymes and bound intermediates involved in photosynthetic sulfate reduction of spinach chloroplasts and *Chlorella*, in *Proc. 3rd Int. Congr. Photosynthesis*, Avron, M., Ed., Elsevier, Amsterdam, 1974, 629.
35. **Humberto de Meio, R.**, Sulfate activation and transfer, in *Metabolic Pathways*, Vol. 3, Academic Press, New York, 1975, chap. 8.
36. **Mayer, A. M.**, Subcellular location of sulphite reductase in plant tissues, *Plant Physiol.*, 42, 324, 1967.
37. **Asada, K., Tamura, G., and Bandurski, R. S.**, Methyl viologen-linked sulfite reductase from spinach leaves, *J. Biol. Chem.*, 244, 4904, 1969.
38. **Schwenn, J. D., Depka, B., and Hennies, H.-H.**, Assimilatory sulfate reduction in chloroplasts: evidence for the participation of both stromal and membrane bound enzymes, *Plant Cell Physiol.*, 17, 165, 1976.
39. **Schmidt, A.**, Distribution of APS-sulfotransferase activity among higher plants, *Plant Sci. Lett.*, 5, 407, 1975.
40. **Schmidt, A.**, The adenosine-5′-phosphosulfate sulfotransferase from spinach (*Spinacea oleracea* L.). Stabilization, partial, purification and properties, *Planta*, 130, 257, 1976.
41. **Tsang, M. L.-S. and Schiff, J. A.**, Studies of sulfate utilization by algae. XXVIII. Identification of glutathione as a physiological carrier in assimilatory sulfate reduction by *Chorella*, *Plant Sci. Lett.*, 11, 177, 1978.
42. **Tsang, M. L.-S. and Schiff, J. A.**, Studies of sulfate utilization by algae. XIV. Distribution of adenosine-3′-phosphate-5′phosphosulfate (PAPS) and adenosine-5′-phosphosulfate (APS) sulfotransferases in assimilatory sulfate reduces, *Plant Sci. Lett.*, 4, 301, 1975.
43. **Schmidt, A.**, Exchange of cysteine-bound sulfide with free sulfide and cysteine synthase activity in *Chlorella pyrenoidos*, *Z. Pflanzenphysiol.*, 84, 435, 1977.
44. **Burnell, J. N. and Anderson, J. W.**, Adenosine 5′-sulfatophosphate kinase activity in spinach leaf tissue, *Biochem. J.*, 134, 565, 1973.
45. **Balharry, J. E. and Nicholas, D. J. D.**, ATP-Sulfurylase in spinach leaves, *Biochim. Biophys. Acta*, 220, 513, 1970.
46. **Hodson, R. C. and Schiff, J. A.**, Preparation of adenosine-3′-phosphate-5′-phosphosulfate (PAPS): an improved enzymatic method using *Chorella pyrenoidosa*, *Arch. Biochem. Biophys.*, 132, 151, 1969.
47. **Tsang, M. L.-S. and Schiff, J. A.**, Properties of enzyme fraction A from *Chorella* and copurification of 3′(2′), 5′-bisphosphonucleoside 3′(2′)-phosphohydrolase, adenosine 5′-phosphosulfate sulfohydrolase and adenosine-5′-phosphosulfate cyclase activities, *Eur. J. Biochem.*, 65, 113, 1976.
48. **Greenberg, D. M.**, Biosynthesis of cysteine and cystine, in *Metabolic Pathways*, Vol. 7, Greenberg, D. M., Ed., Academic Press, New York, 1975, chap. 12.
49. **Giovanelli, J. and Mudd, S. H.**, Sulfuration of O-acetylhomoserine and O-acetylserine by two enzyme fractions from spinach, *Biochim. Biophys. Res. Commun.*, 31, 275, 1968.
50. **Frankhauser, H., Brunold, Chr., and Erismann, K. H.**, Subcellular localization of O-acetylserine sulfhydrylase in spinach leaves, *Experientia*, 32, 1494, 1976.
51. **Schmidt, A.**, Protein-catalyzed isotopic exchange reaction between cysteine and sulfide in spinach leaves, *Z. Naturforsch. Teil C*, 32, 219, 1977.
52. **Hendley, D. D. and Conn, E. E.**, Enzymatic reduction and oxidation of glutathione by illuminated chloroplasts, *Arch. Biochem. Biophys.*, 46, 454, 1953.
53. **Porque, P. G., Baldesten, A., and Reichard, P.**, The envolvement of the thioredoxin system in the reduction of methionine sulphoxide and sulphate, *J. Biol. Chem.*, 245, 2371, 1970.
54. **Wolosiuk, R. A. and Buchanan, B. B.**, Thioredoxin and glutathione regulate photosynthesis in chloroplasts, *Nature (London)*, 266, 565, 1977.
55. **Wolosiuk, R. A., Buchanan, B. B., and Crawford, N. A.**, Regulation of NADP-Malate dehydrogenase by the light-actuated ferredoxin thioredoxin system of chloroplasts, *FEBS Lett.*, 81, 253, 1977.
56. **Tsang, M. L.-S. and Schiff, J. A.**, Sulfate-reducing pathway in *Escherichia coli* involving bound intermediates, *J. Bacteriol.*, 125, 923, 1976.
57. **Schmidt, A. and Christen, U.**, A factor-dependent sulfotransferase specific for 3′-phosphoadenosine-5′-phosphosulfate (PAPS) in the cyanobacterium *Synehococcus* 6301, *Plata*, 140, 239, 1978.

HYDROGEN METABOLISM OF PHOTOSYNTHETIC BACTERIA AND ALGAE

Shuzo Kumazawa and Akira Mitsui

INTRODUCTION

Hydrogenase-catalyzed hydrogen metabolism in photosynthetic organisms can be summarized by the following reactions:

In the light:

Hydrogen photoproduction[1]

$$H_2O \longrightarrow H_2 + \tfrac{1}{2} O_2 \qquad (1)$$

$$RH_2 \longrightarrow H_2 + R \qquad (2)$$

Photoreduction[2-7]
(Carbon dioxide photoreduction)

$$2H_2 + CO_2 \xrightarrow{nATP} (CH_2O) + H_2O \qquad (3)$$

In the dark:

Hydrogen production[1,8,9]

$$RH_2 \longrightarrow H_2 + R \qquad (4)$$

Carbon dioxide reduction[10]

$$2H_2 + CO_2 \xrightarrow{nATP} (CH_2O) + H_2O \qquad (5)$$

Oxyhydrogen reaction[5,10]

$$H_2 + \tfrac{1}{2} O_2 \longrightarrow H_2O \qquad (6)$$

Hydrogen uptake[2,11]

$$R + H_2 \longrightarrow RH_2 \qquad (7)$$

In addition to the above hydrogenase reactions, nitrogenase, the enzyme responsible for dinitrogen fixation, participates in hydrogen production.[12,13] Hydrogenase is a ferredoxin-type nonheme iron-sulfur protein, and nitrogenase is a complex of two proteins, Component II (Fe protein) and Component I (Fe-Mo protein). These enzymes are basically nonheme iron-sulfur proteins.[14,15]

The two different enzyme systems have distinct catalytic properties in hydrogen production. Hydrogenase catalysis is ATP independent, CO sensitive, while nitrogenase catalysis is ATP dependent, CO insensitive.[16-19] The presence of nitrogenase enzyme is restricted to prokaryotic organisms, while hydrogenase is observed in both prokar-

Table 1
OCCURRENCE OF HYDROGENASE ACTIVITY IN ALGAE[a]

Alga	Ref.	Alga	Ref.
Cyanophyceae (Cyanobacteria)			
Anabaena cylindrica	38	*Chlorella protothecoides*	35
Anabaena variabilis	36	*Chlorella vacuolata*	35
Anacystis nidulans	36	*Chlorella vulgaris f. tertia*	48
Aphanothece halophytica	39	*Chlorococcum vacuolatum*	23
Cyanophora paradoxa	36	*Coelastrum* spp.	50, 51
Nostoc muscorum	40	*Coelastrum proboscideum*	35
Oscillatoria limnetica	39	*Crucigenia apiculata*	23
Synechococcus elongatus	41	*Kirchneriella lunaris*	35
Synechocystis sp.	42	*Rhaphidium* spp.	3, 45
		Scenedesmus sp.	3, 45
Euglenophyceae		*Scenedesmus obliquus*	3
Euglena gracilis	43	*Scenedesmus quadricauda*	33
Euglena sp.	44	*Selenastrum gracile*	50
		Selenastrum sp.	35
Chlorophyceae		*Ulva faciata*	40
Ankistrodesmus spp.	10, 45	*Ulva lactuca*	42
Ankistrodesmus braunii	11		
Ankistrodesmus falcatus	35	Phaeophyceae	
Chlamydomonas debaryana	33	*Ascophyllum nodosum*	42
Chlamydomonas dysosmos	33		
Chlamydomonas eugametos	46	Rhodophyceae	
Chlamydomonas intermedia	48	*Ceramium rubrum*	34
Chlamydomonas moewusii	47	*Chondrus crispus*	34
Chlamydomonas reinhardtii	43	*Corallina officinalis*	34
Chlorella fusca	40	*Porphyra umbilicalis*	42
Chlorella homosphaera	49	*Porphyridium aerugineum*	34
Chlorella kessleri	48	*Porphyridium cruentum*	42

[a] This table is based on Kessler's work,[23] with the addition of recent findings.

yotic and eukaryotic organisms.[20] Reviews of hydrogen metabolism have been published on algae[21-25] and on photosynthetic bacteria (anoxygenic phototrophic bacteria).[7,26-32]

OCCURRENCE OF HYDROGENASE IN PHOTOSYNTHETIC ORGANISMS

The occurrence of hydrogenase in algae has been reviewed.[23,25] Kessler classified algae into those with and without hydrogenase activity and observed that 50% of the algae so far tested exhibited activity.[23] However, the presence of hydrogenase in cells may not correspond to the capability of an organism to carry out hydrogen photoevolution.[33-37]

Algae and photosynthetic bacteria containing hydrogenase are listed in Tables 1 and 2, respectively. The presence of hydrogenase activity in photosynthetic bacteria (see Table 2), is based on reports of photoreduction activity (see Equation 3). Since the occurrence of hydrogenase activity can depend on preculture conditions, future additions of organisms to this list can be expected.

Table 2
OCCURRENCE OF HYDROGENASE ACTIVITY IN PHOTOSYNTHETIC BACTERIA (ANOXYGENIC PHOTOTROPHIC BACTERIA)

Bacteria	Ref.
Rhodospirillaceae	
(Purple nonsulfur bacteria)	
Rhodospirillum rubrum	7, 52
Rhodospirillum tenue	163
Rhodomicrobium vannielii	53
Rhodopseudomonas capsulata	54, 55
Rhodopseudomonas gelatinosa	54
Rhodopseudomonas acidophila	56
Rhodopseudomonas palustris	54
Rhodopseudomonas sphaeroides	164
Rhodopseudomonas sulfidophila	165
Chromatiaceae	
(Purple sulfur bacteria)	
Chromatium D	57
Chromatium sp.	58
Chromatium minutissimum	59
Ectothiorhodospira mobilis	166
Ectothiorhodospira shaposhnikovii	164
Thiocapsa floridana	58
Thiocapsa roseopersicina	60
Chlorobiaceae	
(Green sulfur bacteria)	
Chlorobium limicola	61, 62
Chlorobium thiosulfatophilum	61, 62
Chloropseudomonas ethylica	63

[a] Based on hydrogen uptake in the dark.

METHODS USED FOR MEASURING HYDROGENASE ACTIVITY

Methods used for measuring hydrogenase activity are listed in Table 3. Of these methods, manometry and gas chromatography are the most widely used. Amperometric methods or mass spectroscopy can be used for monitoring the change in activity over second to minute intervals. Redox reactions of viologen dyes also are used for measuring hydrogenase activity spectrophotometrically. Short-term measurements can be used as an index of potential capability for hydrogen photoevolution, while long-term measurements can be used as an index of capacity for hydrogen photoevolution for applications.

HYDROGEN PRODUCTION

The capability of algae to produce hydrogen gas and the effects of light on the process (at low light intensities) were first reported by Gaffron and Rubin in 1942.[1] In photosynthetic bacteria, hydrogen production, under dark anaerobic conditions, was first reported by Roelofsen in 1934[8] and Nakamura in 1937.[9] The capacity of photo-

Table 3
METHODS FOR INVESTIGATION OF HYDROGENASE ACTIVITY

Method	Time scale for measurement[a]	Ref.
Manometry	Minutes-hours	64—66
Modified manometry (syringe method)	Minutes-days	67
Gas chromatography	Minutes-days	
Amperometric method		
Clark type electrode	Seconds-minutes	68
Zirconium oxide high temperature electrode	Seconds	69
Enzyme electric cell	Minutes	70
Mass spectroscopy	Seconds-minutes	71
Hydrogen isotope (exchange reaction between hydrogen and deuterated or tritiated water)	Minutes	43, 72
Spectroscopic method	Minutes-hours	72, 73

[a] Time scale indicates that of ordinal use.

synthetic bacteria to carry out photoproduction of hydrogen was first reported by Gest and Kamen in 1949.[12,13]

The potential for hydrogen photoevolution in algae and photosynthetic bacteria are listed in Tables 4 and 5, respectively. The rates of hydrogen photoproduction are dependent on experimental methods and conditions. Some of the rates in the tables are extrapolated from short-term measurements (initial rates: minutes), while others are averages for rates observed over hours or days. Usually, higher rates are obtained in short-term experiments.

Electron Donors and Mechanisms of Hydrogen Photoproduction by Living Cells

Algae have two photosystems (PSI and PSII) and are capable of generating reducing potential from water. On the other hand, bacteria have one photosystem and require exogenous electron donors for photosynthesis.[26] The major difference in hydrogen photoproduction in the two groups is that algae are theoretically capable of producing hydrogen directly through the photolysis of water or through photosynthetic products, while photosynthetic bacteria require external electron donors such as organic compounds or reduced sulfur compounds.

Green Algae

Gaffron and Rubin suggested two possible mechanisms for the generation of electron donors for hydrogen photoproduction.[1] One was a generation from endogenous stored compounds photochemically metabolized, and the other was a generation through photolysis of water. Hydrogen photoproduction coupled to endogenous fermentation has been demonstrated by many researchers.[74,90-96] Embden-Meyerhof pathway,[91] anaerobic Krebs cycle,[90] and pyruvate dehydrogenase reaction (dark production)[96] have been suggested as sources of reductants. On the other hand, many

Table 4
HYDROGEN PHOTOEVOLUTION RATES IN ALGAE

Alga	H_2 produced per 1 hr (μmol)		Method[a]	Light intensity	Ref.
Green Algae					
Ankistrodesmus braunii		0.49/mg Chl	M	16.7 W/m^2	74
		11.1/mg Chl	A		35
Chlamydomonas reinhardtii	0.25/mg dry wt	17/mg Chl[b]	M	300 lx	33
		3.2/mg Chl	M	16.7 W/m^2	74
Chlorella fusca		0.49/mg Chl	M	16.7 W/m^2	74
Chlorella protothecoides		16.4/mg Chl	A		35
Chlorella sp.	0.16/mg dry wt	11/mg Chl[b]	M	400 lx	33
Coelastrum proboscideum		24.6/mg Chl	A		35
Selenastrum sp.		11.8/mg Chl	A		35
Scenedesmus obliquus	0.22/mg dry wt	15/mg Chl[b]	M	500 lx	33
		0.4/mg Chl	M	16.7 W/m^2	74
		20.2/mg Chl	A		35
Kirchneriella lunaris		32.1/mg Chl	A		35
Blue-green algae (cyanobacteria)					
Anabaena cylindrica	1.4/mg dry wt	320/mg Chl	G	400 W/m^2	75
		0.7/mg Chl	G	35,000 lx	76
	0.17/mg dry wt	11/mg Chl[b]	G	7,000 lx	77
	0.58/mg dry wt		G	32 W/m^2	78
	0.10/mg dry wt	6.7/mg Chl[b]	G	4,000 lx	79
Anabaena flos-aquae	0.025/μL pcv[a]		A		80
Calothrix scopulorum	0.13/mg dry wt	8.7/mg Chl[b]	G	4,000 lx	79
Nostoc muscorum	0.005/μL pcv[a]		G	20 W/m^2	81
Oscillatoria brevis	0.17/mg dry wt	11/mg Chl[b]	G	4,000 lx	79
Oscillatoria sp.	0.18/mg dry wt	230/mg Chl	G	63 μ Einstein/m^2/sec	82

[a] Abbreviations used: M, manometry; A, amperometry; G, gas chromatography; pcv, packed cell volume.
[b] Converted to chlorophyll basis with following factor: 67 for dry weight to chlorophyll.

Table 5
HYDROGEN PRODUCTION RATES IN PHOTOSYNTHETIC BACTERIA (ANOXYGENIC PHOTOTROPHIC BACTERIA)

Bacteria	μmoles H_2 produced per 1 hr		Electron donor	Light intensity	Ref.
Rhodospirillum rubrum	10.7/mg N	1.7/mg protein[a]	Malate	10,800 lx	83
		1.68/mg protein	Malate	135 W/m^2	84
		0.96/mg protein	Malate	46 W/m^2	85
R. rubrum mutant C		4.0/mg protein	Formate	Dark	86
Rhodopseudomonas capsulata	5.80/mg dry wt	12/mg protein[a]	Lactate or pyruvate	10,800 lx	67
Rhodopseudomonas palustris		0.78/mg protein	Glucose	29 W/m^2	87
Chromatium D	0.09/mg cells[b]	1.2/mg protein[a]	Thiosulfate	50,000 lx	88
Thiocapsa roseopersisicina		1.84/mg protein	Pyruvate	20 W/m^2	89

[a] Converted to protein basis with following factors: 2 for dry weight to protein, 13 for cells to protein, and 0.16 for nitrogen to protein.

[b] Projected from the figure.[88]

researchers have suggested that the generation of reductants for hydrogen photoproduction is the same system as for photosynthesis, namely biophotolysis of water.[25,35,97]

Blue-Green Algae (Cyanobacteria)

Hydrogen photoproduction by blue-green algae has been investigated for both heterocystous and nonheterocystous forms. Hydrogen production in the hetereocystous blue-green algae is catalyzed by nitrogenase.[76,80,98,99] It has been hypothesized that electrons for hydrogen production are transported from vegetative cells to heterocysts.[100] The oxidative pentose phosphate cycle (glucose-6-phosphate dehydrogenase, phosphogluconate dehydrogenase)[101,102] and isocitrate dehydrogenase reaction[103] coupled to ferredoxin-NADP oxidoreductase and the pyruvate ferredoxin oxidoreductase reaction[104,105] are believed to supply electrons to nitrogenase. PSI in heterocysts may function as a supplier of ATP (through cyclic photophosphorylation)[106] and as a reducer of ferredoxin from endogenous reductant.[103] Hydrogen evolution from sulfide (Na_2S) via a PSI-hydrogenase pathway has been reported in some blue-green algae.[107] Since research on the mechanisms of hydrogen photoproduction in blue-green algae started only during the last few years, a comprehensive understanding of the mechanism is not available.

Photosynthetic Bacteria (Anoxygenic Phototropic Bacteria)

Hydrogen photoproduction by photosynthetic bacteria is stimulated when certain amino acids are used as nitrogen sources for growth or when ammonium ions in culture media limit growth.[12,53,57,84,108] Light-dependent hydrogen production by the photosynthetic bacteria seems to be catalyzed by nitrogenase enzyme. Electron donors used for hydrogen photoproduction are listed in Table 6.

Light-dependent anaerobic Krebs cycle,[7,67,108] pyruvate ferredoxin oxidoreductase,[122,124] and photoreduction of ferredoxin from thiosulfate[88] have been shown to operate in the formation of reductants. The role of light seems to be to supply ATP through cyclic photophosphorylation[27,29] and to reduce ferredoxin, and/or the primary electron acceptor through photochemical reactions.[32,125-127]

Like nonphotosynthetic bacteria,[128] photosynthetic bacteria are able to produce hydrogen fermentatively through formic hydrogenlyase[86,111,121] or pyruvate ferredoxin oxidoreductase reactions[122,124] via hydrogenase.

Hydrogen Photoproduction by Cell-Free System

Hydrogen photoproduction by reconstituted cell components was first demonstrated by Mitsui et al. in 1961 to 1962.[129-131] This cell-free hydrogen photoproduction system consisted of chloroplasts, bacterial hydrogenase, and methylviologen. The electron donor was cysteine plus DPIP (2,6-dichlorophenolindophenol). Chloroplasts and methylviologen could be replaced with blue-green algal lamella (*Nostoc muscurum*) and ferredoxin, respectively. Bacterial hydrogenase was obtained from *Chromatium, Desulfovibrio,* and *Clostridium.* The system demonstrated that hydrogen photoproduction was directly coupled to PSI and hydrogenase when an electron mediator, such as ferredoxin or methylviologen, is present. In 1964, it was suggested that a complete photosystem (PSI and II) could produce hydrogen, using water as an electron donor.[133] In the same year, Abeles[46] demonstrated hydrogen photoproduction from a cell-free preparation of green alga, *Chlamydomonas eugametos,* using reduced pyridine nucleotides as an electron donor.

In the 1970s, the production of hydrogen through biophotolysis was proposed as an alternative energy resource, and thus earlier experiments were reexamined. Kramptiz[134] and Benemann et al.[135] demonstrated hydrogen photoproduction from water. In using similar systems mentioned above (chloroplasts, bacterial hydrogenase and methylviol-

Table 6
ELECTRON DONORS USED FOR HYDROGEN PRODUCTION BY PHOTOSYNTHETIC BACTERIA (ANOXYGENIC PHOTOTROPHIC BACTERIA)

Bacteria	Ref.
Rhodospirillaceae	
(Purple nonsulfur bacteria)	
Rhodospirillum rubrum	
Acetate	108
Lactate	109
Pyruvate	12, 110—115
Succinate	12, 13, 52, 108—110
Fumarate	12, 13, 108—110
Malate	12, 13, 52, 84, 85, 108—110, 116—118
Oxalacetate	12, 110
Rhodopseudomonas acidophila	
Lactate	56
Rhodopseudomonas capsulata	
Propionate	67
Lactate	67, 119, 120
Pyruvate	67, 119
Succinate	67, 119
Fumarate	119
Malate	55, 67, 119
Butyrate	67
Glucose	67
Fructose	67
Sucrose	67
Rhodopseudomonas palustris	
Formate	121
Pyruvate	87
α-Ketoglutarate	87
R. palustris	
Succinate	87
Malate	87
Oxalacetate	87
Glucose	87
Thiosulfate	87
Chromatiaceae	
(Purple sulfur bacteria)	
Chromatium D	
Thiosulfate	88, 117
Pyruvate	122
Chromatium sp.	
Malate + CO_2	123
Thiocapsa roseopersicina	
Thiosulfate	60
Acetate	60
Pyruvate	60, 89
Oxalacetate	60
Chlorobiaceae	
(Green sulfur bacteria)	
Chloropseudomonas ethylica	
Lactate	63
Pyruvate	63
Citrate	63
α-Ketoglutarate	63
Xylose	63
Mannitol	63
Glucose	63
Chloropseudomonas sp.	
Formate	63

ogen or ferredoxin), glucose-glucose oxidase and/or ethanol-catalase were added as an oxygen trap to protect hydrogenase enzyme from oxygen inactivation evolved through the biophotolysis of water. The cell-free systems so far studied are summarized in Table 7. These include the use of dithiothreitol as an electron donor,[92,136] the stabilization of photosystems by microencapsulation[137] or glutaraldehyde treatment,[141] and on the examination of interactions between cell components from different origins.[142,143]

PARTICIPATION OF HYDROGENASE AND NITROGENASE IN HYDROGEN PRODUCTION AND HYDROGEN UPTAKE

Hydrogen metabolism in green algae which lack nitrogenase is ascribed to hydrogenase catalyzed reactions. Depending upon environmental and physiological conditions, these algae can carry out several reactions. In contrast, hydrogen production by blue-green algae and photosynthetic bacteria is complicated by the interaction of hydrogenase and nitrogenase systems. These potential relationships are shown in Figure 1.

Table 7
HYDROGEN PHOTOPRODUCTION BY CELL-FREE SYSTEM

Electron donor	Cell-free system		Rate of H_2 production	Ref.
Cysteine-DPIP	Chloroplast (spinach) + MV (Fd) + CMU	Hydrogenase	5.5 μmol/hr/2 mℓ reaction mixture	129—131
Ascorbate-DPIP	Lamelae (blue-green alga, *Nostoc*) + MV (Fd) + CMU (blue-green alga, *Nostoc*)	*(Chromatium, Desulfovibrio, Clostridium)*		
H_2O	Chloroplast (spinach) + MV	Hydrogenase *(Clostridium)*		133
NADH or NADPH	*Chlamydomonas* cell-free preparation + DCMU		3.2 μmol/hr/2.2 mℓ reaction mixture	46
Dithiothreitol	Chloroplast (spinach) + DCMU	Hydrogenase *(Chlamydomonas)*	4.3 μmol/hr/mg Chl	136
Dithiothreitol or NADH	*Chlamydomonas* cell-free preparation + DCMU		7.4 μmol/hr/mg Chl	92
Ascorbate-DPIP	PS I particle + MV #microencapsulation of the system	Hydrogenase *(Chromatium)*		137
Ascorbate-TMPD	Chloroplast (spinach) + MV	Hydrogenase *(Clostridium)*	125 μmol/hr/mg Chl	138
H_2O	Chloroplast (spinach) + MV #glucose glucose oxidase and ethanol catalase as an oxygen trap	Hydrogenase	2.1 μmol/hr/mg Chl	134
H_2O	Chloroplast (spinach) + Fd #glucose glucose oxidase and catalase as an oxygen trap	Hydrogenase *(Clostridium)*	14 μmol/hr/mg Chl	115
H_2O	Chloroplast (spinach) + Fd (spinach) #glucose glucose oxidase and ethanol catalase as an oxygen trap	Hydrogenase *(Clostridium)*	20 μmol/hr/mg Chl	139, 140
H_2O	Chloroplast (spinach) (lettuce) (tobacco) *(Chenopodium)*, etc. + Fd (spinach) *(Spirulina*, etc.) #glucose glucose oxidase as an oxygen trap, #stabilization of chloroplast by glutaraldehyde fixation	Hydrogenase *(Chlostridium)* *(Escherichia coli)* *(Thiocapsa*, etc.)	94 μmol/hr/mg Chl	141—143

Note: Abbreviations used: CMU, 3-(4-chlorophenyl)-1,1-dimethylurea; DCMU, 3-(3,4-dichlorophenyl)-1,1-dimethylurea; DPIP, 2,6-dichlorophenolindophenol; Fd, ferredoxin; NADH, nicotinamide adenine dinucleotide (reduced form); NADPH, nicotinamide adenine dinucleotide phosphate (reduced form); TMPD, N-tetramethyl-*p*-phenylenediamine.

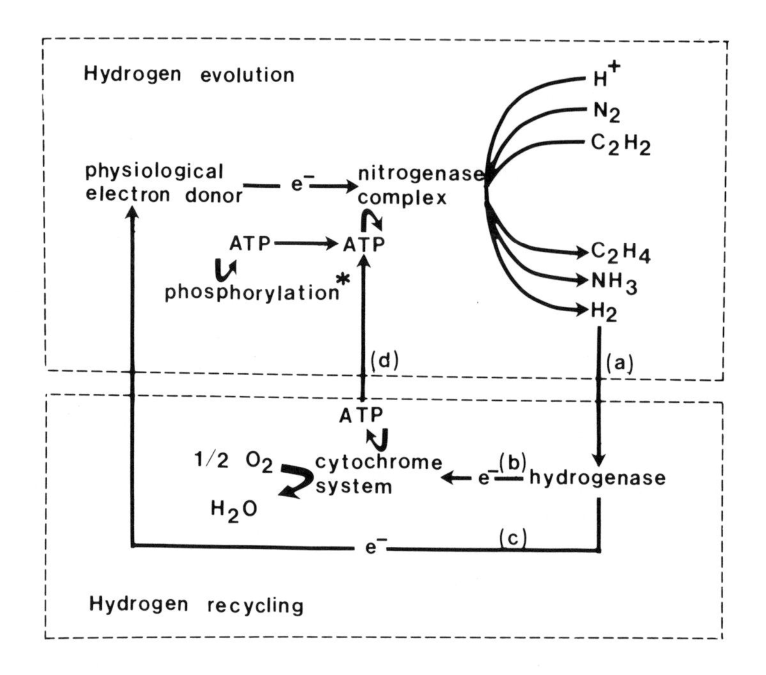

FIGURE 1. Nitrogenase/hydrogenase relationship scheme (working theory). (a) References 76, 144; (b) References 102, 144; (c) References 145, 146; (d) Reference 102. This illustration does not imply stoichiometrical relationship of the compounds or electrons; *, cyclic photophosphorylation and oxidative phosphorylation (cytochrome) are suggested.

Hydrogen photoproduction by the heterocystous blue-green algae, *Anabaena cylindrica,* has been studied by several research groups. In this alga, hydrogen photoproduction seems to be catalyzed by the nitrogenase enzyme. In the study of hydrogenase-nitrogenase interactions, the rate of hydrogen production is lower than acetylene reduction.[76] This low hydrogen photoproduction rate was shown to be attributable to hydrogenase-mediated hydrogen uptake.[76,144] The oxyhydrogen reaction[102,144] and reductant supply (H_2 supported acetylene reduction)[145,146] are suggested as a mechanism of hydrogen uptake in nitrogen fixing photosynthetic microorganisms.[37,147]

The oxyhydrogen reaction has been shown in several species of nitrogen-fixing blue-green algae, including *Anabaena*[36,37,99,102,144] and *Nostoc.*[36,37] The involvement of cytochrome-cytochrome oxidase in the reaction has been suggested.[102] The stimulation of oxyhydrogen reaction under nitrogen-fixing conditions and the localization of major activity in the heterocysts has been reported in recent papers.[102,147,148] The formation of ATP by this oxyhydrogen reaction has been demonstrated.[102] In photosynthetic bacteria, the oxyhydrogen reaction was found in *Rhodopseudomonas capsulata.*[149] The coupling of cytochrome-cytochrome oxidase to hydrogenase in oxyhydrogen reaction was shown in *Rhodopseudomonas palustris.*[150]

The presence of active uptake hydrogenase activity may benefit the efficiency of the nitrogen-fixing microorganisms by recapturing hydrogen produced by nitrogenase.[151]

The relationship between photoreduction and nitrogen fixation in photosynthetic bacteria has been studied by many researchers. The activity of photoreduction appears to be enhanced by increased intracellular concentrations of ammonium ion.[53,56,83,84,119] Photoreduction and nitrogen fixation seem to operate reciprocally, depending on the intracellular supply of carbon and nitrogen. More specifically, when the carbon to nitrogen ratio in cells increases, photoreduction activity is reduced and nitrogen-fixing activity is enhanced and vice versa.

ENHANCEMENT AND STABILIZATION OF HYDROGEN PHOTOPRODUCTION

Several factors affect the stability and efficiency of hydrogen photoproduction. First, hydrogenase and/or nitrogenase are susceptible to oxygen inactivation. Second, there is competition for electron flow to hydrogen production from several biological reactions. It has been shown in cell-free experiments that electron flow to hydrogen production can be diverted to NADP reduction and cyclic phosphorylation.[129,130,132] The apparent Michaelis constant value for ferredoxin, isolated from several organisms, has been reported to be much higher for hydrogen evolution than for NADP reduction.[92] It has also been suggested that nitrite reduction and nitrogen fixation (if nitrogenase is present) also diminishes the high rate of stability of hydrogen photoproduction.[82,152] Finally, the reversible nature of hydrogenase may restrict the accumulation of hydrogen at certain concentrations.

The following treatment or procedures could be employed to enhance or stabilize hydrogen photoproduction:

1. The removal of oxygen and/or hydrogen by the continuous flushing with inert gas[75,78]
2. Balancing light and dark metabolism by intermittent illumination[35,153]
3. Illumination with light below the compensation piont of oxygen production

Living Cell Hydrogen Photoproduction

Green Algae

In some green algae (e.g., *Scenedesmus*), the potential for hydrogen production from water has been demonstrated by Bishop et al.[35] However, high rates of hydrogen photoproduction seem to be short lived (on the order of minutes). Intermittent illumination stabilized hydrogen photoproduction, indicating that continuous light caused the accumulation of inhibitory levels of O_2.[35] The involvement of an O_2-labile, low redox potential, electron carrier in green algae has also been suggested as a cause of inhibition.[97] Production and accumulation of O_2 in reaction systems using green algae are major obstacles in sustained hydrogen photoproduction.

Blue-Green Algae

In blue-green algae, high rates of sustained hydrogen photoproduction by nitrogen-starved cultures of the heterocystous blue-green alga *A. cylindrica* have been reported by several laboratories.[75,78] However, other laboratories have reported conflicting results.[76,99] This difference in yields may come from the experimental methods rather than different catalytic activity. In the former experiments,[75,78] a gas flushing system ($Ar + CO_2$) was used to lower oxygen and hydrogen tensions, thereby limiting the problem of oxyhydrogen reaction.

Sustained hydrogen photoproduction (for a week) has been observed in a closed-flask system by a strain of blue-green algae isolated from subtropical marine environments.[82] This strain is a filamentous blue-green alga without heterocysts (tentatively identified as *Oscillatoria* sp.). When this strain is inoculated into combined nitrogen-free medium, nitrogen chlorosis is observed. Under nitrogen-limiting conditions, this alga adapts to the culture medium by decreasing phycocyanin and chlorophyll (0.1 to 0.2% of dry weight) content and increasing the carbohydrate content, relative to dry weight. Under anaerobic conditions active hydrogen photoproduction is observed, without apparent H_2 uptake activity.[154]

Photosynthetic Bacteria

In photosynthetic purple nonsulfur bacteria, hydrogen photoproduction occurs through light-dependent anaerobic Krebs cycle from organic substrates. In hydrogen photoproduction by *R. capsulata,* Hillmer and Gest[67] reported 5.8 μmol of hydrogen produced per hour on a milligram dry weight basis (or 580 μmol of hydrogen production per hour on a bacterial chlorophyll basis, assuming a 1% chlorophyll content in dry weight). The photoproduction of hydrogen appears to be more efficient than dark or bacterial fermentative hydrogen production.[155] Ammonium ion is known to repress nitrogenase enzyme synthesis when incorporated at the level of glutamine and glutamate.[156] For sustained nitrogen fixation or hydrogen production, genetic studies to derepress nitrogenase enzymes have been conducted.[157]

Cell-Free Hydrogen Photoproduction System

The key to the successful construction of a biophotolytic hydrogen production system will be the stabilization of photosynthetic particles and hydrogenase enzymes against O_2 inactivation. Advanced and inexpensive O_2 trapping systems, other than glucose-glucose oxidase system, will have to be developed. Stabilization of hydrogenase[158-161] and photosynthetic particles,[137,139,141] use of synthetic electron mediator,[162] and use of inorganic or synthetic catalysts instead of hydrogenase[97] have been suggested as solutions.

Biological solar energy conversion to hydrogen gas is being pursued as an alternative energy resource.[167-173] However, many barriers remain before an economical biophotolytic production system is realized.

REFERENCES

1. **Gaffron, H. and Rubin, J.,** Fermentative and photochemical production of hydrogen in algae, *J. Gen. Physiol.,* 26, 219, 1942.
2. **Gaffron, H.,** The reduction of carbon dioxide with molecular hydrogen in green algae, *Nature (London),* 143, 204, 1939.
3. **Gaffron, H.,** Carbon dioxide reduction with molecular hydrogen in green algae, *Am. J. Bot.,* 27, 273, 1940.
4. **Frenkel, A. W. and Lewin, R. A.,** Photoreduction by *Chlamydomonas, Am. J. Bot.,* 41, 586, 1954.
5. **Gingras, G., Goldsby, R. A., and Calvin, M.,** Carbon dioxide metabolism in hydrogen-adapted *Scenedesmus, Arch. Biochem. Biophys.,* 100, 178, 1963.
6. **Senger, H. and Bishop, N. I.,** Changes in the quantum yield and photoreduction during the synchronous life cycle of *Scenedesmus obliquus, Nature (London),* 214, 140, 1967.
7. **Ormerod, H. G. and Gest, H.,** Hydrogen photosynthesis and alternative metabolic pathways in photosynthetic bacteria, *Bacteriol. Rev.,* 26, 51, 1962.
8. **Roelofsen, P. A.,** Metabolism of purple sulfur bacteria, *Proc. Acad. Sci. Amsterdam,* 37, 660, 1934.
9. **Nakamura, H.,** Über das Vorkommen der Hydrogenlyase in *Rhodobacillus palustris* und über ihre Rolle im Mechanismus der Bakteriellen Photosynthese. Beitrage zur Stoffwechselphysiologie der Purpurbakterien. III, *Acta Phytochim.,* 10, 211, 1937.
10. **Gaffron, H.,** Oxyhydrogen reaction in green algae and the reduction of carbon dioxide in the dark, *Science,* 91, 529, 1940.
11. **Kessler, E.,** Reduction of nitrite with molecular hydrogen in algae containing hydrogenase, *Arch. Biochem. Biophys.,* 62, 241, 1956.
12. **Gest, H. and Kamen, M. D.,** Photoproduction of molecular hydrogen by *Rhodospirillum rubrum, Science,* 109, 558, 1949.
13. **Gest. H. and Kamen, M. D.,** Studies on the metabolism of photosynthetic bacteria. IV. Photochemical production of molecular hydrogen by growing cultures of photosynthetic bacteria, *J. Bacteriol.,* 58, 239, 1949.
14. **Orme-Johnson, W. H.,** Iron-sulfur proteins: structure and function, *Annu. Rev. Biochem.,* 42, 159, 1973.
15. **Hardy, R. W. F. and Burns, R. C.,** Comparative biochemistry of iron-sulfur proteins and dinitrogen fixation, in *Iron Sulfur Proteins,* Vol. 1, Lovenberg, W., Ed., Academic Press, New York, 1973, 65.

16. **Bulen, W. A., Burns, R. C., and LeComte, J. R.,** Nitrogen fixation: hydrosulfite as electron donor with cell-free preparations of *Azotobacter vinelandii* and *Rhodospirillum rubrum, Proc. Natl. Acad. Sci. U.S.A.,* 53, 532, 1965.
17. **Burns, R. C. and Bulen, W. A.,** A procedure for the preparation of extracts from *Rhodospirillum rubrum* catalyzing nitrogen reduction and ATP-dependent hydrogen evolution, *Arch. Biochem. Biophys.,* 113, 461, 1966.
18. **Burris, R. H.,** Progress in the biochemistry of nitrogen fixation, *Proc. R. Soc. London Ser. B,* 172, 339, 1969.
19. **Burris, R. H.,** Energetics of biological N_2 fixation, in *Biological Solar Energy Conversion,* Mitsui, A., Miyachi, S., San Pietro, A., and Tamura, S., Eds., Academic Press, New York, 1977, 275.
20. **Kessler, E.,** Effect of anaerobiosis on photosynthetic reactions and nitrogen metabolism of algae with and without hydrogenase, *Arch. Mikrobiol.,* 93, 91, 1973.
21. **Bishop, N. I.,** Partial reactions of photosynthesis and photoreduction, *Annu. Rev. Plant Physiol.,* 17, 185, 1966.
22. **Bishop, N. I. and Jones, L. W.,** Alternate fates of photochemical reducing power generated in photosynthesis: hydrogen production and nitrogen fixation, *Curr. Top. Bioenerg.,* 8 (Part B), 3, 1978.
23. **Kessler, E.,** Hydrogenase, photoreduction and anaerobic growth, in *Algal Physiology and Biochemistry,* Stewart, W. D. P., Ed., Blackwell Scientific, Oxford, 1974, 456.
24. **Kessler, E.,** Hydrogen metabolism of eukaryotic organisms, in *Microbial Production and Utilization of Gases,* Schlegel, H. G., Gottschalk, G., and Pfennig, N., Eds., Akademie der Wissenschaften zu Göttingen, Göttingen, 1976, 247.
25. **Spruit, C. J. P.,** Photoreduction and anaerobiosis, in *Physiology and Biochemistry of Algae,* Lewin, R. A., Ed., Academic Press, New York, 1962, 47.
26. **Gest, H.,** Metabolic aspects of bacterial photosynthesis, in *Bacterial Photosynthesis,* Gest, H., San Pietro, A., and Vernon, L. P., Eds., The Antioch Press, Yellow Springs, Ohio, 1963, 129.
27. **Gest, H.,** Energy conversion and generation of reducing power in bacterial photosynthesis, *Adv. Microbial Physiol.,* 243, 1971.
28. **Gray, C. T. and Gest, H.,** Biological formation of molecular hydrogen, *Science,* 148, 186, 1965.
29. **Keister, D. L. and Fleischman, D. E.,** Nitrogen fixation in photosynthetic bacteria, *Photophysiology,* 8, 157, 1973.
30. **Pfennig, N.,** Photosynthetic bacteria, *Annu. Rev. Microbiol.,* 21, 285, 1967.
31. **van Niel, C. B.,** The present status of the comparative study of photosynthesis, *Annu. Rev. Plant Physiol.,* 13, 1, 1962.
32. **Yoch, D. C. and Arnon, D. I.,** Biological nitrogen fixation by photosynthetic bacteria, in *The Biology of Nitrogen Fixation,* Quispel, A., Ed., Elsevier, New York, 1974, 687.
33. **Healey, F. P.,** Hydrogen evolution by several algae, *Planta,* 91, 220, 1970.
34. **Ben-Amotz, A., Erbes, D. L., Riederer-Henderson, M. A., Peavey, D. G., and Gibbs, M.,** H_2 metabolism in photosynthetic organisms. I. Dark H_2 evolution and uptake by algae and mosses, *Plant Physiol.,* 56, 72, 1975.
35. **Bishop, N. I., Frick, M., and Jones, L. W.,** Photohydrogen production in green algae: water serves as the primary substrate for hydrogen and oxygen production, in *Biological Solar Energy Conversion,* Mitsui, A., Miyachi, S., San Pietro, A., and Tamura, S., Eds., Academic Press, New York, 1977, 3.
36. **Eisbrenner, G., Distler, E., Floener, L., and Bothe, H.,** The occurrence of the hydrogenase in some blue-green algae, *Arch. Microbiol.,* 118, 117, 1978.
37. **Tel-Or, E., Luijk, L. W., and Packer, L.,** Hydrogenase in N_2-fixing cyanobacteria, *Arch. Biochem. Biophys.,* 185, 185, 1978.
38. **Hattori, A.,** Effect of hydrogen on nitrite reduction by *Anabaena cylindrica,* in *Studies on Microalgae and Photosynthetic Bacteria* (Special issue of Plant Cell Physiol.), Tokyo University Press, Tokyo, 1963, 485.
39. **Belkin, S. and Padan, E.,** Hydrogen metabolism in the facultative anoxygenic cyanobacteria (blue-green algae) *Oscillatoria limnetica* and *Aphanothece halophytica, Arch. Microbiol.,* 116, 109, 1978.
40. **Ward, M. A.,** Whole cell and cell-free hydrogenases of algae, *Phytochemistry,* 9, 259, 1970.
41. **Frenkel, A., Gaffron, H., and Battley, E. H.,** Photosynthesis and photoreduction by the blue-green alga, *Synechococcus elongatus,* Näg, *Biol. Bull. (Woods Hole, Mass.),* 99, 157, 1950.
42. **Frenkel, A. W. and Rieger, C.,** Photoreduction in algae, *Nature (London),* 167, 1030, 1951.
43. **Hartman, H. and Krasna, A. I.,** Studies on the 'Adaptation' of hydrogenase in *Scenedesmus, J. Biol. Chem.,* 238, 749, 1963.
44. **Krasna, A. I. and Rittenberg, D.,** The mechanism of action of the enzyme hydrogenase, *J. Am. Chem. Soc.,* 76, 3015, 1954.
45. **Kessler, E. and Czygan, F.-C.,** Physiologische und biochemische Beiträge zur Taxonomie der Gattungen *Ankistrodesmus* und *Scenedesmus.* I. Hydrogenase, Sekundär-Carotinoide und Gelatine-Verflüssigung, *Arch. Mikrobiol.,* 55, 320, 1967.

46. **Abeles, F. B.**, Cell-free hydrogenase from *Chlamydomonas, Plant Physiol.*, 39, 169, 1964.
47. **Frenkel, A.**, A study of the hydrogenase systems of green and blue-green algae, *Biol. Bull., (Woods Hole, Mass.)*, 97, 261, 1949.
48. **Kessler, E.**, Physiologische und biochemische Beiträge zur Taxonomie der Gattung *Chlorella*. III. Merkmale von 8 autotrophen Arten, *Arch. Mikrobiol.*, 55, 346, 1967.
49. **Kessler, E. and Zweier, I.**, Physiologische und biochemische Beiträge zur Taxonomie der Gattung *Chlorella*. V. Die auxotrophen und mesotrophen Arten, *Arch. Microbiol.*, 79, 44, 1971.
50. **Kessler, E. and Maifarth, H.**, Occurrence and activity of hydrogenase in some green algae, *Arch. Mikrobiol.*, 37, 215, 1960.
51. **Sodomkova, M.** Physiologische und biochemische Charakterisierung einiger *Coelastrum* -Arten, *Arch. Protistenkd.*, 111, 223, 1969.
52. **Bose, S. K. and Gest, H.**, Hydrogenase and light stimulated electron transfer reactions in photosynthetic bacteria, *Nature (London)*, 195, 1168, 1962.
53. **Hoare, D. S. and Hoare, S. L.**, Hydrogen metabolism by *Rhodomicrobium vannielii, J. Bacteriol.*, 100, 1124, 1969.
54. **Klemme, J. H.**, Untersuchungen zur Photoautotrophie mit molekularem Wasserstoff bei neuisolierten schwefelfreien Purpurbakterien, *Arch. Mikrobiol.*, 64, 29, 1968.
55. **Weaver, P. F., Wall, J. D., and Gest, H.**, Characterization of *Rhodopseudomonas capsulata, Arch. Microbiol.*, 105, 207, 1975.
56. **Siefert, E. and Pfennig, N.**, Hydrogen metabolism and nitrogen fixation in wild type and Nif^- mutants of *Rhodopseudomonas acidophila, Biochimie*, 60, 261, 1978.
57. **Hendley, D. D.**, Endogenous fermentation in thiorhodaceae, *J. Bacteriol.*, 70, 625, 1955.
58. **Thiele, H. H.**, Sulfur metabolism in thiorhodaceae. IV. Assimilatory reduction of sulfate by *Thiocapsa floridana* and *Chromatium* species, *Antonie van Leeuwenhoek J. Microbiol. Serol.*, 34, 341, 1968.
59. **Nakamura, H.**, Weitere Untersuchungen über den Wasserstoffumsatz bei den Purpurbakterien, nebst einer Bemerkung über die gegenseitige Beziehung zwischen Thio- und Athiorhodaceen. Beiträge zur Stoffwechselphysiologie der Purpurbakterien. V, *Acta Phytochim.*, 11, 109, 1939.
60. **Gogotov, I. N., Zorin, N. A., and Bogorov, L. V.**, Hydrogen metabolism and the ability for nitrogen fixation in *Thiocapsa roseopersicina, Mikrobiologiya*, 43, 5, 1974.
61. **Larsen, H.**, On the culture and general physiology of the green sulfur bacteria, *J. Bacteriol.*, 64, 187, 1952.
62. **Lippert, K. D. and Pfennig, N.**, Die Verwertung von molekularem Wasserstoff durch *Chlorobium thiosulfatophilum, Arch. Mikrobiol.*, 65, 29, 1969.
63. **Kondratieva, E. N. and Gogotov, I. N.**, Production of hydrogen by green photosynthetic bacteria (*Chloropseudomonas*), *Nature (London)*, 221, 83, 1969.
64. **Umbereit, W. W., Burris, R. H., and Stauffer, J. F.**, *Manometric and Biochemical Techniques*, 5th ed., Burgess, Minneapolis, 1972, 387.
65. **Peck, H. D., Jr. and Gest, H.**, A new procedure for assay of bacterial hydrogenases, *J. Bacteriol.*, 71, 70, 1956.
66. **Davis, D. D. and Stevenson, K. L.**, A recording gas microvolumeter, *J. Chem. Educ.*, 54, 394, 1977.
67. **Hillmer, P. and Gest, H.**, H_2 metabolism in the photosynthetic bacterium *Rhodopseudomonas capsulata*: H_2 production by growing cultures, *J. Bacteriol.*, 129, 724, 1977.
68. **Wang, R., Healey, F. P., and Myers, J.**, Amperometric measurement of hydrogen evolution in *Chlamydomonas, Plant Physiol.*, 48, 108, 1971.
69. **Greenbaum, E.**, The molecular mechanisms of photosynthetic hydrogen and oxygen production, in *Biological Solar Energy Conversion*, Mitsui, A., Miyachi, S., San Pietro, A., and Tamura, S., Eds., Academic Press, New York, 1977, 101.
70. **Yagi, T.**, Separation of hydrogenase-catalyzed hydrogen-evolution system from electron-donating system by means of enzymic electric cell technique, *Proc. Natl. Acad. Sci. U.S.A.*, 73, 2947, 1976.
71. **Stuart, T. S., Herold, E. W., Jr., and Gaffron, H.**, A simple combination mass spectrometer inlet and oxygen electrode chamber for sampling gasses dissolved in liquids, *Anal. Biochem.*, 46, 91, 1972.
72. **Krasna, A. I.**, Oxygen-stable hydrogenase and assay, in *Methods in Enzymology*, Vol. 53, Fleischer, S. and Packer, L., Eds., Academic Press, New York, 1978, 296.
73. **Yu, L. and Wolin, M. J.**, Hydrogenase measurement with photochemically reduced methyl viologen, *J. Bacteriol.*, 98, 51, 1969.
74. **Stuart, T. S. and Gaffron, H.**, The mechanism of hydrogen photoproduction by several algae. II. The contribution of photosystem II, *Planta*, 106, 101, 1972.
75. **Weissman, J. C. and Benemann, J. R.**, Hydrogen production by nitrogen-starved cultures of *Anabaena cylindrica, Appl. Environ. Microbiol.*, 33, 123, 1977.
76. **Bothe, H., Tennigkeit, J., Eisbrenner, G., and Yates, M. G.**, The hydrogenase-nitrogenase relationship in the blue-green alga *Anabaena cylindrica, Planta*, 133, 237, 1977.

77. **Daday, A., Platz, R. A., and Smith, G. D.** Anaerobic and aerobic hydrogen gas formation by the blue-green alga *Anabaena cylindrica, Appl. Environ. Microbiol.,* 34, 478, 1977.
78. **Jeffries, T. W., Timourian, H., and Ward, R. L.,** Hydrogen production by *Anabaena cylindrica*: effects of varying ammonium and ferric ions, pH and light, *Appl. Environ. Microbiol.,* 35, 704, 1978.
79. **Lambert, G. R. and Smith, G. D.,** Hydrogen formation by marine blue-green algae, *FEBS Lett.,* 83, 159, 1977.
80. **Jones, L. W. and Bishop, N. I.,** Simultaneous measurement of oxygen and hydrogen exchange from the blue-green alga *Anabaena, Plant Physiol.,* 57, 659, 1976.
81. **Spiller, H., Ernst, A., Kerfin, W., and Böger, P.,** Increase and stabilization of photoproduction of hydrogen in *Nostoc muscorum* by photosynthetic electron transport inhibitors, *Z. Naturforsch. Teil C,* 33, 541, 1978.
82. **Mitsui, A. and Kumazawa, S.,** Hydrogen production by marine photosynthetic organisms as a potential energy source, in *Biological Solar Energy Conversion,* Mitsui, A., Miyachi, S., San Pietro, A., and Tamura, S., Eds., Academic Press, New York, 1977, 23.
83. **Stiffler, H. J. and Gest, H.,** Effects of light intensity and nitrogen growth source on hydrogen metabolism in *Rhodospirillum rubrum, Science,* 120, 1024, 1954.
84. **Schick, H.-J.,** Interrelationship of nitrogen fixation, hydrogen evolution and photoreduction in *Rhodospirillum rubrum, Arch. Mikrobiol.,* 75, 102, 1971.
85. **Gogotov, I. N. and Zorin, N. A.,** Hydrogen metabolism and hydrogenase activity of *Rhodospirillum rubrum, Mikrobiologiya,* 41, 947, 1972.
86. **Gorrell, T. E. and Uffen, R. L.,** Light-dependent and light-independent production of hydrogen gas by photosynthesizing *Rhodospirillum rubrum* mutant C, *Photochem. Photobiol.,* 27, 351, 1978.
87. **Gogotov, I. N., Mitkina, T. V., and Glinskii, V. P.,** Effect of ammonium on hydrogen liberation and nitrogen fixation by *Rhodopseudomonas palustris, Mikrobiologiya,* 43, 586, 1974.
88. **Arnon, D. I., Losada, M., Nozaki, M., and Tagawa, K.,** Photoproduction of hydrogen, photofixation of nitrogen and a unified concept of photosynthesis, *Nature (London),* 190, 601, 1961.
89. **Gogotov, I. N.,** Relationships in hydrogen metabolism between hydrogenase and nitrogenase in phototrophic bacteria, *Biochimie,* 60, 267, 1978.
90. **Healey, F. P.,** The mechanism of hydrogen evolution by *Chlamydomonas moewusii, Plant Physiol.,* 45, 153, 1970.
91. **Kaltwasser, H., Stuart, T. S., and Gaffron, H.,** Light-dependent hydrogen evolution by *Scenedesmus, Planta,* 89, 309, 1969.
92. **King, D., Erbes, D. L., Ben-Amotz, A., and Gibbs, M.,** The mechanism of hydrogen photoevolution in photosynthetic organisms, in *Biological Solar Energy Conversion,* Mitsui, A., Miyachi, S., San Pietro, A., and Tamura, S., Eds., Academic Press, New York, 1977, 69.
93. **Senger, H. and Bishop, N. I.,** Observations on the photohydrogen producing activity during the synchronous cell cycle of *Scenedesmus obliquus, Planta,* 145, 53, 1979.
94. **Stuart, T. S. and Gaffron, H.,** The kinetics of hydrogen photoproduction by adapted, *Scenedesmus, Planta,* 100, 228, 1971.
95. **Stuart, T. S. and Gaffron, H.,** The mechanisms of hydrogen photoproduction by several algae. I. The effect of inhibitors of photophosphorylation, *Planta,* 106, 91, 1972.
96. **Klein U. and Betz, A.,** Fermentative metabolism of hydrogen-evolving *Chlamydomonas moewusii, Plant Physiol.,* 61, 953, 1978.
97. **Krasna, A. I.,** Catalytic and structural properties of the enzyme hydrogenase and its role in biophotolysis of water, in *Biological Solar Energy Conversion,* Mitsui, A., Miyachi, S., San Pietro, S., and Tamura, S., Eds., Academic Press, New York, 1977, 53.
98. **Benemann, J. R. and Weare, N. M.,** Hydrogen evolution by nitrogen-fixing *Anabaena cylindrica* cultures, *Science,* 184, 174, 1974.
99. **Bothe, H., Distler, E., and Eisbrenner, G.,** Hydrogen metabolism in blue-green algae, *Biochimie,* 60, 227, 1978.
100. **Wolk, C. P.,** Movement of carbon from vegetative cells to heterocysts in *Anabaena cylindrica, J. Bacteriol.,* 96, 2138, 1968.
101. **Winkenbach, F. and Wolk, C. P.,** Activities of enzymes of the oxidative and the reductive pentose phosphate pathways in heterocysts of a blue-green alga, *Plant Physiol.,* 52, 480, 1973.
102. **Peterson, R. B. and Burris, R. H.,** Hydrogen metabolism in isolated heterocysts of *Anabaena* 7120, *Arch. Microbiol.,* 116, 125, 1978.
103. **Smith, R. V., Noy, R. J., and Evans, M. C. W.,** Physiological electron donor systems to the nitrogenase of the blue-green alga *Anabaena cylindrica, Biochim. Biophys. Acta,* 253, 104, 1971.
104. **Leach, C. K. and Carr, N. G.,** Pyruvate: ferredoxin oxidoreductase and its activation by ATP in the blue-green alga *Anabaena variabilis, Biochim. Biophys. Acta,* 245, 165, 1971.

105. **Bothe, H., Falkenberg, B., and Nolteernsting, U.,** Properties and function of the pyruvate: ferredoxin oxidoreductase from the blue-green alga *Anabaena cylindrica, Arch. Mikrobiol.,* 96, 291, 1974.
106. **Bothe, H. and Loos, E.,** Effect of far red light and inhibitors on nitrogen fixation and photosynthesis in the blue-green alga *Anabaena cylindrica, Arch. Mikrobiol.,* 86, 241, 1972.
107. **Belkin, S. and Padan, E.,** Sulfide-dependent hydrogen evolution in the cyanobacterium *Oscillatoria limnetica, FEBS Lett.,* 94, 291, 1978.
108. **Gest, H., Ormerod, J. G., and Ormerod, K. S.,** Photometabolism of *Rhodospirillum rubrum*: light-dependent dissimilation of organic compounds to carbon dioxide and molecular hydrogen by an anaerobic citric acid cycle, *Arch. Biochem. Biophys.,* 97, 21, 1962.
109. **Kohlmiller, E. F. and Gest, H.,** A comparative study of the light and dark fermentations of organic acids by *Rhodospirillum rubrum, J. Bacteriol.,* 61, 269, 1951.
110. **Gest, H., Kamen, M. D., and Bregoff, H. M.,** Studies on the metabolism of photosynthetic bacteria. V. Photoproduction of hydrogen and nitrogen fixation by *Rhodospirillum rubrum, J. Biol. Chem.,* 182, 153, 1950.
111. **Gorrell, T. E. and Uffen, R. L.,** Fermentative metabolism of pyruvate by *Rhodospirillum rubrum* after anaerobic growth in darkness, *J. Bacteriol.,* 131, 533, 1977.
112. **Schön, G. and Biedermann, M.,** Growth and adaptive hydrogen production of *Rhodospirillum rubrum* (F_1) in anaerobic dark cultures, *Biochim. Biophys. Acta,* 304, 65, 1973.
113. **Uffen, R. L.,** Growth properties of *Rhodospirillum rubrum* mutants and fermentation of pyruvate in anaerobic, dark conditions, *J. Bacteriol.,* 116, 874, 1973.
114. **Uffen, R. L., Sybesma, C., and Wolfe, R. S.,** Mutants of *Rhodospirillum rubrum* obtained after long-term anaerobic, dark growth, *J. Bacteriol.,* 108, 1348, 1971.
115. **Voelskow, H. and Schön, G.,** Pyruvate fermentation in light-grown cells of *Rhodospirillum rubrum* during adaptation to anaerobic dark conditions, *Arch. Mikrobiol.,* 119, 129, 1978.
116. **Bregoff, H. M. and Kamen, M. D.,** Studies on the metabolism of photosynthetic bacteria. XIV. Quantitative relations between malate dissimilation, photoproduction of hydrogen and nitrogen metabolism in *Rhodospirillum rubrum, Arch. Biochem. Biophys.,* 36, 202, 1952.
117. **Ormerod, J. G., Ormerod, K. S., and Gest, H.,** Light-dependent utilization of organic compounds and photoproduction of molecular hydrogen by photosynthetic bacteria: relationships with nitrogen metabolisms, *Arch. Biochem. Biophys.,* 94, 449, 1961.
118. **Paschinger, H.,** A changed nitrogenase activity in *Rhodospirillum rubrum* after substitution of tungsten for molybdenum, *Arch. Mikrobiol.,* 101, 379, 1974.
119. **Hillmer, P. and Gest, H.,** H_2 metabolism in the photosynthetic bacterium *Rhodopseudomonas capsulata*: production and utilization of H_2 by resting cells, *J. Bacteriol.,* 129, 732, 1977.
120. **Kelley, B. C., Meyer, C. M., Gandy, C. G., and Vignais, P. M.,** Hydrogen recycling by *Rhodopseudomonas capsulata, FEBS Lett.,* 81, 281, 1977.
121. **Qardi, S. M. H. and Hoare, D. S.,** Formic hydrogenlyase and the photoassimilation of formate by a strain of *Rhodopseudomonas palustris, J. Bacteriol.,* 95, 2344, 1968.
122. **Bennett, R., Rigopoulos, N., and Fuller, R. C.,** The pyruvate phosphoroclastic reaction and light-dependent nitrogen fixation in bacterial photosynthesis, *Proc. Natl. Acad. Sci. U.S.A.,* 52, 762, 1964.
123. **Newton, J. W. and Wilson, P. W.,** Nitrogen fixation and photoproduction of molecular hydrogen by Thiorhodaceae, *Antonie van Leeuwenhoek J. Microbiol. Serol.,* 19, 71, 1953.
124. **Bennett, R. and Fuller, R. C.,** The pyruvate phosphoroclastic reaction in *Chromatium.* A probable role for ferredoxin in a photosynthetic bacterium, *Biochem. Biophys. Res. Commun.,* 16, 300, 1964.
125. **Weaver, P., Tinker, K., and Valentine, R. C.,** Ferredoxin linked DPN reduction by the photosynthetic bacteria *Chromatium* and *Chlorobium, Biochem. Biophys. Res. Commun.,* 21, 195, 1965.
126. **Yoch, D. C. and Arnon, D. I.,** The nitrogen fixation system of photosynthetic bacteria. II. *Chromatium* nitrogenase activity linked to photochemically generated assimilatory power, *Biochim. Biophys. Acta,* 197, 180, 1970.
127. **Arnon, D. I. and Yoch, D. C.,** Photosynthetic bacteria, in *The Biology of Nitrogen Fixation,* Quispel, A., Ed., Elsevier, New York, 1974, 168.
128. **Thauer, R. K., Jungermann, K., and Decker, K.,** Energy conservation in chemotrophic anaerobic bacteria, *Bacteriol. Rev.,* 41, 100, 1977.
129. **Mitsui, A. and Arnon, D. I.,** Photoproduction of hydrogen gas by isolated chloroplasts in relation to cyclic and non-cyclic electron flow, *Plant Physiol.,* 37 (Suppl.), IV, 1962.
130. **Arnon, D. I., Mitsui, A., and Paneque, A.,** Photoproduction of hydrogen gas coupled with photosynthetic photophosphorylation, *Science,* 134, 1425, 1961.
131. **Mitsui, A.,** Utilization of solar energy for hydrogen production by cell free system of photosynthetic organisms, in *Hydrogen Energy,* Part A, Veziroglu, T. N., Ed., Plenum Press New York, 1975, 309.
132. **Arnon, D. I.,** Photosynthetic electron transport and phosphorylation in chloroplasts, in *Photosynthetic Mechanism in Green Plants,* Kok, B. and Jagendorf, A. T., Eds., National Academy of Science — National Research Council, Washington, D.C., 1963, 195.

133. **Whatley, F. R. and Grant, B. R.**, Photoreduction of methyl viologen by spinach chloroplasts, *Fed. Proc.*, and oral presentation at the Federation Annual Meeting, Atlantic City, 1964.
134. **Krampitz, L. O.**, Hydrogen Production by Photosynthesis and Hydrogenase Activity, NSF/RANN Report N-73-013, Biophotolysis of Water, NSF/RANN Report N-73-014, 1973.
135. **Benemann, J. R., Berenson, J. A., Kaplan, N. O., and Kamen, M. D.**, Hydrogen evolution by a chloroplast-ferredoxin-hydrogenase system, *Proc. Natl. Acad. Sci. U.S.A.*, 70, 2317, 1973.
136. **Ben-Amotz, A. and Gibbs, M.**, H_2 metabolism in photosynthetic organisms. II. Light-dependent H_2 evolution by preparations from *Chlamydomonas, Scenedesmus* and Spinach, *Biochem. Biophys. Res. Commun.*, 64, 355, 1975.
137. **Kitajima, M. and Butler, W. L.**, Microencapsulation of chloroplast particles, *Plant Physiol.*, 57, 746, 1976.
138. **Hoffmann, D., Thauer, R., and Trebst, A.**, Photosynthetic hydrogen evolution by spinach chloroplasts coupled to a *Clostridium* hydrogenase, *Z. Naturforsch. Teil C*, 32, 257, 1977.
139. **Packer, L.**, Problems in the stabilization of the *in vitro* photochemical activity of chloroplasts used for H_2 production, *FEBS Lett.*, 64, 17, 1976.
140. **Fry, I., Papageorgiou, G., Tel-Or, E. and Packer, L.**, Reconstitution of a system for H_2 evolution with chloroplasts, ferredoxin and hydrogenase, *Z. Naturforsch. Teil C*, 32, 110, 1977.
141. **Rao, K. K., Rosa, L., and Hall, D. O.**, Prolonged production of hydrogen gas by a chloroplast biocatalytic system, *Biochem. Biophys. Res. Commun.*, 68, 21, 1976.
142. **Rao, K. K., Gogotov, I. N., and Hall, D. O.**, Hydrogen evolution by chloroplast-hydrogenase systems: improvements and additional observations, *Biochimie*, 60, 291, 1978.
143. **Reeves, S. G., Rao, K. K., Rosa, L., and Hall, D. O.**, Biocatalytic production of hydrogen, in *Microbial Energy Conversion*, Schlegel, H. G. and Barnea, J., Eds., Pergamon Press, New York, 1977, 235.
144. **Bothe, H., Tennigkeit, J., and Eisbrenner, G.**, The utilization of molecular hydrogen by the blue-green alga *Anabaena cylindrica, Arch. Microbiol.*, 114, 43, 1977.
145. **Wolk, C. P. and Wojciuch, E.**, Photoreduction of acetylene by heterocysts, *Planta*, 97, 126, 1971.
146. **Benemann, J. R. and Weare, N. M.**, Nitrogen fixation by *Anabaena cylindrica*. III. Hydrogen-supported nitrogenase activity, *Arch. Microbiol.*, 101, 401, 1974.
147. **Tel-Or, E., Luijk, L. W., and Packer, L.**, An inducible hydrogenase in cyanobacteria enhances N_2 fixation, *FEBS Lett.*, 78, 49, 1977.
148. **Peterson, R. B. and Wolk, C. P.**, Localization of an uptake hydrogenase in Anabaena, *Plant Physiol.*, 61, 688, 1978.
149. **Meyer, J., Kelley, B. C., and Vignais, P. M.**, Nitrogen fixation and hydrogen metabolism in photosynthetic bacteria, *Biochimie*, 60, 245, 1978.
150. **Izawa, S.**, Hydrogenase reactions in *Rhodopseudomonas palustris, Plant Cell Physiol.* 3, 23, 1962.
151. **Dixon, R. O. D.**, Hydrogenases and efficiency of nitrogen fixation in aerobes, *Nature (London)*, 262, 173, 1976.
152. **Radmer, R. and Kok, B.**, Photosynthesis: limited yields, unlimited dreams, *BioScience*, 27, 599, 1977.
153. **Jeffries, T. W. and Leach, K. L.**, Intermittent illumination increases biophotolytic hydrogen yield by *Anabaena cylindrica, Appl. Environ. Microbiol.*, 35, 1228, 1978.
154. **Kumazawa, S. and Mitsui, A.**, Hydrogenase and nitrogenase participation in the photoproduction of hydrogen by marine blue-green algae, *Plant Physiol.*, 61 (Suppl.), 77, 1978.
155. **Thauer, R.**, Limitation of microbial H_2-formation via fermentation, in *Microbial Energy Conversion*, Schlegel, H. G. and Barnea, J., Eds., Pergamon Press, New York, 1977, 201.
156. **Shanmugam, K. T., O'Gara, F., Andersen, K., and Valentine, R. C.**, Biological nitrogen fixation, *Annu. Rev. Plant Physiol.*, 29, 263, 1978.
157. **Weare, N. M.**, The photoproduction of H_2 and NH_4^+ fixed from N_2 by a derepressed mutant of *Rhodospirillum rubrum, Biochem. Biophys. Acta*, 502, 486, 1978.
158. **Lappi, D. A., Stolzenbach, F. E., Kaplan, N. O., and Kamen, M. D.**, Immobilization of hydrogenase on glass beads, *Biochem. Biophys. Res. Commun.*, 69, 878, 1976.
159. **Yagi, T.**, Use of an enzymic electric cell and immobilized hydrogenase in the study of the biophotolysis of water to produce hydrogen, in *Biological Solar Energy Conversion*, Mitsui, A., Miyachi, S., San Pietro, A., and Tamura, S., Eds., Academic Press, New York, 1977, 61.
160. **Klibanov, A. M., Kaplan, N. O., and Kamen, M. D.**, A rationale for stabilization of oxygen-labile enzymes: application to a clostridial hydrogenase, *Proc. Natl. Acad. Sci. U.S.A.*, 75, 3640, 1978.
161. **Berenson, J. A. and Benemann, J. R.**, Immobilization of hydrogenase and ferredoxins on glass beads, *FEBS Lett.*, 76, 105, 1977.
162. **Adams, M. W. W., Reeves, S. G., Hall, D. O., Christou, C., Ridge, B., and Rydon, H. N.**, Biological activity of synthetic tetranuclear iron-sulphur analogues of the active sites of ferrodoxins, *Biochem. Biophys. Res. Commun.*, 79, 1184, 1977.

163. **Pfennig, N.,** *Rhodospirillum tenue* sp. n., a new species of the purple nonsulfur bacteria, *J. Bacteriol.,* 99, 619, 1969.
164. **Pfennig, N. and Trüper, H. G.,** The phototrophic bacteria, in *Bergey's Manual of Determinative Bacteriology,* 8th ed., Buchanan, R. E. and Gibbons, N. E., Eds., Williams & Wilkins, Baltimore, 1974, 24.
165. **Hansen, T. A. and Veldkamp, H.,** *Rhodopseudomonas sulfidophila,* nov. spec., a new species of the purple nonsulfur bacteria, *Arch. Mikrobiol.,* 92, 45, 1973.
166. **Trüper, H. G.,** *Ectothiorhodospira mobilis* Pelsh, a photosynthetic sulfur bacterium depositing sulfur outside the cells, *J. Bacteriol.,* 95, 1910, 1968.
167. **Calvin, M.,** Solar energy by photosynthesis, *Science,* 184, 375, 1974.
168. **Mitsui, A., Miyachi, S., San Pietro, A., and Tamura, S.,** Eds., *Biological Solar Energy Conversion,* Academic Press, New York, 1977, 454.
169. **Lien, S. and San Pietro, A.,** *An Inquiry into Biophotolysis of Water to Produce Hydrogen,* Department of Plant Science, Indiana University, 1975, 50.
170. **Hall, D. O.,** Photobiological energy conversion, *FEBS Lett.,* 64, 6, 1976.
171. **Zajic, J. E., Kosaric, N., and Brosseau, J. D.,** Microbial production of hydrogen, *Adv. Biochem. Eng.,* 9, 57, 1978.
172. **Mitsui, A.,** Photoproduction of hydrogen via microbial and biochemical processes, in *Symp. Proc. Hydrogen Energy Fundamentals,* S-2, Veziroglu, T. N., Ed., University of Miami Press, Miami, 1975, 31.
173. **Mitsui, A.,** Marine photosynthetic organisms as potential energy resources: research on nitrogen fixation and hydrogen production, in *Proc. 5th Int. Ocean Development Conference,* IODC Organizing Committee, Tokyo, 1978, Bi 29.

PHOTORESPIRATION IN PHOTOSYNTHETIC BACTERIA

Takashi Akazawa, Sumio Asami, and Tetsuko Takabe

INTRODUCTION

Among the photosynthetic bacteria in the biosphere, the following three classes have been used frequently in photosynthesis research:

1. Purple sulfur bacteria (Chromatiaceae)
2. Green sulfur bacteria (Chlorobiaceae)
3. Purple nonsulfur bacteria (Rhodospirillaceae)[1-4]

Intensive investigations have been undertaken in many laboratories to unveil the mechanism of their carbon metabolism as well as photosynthetic electron-transport system. The photosynthetic reactions (light-dependent CO_2 fixation) carried out by autotrophic bacteria proceed under an anaerobic environment. This feature is different from the photosynthetic process in green plants. It is often claimed that the magnitude of the solar energy conversion to biomass production by bacterial photosynthesis is large. Therefore, O_2 effects on the activities of photosynthetic bacteria is an important factor in net carbon assimilation.

Both prokaryotic blue-green algae (cyanophyta) and all other eukaryotic autotrophs have Photosystem I and II, but photosynthetic bacteria lack Photosystem II. In the latter, instead of H_2O, reduced sulfur compounds such as S^{2-}, $S_2O_3^{2-}$ or organic acids, e.g., malate and succinate, are utilized as the electron donor and are equivalent to H_2A in the following reaction. The overall reaction for bacterial photosynthesis is generally expressed as

$$CO_2 + 2H_2A \xrightarrow[\text{bacteriochlorophyll}]{\text{light}} (CH_2O) + 2A + H_2O$$

Clearly there is no O_2 evolution in bacterial photosynthesis, and it will be readily recognized that the effect of O_2 on the process is distinguishable from that of photosynthesis by green plants. By examining the inhibitory effects of O_2 on photosynthesis, the adaptability of microorganisms to O_2 environments, and the mechanism of O_2 toxicity, one may envision the evolutionary origin of photorespiration in chlorophyll-bearing green plants.

INHIBITORY EFFECT OF O_2 ON BACTERIAL PHOTOSYNTHESIS

There are diversities in the mechanism of photosynthetic CO_2 fixation among different types of photosynthetic bacteria which are reflected in different responses of their photosynthetic reactions to O_2 (see Table 1). A variety of experiments indicate the operation of the reductive pentose phosphate cycle (C_3 cycle) in *Chromatium vinosum*.[5] Even though recent investigation using $^{13}CO_2$ supports the double carboxylation pathway containing both RuBP carboxylase and PEP carboxylase in vivo,[6] the role of PEP carboxylase is believed to be anaplerotic in the overall CO_2 fixation reaction. The response of the photosynthetic reactions of *Chromatium* to O_2 is in many ways similar to that observed in green plants.[7] in 100% O_2, the photosynthetic rate is constant as a function of time for a short period of incubation, although the net CO_2

Table 1
OXYGEN EFFECTS ON BACTERIAL PHOTOSYNTHESIS

Bacteria	Growth condition	CO_2 fixation pathway	O_2 effect
Chromatium vinosum	Autotrophic	C_3 cycle[5]	Reversible inhibition[9] Photooxidative damage[20]
Chlorobium thiosulfatophilum	Autotrophic	Reductive carboxylic acid cycle[10]	Complete inhibition (50 μM O_2)[9]
Rhodospirillum rubrum	Heterotrophic (butyrate, malate)	C_3 cycle and reductive tricarboxylic acid cycle[22]	Reversible and/or irreversible inhibition[11]

fixation is one third to one half of that observed under anaerobic conditions;[8] the overall view is quite analogous to the competitive inhibition of photosynthesis by O_2 (Warburg effect) in green plants.[9]

In contrast, a view has prevailed that the reductive carboxylic acid cycle governs carbon assimilation in the green sulfur bacterium, *Chlorobium thiosulfatophilum*, in which the reduced ferredoxin (Fd_{red})-dependent reductive carboxylation reaction occupies the key role.[10] Photosynthesis is strongly inhibited by O_2, even at a very low concentration level, since Fd_{red} is subject to autooxidation under such environmental conditions.[9]

The purple nonsulfur bacterium, *Rhodospirillum rubrum*, grown on butyrate as a carbon source exhibits a high RuBP carboxylase activity, and it was shown that photosynthetic CO_2 fixation is reversibly inhibited by O_2.[11] However, the reaction in the bacterial cells grown on malate which contain much lower carboxylase activity is irreversibly inhibited by O_2.[11] It can be surmised that the reversible-type O_2 inhibition of photosynthesis (Warburg effect) occurs quite ubiquitously in divergent photosynthetic microorganisms in which the C_3 cycle functions.

O_2 EFFECT AND GLYCOLATE PRODUCTION IN PHOTOSYNTHETIC BACTERIA

RuBP Oxygenase

A currently prevailing view is that the competitive inhibition of photosynthetic CO_2 fixation by O_2 and the accompanying production of glycolate in green plants is governed by the RuBP carboxylase/oxygenase reactions.[12] As compiled in Table 2, in several bacterial species in which the C_3 cycle predominates, it was shown that the glycolate molecule is produced and excreted under aerobic or O_2-containing atmospheric environments. Results of a kinetic analysis of glycolate formation in *Chromatium vinosum* show that the underlying mechanism appears to be analogous to that of photorespiratory metabolism in green plants.[8,9] It was found that the RuBP carboxylase reaction exhibited by *Chromatium* is competitively inhibited by O_2 in a similar manner in the higher plant enzyme.[13] Furthermore, the nearly sole labeling with ^{18}O of the carboxyl group of the glycolate molecule produced by both *Chromatium*[14] and *Rhodospirillum rubrum*[11] incubated in an $^{18}O_2$-enriched atmosphere strongly supports that RuBP oxygenase is actively involved in glycolate production. However, other enzymic mechanism(s) such as transketolase-catalyzed reaction utilizing the superoxide anion radical (O_2^-) in glycolate synthesis cannot be totally excluded.[15] It remains to be determined whether O_2^- production coupled to Photosystem I is linked to the photooxidative (photorespiratory) formation of glycolate in *Chromatium* and other photosynthetic bacteria under O_2-containing atmosphere.

Glycolate Formation and Metabolism

In *Chromatium vinosum*, a majority of the glycolate molecules produced are extracellularly excreted, and a part is transformed into glycine via glyoxylate.[16] However, the bacterium is unable to directly utilize O_2 to oxidize glycolate, and the exact enzymic mechanism of the glycolate oxidation remains unknown.[17] Glycine also is excreted extracellularly, and there is experimental evidence that glycine is converted to serine; thus, the overall view of glycolate metabolism is substantially different from that observed in eukaryotic green plants.[18] Under a low O_2 pressure, there is no measurable extracellular excretion of either glycolate or glycine which is derived from products of photosynthetic CO_2 fixation. However, an exogenous addition of either nonradioactive glycolate or glycine to the bacterial suspension medium results in the labeling of these

Table 2
GLYCOLATE EXCRETION BY PHOTOSYNTHETIC BACTERIA

Bacteria	Growth condition	Glycolate excretion
Chromatium vinosum	Autotrophic	9.2 μmol/mg Bchl/hr (100% O_2)[8]
	Heterotrophic	1.2 μmol/mg Bchl/hr (100% O_2)[8]
	(malate, acetate)	0.5 μmol/mg Bchl/hr (100% O_2)[8]
Rhodospirillum rubrum	Heterotrophic (malate)	0.02 μmol/mg protein/hr (10% O_2)[21]
Rhodomicrobium vannielii	Heterotrophic (malate)	0.07 μmol/mg protein/hr (10% O_2)[21]
Alcaligenes eutrophus	Autotrophic	0.02 μmol/mg dry weight/hr (20% O_2)[19]

compounds in the extracellular fraction.[18] These results suggest that in the prokaryotic bacterium, *Chromatium vinosum*, both glycolate oxidation and its subsequent transformation to glycine are the rate-limiting step in the overall process under photorespiratory conditions.

In green plant cells, on the other hand, the glycolate pathway proceeds through the mutual interactions of three cell organelles, i.e., chloroplasts, peroxisome, and mitochondria, and effectively handles glycolate molecules produced under an O_2-containing atmosphere. In this scheme, a partial recovery of 3-PGA into the C_3 cycle in chloroplasts after a turn of the glycolate pathway is considered to be a mechanism for conserving carbon compounds which are drained off from the chloroplastic carbon pathway. In contrast, as typically seen in *Chromatium vinosum*, in the primitive phototrophic bacteria, there is neither the cellular compartmentation nor the labor division among different cell organelles, ensuing in a small metabolic capacity for the glycolate molecules produced.[18] One can visualize that the mechanism in photosynthetic bacteria is a prototype of the photorespiratory glycolate pathway which is more evolutionarily developed in multicellular green plants.

The activities of some enzymes involved in glycolate formation and its metabolism in several photosynthetic bacteria are summarized in Table 3. As presented in the table, the activity of RuBP oxygenase proved to be invariably high enough to sustain glycolate formation in these bacteria. It appears that the autotrophically grown cells of *Alcaligenes eutrophus* contain relatively high activities of P-glycolate phosphatase and glycolate oxidoreductase as compared with those in heterotrophically grown bacterial cells.[19] However, the role of RuBP oxygenase in glycolate formation is hypothetical, and the exact nature of the enzymic machinery participating in bacterial photorespiration remains to be studied.

PHOTOOXIDATIVE DAMAGE OF BACTERIAL PHOTOSYNTHESIS

The irreversible inhibition of O_2 of photosynthetic activity is another important facet of photorespiration, although this phenomenon is not visibly observed in the photosynthetic system of higher plants. On the other hand, the O_2 sensitivity of anaerobic photosynthetic bacteria such as *Chromatium vinosum* is a useful system to elucidate the general mechanism. The irreversible O_2 inhibition of photosynthetic activities in *Chromatium* cells appears to be based on the photodynamic action;[20] the effect is likely to be caused by active O_2 molecules such as O_2^- or $^1O_2$ which are produced in coupling with the electron transport system under a light and O_2 environment. All of these O_2 radicals appear to attack the site(s) of the photophosphorylation reaction.[20] Although the photosynthetic system will be universally affected by the same mechanism, the

Table 3
ACTIVITIES OF ENZYMES INVOLVED IN GLYCOLATE FORMATION AND METABOLISM

Enzyme	Bacteria	Growth condition	Activities
RuBP oxygenase (purified)	*Chromatium vinosum*	Autotrophic	0.09 μmol/mg protein/min (air)[7]
	Rhodospirillum rubrum	Heterotrophic (butyrate)	0.8 μmol/mg protein/min (100% O_2)[23]
			0.1 μmol/mg protein/min (air)[24]
RuBP oxygenase (crude extract)	*Chromatium vinosum*	Autotrophic	5 μmol/mg Bchl/hr (air)[14]
	Rhodospirillum rubrum	Heterotrophic	1.3 μmol/mg Bchl/hr (air)[11]
		(malate)	14.5 μmol/mg Bchl/hr (air)[11]
		(butyrate)	
P-glycolate phosphatase (crude extract)	*Chromatium vinosum*	Autotrophic	27 μmol/mg Bchl/hr[14]
	Rhodospirillum rubrum	Heterotrophic (malate)	1.74 μmol/mg protein/hr[21]
	Rhodomicrobium vannielii	Heterotrophic (malate)	8.39 μmol/mg protein/hr[21]
		Autotrophic	0.73 μmol/mg protein/hr[19]
	Alcaligenes eutrophus	Heterotrophic	0.03 μmol/mg protein/hr[19]
		(malate)	0.16 μmol/mg protein/hr[19]
		(fructose)	
Glycolate oxidoreductase (crude extract)	*Rhodospirillum rubrum*	Heterotrophic (malate)	0.26 μmol/mg protein/hr[21]
	Rhodospirillum vannielii	Heterotrophic (malate)	0.19 μmol/mg protein/hr[21]
	Alcaligenes eutrophus	Autotrophic	0.48 μmol/mg protein/hr[21]
		Heterotrophic	0.19 μmol/mg protein/hr[19]
		(pyruvate)	0.20 μmol/mg protein/hr[19]
		(fructose)	

photooxidative damaging effect is likely to be most prominent in anaerobic photoautotrophs, since their protective device is not highly developed such as that in the chloroplast system. The exogenous addition of $NaHCO_3$ to the *Chromatium* cells results in an effective protection against irreversible O_2 inhibition. It can be hypothesized that photorespiratory CO_2 evolution in *Chromatium* entails a partial protective mechanism which will effectively compensate the irreversible O_2 inhibition in vivo. Thus, one of the physiological roles of photorespiration in green plant cells can be visualized as this effect of CO_2 produced during the photorespiratory process.

ACKNOWLEDGMENTS

The authors express their gratitude for financial support to the Ministry of Education of Japan, the Toray Science Foundation (Tokyo), and the Nissan Science Foundation (Tokyo). Sumio Asami is a recipient of the JSPS Postdoctoral Fellowship (1978).

REFERENCES

1. **Van Niel, C. B.**, The chemoautotrophic and photosynthetic bacteria, *Annu. Rev. Microbiol.*, 8, 105, 1954.
2. **Pfennig, N.**, Photosynthetic bacteria, *Annu. Rev. Microbiol.* 21, 285, 1967.
3. **Hind, G. and Olson, J. M.**, Electron transport pathways in photosynthesis, *Annu. Rev. Plant Physiol.*, 19, 249, 1968.
4. **Gromet-Elhanan, Z.**, Electron transport and photophosphorylation in photosynthetic bacterium, in *Encyclopedia of Plant Physiology*, New Series, Vol. 5, Trebst, A. and Avron, M., Eds., Springer-Verlag, Berlin, 1977, 637.
5. **Fuller, R. C., Smillie, R. M., Sisler, E. C., and Kornberg, H. L.**, Carbon metabolism in *Chromatium*, *J. Biol. Chem.*, 236, 2140, 1961.
6. **Wong, W., Sackett, W. M., and Benedict, C. R.**, Isotope fractionation in photosynthetic bacteria during carbon dioxide assimilation. *Plant Physiol.*, 55, 475, 1975.
7. **Asami, S. and Akazawa, T.**, Oxidative formation of glycolic acid in photosynthesizing cells of *Chromatium*, *Plant Cell Physiol.*, 15, 571, 1974.
8. **Asami, S. and Akazawa, T.**, Biosynthetic mechanism of glycolate in *Chromatium*. I. Glycolate pathway, *Plant Cell Physiol.*, 16, 631, 1975.
9. **Takabe, T. and Akazawa, T.**, A comparative study on the effect of O_2 on photosynthetic carbon metabolism by *Chlorobium thiosulfatophilum* and *Chromatium vinosum*, *Plant Cell Physiol.*, 18, 753, 1977.
10. **Buchanan, B. B.**, Ferredoxin-linked carboxylation reaction, in *The Enzymes*, IV, Boyer, P. D., Ed., Academic Press, New York, 1972, 193.
11. **Takabe, T., Osmond, C. B., Summons, R. E., and Akazawa, T.**, Effect of oxygen on photosynthesis and biosynthesis of glycolate in photoheterotrophically grown cells of *Rhodospirillum rubrum*, *Plant Cell Physiol.*, 20, 233, 1979.
12. **Andrews, T. J. and Lorimer, G. H.**, Photorespiration still unavoidable?, *FEBS Lett.*, 90, 1, 1978.
13. **Takabe, T. and Akazawa, T.**, Oxidative formation of phosphoglycolate from ribulose 1,5-diphosphate catalysed by *Chromatium* ribulo 1,5-diphosphate carboxylase, *Biochem. Biophys. Res. Commun.*, 53, 1173, 1973.
14. **Lorimer, G. H., Osmond, C. B., Akazawa, T., and Asami, S.**, On the mechanism of glycolate synthesis by *Chromatium* and *Chlorella*, *Biochim. Biophys. Acta*, 185, 49, 1978.
15. **Asami, S. and Akazawa, T.**, Enzymic formation of glycolate in *Chromatium*. Role of superoxide radical in a transketolase-type mechanism, *Biochemistry*, 16, 2202, 1977.
16. **Asami, S., Takabe, T., and Akazawa, T.**, Biosynthetic mechanism of glycolate in *Chromatium*. IV. Glycolate-glycine transformation, *Plant Cell Physiol.*, 18, 149, 1977.
17. **Asami, S. and Akazawa, T.**, Biosynthetic mechanism of glycolate in *Chromatium*. III. Effect of α-hydroxy-2-pyridinemethanesulfonate, glycidate and cyanide on glycolate excretion in *Chromatium*, *Plant Cell Physiol.*, 17, 1119, 1976.
18. **Asami S. and Akazawa, T.**, Biosynthetic mechanism of glycolate in *Chromatium*. VI. Glycolate formation and metabolism under low O_2, *Plant Cell Physiol.*, 19, 1353, 1978.
19. **Codd, G. A., Bowien, B., and Schlegel, H. G.**, Glycolate production and excretion of *Alcaligenes eutrophus*, *Arch. Microbiol.*, 110, 167, 1976.
20. **Asami, S. and Akazawa, T.**, Photooxidative damage in photosynthetic activities of *Chromatium vinosum*, *Plant Physiol.*, 62, 981, 1978.
21. **Codd, G. A. and Turnbull, F.**, Enzymes of glycolate formation and oxidation in two members of the *Rhodospirillaceae* (purple non-sulfur bacteria), *Arch. Microbiol.*, 104, 155, 1975.
22. **Anderson, L. and Fuller, R. C.**, Photosynthesis in *Rhodospirillum rubrum*. II. Photoheterotrophic carbon dioxide fixation, *Plant Physiol.*, 42, 491, 1967.
23. **Akazawa, T., Takabe, T., Asami, S., and Kobayashi, H.**, RuBP carboxylases from *Chromatium vinosum* and *Rhodospirillum rubrum* and their role in photosynthetic carbon assimilation, *Brookhaven Symp. Biol.*, 30, 209, 1978.
24. **McFadden, B. A.**, The oxygenase activity of ribulose-1,5-bisphosphate carboxylase from *Rhodospirillum rubrum*, *Biochem. Biophys. Res. Commun.*, 60, 312, 1974.

RESPIRATION AND PHOTORESPIRATION IN PLANTS

George H. Lorimer

INTRODUCTION

Respiration, the uptake of O_2 and release of CO_2 which accompanies the oxidation of carbohydrates, occurs in photosynthetic tissue of all eukaryotes. In the last decade, important differences have emerged between the respiration which occurs in the dark and that which occurs in light (photorespiration). The purpose of this brief review is to summarize the mechanisms of respiration and to point out the differences between dark respiration and photorespiration. For more comprehensive accounts, the reader is referred to other monographs and reviews[1-14] in which references to the original experimental data can be found.

PATHWAYS OF CARBOHYDRATE OXIDATION

Dark Respiration

The major pathways for the oxidation of carbohydrates during dark respiration are shown in Figure 1. Hexose phosphates are converted to triose phosphates either directly via

$$\text{Hexose-6-phosphate} + \text{ATP} \dashrightarrow 2\,(\text{triose phosphate}) + \text{ADP} \quad (1)$$

or via the pentose phosphate pathway, the overall stoichiometry of which is

$$\text{Hexose-6-phosphate} + 6\text{NADP} \dashrightarrow \text{triose phosphate} + 6\text{NADPH} + CO_2 \quad (2)$$

Triose phosphate is converted to pyruvate by the series of reactions shown in Figure 1. Pyruvate then is completely oxidized to CO_2 by the reactions of the tricarboxylic acid cycle. The reduced pyridine nucleotides and flavins so generated are oxidized via the mitochondrial electron transport system, the ultimate electron acceptor being O_2. Energy is conserved by both substrate level and oxidative phosphorylation.

Photorespiration

The major pathway of photorespiration, the C_2 photorespiratory carbon oxidation (PCO) cycle, is shown in Figure 2. The C_2PCO cycle cannot be considered alone, but rather in conjunction with the C_3 photosynthetic carbon reduction (PCR) cycle upon which it is entirely dependent for its capacity to regenerate the substrate ribulose 1,5-bisphosphate. At the CO_2 compensation point, these two cycles balance one another, there being no net respiration or photosynthesis. Below the CO_2 compensation point, CO_2 released by the C_2PCO cycle exceeds that fixed by the C_3PCR cycle. Leaf reserves such as starch can be converted via hexose and triose phosphate to ribulose 1,5-bisphosphate to serve as fuel for the C_2PCO cycle in such circumstances. Under natural conditions (21% O_2 and 0.03% CO_2), the CO_2 fixed by the C_3PCR cycle exceeds that released by the C_2PCO cycle by a factor of 5 to 7.

There are several distinct differences between "dark" respiration and photorespiration. Table 1 provides a comparison.

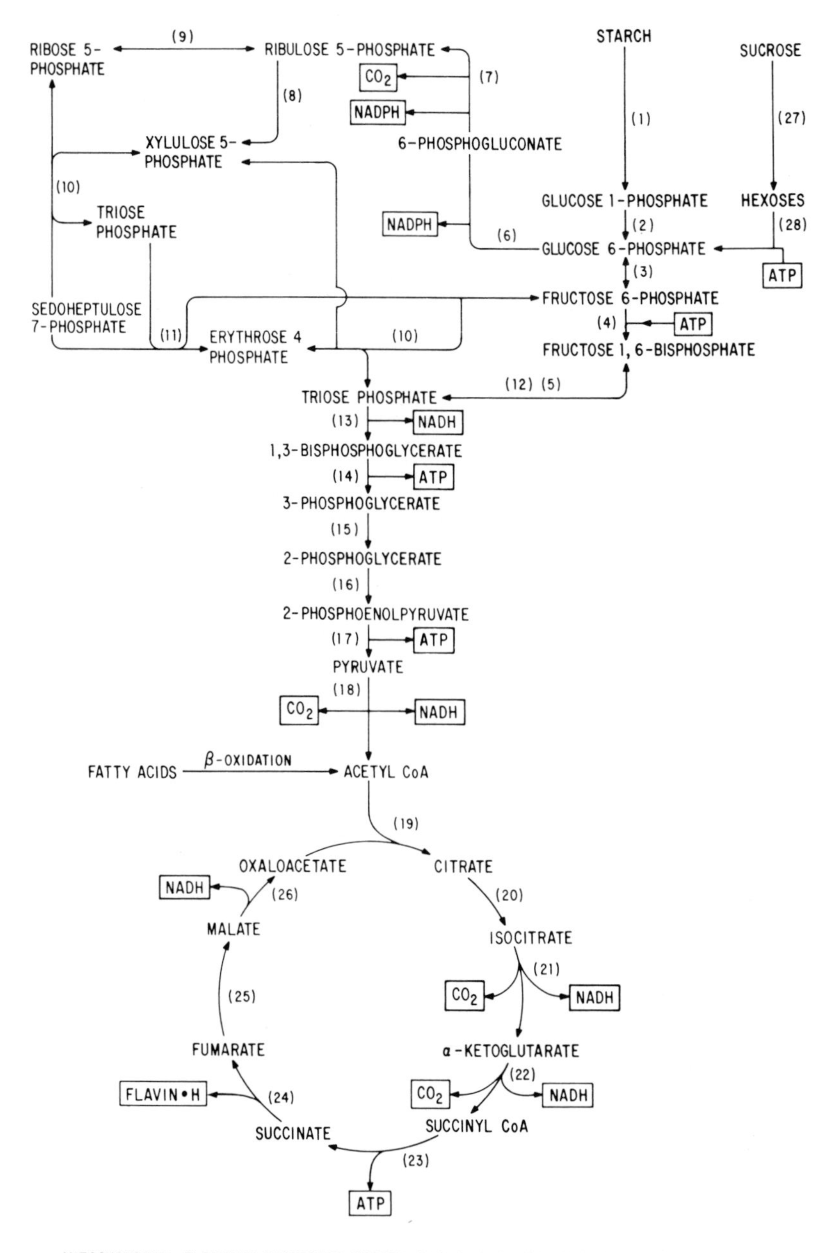

MITOCHONDRIAL ELECTRON TRANSPORT SYSTEM AND OXIDATIVE PHOSPHORYLATION

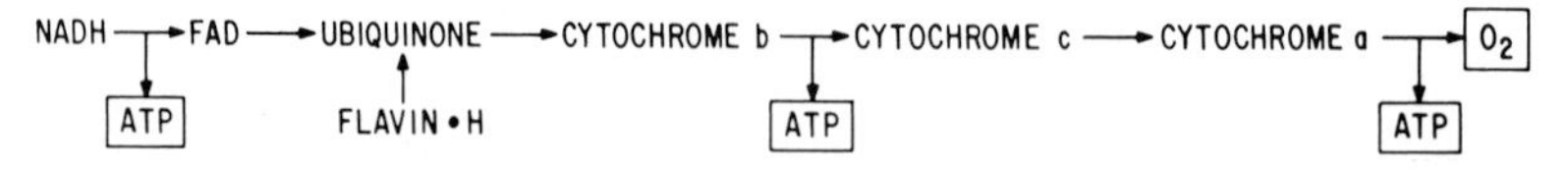

FIGURE 1. Metabolic pathways of carbohydrate oxidation associated with "dark" respiration. Enzymes involved are (1) starch phosphorylase; (2) phosphoglucomutase; (3) phosphohexoisomerase; (4) phosphofructokinase; (5) aldolase; (6) glucose-6-phosphate dehydrogenase; (7) 6-phosphogluconate dehydrogenase; (8) ribulose-5-phosphate epimerase; (9) phosphoriboisomerase; (10) transketolase; (11) transaldolase; (12) triose phosphate isomerase; (13) triose phosphate dehydrogenase; (14) phosphoglycerate kinase; (15) phosphoglyceromutase; (16) enolase; (17) pyruvate kinase; (18) pyruvate dehydrogenase; (19) citrate synthase; (20) aconitase; (21) isocitrate dehydrogenase; (22) α-keto-glutarate dehydrogenase; (23) succinyl CoA synthase; (24) succinate dehydrogenase; (25) fumarase; (26) malate dehydrogenase; (27) invertase; and (28) hexokinase.

Table 1
COMPARISON BETWEEN DARK RESPIRATION AND PHOTORESPIRATION OF PHOTOSYNTHETIC TISSUES OF HIGHER PLANTS

Dark respiration	Photorespiration
Light independent. Whether or not dark respiration continues or is suppressed in the light is debatable. There is evidence for both. It is clear, however, that the mitochondrial electron transport system terminating in cytochrome oxidase does continue to function in the light in order to transfer electrons from glycine to O_2 in photorespiration. (See third item.)	Light dependent. The transient post-illumination CO_2 efflux is due to photorespiration. However, none of the reactions shown in Figure 2 is photochemical. Like the photosynthetic fixation of CO_2, photorespiration requires the regeneration of the common substrate, ribulose 1,5-bisphosphate for which ATP and NADPH are needed. It is the production of these which is light dependent. Note that elements of the PCO cycle (e.g., glycolate synthesis) occur independently of light in chemosynthetic bacteria which fix CO_2 via the C_3 cycle.
Occurs in all photosynthetic tissue.	The potential to photorespire is present in all photosynthetic tissue (exception — the photosynthetic bacterium *Chlorobium*). The expression of this potential is severely suppressed in C_4 plants in which it appears confined to the bundle sheath cells. Glycolate synthesis is a characteristic feature of all organisms which fix CO_2 via ribulose 1,5-bisphosphate carboxylase (hence *Chlorobium* as the exception). All ribulose 1,5-biphosphate carboxylases so far tested have oxygenase activity regardless of the taxonomic origin of the enzyme.
The rate of dark respiration varies considerably depending upon the age and physiological state of the tissue. Rates of 1 to 2 mg CO_2 $dm^{-1}h^{-1}$ (about 5 to 10 μmol CO_2 mg chlorophyll$^{-1}h^{-1}$) are typical for mature, healthy C_3 leaves at 25°C.	Being coupled to the C_3 cycle, the rate of photorespiration is to some extent dependent upon the rate of photosynthesis. Under natural conditions (21% O_2, 0.03% CO_2) the rate of photorespiration is about 15 to 25% of the rate of photosynthesis at 25°C. This corresponds to about 20 to 30 μmol CO_2 mg chlorophyll$^{-1}h^{-1}$. Note that this is 3 to 4 times the rate of dark respiration.
Not markedly influenced by the concentration of CO_2; cytochrome oxidase is essentially saturated by 2 to 3% O_2; no competition between CO_2 and O_2.	Markedly influenced by the concentrations of CO_2 and O_2; photorespiration is not saturated by 21% O_2; competition between CO_2 and O_2 is evident — in general high concentrations (>21%) of O_2 and low concentrations (<0.02%) of CO_2 stimulate photorespiration and vice versa. Ribulose 1,5-bisphosphate carboxylase competitively inhibited by O_2; ribulose 1,5-bisphosphate oxygenase competitively inhibited by CO_2.
Involves several reactions (7,18,21,22 of Figure 1) which produce CO_2 but only one reaction (cytochrome oxidase) which consumes O_2. Ammonia is not involved.	Involves only one (major) reaction (glycine decarboxylase) which produces CO_2 but several (ribulose 1,5-bisphospahte oxygenase, glycolate oxidase and cytochrome oxidase) which consume O_2. In steady state 1 mol of ammonia is released (and presumably refixed) per mole of CO_2 produced.
Oxidation only by transfer of electrons to O_2 (ultimately); no incorporation of molecular O_2. $2AH_2 + {}^*O_2 \rightarrow 2A + H_2{}^*O$	Oxidation both by transfer of electrons to O_2 and by incorporation of molecular O_2 $AH_2 + {}^*O_2 \rightarrow A{}^*O + H_2{}^*O$; $2AH_2 + {}^*O_2 \rightarrow 2A + 2H_2{}^*O$

Table 1 (continued)
COMPARISON BETWEEN DARK RESPIRATION AND PHOTORESPIRATION OF PHOTOSYNTHETIC TISSUES OF HIGHER PLANTS

Dark respiration	Photorespiration
O_2 consumption confined to the mitochondria (cytochrome oxidase).	O_2 consumption occurs in the chloroplasts (ribulose 1,5-bisphosphate oxygenase), peroxisomes (glycolate oxidase) and mitochondria (cytochrome oxidase).
No immediate connection with photosynthetic carbon metabolism.	Intimately associated with photosynthetic carbon metabolism; the C_2 PCO cycle is dependent upon the C_3 cycle.
Conservation of energy via both substrate level and oxidative phosphorylation. Of the 672 kcal available from the complete oxidation of a mole of glucose, about 35 to 40% is conserved in the form of ATP.	No net conservation of energy; indeed an input of energy is required to drive the C_2 PCO cycle. This is energy input and is reflected by an increase in the quantum requirement of net photosynthesis from about 12 to 13 under conditions in which photorespiration is suppressed (high CO_2 and/or low O_2 concentrations) to about 17 to 19 under natural conditions at 25°C.
Function — the provision of energy in the form of ATP with which the cell performs a multitude of endergonic tasks.	Function — the purpose of photorespiration (if indeed it serves any at all) constitutes an enigma.

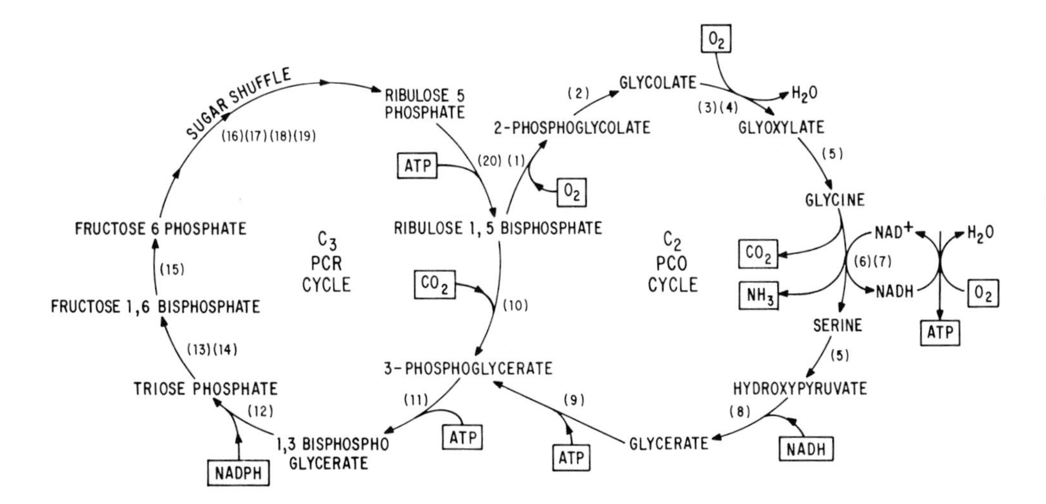

FIGURE 2. Metabolic pathways of carbohydrate oxidation and reduction associated with photorespiration. Enzymes involved are (1) ribulose 1,5-bisphosphate oxygenase; (2) 2-phosphoglycolate phosphatase; (3) glycolate oxidase; (4) catalase; (5) aminotransferase; (6) glycine decarboxylase; (7) serine hydroxymethyl transferase; (8) hydroxypyruvate reductase; (9) glycerate kinase; (10) ribulose 1,5-bisphosphate carboxylase; (11) 3-phosphoglycerate kinase; (12) glyceraldehyde 3 phosphate dehydrogenase; (13) triose phosphate isomerase; (14) aldolase; (15) fructose 1,6-bisphosphate phosphatase; (16) transketolase; (17) pentose phosphate isomerase; (18) sedoheptulose 1,7-bisphosphate phosphatase; (19) xylulose 5-phosphate-3-epimerase; and (20) ribulose 5-phosphate kinase.

REFERENCES

1. **Beevers, H.,** *Respiratory Metabolism in Plants,* Row and Peterson, White Plains, N.Y., 1960.
2. **Palmer, J. M.,** The organization and regulation of electron transport in plant mitochondria, *Annu. Rev. Plant Physiol.,* 27, 133, 1976.
3. **Zelitch, I.,** Organic acids and respiration in photosynthetic tissues, *Annu. Rev. Plant Physiol.,* 15, 121, 1964.
4. **Bonner, J. and Varner, J. E.,** The path of carbon in respiratory metabolism, in *Plant Biochemistry,* 1st ed., Bonner, J. and Varner, J. E., Eds., Academic Press, New York, 1965, 213.
5. **Davies, D. D.,** *The Biochemistry of Plants,* Vol. 2, Academic Press, New York, 1979.
6. **Jackson, W. A. and Volk, R. J.,** Photorespiration, *Annu. Rev. Plant Physiol.,* 21, 385, 1970.
7. **Zelitch, I.,** *Photosynthesis, Photorespiration and Plant Productivity,* Academic Press, New York, 1971.
8. **Chollet, R. and Ogren, W. L.,** Regulation of photorespiration in C_3 and C_4 species, *Bot. Rev.,* 41, 137, 1975.
9. **Schnarrenberger, C. and Fock, H.,** Interactions among organelles involved in photorespiration, in *Encyclopedia of Plant Physiology,* Vol. 3 (New Series), Stocking, C. R. and Heber, U., Eds., Springer-Verlag, Berlin, 1976, 185.
10. **Lorimer, G. H., Woo, K. C., Berry, J. A., and Osmond, C. B.,** The C_2 photorespiratory carbon oxidation cycle in leaves of higher plants: pathway and consequences, in *Photosynthesis '77,* Proc. 4th Int. Congr. Photosynthesis, Hall, D. O., Coombs, J., and Goodwin, T. W., Eds., The Biochemical Society, London, 1977, 311.
11. **Burris, R. H. and Black, C. C.,** *CO_2 Metabolism and Plant Productivity,* University Park Press, Baltimore, 1976.
12. **Andrews, T. J. and Lorimer, G. H.,** Photorespiration — still unavoidable?, *FEBS Lett.,* 90, 1, 1978.
13. **Campbell, W. H. and Black, C. C.,** The relationship of CO_2 assimilation pathways and photorespiration to the physiological quantum requirement of green plant photosynthesis, *BioSystems,* 10, 253, 1978.
14. **Keys, A. J., Bird, I. F., Cornelius, M. J., Lea, P. J., Wallsgrove, R. M., and Miflin, B. J.,** Photorespiratory nitrogen cycle, *Nature (London),* 275, 741, 1978.

Section 3
Major Biosynthetic Pathways

BIOSYNTHESIS OF FRUCTOSE, SUCROSE, AND STARCH

Jack Preiss

BIOSYNTHESIS OF STARCH

Prior to 1962, synthesis of the α-1,4-glucosidic linkage in plants was believed to be catalyzed by phosphorylase (Equation 1).[1]

$$\text{Glucose-1-P} + (\text{glucosyl})_n \rightleftharpoons P_i + (\text{glycosyl})_{n+1} \quad (1)$$

However, it is now generally agreed that the starch synthase reaction (Equation 2) is the main, if not exclusive, reaction involved in the synthesis of the starch α-1,4 glucosidic linkages.[2]

$$\text{ADP(UDP)-glucose} + (\text{glucosyl})_n \rightarrow \text{ADP(UDP)} + (\text{glucosyl})_{n+1} \quad (2)$$

This reaction was first described using UDP-glucose as the sugar nucleotide donor. Subsequently, Recondo and Leloir[3] showed that ADP-glucose was a better substrate both in terms of V_{max} and K_m. Leaf starch synthases[4-6] and bacterial glycogen synthases,[7] however, are relatively inactive with UDP-glucose. The activity with UDP-glucose is only 1% or less than that observed with ADP-glucose. Plant and algal starch synthases can be either bound to the starch granule or can be found in soluble form in the cytoplasm.

Both ADP-glucose and UDP-glucose are synthesized by classical pyrophosphorylase reactions (Equation 3)[8] or by reversal of the sucrose synthase reaction (Equation 4).[9,10]

$$\text{ATP(UTP)} + \text{glucose-1-P} \rightleftharpoons \text{ADP(UDP)-glucose} + PP_i \quad (3)$$

$$\text{Sucrose} + \text{ADP(UDP)} \rightleftharpoons \text{fructose} + \text{ADP(UDP)-glucose} \quad (4)$$

The synthesis of α-1,6 linkages found in amylopectin and phytoglycogen is catalyzed by branching enzyme or Q enzyme.[11,12] A distinction is usually made between Q enzyme, which is able to make a branched production from amylose similar to amylopectin, and branching enzyme, which is able to branch α-glucans to a greater degree to form glycogen-type molecules. However, this distinction will not be made in this review, as there is no clear evidence distinguishing the branching activities presently involved in the synthesis of amylopectin and phytoglycogen.

All the above enzymes appear to be ubiquitous both in plants and in bacteria.[13]

The above reactions indicate at least three possible enzymatic routes from glucose-1-P to the biosynthesis of α-1,4 glucosidic linkages in plant systems: the phosphorylase route, via UDP-glucose or via ADP-glucose. The question to be asked is which pathway is physiologically the more important one. At present there is no definitive answer to this question. However, on the basis of available information, it appears that the ADP-glucose pathway is predominant. This view is based on the findings that in leaves the leaf starch synthase, whether starch-bound or soluble, is solely specific for ADP-glucose.[4-6] Similarly the soluble starch synthases in reserve tissues also appear to be specific for ADP-glucose and inactive with UDP-glucose.[4,14] Reserve tissue starch synthase bound to starch granule can utilize both UDP-glucose and ADP-glucose. However, the rate of glucosyl transfer from UDP-glucose is usually one third to one tenth of that observed from ADP-glucose, and the K_m for UDP-glucose is about 15- to 30-

fold higher than the K_m for ADP-glucose.[2,4] Thus, ADP-glucose would be the preferred sugar nucleotide donor.

A number of observations would also suggest that phosphorylase may not be operative in starch biosynthesis. It can be readily calculated that the P_i to α-glucose-1-P ratio is about 3 to 300 in algae or leaves.[15,16] Since the phosphorylase reaction in vitro has an equilibrium constant of 2.4 at pH 7.3,[17] formation of α-1,4 glucosyl linkages would not occur. Furthermore, the concentration of glucose-1-P in plants is very low and usually cannot be detected. Its concentration can be estimated by assuming that it is in equilibrium with glucose-6-P via the P-glucomutase reaction [K_{eq} = 17.2].[18] Since the concentration of glucose-6-P in *Chlorella pyrenoidosa* is about 0.7 m*M*[15] and can be calculated to be about 1.23 to 1.88 m*M* in spinach chloroplasts,[19,20] the calculated concentration of glucose-1-P would be about 40 μM in *C. pyrenoidosa* and 71.5 to 109 μM in spinach chloroplast. The K_m values of glucose-1-P for phosphorylase in plants ranges from 1 to 50 m*M*[21-24] and are therefore far above the physiological glucose-1-P concentration. The in vitro activity of phosphorylase at saturating glucose-1-P may be 10- to 20-fold in excess of starch synthase activity measured at saturating ADP-glucose concentrations in certain tissues.[25] However, if the physiological concentration of the substrate is taken in account, the phosphorylase synthetic activity may be less than the starch synthase activity in vivo. There is always the possibility that at the site of phosphorylase in the cell that glucose-1-P may be extraordinarily high compared to the rest of the cell, but no evidence for this occurring has ever been reported.

Strong suggestive evidence showing that in maize endosperm 75% of the starch is synthesized via the ADP-glucose pathways has been reported. Two maize endosperm mutants, shrunken-2 (sh-2) and brittle-2 (bt-2), have reduced levels of starch which are 25% of the normal.[26,27] Correlated with this are the considerably reduced levels of the ADP-glucose pyrophosphorylase activity in the mutant endosperm that are about 10% the levels found in the normal starch maize endosperm.[28,29] Thus, at least 75%, if not all, of the starch may be synthesized via the ADP-glucose pyrophosphorylase reaction. Although not entirely conclusive, the above evidence strongly indicates that the ADP-glucose pathway is the more important route to starch biosynthesis. The reader is referred to a recent review that presents more evidence in favor of the ADP-glucose pathway being the dominant pathway in starch synthesis.[30]

Regulation of Starch Biosynthesis

Starch granule accumulation in leaves or algae is dependent on light, and this suggests regulatory mechanisms coordinating increased CO_2 fixation and photophosphorylation with starch biosynthesis. It was of great significance therefore to find that all the leaf and green algal ADP-glucose pyrophosphorylases studied so far are subject to allosteric phenomena and are activated by 3-P-glycerate and inhibited by P_i.[31,34] Other glycolytic intermediates, such as P-enol-pyruvate, fructose-1,6-P_2, and fructose-6-P, activate to lesser extents, and much higher concentrations are required to elicit the activation. Table 1 shows the activation of various leaf enzymes by a number of metabolites. Of the lesser activators fructose-6-P, P-enol-pyruvate, and fructose-P_2 are most active. Phosphoglycolate gives negligible activation. The enzyme studied in the greatest detail is that obtained from spinach leaf.[31,32,35] 3-Phosphoglycerate increases the apparent affinity of the substrates from 2- to 13-fold. All substrate saturation curves are hyperbolic in the presence or absence of 3-P-glycerate at pH 7.5. The $MgCl_2$ saturation curve is sigmoidal in the presence or absence of the activator, and its apparent affinity is not changed by 3-P-glycerate.

Inorganic phosphate is an effective inhbitor of ADP-glucose synthesis for all the leaf and algal enzymes studied. ADP-glucose synthesis catalyzed by the spinach leaf

Table 1
ACTIVATION OF LEAF ADP-GLUCOSE PYROPHOSPORYLASES[a]

	Activator				
Leaf source	None	3-P-glycerate	P-enol-pyruvate	Fructose-P_2	Fructose-6-P
Spinach	1.6	90	23.2	24.6	36.9
Tobacco	5.7	51.5	18.6	32.3	19.0
Sugar beet	2.8	24.0	10.7	7.9	10.5
Sorghum	0.56	4.0	0.8	1.2	1.0
Rice	2.2	32.2	7.9	7.3	6.2
Barley	37	493	26	136	290
Kidney bean	2.2	11.1	2.2	3.1	3.6
Tomato	14.6	79.5	34.2	27.0	30.5

[a] The results are expressed as nanomoles of ADP-glucose formed per minute per milligram protein, with the exception for the data obtained with the spinach leaf enzyme where the results are expressed as micromoles of ADP-glucose formed per minute per milligram protein. The concentration of activators present in the reaction mixtures containing the leaf enzymes were 1 m*M* except for the sorghum enzyme reaction where the concentration was 5 m*M*.[32,33]

enzyme is inhibited 50% by 60 μM P_i in the absence of activator at pH 7.5.[32] In the presence of 1 m*M* 3-P-glycerate, 50% inhibition requires 1.3 m*M* phosphate. Thus, the activator decreases sensitivity to P_i inhibition about 450-fold. As would be expected, P_i at 0.5 m*M* increases the concentration of 3-P-glycerate needed for activation.[32] Phosphate is a noncompetitive or mixed inhibitor with respect to the substrates, ADP-glucose, PP_i, ATP, and glucose-1-P.

The concentration of phosphate required for 50% inhibition of the other leaf ADP-glucose pyrophosphorylases varies from 20 μM for the barley enzyme to 200 μM for the sorghum leaf enzyme (see Table 2). In all cases the concentration of phosphate required for 50% inhibition is increased by the presence of the activator 3-P-glycerate.

The spinach leaf ADP-glucose pyrophosphorylase has been purified to homogeneity.[35] The molecular weight as determined by disc gel electrophoresis is about 210,000. Subunit molecular weight has been determined to be 48,000 and 44,000 for the two subunits found (unpublished results).

The ADP-glucose pyrophosphorylase from the green alga is very similar in properties to the leaf enzymes.[34] Some of the properties of the *Chlorella pyrenoidosa* enzyme are summarized in Table 2. The ADP-glucose pyrophosphorylases of *Chlorella vulgaris, Scenedesmus obliquus,* and *Chlamydomonas reinhardii* are also activated by 3-P-glycerate and inhibited by orthophosphate.[34]

Evidence for 3-P-Glycerate activation and P_i Inhibition of the Leaf and Algal ADP-Glucose Pyrophosphorylases Being Functional In Vivo

The great sensitivity of the leaf ADP-glucose pyrophosphorylases to 3-P-glycerate, the primary CO_2 fixation product of photosynthesis, and to P_i suggests that they play a significant role in the regulation of starch biosynthesis. The level of P_i has been shown to decrease in leaves during photosynthesis because of photophosphorylation.[16] The levels of ATP and reduced pyridine nucleotides are also increased, leading to the formation of sugar phosphates from 3-P-glycerate. These contribute to conditions necessary for optimal starch synthesis via the increased rate of formation of ADP-glucose. In the dark, there is an increase in phosphate concentration with concomitant decreases in the levels of glycolytic intermediates, ATP, and reduced pyridine nucleotides. This leads to inhibition of ADP-glucose synthesis and therefore starch synthesis.

Table 2
3-P-GLYCERATE ACTIVATION AND PHOSPHATE INHIBITION OF ADP-GLUCOSE SYNTHESIS CATALYZED BY LEAF AND ALGAL ADP-GLUCOSE PYROPHOSPHORYLASES

Enzyme source	$A_{0.5}$ (μM)	Fold-activation	$I_{0.5}$ (no 3-P-glycerate) (μM)	$I_{0.5}$ (+3-P-glycerate) (μM)	Conc of 3-P-glycerate (mM)
Chlorella pyrenoidosa	400	18.0	180	1000	2.0
Turkish tobacco leaf	45	9.0	30	1010	1.0
Red cherry tomato leaf	90	5.5	80	880	2.5
Barley leaf	7.0	13.3	20	2300	1.4
Sorghum leaf	370	7.1	190	410	2.2
Sugar beet leaf	190	8.6	50	430	0.87
Rice leaf	180	14.6	60	270	1.0
Spinach leaf	20	9.3	60	1200	1.0

Note: The data are obtained from References 31, 33, and 34. $A_{0.5}$ represents the concentration required for 50% of maximal stimulation of ADP-glucose synthesis. Fold-activation is the maximum stimulation seen at saturating 3-P-glycerate. The assay conditions for measuring ADP-glucose synthesis are indicated in the above references. The $I_{0.5}$ is the concentration of inhibitor required for 50% inhibition under the conditions of the experiment.

The cyanobacteria, in contrast to other photosynthetic bacteria, grow very poorly on organic compounds and depend on CO_2 fixation via the ribulose-P_2 carboxylase reaction for carbon assimilation, as do algae and higher plants, and therefore would be expected to have the same allosteric activator as observed for the plant ADP-glucose pyrophosphorylase.

MacDonald and Strobel[36] reported that wheat leaves infected with the fungus *Puccinia striiformis* accumulated more starch than noninfected leaves. They could correlate starch accumulation with the decrease observed in P_i levels in diseased leaves during the infection process. These data suggested that in diseased leaves the variations in the level of P_i and to a lesser extent variations in the level of activators of the wheat leaf ADP-glucose pyrophosphorylase (3-P-glycerate, fructose-P_2, etc.) regulated the rate of starch synthesis via control of ADP-glucose pyrophosphorylase activity.

Kanazawa et al.[37] have shown in *C. pyrenoidosa* cells that starch and ADP-glucose synthesis occur in the light. Starch synthesis abruptly ceases and the ADP-glucose level drops to below detectable limits when the light is turned off. UDP-glucose levels do not perceptibly change in the light to dark transition. ADP-glucose is not detectable at any time later in the dark, despite the high steady state level of the substrates of the pyrophosphorylase reaction, ATP, and hexose phosphate. Kanazawa et al.[37] indicate that this observation provides strong support for the importance of the regulatory role of ADP-glucose pyrophosphorylase in starch synthesis in vivo.

Recent observations from several laboratories strongly suggest that P_i and 3-P-glycerate play a role in regulating starch biosynthesis and degradation. Steup et al.[38] have shown that during $^{14}CO_2$ fixation in spinach leaf chloroplasts in the light, increasing levels of P_i (0.1 to 0.25 mM) prolong the lag of starch synthesis slightly while having little effect on the rate of $^{14}CO_2$ fixation. Higher concentrations of P_i, 0.5 and 1.0 mM, not only induce a longer lag in starch synthesis, but also inhibit the maximum rate attained. Steup et al.[38] conclude that inhibition of starch synthesis at high phosphate concentrations can be explained by the allosteric inhibition of ADP-glucose pyrophosphorylase by P_i. P_i also stimulated starch degradation.

In a similar study by Heldt et al.,[20] it was shown that P_i as low as 1 m*M* in the medium completely inhibited starch synthesis in intact chloroplasts. The inhibiting action of P_i was overcome by 3-P-glycerate addition. The authors showed that a high ratio of P-glycerate to P_i in the stroma could be correlated with a high rate of CO_2 incorporated into starch; a low ratio correlated with a low rate. No correlation was seen with stromal hexose monophosphate levels. These authors compared their data on stromal concentrations of P_i and 3-P-glycerate with the in vitro data of P_i and 3-P-glycerate effects on the spinach leaf ADP-glucose pyrophosphorylase. The in vitro ADP-glucose pyrophosphorylase activity at the different 3-P-glycerate/P_i ratios[31,32] agreed very well with the rates of starch synthesis observed in the chloroplast.[20]

Shen-Hwa et al.[39] have shown that photosynthetic starch synthesis in leaf discs of spinach beet, sugar beet, or spinach was increased more than tenfold by the presence of exogenous 10 m*M* mannose. In the above tissues, mannose is converted to mannose-6-P in the cytoplasm, but cannot be incorporated into starch. Shen-Hwa et al.[39] postulated two reasons for the mannose effect. They believe the cytoplasmic P_i levels are decreased because of the sequestration into mannose-6-P via phosphorylation. Thus, troise phosphates and phosphoglycerate could not be translocated from the chloroplast into the cytoplasm during photosynthesis as the translocation process requires P_i. Because of the increased levels of glycolytic intermediates in the chloroplast, it was postulated that starch synthesis would be stimulated. Shen-Hwa et al.[39] also suggested that the increased starch synthesis may also occur because the lower phosphate concentration and high 3-P-glycerate concentration in the chloroplast would increase the activity of the ADP-glucose pyrophosphorylase. Herold et al.[40] also showed that leaf discs of plants grown in phosophate deficient media contained much greater amounts of starch than the leaf discs from control plants.

All these data strongly support the hypothesis that the allosteric phenomena observed in vitro for the leaf and algal ADP-glucose pyrophosphorylases are operative in vivo and play an important role in the biosynthesis of starch. It is important to emphasize that the most important consideration in regulating ADP-glucose and starch synthesis is not the absolute concentrations of activator and inhibitor, but rather the fluctuation of the ratio of the activator 3-P-glycerate to the inhibitor, orthophosphate.

ADP-Glucose Pyrophosphorylases of Nonchlorophyllous Plant Tissue

The ADP-glucose pyrophosphorylases occurring in nonphotosynthetic plant tissues, maize endosperm and embryo,[41-43] wheat germ, etiolated peas, mung bean seedlings, potato tuber,[31,44] carrot roots, and avocado mesocarp[31] are similar to the leaf enzymes in that they are activated by 3-P-glycerate. The stimulation by 3-P-glycerate is about 1.5- to 10-fold for these enzymes. The enzyme of this group studied in most detail is the one purified from maize endosperm[41] and from sweet corn endosperm.[45] The stimulation by 3-P-glycerate is only about three- to fourfold, and the concentration needed for activation is very high (2.2 to 10 m*M*) compared to that of the leaf enzymes. Fructose-6-P also stimulates the enzyme activity about threefold at 10 m*M*. P_i at 3 m*M* causes 50% inhibition in the absence of activator.

The lesser sensitivity of the maize enzyme to allosteric effects as compared to the leaf enzymes may reflect differences between leaf and endosperm cells with respect to intracellular levels of metabolites. It is more probable that nonphotosynthetic tissues such as endosperm have no need for the allosteric type regulation. The starch synthetic and degradative rates in leaves appear to be a more dynamic phenomena than what is observed in a reserve tissue, such as endosperm or seed, where the processes of synthesis and degradation are temporally separated during the development and germination of the tissue. Thus, the endosperm enzyme may have evolved to a form which is comparatively insensitive to activation and inhibition. The most important aspect in con-

Table 3
ACTIVATORS OF THE PHOTOSYNTHETIC BACTERIAL ADP-GLUCOSE PYROPHOSPHORYLASES

Organism	Activators
Rhodospirillum fulvum, Rhodospirillum molischianum, Rhodospirillum photometricum, Rhodospirillum rubrum, Rhodospirillum tenue, Rhodocyclus purpureus	Pyruvate
Rhodopseudomonas acidophila, Rhodopseudomonas capsulata, Rhodopseudomonas palustris, Rhodomicrobium vanniellii, Chlorobium limicola, Chromatium vinosum	Pyruvate, fructose-6-P
Rhodospeudomonas viridis	Fructose-6-P, fructose-P_2
Rhodopseudomonas gelatinosa, Rhodopseudomonas globiformis, Rhodopseudomonas spheroides	Fructose-6-P, fructose-P_2, pyruvate
Synechococcus 6301, *Aphanocapsa* 6308	3-P-glycerate

trolling starch biosynthesis in endosperm may be the regulation of synthesis of the starch biosynthetic enzymes, ADP-glucose pyrophosphorylase and ADP-glucose: α-glucan-4-glycosyltransferase.[25,29,44,46-48]

Biosynthesis of Glycogen in Photosynthetic Bacteria

During growth in the light or dark, many photosynthetic bacteria synthesize an α-glucan of the glycogen type. It has been reported that the bacterial glycogen synthase is specific for ADP-glucose.[7,49] Similarly the bacterial ADP-glucose pyrophosphorylase has been found to be an allosteric enzyme.[50] However, in most cases they are activated by glycolytic intermediates other than 3-P-glycerate. Table 3 shows the activator specificity of the ADP-glucose pyrophosphorylases isolated from various photosynthetic bacteria. The ADP-glucose pyrophosphorylases can be classified into five different groups based on their allosteric activator specificity. The enzyme isolated from the bacteria belonging to the genus *Rhodospirillum* is solely activated by pyruvate.[50] However, other phototropic bacteria belonging the families Rhodospirillaceae, Chromatiaceae, and Chlorobiaceae contain ADP-glucose pyrophorylases activated by both fructose-6-P and pyruvate [*Rhodopseudomonas capsulata, Rhodopseudomonas palustris, Rhodopseudomonas acidophila, Rhodomicrobium vanniellii, Chromatium vinosum, and Chlorobium limicola.*[50] ADP-glucose pyrophosphorylases isolated from three Rhodopseudomonads [*Rhodopseudomonas spheroides, Rhodopseudomonas gelatinosa, and Rhodopseudomonas globiformis*] (unpublished results)[50] are activated by fructose-P_2 in addition to fructose-6-P and pyruvate and thus represent a class distinct from those observed in the other photosynthetic bacteria. The *Rhodopseudomonas viridis* ADP-glucose pyrophosphorylase is not activated by pyruvate which makes it distinct from the other purple sulfur and nonsulfur bacteria. It is, however, activated by fructose-P_2 and fructose-6-P. The unicellular blue-green bacteria (Cyanobacteria) *Synechococcus* 6301 and *Aphanocapsa* 6308 contain ADP-glucose pyrophosphorylases activated by 3-P-glycerate and in this respect are similar to the ADP-glucose pyrophosphorylases found in green algae and in higher plants.[51] The activators of the photosynthetic bacteria appear to have the same effects observed for 3-P-glycerate activation of the ADP-glucose pyrophosphorylase of the plant leaves or green algae in Table 1.

They increase V_{max} of ADP-glucose synthesis, increase the affinity of the substrates, and antagonize or reverse the inhibition caused by the allosteric inhibitor.

The *Rhodospirillum tenue* enzyme is inhibited by AMP, but the ADP-glucose pyrophosphorylases of the other *Rhodospirillum* species are not inhibited by either AMP, ADP, or P_i. The *Rhodospirillum fulvum* enzyme is inhibited by ADP, but only at concentrations higher than 1 m*M*. The enzymes from *R. spheroides, R. capsulata*, and *R. viridis* are inhibited most effectively by orthophosphate, while the *R. gelatinosa* enzyme is inhibited effectively by either P_i or AMP. The enzymes from other photosynthetic anaerobic bacteria have not been fully studied with respect to their sensitivity to inhibition by either AMP, ADP, or P_i.

The cyanobacterial ADP-glucose pyrophosphorylases are inhibited by P_i, but not by ADP nor AMP, and thus their activator and inhibitor specificities resemble most closely those of the enzyme found in the leaf and in green algae.

Table 4 shows that the photosynthetic bacterial ADP-glucose pyrophosphorylases are highly sensitive to activation and that in the presence of activator, much higher concentrations of inhibitor are required to elicit 50% inhibition. Alternatively, it has been shown that the presence of P_i reduces the apparent affinity of the activator as it increases the activator concentration necessary for half-maximal activation.[50]

Thus, as shown previously for the leaf ADP-glucose pyrophosphorylases interaction between activator and inhibitor can modulate the activity of the photosynthetic bacterial ADP-glucose pyrophosphorylase. The ratio of activator concentration to inhibitor concentration may therefore determine the activity of the enzyme. A high ratio would permit a high rate of synthesis of ADP-glucose. A low ratio would repress ADP-glucose synthesis and therefore glycogen synthesis.

Possible Reasons for the Variation of Activator Specificity of the Photosynthetic Bacterial ADP-Glucose Pyrophosphorylases

Upon examining Tables 1 and 3, an apparent question arises. Why is there a variation in the activator specificity in the various photosynthetic systems?

Table 3 shows that pyruvate is the only glycolytic intermediate capable of activating the ADP-glucose pyrophosphorylases of *Rhodospirillum* species and *Rhodocyclus purpureus*. These photosynthetic organisms are capable of growth under heterotrophic conditions in the light or dark and under autotrophic conditions in the light. Generally they grow very well on pyruvate and TCA cycle intermediates as carbon sources and photosynthetic electron donors. These organisms cannot utilize glucose or fructose for these purposes. Some *Rhodospirillum rubrum* strains, however, may utilize fructose poorly for growth. Activation of ADP-glucose pyrophosphorylase activity by pyruvate in *R. rubrum* is seen whether the cells are grown aerobically in the dark with malate or anaerobically in the light with either malate, acetate, acetate + CO_2, or CO_2 + H_2.[52] Stanier et al.[53] have pointed out the relation of pyruvate formation with glycogen accumulation and this has been reviewed elsewhere.[50]

Since several mechanisms for the synthesis of pyruvate are available when these organisms are grown under various nutritional conditions giving rise to accumulation of glycogen, pyruvate may be considered as the first glycolytic intermediate in glycogen synthesis in *Rhodospirillum*.

The rationale for fructose-6-P or fructose-P_2 being activators is that some of the photosynthetic organisms can metabolize either fructose or glucose or both via Entner-Doudoroff or glycolytic pathways. Besides the Rhodopseudomonads, *R. gelatinosa, R. spheroides, R. capsulata*, and *R. viridis, Chlorobium limicola* and *Chlormatium vinosum* are other organisms listed in Table 4 able to utilize fructose or glucose. The metabolism of glucose by *R. spheroides* has been shown to occur via the Entner-Doudoroff pathway, while degradation of fructose occurs mainly via the Entner-Doudo-

Table 4
KINETIC PROPERTIES OF THE ACTIVATORS AND INHIBITORS OF PHOTOSYNTHETIC BACTERIAL ADP-GLUCOSE PYROPHOSPHORYLASES

	$A_{0.5}$ (μM)				$I_{0.5}$ (μM)					
Source of enzyme	3-P-glycerate	Pyruvate	Fructose-P_2	Fructose-6-P	P_i (−)	P_i (+)	AMP (−)	AMP (+)	ADP (−)	ADP (+)
Rhodospirillum fulvum	—	19	—	—	—	—	—	—	~1000	>2500
Rhodospirillum molischianum	—	11	—	—	—	—	—	—	—	—
Rhodospirillum rubrum	—	280	—	—	—	—	—	—	—	—
Rhodopseudomonas tenue	—	25	—	—	—	—	—	—	—	—
Rhodopseudomonas acidophila	—	500	—	60	760	>5000	—	—	—	—
Rhodopseudomonas capsulata	—	55	—	77	260	5800	640	7300	410	3800
Rhodopseudomonas gelatinosa	—	150	54	200	145[a]	1200	170[b]	1040	—	—
Rhodopseudmonas spheroides	—	122	16	9	250	2400[b]	—	—	—	—
Rhodopseudomonas viridis	—	—	180	180	380	1350	—	—	—	—
Synechococcus 6301	112	—	—	—	72	1800	—	—	—	—

Note: $A_{0.5}$ is the concentration of activator required for 50% of maximal activation. $I_{0.5}$ is the concentration needed for 50% inhibition; + and − signify whether 1 mM activator was present (+) or not (−). The activator used was the one with the lowest $A_{0.5}$ value.

[a] Value obtained in the presence of 100 μM fructose-6-P_2.
[b] Value obtained in the presence of 100 μM fructose-6-P.

roff pathway in dark-aerobic conditions and via the glycolytic pathway under anaerobic phototrophic conditions.[54] In *R. capsulata*, glucose is metabolized via the Entner-Doudoroff pathways,[55,56] and fructose is metabolized via the glycolytic pathway.[56]

The ADP-glucose pyrophosphorylases of these organisms may therefore exhibit multiple activation specificity because of the utilization of carbohydrates via a number of metabolic pathways. Pyruvate would be principal activator for phototrophic or dark-aerobic growth of the organisms on pyruvate and TCA cycle intermediates. Fructose-6-P and fructose-P_2 would be most important for the time where the Entner-Doudoroff pathway or glycolysis are the predominant carbon assimilatory pathways. However, the predominant metabolic pathways occurring under different nutritional and physiological conditions for many of the anaerobic photosynthetic organisms have not been fully explored. Thus, the above remains a tentative hypothesis.

The blue-green bacteria grow very poorly if at all on organic compounds and thus must depend on CO_2 fixation for carbon assimilation. This is done via the ribulose-P_2 carboxylase reaction which has been demonstrated in blue-green bacterial extracts.[57] Furthermore, the obvious evolutionary relationship of the blue-green bacteria to the green algae and higher plants has been pointed out previously, with respect to ability to evolve O_2 during photosynthesis and with respect to having similar photosynthetic pigments.[58] Because of these similarities as well as the similar CO_2 fixation pathway used by the green algae, plant chloroplast, and blue-green bacteria (Calvin-Bassham cycle), it would be anticipated that they would have a similar activator for their ADP-glucose pyrophosphorylase. In this case, the primary CO_2 fixation product, 3-P-glycerate, is an activator for the first unique enzyme in the biosynthetic pathway of one of the end-products of photosynthesis, starch.

Properties of the ADP-Glucose:1,4-α-D-Glucan 4-α-Glucosyltransferase

Starch-Bound and Soluble Starch Synthases

As indicated previously the formation of the α-1,4-glucosidic linkages of starch was first observed by Leloir et al.[2] Synthesis occurred by the transfer of glucose from UDP-glucose to intact starch granules (Equation 2). The starch granules prepared from dwarf beans, young potatoes, and sweet corn were used. Subsequently, Recondo and Leloir[3] reported that ADP-glucose was far superior to UDP-glucose as a glucosyl donor. Other glucosyl nucleotides were found to be inactive. Subsequently, many other 4-glucosyl transferases adsorbed to the starch granules have been found.[4] In all cases, the rate of transfer of glucose from ADP-glucose was three to ten times higher than from UDP-glucose. Since the enzyme was associated with the starch granules, it was not possible to show primer requirements. However, if oligosaccharides of the maltodextrin series were added to the reaction mixtures containing ADP-glucose-^{14}C (or UDP-glucose-^{14}C) and the active starch granules, glucose transfer to these oligosaccharides could be observed.

Recent studies with the α-1,4-glucan synthases from a number of starch grains have shown that they can vary with respect to specificity of glucosyl donor. Both TDP-glucose and GDP-glucose could act as glucosyl donors in certain α-1,4-glucan synthesizing systems.[4,59] However, with all starch granule systems, ADP-glucose was the superior glucosyl donor. Deoxy-ADP-glucose also appears to be an effective donor.[6,59] However, this sugar nucleotide is not a naturally occurring compound.

In addition to the α-1,4-glucan synthase associated with the starch granules, there is also present in the same tissues a soluble α-1,4-glucan synthase. The soluble synthase systems are active with only ADP-glucose and not with any other naturally occurring glucosyl nucleotides in contrast to the particulate systems.[4,6,60] It has been suggested that perhaps the adsorption or entrapment of the α-1,4-glucan synthase in the starch granule may change its properties with respect to glucosyl donor specificity. It has been shown that the properties of starch granule enzymes from a number of sources could be changed by mere mechanical disruption.[59,61] First the activity of the starch synthase with ADP-glucose is increased, and second, in the case of the potato enzyme, the K_m of ADP-glucose is decreased from 40 to 3.3 m*M*. Moreover, the activity with UDP-glucose is drastically diminished. Not all starch granule-bound starch synthases are active with UDP-glucose. Murata and Akazawa[5] showed that the bound starch synthase from soybean leaves is solely specific for ADP-glucose. This has been shown to be true for other leaf granule-bound starch synthases.[62] It has been shown that in certain varieties of rice or maize where the granules are devoid of the amylose portion of the starch granule, the α-1,4-glucan synthase is not present in the granule, but present only in the soluble portion of the cell.[63,64] It appears therefore that the adsorption of the α-1,4-glucan synthase by the starch granule is because of the presence of amylose. In a number of cases, it has been shown that amylose is capable of absorbing the synthase and thus transforming it into a particulate enzyme.[65]

The granule-bound starch synthase is diminished in the endosperm of the maize mutant, waxy.[64,66] Waxy starch granules contain only amylopectin and no amylose. This may suggest that amylose synthesis is catalyzed by the granule-bound starch synthase and not by the soluble starch synthases. It is possible that normal starch synthase has some binding affinity for amylose or linear α-1,4-oligosaccharides. The waxy mutation may decrease this affinity, and therefore the starch synthase is no longer entrapped in the granule. Another equally viable possibility is that the starch-bound synthase is an enzyme distinct from the soluble form(s) and is missing from waxy endosperm because of the mutation. Further research is required to determine which possibility is correct.

Requirements for Starch Synthase Activity

The requirements for starch synthase activity are sugar nucleotides and primers. The primer requirement may be satisfied by either amylose, amylopectin, glycogen, or starch granules.[4] Various oligosaccharides of the maltodextrin series starting with maltose can also be utilized. The immediate product is the next higher oligosaccharide, i.e., the product with maltose as the acceptor is maltotriose.[6] Longer incubations, however, will convert the maltotriose product to maltotetraose and maltopentaose. A number of reports have indicated the requirement of univalent cations on the activity of starch synthases associated with the granule.[67-69] In general a univalent cation stimulates starch synthase activity, but the effect varies from a minimal to an absolute effect, depending on the source of the enzyme. The starch synthases are not affected by glycolytic intermediates. ADP is a competitive inhibitor with ADP-glucose.[6] The enzyme also appears to have a sulfhydryl group requirement for activity, as it is quite sensitive to sulfhydryl group reagents.[4,6]

Multiple Forms of the Soluble Starch Synthase

Many reports since 1971 have reported the presence of multiple forms of the soluble starch synthases of leaves and reserve tissues.[70-74] In most cases, these enzymes forms were separated by DEAE-cellulose chromatography. The number of forms found in the systems ranged from two to four, with most of the activity residing in two peaks. The multiple enzyme forms studied in greatest detail are from spinach leaf[72] and waxy maize endosperm[71] and from starchy maize endosperm.[75] The isozymes can be distinguished from each other with respect to their reaction rates for different primers. The most distinct difference is that one fraction in each of the plant extracts (starch synthase I in maize) is able to catalyze the synthesis of an α-1,4-glucan with the slight amount of endogenous primer associated with the enzyme in the presence of 0.5 *M* citrate.[70-75] The molecular weight of the starch synthase carrying out this "unprimed" synthesis is about 70,000 in sucrose density ultracentrifugation for all tissues studied,[76] while the molecular weight of the second starch synthase fraction is 92,000 to 95,000.

These data suggest the presence of at least two different classes of soluble starch synthases. The 70,000 mol wt starch synthase fraction which has high affinity for the endogenous α-1,4-glucan primer in the presence of citrate (the K_m for amylopectin in the spinach leaf system is reduced from 530 to 1.9 μg/mℓ in the presence of citrate) is associated with branching enzyme activity.[71,72,74,78] In spinach leaf extracts, the branching enzyme activity is only resolved from the synthase via chromatography on the affinity resin ADP-hexanolamine-Sepharose® 4B.[76] Under certain conditions, the branching enzyme can stimulate the "unprimed" synthesis catalyzed by the starch synthase about 11- to 14-fold. The glucan product formed contains α-1,6 linkages as well as α-1,4 linkages, and the molecular weight and absorption spectra of the I_2-glucan complex was typical of amylopectin.

α-1,4-Glucan:α-1,4-Glucan 6-Glycosyl Transferase (Branching or Q Enzyme)

Properties of Potato Tuber Branching Enzyme

Although publications on branching enzyme or Q enzyme date from the 1940s, very little is known about the mechanism of action and the nature of the glucan substrate. The enzyme that has been studied in most detail is that isolated from potato tubers.[79,80] The enzyme has been purified to near homogeneity, has a molecular weight of 85,000, and is composed of only one subunit.[80] The enzyme can utilize as acceptors amyloses of chain lengths of 40 glucose units or greater.[81-84] Experiments that showed that radioactive label could be transferred from one amylose chain to another amylose chain of disparate length indicated that formation of the α-1,6 branch linkage occurred by interchain transfer.[84] The possibility of intrachain transfer occurring, however, was not eliminated.

Two possible models for the branching enzyme catalytic mechanism have been postulated[84] and invoke complexing of amylose chains in double helices. In the first model, branching enzyme reacts with the amylose molecule to form a covalent bond with a fragment from the donor chain. The branching enzyme oligosaccharide complex then interacts with acceptor amylose chain to form the branching linkage. In the second model, the two amylose chains form a double helix first which facilitates the branching enzyme action of transferring a portion of the oligosaccharide chain from one amylose molecule to another. Borovsky et al.[84] believe that the minimum chain length requirement of 40 for branching enzyme would be explained by the need for a double helix formation. They refer to the postulation by French[85] that amylose chains which can adopt a helical configuration can also form double helices that are stabilized via hydrogen bonding. Some suggestive evidence of double helix formation of amylose chains is seen in the experiments that branching action can occur at a lower temperature (4°C) with amyloses of chain lengths lower than 40.[83]

Spinach Leaf Branching Enzymes

The most detailed report on leaf branching enzymes is by Hawker et al.[76] They have shown that on DEAE-cellulose chromatography, spinach leaf extract contains multiple peaks of branching enzyme activity. One of these peaks is coincident with one of the starch synthase activity peaks. Both branching enzyme activities stimulate starch synthase activity when low or unsaturating concentrations of primer are used. The product formed resembles amylopectin with respect to chain length, I_2 spectra, molecular weight, and resistance to α-amylase and β-amylase digestion.

The molecular weight of the spinach leaf enzymes is about 80,000 and therefore is similar to the potato tuber enzyme in this respect. The only different property noted between the two enzyme forms in spinach leaf is their pH optima in citrate buffer.

The in vitro stimulation of starch synthase activity by branching enzyme has been observed in many other systems[59,74,86] and could be the in vivo mechanism for amylopectin synthesis, i.e., the continuous enlargement of oligosaccharide chains occurs via the starch synthase reaction. After the chain increases to a required length (40?), branching enzyme catalyzes transfer of a portion of the enlarged chain to form the new branch linkage. The stimulation of starch synthase activity by branching enzyme is believed because of the creation of an increased number of nonreducing ends able to accept glycosyl residues from ADP-glucose.

Multiple Forms of Branching Enzyme Activity in Maize Endosperm

Multiple activities of branching enzyme has been observed in a number of systems.[74,76-78] In some maize endosperm mutant systems where in addition to amylopectin a more highly branching α-glucan, phytoglycogen, is formed, two branching enzyme activities have been distinguished.[87-89] One is able to catalyze the branching of amylose to form an amylopectin type product and is called Q enzyme. The other enzyme could catalyze the branching of amylose or amylopectin to form a glycogen type product and is called branching enzyme. However, the occurrence of phytoglycogen in plants is limited[90] and therefore cannot explain the presence of multiple forms of branching enzyme in the many tissues where it is not present.

The branching enzyme isozymes of maize endosperm have recently been studied in detail in normal starchy maize, as well as in endosperm mutants affected in the nature of starch accumulated. Some correlations have been noted with branching enzyme deficiency or alteration and the structure of the starch or glucan accumulated.[75,77]

Remaining Problems in Starch Synthesis

Despite considerable efforts in studying various aspects of starch biosynthesis, the complete formation of a starch granule in vitro has never been demonstrated. The formation of the particular submicroscopic structure of the native starch granule which is unique for each plant source remains unsolved. Likewise, the various functions and interactions of the multiple forms of starch synthase and branching enzyme discussed above remain to be characterized. Neither are the relative importances of either the starch-bound or soluble starch synthase known. In this connection, it is not clear how each plant tissue produces in constant proportions of the two polymers amylose and amylopectin. Theories, such as compartmentation,[91] or that excess amylose is produced because branching enzyme activity is insufficient to branch all the α-1,4-polyglucose chains produced,[92] or that amylopectin is synthesized by a starch synthase-Q enzyme complex and amylose is synthesized by a starch synthase per se[93] have been presented. However, there is presently no evidence to lend support to any of these theories.

Finally, another possible perplexing problem is the *de novo* synthesis of starch in the absence of primer, i.e., the initiation of starch synthesis. Some recent reports claim evidence for the formation of a glucoprotein via transfer of glucose from UDP-glucose to an acceptor protein.[94,95] The presumptive glucoprotein which is thought to contain oligosaccharide chains in α-1,4 linkage (the nature of these glucosyl linkages has never been demonstrated) is believed to function as primer for synthesis of starch. The suggestive evidence is that the glucose transferred from UDP-glucose is precipitable by 5% trichloroacetic acid and that the presumed glucoprotein is no longer precipitated after pronase action. However, it is quite possible that amylose or similar glucan can coprecipitate with protein in 5% trichloroacetic acid. This has been shown to occur in at least two purported *de novo* systems studied in detail.[76,96] Moreover, the studies on the above *de novo* systems have not eliminated the possibility that endogenous maltodextrin or glucan primers are present. Indeed the presence of endogenous primer molecules has been found in some of the purported *de novo* systems.[76,96,97] At present no direct transfer of a glucosyl group to a specific amino acid residue in protein has been demonstrated in the above *de novo* systems. This type of proof is required to definitively establish the formation of the presumed glucoprotein intermediate. Many other postulations have been made to explain the formation of new primer molecules, such as amylases[98] or D enzyme[99] or even by unprimed synthesis by phosphorylase.[22,100] The objections of phosphorylase involvement, however, in starch synthesis have been previously mentioned. The mechanism for generating new primer molecules remains to be shown whether it be via glucoprotein formation or via other routes.

Biosynthesis of Sucrose and Fructose

Sucrose may be synthesized by either sucrose synthase[9,10] (see Equation 4) or by the combined action of sucrose-P synthase[101] (see Equation 5) and sucrose-P phosphatase[102-104] (see Equation 6). All of these enzymes seem to be ubiquitous

$$\text{UDP-glucose} + \text{fructose-6-P} \rightleftharpoons \text{UDP} + \text{sucrose-6-P} \qquad (5)$$

$$\text{Sucrose-6-P} \rightarrow \text{sucrose} + P_i \qquad (6)$$

in plants.[105] Fructose-6-P can be synthesized either from glucose-6-P via phosphoglucoisomerase action or from fructose-bis-P via fructose-bis-P-phosphatase.

Properties of Sucrose Synthase

The enzyme appears to catalyze a reversible reaction with a K'_{eq} [(sucrose) (UDP)/ UDP-glucose)(fructose-6-P)] determined to be 1.3 to 6.7 in a range of 25 to 37°C and pH 7.2 to 7.6.[9,106-109] The variation in K'_{eq} is most probably because of the fact that H^+ participates in the stoichiometry of the reaction. Thus, pH should affect the equilibrium of the reaction. Because of the free reversibility of the enzyme catalyzed reaction, it has been proposed that the enzyme may not only function in the synthesis of sucrose, but also in the synthesis of sugar nucleotides.[110] The view that catalysis of sucrose cleavage is the physiological function of synthase is supported by the finding that its activity is high in some developing plant seeds where sucrose synthesis is relatively unimportant and low in germinating seeds and photosynthetic tissues where sucrose synthesis is high.[110-113]

Other glucosyl nucleotides, besides UDP-glucose (ADP-glucose, dTDP-glucose, CDP-glucose, and GDP-glucose), are active in the sucrose synthase reaction.[105,114,115] However, the uridine nucleotide is always reported to have a lower K_m (higher apparent affinity) for the enzyme. Moreover, V_{max} is always greater with UDP-glucose than with the other sugar nucleotides.[105] The K_m for sucrose can vary from 10 to 400 m*M*, depending on the enzyme source. The K_m for fructose ranges about 5 m*M*.[105] The kinetic mechanism has been determined to be sequential.[105,116] Mg^{2+} has been shown to activate the enzyme in sucrose synthesis while inhibiting cleavage.[109,117,118]

The enzyme has been purified to homogeneity from a number of plant sources and is usually found to be tetrameric in structure with similar subunits of molecular weight of 88,000 to 100,000.[108,113,119] Higher aggregates of the enzyme have been found and the formation of these forms depends on the ionic strength of the solution.[119] Other properties of the enzyme such as amino acid composition have been reported.[108,119]

Properties of Sucrose-P Synthase

Many observations showing that incorporation of $^{14}CO_2$[120-122] or ^{14}C-glucose[123] into sucrose in leaf tissue yielded highly radioactive moieties of glucose and fructose before substantial labeling of the free hexose. This suggested that free fructose was not an intermediate in sucrose synthesis. Fructose-6-P was also found to be labeled before sucrose, and small amounts of sucrose-P were found suggesting that sucrose-P was formed first. This implicated the sucrose-P synthase reaction as the major reaction in the synthesis of sucrose in plants. The K'_{eq} [(sucrose-6-P)(UDP)/(UDP-glucose)(fructose-6-P)] of sucrose-P synthase was determined to be 3250 at pH 7.5.[102] Sucrose-P synthesis is thus favored. The molecular weight of the enzyme is estimated to be about 360,000 to 400,000.[124] No determination of subunit weight has been made as the enzyme has not been purified to homogeneity. Higher molecular weight forms have also been detected. In contrast to the sucrose synthase reaction, the sucrose-P synthase reaction is virtually absolutely specific for UDP-glucose[102] and fructose-6-P.[125] For most enzymes, the K_m for UDP-glucose ranges from 2 to 7.5 m*M*, while for fructose-6-P, the K_m ranges from 0.6 to 4 m*M*.[105]

The leaf enzyme has been shown to be located almost exclusively in the cytosol and not in the chloroplast.[126]

Sucrose-6-Phosphate Phosphatase

Very little is known about sucrose-6-P phosphatase. The enzyme is just as ubiquitous in occurrence as sucrose synthase and sucrose-6-P synthase.[105] The enzyme has been partially purified from carrot roots and immature tissue of sugar cane.[103,104] These enzyme preparations have been shown to be quite specific for sucrose phosphate. Phosphoenolpyruvate was the next best substrate for the carrot enzyme and was hydrolyzed at 5%, the rate observed for sucrose-6-P.[103] Fructose-6-P was hydrolyzed at about 2%, the rate observed for sucrose-6-P for both the carrot and sugar cane preparations.[103,104] The enzyme reaction required Mg^{2+} and was completely inactive in the presence of EDTA.

The K_m for sucrose-6-P for the sugar cane[103] and carrot enzymes[104] range between 130 to 170 μM. However, the enzyme studied from broad bean, maize, and castor bean had a K_m from 45 to 65 μM.[112] Sucrose inhibits the sugar cane enzyme with a K_i of 10 m*M*.[127] Thus, sucrose concentrations in vivo may regulate sucrose phosphatase activity. The inhibition has been shown to be competitive with sucrose-6-P.[127]

Regulation of Sucrose Synthesis

At present the regulation of sucrose synthesis is not understood. However, some allosteric properties of the sucrose synthase[128] and the sucrose-P synthase[129,130] have been reported. Sigmoidal saturation curves for fructose-6-P and UDP-glucose have been demonstrated for the wheat germ sucrose-P synthase.[132] Sucrose synthase from potato tubers and rice grains exhibits sigmoidal saturation curves with sucrose and UDP.[128] However, it is not clear what these observations mean with respect to the in vivo regulation of sucrose formation.

Recently Pontis and Salerno[124,131] have noted that Mg^{2+} enhances the activity of both sucrose synthase and sucrose-6-P synthase in the direction of sucrose synthesis. This had been originally noted previously for the wheat germ enzyme.[129] Thus, any alteration in the level of Mg^{2+} ions in the cell is postulated to modulate rate of sucrose synthesis.[124] Moreover Salerno and Pontis have shown that sucrose inhibits sucrose-P synthase activity.[124,132] Sucrose also decreases the apparent affinity of the enzyme for Mg^{2+}, thus reducing the activity induced by Mg^{2+}. Whether these phenomena are functional in the in vivo system remains to be proven. Regulation of the enzymes involved in sucrose synthesis remains an outstanding problem to be elucidated. Furthermore, the precise function of the enzyme catalyzing Reactions 4 and 5 in vivo is still unresolved.

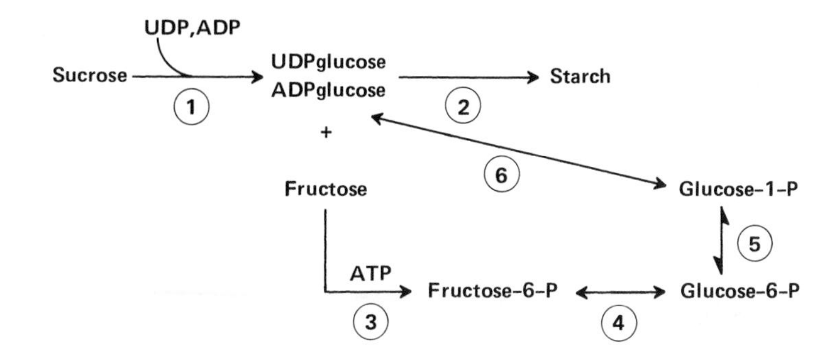

FIGURE 1. Postulated pathway for conversion of sucrose to starch. The numbered reactions are respectively catalyzed by ① sucrose synthase, ② starch synthase, ③ fructokinase (hexokinase), ④ phosphoglucoisomerase, ⑤ phosphoglucomutase, and ⑥ ADP-glucose pyrophosphorylase or UDP-glucose pyrophosphorylase.

Sucrose-Starch Transformation

In developing starchy seeds, sucrose can be directly converted to starch. The pathway in Figure 1 shows that adenosine diphosphate combined with the action of sucrose synthase and starch synthase could convert half of the sucrose molecule to starch. The released fructose portion could be also converted by the combined action of hexokinase, phosphoglucoseisomerase, phosphoglucomutase, ADP-glucose pyrophosphorylase, and starch synthase. Some evidence in support for this pathway has come from experiments where ^{14}C-sucrose was fed to developing seed and was converted to starch and from in vitro studies where sucrose synthase and starch synthase activities were present.[10,133,134] However, it should be noted that the apparent affinity of sucrose synthase is preferentially for UDP rather than ADP[105] and that UDP drastically inhibits ADP-glucose formation from sucrose in the sucrose synthase catalyzed reaction.[105,114,115,134] In contrast starch synthase utilizes ADP-glucose as a glucosyl donor in greater preference than UDP-glucose. The exact in vivo mechanism for conversion of sucrose to starch is still uncertain.

Chourey and Nelson[135] have shown that sh-1 endosperm possesses less than 10% sucrose synthase activity (see Equation 4) as compared to normal endosperm. The sh-1 mutant endosperm is somewhat deficient in its starch content containing about 62% of the normal starch content on a milligram per endosperm or 70% on a percent dry weight basis. This evidence strongly suggests the participation of sucrose synthase in starch synthesis in maize endosperm. How sucrose synthase plays a role is not known. As indicated in Equation 4 and Figure 1, the enzymic reaction is reversible and can either form sucrose or sugar nucleotide (UDP-glucose or ADP-glucose). There is a tendency to believe that the catalysis of sugar nucleotide synthesis is the physiological reaction, but that remains to be proven.

It is important to note that the sh-1 mutation (sucrose synthase deficiency) decreases starch accumulation by 30%, while the sh-2 and bt-2 mutations (ADP-glucose pyrophosphorylase deficiency)[26,29] decrease starch accumulation by 70 to 75%, strongly suggesting that almost all starch synthesis occurs through the sugar nucleotide or ADP-glucose pathway via starch synthase action.

REFERENCES

1. **Hanes, C. S.,** The reversible formation of starch from glucose-1-phosphate catalyzed by potato phosphorylase, *Proc. R. Soc. London Ser. B,* 129, 174, 1940.
2. **Leloir, L. F., de Fekete, M. A. R., and Cardini, C. E.,** Starch and oligosaccharide synthesis from uridine diphosphate glucose, *J. Biol. Chem.,* 236, 636, 1961.
3. **Recondo, E. and Leloir, L. F.,** Adenosine diphosphate glucose and starch synthesis, *Biochem. Biophys. Res. Commun.,* 6, 85, 1961.
4. **Cardini, C. E. and Frydman, R. B.,** ADP-glucose:α-1,4 glucan glucosyl transferases (starch synthetases and related enzymes) from plants, *Methods Enzymol.,* 8, 387, 1966.
5. **Murata, T. and Akazawa, T.,** The role of adenosine diphosphate glucose in leaf starch formation, *Biochem. Biophys. Res. Commun.,* 16, 6, 1964.
6. **Ghosh, H. P. and Preiss, J.,** Biosynthesis of starch in spinach chloroplasts, *Biochemistry,* 4, 1354, 1965.
7. **Greenberg, E. and Preiss, J.,** The occurrence of adenosine diphosphate glucose:glycogen transglucosylase in bacteria, *J. Biol. Chem.,* 239, 4314, 1964.
8. **Espada, J.,** Enzymic synthesis of adenosine diphosphate glucose from glucose-1-phosphate and adenosine triphosphate, *J. Biol. Chem.,* 237, 3577, 1962.
9. **Cardini, C. E., Leloir, L. F., and Chiriboga, J.,** The biosynthesis of sucrose, *J. Biol. Chem.,* 214, 149, 1955.
10. **de Fekete, M. A. R. and Cardini, C. E.,** Mechanism of glucose transfer from sucrose into the starch granule of sweet corn, *Arch. Biochem. Biophys.,* 104, 173, 1964.
11. **Bourne, E. J. and Peat, S.,** The enzymic synthesis and degradation of starch. I. The synthesis of amylopectin, *J. Chem. Soc.,* 1945, 877, 1945.
12. **Hobson, P. N., Whelan, W. J., and Peat, S.,** The enzymic synthesis and degradation of starch. X. The phosphorylase and Q enzyme of broad bean. The Q enzyme of wrinkled pea, *J. Chem. Soc.,* 1950, 3566, 1950.
13. **Preiss, J.,** Regulation of the biosynthesis of α1,4-glucans in bacteria and plants, in *Current Topics in Cellular Regulation,* Vol. 1, Horecker, B. L. and Stadtman, E. R., Eds., Academic Preiss, New York, 1969, 125.
14. **Frydman, R. B., DeSouza, B. C., and Cardini, C. E.,** Distribution of adenosine diphosphate D-glucose α1,4-glucan α4-glucosyl transferase in higher plants, *Biochim. Biophys. Acta,* 113, 620, 1966.
15. **Bassham, J. A. and Krause, G. H.,** Free energy changes and metabolic regulation in steady state photosynthetic carbon reduction, *Biochim. Biophys. Acta,* 189, 207, 1969.
16. **Heber, U. and Santarius, K. A.,** Compartmentation and reduction of pyridine nucleotides in relation to photosynthesis, *Biochim. Biophys. Acta,* 109, 390, 1965.
17. **Cohn, M.,** Phosphorylases (survey), in *The Enzymes,* Vol. 5, 2nd ed., Boyer, P. D., Lardy, H., and Myrback, K., Eds., Academic Press, New York, 1961, 179.
18. **Colowick, S. P. and Sutherland, E. W.,** Polysaccharide synthesis from glucose by means of purified enzymes, *J. Biol. Chem.,* 144, 423, 1942.
19. **McC. Lilley, R., Chon, C. J., Moobach, A., and Heldt, H. W.,** The distribution of metabolites between spinach chloroplasts and medium during photosynthesis *in vitro, Biochim. Biophys. Acta,* 460, 259, 1977.
20. **Heldt, H. W., Chon, C. J., Maronde, D., Herold, A., Stankovic, Z. S., Walker, D. A., Kraminer, A., Kirk, M. R., and Heber, U.,** Role of orthophosphate and other factors in the regulation of starch formation in leaves and isolated chloroplasts, *Plant Physiol.,* 59, 1146, 1977.
21. **Alexander, A. G.,** Polyglucoside synthesis in *Saccharum* species, *Ann. N.Y. Acad. Sci.,* 210, 64, 1973.
22. **Frydman, R. B. and Slabnik, E.,** The role of phosphorylase in starch biosynthesis, *Ann. N.Y. Acad. Sci.,* 210, 153, 1973.
23. **Burr, B. and Nelson, O. E.,** Maize α-glucan phosphorylase, *Eur. J. Biochem.,* 56, 539, 1975.
24. **Chen, M. and Whistler, R. L.,** Use of structural analog in the study of potato phosphorylase, *Int. J. Biochem.,* 7, 433, 1976.
25. **Ozbun, J. L., Hawker, J. S., Greenberg, E., Lammel, C., Preiss, J., and Lee, E. Y. C.,** Starch synthetase, phosphorylase, ADPglucose pyrophosphorylase and UDPglucose pyrophosphorylase in developing maize kernels, *Plant Physiol.,* 51, 1, 1973.
26. **Cameron, J. W. and Teas, H. J.,** Carbohydrate relationships in developing and mature endosperms of brittle and related maize genotypes, *Am. J. Bot.,* 41, 50, 1954.
27. **Creech, R. G.,** Genetic control of carbohydrate synthesis in maize endosperm, *Genetics,* 52, 1175, 1965.
28. **Dickinson, D. B. and Preiss, J.,** Presence of ADPglucose pyrophosphorylase in shrunken-2 and brittle-2 mutants of maize endosperm, *Plant Physiol.,* 44, 1058, 1969.

29. **Tsai, C.-Y. and Nelson, D. E.,** Starch deficient mutants lacking adenosine diphosphate glucose pyrophosphorylase activity, *Science,* 151, 341, 1966.
30. **Preiss, J. and Levi, C.,** Metabolism of starch in leaves, in *Handbuch der Pflanzen Physiologie,* Springer-Verlag, Berlin, Vol. 6, 282, 1979.
31. **Preiss, J., Ghosh, H. P., and Wittkop, J.,** Regulation of the biosynthesis of starch in spinach leaf chloroplasts, in *Biochemistry of Chloroplasts,* Vol. 2, Academic Press, London, 1967, 131.
32. **Ghosh, H. P. and Preiss, J.,** Adenosine diphosphate glucose pyrophosphorylase: a regulatory enzyme in the biosynthesis of starch in spinach leaf chloroplasts, *J. Biol. Chem.,* 241, 4491, 1966.
33. **Sanwal, G. G., Greenberg, E., Hardie, J., Cameron, E., and Preiss, J.,** Regulation of starch biosynthesis in plant leaves: activation and inhibition of ADPglucose pyrophosphorylase, *Plant Physiol.,* 43, 417, 1968.
34. **Sanwal, G. G. and Preiss, J.,** Biosynthesis of starch in *Chlorella pyrenoidosa.* II. Regulation of ATP:α-D-glucose-1-phosphate adenyl transferase (ADP-glucose pyrophosphorylase) by inorganic phosphate and 3-phosphoglycerate, *Arch. Biochem. Biophys.,* 119, 454, 1967.
35. **Ribereau-Gayon, G. and Preiss, J.,** Adenosine diphosphate glucose pyrophosphorylase from spinach leaf, *Methods Enzymol.,* 23, 618, 1971.
36. **MacDonald, B. M. and Strobel, G. A.,** Adenosine diphosphate glucose pyrophosphorylase control of starch accumulation in rust-infected wheat leaves, *Plant Physiol.,* 46, 126, 1970.
37. **Kanzawa, T., Kanazawa, K., Kirk, M. R., and Bassham, J. A.,** Regulatory effects of ammonia on carbon metabolism in *Chlorella pyrenoidosa* during photosynthesis and respiration, *Biochim. Biophys. Acta,* 256, 656, 1972.
38. **Steup, M., Peavey, D. G., and Gibbs, M.,** The regulation of starch metabolism by inorganic phosphate, *Biochem. Biophys. Res. Commun.,* 72, 1554, 1976.
39. **Shen-Hwa, C.-S., Lewis, D. H., and Walker, D. A.,** Stimulation of photosynthetic starch formation by sequestration of cytoplasmic orthophosphate, *New Phytol.,* 74, 383, 1975.
40. **Herold, A., Lewis, D. H., and Walker, D. A.,** Sequestration of cytoplasmic orthophosphate by mannose and its differential effect on photosynthetic starch synthesis in C3 and C4 species, *New Phytol.,* 76, 397, 1976.
41. **Dickinson, D. B. and Preiss, J.,** ADPglucose pyrophosphorylase from maize endosperm, *Arch. Biochem. Biophys.,* 130, 119, 1969.
42. **Hannah, L. C. and Nelson, O. E.,** Characterization of adenosine diphosphate glucose pyrophosphorylases from developing maize seeds, *Plant Physiol.,* 55, 297, 1975.
43. **Preiss, J., Lammel, C., and Sabraw, A.,** A unique adenosine diphosphoglucose pyrophosphorylase associated with maize embryo tissue, *Plant Physiol.,* 47, 104, 1971.
44. **Sowokinos, J. R.,** Pyrophosphorylases in *Solanum tuberosum.* I. Changes in ADPglucose and UDPglucose pyrophosphorylase activities associated with starch biosynthesis during tuberization, maturation and storage of potatoes, *Plant Physiol.,* 57, 63, 1976.
45. **Amir, J. and Cherry, J. H.,** Purification and properties of adenosine diphosphoglucose pyrophosphorylase from sweet corn, *Plant Physiol.,* 49, 893, 1972.
46. **Turner, J. F.,** Physiology of pea fruits. VI. Changes in uridine diphosphate glucose pyrophosphorylase and adenosine diphosphate glucose pyrophosphorylase in the developing seed, *Aust. J. Biol. Sci.,* 22, 1145, 1969.
47. **Tsai, C. Y., Salamini, F., and Nelson, O. E.,** Carbohydrate metabolism in the developing endosperm of maize, *Plant Physiol.,* 46, 299, 1970.
48. **Baxter, E. D. and Duffus, C. M.,** Starch synthetase in developing barley chloroplasts, *Phytochemistry,* 10, 2641, 1971.
49. **Krebs, E. G. and Preiss, J.,** Regulatory mechanisms in glycogen metabolism, in *Biochemistry of Carbohydrates,* Vol. 5, Whelan, W., Ed., MTP International Review of Science, Biochemistry Series One, Butterworths, London, 1975, 337.
50. **Preiss, J.,** Regulation of adenosine diphosphate glucose pyrophosphorylase, in *Advances in Enzymology and Related Areas of Molecular Biology,* Meister A., Ed., John Wiley & Sons, New York, 1978, 317.
51. **Levi, C. and Preiss, J.,** Regulatory properties of the blue-green bacterium *Synechococcus* 6301, *Plant Physiol.,* 58, 753, 1976.
52. **Furlong, C. E. and Preiss, J.,** The regulation of the biosynthesis of α-1,4-glucans in photosynthetic systems, in *Progress in Photosynthetic Research,* Vol. 3, Metzner, H., Ed., Tubinger, Germany, 1969, 1604.
53. **Stanier, R. Y., Doudoroff, M., Kunisawa, R., and Coutopoulou, R.,** The role of organic substrates in bacterial photosynthesis, *Proc. Natl. Acad. Sci. U.S.A.,* 45, 1246, 1959.
54. **Conrad, R. and Schlegel, H. G.,** Influence of aerobic and phototrophic growth conditions on the distribution of glucose and fructose carbon into the Entner-Doudoroff and Embden-Meyerhof pathways in *Rhodopseudomonas spheroides, J. Gen. Microbiol.,* 101, 277, 1977.

55. **Eidels, L. and Preiss, J.**, Carbohydrate metabolism in *Rhodopseudomonas capsulata*; enzyme titers, glucose metabolism and polyglucose polymer synthesis, *Arch. Biochem. Biophys.*, 140, 75, 1970.
56. **Conrad, R. and Schlegel, H. F.**, Different degradation pathways for glucose and fructose in *Rhodopseudomonas capsulata*, *Arch. Microbiol.*, 112, 39, 1977.
57. **Smith, A. J.**, Synthesis of metabolic intermediates, in *The Biology of Blue Green Algae*, Carr, N. G. and Whitton, B. A., Eds., University of California Press, Berkeley, 1973, 1.
58. **Smith, A. J.**, Blue-green bacteria: status and photoautotrophic metabolism, *Biochem. Soc. Trans.*, 3, 12, 1975.
59. **Frydman, R. B. and Cardini, C. E.**, Studies on the biosynthesis of starch. II. Some properties of the adenosine diphosphate glucose:starch glucosyl transferase, *J. Biol. Chem.*, 242, 312, 1967.
60. **Frydman, R. B. and Cardini, C. E.**, Studies of the biosynthesis of starch. I. Isolation and properties of the soluble adenosine diphosphate glucose: starch glucosyl transferase of *Solanum tuberosum*, *Arch. Biochem. Biophys.*, 116, 9, 1966.
61. **Chandorkar, K. R. and Badenhuizen, N. P.**, How meaningful are determinations of glucosyl transferase activities in starch enzyme complexes?, *Staerke*, 18, 91, 1966.
62. **Nomura, T., Nakayama, N., Murata, T., and Akazawa, T.**, Biosynthesis of starch in chloroplasts, *Plant Physiol.*, 42, 327, 1967.
63. **Murata, T., Sugiyama, T., and Akazawa, T.**, Enzyme mechanism of starch synthesis in glutinous rice grains, *Biochem. Biophys. Res. Commun.*, 18, 371, 1965.
64. **Frydman, R. B.**, Starch synthetase of potatoes and waxy maize, *Arch. Biochem. Biophys.*, 102, 242, 1963.
65. **Akazawa, T., and Murata, T.**, Adsorption of ADPG-starch transglucosylase by amylose, *Biochem. Biophys. Res. Commun.*, 19, 21, 1965.
66. **Nelson, O. E., Chourey, P. S., and Chang, M. T.**, Nucleoside diphosphate sugar-starch glucosyl transferase activity of wx starch granules, *Plant Physiol.*, 62, 383, 1978.
67. **Murata, T. and Akazawa, T.**, Enzyme mechanisms of starch synthesis in sweet potato roots. I. Requirement of potassium ions for starch synthetase, *Arch. Biochem. Biophys.*, 126, 873, 1968.
68. **Nitsos, R. E. and Evans, H. J.**, Effects of univalent cations on the activity of the particulate starch synthetase, *Plant Physiol.*, 44, 1260, 1969.
69. **Hawker, J. S., Marschner, H., and Downton, W. J. S.**, Effects of sodium and potassium on starch synthesis in leaves, *Aust. J. Physiol.*, 1, 491, 1974.
70. **Ozbun, J. L., Hawker, J. S., and Preiss, J.**, Multiple forms of α-1,4-glucan synthetase from spinach leaves, *Biochem. Biophys. Res. Commun.*, 43, 631, 1971.
71. **Ozbun, J. L., Hawker, J. S., and Preiss, J.**, Adenosine diphosphoglucose-starch glucosyltransferases from developing kernels of waxy maize, *Plant Physiol.*, 48, 765, 1971.
72. **Ozbun, J. L., Hawker, J. S., and Preiss, J.**, Soluble adenosine diphosphate glucose α-1,4-glucan α-4-glucosyltransferases from spinach leaves, *Biochem. J.*, 126, 953, 1971.
73. **Hawker, J. S., Ozbun, J. L., and Preiss, J.**, Unprimed starch synthesis by soluble ADPglucose-starch glucosyltransferase from potato tubers, *Phytochemistry*, 11, 1287, 1972.
74. **Hawker, J. S. and Downton, W. J. S.**, Starch synthetases from *Vitis vinifera* and *Zea mays*, *Phytochemistry*, 13, 893, 1974.
75. **Preiss, J. and Boyer, C. D.**, Evidence for independent genetic control of the multiple forms of maize endosperm branching enzymes and starch synthases, in *Mechanisms of Polysaccharide Polymerization and Depolymerization*, Marshall, J. J., Ed., Academic Press, New York, 1980, 161.
76. **Hawker, J. S., Ozbun, J. L., Ozaki, H., Greenberg, E., and Preiss, J.**, Interaction of spinach leaf adenosine diphosphate glucose α1,4-glucan α-4-glucosyl transferase and α1,4-glucan α1,4-glucan-6-glycosyl transferase in synthesis of branched α-glucan, *Arch. Biochem. Biophys.*, 160, 530, 1974.
77. **Boyer, C. and Preiss, J.**, Multiple forms of starch branching enzyme of maize: evidence for independent genetic control., *Biochem. Biophys. Res. Commun.*, 80, 169, 1978.
78. **Boyer, C. and Preiss, J.**, Multiple forms of (1 → 4)-α-D-glucan, (1 → 4)-α-D-glucan-6-glycosyl transferase from developing *Zea mays* L. kernels, *Carbohydr. Res.*, 61, 321, 1978.
79. **Drummond, G. S., Smith, E. E., and Whelan, W. J.**, Purification and properties of potato α1,4-glucan, α1,4-glucan-6-glycosyltransferase (Q enzyme), *Eur. J. Biochem.*, 26, 168, 1972.
80. **Borovsky, D., Smith, E. E., and Whelan, W. J.**, Purification and properties of potato 1,4-α-D-glucan:1,4-α-D-glucan 6α(1,4-α-glucan) transferase, *Eur. J. Biochem.*, 59, 615, 1975.
81. **Peat, S., Whelan, W. J., and Bailey, J. M.**, The enzyme synthesis and degradation of starch. XVIII. The minimum chain length for Q enzyme action, *J. Chem. Soc.*, 1953, 1422, 1953.
82. **Nussenbaum, S. and Hassid, W. Z.**, Mechanism of amylopectin formation by the action of Q enzyme, *J. Biol. Chem.*, 196, 785, 1952.
83. **Borovsky, D., Smith, E. E., and Whelan, W. J.**, Temperature-dependence of the action of Q enzyme and the nature of the substrate for Q enzyme, *FEBS Lett.*, 54, 201, 1975.
84. **Borovsky, D., Smith, E. E., and Whelan, W. J.**, On the mechanism of amylose branching by potato Q enzyme, *Eur. J. Biochem.*, 62, 307, 1976.

85. **French, D.**, Fine structure of starch and its relationship to the orga nization of starch granules, *J. Jpn. Soc. Starch Sci.*, 19, 8, 1972.
86. **Doi, A.**, ADP-D-glucose:α1,4-glucan α-4-glucosyltransferase of spinach leaves. Enzymatic synthesis of amylopectin type polysaccharide in a two-enzyme system, *Biochim. Biophys. Acta*, 184, 477, 1969.
87. **Manners, D. J., Rowe, J. J. M., and Rowe, K. L.**, Studies on carbohydrate-metabolizing enzymes. XIX. Sweet corn branching enzymes, *Carbohydr. Res.*, 8, 72, 1968.
88. **Lavintman, N.**, The formation of branched glucans in sweet corn, *Arch. Biochem. Biophys.*, 116, 1, 1966.
89. **Hodges, H. F., Creech, R. G., and Loerch, J. D.**, Biosynthesis of phytoglycogen in maize endosperm. The branching enzyme, *Biochim. Biophys. Acta*, 185, 70, 1969.
90. **Black, R. C., Loerch, J. D., McArdle, F. J., and Creech, R. G.**, Genetic interactions affecting maize phytoglycogen and the phytoglycogen-forming branching enzyme, *Genetics*, 53, 661, 1966.
91. **Whelan, W. J.**, Starch and similar polysaccharides, in *Encyclopedia of Plant Physiology*, Vol. 6, Ruhland, W., Ed., Springer-Verlag, Berlin, 1958, 155.
92. **Geddes, R. and Greenwood, C. T.**, Observations on the biosynthesis of the starch granule, *Staerke*, 21, 148, 1969.
93. **Schiefer, S., Lee, E. Y. C., and Whelan, W. J.**, Multiple forms of starch synthetase in maize varieties as revealed by disc-gel eletrophoresis and activity staining, *FEBS Lett.*, 30, 129, 1973.
94. **Lavintman, N. and Cardini, C. E.**, Particulate UDPglucose: protein transglucosylase from potato tuber, *FEBS Lett.*, 29, 43, 1973.
95. **Lavintman, N., Tandecarz, J., Carceller, M., Mendiara, S., and Cardini, C. E.**, Role of uridine diphosphate glucose in the biosynthesis of starch, *Eur. J. Biochem.*, 50, 145, 1974.
96. **Kawaguchi, K., Fox, J., Holmes, E., Boyer, C., and Preiss, J.**, *De novo* synthesis of *Escherichia coli* glycogen is due to primer associated with glycogen synthase and activation by branching enzyme, *Arch. Biochem. Biophys.*, 190, 385, 1978.
97. **Schiefer, S., Lee, E. Y. C., and Whelan, W. J.**, The requirement for a primer in the *in vitro* synthesis of polysaccharide by sweet corn (1 → 4)-α-D-glucan synthase, *Carbohydr. Res.*, 61, 239, 1978.
98. **Parodi, A. J.**, Factors affecting the molecular weight distribution of liver glycogen, *Arch. Biochem. Biophys.*, 120, 547, 1967.
99. **Whelan, W. J.**, The action patterns of starch metabolizing enzymes, *Staerke*, 9, 74, and 98, 1957.
100. **Slabnik, E. and Frydman, R. B.**, A phosphorylase involved in starch biosynthesis, *Biochem. Biophys. Res. Commun.*, 38, 709, 1970.
101. **Leloir, L. F. and Cardini, C. E.**, The biosynthesis of sucrose phosphate, *J. Biol. Chem.*, 214, 157, 1955.
102. **Mendicino, J.**, Sucrose phosphate synthesis in wheat germ and green leaves, *J. Biol. Chem.*, 235, 3347, 1960.
103. **Hawker, J. S. and Hatch, M. D.**, A specific surcrose phosphatase from plant tissues, *Biochem. J.*, 99, 102, 1966.
104. **Hawker, J. S.**, Studies on the location of sucrose phosphatase in plant tissues, *Phytochemistry*, 5, 1191, 1966.
105. **Pontis, H. G.**, Riddle of sucrose, in *Plant Biochemistry II*, Vol. 13, Northcote, D. H., Ed., University Park Press, Baltimore, 1977, 79.
106. **Avigad, G. and Milner, Y.**, UDPglucose:fructose transglucosylase from sugar beet roots, *Methods Enzymol.*, 8, 341, 1966.
107. **Akazawa, T., Minamikawa, T., and Murata, T.**, Enzyme mechanism of starch synthesis in ripening rice grains, *Plant Physiol.*, 39, 371, 1964.
108. **Delmer, D. P.**, The purification and properties of sucrose synthetase from etiolated *Phaseolus aureus* seedlings, *J. Biol. Chem.*, 247, 3822, 1972.
109. **Avigad, G.**, Sucrose-uridine diphosphate glucosyltransferase from Jerusalem artichoke tubers, *J. Biol. Chem.*, 239, 3613, 1964.
110. **Delmer, D. P. and Albersheim, P.**, The biosynthesis of sucrose and nucleoside diphosphate glucoses in *Phaseolus aureus*, *Plant Physiol.*, 45, 782, 1970.
111. **de Fekete, M. A. R., Zum stoffwecksel der starke.** I. Die umwandlung von saccharose in starke in den kotyledonen von *Vicia fabia*, *Planta*, 87, 311, 1969.
112. **Hawker, J. S.**, Enzymes concerned with sucrose synthesis and transformation in seeds of maize, broad bean and castor bean, *Phytochemistry*, 10, 2313, 1971.
113. **Nomura, T. and Akazawa, T.**, Enzymatic mechanism of starch synthesis in ripening rice grains. VII. Purification and properties of sucrose synthase, *Arch. Biochem. Biophys.*, 156, 644, 1973.
114. **Grimes, W. J., Jones, B. L., and Albersheim, P.**, Sucrose synthetase from *Phaseolus aureus* seedlings, *J. Biol. Chem.*, 245, 188, 1970.
115. **Cardini, C. E. and Recondo, E.**, Specificity of nucleoside diphosphate sugars in sucrose biosynthesis, *Plant Cell Physiol.*, 3, 313, 1962.

116. **Wolusiuk, R. A. and Pontis, H. G.,** Studies on sucrose synthetase: kinetic mechanism, *Arch. Biochem. Biophys.,* 165, 140, 1974.
117. **Pressy, R.,** Potato sucrose synthase: purification, properties and changes in activity associated with maturation, *Plant Physiol.,* 44, 759, 1969.
118. **Delmer, D. P.,** The regulatory properties of purified *Phaseolus aureus* sucrose synthetase, *Plant Physiol.,* 50, 469, 1972.
119. **Su, J.-C. and Preiss, J.,** Purification and properties of sucrose synthase from maize kernels, *Plant Physiol.,* 61, 389, 1978.
120. **Buchanan, J. G., Lynch, V. H., Benson, A. A., Bradley, D. F., and Calvin, M.,** The path of carbon in photosynthesis. XVIII. The identification of nucleotide coenzymes, *J. Biol. Chem.,* 203, 935, 1953.
121. **Edelman, J., Ginsburg, V., and Hassid, W. Z.,** Conversion of monosaccharides to sucrose and cellulose in wheat seedlings, *J. Biol. Chem.,* 213, 843, 1955.
122. **Bean, R. C., Ban, B. K., Welch, H. V., and Porter, G.,** Carbohydrate metabolism of the avocado: relations between sugars in leaves during photosynthesis and subsequent dark periods, *Arch. Biochem. Biophys.,* 96, 524, 1962.
123. **Putnam, E. W. and Hassid, W. Z.,** Sugar transformation in leaves of *Canna indica.* I. Synthesis and inversion of sucrose, *J. Biol. Chem.,* 207, 885, 1954.
124. **Salerno, G. L. and Pontis, H. G.,** Regulation of sucrose levels in plant cells, in *Mechanisms of Polysaccharide Polymerization and Depolymerization,* Marshall, J. J., Ed., Academic Press, New York, 1979.
125. **Slabnik, E., Frydman, R. B., and Cardini, C. E.,** Some properties of potato tuber UDPG:D-fructose-2-glycosyltransferase (E.C. 2.4.1.14) and UDPG:D-fructose-6-phosphate-2-glucosyltransferase (E.C. 2.4.1.13), *Plant Physiol.,* 43, 1063, 1968.
126. **Bird, I. F., Cornelius, M. J., Keys, J., and Whittingham, C. P.,** Intracellular site of sucrose synthesis in leaves, *Phytochemistry,* 13, 59, 1974.
127. **Hawker, J. S.,** Inhibition of sucrose phosophatase by sucrose, *Biochem. J.,* 102, 401, 1967.
128. **Murata, T.,** Sucrose synthetase of rice grains and potato tubers, *Agric. Biol. Chem.,* 36, 1815, 1972.
129. **Preiss, J. and Greenberg, E.,** Allosteric regulation of urdine diphosphoglucose:D-fructose-6-phosphate-2-glucosyl transferase (E.C. 2.4.1.14), *Biochem. Biophys. Res. Commun.,* 36, 289, 1969.
130. **de Fekete, M. A. R.,** The regulative properties of UDPglucose:D-fructose-6-phosphate 2 glucosyl transferase (sucrose phosphate synthetase) from *Vicia fabia* cotyledons, *Eur. J. Biochem.,* 19, 73, 1971.
131. **Salerno, G. L. and Pontis, G. L.,** Studies on sucrose phosphate synthetase. Reversal of UDP inhibition by divalent ions, *FEBS Lett.,* 64, 415, 1976.
132. **Salerno, G. L. and Pontis, G. L.,** Studies on sucrose-phosphate synthetase. The inhibitory action of sucrose, *FEBS Lett.,* 86, 263, 1978.
133. **Murata, T., Sugiyama, T., and Akazawa, T.,** Enzymic mechanism of starch synthesis in ripening rice grains. II. Adenosine diphosphate glucose pathway, *Arch. Biochem. Biophys.,* 107, 92, 1964.
134. **Murata, T., Sugiyama, T., Minamikawa, T., and Akazawa, T.,** Enzyme mechanism of starch synthesis in ripening rice grains. III. Mechanism of the sucrose-starch conversion, *Arch. Biochem. Biophys.,* 113, 34, 1966.
135. **Chourey, P. S. and Nelson, O. E.,** The enzymatic deficiency conditioned by the shrunken-1 mutations in maize, *Biochem. Genet.,* 14, 1041, 1976.

CELLULOSE SYNTHESIS

Deborah Delmer

In contrast to the situation for starch biosynthesis, so little is known about cellulose biosynthesis in photosynthetic organisms that it is not possible even to diagram the pathway of biosynthesis with any degree of certainty. This situation is particularly unfortunate, since cellulose represents the most abundant organic compound of the world; approximately 10^{11} tons of cellulose are synthesized per year in the biosphere.[1,2]

With the exception of a few specialized cell types, nearly all cells of higher plants accumulate cellulose as a major structural component of their cell walls. Undifferentiated plant cells which are still growing are characterized by having only a thin primary cell wall; in such cells, cellulose usually accounts for between 5 to 30% of the dry weight of the cell wall. Cells which have stopped or are stopping growth and which are differentiating into cells with specialized functions, deposit, inside the primary wall, a much thicker secondary cell wall which is often heavily cellulosic. For example, in the mature cells of normal wood, cellulose usually accounts for between 40 to 50% of the total dry weight and is the major factor responsible for the tensile strength of such tissues.[3] In some specialized cells like cotton fibers, cellulose can represent 90 to 95% of the dry weight of the secondary cell wall.[4] In contrast to starch, once synthesized, cellulose does not seem to ever be significantly degraded by the plant, and thus it represents a stable end product of carbon metabolism for the plant. Many algae also contain cellulose as a major structural component of their cell walls, although some types do substitute mannans, xylans, or other polysaccharides as the microfibrillar framework.[5] The most widely studied of the algae with cellulosic walls have been the genera *Valonia, Pleurochrysis, Oocystis, Glaucocystis, Cladophora*, and *Chaetomorpha*. Few bacteria contain significant amounts of cellulose; one notable exception is the nonphotosynthetic bacterium *Acetobacter xylinium* which synthesizes ribbons of cellulose into the medium in which it is grown.[6]

Consideration of the structure and orientation of cellulose microfibrils is essential for any discussion of biosynthesis. Cellulose is composed of many extended chains of β-(1→4)-glucan which are tightly hydrogen bonded to each other. Such an array is termed a fibril and is highly insoluble and usually also highly crystalline. The current, most widely accepted model for the crystalline structure of native cellulose (termed Cellulose I) is one in which the chains are proposed to be in parallel alignment with both intramolecular hydrogen bonding between the adjacent glucose residues and intermolecular hydrogen bonding between the chains.[7] Less current and less favored models proposed that the chains were folded[8] or extended, but antiparallel.[9] If the parallel model is the correct one, it simplifies our concepts of biosynthesis. Since the microfibrils seem to grow from one end only, an antiparallel arrangement would necessitate two separate polymerization mechanisms whereby half the chains were polymerized from the reducing end of the molecule and the other half from the nonreducing end.

Controversy still exists as to what represents a basic fibril. In the electron microscope, large, so-called microfibrils (often 10 to 30 nm) are commonly observed in many cell walls (the size will vary with the organism under study); however, smaller, "elementary fibrils" of approximately 3.5 nm diameter have also been observed, although they have been dismissed by some as an artifact. However, according to a recent model of Blackwell and Kolpack,[10] the elementary fibril is a real structural entity, one plausible model for which was considered to be an array of 6×6 (= 36) chains, with the so-called microfibril representing an imperfect array of elementary fibrils.

The length (degree of polymerization) of the individual glucan chains is also open to some question, since measurement can only be made following solubilization of the chains, a procedure which may result in some chain breakage. One of the more reliable estimates which is often quoted is the report of Marx-Figini and Schulz, who used rather mild solubilization techniques to determine by viscosity the degree of polymerization of cotton fiber cellulose.[11] Their data indicate that the cellulosic chains of the primary wall are of rather heterogeneous length, ranging from 2000 to 6000 glucose residues per chain, whereas those of the secondary cell wall cellulose are longer and rather homogeneous at about 14,000 residues per chain. It has also been speculated[12] that the chains may really have no end, but are synthesized as one continuous structure in the cell wall.

For the discussion which is to follow, the reader may use as an aid the stylized drawing (see Figure 1) of this author's conception of a probable pathway and mechanism for the biosynthesis of fibrillar cellulose. The model is hypothetical and has many points unproven, however, it represents an attempt to synthesize and interpret our current state of knowledge.

For the purpose of discussing biosynthesis, the most relevant features of the structure of cellulose would seem to be the following:

1. Each glucose residue is rotated approximately 180° with respect to its neighbor; thus, the basic repeating unit of cellulose is cellobiose and not glucose, and, although it may not be necessary, perhaps some activated form of cellobiose should be considered as a possible precursor.
2. Many glucan chains compose one microfibril; thus, it might be expected that the cellulose synthetase would be a multisubunit protein complex, with each subunit responsible for catalyzing the polymerization of one, or perhaps two, glucan chains.
3. Particularly in secondary cell walls, the microfibrils are precisely oriented and are usually deposited in sequential layers of opposing orientation; any ultimate understanding of the synthesis mechanism must consider how the events of synthesis and orientation of microfibrils are coupled.
4. In higher plants, the orientation of primary wall fibrils often appear to be random; furthermore, the degree of polymerization may be considerably lower than that of secondary wall cellulose, and the onset of secondary wall synthesis appears to be a precisely regulated event.

All of these factors suggest that a separate enzyme system, or at least a separate regulatory system, may be involved for primary and secondary wall cellulose synthesis; however, to date there is essentially no specific information to support or refute this concept.

It is now generally agreed that, in most algae and all higher plants, cellulose is synthesized at the cell surface. One example of an exception is represented by the alga *Pleurochrysis* which synthesizes scales of cellulose within a large, intracellular Golgi complex; the scales are then transported and deposited in the cell wall after assembly.[13] The conclusion that most other photosynthetic organisms synthesize cellulose at the cell surface is based partly on the fact that intracellular cellulosic fibrils have not been observed cytologically nor demonstrated chemically. This conclusion is also supported by freeze-fracture studies of plasma membranes such as those by Brown and Montezinos[14] which show what appear to be multisubunit complexes on the interior face of the plasma membrane of *Oocystis*. These particles appear to be associated with the ends of cellulose microfibrils, and it has been suggested (without any biochemical evidence yet) that these particles represent a cellulose-synthesizing complex. Similar, but

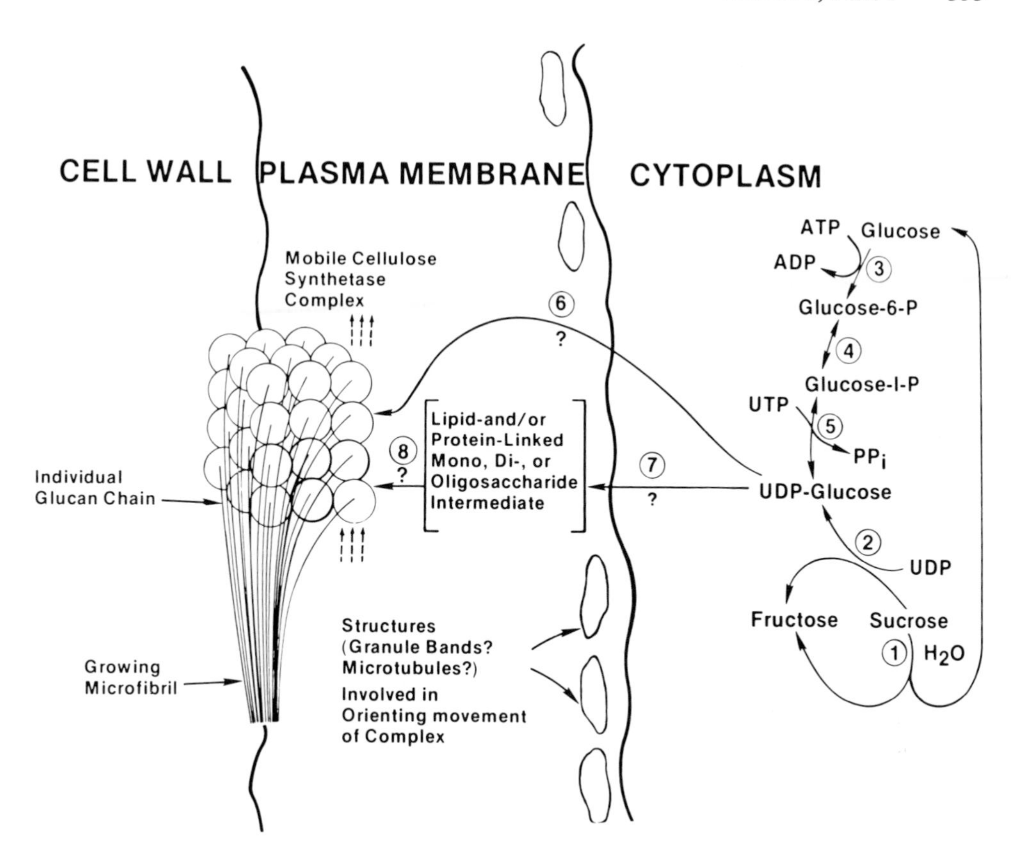

FIGURE 1. Hypothetical model for the biosynthesis of cellulose. Numbers refer to reactions catalyzed by the following enzymes: 1, invertase; 2, sucrose synthetase; 3, hexokinase; 4, phosphoglucomutase; 5, UDP-glucose pyrophosphorylase; 6, 7, and 8, hypothetical reactions on the pathway to cellulose.

smaller, globular particles have been observed on the plasma membrane of corn root cells and have been seen associated with presumably cellulosic fibrils.[15] Since each microfibril consists of a large number of glucan chains, a multisubunit enzyme complex would be expected to participate in cellulose synthesis (see Figure 1).

The manner by which the cellulose microfibrils are oriented within the cell wall is poorly understood. The fact that colchicine interferes with orientation of newly synthesized microfibrils has been interpreted to indicate that microtubules play some role in the orientation process.[16,17] This possibility is further supported by the observation that microtubule orientation itself is often (but not always) correlated with the orientation of the microfibrils of the wall.[17] On the other hand, ordered granules present in the plasma membrane of algae have been suggested to play a role in orientation of microfibril deposition in these organisms.[14]

The area of least understanding concerns the path of carbon into cellulose. This is so because, for obscure reasons, scientists have been unable to demonstrate convincingly in vitro synthesis of cellulosic microfibrils using any higher plant system. As a consequence of this, it is not even possible to state with certainty what the activated form(s) of glucose is which is used as substrate for polymerization. By analogy with other polysaccharide-synthesizing enzymes, it has usually been assumed that a nucleoside diphosphate glucose is the most logical precursor; of the various forms known, most attention has centered upon GDP-glucose and UDP-glucose as potential substrates. In this author's opinion, a role for GDP-glucose as substrate in higher plants seems now rather unlikely; the arguments upon which this conclusion is based have been summarized in detail elsewhere,[12,18] and therefore will not be repeated here. UDP-glucose, however, is still considered by most workers to be a likely intermediate on the pathway to cellulose. The reasons for this are

1. In the nonphotosynthetic bacterium *Acetobacter xylinium*, radioactivity from UDP-^{14}C-glucose has been shown to be incorporated into cellulose.[19]
2. The capacity for synthesis of this nucleotide sugar via either sucrose synthetase or UDP-glucose pyrophosphorylase (see Figure 1) is quite high in those plants where it has been studied.[18]
3. In the few cases investigated, UDP-glucose is the only nucleoside diphosphate glucose found in significant quantity in plant cells[18] and is particularly abundant, for example, in cotton fibers at the time of secondary wall cellulose deposition.[20]

There are a number of reports in the literature which claim that UDP-glucose can serve as a substrate in vitro for cellulose synthesis using extracts derived from tissues of higher plants. Unfortunately, many of these reports are of questionable accuracy, and the product synthesized may in fact have been β-(1→3)-glucan or the backbone of xyloglucan chains which contain β(→4)-glucose residues.[12,18]

If UDP-glucose is involved as precursor, by analogy with recent elegant studies on chitin synthesis in yeast,[21] it is possible that polymerization involves direct enzyme-catalyzed transfer of glucose from UDP-glucose to the growing glucan chain (Reaction 6 of Figure 1). However, by analogy with the synthesis of cell wall components in bacteria,[22] transfer of activated glycosyl units through the hydrophobic region of the plasma membrane may necessitate the use of so-called "lipid-intermediates". In these types of reactions, sugars are transferred from nucleotide sugars to a membrane-associated phosphorylated long-chain polyisoprenol. In this way, a phosphorylated lipid oligosaccharide can be synthesized which serves as donor for final polymerization of the polysaccharide product (see Reactions 7 and 8, Figure 1). Interestingly, it was Colvin, while studying cellulose synthesis in *Acetobacter*, who first evolved the concept of a lipid intermediate. Evidence for such an intermediate in this organism has accumulated, but to date such a compound has never been purified from *Acetobacter* or completely structurally characterized.[23] In higher plants, recent data support the role for such intermediates in the glycosylation of N-glycosidically linked glycoproteins,[12] but no strong evidence exists for (or against) their role in cellulose synthesis. One very recent report (yet to be confirmed in other laboratories) presented data which indicate a role for both UDP-glucose and GDP-glucose, as well as lipid-linked and protein-linked glucan oligosaccharides, as intermediates in cellulose synthesis in the nonphotosynthetic alga *Prototheca*.[24] Although quite interesting, the evidence for these intermediates must still be considered preliminary.

The inability to achieve significant synthesis of cellulose could be for any one of a variety of reasons, such as inadequate knowledge concerning the appropriate substrates and/or an extreme lability of the synthesizing complex. Since cellulose is the most abundant molecule in the biomass, it is hoped that new information concerning its biosynthesis will be forthcoming in the near future.

REFERENCES

1. **Hess, K.,** *Die Chemie der Zellulose un Ihrer Begleiter,* Akademische Verlagsgesellschaft, M. B. H., Leipzig, 1928, 5.
2. **Hall, D. O.,** Solar energy use through biology — past, present and future, *Sol. Energy,* 22, 307, 1979.
3. **Côté, W. A.,** Wood ultrastructure in relation to chemical composition, in *The Structure, Biosynthesis, and Degradation of Wood,* Vol. 11, Loewus, F. A. and Runeckles, V. C., Eds., Plenum Press, New York, 1974, chap. 1.
4. **Meinert, M. and Delmer, D. P.,** Changes in biochemical composition of the cell wall of the cotton fiber during development, *Plant Physiol.,* 59, 1088, 1977.
5. **Preston, R. D.,** *The Physical Biology of Plant Cell Walls,* Chapman and Hall, London, 1974.
6. **Brown, R. M., Jr., Willison, J. H., and Richardson, C. L.,** Cellulose biosynthesis in *Acetobacter xylinium*: visualization of the site of synthesis and direct measurement of the *in vivo* process, *Proc. Natl. Acad. Sci. U.S.A.,* 73, 4565, 1976.
7. **Gardner, K. H. and Blackwell, J.,** The structure of native cellulose, *Biopolymers,* 13, 1975, 1974; **Sarko, A. and Muggli, R.,** Packing analysis of carbohydrates and polysaccharides. III. *Valonia* cellulose and cellulose II, *Macromolecules,* 7, 486, 1974.
8. **Manley, R. St. J.,** Fine structure of native cellulose microfibrils, *Nature (London),* 204, 1155, 1964.
9. **Muhlethaler, K.,** Fine structure of natural polysaccharide systems, *J. Polym. Sci. Part C,* 28, 305, 1969.
10. **Blackwell, J. and Kolpak, F. J.,** The cellulose microfibril as an imperfect array of elementary fibrils, *Macromolecules,* 8, 322, 1975.
11. **Marx-Figini, M. and Schulz, G. V.,** Uber die Kinetik und den Mechanisms der Biosynthese der Cellulose in den hoheren Pflanzen (nach Versuchen an den Samenhaaren der Baumwolle), *Biochim. Biophys. Acta,* 112, 74, 1966.
12. **Darvill, A., McNeil, M., Albersheim, P., and Delmer, D. P.,** The primary cell walls of flowering plants, in *The Biochemistry of Plants,* Vol. 1, Tolbert, N. E., Ed., Academic Press, New York, 1980 chap. 3.
13. **Brown, R. M., Jr., Franke, W. W., Kleinig, H., Falk, H., and Sitte, P.,** Cellulosic wall component produced by the Golgi apparatus of *Pleurochrysis Scherffeli* Pringsheim, *Science,* 166, 894, 1969.
14. **Brown, R. M., Jr. and Montezinos, D.,** Cellulose microfibrils: visualization of biosynthetic and orienting complexes in association with the plasma membrane, *Proc. Natl. Acad. Sci. U.S.A.,* 73, 143, 1976.
15. **Mueller, S. C., Brown, R. M., Jr., and Scott, T. K.,** Cellulosic microfibrils: nascent stages of synthesis in a higher plant cell, *Science,* 194, 949, 1976.
16. **Pickett-Heaps, J. D.,** The effects of colchicine on the ultrastructure of dividing plant cells, xylem wall differentiation and distribution of cytoplasmic microtubules, *Dev. Biol.,* 15, 206, 1967.
17. **Robinson, D. G.,** Plant cell wall synthesis, *Adv. Bot. Res.,* 5, 89, 1977.
18. **Delmer, D. P.,** The biosynthesis of cellulose and other cell wall polysaccharides, in *The Structure, Biosynthesis, and Degradation of Wood,* Vol. 11, Loewus, F. A. and Runeckles, V. C., Eds., Plenum Press, New York, 1974, chap. 2.
19. **Glaser, L.,** The synthesis of cellulose in cell-free extracts of *Acetobacter xylinum, J. Biol. Chem.,* 232, 627, 1958.
20. **Carpita, N. and Delmer, D. P.,** Changes in nucleotide sugar pools in cotton fibers, *Plant Physiol.,* 61 S-639, 1978.
21. **Ruiz-Herrera, J., Lopez-Romero, E., and Bartnicki-Garcia, S.,** Properties of chitin synthetase in isolated chitosomes from yeast cells of *Mucor rouxii, J. Biol. Chem.,* 252, 338, 1977.
22. **Lennarz, W. J. and Scher, M. G.,** The role of lipid-linked activated sugars in glycosylation reactions, in *Membrane Structure and Mechanisms of Biological Energy Transduction,* Avery, J., Ed., Plenum Press, New York, 1973, 441.
23. **Kjozbakken, J. and Colvin, J. R.,** Biosynthesis of cellulose by a particulate enzyme system from *Acetobacter xylinium,* in *Biogenesis of Plant Cell Wall Polysaccharides,* Loewus, F., Ed., Academic Press, New York, 1973, 361.
24. **Hopp, H. E., Romero, P. A., Daleo, G. R., and Pont-Lezica, R.,** Synthesis of cellulose precursors. The involvement of lipid-linked sugars, *Eur. J. Biochem.,* 84, 561, 1978.

HEMICELLULOSE SYNTHESIS

Deborah Delmer

Some years ago, Meier[1] pointed out that the term hemicellulose (half cellulose) is not a very good one. In general, this term has come to refer to the noncellulosic polysaccharides found in the cell walls of plants.* However, the so-called pectic polysaccharides are usually not considered to be hemicelluloses. These distinctions are really becoming rather archaic; they were based originally on the criteria of solubility and extractability of the polysaccharides from the cell wall. Those polysaccharides which are extracted from walls by hot water, chelating agents, or weak acid have been grouped together as the pectic substances; the remaining noncellulosic polysaccharides which require more drastic procedures, such as alkali for extraction, are classed as the hemicelluloses. However, another definition of hemicelluloses is sometimes that of the "matrix substances", and by this terminology would include all the noncellulosic polysaccharides which represent the total matrix in which cellulosic microfibrils are embedded. This brief review considers the biosynthesis of all the noncellulosic polysaccharides of the cell wall. In any case, the review can be kept brief because we know so little about the biosynthesis of any of these polymers. Other recent reviews further document this situation.[2-4]

Table 1 summarizes the general types of noncellulosic polysaccharides found in the cell walls of plants. The data were collectively derived from a variety of sources, the best reviews of which are those of Darvill et al.,[4] Timell,[5] Côté,[6] Aspinall,[7] and Preston.[8] For the purpose of discussing biosynthesis, several points about these polymers should be emphasized:

1. Except for a few cases, almost all of the polymers contain more than one type of monosaccharide residue. Furthermore, the names in themselves are oversimplifications; to cite just one example, xyloglucans are characterized by having a β-1,4-glucan backbone to which terminal xylose residues are α-linked at carbon-6 of the glucosyl units. However, lesser amounts of galactosyl residues may also be β-1,2-linked to the xylosyl residues, and fucosyl residues may be further α-1,2-linked to the galactosyl residues.[4] Clearly, the biosynthesis of such polymers requires a delicate coordination of a variety of substrates and enzymes acting in concert. In attempts to synthesize polymers in vitro using plant extracts, it is conceivable that a loss of even one of these activities may render the entire synthesis inoperable. In view of this, it is unfortunate how few studies have been carried out which attempt in vitro synthesis using more than one type of substrate.
2. We still know very little about how these polymers may be interconnected in the cell wall. This is particularly true for the matrix polymers of primary cell wall where the heterogeneity of polymers appears to be more extensive. In fact, one recent model has proposed that all of the matrix components are covalently linked to each other, forming one great macromolecule;[9] this, however, is probably something of an overstatement, although it is clear that many polymers are interconnected.
3. The distribution of polymers, both qualitative and quantitative, is a function of the species examined and of the cell type within that species. In terms of contribution to the biomass of the world, the hemicellulosic components of wood are

* The biosynthesis of algal polysaccharides is discussed elsewhere in this volume.

Table 1
NONCELLULOSIC POLYSACCHARIDES OF PLANT CELL WALLS

Pectic polymers of primary cell walls	Hemicellulosic polymers of primary cell walls	Hemicellulosic polymers of secondary cell walls
Rhamnogalacturonan	Xyloglucan[a]	*O*-Acetyl-4-*O*-methyl-glucurono-xylan[b]
Apiogalacturonan	Glucuronoarabinoxylan[c]	Arabino-4-*O*-methyl-glucurono-xylan[d]
Homogalacturonan	Mixed-linked glucan[c]	Glucomannan[b]
Araban		*O*-acetyl-galactoglucomannan[d]
Galactan		β-1,3-glucan[e]
Arabinogalactan[f]		

[a] More predominant in the primary walls of dicotyledonous plants. Also found as an amyloid component of seeds.[4]
[b] Predominantly a hemicellulosic component of the wood of angiosperms.[1,5,6]
[c] More predominant in the primary walls of monocotyledonous plants.[4]
[d] Predominantly a hemicellulosic component of the wood of conifers.[1,5,6]
[e] Found in secondary walls of cotton fibers[13,14] and primary walls of pollen.[15] Its varying solubility makes it difficult to classify as a pectic or hemicellulosic component; it is more commonly found as a wound polymer in plants.
[f] Considered by some not to be a true cell wall constituent.[4,6]

certainly the most abundant of the noncellulosic polymers. Unfortunately, because of the nature of the tissues, studies on in vitro biosynthesis in these cell types are almost nonexistent. Because of their rapid growth rate, relative homogeneity of cell type, and ease of breakage, the cells of growing plant parts, such as hypocotyls, epicotyls, and roots of seedlings, pollen tubes, cotton fibers, and undifferentiated cells in tissue culture, have been used most often for biosynthetic studies.

As regards the site of synthesis of the matrix components, most workers believe that, in contrast to cellulose, synthesis of these polymers occurs intracellularly, probably within the Golgi apparatus. From this site, the polymers are secreted via Golgi-derived vesicles which release their contents to the cell wall through fusion of the vesicles with the plasma membrane. These conclusions are based partly on autoradiographic and partly on biochemical studies, and the literature supporting these conclusions has recently been reviewed in some detail elsewhere.[3,4]

It is also generally agreed now that nucleotide sugars are the most likely substrates, serving as glycosyl donors via transglycosylation reactions. Figure 1 outlines what are now thought to be the major pathways for synthesis and interconversions of those sugar nucleotides most likely to be involved in cell wall synthesis. The literature on this topic has been reviewed recently in detail elsewhere.[10] It should be noted that two alternate pathways are given for the synthesis of the uronosyl and pentosyl nucleotide sugars. The pathway through UDP-glucose (the nucleotide sugar oxidation pathway) was originally thought to be the only pathway; however, careful studies by Loewus and co-workers[11,12] have established that another pathway (the so-called myo-inositol oxidation pathway) is also operative in plants. It is not yet clear how widespread is the occurrence of this latter pathway, but in some tissues such as lily pollen, 70% of the flow of carbon into the uronosyl and pentosyl units of the wall can be derived from this pathway.[12]

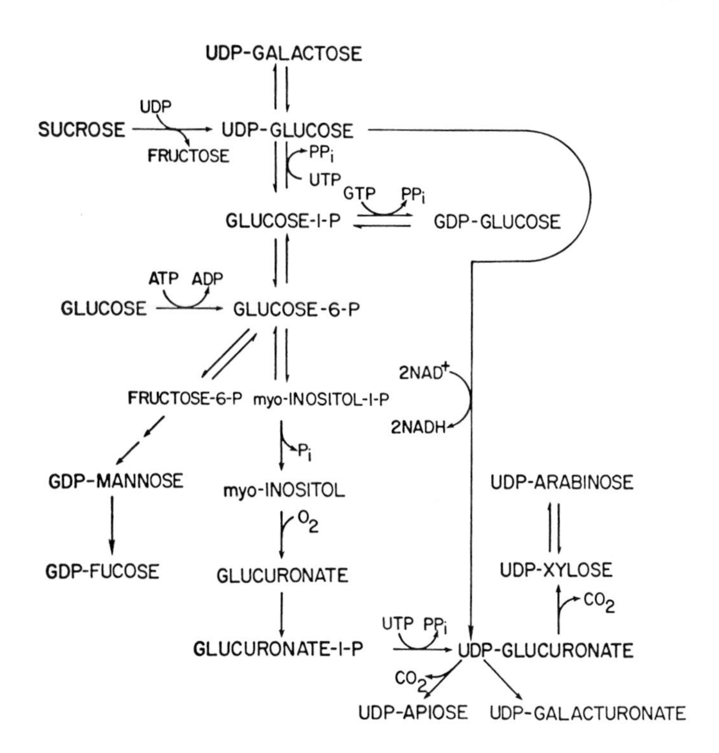

FIGURE 1. Pathways for the synthesis and interconversions of nucleotide sugars in plants.

As discussed in the article for cellulose synthesis, the question of whether lipid intermediates play a role in cell wall synthesis in plants is still unresolved. To date, there is no strong evidence to support or refute a role for these compounds in the synthesis of the noncellulosic polysaccharides.[2,4]

Table 2 summarizes the current studies on attempts at in vitro synthesis of noncellulosic cell wall polysaccharides. Most of these studies were performed by supplying radioactive nucleotide sugars to isolated membrane preparations of tissues studied and monitoring for incorporation of radioactivity into ethanol-insoluble material. The reader should consult the original reference cited for details of the characterization of the products synthesized. In many cases, only limited characterization was performed of the products with respect to purification or analyses of linkages and anomeric configurations of the glycosyl residues. In no case was it established whether endogenous components in the preparations served as primers. In view of these shortcomings, only limited conclusions can be drawn about the final polymerization steps involved in the synthesis of the noncellulosic polysaccharides. They are simply that the enzymes are usually membrane bound (probably Golgi membranes) and that nucleotide sugars can serve as substrates for polymer synthesis with the uracil derivative being the most frequently active. Clearly much more exploration is warranted for this area of polysaccharide biosynthesis.

Table 2
IN VITRO SYNTHESIS OF NONCELLULOSIC CELL WALL POLYSACCHARIDES

Substrate used	General type of product produced	Tissue studied	Ref.
UDP-galacturonic acid or TDP-galacturonic acid	Polygalacturonic acid	Hypocotyls of etiolated *Phaseolus aureus*	16,17
S-adenosylmethionine	Methyl esters of polyuronides	Hypocotyls of etiolated *P. aureus*	18
UDP-arabinose and/or UDP-xylose	Araban, xylan	Seedlings of *Avena sativa*	19
		Hypocotyls of etiolated *P. aureus*	20
		Immature cobs of *Zea mays*	21
UDP-galactose	Galactan	Hypocotyls of etiolated *P. aureus*	22
UDP-xylose and UDP-glucose	Xyloglucan	Epicotyls of etiolated *Pisum sativum*	23,24
GDP-mannose	Mannan	Tubers of *Orchis*	25
		Suspension cultures of *Acer pseudoplatanus*	26
GDP-mannose and GDP-glucose	Glucomannan	Hypocotyls of etiolated *Phaseolus aureus*	27,28
UDP-apiose and UDP-galacturonic acid	Apiogalacturonan	*Lemna*	29,30
UDP-glucose	Mixed linked glucan and/or β-1,3-glucan	Epicotyls of etiolated *Pisum sativum*	31,32
		Cotton fibers	33,34
		Hypocotyls of etiolated *Phaseolus aureus*	35
		Avena coleoptiles	36
		Endosperm of *Lolium*	37
		Seedlings of *Triticum*	38

REFERENCES

1. **Meier, H.,** General chemistry of cell walls and distribution of the chemical constituents across the walls, in *The Formation of Wood in Forest Trees,* Zimmerman, M. H., Ed., Academic Press, New York, 1964, 137.
2. **Delmer, D. P.,** The biosynthesis of cellulose and other plant cell wall polysaccharides, in *The Structure, Biosynthesis, and Degradation of Wood,* Vol. 11, Loewus, F. and Runeckles, V. C., Eds., Plenum Press, New York, 1974, chap. 2.
3. **Robinson, D. G.,** Plant cell wall synthesis, *Adv. Bot. Res.,* 5, 89, 1977.
4. **Darvill, A., McNeil, M., Albersheim, P., and Delmer, D. P.,** The primary cell walls of flowering plants, in *The Biochemistry of Plants,* Vol. 1, Tolbert, N. E., Ed., Academic Press, New York, 1979, chap. 3.
5. **Timell, T. E.,** Wood and bark polysaccharides, in *Cellular Ultrastructure of Woody Plants,* Côté, W. A., Ed., Syracuse University Press, Syracuse, N. Y., 1965, 127.
6. **Côté, W. A.,** Wood ultrastructure in relation to chemical composition, in *The Structure, Biosynthesis and Degradation of Wood,* Vol. 11, Loewus, F. and Runeckles, V. C., Eds., Plenum Press, New York, 1974, chap. 1.
7. **Aspinall, G.,** Carbohydrate polymers of plant cell walls, in *Biogenesis of Plant Cell Wall Polysaccharides,* Loewus, F., Ed., Academic Press, New York, 1973, 95.
8. **Preston, R. D.,** *The Physical Biology of Plant Cell Walls,* Chapman and Hall, London, 1974.
9. **Keegstra, K., Talmadge, K. W., Bauer, W. D., and Albersheim, P.,** The structure of plant cell walls. III. A model of the walls of suspension-cultured sycamore cells based on the interconnections of the macromolecular components, *Plant Physiol.,* 51, 188, 1973.

10. **Gander, J. E.,** Mono- and oligosaccharides, in *Plant Biochemistry,* 3rd ed., Bonner, J. and Varner, J. E., Eds., Academic Press, New York, 1976, chap. 11.
11. **Loewus, F., Chen, M. S., and Loewus, M. W.,** The myo-inositol oxidation pathway to cell wall polysaccharides, in *Biogenesis of Plant Cell Wall Polysaccharides,* Loewus, F., Ed., Academic Press, New York, 1973, 1.
12. **Maiti, I. B. and Loewus, F. A.,** Evidence for a functional myo-inositol oxidation pathway in *Lilium longiflorum* pollen, *Plant Physiol.,* 62, 280, 1978.
13. **Maltby, D., Carpita, N., Montezinos, D., and Delmer, D. P.,** β-1,3-glucan in developing cotton fibers: structure, localization, and relationship of synthesis to that of secondary wall cellulose, *Plant Physiol.,* 63, 1158, 1979.
14. **Huwyler, H. R., Franz, G., and Meier, H.,** β-1,3-Glucans in the cell walls of cotton fibers, *Plant Sci. Lett.,* 12, 55, 1978.
15. **Herth, W., Franke, W. W., Bittiger, H., Kuppel, A., and Keilich, G.,** Alkali-resistant fibrils of β-1,3- and β-1,4-glucans: structural polysaccharides in the pollen tube wall of *Lilium longiflorum, Cytobiologie,* 9, 344, 1974.
16. **Liu, T. Y., Elbein, A. D., and Su, J. C.,** Substrate specificity in pectin synthesis, *Biochem. Biophys. Res. Commun.,* 22, 650, 1966.
17. **Villemez, C. L., Liu, T. Y., and Hassid, W. Z.,** Biosynthesis of the polygalacturonic acid chain of pectin by a particulate enzyme preparation from *Phaseolus aureus* seedlings, *Proc. Natl. Acad. Sci. U.S.A.,* 54, 1626, 1965.
18. **Kauss, H.,** Biosynthesis of pectin and hemicelluloses, in *Plant Carbohydrate Chemistry,* Pridham, J. B., Ed., Academic Press, New York, 1974, 191.
19. **Ben-Arie, R., Ordin, L., and Kindinger, J. L.,** A cell-free xylan synthesizing enzyme from *Avena sativa, Plant Cell Physiol.,* 14, 427, 1973.
20. **Ozduck, W. and Kauss, H.,** Biosynthesis of pure araban and xylan, *Phytochemistry,* 11, 2489, 1972.
21. **Bailey, R. W. and Hassid, W. Z.,** Xylan synthesis from uridine-diphosphate-D-xylose by particulate preparations from immature corncobs, *Proc. Natl. Acad. Sci. U.S.A.,* 56, 1586, 1966.
22. **Panayotatos, N. and Villemez, C. L.,** The formation of a β-(1→4)-D-galactan chain catalyzed by a *Phaseolus aureus* enzyme, *Biochem. J.,* 133, 263, 1973.
23. **Ray, P. M.,** Golgi membranes form xyloglucan from UDPG and UDP-xylose, *Plant Physiol.,* 56, S-84, 1975.
24. **Villemez, C. L. and Hinman, M.,** UDP-glucose stimulated formation of xylose containing polysaccharides, *Plant Physiol.,* 56, S-79, 1975.
25. **Franz, G.,** Biosynthesis of salep mannan, *Phytochemistry,* 12, 2369, 1973.
26. **Smith, M. M., Axelos, M., and Péaud-Lenoël, C.,** Biosynthesis of mannan and mannolipids from GDP-man by membrane fractions of sycamore cell cultures, *Biochimie,* 58, 1195, 1976.
27. **Elbein, A. D.,** Biosynthesis of a cell wall glucomannan in mung bean seedlings, *J. Biol. Chem.,* 244, 1608, 1969.
28. **Villemez, C. L.,** Rate studies of polysaccharide biosynthesis from guanosine diphosphate-α-D-glucose and guanosine diphosphate-α-D-mannose, *Biochem. J.,* 121, 151, 1971.
29. **Mascaro, L. J. and Kindel, P. K.,** Characterization of [^{14}C]apiogalacturonans synthesized in a cell-free system from *Lemna minor, Arch. Biochem. Biophys.,* 183, 139, 1977.
30. **Pan, Y. T. and Kindel, P. K.,** Characterization of particulate D-apiosyl- and D-xylosyltransferase from *Lemna minor, Arch. Biochem. Biophys.,* 183, 131, 1977.
31. **Anderson, R. L. and Ray, P. M.,** Labeling of the plasma membrane of pea cells by a surface localized glucan synthetase, *Plant Physiol.,* 61, 723, 1978.
32. **Raymond, Y., Fincher, G. B., and MacLachlan, G. A.,** Tissue slice and particulate β-glucan synthetase activities from *Pisum* epicotyls, *Plant Physiol.,* 61, 938, 1978.
33. **Delmer, D. P., Heiniger, U., and Kulow, C.,** UDP-glucose: glucan synthetase from developing cotton fibers. I. Kinetic and physiological properties, *Plant Physiol.,* 59, 713, 1977.
34. **Heiniger, U. and Delmer, D. P.,** UDP-glucose: glucan synthetase in developing cotton fibers. II. Structure of the reaction product, *Plant Physiol.,* 59, 719, 1977.
35. **Chambers, J. and Elbein, A. D.,** Biosynthesis of glucans in mung bean seedlings. Formation of β-(1→4)-glucans from GDP-glucose and β-(1→3)-glucans from UDP-glucose, *Arch. Biochem. Biophys.,* 138, 620, 1970.
36. **Ordin, L. and Hall, M. A.,** Cellulose synthesis in higher plants from UDP-glucose, *Plant Physiol.,* 43, 473, 1968.
37. **Smith, M. M. and Stone, B. A.,** β-Glucan synthesis by cell free extracts from *Lolium multiflorum* endosperm, *Biochim. Biophys. Acta,* 313, 72, 1973.
38. **Pêaud-Lenoël, C. and Axelos, M.,** Structural features of the β-glucans enzymatically synthesized from uridine diphosphate glucose by wheat seedlings, *FEBS Lett.,* 8, 224, 1970.

BACTERIAL POLYSACCHARIDE SYNTHESIS*

I. W. Sutherland

INTRODUCTION

Three distinct structural types of bacterial extracellular polysaccharides can be recognized. Depending on whether these polymers are composed of a single type of monosaccharide unit or of two or more monosaccharides, they are termed homopolysaccharides and heteropolysaccharides, respectively.[1] The heteropolysaccharides may either be formed from regular oligosaccharide subunits or, more rarely, may be a random copolymer of two constituent monosaccharides, as exemplified by bacterial alginate synthesized by *Azotobacter vinelandii* [2] and by *Pseudomonas aeruginosa*.[3] The monosaccharides found in microbial polysaccharides commonly are neutral sugars — hexoses or 6-deoxyhexoses — and uronic acids or acetylamino-sugars. Many bacterial polysaccharides also contain acyl groups, most frequently *O*-acetyl groups or pyruvate linked as a ketal (see Figure 1). Other acyl groups which have been detected include formate[4] and succinate,[5] and more than one type of acyl group may be present in the same polysaccharide.

Physically, the polysaccharides may form a capsule surrounding the bacterial cell and attached to it under normal growth conditions, or it may all be released into the environment in the form of soluble slime unattached to the microorganisms.

Most polysaccharides are strain specific, i.e., each strain of any particular bacterial species produces a polysaccharide of distinct serological type and chemical structure. This has been most intensively studied in *Enterobacter (Klebsiella) aerogenes*, in which species 80 different polysaccharide serological types and a rather smaller number of chemotypes have been identified. The structure of many of these polymers have now been determined.[1] Alternatively, the same polymer may be synthesized by more than one bacterial species. This is seen in the polysaccharide colanic acid, which is formed by several closely related species of the Enterobacteriaceae, including *Escherichia coli, Enterobacter cloacae*, and *Salmonella typhimurium*.[6] Although the carbohydrate structure of the product from these different sources was constant, variations were discovered by Garegg et al. in the acyl substituents.[7] The production of the same polysaccharide by an *E. coli* strain and an *E. aerogenes* strain was also noted.[8] Of less common occurrence is the production of the same polysaccharide by bacterial strains which are less closely related to one another, as already mentioned with bacterial alginate production in cultures of *A. vinelandii* and *P. aeruginosa*.[2,3] In some bacterial species studied, all thoroughly tested strains yield the same polysaccharide. This is the case for *A. vinelandii* (alginate) and *Acetobacter xylinum* (cellulose).

FACTORS AFFECTING POLYSACCHARIDE SYNTHESIS

Synthesis by whole cells may be affected by various physiological factors (see Table 1), so the conditions employed for cell cultivation are important. Polysaccharide production may occur both during and after growth or, in a few bacteria, only after growth has ceased. As a result, optimal yields of the enzymes and precursors involved in polysaccharide production may depend on both the physical and chemical conditions applied to the culture. It also has been noted frequently that strains producing

* There is no definitive study on either the structure or synthesis of polysaccharides in photosynthetic bacteria. This section deals with bacteria in general.

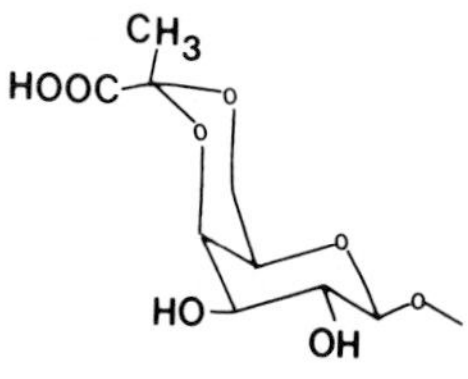

FIGURE 1. Ketals attached to the terminal reducing galactose residues of colanic acids from different strains of *Escherichia coli* and *Salmonella typhi.*[8]

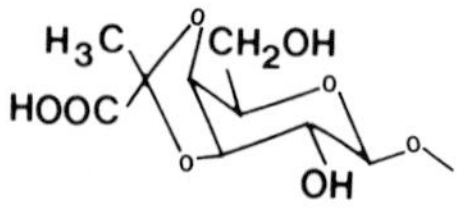

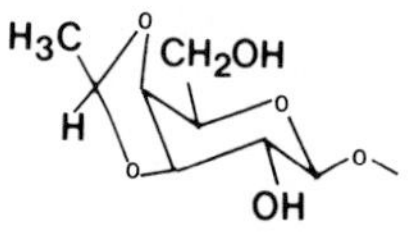

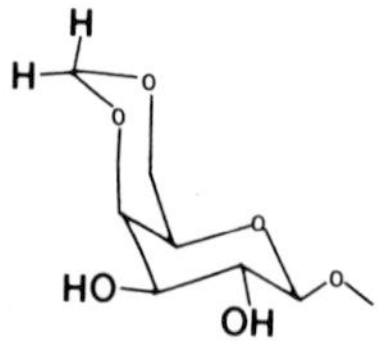

copious polysaccharide on solid media either fail to form polymer or do so with much lower yields in liquid media. Consequently, studies are needed on the conditions favoring and inhibiting polysaccharide production in batch and continuous culture. Few generalizations can be made about polysaccharide production by bacteria.[15]

BIOSYNTHETIC MECHANISMS

Two distinct mechanisms for the biosynthesis of bacterial extracellular polysaccharides are known. One of these is limited to certain homopolysaccharides, specifically to linear polyfructoses (levans) and branched polyglucoses (dextrans), while the other is apparently utilized for the formation of both homopolysaccharides and heteropolysaccharides. These two mechanisms have a number of distinguishing features; levan and dextran synthesis is essentially extracellular and requires a specific carbohydrate substrate, while other polysaccharides are produced by an intracellular process involving cytoplasmic and membrane-bound enzymes and precursors. A modification of this latter mechanism enables bacteria to form one heteropolysaccharide lacking repeating units, alginate.

Levan and Dextran Synthesis

The production of levans is found in *Bacillus* spp., *Streptococcus salivarius,* and *Acetobacter* spp. Dextrans are synthesized by *Streptococcus sanguis, Streptococcus mutans,* and *Leuconostoc* spp. as well as a few other bacteria. Both types of polymer are formed from the specific substrate sucrose by extracellular enzymes found in culture supernatant obtained from the bacteria which produce them. Dextran synthesis has been most thoroughly studied, but levan synthesis appears to be very similar. Although a few strains of *Leuconostoc mesenteroides* produce unbranched dextrans, the

Table 1
SOME FACTORS AFFECTING BACTERIAL POLYSACCHARIDE PRODUCTION

Bacterial species	Factors Favoring synthesis	Inhibitory factors	Ref.
Enterobacter aerogenes	High C:N ratio; high C:P ratio; high C:S ratio	Acid pH (4.5 to 6); anaerobiosis	9
	Lower incubation temperature (15 to 20°C) Ca^{2+}	Lack of Mg^{2+}, K^+, Ca^{2+}	10
Chromobacterium violaceum	High C:N ratio Ca^{2+}	Fe^{2+}	11
Rhizobium meliloti	High C:N ratio		12
Azotobacter vinelandii	Limitation of MoO_4^{2-}, PO_4^{3-}, SO_4^{2-} or Fe in presence of excess C source	Oxygen limitation	13
Pseudomonas sp.	High C:N ratio; pH 7 to 9	Acid pH; low P (1 m*M*)	14

Table 2
CELL-FREE SYSTEMS FOR BACTERIAL POLYSACCHARIDE SYNTHESIS

Bacteria	Polysaccharide structure	Ref.
Klebsiella (Enterobacter) aerogenes	— [Gal 1 ⟶ 3 Man 1 ⟶ 3 Gal] — ↑ 2 (on Man) 1 | β-GlcA	21
Klebsiella aerogenes type 8	α—Glc A 1 ↓ 4 $\xrightarrow{\beta}$ 3 [Gal 1 $\xrightarrow{\beta}$ 3 Gal 1 $\xrightarrow{\beta}$ 3 Glc]	22
Escherichia coli K12	PYRUV $=^{4}_{6}$ β Gal 1 ↓ 4 β GlcA 1 ↓ 3 Gal ↓ — 3 [β Glc 1 ⟶ Fuc 1 ⟶ 4 Fuc] 1 — ↑ (on Fuc) Ac	23
Acetobacter xylinum	Cellulose	24
E. coli	Sialic acid	25
Neisseria meningitidis	Sialic acid	26

products most frequently formed are branched polymers with 60 to 90% of $\alpha(1 \rightarrow 6)$ linkages and 5 to 30% of $\alpha(1 \rightarrow 4)$ linkages,[16] while other types found in oral *Streptococci* such as *S. mutans* may contain either $\alpha(1 \rightarrow 3)$ and $\alpha(1 \rightarrow 6)$ linkages or $\alpha(1 \rightarrow 3)$ linkages only.[17]

The mechanism of biosynthesis of dextrans requires an acceptor (i.e., preformed dextran of lower molecular weight). The overall synthetic process is thus a chain elongation of the polyglucose together with the liberation of fructose from the substrate:

$$\text{Sucrose} + \text{glucose}_n \xrightarrow{\text{dextransucrase}} \text{glucose}_{n+1} + \text{fructose}$$

or

$$\text{x Sucrose} + \text{glucose}_n \longrightarrow \text{glucose}_{n+x} + \text{x fructose}$$

Other substances as well as preformed dextran may act as acceptor, resulting in glucosyl transfer and chain termination. Dextran synthesis is favored by pH values near 8.0, and glucosyl transferase activity is inhibited by various compounds, including quaternary ammonium ions, aliphatic amines, and anionic detergents.[18] From results with partially purified enzyme systems, it is probable that branching is caused by specific enzymes, although mechanisms analogous to those found in glycogen synthesis have also been suggested.[19] The specific requirement for sucrose as a substrate for polysaccharide synthesis as is found in both levan and dextran production is not in itself unique. However, no other bacterial polysaccharides are known to be formed by extracellular processes in this way.

Levan synthesis closely resembles the production of dextran, and, indeed, formation of the two polymers may occur occasionally in the same organism. The polymer differs from dextran in being linear, and the enzyme levansucrase shows a broad pH optimum from 6.0 to 8.0.[20]

$$\text{Sucrose} + \text{fructose}_n \xrightarrow{\text{levansucrase}} \text{fructose}_{n+1} + \text{glucose}$$

Synthesis of Polysaccharides with Repeating Unit Structures

Most bacterial extracellular polysaccharides are composed of repeating units, and these polymers are formed by an entirely different mechanism from that used for levan and dextran synthesis. The first stage involves the activation of the component monosaccharides as "sugar nucleotides", i.e., usually as nucleoside diphosphate monosaccharides. These are normally formed by interconversions, starting with compounds, such as either UDP-glucose or GDP-mannose. These reactions may involve epimerization, oxidation, decarboxylation, reduction, or rearrangement. A typical series of reactions yields galactose and glucuronic acid (as their UDP derivatives from UDP-glucose):

$$\begin{array}{ccccc} \text{Glucose-1-P} & \longrightarrow & \text{UDP-glucose} & \longrightarrow & \text{UDP-glucuronic acid} \\ & & \updownarrow & & \\ & & \text{UDP-galactose} & \longrightarrow & \text{UDP-galacturonic acid} \end{array}$$

In all the systems which have been exhaustively studied (see Table 2), the monosaccharides are transferred by specific enzymes to a lipid carrier, and the repeating unit accumulates in the form of a lipid-pyrophosphate-oligosaccharide. The lipid carrier has been identified in the system studied by Troy et al.[21] as the same C_{55} isoprenoid alcohol as is utilized for the synthesis of peptidoglycan and lipopolysaccharide components of the bacterial wall. As the amount of this lipid is thought to be limited to about 10^5 molecules per bacterium,[27] there may be circumstances under which there is competition for this lipid between the systems which require it. The possible priorities in the utilization of isoprenoid lipid are indicated in Figure 2.

The first stage in the synthesis of polysaccharide from sugar nucleotides involves the transfer of a sugar-1-phosphate to the isoprenoid lipid phosphate; subsequent steps are achieved by the transfer of sugars with release of the nucleoside diphosphates. A scheme for the biosynthesis of *E. (Klebsiella) aerogenes* type 8 polysaccharide is indicated in Figure 3. This polymer and another from an *E. aerogenes* strain[21] were devoid of acyl groups. These substituents could be added either at the lipid intermediate stage or after polymerization (see Figure 4). Although an enzyme catalyzing the acetylation of preformed polysaccharide was detected in spheroplast membranes of Meningococci,[26] it seems most probable that acetylation occurs at the lipid intermediate stage as the presumed precursor of the acetyl groups; acetyl CoA would then be available in the cytoplasmic membrane, as would phosphoenol pyruvate which is thought to provide the pyruvate groups.

Bacterial homopolysaccharides such as cellulose and glucans other than dextran probably are synthesized in the same manner, as is indicated in Figure 3. The mode of

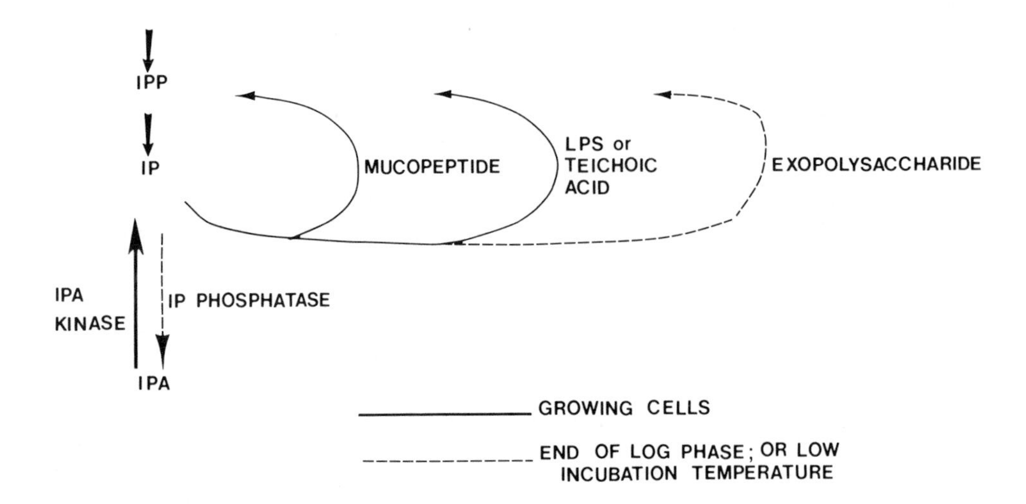

FIGURE 2. Carrier lipid utilization.

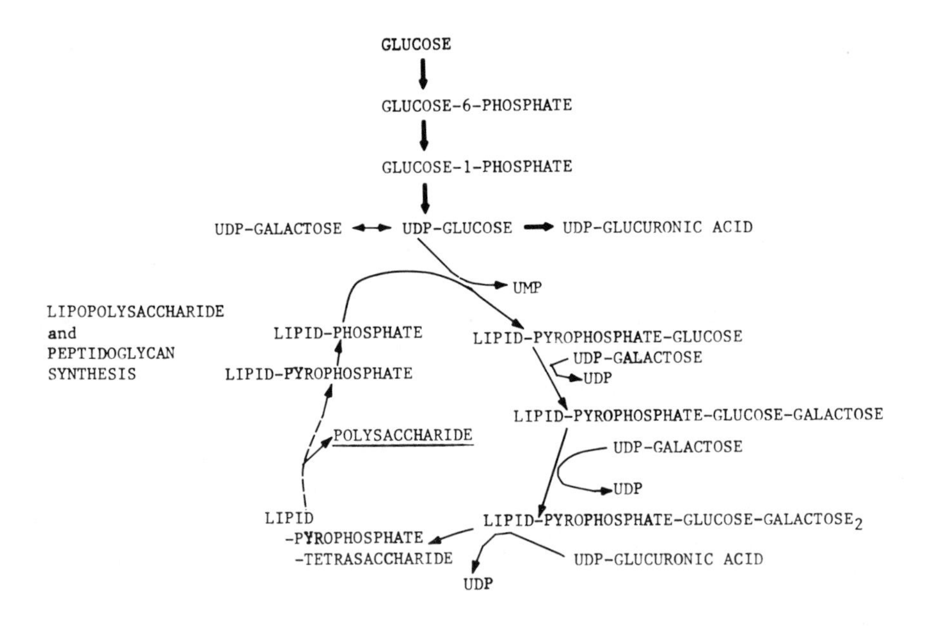

FIGURE 3. The biosynthesis of an exopolysaccharide.

polysaccharide chain termination is still uncertain, although mutants with polysaccharide thought to be of increased chain length have now been isolated from a number of polysaccharide-producing bacterial species. Such mutants are stable, and the polysaccharides derived from them have greatly increased viscosity in solution.[28]

Bacterial Alginate Synthesis

The mechanism of biosynthesis outlined in Figure 3 clearly could not account for the formation of a polymer such as alginate which lacks regular repeating units. Instead, a homopolymer of D-mannuronic acid is synthesized from GDP-mannuronic acid. The homopolymer is acetylated at this stage, as all the *O*-acetyl groups are asso-

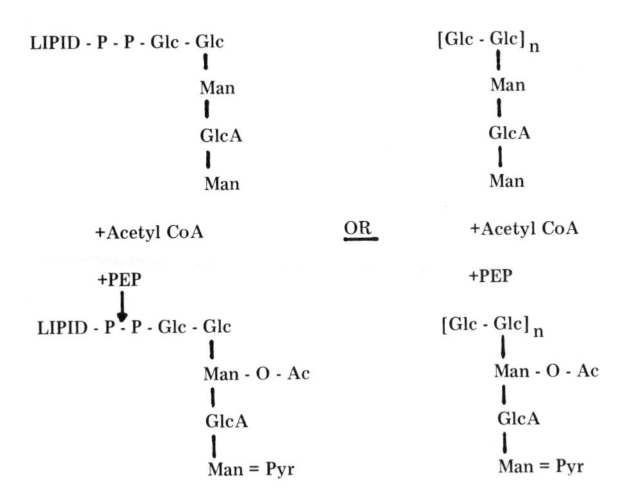

FIGURE 4. Possible exopolysaccharide acylation mechanisms.

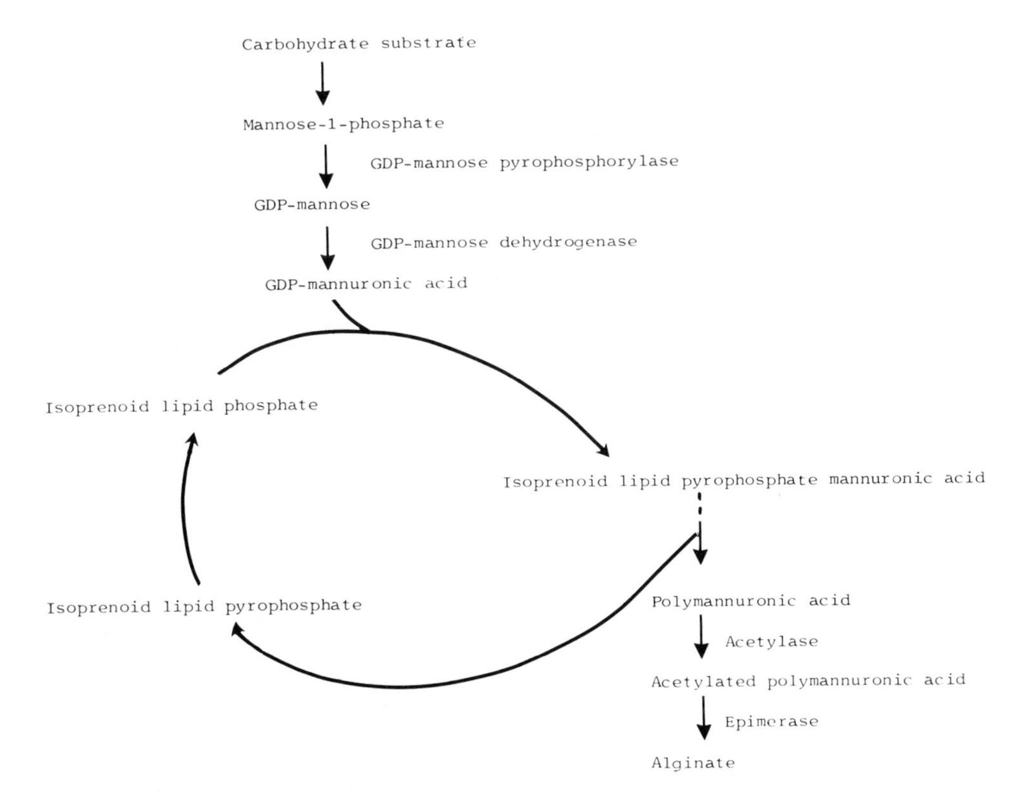

FIGURE 5. Bacterial alginate biosynthesis.

ciated with mannuronic acid residues.[29] The L-glucuronic acid residues are introduced after polymerization by an extracellular epimerase,[30] the sequence of reactions being indicated in Figure 5. Whether the postpolymerization epimerization observed in this mechanism is unique for bacterial extracellular polysaccharides is not clear, but similar enzymes for the epimerization of D-glucuronic acid to L-iduronic acid have been reported in eukaryotic systems.

REFERENCES

1. **Sutherland, I. W.,** Bacterial exopolysaccharides — their nature and production, in *Surface Carbohydrates of the Prokaryotic Cell,* Sutherland, I. W., Ed., Academic Press, London, 1977, chap. 3.
2. **Gorin, P. A. J. and Spencer, J. F. T.,** Exocellular alginic acid from *Azotobacter vinelandii, Can. J. Chem.,* 44, 993, 1966.
3. **Linker, A. and Jones, R. S.,** A new polysaccharide resembling alginic acid isolated from Pseudomonads, *J. Biol. Chem.,* 241, 3845, 1966.
4. **Sutherland, I. W.,** Formate, a new component of bacterial exopolysaccharides, *Nature (London),* 228, 280, 1970.
5. **Hisamatsu, M., Sano, K., Amemura, A., and Harada, T.,** Acidic polysaccharides containing succinic acid in various strains of *Agrobacterium, Carbohydr. Res.,* 61, 89, 1978.
6. **Sutherland, I. W.,** Structural studies on colanic acid, the common exopolysaccharide found in the Enterobacteriaceae, by partial acid hydrolysis, *Biochem. J.,* 115, 935, 1969.
7. **Garegg, P. J., Lindberg, B., Onn, T., and Sutherland, I. W.,** Comparative structural studies on the M-antigen from *Salmonella typhimurium, Escherichia coli* and *Aerobacter cloacae, Acta Chem. Scand.,* 25, 2103, 1971.
8. **Niemann, H., Chakraborty, A. K., Friebolin, H., and Stirm, S.,** Primary structure of the *Escherichia coli* serotype K42 capsular polysaccharide and its serological identity with the *Klebsiella* K63 polysaccharide, *J. Bacteriol.,* 133, 390, 1978.
9. **Duguid, J. P. and Wilkinson, J. F.,** The influence of cultural conditions on polysaccharide production by *Aerobacter aerogenes, J. Gen. Microbiol.,* 9, 174, 1953.
10. **Wilkinson, J. F. and Stark, G. H.,** The synthesis of polysaccharide by washed suspensions of *Klebsiella aerogenes, Proc. R. Phys. Soc. Edinburgh,* 25, 35, 1956.
11. **Corpe, W. A.,** Factors influencing growth and polysaccharide formation by strains of *Chromobacterium violaceum, J. Bacteriol.,* 88, 1433, 1964.
12. **Dudman, W. F.,** Growth and extracellular polysaccharide production by *Rhizobium meliloti* in defined medium, *J. Bacteriol.,* 88, 640, 1964.
13. **Jarman, T. R., Deavin, L., Slocombe, S., and Righelato, R. C.,** Investigation of the effect of environmental conditions on the rate of exopolysaccharide synthesis in *Azotobacter vinelandii, J. Gen. Microbiol.,* 107, 59, 1978.
14. **Williams, A. G. and Wimpenny, J. W. T.,** Exopolysaccharide production by *Pseudomonas* NCIB 11264 grown in batch culture, *J. Gen. Microbiol.,* 102, 13, 1977.
15. **Sutherland, I. W. and Ellwood, D. C.,** Microbial exopolysaccharides — industrial polymers of current and future potential, in *Microbial Technology: Current State, Future Prospects,* Bull, A. T., Ellwood, D. C., and Ratledge, C., Eds., Cambridge University Press, Cambridge, 1979, 107.
16. **Jeanes, A., Haynes, W. C., Wilham, C. A., Rankin, J. C., Melvin, E. H., Austin, M., Cluskey, J. E., Fisher, B. E., Tuchiya, H. M., and Rist, C. E.,** Characterization and classification of dextrans from ninety six strains of bacteria, *J. Am. Chem. Soc.,* 76, 5041, 1954.
17. **Baird, J. K., Longyear, V. M. C., and Ellwood, D. C.,** Water insoluble and soluble glucans produced by extracellular glycosyltransferases from *Streptococcus mutans, Microbios,* 8, 143, 1973.
18. **Ciardi, J. E., Bowen, W. H., and Rölla, G.,** The effect of antibacterial compounds on glucosyltransferase activity from *Streptococcus mutans, Arch. Oral Biol.,* 23, 301, 1978.
19. **McCabe, M. M. and Smith, E. E.,** The dextran acceptor reaction of dextransucrase from *Streptococcus mutans* K1 R, *Carbohydr. Res.,* 63, 223, 1978.
20. **Gibbons, R. J. and Nygaard, M.,** Synthesis of insoluble dextran and its significance in the formation of gelatinous deposits by plaque-forming Streptococci, *Arch. Oral Biol.,* 13, 1249, 1968.
21. **Troy, F. A., Frerman, F. E., and Heath, E. C.,** The biosynthesis of capsular polysaccharide, *J. Biol. Chem.,* 246, 118, 1971.
22. **Sutherland, I. W. and Norval, M.,** The synthesis of exopolysaccharide by *Klebsiella aerogenes* membrane preparations and the involvement of lipid intermediates, *Biochem. J.,* 120, 567, 1970.
23. **Johnson, J. G. and Wilson, D. B.,** Role of a sugar-lipid intermediate in colanic acid synthesis by *Escherichia coli,* J. Bact., 129, 225 (1977).
24. **Garcia, R. C., Recondo, E., and Dankert, M.,** Polysaccharide biosynthesis in *Acetobacter xylinum, Eur. J. Biochem.,* 43, 93, 1974.
25. **Vijay, I. K. and Troy, F. A.,** Properties of membrane-associated sialyltransferase of *Escherichia coli, J. Biol. Chem.,* 250, 164, 1975.
26. **Vann, W. F., Liu, T. Y., and Robbins, J. B.,** Cell-free biosynthesis of the *O*-acetylated *N*-Acetylneuraminic acid capsular polysaccharide of Group C Meningococci, *J. Bacteriol.,* 133, 1300, 1978.
27. **Wright, A.,** Mechanism of conservation of the Salmonella O-antigen by bacteriophage ε^{34}, *J. Bacteriol.,* 105, 927, 1971.

28. **Sutherland, I. W.,** Enhancement of polysaccharide viscosity by mutagenesis, *J. Appl. Biochem.*, 1, 60, 1979.
29. **Davidson, I. W., Sutherland, I. W., and Lawson, C. J.,** Localization of *O*-acetyl groups of bacterial alginate, *J. Gen. Microbiol.*, 98, 603, 1977.
30. **Haug, A. and Larsen, B.,** Biosynthesis of alginate. II. Poly mannuronic acid C-5 epimerase from *Azotobacter vinelandii*, *Carbohydr. Res.*, 17, 297, 308.

ALGAL POLYSACCHARIDE SYNTHESIS

Kazutosi Nisizawa

INTRODUCTION

There are a variety of complex polysaccharides whose chemical properties have not been fully elucidated. Few studies on the biosynthesis of algal polysaccharides had been carried out by the 1960s. Studies have rapidly developed during the last decade, particularly on their storage and on some mucilaginous intercellular polysaccharides useful for industrial purposes.

STARCH AND RELATED POLYSACCHARIDES

Fredrick, in a series of works on the storage polysaccharides of blue-green, red, and green algae found that they are synthesized by at least three kinds of enzymes. These are (1) phosphorylase, (2) branching enzyme or Q-enzyme, and (3) starch synthetase. These also have been investigated in animals and higher plants. He carried out these studies from a phylogenetic or taxonomic point of view using members of the three algal phyla. Specifically, he examined *Oscillatoria princeps* [1] and *Cyanidium caldarium*,[2] the latter often being debated for its taxonomic position as the test sample for blue-green algae, *Rhodymenia pertusa* [3] for red algae, and *Spirogyra setiformis* [1] and *Chlorella pyrenoidosa* [3] for green algae.

Phosphorylase

The phosphorylases of blue-green and red algae have two isozymes (a_1 and a_2). The a_2 is activated by AMP and Mn^{2+}, as is phosphorylase b in animals.[4] In *Rhodymenia* the amount of a_1, AMP-independent phosphorylase tends to diminish as compared to that in blue-green algae.[1] The a_2 requires no primer for its activity.[5]

Branching Enzymes

The branching enzymes (a_5) of *Spirogyra* [1] and *Chlorella* [3] can be separated on polyacrylamide gel electrophoresis into three isozymes as in *R. pertusa* [1] and into two in blue-green algae.[1] The synthetase had two isozymes (a_3 and a_4) in all algae tested, and they use UDPG and/or ADPG as the glucosyl donor. The a_3 forms a hybrid with a_2, so that the hybrid enzyme is capable of utilizing not only glucose-1-phosphate, but also the nucleotide sugars.[1] Shortly before these investigations, Kauss and Kandler[6] found that ADPG or UDPG is used in *C. pyrenoidosa* for its starch synthesis, and that the predominant use of nucleotide sugars is dependent on culture temperature. Using a 25-fold purified starch synthetase from this green alga, Preiss and Greenberg[7] demonstrated that deoxy-ADPG as well as ADPG can serve as the glucosyl donor of this transferase. Also amylose and amylopectin were found to be more effective than glycogen as the primer.[7] The presence of synthetases of both bound and soluble forms was also found in *Chlorella*. In contrast to UDPG-pyrophosphorylase, the synthetase was activated about 20-fold by 3-phosphoglyceric acid, the first product of photoassimilation of CO_2 and inhibited by phosphate and ADP.[8]

More recently, it was reported that a similar enzyme system to that found in *Spirogyra* is present in the cyanelle of a lower green alga, *Glaucocystis nostochinearum*, and that the polysaccharide synthesized resembles amylopectin rather than cyanophycean starch.[9] Based on a comparison of the properties of these enzymes with those from the typical blue-green alga, *Anacystis nidulans*, Fredrick[10] suggested that *Glaucocystis* (which involves various toxonomic obscurities) should be placed between green and blue-green algae.

Starch Synthetase

Few investigations of starch synthesis in green macroalgae have been carried out, but the synthesis seems to be performed by a similar enzyme to that of *C. pyrenoidosa.*[1] In fact, the presence of phosphorylase in *Acetabularia mediterranea* has been reported by Clauss[11] and in *Ulva pertusa* by Kashiwabara et al.[12]

The starch synthetase, as found in *R. pertusa,* was also investigated in *Serraticardia maxima* by Nagashima et al.,[13] but unlike the synthetase of *Chlorella,* only a starch granule-bound form was detected. The authors failed to separate it from the granule despite various attempts. By isotopic techniques, it was found that the synthetase uses not only ADPG but also UDPG and GDPG as the glucosyl donor with their respective effectiveness in that order.

The iodine complex of *Serraticardia* starch showed a maximum absorption at 550 nm, and its average chain length was estimated to be 17 to 18. The specific optical rotation was +199.5°. Thus, the floridean starch of this red alga seemed to be of an intermediate type, with chemical properties between amylopectin and glycogen.[14]

LAMINARAN AND RELATED POLYSACCHARIDES

In contrast to higher plants,[15] few biosynthetic studies on β-1,3-glucan in algae have been done, except for paramylon, the storage β-1,3-glucan of *Euglena.* Using isotopic techniques, Goldemberg and Marechal[16] found that the β-1,3-glucan of *Euglena gracilis* is synthesized from nucleotide ^{14}C-glucoses by extracts from dark-cultured cells. The enzyme was present in a bound form, but could be solubilized by sodium deoxycholate. Of the ^{14}C-nucleotide glucoses tested, UDPG was most efficient. The synthesis of the glucan took place regardless of the addition of the primer paramylon. The enzymatic transformation of UDPG to the β-1,3-glucan attained 20 to 30% yields.[17] The enzyme which carried out the synthesis of paramylon was a UDPG glucotransferase, but degradation seemed to be taken over by other enzymes, such as β-1,3-glucan phosphorylase and laminarase (β-1,3-glucanase).[18,19]

A seawater culture of fronds of the brown alga *Eisenia bicyclis* was used to study in vivo biosynthesis of β-1,3-glucan. Using $H^{14}CO^{-}_3$, Yamaguchi et al.[20] observed that mannitol and then laminaran was rapidly photosynthesized. Upon a chase culture of these fronds, the laminaran synthesized was rapidly consumed in the dark in concert with a rapid decrease of ^{14}C-mannitol.

ALGINIC ACID

Biosynthetic studies on alginic acid have developed in two ways owing to its particular structural feature: (1) the synthesis of each block consisting only of β-1,4-linked D-mannuronic acid (MA) residue or only of α-1,4-linked L-guluronic acid (GA) residue and (2) the synthesis of a block consisting of both residues.[21]

As to the former direction of studies, the presence of GDPMA and GDPGA in *Fucus gardneri* was first reported by Lin and Hassid.[22] The authors also found that the radioactivity of ^{14}C-MA or ^{14}C-GA incubated with the disks of these algal fronds is incorporated into alginic acid and fucoidan. Subsequently, they detected the following enzyme activities in the algal extracts: hexokinase, phosphomannomutase, D-mannose-1-phosphate guanyltransferase, GDPM dehydrogenase, and GDPMA transferase. The ^{14}C label of GDPMA was incorporated into the primer, alginic acid by a particulate enzyme preparation from algal fronds. Based on these findings, the authors postulated that GDPGA is formed from GDPMA, possibly by the action of 5-epimerase. This is similar to rabbit skin which can convert UDP-D-glucuronic acid to UDP-L-iduronic acid.[23] MA and GA residues are transferred to some precursors to form alginic acid, although the existence of GDPGA transferase could not be verified.[24]

As to the second direction of study, Larsen and Haug[21,25,26] first reported the presence of 5-epimerase in the culture medium of *Azotobacter vinelandii*. It produces alginic acid containing acetoester residues, the enzyme being capable of epimerizing MA to GA on the polymer level. This was quite similar to the polymerlevel epimerization of D-glucuronic acid residue of heparin to L-iduronic acid residue by mouse mastocytoma microsomal fraction.[27] A similar polymannuronic acid-5-epimerase found in *Azotobacter* was also found in extracts from a brown alga, *Pelvetia canaliculata*.[28] Strong evidence for the existence of this enzyme in both bacterium and alga was found when tritium (in tritiated water used as incubation medium) was incorporated into GA residues which had been newly formed from MA residues of the substrate used. The algal epimerase required Mg^{2+} in addition to Ca^{2+}.[29]

The polymannuronic acid-5-epimerase activity was higher in the growing part than in other parts of *E. bicyclis* fronds.[30] The result also suggests that at least some of the GA of the alginic acid molecule is formed secondarily from their MA residues during the biosynthesis of the acid. Thus, the 5-epimerase action can explain the existence of a block structure consisting of the mixture of MA and GA residues. However, the occurrence of blocks consisting exclusively of MA or GA is hardly conceivable from only the action of this enzyme. The possible existence of a GDPGA transferase[24] seems to be important. Pinder and Bucke,[31] however, found in experiments of *A. vinelandii* that the GA residues of the alginate molecule are formed from MA by the 5-epimerase and not from GDPGA by the possible glycosyltransfer reaction, in contrast to the concept of Lin and Hassid.[24]

On the other hand, several studies of alginate biosynthesis in vivo have already been carried out. Incorporation of radioactivity into alginate of *E. bicyclis* cultured in $H^{14}CO^{-}_{3}$-containing seawater proceeded at almost one third the rate of that into laminaran[20] and was greater into the alginates of the cortex and medulla of stipe sections in *Laminaria digitata*[32] and *Laminaria hyperborea*[33] than into the alginate of their barks. Moreover, the rate of incorporation into alginate of blades in the dark was very low, while it was greater in the stipe. Incorporation occurred predominantly into MA residues in short-term culture in the light, while GA residues were more rapidly labeled in dark culture.[34]

The alginate of *Ishige okamurai* was commonly higher in GA content than in MA; photoassimilated radiocarbon was incorporated into each fraction of alginate in the order of MA-rich>MAGA-rich>GA-rich. A chase culture showed that the radiocarbon incorporated into alginate is increased at earlier stages of culture in both light and dark. The rate of increase of each fraction was in the same order as above. Decreasing rates observed at later stages of culture occurred most rapidly with MA-rich alginate. No decreasing rate was observed with GA-rich alginate in which only a gradual increase in the incorporation was observed throughout the chase period.[35] The results suggest that the synthesis of GA residues is carried out following the synthesis of MA residues in vivo, possibly by 5-epimerase.

SULFATED POLYSACCHARIDES

Few investigations on the biosynthesis of algal sulfated polysaccharides have been conducted at the enzymatic level; most have been done in vivo. Porphyran was found by Rees and Conway[36] to be highly heterogeneous in the proportion of constituent sugar residues. Based on the analysis of porphyran samples from *Porphyra umbilicalis* harvested in different seasons, the authors reported that the enzyme system synthesizing these porphyrans does not seem to assemble the units in a rigidly predetermined manner and that the synthesis is flexible to some extent because the molar ratio of the sugar residues vary seasonally. However, the ratio of 3,6-anhydro-L-galactose to 6-*O*-methylgalactose and of galactose to sulfate were almost constant in any sample.

Rees[37] also reported that the formation of 3,6-anhydrogalactose residues in porphyran occurs on enzymatic release of the sulfate at the polysaccharide level. This enzymic reaction has been thought to be analogous to the alkaline elimination of sulfate from hexose 6-sulfate. The alkyltransferase of *Porphyra* diminished its activity by partial acid and enzymatic hydrolysis products of porphyran, and the reaction is irreversible. Later, a similar alkyltransferase which was capable of forming the 3,6-anhydro-L-galactose residue of carrageenan was extracted from *Gigartine stellata* by Lawson and Rees.[38]

The 6-*O*-methylation of galactose residues of porphyran was investigated by Su and Hassid.[39] The following mechanism was postulated:

1. D- and L-galactose residues which had been incorporated into porphyran from UDP-D-galactose and GDP-L-galactose, respectively, are first esterified with sulfate (and then)
2. A part of D-isomer undergoes etherification to give 6-*O*-methyl-D-galactose, while a part of the L-isomer is dehydrated by the alkyl-transferase as described above.

However, methylation might occur in mediation of S-adenosyl-L-menthionine as has been observed for pectin synthesis in *Phaseolus aureus*[40] and for the formation of 4-*O*-methyl-D-glucuronic acid residue of hemicellulose.[41]

In some in vivo experiments using $^{14}CO_2$ in the culture medium, Percival et al.[42] found that an acid-soluble cell wall polysaccharide, glucuronoxylofucan, from *Fucus vesiculosus* was metabolically active. Bidwell et al.[43] assumed that fucoidan may be synthesized from the xylogalactofucoglucuronan (the same polysaccharide as the glucuronoxylofucan mentioned above) via xyloglucuronogalactofucan).

The biosynthesis site of algal sulfated polysaccharides in some brown algae, such as *P. canaliculata, L. digitata, Laminaria saccharina,* and *L. hyperborea,* was investigated histochemically by Evans et al.[44] It was found that the synthesis is performed mainly by epidamal cells in *Pelvetia,* while confined to specialized secretory cells in *Laminaria,*[44] and that the sulfation of the polysaccharides appears to occur in the Golgi-rich perinuclear region.[44] The same conclusion for the sulfation of capsular polysaccharide of the red alga *Porphyridium aerugineum* cells has been reached by Ramus and Groves.[45] The sulfation of carragenans in *Chondrus crispus* was found to occur simultaneously at both KCl-soluble (λ-) fraction and KCl-insoluble (κ-) fraction.[46]

REFERENCES

1. **Fredrick, J. F.**, Biochemical evolution on glucosyltransferase isozymes in algae, *Ann. N.Y. Acad. Sci.*, 151, 413, 1968.
2. **Fredrick, J. F.**, Glucosyltransferase isozymes in algae. III. The polyglucoside and enzymes of *Cyanidium caldarium Phytochemistry*, 7, 1573, 1968.
3. **Fredrick, J. F.**, Effect of amitrole on biosynthesis of phosphorylases in different algae, *Phytochemistry*, 8, 725, 1969.
4. **Fredrick, J. F.**, An algal α-glucan phosphorylase which requires adenosine-5-phosphate as coenzyme, *Phytochemistry*, 2, 413, 1963.
5. **Fredrick, J. F.**, A non-primer requiring α-1,4-glucan phosphorylase of *Cyanidium caldarium, Phytochemistry*, 11, 3259, 1972.
6. **Kauss, H. and Kandler, O.**, Adenosinediphosphate glucose aus *Chlorella*, *Z. Naturforsch. Teil B*, 17, 858, 1962.
7. **Preiss, J. and Greenberg, E.**, Biosynthesis of starch in *Chlorella pyrenoidosa*. I. Purification and properties of the adenosine diphosphoglucose: α-1,4-glucan, α-4-glucosyl transferase from *Chlorella*, *Arch. Biochem. Biophys.*, 118, 702, 1967.
8. **Sanwal, G. G. and Preiss, J.**, Biosynthesis of starch in *Chlorella pyrenoidosa*. II. Regulation of ATP: α-D-glucose 1-phosphate adenyl transferase (ADP-glucose pyrophosphorylase) by inorganic phosphate and 3-phosphoglycerate, *Arch. Biochem. Biophys.*, 119, 454, 1967.
9. **Fredrick, J. F.**, Glucan and glucosyltransferase isozymes of *Glaucocystis nostochinearum*, *Phytochemistry*, 16, 1489, 1977.
10. **Fredrick, J. F.**, Protein and isozyme patterns of the cyanebles of *Glaucocystis nostochinearum* compared with *Anacystis nidulans*, *Phytochemistry*, 16, 1571, 1977.
11. **Clauss, H.**, Das Verhalten der Phosphorylase in kernhaltigen und kernlosen Teilen von *Acetabularia mediterranea*, *Planta*, 52, 534, 1959.
12. **Kashiwabara, Y., Suzuki, H., and Nisizawa, K.**, Some enzymes relating to the metabolism of starch in a sea-lettuce, *Plant Cell Physiol.*, 6, 537, 1965.
13. **Nagashima, H., Nakamura, S., Nisizawa, K., and Hori, T.**, Enzymic synthesis of floridean starch in a red alga, *Serraticardia mxima*, *Plant Cell Physiol.*, 12, 243, 1971.
14. **Ozaki, H., Maeda, M., and Nisizawa, K.**, Floridean starch of calcareous red alga, *Joculator maximus* (*Serraticardia maxima*), *J. Biochem. (Tokyo)*, 61, 497, 1967.
15. **Nikaido, H. and Hassid, W. Z.**, Biosynthesis of saccharides from glycopyranosyl esters of nucleoside pyrophosphates ("sugar nucleotides"), *Adv. Carbohydr. Chem. Biochem.*, 26, 351, 1971.
16. **Goldemberg, S. H. and Marechal, L. R.**, Biosynthesis of paramyron in *Euglena gracilis*, *Biochim. Biophys. Acta*, 71, 743, 1963.
17. **Marechal, L. R. and Goldenberg, S. H.**, Uridine diphosphate glucose-β-1,3-glucan β-3-glucosyltransferase from *Euglena gracilis*, *J. Biol. Chem.*, 239, 3163, 1964.
18. **Dwyer, M. R., Smydzuk, J., and Smillie, R. M.**, Synthesis and break down of β-1,3-glucan in *Euglena gracilis* during growth and carbon depletion, *Aust. J. Biol. Sci.*, 23, 1005, 1970.
19. **Dwyer, M. R. and Smillie, R. M.**, A light-induced β-1,3-glucan breakdown associated with the differentiation of chloroplasts in *Euglena gracilis*, *Biochim. Biophys. Acta*, 216, 392, 1970.
20. **Yamaguchi, T., Ikawa, T., and Nisizawa, K.**, *Plant Cell Physiol.*, 7, 217, 1966.
21. **Haug, A. and Larsen, B.**, Biosynthesis of algal polysaccharides, in Plant Carbohydrate Biochemistry, Academic Press, London, 1973, 207.
22. **Lin, T.-Y. and Hassid, W. Z.**, Isolation of guanosine diphosphate uronic acids from a marine brown alga, *Fucus gardneri*, *J. Biol. Chem.*, 241, 3282, 1966.
23. **Jacobson, B. and Davidson, E. A.**, *J. Biol. Chem.*, 237, 638, 1962.
24. **Lin, T.-Y. and Hassid, W. Z.**, *J. Biol. Chem.*, 241, 5284, 1966.
25. **Larsen, B. and Haug, A.**, Biosynthesis of alginate. I. Composition and structure of alginate produced by *Azotobacter vinelandii* (Lipmann), *Carbohydr. Res.*, 17, 287, 1971.
26. **Haug, A. and Larsen, B.**, Biosynthesis of alginate. II. Polymannuronic acid C-5-epimerase from *Azotobacter vinelandii* (Lipmann), *Carbohydr. Res.*, 17, 297, 1971.
27. **Lindahl, U. and Bäckström, G.**, Biosynthesis of L-iduronic acid in heparin: epimerization of D-glucuronic acid on the polymer level, *Biochem. Biophys. Res. Commun.*, 46, 985, 1972.
28. **Madgwick, J., Haug, A., and Larsen, B.**, Polymannuronic acid 5-epimerase from the marine alga, *Pelveti canaliculata* (L.) Dcne. et Thur., *Acta Chem. Scand.*, 27, 3592, 1973.
29. **Madgwick, J., Haug, A. (Deceased), and Larsen, A.**, Ionic requirements of alginate-modifying enzymes in the marine alga *Pelvetia canaliculata* (L.) Dcne et Thur., *Bot. Mar.*, 21, 1, 1978.
30. **Nisizawa, K. and Ishikawa, M.**, Some enzymic properties of algal polymannuronate 5-epimerase, 10th Int. Seaweed Symp. Göteborg, Sweden, 1980. Abstract B78, *Bull. Jpn. Soc. Sci. Fish.*, 47, 1981, in press.

31. **Pinder, D. F. and Bucke, C.**, The biosynthesis of alginic acid by *Azotobacter venelandii, Biochem. J.*, 152, 617, 1975.
32. **Hellebust, J. A. and Haug, A.**, Alginic acid synthesis in *Laminaria digitata* (L.) Lamour, in Proceedings of 6th Int. Seaweed Symp., Madrid, 1969, 463.
33. **Hellebust, J. A. and Haug, A.**, Photosynthesis, translocation, and alginic acid synthesis in *Laminaria digitata* and *Laminaria hyperborea, Can. J. Bot.*, 50, 169, 1972.
34. **Hellebust, J. A. and Haug. A.**, *In situ* studies on alginic acid synthesis and other aspects of the metabolism of *Laminaria digitata, Can. J. Bot.*, 50, 177, 1972.
35. **Abe, K., Sakamoto, T., Sasaki, S. F., and Nisizawa, K.**, *In vivo* studies on the synthesis of alginic acid in *Ishige okamurai, Bot. Mar.*, 16, 229, 1973.
36. **Rees, D. A. and Conway, E.**, The structure and biosynthesis of porphyran: a comparison of some samples, *Biochem. J.*, 84, 411, 1962.
37. **Rees, D. A.**, Enzymatic synthesis of 3:6-anhydro-L-galactose within porphyran from L-galactose 6-sulfate units, *Biochem. J.*, 81, 347, 1961.
38. **Lawson, C. J. and Rees, D. A.**, An enzyme for the metabolic control of polysaccharide conformation and function, *Nature (London)*, 227, 392, 1970.
39. **Su, J.-C. and Hassid, W. Z.**, Carbohydrates and nucleotides in the red alga *Porphyra perforata.* II. Separation and identification of nucleotides, *Biochemistry*, 1, 474, 1962.
40. **Kauss, H. and Hassid, W. Z.**, Enzymic introduction of the methyl ester groups of pectin, *J. Biol. Chem.*, 242, 3449, 1967.
41. **Kauss, H. and Hassid, W. Z.**, Biosynthesis of the 4-O-methyl-D-glucuronic acid unit of hemicellulose B by transmethylation from S-adenosyl-L-methionine, *J. Biol. Chem.*, 242, 1680, 1967.
42. **Percival, E., Bourne, E. J., and Brush, P.**, The active carbohydrate metabolites of *Fucus vesiculosus*, Proc. 6th Int. Seaweed Symp., Madrid, 1969, 575.
43. **Bidwell, R. G. S., Percival, E., and Smestad, B.**, Photosynthesis and metabolism of marine algae. VIII. Incorporation of ^{14}C into the polysaccharides metabolized by *Fucus vesiculosus* during pulse labeling experiments, *Can. J. Bot.*, 50, 191, 1972.
44. **Evans, L. V., Simpson, M., and Callow, E.**, Sulfated polysaccharide synthesis in brown algae, *Planta*, 110, 237, 1973.
45. **Ramus, J. and Groves, S. T.**, Incorporation of sulfate into the capsular polysaccharide of the red alga *Porphyridium, J. Cell Biol.*, 54, 399, 1972.
46. **Loewus, F. and Wagner, G.**, The incorporation of ^{35}S-labelled sulfate into carrageenan in *Chondrus crispus, Plant Physiol.*, 48, 373, 1971.

LIPIDS AND FATTY ACIDS OF PHOTOSYNTHETIC BACTERIA

Mario Snozzi and Reinhard Bachofen

LIPIDS

The lipid composition of photosynthetic bacteria, and bacteria in general, is relatively simple. Neutral lipids are rare, and sterols are absent.[1] Phospholipids are localized in the cell membrane, in intracytoplasmic membranes (chromatophores), and in cell envelopes. According to the method of lipid determination, the phospholipid composition may reflect that of whole cells or of cell fragments with various impurities. Little is known about the specific functions of these compounds. Birrell et al.[2] demonstrated by ESR of spin labeled compounds in chromatophores of *Rhodopseudomonas spheroides* that part of the phospholipids were immobilized by interactions with the membrane proteins. Mild extraction of various lipid components from freeze-dried chromatophores of *Rhodospirillum rubrum* with solvents of increasing polarity exhibited different solubilities.[3]

The composition of phospholipids of bacteria depends on the culture medium and culture conditions.[4,5] Therefore, variable lipid composition at different stages of growth has been detected.

Phosphatidylethanolamine is one of the major lipids in photosynthetic bacteria. As shown by Randle et al.,[6] the relative amounts of phosphatidylethanolamine and phosphatidylcholine in *Agrobacterium tumefaciens* varies during growth in batch cultures. In the log phase, phosphatidylethanolamine was predominant; in the stationary phase, the ratio was reversed. Similar results were obtained by Shively and Benson[7] with *Thiobacillus thiooxidans*. Hence, one might also expect such changes in photosynthetic bacteria which contain these two phospholipids.

Phosphatidylglycerol and diphosphatidylglycerol (cardiolipin) have also been found in photosynthetic bacteria. It is known from *Escherichia coli* that cardiolipin is derived from phosphatidylglycerol.[8] When batch cultures of Gram-negative cells go from the logarithmic phase of growth to the stationary phase, the portion of phosphatidylglycerol tends to fall while that of cardiolipin will rise.[6-8] Therefore, the relative amount of phospholipids given in the tables differs greatly, depending on growth conditions. Moreover, some results cover the composition of the intracytoplasmic membranes only, and others cover the whole cell, including the cell envelope. Even with the same species, the lipid composition may vary because of strain differences.

Rothman and Kennedy[9] have reported that the distribution of phosphatdylethanolamine is asymmetric in the membranes of *Bacillus megaterium*. The results indicate that bacteria as well as eucariotic cells have membranes with an asymmetric lipid bilayer.

Cronan[10] has recently reviewed the function of lipids in bacterial membranes, the mechanisms of lipid synthesis and degradation, and the regulation of lipid synthesis. The composition of bacterial lipids and the comparison of lipids among different bacterial strains were reviewed by Goldfine[11] and Ikawa.[12] (Table 1)

FATTY ACIDS

The fatty acids of bacteria have predominantly 16 and 18 carbon atoms generally arranged as straight chains (saturated or mono-unsaturated) or as branched chains (iso, anteiso, and cyclopropane). The composition of fatty acids also varies considerably, depending on culture conditions, age, and strain.[13] (Table 2)

Table 1
LIPID COMPOSITION OF PHOTOSYNTHETIC BACTERIA[a]

Organism	PC	PEA	PG	DPG	Ornithin lipid	Others	Ref.
Rhodopseudomonas							
R. spheroides anaerobic	8	36	54	2			14
R. spheroides aerobic	4	36	54	7			14
R. spheroides[b]	+	+	+	+	+	+	17
R. spheroides CPH	22.6	35.8	28.7		9.3	3.6	15
CPH	23.3	35.3	34	3.3	+	3.9 (PA)	16
R. spheroides[c]	22	40	14			24 (PA)	18
						2.6[d]	30
R. capsulata[b]	+	+	+		+		17
	+	+ +	+ +				5
R. capsulata		54.4	15.8	1.8		14 (PA) + 14	19
R. gelatinosa[b]		+	+		+		17
R. palustris[b]	+	+	+	+	trace		17
Rhodospirillum rubrum							
R. rubrum aerobic	12	62	26	+			14
R. rubrum anaerobic	6	57	29	8			14
					+		20
		29.2	15.4	7.7	26.1	9.2(PA) + 12.4	19
R. rubrum[b]		+	+	+	+		17
R. rubrum CPH		43.7	32.9	17.1	6.3		3
R. rubrum CPH[c]	30	12	42			15 (PI)	21
		65	22	13			22
Chromatium D		47	53	+			14
		57	17.8	3.75		+ Glycolipids	23
		55	39	6			4
Rhodomicrobium vannielii	26.5	4.5	9.7		50	8.5 (PA)	24

[a] Figures given are the percentage of total lipid of the analyzed material. Differences may be caused by culture conditions, strain, or method of determination. Abbreviations: PA = phosphatidic acid, PC = phosphatidylcholine, PEA = phosphatidylethanolamine, PG = phosphatidylglycerol, DPG = diphosphatidylglycerol (cardiolipin), PI = phosphatidylinositol; and CPH = chromatophores;

[b] +, present.

[c] Percentage of total lipid phosphate.

[d] Sulphoquinovosyl-diacyl glycerol.

Table 2
FATTY ACID COMPOSITION OF PHOTOSYNTHETIC BACTERIA[a]

Organism	14:0	14:1	16:0	16:1	18:0	18:1	Others	Ref.
Rhodopseudomonas								
R. spheroides aerobic	0.3	1.1	9.6	3.5	13.0	72.5		14
			Trace	Trace	Trace	99.0		17
R. spheroides anaerobic	1.2	0.7	8.9	7.4	11.4	70.4		14
			1.7	1.0	4.8	90.8		17
	Trace	Trace	5.0	2.4	9.7	76.6	6.4 (19:0), trace (17:0)	25
			7	2	5	86		26
R. spheroides CPH	Trace	Trace	4.6	2.1	9.7	76.8	6.8 (19:0), trace (17:0)	25
R. capsulata aerobic		2.4	0.9	11.8	0.2	84.0		17
R. capsulata anaerobic			2.3	2.4	4.2	90.3		17
			10	2		28	14 (17:0), 33 (17:1) 8 (19:1)	19
R. gelatinosa aerobic	2.2		13.8	58	Trace	24.4		17
R. gelatinosa anaerobic	2.9		33.4	51	Trace	6.2		17
R. palustris aerobic			3.5	4.8	Trace	91.5		17
R. palustris anaerobic			12.7	3.1	6.2	72.3		17
Strain 1-7	0.17		5.17	1.57	7.10	78.0		27
Strain 1-23	0.11		6.6	1.97	7.8	75.0		27
Rhodospirillum								
R. rubrum aerobic	2.9		17.4	30.0	+	49.7		14
	1.6	1.4	8.3	51.0	Trace	35.3		17
R. rubrum anaerobic	3.0		19	26.1	+	51.9		14
	1.6		16.3	37.6	0.9	37.3		17
	5		18	37	0.5	35	+ (15:0, 17:0, 17:1, 19:1)	19
	2.0		25.5	16.4	1.6	54		28
R. rubrum CPH[b]	0.9		16.8	38.2	0.7	39.2	0.5 (15:0), 3.6 (17:1)	3
Chromatium D	+		28.7	32.6	+	38.7		14
Chromatium D CPH	2.8		28.5	30.3	3.1	35.3		14
Rhodomicrobium vannielii	1.75	0.5	5.40	1.08	0.11	88.20	2.7 (18:2), trace (10:0, 12:0)	29

[a] The figures reported are the percentage of each length of the total of the analyzed material.
[b] Percentage of chromatophore phospholipids.

REFERENCES

1. **Kamio, Y., Kanegasaki, S., and Takahashi, H.,** Occurrence of plasmalogens in anaerobic bacteria, *J. Gen. Appl. Microbiol.,* 15, 439, 1969.
2. **Birrell, G. B., Sistrom, W. R., and Griffith, O. H.,** Lipid-protein associations in chromatophores from the photosynthetic bacterium *Rhodopseudomonas spheroides, Biochemistry,* 17, 3768, 1978.
3. **Costes, C., Bazier, R., Baltscheffsky, H., and Hallberg, C.,** Mild extraction of lipids and pigments from *Rhodospirillum rubrum* chromatophores, *Plant Sci. Lett.,* 12, 241, 1978.
4. **Steiner, S., Burnham, J. C., Conti, S. F., and Lester, R. L.,** Polar lipids of *Chromatium* strain D grown at different light intensities, *J. Bacteriol.,* 103, 500, 1970.
5. **Steiner, S., Sojka, G. A., Conti, S. F., Gest, H., and Lester R. L.,** Modification of membrane composition in growing photosynthetic bacteria, *Biochim. Biophys. Acta,* 203, 571, 1970.
6. **Randle, C. L., Albro, P. W., and Dittmer, J. C.,** The phosphoglyceride composition of Gram-negative bacteria and the changes in composition during growth, *Biochim. Biophys. Acta,* 187, 214, 1969.
7. **Shively, J. M. and Benson A. A.,** Phospholipids of *Thiobacillus thiooxidans, J. Bacteriol.,* 94, 1979, 1967.
8. **Cronan, J. E., Jr.,** Phospholipid alterations during growth of *Escherichia coli, J. Bacteriol.,* 95, 2054, 1968.
9. **Rothman, J. E. and Kennedy, E. P.,** Asymmetrical distribution of phospholipids in the membrane of *Bacillus megaterium, J. Mol. Biol.,* 110, 603, 1977.
10. **Cronan, J. E., Jr.,** Molecular biology of bacterial membrane lipids, *Annu. Rev. Biochem.,* 47, 163, 1978.
11. **Goldfine, H.,** Comparative aspects of bacterial lipids, *Adv. Microb. Physiol.,* 8, 1, 1972.
12. **Ikawa, M.,** Bacterial phosphatides and natural relationships, *Bacteriol. Rev.,* 31, 54, 1967.
13. **O'Leary, W. M.,** *The Chemistry and Metabolism of Microbial Lipids,* World, Cleveland, Ohio, 1967.
14. **Haverkate, F., Teulings, F. A. G., and van Deenen, L. L. M.,** Studies on the phospholipids of photosynthetic microorganisms. II, *K. Ned. Akad. Wet. Proc. Ser B,* 68, 154, 1965.
15. **Gorchein, A.,** Ornithine in *Rhodopseudomonas spheroides, Biochim. Biophys. Acta,* 84, 356, 1964.
16. **Gorchein, A.,** The separation and identification of the lipids of *Rhodopseudomonas spheroides, Proc. R. Soc. London Ser. B,* 170, 279, 1968.
17. **Wood, B. J. B., Nichols, B. W., and James, A. T.,** The lipids and fatty acid metabolism of photosynthetic bacteria, *Biochim. Biophys. Acta,* 106, 261, 1965.
18. **Lascelles, J. and Szilagyi, J. F.,** Phospholipid synthesis by *Rhodopseudomonas spheroides* in relation to the formation of photosynthetic pigments, *J. Gen. Microbiol.,* 38, 55, 1965.
19. **Hirayama, O.,** Lipids and lipoprotein complex in photosynthesic tissues. IV. Lipids and pigments of photosynthetic bacteria, *Agric. Biol. Chem.,* 32, 34, 1968.
20. **Depinto, J. A.,** Ornithine-containing lipid in *Rhodospirillum rubrum, Biochim. Biophys. Acta,* 144, 113, 1967.
21. **Benson, A. A., Wintermans, J. F. G. M., and Wiser, R.,** Chloroplast lipids as carbohydrate reservoirs, *Plant Physiol.,* 34, 315, 1959.
22. **Snozzi, M. and Bachofen, R.,** Characterisation of reaction centers and their phospholipids from *Rhodospirillum rubrum, Biochim. Biophys. Acta,* 546, 236, 1979.
23. **Steiner, S., Conti, S. F., and Lester, R. L.,** Separation and identification of the polar lipids of *Chromatium* strain D, *J. Bacteriol.,* 98, 10, 1969.
24. **Park, Ch. and Berger, L. R.,** Complex lipids of *Rhodomicrobium vannielii, J. Bacteriol.,* 93, 221, 1967.
25. **Schmitz, R.,** Ueber die Zusammensetzung der pigmenthaltigen Strukturen des Purpurbakteriums *Rhodopsueodmonas spheroides, Z. Naturforsch. Teil B,* 22, 645, 1967.
26. **Gorchein, A.,** Studies on the structure of an ornithin-containing lipid from non-sulphur purple bacteria, *Biochim. Biophys. Acta,* 152, 358, 1968.
27. **Constantopoulos, G. and Bloch, K.,** Isolation and characterisation of glycolipids from some photosynthetic bacteria, *J. Bacteriol.,* 93, 1788, 1967.
28. **Cho, K. Y. and Salton, M. R. J.,** Fatty acid composition of bacterial membrane and wall lipids, *Biochim. Biophys. Acta,* 116, 73, 1966.
29. **Park, Ch. and Berger, L. R.,** Fatty acids of extractable and bound lipids of *Rhodomicrobium vannielii, J. Bacteriol.,* 93, 230, 1967.
30. **Radunz, A.,** Ueber das Sulfochinovosyl-diacylglycerin aus höheren Pflanzen, Algen und Purpurbakterien, *Hoppe-Seylers Z. Physiol. Chem.,* 350, 411, 1969.

LIPIDS AND FATTY ACIDS OF MICROALGAE

Peter Pohl

INTRODUCTION

Lipids and fatty acids are constituents of plant cells, where they serve as membrane components and as a source of energy and plant metabolites. Lipids can be extracted from plant materials with lipophilic organic solvents, such as ether, petroleum ether, and chloroform, or with solvent mixtures (e.g., chloroform-methanol). Thus, one obtains the so-called "total lipids". Most of the lipids are acyl lipids containing fatty acids bound to hydrophilic moieties, such as alcohols (mainly glycerol), sugars, and bases. Frequently the "total lipids" contain some other lipid-like compounds, such as chlorophylls, carotenoids, sterols, and sterol esters.

Lipids

These can be subdivided according to their polarity which depends on the nonpolar (lipophilic) carbon chains (fatty acids, long-chain alcohols) and on the polar (hydrophilic) moieties, such as phosphate and carboxylic groups, alcohols, sugars, or bases in each lipid. The main nonpolar lipids ("neutral lipids") of microalgae are triglycerides, free fatty acids, hydrocarbons, and wax esters (only in few algae). The polar lipids all are acyl lipids and mainly consist of phospholipids and glycolipids (see References 1 to 3 for structures). The main phospholipids of algae are phosphatidylcholine (PC), phosphatidylethanolamine (PE), phosphatidylinositol (PI), phosphatidylserine (PS), phosphatidylglycerol (PG), phosphatidic acid (PA), and diphosphatidylglycerol (cardiolipin, DPG). The major algal glycolipids are monogalactosyldiglyceride (MGDG), digalactosyldiglyceride (DGDG), and sulphoquinovosyldiglyceride (sulpholipid, SL).

A novel class of algal lipids are chlorosulpholipids which are derivatives of *N*-docosane-1,14-diol and of *N*-tetracosane-1,15-diol disulphates and which have been found in Chrysophyceae, Xanthophyceae, Chlorophyceae, and Cyanophyceae.

Fatty acids

Most of the fatty acids of algae are bound to the above lipids. In some algae, however, minor quantities of free fatty acids can be found.

1. Saturated fatty acids. Algae produce mostly larger amounts of even-numbered fatty acids (12:0, 14:0, 16:0, 18:0)* with minor quantities of odd-numbered acids (13:0 to 19:0). Small amounts of branched-chain (iso and anteiso) fatty acids have also been found frequently.
2. Unsaturated fatty acids.** Algae produce a great variety of unsaturated fatty acids mostly with C_{12} to C_{22} carbon chains and 1 to 6 double bonds. Normally, the double bonds are in the *cis*-configuration (exception: *trans* 3-16:1 in the phosphatidylglycerol of photosynthesizing algae). Unlike bacteria, fungi, and terrestrial plants, algae do not appear to synthesize fatty acids with "unusual" structures, such as acetylenic, hydroxy, epoxy, oxo, cyclopropanoic, and cyclopropenoic acids.[4]

* 12:0 is a saturated straight-chain fatty acid with 12 carbon atoms.

** Unsaturated fatty acids: 18:2 (9,12) describes an unsaturated fatty acid with 18 carbon atoms and two double bonds in the positions 9/10 and 12/13 (linoleic acid). The carboxylic group is carbon atom 1.

OCCURRENCE OF LIPIDS AND FATTY ACIDS IN MICROALGAE

So far, only relatively few algae have been analyzed for their lipid content. As to the fatty acids, however, our knowledge is much better. Tables 1 to 7 show that there are significant differences between marine and freshwater microalgae as well as between the individual classes of algae with regard to their fatty acid compositions. The formation of algal fatty acids and lipids is controlled by environmental factors, such as light, temperature, and the composition of the nutrient media. Thus, for correct comparisons, all algae analyzed should have been grown under comparable growth conditions (which, of course, is not the case). Nevertheless, the data in Tables 1 to 7 show that there is still considerable correspondence among the algae of each class.

Chrysophyceae (see Table 1)

These organisms have a great variety of saturated and unsaturated C_{16} to C_{22} fatty acids. Three Chrysophyceae presented in Table 1 (*Cricosphaera carterae, Isochrysis galbana, Prymnesium parvum*) were also analyzed by Lee and Loeblich,[5] who found much larger amounts of 14:0 (68.9%) in *P. parvum* and the 22:6 fatty acid in all of the three algae. There are no reports of lipids in marine Chrysophyceae. The lipids of *Ochromonas danica* (freshwater alga) were composed of MGDG, DGDG, SL, PG, PC, PE, PI, DPG, and triglycerides[6] and of chlorosulpholipids (chloro derivatives of docosane-1,14-diol-disulphate.[7-9] In several species of the Chrysophyceae (by some authors classified as "Haptophyceae"), a highly unsaturated straight chain hydrocarbon (3,6,9,12,15,18—21:6) has been detected.[5,10] In *Syracosphaera carterae*, the main hydrocarbon was a C_{17} *n*-alkane.[11]

Xanthophyceae (see Table 1)

Of the few algae so far investigated, the marine species and one freshwater organism (*Monodus subterraneus*) preferably synthesized the 16:0, 16:1, and 20:5 fatty acids, whereas another freshwater alga (*Tribonema aequale*) made high proportions of 18:1. The lipids of *M. subterraneus* were MGDG, DGDG, SL, PG, PE, PI, and triglycerides.[6] Of these, the two galactolipids had high concentrations of the 20:5 acid. In *Tribonema aequale*,[12] *M. subterraneus*,[13] and another freshwater species, *Botrydium granulatum*,[13] chlorosulpholipids (docosane-1,14-diol-disulphate and several chlorinated derivatives) have been detected. The hydrocarbons of *T. aequale* mainly contain short-chain alkanes and alkenes (14:0, 15:0, 15:1, 15:2).[10]

Bacillariophyceae (Diatoms) (see Table 2)

The major fatty acids are 16:0, 16:1, and 20:5. Most of the Bacillariophyceae produce rather low proportions of C_{18} fatty acids. The lipids of several diatoms have been analyzed by Kates[14] and Opute.[15] They consisted of triglycerides, MGDG, DGDG, SL, PG, PC, and PI. Four sulpholipids (phosphatidylsulphocholine, lysophosphatidylsulphocholine, deoxycertmide sulphonic acid, and sterol sulphate) were detected in *Nitzschia alba* (marine, nonphotosynthetic) by Anderson et al.[16,17] Many diatoms produce hydrocarbons.[5,11,18,19] Among these, a polyunsaturated hydrocarbon (3,6,9,12,15,18 to 21:6) often predominates.[5,10,19]

Dinophyceae (see Table 3)

In most of these algae, the main fatty acids are 16:0, 18:4, 20:5, and 22:6. In some natural and laboratory-cultured Dinophyceae, a very unusual polyunsaturated fatty acid (18:5[3,6,9,12,15]) was discovered.[20,21] The fatty acids of two more Dinophyceae not listed in Table 3 (*Gloedinium montanum* and Peridinium sociale) were analyzed by Lee and Loeblich,[5] who could not detect the 18:4 fatty acid. There is little infor-

Table 1
FATTY ACIDS OF CHRYSOPHYCEAE AND XANTHOPHYCEAE (EXPRESSED AS PERCENT OF TOTAL FATTY ACIDS)

	Chrysophyceae										Xanthophyceae			
Fatty acids Position of double bonds	*Monochrysis lutheri* (marine) (100/97)	*Syracosphaera carterae* (marine) (97/101)	*Pseudopedinella* sp. (marine) (100)	*Prymnesium parvum* (marine) (100)	*Isochrysis galbana* (marine) (100)	*Discrateria inornata* (marine) (100)	*Coccolithus huxley* (marine) (100)	*Cricosphaera carterae* (marine) (100)	*Cricosphaera elongata* (marine) (100)	*Ochromonas danica* (freshwater) (6)	*Olisthodiscus* sp. (marine) (97/100)	Undescribed flagellate (marine) (100)	*Monodus subterraneus* (freshwater) (6)	*Tribonema aequale* (freshwater) (5)
12:0	0.6/ 0.2	0.1/ 0.5	0.6	—	0.1	0.4	0.4	0.3	0.2	—	0.5/ 0.6	0.4	—	tr
14:0	9.2/11.4	0.6/ 1.4	4.9	5.6	0.6	1.3	6.4	8.6	6.6	13	7.4/ 9.1	6.4	2	4.7
14:1	— / 0.2	0.2/ tr	—	—	—	—	—	—	—	—	0.4/ —	—	—	tr
15:0	0.6/ 0.2	0.1/ —	0.4	—	0.3	0.3	0.4	0.2	1.3	—	0.4/ 0.3	0.2	—	0.5
16:0	10.1/15.1	21.8/49.1	7.9	15.7	15.6	15.9	16.9	8.8	9.5	12	21.8/ 6.0	12.3	24	7.9
16:1 (Various isomers)	20.2/25.4	3.0/ 9.9	23.6	10.1	5.1	7.9	27.8	21.4	20.8	1	19.9/ 0.4	18.1	24	9.5
16:2 (Various isomers)	7.1/ 1.9	3.8/ 6.1	7.4	2.3	2.9	1.7	3.8	9.8	7.6	—	2.2/ 1.0	4.0	—	tr
16:3 (Various isomers)	14.8/ 0.6	0.8/ 0.3	16.7	0.9	2.1	2.1	7.1	14.8	13.8	—	0.3/ 2.5	12.3	—	0.2
16:4 (Various isomers)	1.5/ 0.3	0.6/ —	1.3	0.5	0.2	0.4	1.2	0.4	1.1	—	0.2/ 0.9	1.2	—	—
17:0	0.4/ —	0.1/ —	0.4	—	0.3	0.4	—	0.1	—	—				tr
18:0	0.4/ —	0.7/ 3.3	—	—	—	0.1	0.7	0.4	—	2	0.2/ —	0.2	1	0.4
18:1 (9)	5.7/ 0.8	7.3/ 7.9	2.6	24.9	14.6	16.6	10.0	3.3	1.8	14	1.1/ 7.3	0.6	9	73.1
18:2 (9,12)	1.6/ 0.7	9.5/ 8.8	1.9	17.8	10.7	5.0	2.1	2.8	2.3	20	0.9/ 6.8	0.8	4	0.6
18:3 (6,9,12)	— / 0.1	— / —	—	0.4	0.8	0.4	—	—	0.2	14	0.3/ 0.4	—	tr }	
18:3 (9,12,15)	— / 0.2	8.0/11.6	—	11.2	14.3	12.8	1.3	—	0.7	3	6.8/ 6.3	0.2	tr }	0.2
18:4 (6,9,12,15)	0.6/ 4.0	24.1/ —	—	2.0	17.2	20.3	1.1	1.7	0.9	4	7.9/29.3	0.2	—	—
20:1 (Various isomers)	— / 0.6	0.5/ —	0.5	—	—	—	0.2	—	0.3	—				1.0
20:3 (8,11,14)	1.7/ 0.1	— / —	0.7	1.2	0.5	1.4	0.4	—	0.2	5	1.7/ —	—	1	1
20:4 (Various isomers)	0.5/ 0.4	0.1/ —	1.3	1.4	0.4	0.6	0.5	3.2	2.0	6	2.2/ 2.1	5.8	5	0.8
20:5 (5,8,11,14,17)	18.9/16.3	3.9/ —	26.5	4.2	3.1	7.8	17.1	19.8	27.9	—	21.8/ 1.8	31.4	29	
22:5 (4,7,10,13,16)	0.5/ 1.2	1.1/ —	—	0.2	—	—	0.1	—	— }	4				
22:5 (7,10,13,16,19)	3.3/ 0.4	0.1/ —	1.1	0.7	0.3	3.4	0.1	1.8	0.3 }		0.1/13.2	2.5	—	0.4
22:6 (4,7,10,13,16,19)	— /13.1	8.6/ —	—	—	—	—	—	—	—	—	3.0/ —	—	—	

Note: tr = trace.

Table 2
LIPIDS AND FATTY ACIDS OF BACILLARIOPHYCEAE (DIATOMS)

Lipids or fatty acids	Position of double bonds	Freshwater						Marine									
		Navicula muralis (15)	Navicula pelliculosa (14)	Nitzschia palea (15)	Navicula incerta (15)	Nitzschia closterium (14)	Nitzschia angularis (14)	Nitzschia thermalis (14)	Cyclotella cryptica (97)	Phaeodactylum tricornutum (97)	Sceletonema costatum (97)	Ditylum brightwelli (100)	Bidulphia sinensis (100)	Chaetoceros septentrionale (100)	Lauderia borealis (100)	Asterionella japonica (100)	Thalassiosira fluviatilis (97)
MGDG		+	+	+	+	+	+	+	+	+							
DGDG		+	+	+	+	+	+	+	+	+							
SL		+	+	+	+	+	+	+	+	+							
PG		+	+	+	+	+	+	+	+	+							
PC		+	+	+	+	+	+	+	+	+							
PE		+	+ (?)	+	+	+	+	+	+	+							
PI		+	+	+	+	+	+	+	+	+							
DPG			+			+	+	+	+	+							
PA			+			+	+	+	+	+							
N-methyl-PE			+			+	+	+	+	+							
glycerides		+	+	+	+	+	+	+	+	+							
hydrocarbons[a]																	
12:0		—	—	—	—	—	—	—	0.1	0.1	0.2	0.9	—	0.6	—	0.5	0.1
14:0		5.0	2.8	6.2	5.5	5.5	5.3	2.2	7.2	7.8	32.7	7.8	11.4	6.3	7.3	6.3	7.9
14:1		tr	tr	tr	tr	0.4	0.5	0.3	0.1	0.2	1.0	—	—	—	—	—	0.1

15:0	—	tr	—	—	0.4	1.3	0.5	1.3	0.2	1.4	0.7	0.4	0.5	0.5	0.4	1.0
16:0	27.4	9.1	22.8	30.1	25.4	23.6	30.4	27.8	26.6	6.9	13.4	14.1	11.0	11.7	11.3	23.2
16:1 (Various isomers)	45.7	30.8	44.7	38.3	24.9	30.5	45.5	42.7	46.9	22.4	33.0	27.6	21.2	21.3	21.7	44.8
16:2 (Various isomers)	3.8	3.2	3.6	4.4	1.8	2.6	2.3	2.6	2.6	7.4	2.7	5.4	2.9	2.9	2.8	2.8
16:3 (Various isomers)	3.8	18.3	1.6	2.7	3.7	3.6	2.4	4.1	2.9	2.3	9.7	4.0	8.7	12.1	9.4	6.5
16:4 (Various isomers)	—	—	—	—	—	—	—	0.2	0.3	6.6	0.4	9.4	1.2	0.9	1.2	0.1
18:0	tr	tr	tr	tr	0.6	1.7	0.8	0.3	0.5	0.1	0.2	0.7	0.1	—	0.2	0.3
18:1 (Various isomers)	3.0	6.2	2.5	2.7	2.6	3.2	1.2	0.5	1.1	0.3	6.3	2.5	3.4	1.8	2.9	0.4
18:2 (9,12)	tr	3.9	tr	tr	2.5	2.2	0.8	0.3	1.0	1.1	2.1	1.1	0.5	1.1	0.8	0.4
18:3 (Various isomers)	tr	2.6	tr	tr	—	1.6	1.3	0.1	0.6	0.6	1.0	tr	0.5	0.3	0.2	0.2
18:4 (6,9,12,15)		—		—	—	—	—	2.6	0.1	2.2	0.2	2.8	0.1	—	0.4	0.2
20:4 (5,8,11,14)				—				—	0.2	—	—	0.4	0.5	—	6.5	0.5
20:4 (8,11,14,17)	} 6.3	} 4.5	} 6.3	—	9.4[c]	—	3.7[b]	0.1	—	—	1.4	—	1.2	1.3	4.3	0.1
20:5 (5,8,11,14,17)	5.0	14.5	12.0	16.4	17.2	21.1	6.9	7.5	8.6	13.8	10.9	15.1	20.8	30.3	19.8	8.0
22:5 (Various isomers)	—	—	—	—	—	—	—	0.1	—	0.4	1.2	1.5	3.7	1.1	3.2	0.4
22:6 (4,7,10,13,16.19)	—	—	—	—	—	—	—	0.9	0.8	1.7	—	—	—	—	—	2.2

Note: Numbers in lower column indicate percent of fatty acids; tr = trace; +, present.

[a] See text.
[b] Positions of double bonds not reported.

Table 3
FATTY ACIDS OF DINOPHYCEAE (MARINE) (EXPRESSED AS PERCENT OF TOTAL FATTY ACIDS)

Fatty acids	Position of double bonds	Species																		
		Amphidinium carteri (22/97)	*Ceratium furca* (20)	*Cochlodinium heteroloblatum* (20)	*Exuviella* sp. (22)	*Glenodinium* sp. (22)	*Gonyaulax catenella* (20)	*G. digensis* (20)	*G. digitale* (20)	*G. polyedra* (102)	*G. tamarensis* (22)	*Gymnodinium nelsoni* (22)	*G. simplex* (20)	*G. splendens* (20)	*Massartia rotundata* (20)	*Peridinium triquetum* (20)	*P. trochoideum* (22/100)	*Prorocentrum micans* (22)	*P. minimum* (22)	*Cryptecodinium cohnii* (heterotrophic) (103)
12:0		—/0.2	—	—	—	—	—	—	—	—	—	—	—	—	—	—	—/—	—	—	8
14:0		2/3.3	4.4	1.0	3	6	6.5	9.7	4.1	2	11	6	17.5	1.1	12.1	2.6	14/17	1.9	2.2	19
14:1		—/—	—	—	—	1	—	—	—	—	1	—	—	—	—	—	—/—	—	—	1
16:0		12/36	26.1	32.9	26	23	13.9	21.8	28.3	36	28	35	17.6	28.8	15.5	26.6	27/36	27.5	23.1	20
16:1	(Various isomers)	1/0.5	4.6	3.8	2	4	2.6	1.8	—	1	4	5	4.4	—	11.1	1.5	8/4	2.1	1.4	1
16:2	(Various isomers)	2/0.2	—	—	1	—	—	—	—	—	2	—	—	—	—	—	2/1	—	—	—
16:3	(Various isomers)	—/0.1	—	—	—	—	—	—	—	—	—	—	—	—	—	—	—/3	—	—	—
18:0		2/5.7	0.5	1.0	10	3	1.2	1.8	1.5	—	3	1	2.9	2.7	2.4	1.2	4/—	1.7	1.3	1
18:1	(Various isomers)	2/7.5	4.8	6.4	5	5	2.7	7.5	7.1	3	5	16	7.6	9.8	5.6	4.8	9/7	6.7	4.3	14
18:2	(9,12)	1/0.1	3.1	2.4	9	5	1.9	0.3	0.5	2	4	1	2.1	2.8	2.2	4.8	3/—	5.1	5.4	—
18:3	(9,12,15)	3/0.1	0.4	0.1	2	6	1.8	2.1	—	3	5	1	0.4	0.1	3.9	0.3	2/3	0.2	0.3	—
18:4	(6,9,12,15)	19/10.1	1.0	1.4	8	23	7.3	1.4	7.0	14	12	4	2.4	2.1	11.6	4.3	9/7	4.5	2.4	—
18:5	(3,6,9,12,15)	—/—	5.9	8.2	—	—	11.1	8.3	3.8	—	—	—	19.4	11.0	17.7	22.6	—/—	12.5	23.2	—
20:0		1/2.2	—	—	1	—	—	—	—	—	—	—	—	—	—	—	—/—	—	—	—
20:5	(5,8,11,14,17)	20/7.4	7.4	11.4	4	2	11.2	8.2	16.2	14	3	5	1.1	8.0	0.3	5.1	1/13	2.7	5.3	—
22:5	(7,10,13,16,19)	—/0.6	—	—	—	1	—	—	—	—	1	—	—	—	—	—	1/3	—	—	1
22:6	(4,7,10,13,16,19)	24/25.4	20.7	28.0	12	19	33.9	24.5	23.8	23	18	20	20.4	30.3	14.1	21.3	15/—	32.1	24.5	30

mation about the lipids: *Glenodinium* sp. (photoauxotroph) produced MGDG, DGDG, PC, and triglycerides, and *Gyrodinium cohnii* (heterotroph) produced PC and triglycerides.[22] The hydrocarbons of several Dinophyceae were analyzed by Lee and Loeblich[5] and Blumer et al.[10,19] The major hydrocarbon in the photosynthetic organisms was 3,6,9,12,15,18—21:6.

Rhodophyceae

The fatty acids and lipids of one marine unicellular red alga, *Porphyridium cruentum* have been analyzed by Nichols and Appleby.[6] The main fatty acids were 16:0 (23%), 18:2 (16%), 20:4 (36%), and 20:5 (17%). The lipids were composed of MGDG, DGDG, SL, PG, PC, and triglycerides.

Chlorophyceae and Prasinophyceae (see Tables 4 and 5)

Some of the marine microalgae (see Table 4) have formerly been regarded as Chlorophyceae and are now classified as Prasinophyceae.[23-25] All of these algae contain a great number of saturated and unsaturated fatty acids, mostly with higher proportions of C_{16} and C_{18} compounds. There are no reports on lipids of these marine algae. The main hydrocarbons were 7-heptadecene in *Dunaliella tertiolecta* and heptadecane and pentadecane in *Derbesia tenuissima*.[10] The fatty acid compositions of freshwater Chlorophyceae are shown in Table 5. These algae produce no or only small amounts of the 18:4 and of polyunsaturated C_{20} and C_{22} fatty acids. Only few freshwater Chlorophyceae have so far been analyzed for their lipids. The main polar lipids seem to be MGDG, DGDG, PG, SL, PE, PC, and PI.[26-29] The presence of chlorosulpholipids has been detected in *Elakatothrix viridis* and *Zygnema* sp.,[13] and an inositol-free phytosphingoglycolipid was found in *Scenedesmus obliquus*.[30] Freshwater Chlorophyceae also seem to produce small amounts of hydrocarbons with C_{15} to C_{36} carbon chains. The main hydrocarbon was *n*-C_{17} in *Chlorella pyrenoidosa* and *Coelastrum microsporum*,[31] *n*-C_{23} in *Tetraedron* sp.,[31] *n*-C_{27} in *Scenedesmus quadricauda*,[31] a series of saturated *n*-paraffins (C_{17} to C_{36}) in autotroph *Chlorella vulgaris*,[32] and 1-pentacosene and 1-heptacosene in *C. vulgaris* grown heterotrophically.[32]

A unique fatty acid and lipid composition has been observed in *Botryococcus braunii*. This alga produces only very small amounts of fatty acids (0.014%), the major compounds being 16:0, 18:1, and 28:1.[33] By far the major portion of the dry weight (approximately 76%) was made up of hydrocarbons. According to Gelpi et al.[34] and Knights et al.,[35] the main hydrocarbons were C_{27} (1,18) diene, C_{29} (1,20) diene, and C_{31} (1,22) diene, in addition to the commonly found *n*-C_{17} hydrocarbon. Maxwell et al.,[36] however, found no saturated hydrocarbons. Two unsaturated isomeric C_{34} hydrocarbons (botryococcene and isobotryococcene) were present in a ratio of about 9:1 and constituted the bulk of the hydrocarbons.

Euglenophyceae (see Table 6)

Of the algae shown in Table 6, *Euglena gracilis* can be grown heterotrophically in the dark and photoauxotrophically in the light, leading to considerably different fatty acid compositions. The fatty acids of the light-grown organisms (*Euglena, Astasia, Eutreptia*) are composed of C_{14} to C_{22} fatty acids with larger portions of polyunsaturated C_{16} and C_{18} fatty acids; in dark-grown *E. gracilis*, saturated and unsaturated fatty acids (14:0, 16:0, 16:1, 18:1) predominated. In the dark, *E. gracilis* produced large amounts of unpolar lipids (wax esters) and phospholipids (PC, PE), whereas upon illumination the major lipids were MGDG, DGDG, SL, and PG (chloroplast lipids).[37-50]

In two marine species, *Eutrepiella* sp.[10] and *Eutreptia viridis*,[5] a polyunsaturated hydrocarbon (21:6) was detected. *Eutrepiella* sp. also contained the *n*-C_{15} and *n*-C_{17} hydrocarbons.[10]

Table 4
FATTY ACIDS OF MARINE MICROSCOPIC CHLOROPHYCEAE AND PRASINOPHYCEAE (EXPRESSED AS PERCENT OF TOTAL FATTY ACIDS)

Fatty acids	Position of double bonds	Chlorophyceae				Prasinophyceae			
		Dunaliella primolecta (100)	*Dunaliella tertiolecta* (100/97)	*Dunaliella salina* (101)	*Chlamydomonas* sp. (100)	*Heteromastix rotunda* (100)	*Tetraselmis (Platymonas)* sp. (97)	*Platymonas tetrathele* (104)	*Halosphaera viridis* (25)
12:0		0.8	0.9/0.2	0.2	0.8/0.9	0.2	0.1	0.4	0.12
14:0		4.7	5.9/0.5	0.4	1.3/4.5	9.2	0.8	2.0	4.95
15:0		1.7	0.7/1.3	—	0.3/1.3	0.5	0.1	—	0.02
16:0		11.3	12.9/17.6	41.3	16.8/15.1	10.6	19.7	11.2	17.63
16:1	(Various isomers)	9.8	9.5/3.4	14.8	6.8/2.4	15.7	2.9	4.0	1.25
16:2	(Various isomers)	8.3	3.0/2.2	0.4	0.8/6.6	2.9	0.5	2.4	0.13
16:3	(Various isomers)	7.4	4.9/4.3	—	0.3/6.6	3.1	1.8	3.5	0.07
16:4	(Various isomers)	5.9	7.1/16.5	—	0.5/0.3	2.0	14.7	14.0	3.42
18:0		0.1	—/0.4	—	0.2/—	—	0.3	9.2	0.07
18:1	(Various isomers)	5.8	7.7/3.1	11.2	7.1/16.7	2.3	14.8	12.9	44.75
18:2	(9,12)	5.8	5.8/5.1	7.8	28.4/10.1	2.8	3.3	12.4	7.38
18:3	(6,9,12)	2.1	0.8/5.8	—	0.6/3.1	—	0.6	—	0.91
18:3	(9,12,15)	10.4	7.7/37.4	18.5	0.2/28.4	3.7	18.9	16.0	7.28
18:4	(6,9,12,15)	6.7	8.3/1.2	—	18.6/0.8	9.3	7.5	7.6	6.55
20:0 + 20:1		0.3	—/0.1	—	15.8/1.0	—	1.6	5.0	0.09
20:2	(8,11)	0.8	1.0/—	—	—/0.5	—	—	—	—
20:3	(8,11,14)	1.3	1.9/0.1	—	—/—	—	—	0.1	—
20:4	(5,8,11,14)	0.4	0.4/—	—	—/—	0.6	1.3	2.0	—
20:4	(8,11,14,17)	1.7	3.5/—	—	—/—	0.2	—	—	0.39
20:5	(5,8,11,14,17)	9.7	10.2/—	—	—/—	27.9	7.5	3.8	tr[a]
22:5	(Various isomers)	3.9	6.2/—	—	0.3/—	6.8	—	—	0.43
22:6	(4,7,10,13,16,19)	—	—/—	—	—/—	—	1.4 (?)	—	3.83

[a] tr = trace.

Cryptophyceae (see Table 6)

These algae contain a wide range of saturated and unsaturated fatty acids ranging from C_{12} to C_{22}. Interestingly, some of the marine species produce larger portions of 20:1. The lipids of one alga (*Cryptomonas* sp.) were analyzed by Beach et al.[51] and contained MGDG, DGDG, SL, PG, and PC. Among the hydrocarbons of two species of *Cryptomonas*[5,10] and of *Rhodomonas lens*,[5] the major compound was a 21:6 polyunsaturated alkene.

Cyanophyceae (see Table 7)

These algae produce C_{12} to C_{18} fatty acids with zero to four double bonds. They do not appear to be able to synthesize the *trans* 3-16:1, 16:3, 16:4 and unsaturated C_{20} and C_{22} fatty acids. The data of Table 7 indicate that the capability to produce unsaturated fatty acids varies widely among these organisms; some are able to make 18:1, and some produce 18:2, 18:3 (α), 18:3 (γ), or even 18:4, respectively. These great differences have even been observed within strains of the same genus (cluster) as was shown by Kenyon et al.[52,53] (11 strains of *Aphanocapsa*, 20 strains of *Synechococcus*). So far it seems to be quite problematical to make taxonomic subdivisions on the basis of the fatty acid compositions. Possibly the above differences are related to differences in the DNA base compositions.[54] The main lipids in Cyanophyceae appear to be mono- and digalactosyl diglycerides, sulphoquinovosyldiglyceride, and phosphatidylglycerol.[55-58] Zepke et al.[58] found trigalactosyldiglyceride in *Tolypothrix* and *Oscillatoria. Cyanidium caldarium* which cannot clearly be classified as a Cyanophycea[59] also contained other phospholipids such as phosphatidylcholine, phosphatidylethanolamine, and phosphatidylinositol.[59] In many Cyanophyceae, small quantities of hydrocarbons in the range of C_{14} to C_{29} have been detected. Mostly, the dominating hydrocarbon was n-C_{17}; in some species, however, it was C_{25} to C_{29}.[31,34,60-64] Mercer and Davies[13] found chlorosulpholipids (probably chloro derivatives of docosane-1,14-diol disulphate) in *Nostoc* sp.

Under nitrogen-fixing conditions, filamentous heterocystous and unicellular Cyanophyceae synthesize a unique group of glycolipids which, in the heterocystous organisms, are exclusively localized in the heterocyst envelopes forming "laminated layers".[57,65-70] These glycolipids are long-chain hydroxy fatty acids or alcohols attached to sugars, such as the glucopyranosyl esters of 25-hydroxyhexacosanoic acid and of 25,27- or 26,27-dihydroxyoctacosanoic acid, glucopyranosyl-3,25- hexacosanediol, and glucopyranosyl-3,25,27-octacosanetriol.[68-70] Lesser amounts of the corresponding galactosyl esters were also detected.

PATHWAYS OF UNSATURATED FATTY ACID BIOSYNTHESIS IN MICROALGAE

The biosynthesis of unsaturated fatty acids in all of the algae is O_2 dependent. Most of the algae are able to synthesize unsaturated C_{16} to C_{22} fatty acids. An exception to this are the Cyanophyceae and photoautotroph freshwater Chlorophyceae which synthesize no or only very small amounts of unsaturated C_{20} and C_{22} fatty acids. One of the main pathways appears to be

$$C_2 \longrightarrow\longrightarrow 16{:}0 \begin{smallmatrix}\nearrow 16{:}1(9)\\ \longrightarrow\end{smallmatrix} 18{:}0 \longrightarrow 18{:}1(9) \longrightarrow 18{:}2(9,12) \longrightarrow 18{:}3(6,9,12)$$

$$\longrightarrow 20{:}3(8,11,14) \longrightarrow 20{:}4(5,8,11,14) \qquad (1)$$

A characteristic feature of Pathway 1 is the formation of arachidonic acid 20:4(5,8,11,14) via γ-linolenic acid 18:3(6,9,12). This pathway (also called "animal

Table 5
FATTY ACIDS OF FRESHWATER MICROSCOPIC CHLOROPHYCEAE

Lipids or fatty acids	Position of double bonds	*Bracteacoccus minor* (99)[d]	*Chlamydomonas reinhardi* (29)	*Chlorella fusca* (99)[d]	*Chlorella pyrenoidosa* (105)[f]	*Chlorella variegata* (28)	*Chlorella vulgaris* (26)[c]	*Coelastrum microsporum* (106)	*Halochlorella rubescens* (99)[d]	*Polytoma oviforme* (colorless) (28)	*Prototheca portoricensis* (colorless) (28)	*Scenedesrus obliquus* (107)	*Scenedesmus quadricauda* (106)	*Tetraedron* sp. (106)	*Ulothrix aequalis* (108)
Lipids															
MGDG			+			+	+			tr	–				+
DGDG			+			+	+			tr	tr				+
SL			+			+	+			—	tr				
PG			+			+	+			+	+				
PC			+			+	+			+	+				
PE			+			+	+			+	+				
PI						+	+			+	+				
DPG						+	+								
Glycerides						+				+	+				
Hydrocarbons					+[b]		+[a]	+[b]					+[b]	+[b]	
Sterol esters										+	+				
Fatty Acids															
12:0		tr	—	tr	tr	tr	tr	—	tr	0.2	0.5	—	—	—	0.1
14:0		0.5	tr	0.3	0.2	1.2	1.9	0.7	0.5	2.5	0.6	0.9	—	—	1.1

14:1	(?)	—	—	—	tr	—	—	—	—	—		0.3	—	—	0.3
15:0		—	—	—	0.2	—	—	—	—	—		—	—	—	0.5
16:0		15.7	19.2	21.2	13.6	13.5	25.3	13.1	22.7	39.3	24.2	35.4	13.0	18.3	15.1
16:1	(Various isomers)	3.6	3.4	4.9	3.2	2.0	8.2	1.3	2.7	1.9	0.9	1.6	4.0	2.6	3.4
16:2	(Various isomers)	7.8	2.4	4.6	7.0	1.0	5.4	4.6	4.3	0.8	0.2	tr	—	2.7	0.1
16:3	(Various isomers)	0.7	6.7	5.0	5.1	tr	tr	—	5.4	0.9	—	0.2	—	—	2.8
16:4	(Various isomers)	6.3	21.1	—	—	—	—	—	—	—	—	15.0	—	—	30.5
17:0		—	—	—	tr	—	—	—	—	—	—	—	—	—	0.1
18:0		2.6	tr	tr	3.5	1.9	2.9	—	tr	3.2	2.1	—	—	1.7	0.7
18:1	(9)	8.7	5.4	25.4	34.7	13.4	2.9	37.0[g]	24.8	31.4[h]	18.0	7.8	32.0[i]	33.5	2.9
18:2	(9,12)	20.9	10.5	17.4	17.7	53.0	42.4	13.0	18.4	4.9	42.2	6.4	7.0	11.1	4.7
18:3	(Various isomers)	33.2	31.7[e]	21.2	14.6	16.0	12.8	16.0	21.1	8.2	3.2	29.5	29.0	18.3	36.3
18:4	(6,9,12,15)	—	tr	—	—	—	—	7.0	—	—	—	1.9	—	6.3	0.1
C_{20}		—	—	—	tr	—	—	—	—	0.8	2.1	0.65	—	—	0.8
C_{22}		—	—	—	—	—	—	—	—	—	2.0	0.31	—	—	0.4
C_{24}		—	—	—	—	—	—	—	—	—	1.8	—	—	—	—

Note: Numbers in lower column indicate percent of total fatty acids; tr = trace; +, present.

[a] See Reference 32.
[b] See Reference 31.
[c] Grown in the light (inorganic).
[d] Grown at high levels of KNO_3.
[e] 8.7% 18:3 (Δ6,9,12) + 22.4% 18:3 (Δ9,12,15).
[f] + traces of 14:2 (?), 14:3, and of C_{19}, C_{20}, C_{22}, and C_{24} acids.
[g] 30% 18:1 (*cis*9) (oleic acid) + 7% 18:1 (*trans*9) (elaidic acid).
[h] 9.3% 18:1 (Δ5) + 22.1% 18:1 (Δ9).
[i] 18% 18:1 (*cis*9) (oleic acid) + 14% 18:1 (*trans*9) (elaidic acid).

Table 6
FATTY ACIDS OF EUGLENOPHYCEAE AND CRYPTOPHYCEAE (EXPRESSED AS PERCENT OF TOTAL FATTY ACIDS)

		Euglenophyceae				Cryptophyceae										
		Freshwater			Marine	Freshwater			Marine							
Fatty acids	Position of double bonds	*Euglena gracilis* (dark-grown) (50)	*E. gracilis* (light-grown) (50)	*Astasia longa* (37)	*Eutreptia viridis* (5)	*Cryptomonas ovata* (51)	*C. ovata* var. *palustris* (109)	*Chilomonas paramecium* (51)	*Hemiselmis rufescens* (100)	*H. brunescens* (100)	*H. virescens* (51)	*Cryptomonas appendiculata* (100)	*C. maculata* (100)	*C. sp.* (51)	*Chroomonas* sp. (51)	*Rhodomonas lens* (51)
12:0		tr	tr	—	1.2	—	—	—	0.5	0.6	—	—	0.8	—	—	
13:0		—	—	—		—	—	—	0.3	0.9	—	—	—	—	—	
14:0		9.2	3.4	10.4	0.2	1	5	18	0.9	1.2	8	1.7	4.5	6	2	18
15:0		—	—	—	tr	—	—	—	0.5	0.6	—	0.3	1.0	—	—	
16:0		27.6	7.5	17.9	33.4	10	10	18	21.3	12.5	23	16.1	14.7	4	16	13
16:1	(Various isomers)	9.1	4.9	1.4	3.1	5	9	—	10.0	3.4	7	11.8	6.5	2	2	6
16:2	(Various isomers)	1.1	25.8	—	0.1		—		2.9	3.1	—	1.4	2.7			
16:3	(Various isomers)	tr	4.4	—	0.4	2	—	1	0.3	0.2	—	1.5	1.1	2	1	
16:4	(Various isomers)	tr	14.4	—	—		—		1.0	0.1	—	0.4	—			
17:0		—	—	—	tr	—	—	—	0.7	0.4	—	0.5	0.3	—	—	
18:0		1.2	2.4	2.2	—	—	8	1	—	0.1	—	0.4	0.3	—	1	1
18:1	(Various isomers)	24.1	1.9	4.4	8.9	2	16	9	2.3	2.3	7	2.5	3.5	5	5	10
18:2	(9,12)	1.9	8.5	0.9	9.9	—	26	12	1.3	0.1	3	4.2	—	—	3	2
18:3	(6,9,12)	—	—	—	1.4	5	—	—	2.0	0.2	—	—	—	—	1	
18:3	(9,12,15)	1.8	22.5	—		17	5	27	6.8	8.4	22	12.0	6.3	7	23	16
18:4	(6,9,12,15)	3.4	tr	—		34	—	—	17.4	30.5	16	13.3	15.6	44	23	13

20:0		tr	tr		—	—	—	1.8	0.2	—	0.9	—	—	—	
20:1	(Various isomers)	—	—	9.4	—	—	—	14.4	18.0	—	10.3	16.7	—	—	
20:4	(5,8,11,14)	9.2	1.3		—	—	—	—	0.2	—	2.7	0.7	—	—	
20:4	(8,11,14,17)	—	—	0.7	—	—	—	2.6	0.3	—	5.8	1.0	—	—	
20:5	(5,8,11,14,17)	1.5	2.8	12.3	12	—	6	8.2	13.9	7	9.8	17.2	16	14	13
22:4	(7,10,13,16)	—	—		2	—	—	0.9	—	2	—	—	—	1	1
22:5	(7,10,13,16,19)	—	—		—	—	—	3.4	1.3	—	2.5	3.4	—	—	
22:6	(4,7,10,13,16,19)	0.3	tr	6.7[a]	7	—	3	—	—	2	—	—	10	6	5

Note: tr = trace.

[a] + 20:2 (11.2%) + 22:2 (1, 2%).

Table 7
FATTY ACIDS OF CYANOPHYCEAE (CYANOBACTERIA) (EXPRESSED AS PERCENT OF TOTAL FATTY ACIDS)

Species	Fatty acids													
	12:0	14:0	14:1	15:0	16:0	16:1	16:2	18:0	18:1	18:2	18:3[a] (α)	18:3[a] (γ)	18:4	20:0
Unicellulars														
Aphanocapsa (52)	(11 strains[b])													
Agmenellum quadruplicatum (110)	tr	2.0	1.2	1.8	34	15	3.8	2.6	16	14	5.2 (α + γ)			
Chlorogloea fritschii[c] (71)	—	—	—	—	42.3	4.9	tr	5.4	14.3	17.2	15.8	—	—	—
Coccochloris elabens (110)	tr	1.1	0.5	0.8	49	12	—	1.2	13	17	2.6 (α + γ)		—	—
Gloeocapsa sp. (52)	—	11	3	—	19	46	—	3	4	3	—	—	—	—
Microcystis aeruginosa (52)	—	1	tr	—	40	3	—	2	6	15	—	27	—	—
Myxosarcina chroococcoides (71)	—	—	—	—	38.2	8.6	1.2	4.0	6.8	9.2	33.3	—	—	—
Synechococcus cedrorum[d] (111)	0.2	0.5	1.0	—	47.0	38.8	—	1.4	10.0	—	—	—	—	—
Anacystis nidulans[e] (110)	tr	0.5	0.6	tr	49.0	34.0	—	1.7	13.0	—	—	—	—	0.5
Anacystis marina (110)	tr	21	1.4	1.4	32	36	tr	1.6	4.1	—	—	—	—	
Cyanidinium caldarium[f] (92)	tr	tr	—	—	18	tr	—	7	34	11	30 (α + γ)		—	—
Filamentous (without heterocysts)														
Lyngbya lagerheimii[g] (53)	—	tr	—	1.1	33	2	tr	4	7	12	27	—	—	—
Microcoleus chtonoplastes (110)	1.5	5.0	tr	2.5	37	13	—	3.7	14	5.0	18 (α + γ)		—	—
Microcoleus vaginatus (53)	—	tr	tr	—	26	6	—	1	6	8	15	—	35	—
Oscillatoria williamsii[h] (110)	0.6	2.0	1.2	1.7	36	24	14	1.9	11	3.9	—	—	—	—
Oscillatoria agardhii var. *suspensa* (53)	—	1	tr	—	12	14	—	2	7	18	21	4	—	—
Phormidium sp. (53)	—	1	tr	—	18	9	—	3	21	32	1	—	2	—
Plectonema boryanum[i] (53)	—	1	—	—	21	30	—	2	16	25	1	—	—	—
Plectonema terebrans (110)	tr	1.4	0.8	1.1	35	13	5.1	2.5	20	11	6.0 (α + γ)		—	—
Schizothrix calcicola (53)	—	1	tr	—	16	20	—	3	16	16	16	—	—	—
Spirulina platensis[j] (71)	—	—	—	—	43.4	9.7	tr	2.9	5.0	12.4	tr	21.4	—	—
Trichodesmium erythaeum (110)	2.2	21	tr	1.4	17	3.7	—	2.6	2.8	4.2	19 (α + γ)		—	—
Filamentous (with heterocysts)														
Anabaena cylindrica[k] (71)	—	—	—	—	46.0	6.4	5.6	3.6	6.0	24.0	11.2	—	—	—

Anabaena flos-aquae (71)	—	—	—	—	39.5	5.5	4.3	1.0	5.2	36.5	10.7	—	—	—
Anabaena spiroides (53)	—	tr	—	—	18	15	—	1	12	18	40	tr	—	—
Anabaena variabilis[l] (110)	tr	1.8	1.1	0.9	32	15	—	4.4	14	14	17 (α + γ)		—	—
Anabaenopsis circularis (53)	—	1	1	—	19	21	—	2	10	24	11	—	—	—
Calothrix parietina (53)	—	1	—	—	34	6	—	1	4	14	7	2	17	—
Calothrix desertica (53)	—	1	—	—	24	2	—	5	16	22	5	—	6	—
Tolypothrix tenuis (53)	—	1	2	—	22	3	—	3	12	15	6	13	11	—
Nodularia sphaerocarpa (53)	—	1	tr	—	23	3	—	7	14	17	5	—	11	—
Hapalosiphon laminosus[m] (111)	0.4	1.0	—	—	53.7	23.8	—	2.9	18.2	—	—	—	—	—
Mastigocladus laminosus[m] (71)	—	—	—	—	38.5	42.5	—	tr	16.8	2.1	—	—	—	—
Nostoc muscorum[n] (110)	tr	2.7	2.1	2.2	27	20	—	3.1	16	14	11 (α + γ)		—	—

Note: tr = trace.

[a] 18:3 (α) = 18:3 (9,12,15); 18:3 (γ) = 18:3 (6,9,12).

[b] Kenyon[52] investigated 11 strains of *Aphanocapsa*; 5 strains had high percentages of 16:0 (28 to 35%), 18:2 (17 to 21%) and 18:3 (γ) (18 to 31%) and low concentrations of 14:0 (1 to 2%), 14:1 (1%), 16:1 (4 to 13%), and 18:3 (α) (0 to 2%). The other six strains had high percentages of 14:0 (18 to 34%), 14:1 (5 to 13%), and 16:1 (34 to 50%) and low concentrations of 16:0 (6 to 10%), 18:2 (1 to 4%), and no 18:3.

[c] Two species of *Chlorogloea fritschii* were analyzed by Holton et al.,[111] with comparable results. One species of *Chlorogloea* investigated by Kenyon[52] contained higher amounts of 14:0 and 16:1, lower amounts of 16:0 and 18:1, and no 18:3.

[d] *Synechococcus* (20 strains) were investigated by Kenyon,[52] indicating that there are several groups with varying fatty acid contents: (1) high content of 18:2, 18:3 (α), partly 18:3 (γ) and 16:2; and (2) low or undetectable contents of 18:2 and 18:3.

[e] *A. nidulans* was also investigated with similar results by Holton et al.[96] and Orő et al.[60]

[f] Unclear botanical classification.

[g] *L. lagerheimii* was also investigated by Parker et al.,[110] who could not detect 18:3.

[h] One species of *Oscillatoria* was investigated by Holton et al.[111] showing no 16:2, but 6.8% 18:3.

[i] One species of *Plectonema* was investigated by Kenyon et al.,[53] with similar results.

[j] Three strains of *Spirulina* were also investigated by Kenyon et al.,[53] showing a similar fatty acid composition in one strain. The other two strains had no 18:2 and 18:3.

[k] One species of *Anabaena* investigated by Kenyon et al.[53] had no 16:2.

[l] Kenyon et al.[53] found a similar fatty acid composition in *A. variabilis*, too.

[m] *H. laminosus* and *M. laminosus* are by some authors regarded as being identical.

[n] *N. muscorum* and *N.* sp. were also investigated with similar results by Holton et al.[111] and Kenyon et al.,[53] respectively.

pathway" or "γ-linolenic pathway") operates in animals, but has also been found in microalgae, such as *Ochromonas danica* (Chrysophyceae), *Porphyridium cruentum* (Rhodophyceae),[6] and *Spirulina platensis* (Cyanophyceae).[71] There is evidence that in *P. cruentum* arachidonic acid is synthesized in the chloroplast, but in *O. danica* (and in *M. subterraneus* Xanthophyceae) it is synthesized outside the chloroplast.[6]

A variant of the above pathway seems to exist in heterotroph (darkgrown) *Euglena gracilis* (Euglenophyceae).[6,41]

$$C_2 \longrightarrow\longrightarrow 16{:}0 \longrightarrow 18{:}0 \longrightarrow 18{:}1(9) \longrightarrow 18{:}2(9,12) \longrightarrow$$

$$\longrightarrow 20{:}2(11,14) \longrightarrow 20{:}3(8,11,14) \longrightarrow 20{:}4(5,8,11,14) \quad (2)$$

Pathway 2 seems to operate outside the chloroplasts and employs 20:2(11,14) instead of γ-linolenic acid as an intermediate.

One of the most common fatty acids in higher plants and algae is α-linolenic acid 18:3(9,12,15). The formation of this compound by direct conversion of 18:1 to 18:2(9,12) and then to 18:3(9,12,15) was described in several Chlorophyceae[72-75] and in photoauxotroph *E. gracilis.*[41]

Stumpf and co-workers[76-78] have proposed Pathway 3 for the formation of α-linolenic acid in green algae (*Chlorella pyrenoidosa*), as well as in yeast (*Candida*) and for leaves of higher plants (spinach chloroplasts):

$$C_2 \longrightarrow\longrightarrow 12{:}0 \longrightarrow\longrightarrow 12{:}3(3,6,9) \longrightarrow 14{:}3(5,8,11) \longrightarrow 16{:}3(7,10,13)$$

$$\longrightarrow 18{:}3(9,12,15) \quad (3)$$

As for the formation of mono-unsaturated fatty acids (16:1, 18:1), there seem to exist several pathways; there are reports on the direct dehydrogenation (as part of the γ-linolenic pathway?) on exogenously supplied 16:0 and 18:0 to 16:1 and 18:1 in *C. vulgaris* (grown in an inorganic medium (phosphate buffer),[74] *P. cruentum* (Rhodophyceae), *Astasia longa* (Euglenophyceae, colorless), *Ochromonas malhamensis* and *Poteriochromonas stipitata* (Chrysophyceae),[72] and *E. gracilis.*[74] On the other hand, a direct conversion could not be demonstrated in *E. gracilis, Chlamydomonas reinhardii* (Chlorophyceae), *Polytoma uvella* (Chlorophyceae, colorless) and several Chlorophyceae,[72] and also not in *C. vulgaris* grown in media containing carbohydrate.[74] Possibly, dehydrogenations are not carried out when the medium contains a sufficiently high amount of organic carbon.

A new pathway was postulated by Nagai and Bloch[79] for the formation of oleic acid in photoauxotroph *E. gracilis:*

$$C_2 \longrightarrow\longrightarrow 10{:}0 \longrightarrow 12{:}0 \xrightarrow{+0_2?} 12{:}1(3) \longrightarrow 14{:}1(5) \longrightarrow 16{:}1(7) \longrightarrow 18{:}1(9) \quad (4)$$

The 16:1 (*trans* 3) fatty acid is found in all photosynthesizing algae except the Cyanophyceae.[55,71] In *C. vulgaris* (grown in an inorganic medium), this compound was synthesized by direct dehydrogenation of 16:0, the reaction requiring light and probably oxygen.[80] This fatty acid, however, was not synthesized when *C. vulgaris* was grown in organic media.[26]

Probably, complex lipids are involved as intermediates or substrates in the formation of unsaturated fatty acids. So, for example, two pathways were proposed for the conversion of 18:1 to 18:2 in *C. vulgaris,* one mediated by PC,[81-83] the other by MGDG. The pathway mediated by MGDG seems to exist in Cyanophyceae, too.[83,86] In *E. gracilis,* two pathways were proposed for fatty acid biosynthesis.[50] One pathway led to the formation of polyunsaturated C_{20} and C_{22} fatty acids and was independent of light.

This pathway was associated with nonpolar lipids (wax esters) and phospholipids. The other pathway led to polyunsaturated C_{16} and C_{18} fatty acids and required light and ammonium. This pathway was associated with chloroplast lipids (MGDG, DGDG, SL, PC).

As the data of Table 7 show, the Cyanophyceae are extremely versatile in their ability to synthesize unsaturated C_{18} fatty acids. In some of these algae, the end products of fatty acid biosynthesis are 18:1, 18:2, 18:3(6,9,12), 18:3(9,12,15), or even 18:4, respectively. This might indicate that several different pathways (probably similar to one or the other of the pathways mentioned above) are present within individual strains or species of the Cyanophycese. Possibly the other classes of algae evolved from blue-green algae with different fatty acid synthesizing abilities.[71,72]

LIPID METABOLISM IN MICROALGAE

To date, our knowledge about lipid metabolism in microalgae is still incomplete. The reader is referred to the following publications for details.[14,27,50,85,87-91]

LIPIDS AND FATTY ACIDS OF MICROALGAE AND ENVIRONMENTAL FACTORS

The formation of lipids and fatty acids is influenced by several environmental factors, most of all by light and light intensity, temperature, and the chemical composition of the nutrient media. Experiments carried out by various authors showed that in *E. gracilis* (Euglenophyceae) and *C. vulgaris* (Chlorophyceae), the formation of polyunsaturated fatty acids (mainly 16:2, 16:3, 16:4, 18:2, 18:3) and of MGDG, DGDG, SL, and PG (chloroplast lipids) is induced or increased by light.[26,28,37,38,40,45,47,48,50] Low temperatures seem to favor the formation of unsaturated fatty acids, as was shown with *C. caldarium,*[92-95] *Anacystis nidulans,*[96] *E. gracilis,*[39] and *Monochrysis lutheri,*[97] as well as with macroscopic marine algae.[98] In *C. caldarium,* the influence of low temperatures has been attributed to an induction of desaturases (dehydrogenases).[95]

The nitrogen content (NO_3^-, NH_4^+) in the media is important: high levels of nitrogen (more than about 100 μmol N/ℓ) lead to an increase in the formation of polyunsaturated fatty acids and of polar lipids (MGDG, DGDG, SL, PG). This has been demonstrated not only in microalgae such as *E. gracilis*[50] and several species of *Chlorella* and *Bracteacoccus* minor,[99] but also in macroscopic marine algae.[98] The formation of the above fatty acids and lipids seems also to be controlled in a similar way by varying concentrations of Mn^{2+}.[88]

TOTAL LIPID AND FATTY ACID CONTENT OF MICROALGAE

As for the total lipid and fatty acid content, there are few quantitative data because both the dry weight of the algae and the lipid and fatty acid content vary according to the growth conditions. In particular, the nitrogen content (nitrate, ammonia, etc.) of the nutrient media seems to have a strong influence. As shown in Table 8, the dry weight of *C. vulgaris* (freshwater green alga) decreases with decreasing concentrations of NH_4Cl in the medium. The percentage of total lipids at first decreases from 33.1 to 26.9% (at 0.0005% NH_4Cl), but increases again at still lower nitrogen concentrations. This increase, however, is only relative (percent of dry weight), since the alga produces large amounts of polar lipids (mono- and digalactosyl diglycerides, sulpholipid, phospholipids) at higher nitrogen levels. At low nitrogen levels, the alga synthesizes only small amounts of polar lipids, but relatively high amounts of nonpolar lipids (mostly triglycerides and hydrocarbons).

Table 8
DRY WEIGHT AND TOTAL LIPID CONTENT OF *CHLORELLA VULGARIS* GROWN AT VARYING CONCENTRATIONS OF NH_4Cl IN THE NUTRIENT MEDIUM[99]

NH_4Cl (%)	0.025	0.005	0.001	0.0005	0.0001	0.00005
Dry weight (mg)	76.91	73.80	65.55	27.85	26.50	16.15
Total lipids (mg)	25.4	22.6	18.74	7.48	8.20	8.76
Total lipids (%) of dry weight	33.1	30.6	28.6	26.9	31.0	54.2

Similar results have been obtained with total lipid content in other green algae, *Bracteacoccus minor, Chlorella pyrenoidosa, Chlorella fusca,* and *Halochlorella rubescens.*[99]

In some colorless algae (*Prototheca portoricensis, Polytoma ovigorme* — Chlorophyceae, and *Chilomonas paramecium* — Cryptophyceae), total lipid content ranged from 9.8 to 12.9% of the dry weight.[28]

The total fatty acid content (percent of the total lipids) also varied (4.7, 8.5, and 11.8%) in the above-mentioned colorless algae. *Chlorella variegata* grown in the light contained 19.5% total fatty acids (percent of the total lipids) and contained 36.6% when grown in the dark.[28]

REFERENCES

1. **Hitchcock, C.,** Structure and distribution of plant acyl lipids, in *Recent Advances in the Chemistry and Biochemistry of Plant Lipids,* Galliard, T. and Mercer, E. I., Eds., Academic Press, London, 1975, 1.
2. IUPAC-IUB, Nomenclature of phosphorus-containing compounds of biochemical importance, IUPAC-IUB Commission on Biochemical Nomenclature (CBN), *Eur. J. Biochem.,* 79, 1, 1977.
3. IUPAC-IUB, The nomenclature of lipids, IUPAC-IUB Commission on Biochemical Nomenclature (CBN), *Eur. J. Biochem.,* 79, 11, 1977.
4. **Pohl, P. and Wagner, H.,** Fettsäuren im Pflanzen- und Tierreich (eine Übersicht). II. Trans-ungesättigte, Alkin-, Hydroxy-, Epoxy-, Oxo-, Cyclopropan- und Cyclopropen-Fettsäuren, *Fette, Seifen, Anstrichmittel,* 74, 541, 1972.
5. **Lee, R. F. and Loeblich, A. R., III,** Distribution of 21:6 hydrocarbon and its relationship to 22:6 fatty acid in algae, *Phytochemistry,* 10, 593, 1971.
6. **Nichols, B. W. and Appleby, R. S.,** The distribution and biosynthesis of arachidonic acid in algae, *Phytochemistry,* 8, 1907, 1969.
7. **Haines, T. H., Pousada, M., Stern, B., and Mayers, G. L.,** Microbial sulpholipids: (R)-13-Chloro-1-(R)-14-docosanediol disulphate and polychlorosulpholipids in *Ochromonas danica, Biochem. J.,* 113, 565, 1969.
8. **Mayers, G. L., Pousada, M., and Haines, T. H.,** Microbial sulpholipids. III. The disulfate of (+)-1,14-docosanediol in *Ochromonas danica, Biochemistry,* 8, 2981, 1969.
9. **Elovson, J. and Vagelos, P. R.,** A new class of lipids: chlorosulfolipids, *Proc. Natl. Acad. Sci. U.S.A.,* 62, 957, 1969.
10. **Blumer, M., Guillard, P. R. L., and Chase, T.,** Hydrocarbons of marine phytoplankton, *Mar. Biol.,* 8, 183, 1971.
11. **Clark, P. C., Jr., and Blumer, M.,** Distribution of *n*-paraffins in marine organisms and sediment, *Limnol. Oceanogr.,* 12, 79, 1967.
12. **Mercer, E. I. and Davies, C. L.,** Chlorosulpholipids of *Tribonema aequale, Phytochemistry,* 13, 1607, 1974.
13. **Mercer, E. I. and Davies, C. L.,** Chlorosulpholipids in algae, *Phytochemistry,* 14, 1545, 1975.
14. **Kates, M. and Volcani, B. E.,** Lipid components of diatoms, *Biochem. Biophys. Acta,* 116, 264, 1966.

15. **Opute, F. I.**, Lipid and fatty acid composition of diatoms, *J. Exp. Bot.*, 25, 823, 1974.
16. **Anderson, R., Livermore, B. P., Kates, M., and Volcani, B. E.**, The lipid composition of the non-photosynthetic diatom Nitzschia alba, *Biochim. Biophys. Acta*, 528, 77, 1978.
17. **Anderson, R., Kates, M., and Volcani, B. E.**, Identification of the sulfolipids in the non-photosynthetic diatom *Nitzschia alba*, *Biochim. Biophys. Acta*, 528, 89, 1978.
18. **Lee, R. F., Nevenzel, J. C., Paffenhöfer, G. -A., Benson, A. A., Patton, S., and Kavanagh, T. E.**, A unique hexaene hydrocarbon from a diatom (*Skeletonema costatum*), *Biochim. Biophys. Acta*, 202, 386, 1970.
19. **Blumer, M., Mullin, M. M., and Guillard, P. R. L.**, A polyunsaturated hydrocarbon (3,6,9,12,15,18-heneicosahexaene) in the marine food web, *Mar. Biol.*, 6, 226, 1970.
20. **Joseph, J. D.**, Identification of 3,6,9,12,15-octadecapentaenoic acid in laboratory-cultured photosynthetic dinoflagellates, *Lipids*, 10, 395, 1975.
21. **Mayzaud, P., Eaton, C. A., and Ackman, R. G.**, The occurrence and distribution of octadecapentaenoic acid in a natural plankton population. A possible food chain index, *Lipids*, 11, 858, 1976.
22. **Harrington, G. W., Beach, D. H., Dunham, J. E., and Holz, G. G., Jr.**, The polyunsaturated fatty acids of marine dinoflagellates, *J. Protozool.*, 17, 213, 1970.
23. **Parke, M. and Manton, I.**, Preliminary observations on the fine structure of *Prasinocladus marinus*, *J. Mar. Biol. Assoc. U.K.*, 45, 525, 1965.
24. **Craigie, J. S., McLachlan, J., Majak, W., Ackman, R. G., and Tocher, C. S.**, Photosynthesis in algae. II. Green algae with special reference to *Dunaliella* spp. and *Tetraselmis* spp., *Can. J. Bot.*, 44, 1247, 1966.
25. **Ackman, R. G., Addison, R. F., Hooper, S. N., and Prakash, A.**, *Halosphaera viridis:* Fatty acid composition and taxonomical relationships, *J. Fish. Res. Board Can.*, 27, 251, 1970.
26. **Nichols, B. W.**, Light induced changes in the lipids of *Chlorella vulgaris*, *Biochim. Biophys. Acta*, 106, 274, 1965.
27. **Sastry, P. S. and Kates, M.**, Biosynthesis of lipids in plants. I. Incorporation of orthophosphate-^{32}P and glycerophosphate-^{32}P into phosphatides of *Chlorella vulgaris* during photosynthesis, *Can. J. Biochem.*, 43, 1445, 1965.
28. **Glasl, H. and Pohl, P.**, Fettsauren und Lipide in farblosen Algen, *Z. Naturforsch. Teil C*, 29, 399, 1974.
29. **Eichenberger, W.**, Lipids of *Chalmydomonas reinhardii* under different growth conditions, *Phytochemistry*, 15, 459, 1976.
30. **Wagner, H., Pohl, P., and Münzing, A.**, Sphingolipide und Glykolipide von Pilzen und höheren Pflanzen. IV. Mitt.: Isolierung eines inositfreien Phytosphingoglykolipids aus der Grünalge *Scenedesmus obliquus*, *Z. Naturforsch. Teil C*, 24, 360, 1969.
31. **Gelpi, E., Schneider, H. J., Mann, J., and Orő, J.**, Hydrocarbons of geochemical significance in microscopic algae, *Phytochemistry*, 9, 603, 1970.
32. **Patterson, G. W.**, The effects of culture conditions on the hydrocarbon content of *Chlorella vulgaris*, *J. Phycol.*, 3, 22, 1967.
33. **Douglas, A. G., Douraghi-Zadeh, K., and Eglington, G.**, The fatty acids of the alga *Botryococcus braunii*, *Phytochemistry*, 8, 285, 1969.
34. **Gelpi, E., Orő, J., Schneider, H. J., and Bennett, E. O.**, Olefins of high molecular weight in two microscopic algae, *Science*, 161, 700, 1968.
35. **Knights, B. A., Brown, A. C., Conway, E., and Middleditch, B. S.**, Hydrocarbons from the green form of the freshwater alga *Botryococcus braunii*, *Phytochemistry*, 9, 1317, 1970.
36. **Maxwell, J. R., Douglas, A. G., Eglington, G., and McCormick, A.**, The Botryococcenes—hydrocarbons of novel structure from the alga *Botryococcus braunii*, Kützing, *Phytochemistry*, 7, 2157, 1968.
37. **Erwin, J. and Bloch, K.**, Polyunsaturated fatty acids in some photosynthetic microorganisms, *Biochem. Z.*, 338, 496, 1963.
38. **Rosenberg, A.**, A comparison of lipid patterns in photosynthesizing and nonphotosynthesizing cells of *Euglena gracilis*, *Biochemistry*, 2, 1148, 1963.
39. **Blee, E. and Schantz, P.**, Biosynthesis of galactolipids in *Euglena gracilis*. II. Changes in fatty acid composition of galactolipids during chloroplast development, *Plant Sci. Lett.*, 13, 257, 1978.
40. **Rosenberg, A. and Pecker, M.**, Lipid alterations in *Euglena gracilis* cells during light-induced greening, *Biochemistry*, 3, 254, 1964.
41. **Hulanicka, D., Erwin, J., and Bloch, K.**, Lipid metabolism of *Euglena gracilis*, *J. Biol. Chem.*, 230, 2778, 1964.
42. **Guehler, P. F., Peterson, L., Tsuchiya, H. M., and Dodson, R. M.**, Microbiological transformations. XIII. Composition of the wax formed by *Euglena gracilis*, *Arch. Biochem. Biophys.*, 106, 294, 1964.
43. **Korn, E. D.**, The polyunsaturated 20-carbon and 22-carbon fatty acids of *Euglena*, *Biochem. Biophys. Res. Commun.*, 14, 1, 1964.

44. **Korn, E. D.**, The fatty acids of *Euglena gracilis*, *J. Lipid Res.*, 5, 352, 1964.
45. **Rosenberg, A.**, Galactosyl diglycerides: Their possible function in *Euglena chloroplasts*, *Science*, 157, 1191, 1967.
46. **Rosenberg, A.**, *Euglena gracilis:* A novel lipid energy reserve and arachidonic acid enrichment during fasting, *Science*, 157, 1189, 1967.
47. **Rosenberg, A. and Gouaux, J.**, Quantitative and compositional changes in monogalactosyl and di-agalactosyl diglycerides during light-induced formation of chloroplasts in *Euglena gracilis*, *J. Lipid Res.*, 8, 80, 1967.
48. **Constantopoulos, G. and Bloch, K.**, Effect of light intensity on the lipid composition of *Euglena gracilis*, *J. Biol. Chem.*, 242, 3538, 1967.
49. **Kolattukudy, P. E.**, Reduction of fatty acids to alcohols by cell-free preparations of *Euglena gracilis*, *Biochemistry*, 9, 1095, 1970.
50. **Pohl, P. and Wagner, H.**, Control of fatty acid and lipid biosynthesis in *Euglena gracilis* by ammonia, light and DCMU, *Z. Naturforsch. Teil B*, 27, 53, 1972.
51. **Beach, D. H., Harrington, G. W., and Holz, G. G., Jr.**, The polyunsaturated fatty acids of marine and freshwater cryptomonads, *J. Protozool.*, 17, 501, 1970.
52. **Kenyon, C. N.**, Fatty acid composition of unicellular strains of blue-green algae, *J. Bacteriol.*, 109, 827, 1972.
53. **Kenyon, C. N., Rippka, R., and Stanier, R. Y.**, Fatty acid composition and physiological properties of some filamentous blue-green algae, *Arch. Mikrobiol.*, 83, 216, 1972.
54. **Kenyon, C. N. and Stanier, R. Y.**, Possible evolutionary significance of polyunsaturated fatty acids in blue-green algae, *Nature (London)*, 227, 1164, 1970.
55. **Nichols, B. W., Harris, R. V., and James, A. T.**, The lipid metabolism of blue-green algae, *Biochem. Biophys. Res. Commun.*, 20, 256, 1965.
56. **Hirayama, O.**, Pigments and lipids in blue-green alga *Anacystis nidulans*, *J. Biochem.*, 61, 179, 1967.
57. **Winkenbach, F., Wolk, C. P., and Jost, M.**, Lipids of membranes and of the cell envelope in heterocysts of a blue-green alga, *Planta*, 107, 69, 1972.
58. **Zepke, H. D., Heinz, E., Radunz, A., Linscheid, M., and Pesch, R.**, Combination and positional distribution of fatty acids in lipids from blue-green algae, *Arch. Microbiol.*, 119, 157, 1978.
59. **Allen, C. F., Good, P., and Holton, R. W.**, Lipid composition of Cyanidium, *Plant Physiol.*, 46, 748, 1970.
60. **Orô, J., Tornabene, T. G., Nooner, D. W., and Gelpi, E.**, Aliphatic hydrocarbons and fatty acids of some marine and freshwater microorganisms, *J. Bacteriol.*, 93, 1811, 1967.
61. **Han, J., McCarthy, E. D., Calvin, M., and Benn, M. H.**, Hydrocarbon constituents of the blue-green algae Nostoc muscorum, *Anacystis nidulans*, *Phormidium luridum* and *Chlorogloea fritschii*, *J. Chem. Soc. C*, 2785, 1968.
62. **Han, J., Chan, H. W.-S., and Calvin, M.**, Biosynthesis of alkanes in *Nostoc muscorum*, *J. Am. Chem. Soc.*, 91, 5156, 1969.
63. **Winters, K., Parker, P. L., and Van Baalen, C.**, Hydrocarbons of blue-green algae: geochemical significance, *Science*, 163, 467, 1969.
64. **Paoletti, C., Pushpararaj, B., Florenzano, G., Capella, P., and Lercker, G.**, Unsaponafiable matter of green and blue-green algal lipids as a factor of biochemical differentiation of their biomasses. I. Total unsaponifiable and hydrocarbon fraction, *Lipids*, 11, 258, 1976.
65. **Nichols, B. W. and Wood, B. J. B.**, New glycolipid specific to nitrogen-fixing blue-green algae, *Nature (London)*, 217, 767, 1968.
66. **Walsby, A. E. and Nichols, B. W.**, Lipid composition of heterocysts, *Nature (London)*, 221, 673, 1969.
67. **Bryce, T. A., Welti, D., Walsby, A. E., and Nichols, B. W.**, Monohexoside derivatives of long chain polyhydroxy alcohols; a novel class of glycolipid specific to heretocystous algae, *Phytochemistry*, 11, 295, 1972.
68. **Lambein, F. and Wolk, C. P.**, Structural studies on the glycolipids from the envelope of the heterocyst of *Anabaena cylindrica*, *Biochemistry*, 12, 791, 1973.
69. **Lorch, S. K. and Wolk, C. P.**, Application of gas-liquid chromatography to study of the envelope lipids of heterocysts, *J. Phycol.*, 10, 352, 1974.
70. **Abreu-Grobois, F. A., Billyard, T. C., and Walton, T. J.**, Biosynthesis of heterocyst glycolipids of *Anabaena cylindrica*, *Phytochemistry*, 16, 351, 1977.
71. **Nichols, B. W. and Wood, B. J. B.**, The occurrence and biosynthesis of gamma-linolenic acid in a blue-green alga, *Spirulina platensis*, *Lipids*, 3, 46, 1968.
72. **Erwin, J., Hulanicka, D., and Bloch, K.**, Comparative aspects of unsaturated fatty acid synthesis, *Comp. Biochem. Physiol.*, 12, 191, 1964.
73. **Harris, R. V. and James, A. T.**, Linoleic and α-linolenic acid biosynthesis in plant leaves and a green alga, *Biochim. Biophys. Acta*, 106, 456, 1965.

74. **Harris, R. V., Harris, P., and James, A. T.**, The fatty acid metabolism of *Chlorella vulgaris*, *Biochim. Biophys. Acta*, 106, 465, 1965.
75. **Cherif, A., and Dubacq, J. P., Mache, R., Oursel, A., and Tremolieres, A.**, Biosynthesis of α-linolenic acid by desaturation of oleic and linoleic acids in several organs of higher and lower plants and in algae, *Phytochemistry*, 14, 703, 1975.
76. **Jacobson, B. S., Kannangara, C. G., and Stumpf, P. K.**, Biosynthesis of α-linolenic acid by disrupted spinach chloroplasts, *Biochem. Biophys. Res. Commun.*, 51, 487, 1973.
77. **Kannangara, C. G., Jacobson, B. S., and Stumpf, P. K.**, *In vitro* biosynthesis of α-linolenic acid in plants, *Biochem. Biophys. Res. Commun.*, 52, 648, 1973.
78. **Jacobson, B. S., Kannangara, C. G., and Stumpf, P. K.**, The elongation of medium chain trienoic acids to α-linolenic acid by a spinach chloroplast stroma system, *Biochem. Biophys. Res. Commun.*, 52, 1190, 1973.
79. **Nagai, J. and Bloch, K.**, Synthesis of oleic acid by *Euglena gracilis*, *J. Biol. Chem.*, 240, PC 3702, 1965.
80. **Nichols, B. W., Harris, P., and James, A. T.**, The biosynthesis of trans 3-hexadecenoic acid by *Chlorella vulgaris*, *Biochem. Biophys. Res. Commun.*, 21, 473, 1965.
81. **Gurr, M. I., Robinson, M. P., and James, A. T.**, The mechanism of formation of polyunsaturated fatty acids by photosynthetic tissue. The tight coupling of oleate desaturation with phospholipid synthesis in *Chlorella vulgaris*, *Eur. J. Biochem.*, 9, 70, 1969.
82. **Gurr, M. I. and Brawn, P.**, The biosynthesis of polyunsaturated fatty acids by photosynthetic tissue. The composition of phosphatidyl choline species in *Chlorella vulgaris* during the formation of linoleic acid, *Eur. J. Biochem.*, 17, 19, 1970.
83. **Appleby, R. S., Safford, R., and Nichols, B. W.**, The involvement of lecithin and monogalactosyl diglyceride in linoleate synthesis by green and blue-green algae, *Biochim. Biophys. Acta*, 248, 205, 1971.
84. **Nichols, B. W. and Moorhouse, R.**, The separation, structure and metabolism of monogalactosyl diglyceride species in *Chlorella vulgaris*, *Lipids*, 4, 311, 1969.
85. **Safford, R. and Nichols, B. W.**, Positional distribution of fatty acids in monogalactosyl diglyceride fractions from leaves and algae. Structural metabolic studies, *Biochim. Biophys. Acta*, 210, 57, 1970.
86. **Nichols, B. W.**, Fatty acid metabolism in the chloroplast lipids of green and blue-green algae, *Lipids*, 3, 354, 1968.
87. **Kates, M. and Wassef, M. K.**, Lipids chemistry, *Annu. Rev. Biochem.*, 39, 323, 1970.
88. **Constantopoulos, G.**, Lipid metabolism of manganese-deficient algae. I. Effect of manganese deficiency on the greening and the lipid composition of *Euglena gracilis* Z, *Plant Physiol.*, 45, 76, 1970.
89. **Mazliak, P.**, Lipid metabolism in plants, *Annu. Rev. Plant Physiol.*, 24, 287, 1973.
90. **Nichols, B. W.**, Lipid Composition and Metabolism, in *The Biology of Blue-green Algae*, Carr, N. G. and Whitton, B. A., Eds., University of California Press, Berkeley, 1973, 144.
91. **Wood, B. J. B.**, Fatty acids and saponifiable lipids, in *Algal Physiology and Biochemistry*, Stewart, W. D. P., Ed., Blackwell Scientific, Oxford, 1974, 236.
92. **Kleinschmidt, M. G. and McMahon, V. A.**, Effect of growth temperature on the lipid composition of *Cyanidium caldarium*. I. Class separation of lipids, *Plant Physiol.*, 46, 286, 1970.
93. **Kleinschmidt, M. G. and McMahon, V. A.**, Effect of growth temperature on the lipid composition of *Cyanidium caldarium*, *Plant Physiol.*, 46, 290, 1970.
94. **Adams, B. L., McMahon, V., and Seckbach, J.**, Fatty acids in the thermophilic alga, *Cyanidium caldarium*, *Biochem. Biophys. Res. Commun.*, 42, 359, 1971.
95. **Bedord, C. J., McMahon, V., and Adams, B.**, α-linolenic acid biosynthesis in *Cyanidium caldarium*, *Arch. Biochem. Biophys.*, 185, 15, 1978.
96. **Holton, R. W., Blecker, H. H., and Onore, M.**, Effect of growth temperature on the fatty acid composition of a blue-green alga, *Phytochemistry*, 3, 595, 1964.
97. **Ackman, R. G., Tocher, C. S., and McLachlan, J.**, Marine phytoplankter fatty acids, *J. Fish. Res. Board Can.*, 25, 1603, 1968.
98. **Zurheide, F. and Pohl, P.**, Lipids and Fatty Acids in Macroscopic Marine Algae and Their Control by Environmental Factors, papers in preparation.
99. **Pohl, P., Passig, T., and Wagner, H.**, Über den Einfluss von anorganischem Stickstoff-Gehalt in der Nährlosung auf die Fettsäure-Biosynthese in Grünalgen, *Phytochemistry*, 10, 1505, 1971.
100. **Chuecas, L. and Riley, J. P.**, Component fatty acids of the total lipids of some marine phytoplankton, *J. Mar. Biol. Assoc. U.K.*, 49, 97, 1969.
101. **Williams, P. M.**, Fatty acids derived from lipids of marine origin, *J. Fish. Res. Board Can.*, 22, 1107, 1965.
102. **Patton, S., Fuller, G., Loeblich, A. R., III, and Benson, A. A.**, Fatty acids of the "red tide" organism, *Gonyaulax polyedra*, *Biochim. Biophys. Acta*, 116, 577, 1966.
103. **Harrington, G. W. and Holz, G. G., Jr.**, The monoenoic and docosahexaenoic fatty acids of a heterotrophic dinoflagellate, *Biochim. Biophys. Acta*, 164 ,137, 1968.

104. **Pohl, P., Wagner, H., and Passig, T.,** Inhaltsstoffe von Algen. II. Über die unterschiedliche Fettsäurezusammensetzung von Salz- und Süßwasseralgen, *Phytochemistry,* 7, 1565, 1968.
105. **Schlenk, H., Mangold, H. K., Gellermann, J. L., Link, W. E., Morrisette, R. A., Holman, R. T., and Hayes, H.,** Comparative analytical studies of fatty acids of the alga *Chlorella pyrenoidosa, J. Am. Oil Chem. Soc.,* 37, 547, 1960.
106. **Schneider, H., Gelpi, E., Bennett, E. O., and Orô, J.,** Fatty acids of geochemical significance in microscopic algae, *Phytochemistry,* 9, 613, 1970.
107. **Klenk, E., Knipprath, W., Eberhagen, D., and Koof, H. P.,** Über die ungesättigten Fettsäuren der Fettstoffe von Süßwasser-und Meeresalgen, Hoppe-Seylers *Z. Physiol. Chem.,* 334, 44, 1963.
108. **Jamieson, G. R. and Reid, E. H.,** The fatty acid composition of Ulothrix aequalis lipids, *Phytochemistry,* 15, 795, 1976.
109. **Collins, R. P. and Kalnins, K.,** The fatty acids of *Cryptomonas ovata* var. *palustris, Phyton,* 26, 47, 1969.
110. **Parker, P. L., Van Baalen, C., and Maurer, L.,** Fatty acids in eleven species of blue-green algae: geochemical significance, *Science,* 155, 707, 1967.
111. **Holton, R. W., Blecker, H. H., and Stevens, T. S.,** Fatty acids in blue-green algae: possible relation to phylogenetic position, *Science,* 160, 545, 1968.

SYNTHESIS OF PRENYLLIPIDS IN VASCULAR PLANTS (INCLUDING CHLOROPHYLLS, CAROTENOIDS, AND PRENYLQUINONES)

Hartmut K. Lichtenthaler

Plant prenyllipids comprise those lipid-soluble compounds which arise either totally or in part from the isoprenoid (terpenoid) pathway. Among these one has to differentiate: (1) the pure or simple prenyllipids, the carbon skeleton of which is made up solely from isoprene units such as sterols, prenols, and carotenoids, and (2) the mixed prenyllipids (chlorophylls, prenylquinones, prenylchromanols) which possess an isoprenoid side chain (prenyl chain) bound to an aromatic nucleus (e.g., porphyrin ring in the case of chlorophylls, a benzo- or a naphthoquinone ring in the case of prenylquinones). The fat-soluble chromanols (e.g., tocopherols, tocotrienols) represent the cyclic forms of reduced prenylhydroquinones (tocoquinones). The fat-soluble vitamins are also derived from the prenyl pathway and therefore have been named prenyl vitamins. Plants contain provitamin A (β-carotene), vitamin E (α-tocopherol), and vitamin K_1 (phylloquinone) as regular components of their chloroplasts. The two other prenyl vitamins, vitamin A and vitamin D, do not occur in green or nongreen plant tissue. This chapter deals with the biosynthesis, localization, concentration, and metabolism of plant prenyllipids with special emphasis on chlorophylls, carotenoids, and prenylquinones. The tables include provitamin A as well as the vitamins E and K_1. The three plant prenyl vitamins are treated in more detail by Lichtenthaler in this volume. The sterols, which also are vascular plant prenyllipids, are treated by Eichenberger in this volume.

A basic introduction to the prenyllipids of higher plants is given by Goodwin[1] and by Lichtenthaler.[2] The chromatographic methods for the separation of the different prenyllipid classes have recently been described in detail.[3] Information on the isolation and quantitative determination can be found in the following references: chlorophylls,[4,5] carotenoids,[6,7] prenylquinones,[8,9] and polyprenols.[29]

BIOSYNTHESIS AND OCCURRENCE

The biogenetic relationship of the different plant prenyllipids is indicated in Figure 1. Biosynthesis starts from the active C_5-unit, isopentenyl pyrophosphate, which derives from acetyl-CoA via mevalonic acid. Two active C_5 units join head to tail to yield a C_{10} unit (monoterpene) which will be elongated by further C_5 units to yield sesquiterpenes (C_{15}) and diterpenes (C_{20}). Sterols represent triterpenes (C_{30}) and carotenoids are tetraterpenes (C_{40}), both of which arise by dimerization of their sesquiterpene (farnesylpyrophosphate) or monoterpene (geranylgeranylpyrophosphate) precursors, respectively. The mixed prenyllipids, chlorophylls, vitamin K_1, vitamin E, and its oxidized form tocoquinone, contain a diterpene (C_{20}-prenyl) chain. Plastoquinone-9 (PQ-9) and ubiquinone-9 (UQ-9) contain a polyprenyl side chain made up from 9 C_5 units and the ubiquinone-10 possesses a decaprenyl chain consisting of 10 isoprene units. Free polyprenols (prenyl-alcohols) also may occur in green plant tissue, e.g., solanesol, a nonaprenol, or the group of castaprenols which contain 10 to 16 isopentenyl units.[28]

The chemical structures of different plant prenylquinones are shown in Figure 2. Plastoquinones-9, α-tocopherol, α-tocoquinone, and phylloquinone K_1 are natural constituents of chloroplasts (together with chlorophylls and carotenoids) and other plastid forms in vascular plants as well as in the eukaryotic algae such as green, red, and brown algae.[8,10] The prokaryotic blue-green algae (cyanobacteria) have the same

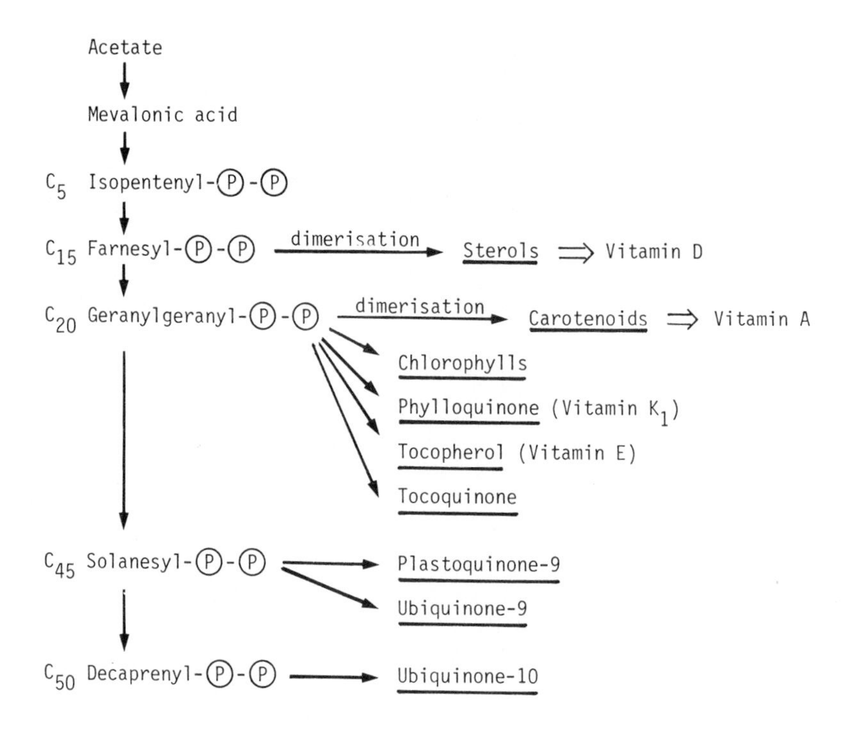

FIGURE 1. Biogenetic relationship of the major plant prenyllipids. The origin of vitamin A (from carotenoids) and of vitamin D (from the steroid pathway) is also indicated; both vitamins occur, however, only in animal tissue.

prenylquinone composition as higher plants.[10,12] Thus, the four plastid prenylquinones first appear during evolution as simultaneous components of the photosynthetic apparatus together with the second photosynthetic light reaction, the photolysis of water and the evolution of O_2. In contrast to chloroplasts, mitochondria possess only one prenylquinone, the ubiquinone. In the case of algae and vascular plants, it possesses a prenyl side chain of either nine or ten isoprene units.[11]

Besides these major plant prenylquinones, some prenylquinone derivatives may occur in minor concentration. The latter (e.g., the group of plastoquinone B and C; β- and δ-tocopherol, monodehydrovitamin K_1) mainly represent biosynthetic precursors or oxidation products of the main components. Some compounds, e.g., plastochromanol or ubichromenol, are apparently artifacts which arise during column chromatography from the corresponding quinones (plastohydroquinone, ubiquinone).

The aromatic nucleus of the different benzo- and naphthoprenylquinones is derived from intermediates of the shikimic acid pathway (Figure 3). Phylloquinone formation branches off from shikimic acid. Homogentisic acid is the common precursor for plastoquinone-9, α-tocopherol and α-tocoquinone.[13] There is some evidence that α-tocoquinone does not arise from oxidation of α-tocopherol but rather from a common intermediate.[14] The benzoquinone ring of ubiquinone is formed via cinnamic acid. Figure 3 also shows the possible interrelationship of prenylquinone formation and aromatic amino acid synthesis. In fact, within the plant cell protein formation (aromatic amino acids) and prenylquinone formation compete for intermediates of the shikimic acid pathway. Thus, when protein synthesis is strongly reduced, e.g., by a latent nitrogen deficiency,[15] there is an increased accumulation of prenylquinones in the chloroplasts of algae and higher plants.

Though the basic pathway for the formation of chlorophylls, carotenoids, and prenylquinones is well established,[13,16,17] there are still some uncertainties. This mainly re-

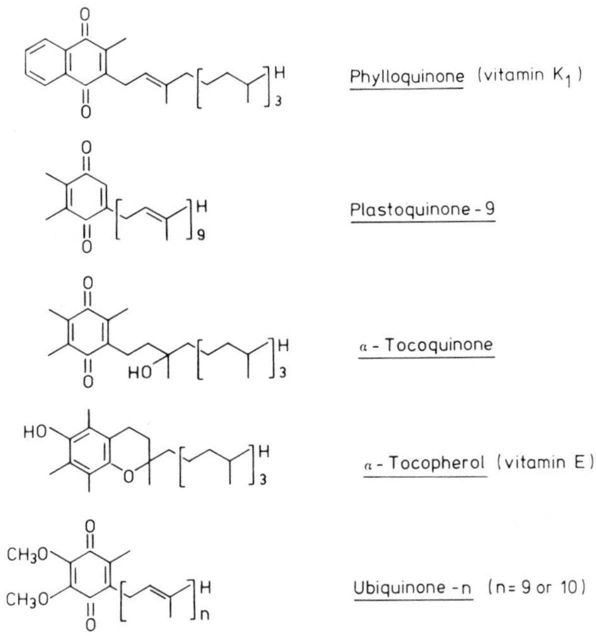

FIGURE 2. Chemical structure of the major plant prenylquinones.

fers to the formation of mixed prenyllipids with either a phytyl chain (1 double bond in the first isoprene unit) such as chlorophylls and phylloquinone K_1 or with fully saturated prenyl chains (α-tocoquinone and its chromanol α-tocopherol). Under certain growth conditions (e.g., delayed illumination of etiolated seedlings or after application of herbicides which block prenyl pigment formation, such as amitrole or the fluor-containing pyridazinone derivatives) one can detect chlorophylls and prenylquinones with 1, 2, or 3 additional double bonds in the prenyl chain, which are regarded to be their biosynthetic intermediates.[14] The question is whether the geranylgeraniol pyrophosphate is bound directly to the porphyrin, benzo- or naphthoquinone nucleus or first becomes reduced to phytylpyrophosphate. This problem does not apply to the biosynthesis of plastoquinone-9 or ubiquinone-9/10, the prenyl chain of which remains unsaturated and always exhibits 1 double bond per each C_5 unit of the chain.

LOCALIZATION AND FUNCTION

Chlorophylls, carotenoids, and prenylquinones are functional constituents of the photochemically active thylakoids of chloroplasts (Table 1) and make up about 25 to 30% of the thylakoid lipids.

Prenyl Pigments

The prenyl pigments (chlorophylls, carotenoids) are quantitatively bound to the thylakoids; their function in the absorption and/or conversion of light energy is evident. Within the photosynthetic membrane, they are organized and distributed in a different way as indicated in the modified tripartite model of the thylakoids (Figure 4), which was originally developed by Butler.[18] β-Carotene (and perhaps other carotenoids) also functions in the protection of chlorophylls and other membrane components against photooxidation at higher light intensities. This is particularly evident in unicellular

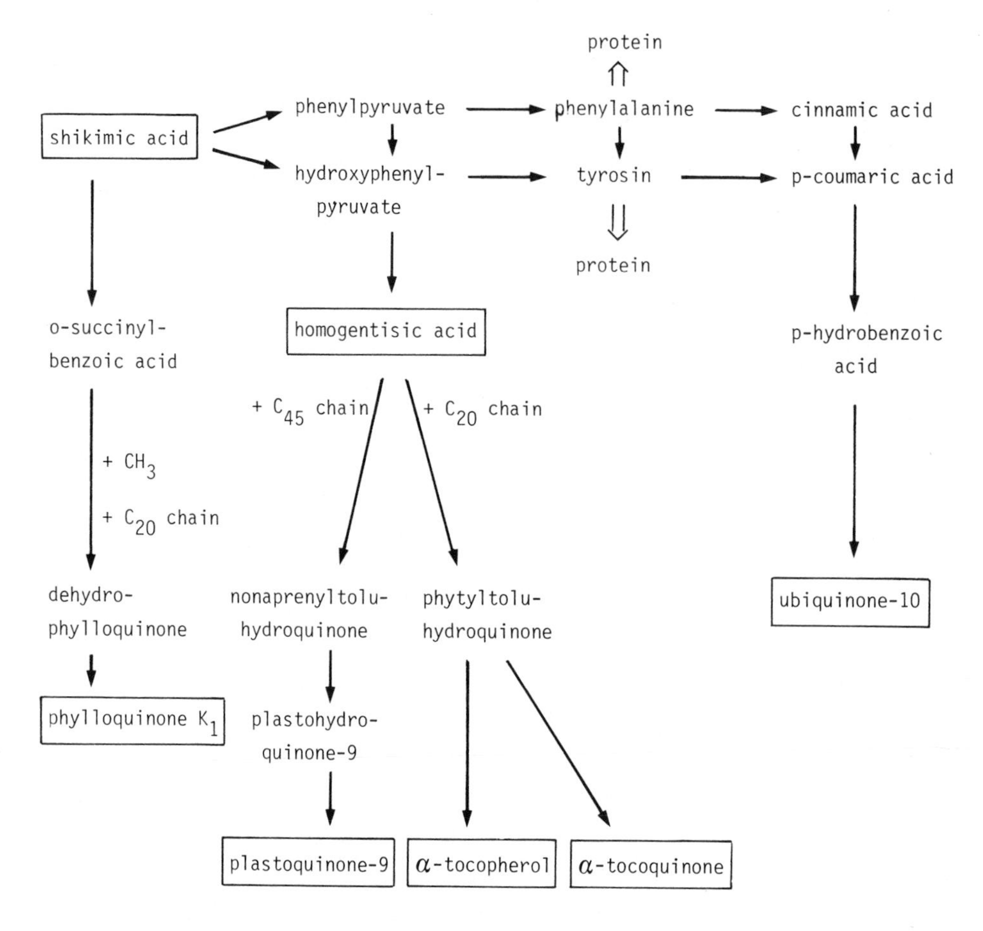

FIGURE 3. Scheme for the biosynthesis of the different plant prenylquinones from intermediates of the shikimic acid pathway.

green algae which considerably increase their total carotenoid content on a chlorophyll and on a cell basis when grown at higher light intensities.

Sun-exposed leaves of trees and plants, grown at a high light intensity, also possess a higher proportion of β-carotene than shade leaves or plants grown under weak light conditions. The mechanism of this light protection is not known. Since the labeling degree of β-carotene is much higher than that of other prenyl pigments,[19,20] it is clear that it undergoes fast degradation.

Prenylquinones

The chloroplast prenylquinones, plastoquinone-9, phylloquinone K_1, and α-tocoquinone, are potential electron carriers of the photosynthetic electron transport chain. The function of plastoquinone-9 as terminal electron acceptor of pigment system II (PQ-shuttle) is well established.[21] Recent inhibitor studies point to a functional site for the phylloquinone K_1 just before that of plastoquinone (Figure 4).[22] It is possible that it controls the electron flow into the large plastoquinone pool and also represents the reentry point for cyclic electrons into the chain. The role of α-tocoquinone is not known. Its concentration is lower than that of plastoquinone and lies in the range of phylloquinone, cytochromes, and other electron carriers.[23] The hitherto hypothetical possibility that it controls the outflow of electrons from the large plastoquinone pool to cytochrome f and plastocyanin should be investigated. α-Tocopherol functions as

Table 1
LIPID COMPOSITION OF THYLAKOIDS FROM SPINACH CHLOROPLASTS BASED ON 1 GRAM ATOM OF MANGANESE

Mol	Lipid	Mol wt
115	Chlorophylls	103,200
	80 Chlorophyll *a*	
	35 Chlorophyll *b*	
24	Carotenoids	13,700
	7—10 β-Carotene (provitamin A)	
	14—17 Xanthophylls	
23	Prenylquinones	15,900
	8—10 Plastoquinone-9	
	2—3 Phylloquinone (vitamin K_1)	
	2—3α-Tocoquinone	
	8—10 α-Tocopherol (vitamin E)	
58	Phospholipids	45,400
	26 Glycerophosphatidylglycerin	
	21 Glycerophosphatidylcholin and others	
173	Monogalactosyldiglyceride	134,100
72	Digalactosyldiglyceride	67,000
24	Sulfolipid	20,500
	Remaining lipids	95,300
	Total lipids	495,000
Gram atoms		464,000
4690	Nitrogen atoms as protein	
1	Manganese (in the water splitting protein Y)	
6	Iron (e.g., ferredoxin, cytochromes)	
3	Copper (e.g., in plastocyanin)	
	Total protein (rounded)	465,000

Note: The thylakoid membrane consists of approximately 50% lipids and 50% protein (data based on Lichtenthaler and Park[34]). In contrast to other membranes, sterols are present only in trace amounts.

lipid antioxidant in the photosynthetic membrane and helps to protect the thylakoid membrane lipids from photooxidation.[24] Its interaction with unsaturated phospholipids also points to a structural function in the photosynthetic membrane, which may be similar to that of sterols in other membranes.[25]

Within the chloroplast, phylloquinone is quantitatively associated with the thylakoids. Plastoquinone, α-tocopherol and α-tocoquinone are bound to the photosynthetic membrane, but are synthesized in excess amounts and deposited in the osmiophilic plastoglobuli of the plastid stroma (Figures 5 and 6).[26,27] Plastoglobuli mainly contain the reduced quinone forms plastohydroquinone and α-tocopherol. Other excess chloroplast lipids are present too in plastoglobuli, e.g., in *Ficus* chloroplasts, also galacto- and phospholipids, and free polyprenols in *Aesculus* chloroplasts.[28,29] The oxidized forms plastoquinone-9 and α-tocoquinone, in turn, are preferentially associated within the thylakoids.

Ubiquinone-9/10 is bound to the internal membrane of mitochondria and functions as electron carrier in the respiratory chain. In contrast to plastoquinone and α-tocopherol, ubiquinone is never accumulated in excess amounts. Its concentration seems to be directly regulated by mitochondria activity and is usually higher on a plant tissue dry weight basis in green tissues and influorescences than in white plant tissue (see Table 7).

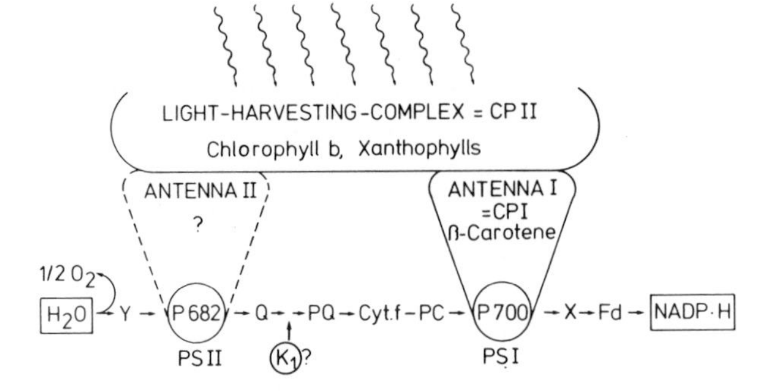

FIGURE 4. Tripartite model of the photosynthetic membrane with the two photosystems (PSI and PSII), the light harvesting complex and the electron transport chain. The scheme shows the differential distribution of prenylpigments in the thylakoids. Chlorophyll *b* and xanthophylls appear to be primarily associated with the light harvesting complex (chlorophyll a/b protein CPII = LHCP). β-Carotene, in turn, is predominantly localized in the antenna (CPI) of pigment system I. The possible functional site of phylloquinone K_1 between the quencher Q and the plastoquinone pool (PQ) is indicated.

REGULATION OF PRENYLLIPID ACCUMULATION BY LIGHT

The light-induced synthesis of thylakoids and their lipids is initiated and regulated by phytochrome.[2,30] The prenyllipid composition of thylakoids and chloroplasts is variable within a certain range and can be modified by light intensity and light quality, which will result in the formation of either "sun type" or "shade type" chloroplasts. Sun leaves of trees and plants grown at a high light intensity possess a higher proportion of chlorophyll a, β-carotene, and prenylquinone than do shade leaves or plants grown at weak light conditions.[2,30] This is documented by a higher value for the ratios chlorophyll a/b, and lower values for xanthophylls/β-carotene (x/c) and chlorophyll a/prenylquinones (a/K_1, a/PQ-9, a/α-TQ, a/α-T). These changes in prenyllipid composition are correlated with changes in ultrastructure and photochemical activity (Table 2). The modification "sun type" chloroplast can be simulated by blue light and by the application of cytokinins. The "shade type" situation can be triggered by weak red light and by the application of photosystem II herbicides such as bentazone, diuron (DCMU), or metabenzthiazuron (MTB).[2,30,31]

Besides the modification of the prenyllipid composition of thylakoids described before, there also exists a continuous accumulation of plastohydroquinone and α-tocopherol with increasing age of leaves, which is strongly light dependent (Table 3). These excess prenylquinones are deposited in the osmiophilic plastoglobuli (Figures 5 and 6), the size and number of which are increased parallel to the continuous accumulation of plastohydroquinone and α-tocopherol.[2,32] It depends on the plant species whether a higher proportion of excess plastohydroquinone (e.g., *Fagus sylvatica* sun leaves, Tables 4 and 5) or higher amounts of excess α-tocopherol (e.g., sun leaves of *Tilia cordata,* Table 3) are accumulated.

Table 2
DEPENDENCE OF THE PRENYLLIPID COMPOSITION, ULTRASTRUCTURE, AND PHOTOSYNTHETIC ACTIVITY OF CHLOROPLASTS ON THE GROWTH CONDITION

Growth condition I	Growth condition II
Shade, shade leaf	Sunlight, sun leaf
Low light intensity	High light intensity
Red light	Blue light
Application of photosystem II-herbicides	Application of cytokinins
Results	
"Shade-type" chloroplast	"Sun-type" chloroplast
Lower values for *a/b*	Higher values for *a/b*
Higher values for *x/c*	Lower values for *x/c*
Higher values for *a*/K_1, a/PQ-9 a/α-T	Lower values for *a*/K_1, a/PQ-9 a/α-T
Many thylakoids per chloroplast section	Less thylakoids per chloroplast section
Higher grana stacks	Low grana stacks
Few and small plastoglobuli	Many and larger plastoglobuli
Lower Hill-activity	Higher Hill-activity
Lower CO_2-fixation capacity	Higher CO_2-fixation capacity

Table 3
MOLAR LEVELS OF CAROTENOIDS AND PRENYLQUINONES IN DIFFERENT LEAVES OF THE LINDEN TREE *(TILIA CORDATA)* BASED ON 100 MOL OF CHLOROPHYLL A

	Chlorophylls							
Type of leaves	*a*	*b*	Carotenoids	Prenylquinones	PQ-9	α-T	α-TQ	K_1
Shade leaves[a]	100	36	54	25	15	6	2	2
Half-shaded leaves[b]	100	36	55	57	34	16	5	2
Sun leaves[c]	100	32	49	128	46	72	7	3
Extreme sun leaves[d]	100	28	48	227	82	130	12	3

Note: The data demonstrate the enhanced accumulation of excess prenylquinones (mainly plastohydroquinone and α-tocoperol) with increasing light exposure of leaves. PQ-9 = total plastoquinone-9 (quinone + hydroquinone); α-T = α-tocopherol (vitamin E); α-TQ = α-tocoquinone, K_1 = phylloquinone (vitamin K_1).

[a] In the inside of the tree canopy without direct sunlight, thin leaves with 30 to 50% more leaf area than sun leaves.
[b] "Blue shade" leaves from the north side of the tree mainly receive the blue sky light and never direct sunlight.
[c] South side of tree, at a height of 2 m, receive full sunlight, thick, xeromorph, and smaller leaves.
[d] Leaves from tree top, at a height of 12 m, thick, xeromorph, and smaller leaves.

PRENYLLIPID CONTENT OF DIFFERENTLY PIGMENTED PLANT TISSUES

Green and Etiolated Leaf Tissue

The prenyllipid content of green leaf tissues can be expressed (a) with respect to the incident light as μg/100 cm² leaf area (Tables 4 and 6) or (b) as μg/g dry weight (Table 7). Sun leaves possess about 1.5 to 2 times higher levels of chlorophylls, carotenoids, α-tocoquinone, phylloquinone, phospholipids, and galactolipids on a leaf area basis

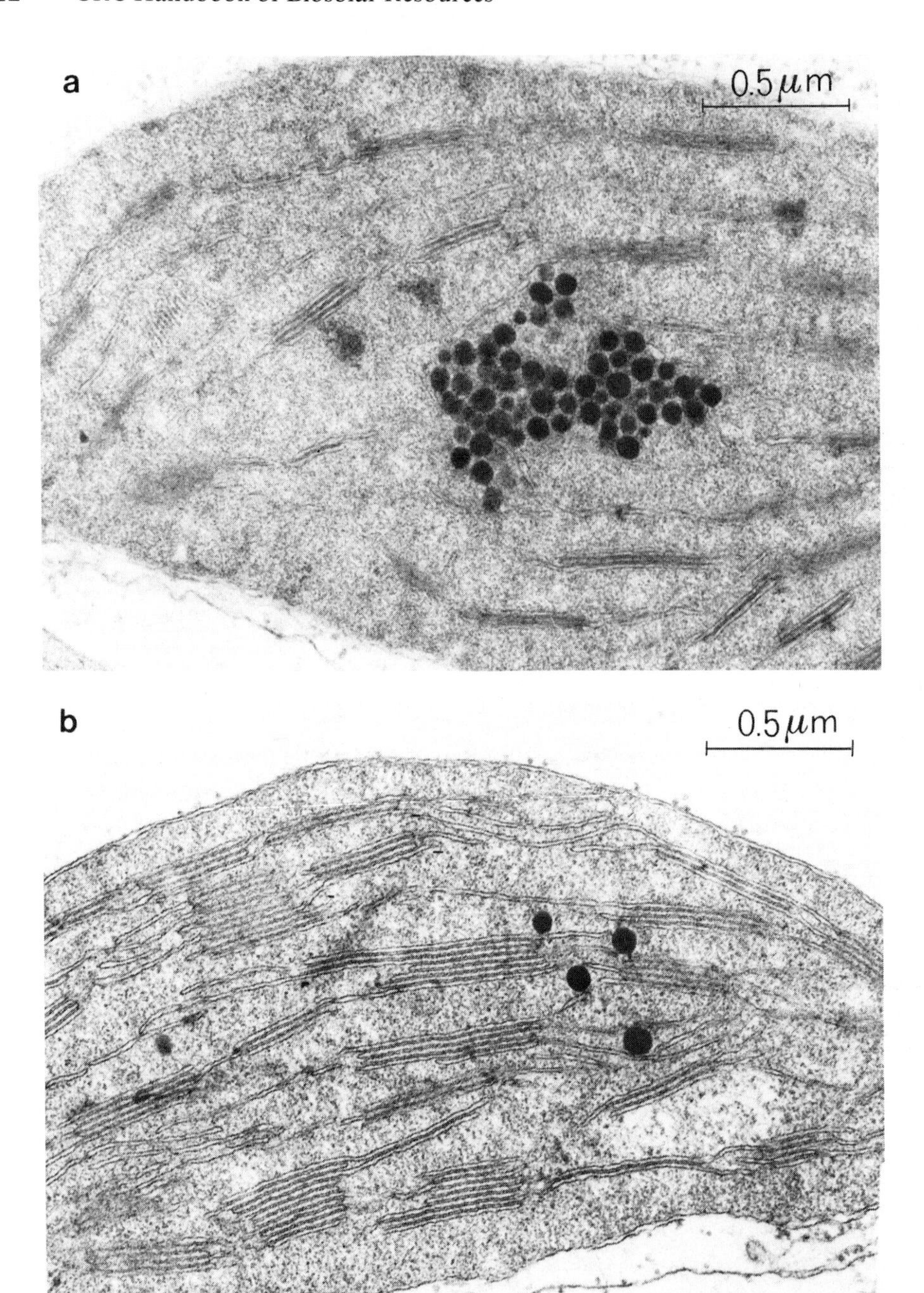

FIGURE 5. Section through chloroplasts from barley seedlings. (a) Greened in weak blue light with many plastoglobuli and few thylakoids; (b) greened in weak red light with high grana stacks and only few plastoglobuli. The ultrastructure of "blue light" chloroplasts resembles that of plants grown at high light intensity. The structure of "red light" chloroplasts corresponds to that of plants grown at low light intensity. (Photographs taken by D. Meier, 1978).

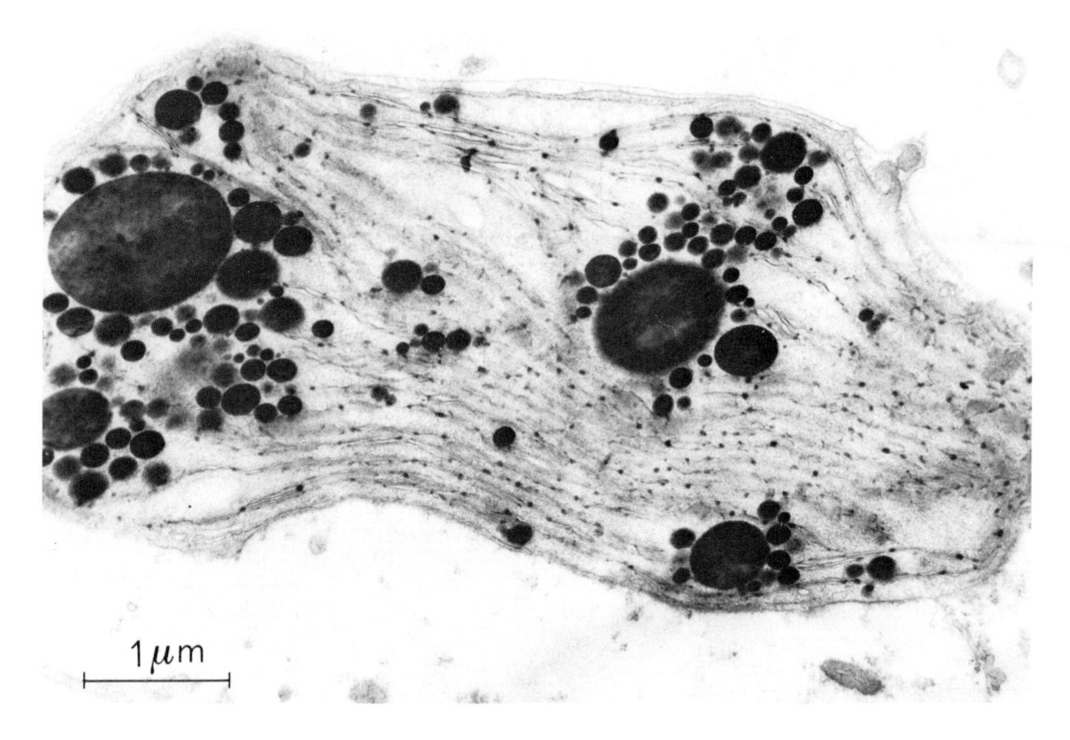

FIGURE 6. Chloroplasts from the 5- to 6-year-old green assimilation tissue of the stem succulent *Cereus peruvianus* with many smaller and some larger osmiophilic plastoglobuli.

than shade leaves (Table 4). The levels of α-tocopherol and plastoquinone-9 are augmented to a higher degree because of the accumulation of excess amounts. These differences, found in older leaves (August), are much less pronounced in the young leaves (June) shortly after leaf unfolding (e.g., May/June, Table 6).

On a dry weight basis the shade leaves exhibit a higher prenylpigment content than sun leaves, which is due to the higher dry weight of the sun leaves. The level of plastoquinone and α-tocopherol of sun leaves is, however, higher even on a dry weight basis (Table 5). Since herbs contain a higher percentage of water in their leaf tissues than tree leaves, they exhibit higher chlorophyll and carotenoid contents on a dry weight basis than leaves of trees (Table 7).

The prenyllipid level of aurea mutants is lower than that of the normal green plants (e.g., *Nicotiana*). The white parts of variegated leaves contain low amounts of pigments. The level of prenylquinones is within the range of that of etiolated tissue (Table 7). The carotenoid and prenylquinone content of etiolated leaves depends on the age of seedlings and can be higher by about 50% than the values given for barley in Table 7.

Other Plant Tissues (Table 7)

Those plant tissue cultures which possess fair amounts of chlorophylls and carotenoids also possess good levels of prenylquinones. Green fruit tissues have much lower chlorophyll and carotenoid (primary carotenoid) levels than green leaf tissues. During fruit ripening, the chloroplasts develop into chromoplasts, a process which is accompanied by a thylakoid and chlorophyll breakdown and the appearance of secondary carotenoids.[26] The level of phylloquinone is reduced and that of plastoquinone and α-tocopherol raised by new accumulations of these compounds. This also applies to chlo-

Table 4
DIFFERENCES IN THE SIZE, WEIGHT, AND WATER AND LIPID CONTENT OF THE TWO MORPHOLOGICALLY DIFFERENT SUN AND SHADE LEAVES OF THE BEECH TREE *(FAGUS SYLVATICA)*

	Shade leaf	Sun leaf	Sun/shade leaf
Leaf area (cm^2)	38.4	26.1	
Dry weight (mg)	126	257	
Water content (%)	61	49	
Lipid content per 100 cm^2 leaf area (μg)			
1. Prenylpigments[a]			
Chlorophyll *a*	590	1230	
Chlorophyll *b*	212	370	
Total Chlorophylls	802	1600	∿ 2.0 ×
β-Carotene (provitamin A)	71	119	
Xanthophylls	125	218	
Total Carotenoids	196	337	∿ 1.7 ×
2. Prenylquinones[a]			
Total plastoquinone-9	85	850	10 ×
α-Tocopherol (vitamin E)	95	364	3.8 ×
α-Tocoquinone	8	19	2.4 ×
Phylloquinone (vitamin K_1)	7	14	2.0 ×
Ubiquinone-9/10	5	18	3.6 ×
Total prenylquinones	200	1265	∿ 6 ×
3. Diglycerides[b]			
Phospholipids	570	1070	∿ 1.9 ×
Galactolipids	3320	5470	∿ 1.6 ×
MGD[c]	1890	2770	∿ 1.5 ×
DGD[c]	1430	2700	∿ 1.9 ×

Note: Similar differences are found between all sun and shade leaves of deciduous trees and other woody plants.

[a] Values of August 10 from Lichtenthaler.[2]
[b] Mean values of August 5 and August 12 from Frey and Tevini 1972 (personal communication).
[c] MGD = monogalactosyldiglyceride; DGD = digalactosyldiglyceride.

roplast degeneration in green leaves where plastoquinone and α-tocopherol accumulation may proceed until the green-yellow transition stage.[2]

The chromoplast-bearing cells of plasmochromous flower petals contain high amounts of prenylquinones which can reach the levels of green leaf tissue. The white, red, or blue flower petals (chymochromous flowers), which mainly possess leucoplasts, exhibit a low pigment and prenylquinone level. The prenylquinone and carotenoid content of chlorophyll-free white plant tissues (roots, tubers, white onion scales) is extremely low and often hard to determine.

In summary, the prenyllipid content of plant tissues is strongly correlated with the differentiation stage of plastids. As a general rule, the highest prenyllipid levels are found in green tissues which contain chloroplasts. The levels decrease from tissues with chromoplasts to that of etioplasts and leucoplasts.

Table 5
PRENYLLIPID CONTENT (μg/1g DRY WEIGHT) OF SUN AND SHADE LEAVES OF THE BEECH TREE *(FAGUS SYLVATICA)* DURING LEAF GROWTH (MAY 2) AND IN THE FULLY GROWN LEAVES BEFORE THE ONSET OF SENESCENCE (AUGUST 10)

	Shade leaf		Sun leaf	
	May 2	August 10	May 2	August 10
Leaf area (cm^2)	22.4	38.4	21.8	26.1
Dry weight (mg)	54	126	87	257
Water content (%)	76	61	71	49
Lipid content (μg) per 1 g dry weight				
1. Prenyl pigments[a]				
Chlorophyll *a*	2,420	1,800	1,630	1,240
Chlorophyll *b*	1,000	650	524	381
Total chlorophylls	3,420	2,450	2,127	1,621
β-Carotene (provitamin A)	340	220	163	121
Xanthophylls	760	380	315	222
Total Carotenoids	1,100	600	478	343
2. Prenylquinones[a]				
Total plastoquinone	283	260	255	860
α–Tocopherol (vitamin E)	146	290	163	370
α–Tocoquinone	96	25	100	19
Phylloquinone (vitamin K_1)	28	21	22	15
Ubiquinone-9/10	17	16	21	18
Total prenylquinones	570	612	561	1,282
3. Diglycerides[b]				
Phospholipids	+	3,484	+	4,352
Galactolipids	+	13,290	+	19,098
MGD[c]	+	7,560	+	9,638
DGD[c]	+	5,730	+	9,460

Note: On a weight basis, the pigment content is highest in the young, not yet fully expanded, leaf and then decreases somewhat parallel to the augmentation of the dry weight. The pigment levels and water content of shade leaves are higher than in sun leaves.

[a] Values from Lichtenthaler.[2]

[b] Mean values of August 5 and August 12 from Frey and Tevini, 1972 (personal communication).

[c] MGD = monogalactosyldiglyceride; DGD = digalactosyldiglyceride.

Table 6
AMOUNTS OF CHLOROPHYLLS, CAROTENOIDS, AND PRENYLQUINONES PER 100 CM^2 LEAF AREA IN GREEN LEAF AND THALLUS TISSUE OF DIFFERENT PLANTS

Plant	Chlorophylls		Carotenoids		Prenylquinones				
	a	*b*	*x*	*c*	PQ-9	α-T	α-TQ	K_1	UQ 9/10
Fagus sylvatica (beech)									
Sun leaves (June 4)	1260	380	225	125	200	220	20	13	17
Sun leaves (August 10)	1230	370	218	119	850	364	19	14	18
Shade leaves (May 11)	768	310	205	100	78	35	19	6.5	4
Shade leaves (August 10)	590	212	125	71	85	95	8	7	5
Ficus elastica									
Light green leaf (20 days old)	1000	330	300	100	125	100	10	16	+
Dark green leaf (1 yr old)	2600	900	630	230	750	1100	34	36	+
Dark green leaf (5 yr old)	2700	950	720	250	1900	1850	36	38	+
Spinacia oleracea (spinach)	2610	825	504	232	302	165	10	12	13
Lunularia cruciata (liver moss)	1900	600	390	170	148	173	15	23	+
Pellia epiphylla (liver moss)	1020	382	280	110	27	21	3	2	+
Buxus japonica									
(Normal green form)	1260	450	210	102	380	192	14	9	+
(Aurea mutant)	270	51	115	40	290	190	5	10	+
Acer negundo									
(Normal green form)	2280	704	575	340	370	540	45	40	+
(Aurea mutant)	856	192	418	162	172	324	25	27	+
(Variegated form)									
Green leaf parts	1148	400	288	80	83	260	26	17	+
White leaf parts	14	6	8	2	5	16	7	0.3	+

Note: In contrast to the prenylpigments, the levels of plastoquinone-9 and α-tocopherol are continuously raised with increasing age of leaf tissue (e.g., *Ficus*). This process depends on the light density and is much higher in sun leaves than in shade leaves (*Fagus*).

[a] a and b = chlorophyll a and b; x and c = xanthophylls and β-carotene; PQ-9 = total plastoquinone-9 (quinone + hydroquinone); α-T = α-tocopherol (vitamin E); α-TQ = α-tocoquinone; K_1 = phylloquinone (vitamin K_1); UQ 9/10 = ubiquinone with prenylchain length of 9 or 10 isoprene units.

Table 7
AMOUNTS OF CHLOROPHYLLS, CAROTENOIDS, PRENYLQUINONES, AND FAT-SOLUBLE VITAMINS (μg/1 g DRY WEIGHT) IN DIFFERENT GREEN AND NONGREEN PLANT TISSUES

	Chlorophylls		Carotenoids		Prenylquinones				
Type of plant tissue	*a*	*b*	Xanthophylls	β-carotene	PQ-9	α-T	α-TQ	K_1	UQ-9/10
1. Green leaf tissue									
Fagus sylvatica									
Sun leaves (August)	1240	381	222	121	860	370	19	15	18
Shade leaves (August)	1800	650	380	220	260	290	25	21	16
Spinacia oleracea (spinach)	6540	2060	1260	580	755	413	25	30	32
Hordeum vulgare (barley)									
Young primary leaves	9200	3400	1120	515	400	256	12	56	19
Nicotiana tabacum									
Normal green plant	8250	2650	788	552	1700	212	177	28	+
Aurea mutant Su/su	1086	214	170	90	370	83	61	31	+
2. Etiolated leaf tissue									
Hordeum vulgare (barley)	—	—	151	15	90	52	3	3	27
3. Variegated leaf tissue									
Acer negundo									
Green parts	1870	800	720	200	450	720	66	32	+
White parts	142	72	76	20	74	155	8	2	+
4. Plant tissue culture									
Petunia hybrida									
Full green callus	990	340	170	110	310	493	26	14	
+									
Yellow-green callus	141	50	35	22	406	241	40	6	
+									
Nicotiana tabacum									
Faint green callus	106	20	13	6	83	40	7	3.8	
+									
Hordeum vulgare									
Faint yellow callus	—	—	4	1	88	32	12	< 1	
+									

Table 7 (continued)
AMOUNTS OF CHLOROPHYLLS, CAROTENOIDS, PRENYLQUINONES, AND FAT-SOLUBLE VITAMINS (μg/1 g DRY WEIGHT) IN DIFFERENT GREEN AND NONGREEN PLANT TISSUES

Type of plant tissue	Chlorophylls		Carotenoids		Prenylquinones			
	a	*b*	Xanthophylls UQ-9/10	β-carotene	PQ-9	α-T	α-TQ	K_1
5. Fruit tissue								
Solanum lycopersicum								
Green tomato tissue	192	73	30	21	56	70	0.4	0.3
+								
Red tomato tissue	—	—	602[a]	74	173	118	1.4	0.1
+								
Capsicum annuum (variety with red fruits)								
Green fruit tissue	273	112	82	25	44	86	0.4	0.5
+								
Young red tissue	—	—	960[a]	115	263	261	1.4	0.5
+								
Full red tissue	—	—	1400[a]	103	220	330	8.0	0.2
+								
Capsicum annuum (variety with yellow fruits)								
Green fruit tissue	492	204	132	32	36	66	3	3.6
+								
Yellow-green tissue	70	33	128	8	51	82	2	1.9
+								
Yellow fruit tissue	—	—	310[a]	50	139	298	7	1.2
+								
6. Flower petals: plasmochromous flowers (colored by secondary carotenoids)								
Rhododendron flavum	25	11	2367[a]	66	890	433	140	15
+								
Escholtzia californica	7	3	14940[a]	35	733	800	100	20
+								
Viola tricolor	77	36	3955[a]	16	325	273	4	9
+								

Chymochromous flowers (white petals or colored by anthocyanus)									
Saponaria officinalis (white petals)	33	14	12	5	82	120	8	4	+
Pelargonium zonale (red petals)	38	19	8	3	117	121	11	8	+
Dahlia variabilis (red petals)	12	5	25	5	56	83	5	1	+
7. Chlorophyll-free tissue									
Solanum tuberosum (potato tubers)	—	—	0.4	—	0.33	0.6	0.07	< 0.03	0.2
Scorzonēra hispanica (roots)	—	—	0.5	—	0.16	1.5	0.08	< 0.04	0.4
Allium cepa (onion scales)	—	—	0.4	0.1	6.5	4.2	0.4	0.2	0.7
Brassica oleracea (cauliflower inflorescence)	—	—	2.6	0.6	13.8	3.0	0.9	0.8	9.6
Daucus carota (carrot roots)	—	—	—	614	31	32	1.9	0.6	0.8

Note: β-carotene = provitamin A; α-tocopherol = vitamin E (α-T); phylloquinone K_1 = vitamin K_1; PQ-9 = plastoquinone-9 (quinone + hydroquinone); α-TQ = α-tocoquinone; UQ-9/10 = ubiquinone-9 or 10.

[a] Xanthophylls + secondary carotenoids and xanthophyllesters.

REFERENCES

1. **Goodwin, T. W.,** The prenyllipids of the membranes of higher plants, in *Lipids and Lipid Polymers in Higher Plants,* Tevini, M. and Lichtenthaler, H. K., Eds., Springer-Verlag, Berlin, 1977, 28.
2. **Lichtenthaler, H. K.,** Regulation of prenylquinone synthesis in higher plants, in *Lipids and Lipid Polymers in Higher Plants,* Tevini, M. and Lichtenthaler, H. K., Eds., Springer-Verlag, Berlin, 1977, 231.
3. **Lichtenthaler, H. K.,** Separation of prenyllipids including chlorophylls, carotenoids, prenylquinones and fat-soluble vitamins, in *CRC Handbook Series in Chromatography,* Vol. 8, Mangold, H., Ed., in press.
4. **Strain, H. H., Cope, B. T., and Scec, W. A.,** Analytical procedures for the isolation, identification, estimation, and investigation of the chlorophylls, in *Methods in Enzymology,* Vol. XXIII, San Pietro, A., Ed., Academic Press, New York, 1971, 452.
5. **Jones, O. T. G., Chlorophyll,** in *Phytochemistry,* Vol. I, Miller, L. P., Ed., Van Nostrand Reinhold, New York, 1973, 75.
6. **Liaaen-Jensen, S., III.** Isolation reactions, in *Carotenoids,* Isler, Q., Ed., Birkhauser Verlag, Basel, 1971, 61.
7. **Davies, B. H.,** Carotenoids, in *Chemistry and Biochemistry of Plant Pigments,* 2nd ed., Vol. 2, Goodwin, T. W., Ed., Academic Press, London, 1976, 38.
8. **Barr, F. and Crane, F. L.,** Quinones and algae in higher plants, in *Methods in Enzymology,* Vol. XXIII, San Pietro, A., Ed., Academic Press, New York, 1971, 372.
9. **Lichtenthaler, H. K., Karunen, P., and Grumbach, K. H.,** Determination of prenylquinones in green photosynthetically active moss and liver moss tissues, *Physiol. Plant.,* 40, 105, 1977.
10. **Lichtenthaler, H. K.,** Verbreitung und relative Konzentration der lipophilen Plastidenchinone in grünen Pflanzen, *Planta* (Berl.), 81, 140, 1968.
11. **Ramasarma, T.,** Lipid quinones, *Adv. Lipid Res.,* 6, 107, 1968.
12. **Carr, N. G. and Hallaway, M.,** Quinones of some blue-green algae, in *Biochemistry of Chloroplasts,* Vol. 1, Goodwin, T. W., Ed., Academic Press, New York, 1966, 466.
13. **Threlfall, D. R. and Whistance, G. R.,** Biosynthesis of isoprenoid quinones and chromanols, in *Aspects of Terpenoid Chemistry and Biochemistry,* Goodwin, T. W., Ed., Academic Press, New York, 1971, 357.
14. **Lichtenthaler, H. K.,** Occurrence and function of prenyllipids in the photosynthetic membrane, in *Recent Advances in the Biochemistry and Physiology of Plant Lipids,* Appelquist, L. -A. and Liljenberg, C., Eds., Elsevier, Amsterdam, 1979, in press.
15. **Verbeek, L. and Lichtenthaler, H. K.,** Der Einflubvon Stickstoffmangel auf die Lipochinon- und Isoprenoidsynthese der Chloroplasten von *Hordeum vulgare* L., *Z. Pflanzenphysiol.,* 70, 245, 1973.
16. **Schneider, H. A. W.,** Chlorophylls. Aspects of enzymology and regulation of biosynthesis, *Ber. Dtsch. Bot. Ges.,* 88, 83, 1975.
17. **Davies, B. H.,** Carotenoids in higher plants, in *Lipids and Lipid Polymers in Higher Plants,* Tevini, M. and Lichtenthaler, H. K., Eds., Springer-Verlag, Berlin, 1977, 199.
18. **Butler, W. L. and Kitojima, M.,** A tripartite model for chloroplast fluorescence, in *Proceedings 3rd Int. Congr. Photosynthesis,* Vol. 1, Avron, M., Ed., Elsevier, Amsterdam, 1974, 13.
19. **Grumbach, K. H., Lichtenthaler, H. K., and Erismann, K. H.,** Incorporation of $^{14}CO_2$ in photosynthetic pigments of *Chlorella pyrenoidosa, Planta,* 140, 37, 1978.
20. **Grumbach, K. H. and Lichtenthaler, H. K.,** Incorporation of $^{14}CO_2$ in prenylquinones of *Chlorella pyrenoidosa, Planta,* 141, 253, 1978.
21. **Trebst, A.,** Energy conservation in photosynthetic electron transport of chloroplasts, *Ann. Rev. Plant Physiol.,* 25, 423, 1974.
22. **Lichtenthaler, H. K. and Pfister, C.,** New aspects on the function of naphthoquinones in photosynthesis, in *Photosynthetic Oxygen Evolution,* Metzner, H., Ed., Academic Press, London, 1978, 171.
23. **Lichtenthaler, H. K.,** Localization and functional concentration of lipoquinones in chloroplasts, in *Progress in Photosynthesis Research,* Vol. I, 1969, 304.
24. **Lichtenthaler, H. K. and Tevini, M.,** Die Wirkung von UV-Strahlen auf die Lipochinon-Pigment-Zusammensetzung isolierter Spinatchloroplasten, *Z. Naturforsch.,* 24b, 764, 1969.
25. **Lucy, J. A.,** Structural interaction between vitamin E and polyunsaturated phospholipids, in *Tocopherol, Oxygen and Biomembranes,* de Duve, C. and Hayaishi, O., Eds., Elsevier, Amsterdam, 1978, 109.
26. **Lichtenthaler, H. K.,** Plastoglobuli and the fine structure of plastids, *Endeavour,* 27, 144, 1968.
27. **Lichtenthaler, H. K. and Sprey, B.,** Über die osmiophilen globulären Lipideinschlüsse der Chloroplasten, *Z. Naturforsch.,* 21b, 690, 1966.

28. **Wellburn, A. R. and Hemming, F. W.**, The subcellular distribution and biosynthesis of castaprenols and plastoquinone in the leaves of *Aesculus hippocastanum, Biochem. J.*, 104, 173, 1967.
29. **Hemming, F. W.**, Polyisoprenoid alcohols (prenols), in *Terpenoids in Plants,* Pridham, J. B., Ed., Academic Press, London, 1967, 223.
30. **Lichtenthaler, H. K. and Buschmann, C.**, Control of chloroplast development by red light, blue light and phytohormones, in *Chloroplast Development,* Akoyunoglou, G., et al., Eds., Elsevier-North Holland, Amsterdam, 1978, 801.
31. **Fedtke, C.**, Changed physiology in wheat plants treated with the herbicide methabenzthiazuron, *Naturwissenschaften,* 61, 272, 1974.
32. **Lichtenthaler, H. K. and Weinert, H.**, Die Beziehungen zwischen Lipochinosynthese und Plastoglobulibildung in den Chloroplasten von *Ficus elastica* Roxb., *Z. Naturforsch.,* 25b, 619, 1970.
33. **Lichtenthaler, H. K.**, Zur Synthese der lipophilen Plastidenchinone und Sekundärcarotinoide während der Chromoplastenentwicklung, *Ber. Dtsch. Bot. Ges.,* 82, 483, 1969.
34. **Lichtenthaler, H. K. and Park, R. B.**, Chemical composition of chloroplast lamellae from spinach, *Nature (London),* 198, 1070, 1963.

SYNTHESIS OF GLYCOLIPIDS AND PHOSPHOLIPIDS IN VASCULAR PLANTS

M. Tevini

CHEMICAL STRUCTURE AND DISTRIBUTION

Glycolipids and phospholipids of vascular plants are mainly diglycerides which show high variations in their fatty acid chains esterified with glycerol. Other glyco- and phospholipids not of the glycerolipid type also are present in plant tissues but in minor amounts. Chemical structures and brief descriptions are given in Table 1, and the percentage of fatty acid composition of several tissues is shown in Table 2.

Glyceroglyco- and phospholipids are distributed in all plant tissues because they are structural components of cell membranes (Table 3). Glycolipids are more concentrated in plastids, especially in chloroplasts, whereas phospholipids occur in all plant cell organelles and their membranes (Table 4). The lipids can be extracted from plant tissue with hot chloroform/alcohol mixtures and separated by two-dimensional thin-layer chromatography.[1]

BIOSYNTHESIS

The biosynthesis of plant glycolipids and phospholipids is described in some recent papers,[2-4] but the biosynthetic pathways and subcellular localization of phospho- and glycolipid synthesis are not firmly established. Figure 1 summarizes the main synthetic pathways in vascular plants. Phosphatidic acid (PA), which is synthesized by acylation of sn-glycerol-3-P, is the key substance for the synthesis of glycerolipids leading to 1,2-diglycerides and to nucleotide diglycerides. 1,2-Diglycerides are the precursors for the synthesis of the phospholipids PE and PC, and of the glycolipids MGDG, DGDG, and SL. The phospholipids PS, PI, PG, and DPG are synthesized via the CDP-diglyceride pathway. PE can also be derived from PS by decarboxylation and PC by transmethylation of PE.[3]

The biosynthesis of glyceroglyco- and phospholipids is not light dependent because the lipids are always found in dark-grown plants or in roots. During chloroplast formation, the biosynthesis of the galactolipids and some phospholipids is stimulated by white or red light, indicating the participation of the phytochrome system in regulating lipid biosynthesis.[5]

Sn-glycerol-3-phosphate

Especially for synthesis of the glycerol backbone of glycerolipid, two pathways are possible in vascular plant tissues. In germinating seeds, glycerol-3-phosphate is derived from glycolytic intermediates by the action of glycerol phosphate dehydrogenase.[6] In leaf cytoplasma, all enzymes are present to follow the pathway from glyceraldehyde phosphate (GAP) via glycerolaldehyde and glycerol (G). The enzymes involved are a phosphatase, which dephosphorylates GAP, and NADPH-dependent aldose reductase to yield glycerol. Finally, the glycerol is phorphorylated by glycerol kinase.[7]

Phosphatidic Acid

Glycerol is stepwise acylated to monoacyl-sn-glycero-3-phosphates and to phosphatidic acid. The acylation occurs in microsomes, mitochondria, fat droplets, and chloroplasts. The synthesis site in plastids is the envelope.[8]

Table 1
GLYCOLIPIDS AND PHOSPHOLIPIDS IN VASCULAR PLANTS

Common name	Structure and chemical name	Source and fatty acid composition
Glycolipids Monogalactosyldiglyceride (MGDG) 1,2-diacyl-[β-D-galactopyranosyl (1′ 3)]-sn-glycerol	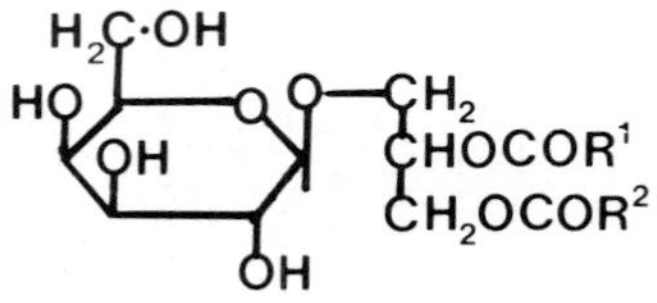	Major component in leaves, especially of chloroplasts, also present in all tissues with special plastid types. Main fatty acids (R^1, R^2) in leaves: $C_{18:3}$, $C_{16:0}$, and $C_{16:3}$ in 16:3 plants (spinach), in roots: $C_{18:2}$, $C_{16:0}$
Digalactosyldiglyceride (DGDG) 1,2 diacyl[α-D-galactopyranosyl-(1′ 6′)-β-D-galactopyranosyl (1′ 3)]-sn-glycerol	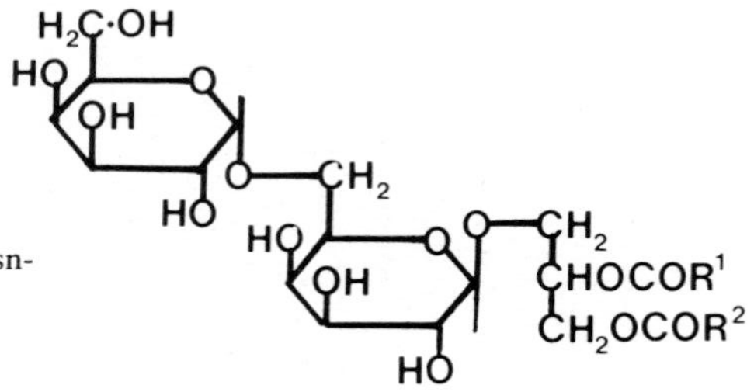	Always together with MGDG in chloroplasts. Also with high proportions of polyene fatty acids, but usually less 18:3 than in MGDG
Sulphoquinovosyldiglyceride (SL) 1,2 diacyl-[6-sulpho-α-D-quinovopyranosyl-(1′ 3)]-sn-glycerol	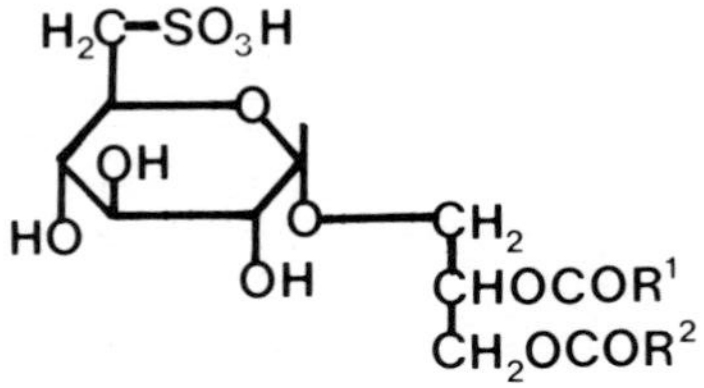	Widely distributed in plant tissue especially in leaves. Contains more saturated fatty acids (mainly $C_{16:0}$) sugar moiety 6-deoxyglucose

Sterylglucoside (SG)	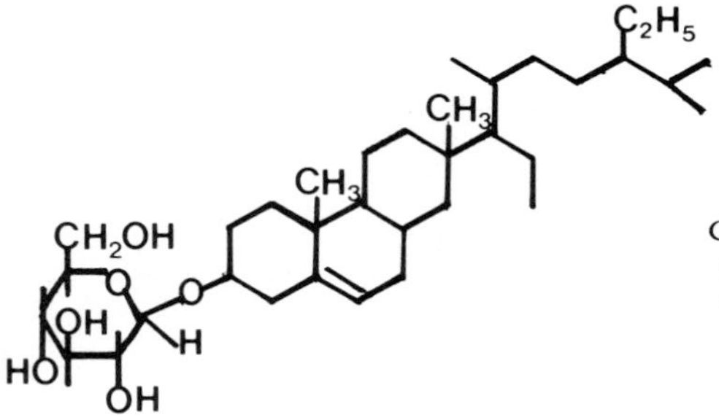	Containing D-glucose and several types of sterols such as sitosterol, stigmasterol, campesterol, and cholesterol
Acylated sterylglucoside (ASG)	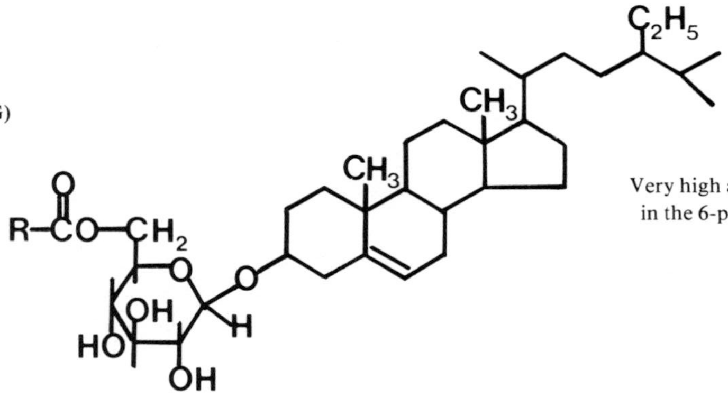	Very high amounts in potato tubers. Glucose is acylated in the 6-position with $C_{16:0}$ and the C_{18}-series
Glycosylceramides	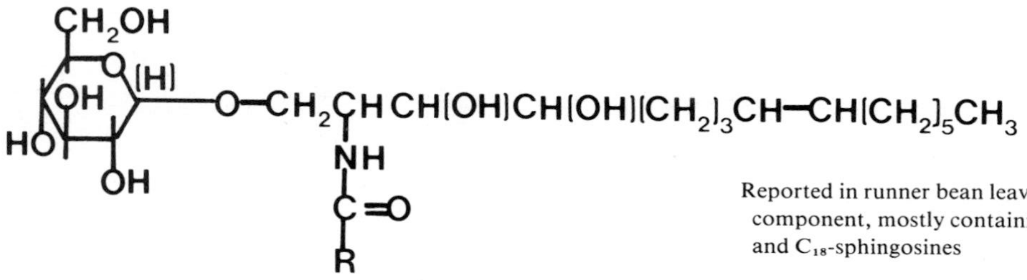 $O-CH_2CH(NH-CO-R)CH(OH)CH(OH)(CH_2)_3CH{=}CH(CH_2)_5CH_3$	Reported in runner bean leaves (wheat flour). Minor component, mostly containing glucose, -hydroxy acids and C_{18}-sphingosines

Glycerophospholipids		
Phosphatidic acid (PA)	X = H	Minor amounts in all tissues. Important as biosynthetic intermediate. Main fatty acids: $C_{16:0}$, $C_{18:0}$, $C_{18:1}$, $C_{18:2}$, but also $C_{18:3}$
Phosphatidylcholine (PC) (Lecithin)	X = OH · $CH_2CH_2N^+(CH_3)_3$ Choline	Main component, in all tissues, especially located in microsomal and mitochondrial fractions, but also in plastids. Main fatty acids $C_{18:2}$, and $C_{16:0}$
Phosphatidylethanolamine (PE)	X = OH · $CH_2CH_2NH_2$ Ethanolamine	Mostly found together with PC, nearly the same fatty acid composition as in PC, fatty amides found in flours
Phosphatidylserine (PS)	X = OH · CH_2CH · NH_2,Serine COOH	Minor component, absent in cnloroplasts
Phosphatidylglycerol (PG)	X = CH_2OH · CHOH · CH_2OH Glycerol	Main component in green tissues, there connected with the unusual fatty acid Δ 3 trans $C_{16:1}$, high amounts of $C_{18:3}$
Phosphatidylinositol (PI)	X = [structure: myo-inositol ring, positions 1–6, OH groups] Myo-inositol	Minor component in leaves, higher amounts in seeds Normal phospholipid fatty acids
Diphosphatidylglycerol (DPG)	X = $-CH_2$ $CHOCOR^1$ CH_2OCOR^2	Minor component, mainly found in mitochondria, microsomes of roots, and cauliflower buds

Table 2
PERCENTAGE OF FATTY ACID COMPOSITION OF GLYCOLIPIDS AND PHOSPHOLIPIDS IN SOME PLANT MATERIAL

	$< C_{14}$	14:0	16:0	16:1	18:0	18:1	18:2	18:3
				Carrot leaves				
MGDG	—	0.1	1.8	0.2	0.6	38	3.4	55.6
DGDG	—	—	20.4	2.6	0.3	4.1	7.1	65.4
SL	—	2.0	57.3	—	0.6	1.4	8.8	29.5
PG	—	1.7	28.6	26.0[a]	0.2	1.3	10.8	31.4
PC	—	11.8	36.4	—	3.1	6.3	40.7	1.6
PE	—	1.7	39.7	—	1.4	1.7	53.5	2.5
PI	4.7	10.6	27.3	1.3	6.4	5.2	43.2	1.0
PA	—	5.8	39.1	—	1.5	5.0	11.5	36.6
				Carrot roots				
MGDG	4.0	3.5	30.8	2.5	2.2	9.3	41.8	5.9
DGDG	2.3	2.2	29.6	0.5	1.8	7.3	48.8	4.8
SL	5.9	8.6	49.9	0.8	5.8	9.0	16.8	—
PG	4.0	7.3	43.9	2.8[a]	1.1	6.4	32.0	0.8
PC	4.9	7.3	53.2	2.7	1.9	10.4	12.4	4.9
PE	3.0	8.3	36.1	3.8	2.3	4.4	38.5	1.6
PI	3.1	8.7	48.3	3.6	5.3	7.3	19.7	2.7
PA	3.4	26.1	29.9	4.1	6.3	7.8	20.9	—
				Barley chloroplast				
MGDG	3.7	4.9	4.1	—	—	3.7	8.1	79.3
DGDG	—	1.2	19.6	—	0.4	1.3	3.9	73.7
SL	—	0.7	34.4	—	—	6.7	16.3	42.0
PG	—	0.1	32.6	14.7[a]	3.3	3.2	8.1	37.2
PC	—	3.9	36.8	—	2.9	2.8	41.0	12.4

[a] 3-trans- $C_{16:1}$

1,2-Diglycerides Pathway

Phosphatidic acid phosphatase splits PA into diglycerides and P_1. The enzyme is found in membranes, in soluble cytoplasmic proteins, and in plastid envelope fractions. The envelopes of spinach chloroplast contain high amounts of diglyceride species with polyenoic acids, which are also found in galactolipids.[9] The fatty acid distribution in the diglycerides may be of great importance during the enzymatic reaction of galactolipid synthesis .

Galactolipids

Galactolipids are synthesized in or at the plastid envelope of the plant cells by galactosylation of 1,2-diglycerides. The first galactosyltransferase uses UDP-Gal to form MGDG. A second and third galactosyltransferase catalyses the synthesis of DGDG and TGDG.[10] Since in many tissues DGDG does not have the same fatty acid distribution as MGDG, several pools of MGDG may exist which are selectively galactosylated.

Sulfo(no) lipid (SL)

The sulfolipid is present in all plant tissues but in different amounts. The pathway of the sugar moiety 6-sulfo-quinovose is rather complex. Its precursor is cysteic acid

Table 3
PHOSPHOLIPID AND GLYCOLIPID COMPOSITION OF SEVERAL PLANT TISSUES[a]

	MGDG	DGDG	SL	PC	PG	PE	PI	PS	PA	DPG
Impatiens balsamina										
Green leaves	373	210	23.7	41.1	19.4	30.5	7.1	0.4	—	1.3
Stems	50	32.5	2.5	5.3	2.4	9.7	3.7	1.8	—	0.3
Roots	20.6	38.6	5.2	16.2	3.1	26.3	3.0	2.9	—	—
Red floral leaves	19.3	18.5	9.6	25.0	4.7	19.8	0.5	1.7	—	—
Daucus carota										
Green leaves	263	173	21.5	80	41	40	18	—	9	—
Orange roots	22.7	24.0	7.7	34.1	6.9	19.3	11.5	—	10.9	—
Young white roots	90.8	63.2	19.7	65.6	6.6	29.1	8.4	—	31.6	—
Gingko biloba leaves[b]	365	265	25	145	68	43	22	—	—	25
Pinus radiata needles[b]	220	190	44	65	44	22	33	—	—	27
Hordeum vulgare										
Etiolated leaves	80.5	71.6	11.6	61.7	7.1	25.1	11.6	8.0	—	9.8
Green leaves (5′ red ill.)	125	108	13.9	73.3	15.9	30.6	17.6	8.3	—	11.1
Tomato fruits[c]	550	301	4	96	47	132	45	10	—	—

[a] mg/100 g fresh weight, except as specified.
[b] Adapted from Ref. 14.
[c] mg/100 g dry weight.

Table 4
LIPID COMPOSITION OF PLANT CELL ORGANELLES[a]

	MGDG	DGDG	SL	PC	PG	PE	PI	PS	Lyso	PA	DPG	Ref.
Spinach microsomes	12	8	2	35	3	30	7	—	—	—	—	9
Spinach chloroplasts	55	31.6	5.2	2.6	4.5	0.9	0.2	—	—	—	—	—
Barley etioplasts	42.3	25.3	10.3	9.0	7.2	5.9	tr.	—	—	—	—	—
Barley chloroplasts	43.4	23.7	14.5	5.9	10.4	1.9	tr.	—	—	—	—	—
Mespilus chromoplasts	33.9	46.3	8.6	2.9	1.0	3.2	1.8	—	—	2.2	—	—
Red pepper chromoplasts	41.5	25.8	11.4	9.6	3.0	3.9	2.8	—	—	1.9	—	—
Carrot chromoplasts	25.9	42.9	8.3	3.8	0.7	2.9	2.2	—	—	13.2	—	—
Potato tuber												
Mitochondria	—	—	—	43	3	30	7	3	—	5	3	15
Peroxisomes	—	—	—	61	15.1	19.6	4.3	—	—	—	—	16
Onion root tip nuclei	—	—	—	27.1	5.6	20.4	8.0	3.1	10.1	24.6	1.2	17

[a] % by weight of determined lipids.

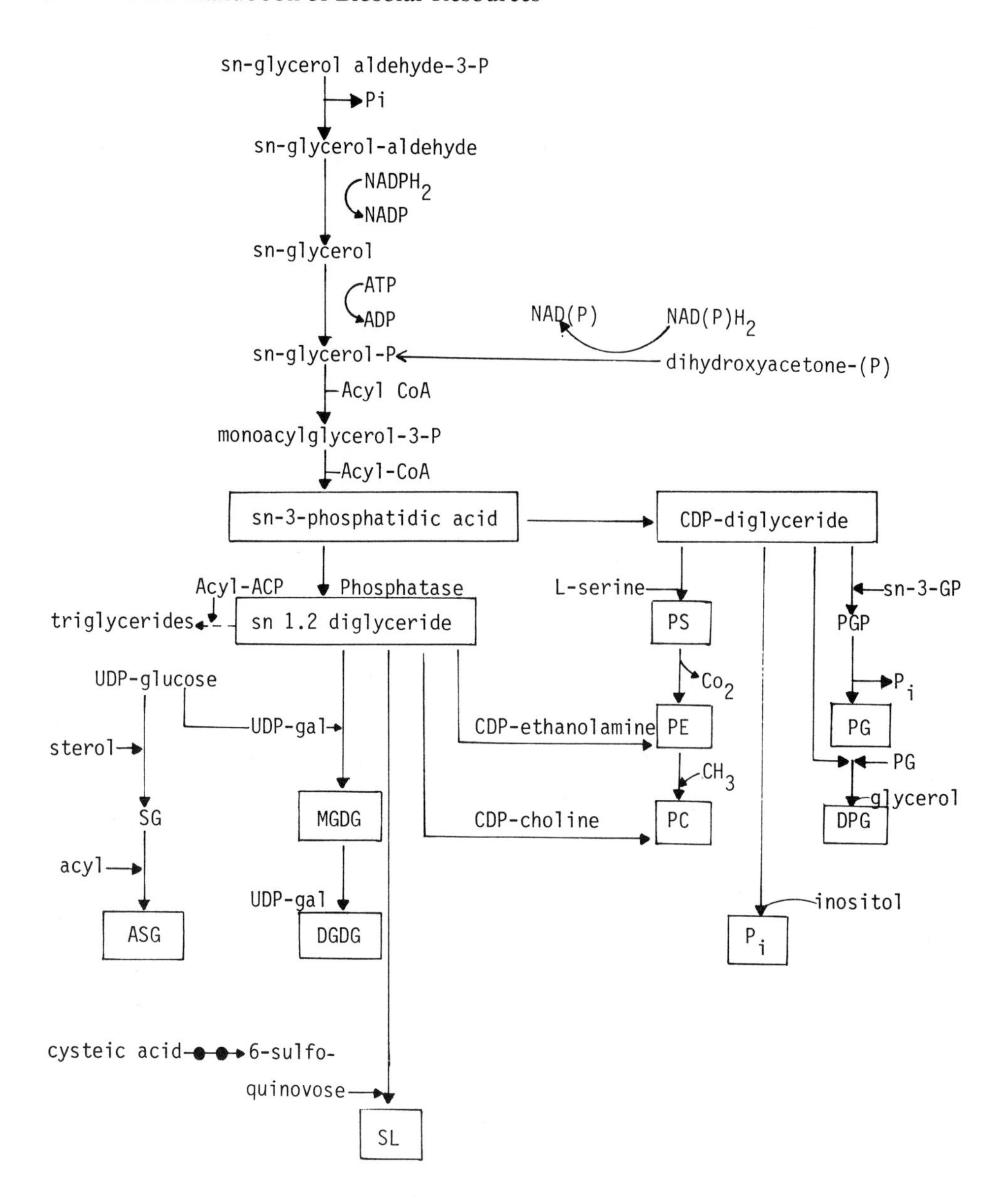

FIGURE 1. Biosynthetic pathways of phospholipids and glycolipids in vascular plants.

followed by several intermediates, e.g., 3-sulfo-lactate and 3-sulfoactaldehyde. The participation of a nucleotide sugar in biosynthesis of sulfolipid has not been demonstrated yet. The substrate for acylation does not seem to be sulfo-quinovosyl glycerol.[11]

Phospholipids

The biosynthesis of PE and PC in the microsomal fraction of spinach leaves was demonstrated with CDP ethanolamine and CDP-choline as substrates.

CDP-diglyceride Pathway

The presence of the CDP-diglyceride pathway was demonstrated by incorporation of ^{32}P from ^{32}P-CTP in spinach leaves, maize, and cauliflower inflorescence. Biosynthetic pathways of the following phospholipids using nucleotide-diglycerides are well summarized by Kates and Marshall.[3]

Phosphatidylglycerol (PG) and Diphosphatidylglycerol (DPG)

CDP-diglyceride is the obligatory substrate for PG synthesis in spinach microsomal membrane fractions. The PG synthesis has an absolute requirement for Mg as a cofactor. DPG (cardiolipin) is shown to be synthesized by incorporation of ^{14}C-GP into mitochondria of cauliflower inflorescences. The formation by the action of phospholipase D exchange reaction may be also possible in vivo.

Phosphatidylserine (PS), Phosphatidylethanolamine (PE), and Phosphatidylcholine (PC)

The biosynthesis of PS has been demonstrated in tomato roots and in spinach leaves. ^{14}C from uniformly labeled L-serine was not only incorporated into PS but also into PE showing that a PS decarboxylase pathway exists in plant leaves. The ethanolamine exchange pathway by the action of phospholipase D using PS and PC as substrates found in spinach leaves may only be operating in vitro. In spinach leaves, PE can be used as precursor of PC biosynthesis but *S*-adenosylmethionine is not the potent methyl donator in this reaction. The 1,2-diglyceride pathway using CDP-choline or CDP-ethanolamine also is possible.

Phosphatidyl Inositol (PI)

Enhanced incorporation of myo-inositol into PI is catalyzed in the presence of CDP-diglyceride by mitochondrial preparations from nonphotosynthetic tissues and spinach leaves as well. In particulate fractions of spinach leaves, the PI synthesis without CDP-diglyceride also is possible.

Sterylglucosides (SG, ASG), Cerebrosides, and Phytoglycolipid

Biosynthetic studies concerning SG and acylated SG have been made with subcellular fractions of several plants.[12] UDP-^{14}C-glucose and labeled sterol served as precursors for steryl ^{14}C-glucose and acylated ^{14}C-glucoside. Either glycosyltransferase or acyltransferase are particle bound. In vivo studies with intact tobacco plants suggest that acylated UDP-glucose may be the precursor for ASG. The origin of fatty acids has to be elucidated. The biosynthesis of cerebrosides as well as phytoglycolipids has not been studied in higher plants. Most of the work on cerebrosides has been done with fungi. Phytoglycolipids are rather complex lipids found in several seeds but have not received much attention.[13]

REFERENCES

1. **Kates, M.,** Techniques of lipidology, in *Laboratory Techniques in Biochemistry and Molecular Biology,* Vol. 3, Part II, Work, T. S. and Work, E., Eds., North-Holland, Amsterdam, 1972.
2. **Mudd, J. B. and Garcia, R. E.,** Biosynthesis of glycolipids, in *Phytochemical Society Symposia,* Series No. 12, Galliard, T. and Mercer, E. J., Eds., Academic Press, London, 1975, 162.
3. **Kates, M. and Marshall, M. O.,** Biosynthesis of phospho-glycerides in plants, in *Phytochemical Society Symposia,* Series No. 12, Galliard, T. and Mercer, E. J., Eds., Academic Press, London, 1975, 115.
4. **Heinz, E.,** Enzymatic reactions in galactolipid biosynthesis, in *Lipids and Lipid Polymers in Higher Plants,* Tevini, M. and Lichtenthaler, H. K., Eds., Springer-Verlag, Berlin, 1977, chap. 6.
5. **Tevini, M.,** Light, function and lipids during plastid development, in *Lipids and Lipid Polymers in Higher Plants,* Tevini, M. and Lichtenthaler, H. K., Eds., Springer-Verlag, Berlin, 1977, chap. 7.
6. **Yamada, M.,** Studies on fat metabolism in germinating castor beans. IV. Glycerol metabolism, *Sci. Pap. Coll. Gen. Educ., Univ. Tokyo,* 10, 283, 1960.
7. **Hippmann, H. and Heinz, E.,** Glycerol kinase in leaves, *Z. Pflanzenphysiol.,* 79, 408, 1976.
8. **Joyard, J. and Douce, R.,** Site of synthesis of phosphatidic acid and diacylglycerol in spinach chloroplasts, *Biochim. Biophys. Acta,* 486, 273, 1977.
9. **Douce, R.,** Site of biosynthesis of galactolipids in spinach chloroplasts, *Science,* 183, 852, 1974.
10. **Van Hummel, H. C.,** Chemistry and biosynthesis of plant galactolipids, in *Progr. Chem. Nat. Org. Prod.,* Vol. 32, Herz, W., Grisebach, H., and Kirby, G. W., Eds., Springer-Verlag, Vienna, 1975, 267.
11. **Haarwood, J. L.,** Synthesis of sulphoquinovosyl diacylglycerol by higher plants, *Biochim. Biophys. Acta,* 398, 224, 1975.
12. **Eichenberger, W.,** Steryl glycosides and acylated steryl glycosides, in *Lipids and Lipid Polymers in Higher Plants,* Tevini, M. and Lichtenthaler, H. K., Eds., Springer-Verlag, Berlin, 1977, chap. 9.
13. **Carter, H. E., Strobach, D. R., and Hawthorne, J. N.,** Biochemistry of the sphingolipids. XVIII. Complete structure of tetrasaccharide phytoglycolipid, *Biochemistry,* 8, 383, 1969.
14. **Roughan, P. G. and Batt, R. D.,** The glycerolipid composition of leaves, *Phytochemistry,* 8, 363, 1969.
15. **Meunier, D. and Mazliak, P.,** Differences de composition lipidique entre les deux membranes des mitochondries de pomme de terre, *C. R. Acad. Sci.,* 275, 213, 1972.
16. **Mazliak, P.,** Glyco- and phospholipids of biomembranes in higher plants, in *Lipids and Lipid Polymers in Higher Plants,* Tevini, M. and Lichtenthaler, H. K., Eds., Springer-Verlag, Berlin, 1977, chap. 3.
17. **Philipp, E. J., et al.,** Characterization of nuclear membranes and endoplasmic reticulum isolated from plant tissues, *J. Cell Biol.,* 68, 11, 1976.

STEROIDS OF ALGAE

Glenn W. Patterson

Steroids are well studied in animal systems because of their importance as hormones. However, in plants, including algae, there have been few identifications of steroids except for the sterols. Sterols are apparently components of all algae.[11,21,55] Sterol composition of algae ranges from 0.001% or less (dry weight basis) in some blue-green algae to 0.5% or more in some green algae and Chrysophytes. The kinds of sterols found in algae vary. Rhodophyta (red algae) are unique among the algae, and other plants as well, in that most of them contain cholesterol as a major sterol (Table 1). Some red algae contain desmosterol or 22-dehydrocholesterol rather than cholesterol (Figure 1). Except for one species, the primary sterol of all red algae contains 27 carbon atoms. The sterols of Phaeophyta (brown algae) are very characteristic of that division. The principal sterol of all brown algae is fucosterol (24-ethylidene-cholesterol), a sterol which is rare elsewhere in the plant kingdom (Table 2). Several biosynthetically related sterols accompany fucosterol in some brown algae, but in others the sterol fraction is essentially pure fucosterol. It has been suggested that fucosterol could be obtained as a useful by-product from commercially utilized brown algae.[69]

The Chlorophyta (green algae) differ markedly in sterol composition from red algae and brown algae which are characterized by one sterol in each division (cholesterol and fucosterol, respectively). The green algae contain complex mixtures of sterols in most species and great variability of sterol structures exists from species to species (Table 3). Some species contain Δ^5-ergostenol, poriferasterol, and clionasterol (Figure 2) which are identical to the sterols of higher plants except for their configuration at C-24 in the sterol side chain. This is additional biochemical evidence linking higher plants with the green algae. Other green algae contain as major sterol components ergosterol, chondrillasterol, and 28-isofucosterol (Figure 3). At present, it appears that ergosterol and chondrillasterol are more typical of the primitive, unicellular green algae, while 28-isofucosterol and other Δ^5-sterols are more common in the larger, more advanced green algae. More data are needed to confirm these observations. Several unusual sterols have also been identified in green algae.[60,64]

Most of the data on sterols of Chrysophyta are very recent (Table 4). The sterols of the Chrysophyceae, Haptophyceae, and Xanthophyceae are primarily Δ^5-sterols similar to those in the green algae (Figure 2). In the the Bacillariophyceae, however, sterol composition is quite variable from species to species. One of the most common sterols is 22-dehydrocampesterol (24α-methyl-22-dehydrocholesterol). The sterols of the Chrysophyta are of special interest due to their importance in the marine food chain.

Only a few species of algae in other divisions have been examined for sterol composition. Originally, the blue-green algae were thought to be devoid of sterols,[44] but more recent work demonstrates their presence, usually in extremely small quantities (Table 5). The Euglenophyta contain sterols similar to those of some of the unicellular green algae. Three species of dinoflagellates have been examined for sterol with very unusual structures being reported.[70,83]

Very little is known about the conditions favoring synthesis of sterols in algae, but recent reports indicate that more total sterol is produced under autotrophic than under heterotrophic conditions.[12,84]

Table 1
STEROLS OF RED ALGAE (RHODOPHYTA)

Order	Organism	Major sterols[a]	Ref.
Porphyridiales	*Porphyridium aerugineum*	chol, erg	10
	P. cruentum	22DC, erg	11
	P. sordidum	22DC	16
	P. violaceum	22DC	16
Bangiales	*Porphyra purpurea*	desmo	28
	P. tenera	chol	20
	P. yamamoto	chol	20
	P. yezoensis	chol	20
	P. sp.	chol	36
Nemalionales	*Liagora distenta*	chol, liago	24
	L. viscida	chol, desmo, liago	25
	Nemalion helminthoides	desmo, chol	25
	Scinaia furcellaria	chol, liago	24
Gelidiales	*Acanthopeltis japonica*	chol	80
	Gelidium amansii	chol, chola	17, 80
	G. japonicum	chol	79
	G. latifolium	chol	24
	G. sesquipedale	chol	46
	G. subcostatum	chol, chola	17, 79
	Pterocladia capillacea	chol	2
	P. pinnata	chol	24
	P. tenuis	chol, chola	17, 79
Cryptonemiales	*Amphiroa beauvoisii*	chol	24
	A. rigida	chol	25
	Callophyllus adhaerens	chola	17
	Carpopeltis cornea	chol	17
	C. divaricata	chol	17
	Corallina granifera	chol, desmo	25, 31
	C. mediterranea	chol	25
	C. officinalis	chol	28
	Cyrtymenia sparsa	chol	81
	Dilsea earnosa	chol	28
	Dumontia incrassata	chol	46
	Gloiopeltis fureata	chol	81
	G. tenax	chol	46
	Goniolithon byssoides	chol	25
	Grateloupia elliptica	chol	81
	G. proteus	chol	24
	Halymenia floresia	chol	25
	Jania rubens	chol, desmo	25
	Polyides caprinus	chol	28
	P. rotundus	chol	36, 46
	Prionitis decipiens	chol	46
	Schimmelmannia schousboea	chol, chol-7	25
	Thureteila schousboei	chol	25
	Tichocarpus crinitus	chol	81
Gigartinales	*Ahnfeltia durvilaei*	chol	46
	A. parakoxa	chol	17
	A. stellata	chol	28
	Caulacanthus ustulatus	chol	24
	Chondrus crispus	chol	4, 67
	C. giganteus	chol	81
	C. ocellatus	chol	81
	Eucheuma cottonii	chol	46
	E. isiforme	chol	46
	E. uncinatum	chol	46
	Furcellaria fastigata	chol, 24M	28, 37, 66

Table 1 (continued)
STEROLS OF RED ALGAE (RHODOPHYTA)

Order	Organism	Major sterols[a]	Ref.
	Gigartina acicularis	desmo, chol	24
	G. chamissoi	chol	33
	G. sckottsbergii	chol	46
	G. stellata	chol	28
	G. teedi	chol, desmo	24
	G. tenella	chol	46
	Gracilaria lemanaeformis	chol	33
	G. texorii	chol	74
	G. verrucosa	chol	23, 25
	Gymnogongrus divaricatus	chol	17
	G. flabelliformis	chol	17
	G. griffithsiae	chol	24
	Hypnea isiformis	chol	46
	H. japonica	22DC	17, 77
	H. musciformis	22DC, chol	24, 46
	Iridaea boryana	chol	46
	I. ciliata	chol	33
	I. laminarioides	chol	33, 34
	Iridophycus cornucopiae	chol	81
	Meristhotheca papulosa	chol	17
	Nemastoma dicotoma	chol	24
	Petroglossum nicaeense	chol	25
	Phyllophora membranifolia	chol	37
	Plocamium coccineum	chol, desmo	25
	P. telfairiae	chol	17
	P. vulgare	chol	28
	Rhodoglossum polchrum	chol	79
	Schizymenia dubyi	desmo, chol	24
	Sphaerococcus coronopifolius	22DC	25
Rhodymeniales	*Coeloseira pacifica*	chol	81
	Halosaccion ramentaceum	chol, desmo	36
	Rhodymenia palmata	desmo, chol	4, 28
Ceramiales	*Acanthophora najadiformis*	chol	24
	Alsidium corrallinum	chol	24
	A. helminthocorton	chol	25
	Caloglossa leprieurii	chol	23
	Centroceras clavulatum	chol	24
	Ceramium ciliatum	chol, desmo	25
	C. rubrum	chol, desmo	23, 25
	Chondria coerulescens	chol	25
	C. dasyphylla	chol	23
	Dasya pedicellata	chol	23
	Grinnellia americana	chol	23
	Halopitys incurvus	chol, desmo	25
	Laurencia obtusa	chol	24
	L. paniculata	chol	24
	L. papillosa	chol	1
	L. pinnatifida	chol	28
	L. undulata	chol, chol-7, 24M	25
	Lophocladia lallemandi	chol	24
	Nitophyllum puncatatium	chol, desmo	25
	Polysiphonia lanosa (fastigata)	chol	23
	P. nigrescens	chol	28
	P. subtilissima	chol	23
	P. subulata	chol	25
	Ptilota serrata	chol	37

Table 1 (continued)
STEROLS OF RED ALGAE (RHODOPHYTA)

Order	Organism	Major sterols[a]	Ref.
	Rhodomelia conferoides	chol	4, 36
	R. larix	chol	81
	Rytiphloea tinctoria	24 M, 24 MC, chol	3, 25
	Spyridia filamentosa	chol	25
	Vidalia volubilis	24 MC, chol	25
	Wrangelia penicillata	chol	25

[a] Chol = cholesterol; chola = cholestanol; erg = ergosterol; 22 DC = 22-dehydrocholesterol; desmo = desmosterol; liago = liagosterol; chol-7 = cholest-7-enol; 24 M = 24-methyl cholesterol; 24MC = 24-methylene cholesterol.

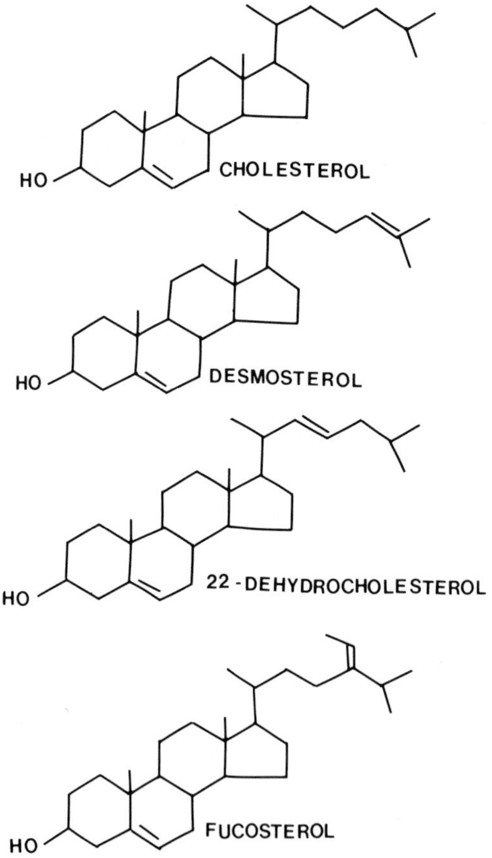

HO CHOLESTEROL

HO DESMOSTEROL

HO 22-DEHYDROCHOLESTEROL

HO FUCOSTEROL

FIGURE 1. Characteristic sterols of red algae and brown algae.

Table 2
STEROLS OF BROWN ALGAE (PHAEOPHYTA)

Order	Organism	Major sterols[a]	Ref.
Ectocarpales	*Pylaiella littoralis*	fuco	32
	Spongonema tomentosum (Ectocarpus spongiosus)	fuco	32
Sphacelariales	*Cladostephus spongiosus*	fuco	32
	C. verticillatus	fuco, chol	32
	Helopteris filicina	fuco, chol	5
	H. scoparia	chol, fuco	5
	Sphacelaria pennata (cirrosa)	fuco	32
Dictyotales	*Dictyopteris divaricata*	fuco	38, 39
	D. membranacea	fuco	5
	Dictyota dichotoma	fuco	32
	D. dichotoma var. *implexa*	fuco	5
	Padina arborescens	fuco	55
Chordariales	*Heterochordaria abietina*	fuco	55
Dictysiphonales	*Myelophycus caespitosus*	fuco	81
Laminariales	*Alaria crassifolia*	fuco	38, 81
	Chorda filum	fuco	32
	Costaria costata	fuco, 24MC	78
	Eisenta bicyclis	fuco	78
	Laminaria angustata	fuco	55
	L. digitata	fuco,24MC	32,53
	L. faeroensis	fuco, 24MC	53
	L. Hyperborea (cloustonii)	fuco	55
	L. japonica	fuco	55
	L. saccharina	fuco	55, 66
Fucales	*Ascophyllum nodosum*	fuco	32, 42
	Cystophyllum hakodatense	fuco	81
	Cystoseira adriatica	fuco	5
	C. corniculata	fuco	30
	C. crinata	fuco	5
	C. fimbricata	fuco	5
	C. stricta	fuco	5
	Fucus gardneri	fuco	61
	F. evanescens	fuco, 24MC	38, 78, 81
	F. divarcarpus	fuco, 24MC	18
	F. ceranoides	fuco	32
	F. serratus	fuco	55
	F. spiralis	fuco	55
	F. vesiculosis	fuco	55
	Halidrys siliquosa	fuco	32
	Macrocystis pyrifera	fuco	69
	Pelvetia canaliculata	fuco	47, 55
	P. wrightii	fuco	78
	Sargassum confuseum	fuco	38
	S. fluitans	fuco, 24MC	71
	S. muticum	fuco	61
	S. ringgoldianum	fuco, 24MC	38
	S. thunbergii	fuco, 24MC	38
	S. vulgare	fuco	5

[a] Chol = cholesterol; fuco = fucosterol; 24MC = 24-methylene cholesterol.

Table 3
STEROLS OF GREEN ALGAE (CHLOROPHYTA)

Order	Organism	Major sterols[a]	Ref.
Volvocales	*Chlamydomonas reinhardii*	porif-5,7,22; erg	9, 57
	Haematococcus pluvialis	erg	82
	Polytoma uvella	erg	82
Chorosphaerales	*Coccomyya elongata*	erg-5, cliona, porif	57
Chlorococcales	*Ankistrodesmus braunii*	chond, chond-7, erg-7	84
	Bracteacoccus cinnabarinus	chond, erg-7, chond-7	84
	Chlorella candida	erg	54
	C. ellipsoidea (TU #246)	erg-5, 8,22; erg; erg-5, 8	60
	C. ellipsoidea (TU #247)	porif, erg-5, cliona	59
	C. emersonii	chond, erg-7	57
	C. fusca	chond, erg-7	57
	C. glucotropha	chond	57
	C. miniata	chond	57
	C. nocturna	erg	54
	C. pringsheimii	porif, erg-5	57
	C. protothecoides var. *mannophila*	erg	55
	C. protothecoides var. *communis*	erg	55
	C. saccharophila	porif, erg-5	59
	C. simplex	erg	54
	C. sorokiniana	erg	54
	C. vulgaris	chond, erg-7	59
	Chlorococcum echinozygotum	erg-7, erg-7,22	84
	Hydrodictyon reticulatum	spin (?)	35
	Neochloris aquatica	chond, erg-7	84
	Oocystis marsonii	porif, erg-5	84
	O. polymorpha	chond, chond-7, erg-7	49
	Prototheca zopfii	erg	82
	Scenedesmus obliquus	chond, erg-7	13, 40
	S. quadricauda	chond, erg-7	51, 84
	Selensastrum gracile	chond-7, erg-7, chond	51
	Trebouxia decolorans	porif, erg-5, cliona	43
	T. sp.	porif, erg-5, cliona	43
Ulotrichales	*Ulotrix flacca*	24MC	22
	Uronema gigas	chond, erg-7, chond-7	51
	U. terrestre	chond, erg-7	51
Ulvales	*Blidingia minima*	24MC	22
	Enteromorpha intestinalis	28-iso	22, 23, 29
	E. linza	28-iso	22, 76
	E. plumosa	28-iso	23
	Monostroma nitidum	hali (?)	76
	M. zostericola	28-iso, 24MC	22
	Ulva fenestrata	28-iso	22
	U. lactuca	28-iso	23, 29
	U. pertusa	chol	38
Prasiolales	*Prasiola meridionalis*	24MC	22
Cladophorales	*Chaetomorpha crassa*	chol, 24MC	38
	C. linum	24EC	22
	C. capillaris	24MC, 24EC	22
	Cladophora flexuosa	28-iso, chol, 24M	57
	C. ruperstris	28-iso, 24MC	22
	Pithophora sp.	24M, chol, 28-iso	57
	Spongomorpha coalita	24MC	22
Codiales	*Bryopsis corticulans*	24EC	22
	B. plumosa	24EC	22
	Codium fragile	clero	64

	C. fragile spp. *tomentosoides*	clero	22
	C. fragile spp. *atlanticum*	clero	22
	C. vermilara	clero	22
	Halimeda incrassata	cliona	57
Zygnematales	*Spirogyra* sp.	cliona, porif	57

[a] Erg = ergosterol; erg-5 = ergost-5-en-3β-ol; porif-5,7,22 = 7-dehydroporiferasterol; cliona = clionasterol; porif = poriferasterol; chond = chondrillasterol; chond-7 = 24β-ethyl cholest-7-en-3β-ol; erg-5, 8,22 = ergosta-5,8,22-trien-3β-ol; erg-5,8 = ergosta-5,7-dien-3β-ol; erg-7,22 = ergosta 7,22-dien-3β-ol; spin = α-spinasterol; 24MC = 24-methylene cholesterol; 28-iso = 28-isofucosterol; 24M = 24-methyl cholesterol; hali = haliclonasterol; chol = cholesterol; 24EC = 24-ethyl cholesterol; clero = clerosterol.

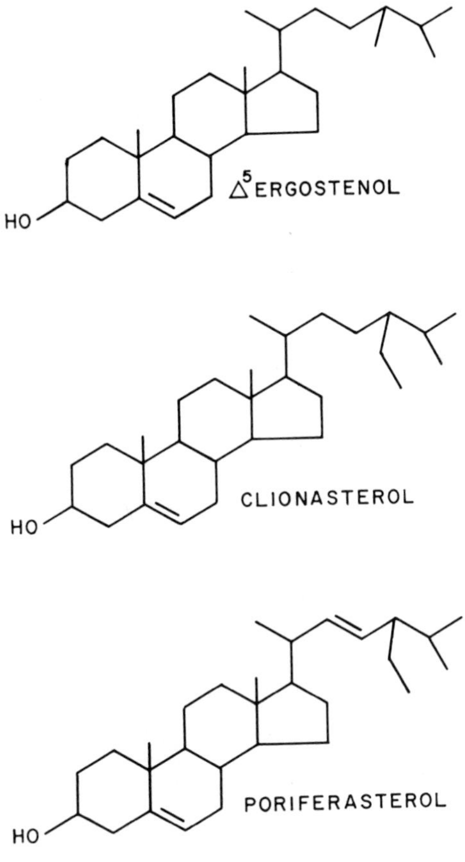

FIGURE 2. Δ^5-Sterols of green algae.

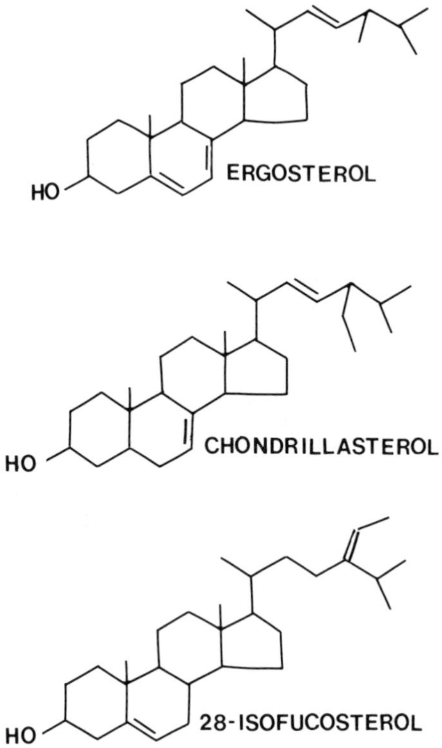

ERGOSTEROL

CHONDRILLASTEROL

28-ISOFUCOSTEROL

FIGURE 3. Principal sterols of green algae.

Table 4
STEROLS OF CHRYSOPHYTA

Class	Organism	Major sterols[a]	Ref.
Chrysophyceae	*Ochromonas danica*	chol, sit (?)	19
	O. malhamensis	porif, brass	27
	O. sociabilis	porif	27
	Synura petersenii	stig (?)	8
Haptophyceae	*Isochrysis galbana*	porif, cliona, erg-5	72
	Pavlova lutheri	porif, cliona, erg-5	72
Xanthophyceae	*Botrydium granulatum*	cliona, chol	45
	Monodus subterranous	cliona, chol	45
	Tribonema aequale	cliona, chol	45
Bacillariophyceae	*Amphora exigua*	stig, chol	50
	A. sp.	stig	50
	Biddulphia aurita	erg-8(9), erg-5,22,erg-5,erg-7,22,fuco	50
	Chaetocerous simplex calcitrans	chol, 24MC	15
	Cyclotella nana[b]	brass (?)	41
	Fragillaria sp.	erg-7,22; erg-5, erg-7,24(29)	50
	Navicula pelliculosa	camp-5,22	50
	Nitzschia alba	brass (?), 24EC	75
	N. closterium	camp-5,22	41, 50
	N. frustulum	camp-5,22; chol	50
	N. longissima	chol	50
	N. ovalis	camp-5,22; 22DC; chol	50
	Phaeodactylum tricornutum	camp-5,22	50, 65
	Thallasiosira pseudonana[b]	erg-7,22; erg-5; erg-7,24(28)	50

[a] Chol = cholesterol; sit = sitosterol; porif = poriferasterol; brass = brassicasterol; stig = stigmasterol; cliona = clionasterol; erg-5 = ergost-5-en-3β-ol; erg-8(9) = ergost-8(9)-en-3β-ol; erg-5,22 = ergosta-5,22-dien-3β-ol; erg-7,22 = ergosta-7,22-dien-3β-ol; fuco = fucosterol; 24MC = 24-methylene cholesterol; erg-7,24(28) = ergosta-7,24(28)-dien-3β-ol; camp-5,22 = 22-dehydrocampesterol; 24EC = 24-ethyl cholesterol; 22DC = 22-dehydrocholesterol.

[b] *Thallasiosira pseudonana* is synonymous with the name *Cyclotella nana* and replaces it in the newer literature.

Table 5
STEROLS OF OTHER ALGAE

Division	Organism	Major sterols[a]	Ref.
Cyanophyta	*Anabaena cylindrica*	brass[b]	73
	Anacystis nidulans	chol, 24EC	62
	Calothrix sp.	chond, brass, 28-iso, chol	51
	Cyanidium caldarium	erg, sit	68
	Fremyella diplosiphon	chol, 24EC	62
	Nostoc commune	chond, erg-7, chond-7, brass, chol	51
	N. muscorum	chol	58
	Phormidium luridum	chond-7, chond	21
	Spirulina maxima	chol, sit	48
	S. platensis	cliona, 28-iso chond-7	51
	S. platensis	chol	26
	S. sp.	cliona, chol	51
Euglenophyta	*Astasia longa*	chol, cliona	63
	A. ocellata	erg	82
	Euglena gracilis	erg, porif-5,7,22	6, 14
	E. gracilis (white)	erg-7, cliona, chol	14
	E. gracilis (green)	chond, erg-7, erg-7,22, chond-7	14
	Peranema tricophorum	erg	82
Pyrrophyta	*Cryptothecodinium cohnii*	dino, dedino	83
	Gonyaulax tamarensis	dino, chol	70
	Pyrocystis lunula	chol	7
Charophyta	*Chara vulgaris*	28-iso, cliona	56
	Nitella flexilis	cliona, 28-iso	56

[a] Brass = brassicasterol; chol = cholesterol; 24EC = 24-ethyl cholesterol; chond = chondrillasterol; 28-iso = 28-isofucosterol; erg = ergosterol; sit = sitosterol; erg-7 = ergost-7-en-3β-ol; chond-7 = 24-β-ethyl cholest-7-en-3β-ol; cliona = clionasterol; porif-5,7,22 = 7-dehydroporiferasterol; erg-7,22 = ergosta-7,22-dien-3β-ol; dino = dinosterol; dedino = dehydrodinosterol.

[b] Although identifications of sterols such as brassicasterol, sitosterol, chondrillasterol, etc. have been made from blue-green algae, in no case has the configuration at C-24 been determined.

REFERENCES

1. **Abdel-Fattah, A. F., Abed, N. M., and Edrees, M.,** Influence of seasonal changes on the constituents of the red seaweed *Laurencia papillosa, Qual. Plant. Mater. Veg.,* XXII(2), 171, 1973.
2. **Abdel-Fattah, A. F., Abed, N. M., and Edreeds, M.,** Seasonal variation in the chemical composition of the agarophyte *Peterocladia capillacea, Aust. J. Mar. Freshwater Res.,* 24, 177, 1973.
3. **Alcaide, A., Barbier, M., Potier, P., Magueur, A. M., and Teste, J.,** Nouveau resultats sur les sterols des algues rouges, *Phytochemistry,* 8, 2301, 1969.
4. **Alcaide, A., Devys, M., and Barbier, M.,** Remarques sur les sterols des algues rouges, *Phytochemistry,* 7, 329, 1968.
5. **Amico, V., Oriente, G., Piatelli, M., Tringali, C., Fattorusso, E., Magno, S., Mayol, L., Santracroce, C., and Sica, D.,** Amino acids, sugars and sterols of some Mediterranean brown algae, *Biochem. Syst.,* 4, 143, 1976.
6. **Anding, C. and Ourisson, G.,** Presence of ergosterol in light-grown and dark-grown *Euglena gracilis, Eur. J. Biochem.,* 34, 345, 1973.
7. **Ando, T. and Barbier, M.,** Sterols from the culture medium of the luminescent dinoflagellate *Pyrocystis lunula, Biochem. Syst.,* 3, 245, 1975.
8. **Avivi, L., Iaron, O., and Halevy, S.,** Sterols of some algae, *Comp. Biochem. Physiol.,* 21, 321, 1967.
9. **Bard, M. and Wilson, K. J.,** Isolation of sterol mutants in *Chlamydomonas reinhardi:* chromatographic analyses, *Lipids,* 13, 533, 1978.
10. **Beastall, G. H., Tyndall, A. M., Rees, H. H., and Goodwin, T. W.,** Sterols in *Porphyridium* species, *Eur. J. Biochem.,* 41, 301, 1974.
11. **Beastall, G. H., Rees, H. H., and Goodwin, T. W.,** Sterols in *Porphyridium cruentum, Tetrahedron Lett.,* 4935, 1971.
12. **Berg, L.,** Fatty Acid and Sterol Composition of *Chlorella emersonii, Chlorella ellipsoidea* and *Chlorella sorokiniana* Grown Autotrophically, Photoheterotrophically and Heterotrophically, M. S. thesis, University of Maryland, College Park, 1976.
13. **Bergmann, W. and Feeney, R. J.,** Sterols of algae. I. The occurrence of chondrillasterol in *Scenedesmus obliquis, J. Org. Chem.,* 15, 812, 1950.
14. **Brandt, R. D., Pryce, R. J., Anding, C., and Ourisson, G.,** Sterol biosynthesis in *Euglena gracilis* Z. (Comparative study of free and found sterols in light and dark-grown *Euglena gracilis* Z.), *Eur. J. Biochem.,* 17, 344, 1970.
15. **Boutry, J. L., Barbier, M., and Ricard, M.,** La Diatomẽe *Chaetoceros simplex* calcitrans Paulson et son environment. II. Effets de la luminiere et des irradiations ultraviolet sur la production primaire et la biosynthese des sterols, *J. Exp. Mar. Biol. Ecol.,* 21, 69, 1976.
16. **Cargile, N. L., Edwards, H. N., and McChesney, J. D.,** Sterols of *Porphyridium* (Rhodophyta). *J. Phycol.,* 11, 457, 1975.
17. **Chardon-Loriaux, I., Morisaki, M., and Ikekawa, N.,** Sterol profiles of red algae, *Phytochemistry,* 15, 723, 1976.
18. **Ciereszko, L. S., Johnson, M. A., Schmidt, R. W., and Koons, C. B.,** Chemistry of coelenterates. VI. Occurrence of gorgosterol, a C_{30} sterol, in coelenterates and their zooxanthellae, *Comp. Biochem. Physiol.,* 24, 899, 1968.
19. **Collins, R. P. and Kalnins, K.,** Sterols produced by *Synura petersenii* (Chrysophyta), *Comp. Biochem. Physiol.,* 30, 779, 1969.
20. **Cook, C. H. and Cho, Y. S.,** Study on sterol component in purple laver, *Seoul Nat. Univ. Fac. Pap., Sci. Technol. Ser.,* 3, 43, 1974.
21. **DeSouza, N. J. and Nes, W. R.,** Sterols: isolation from a blue-green algae, *Science,* 162, 3636, 1968.
22. **Dickson, L. G., Patterson, G. W., and Knights, B. A.,** Distribution of sterols in marine chlorophyceae. *Proc. Int. Seaweed Symp.,* 9, 413, 1978.
23. **Doyle, P. J. and Patterson, G. W.,** Sterols of some Chesapeake Bay alga, *Comp. Biochem. Physiol.,* 413, 355, 1972.
24. **Fattorusso, E., Magno, S., Santacroce, C., Sica, D., Impellizzeri, G., Mangiafico, S., Oriente, G., Piatelli, M., and Sciuto, S.,** Sterols of some red algae, *Phytochemistry,* 14, 1579, 1975.
25. **Fattorusso, E., Magno, S., Santacroce, C., Sica, D., Impellizzeri, G., Mangiofico, S., Piatelli, M., and Sciuto, S.,** Sterols of Mediterranean Florideophyceae, *Biochem. Syst.,* 4, 135,1976.
26. **Forin, M. C., Maume, B., and Baron, C.,** Biochimie cellulair sur les sterols et alcools triterpeniques d'une cyanophycee *Spirulina platensis* Gertler, *C. R. Acad. Sci., Ser. D,* 274, 133, 1972.
27. **Gershengorn, M. C., Smith, A. R. H., Goulston, G., Goad, L. J., Goodwin, T. W., and Haines, T. H.,** The sterols of *Ochromonas danica* and *Ochromonas malhamensis, Biochemistry,* 7, 1706. 1968.
28. **Gibbons, G. F., Goad, L. J., and Goodwin, T. W.,** The sterols of some marine red algae, *Phytochemistry,* 6, 677, 1967.

29. **Gibbons, G. F., Goad, L. J., and Goodwin, T. W.,** The identification of 28-isofucosterol in the marine green algae *Enteromorpha intestinalis* and *Ulva lactuca, Phytochemistry,* 7, 983, 1968.
30. **Guven, K. and Hakyemez, G.,** Sterol of *Cystoseira corniculata* hauck, *Eczacilik Bul.,* 17, 94, 1975.
31. **Guven, K. C. and Bergisadi, N.,** Sterol of *Corallina granifera, Eczacilik Bul.,* 17, 30, 1975.
32. **Heilbron, I. M.,** Some aspects of algal chemistry, *J. Chem. Soc.,* 79, 1942.
33. **Henriquez, P., Trucco, R., and Silva, M.,** Chemical composition of Chilean marine algae. II. *Bot. Mar.,* 15, 117, 1972.
34. **Henriquez, P., Trucco, R., Silva, M., and Sammes, P. G.,** Cholesterol in *Iridaea laminariodes* and *Gracilaria verrucosa, Phytochemistry,* 11, 1171, 1972.
35. **Hunek, S.,** Spinasterin in *Hydrodictyon reticulatum, Phytochemistry,* 8, 1313, 1969.
36. **Idler, D. R., Saito, A., and Wiseman, P.,** Sterols in red algae (Rhodophyceae), *Steroids,* 11, 465, 1968.
37. **Idler, D. R. and Wiseman, P.,** Sterols in red algae (Rhodophyceae): variation in the desmosterol content of dulca *(Rhodymenia palmata), Comp. Biochem. Physiol.,* 35, 679, 1970.
38. **Ikekawa, N., Morisaki, N., Tsuda, K., and Yashida, T.,** Sterol compositions in some green algae and brown algae, *Steroids,* 12, 41, 1968.
39. **Ikekawa, N., Tsuda, K., and Morisake, N.,** Saringosterol: a new sterol from brown algae, *Chem. Ind. (London),* 1179, 1966.
40. **Iwata, J., Nakata, H., Mizuschima, M., and Sakurai, Y.,** Lipids of algae. Part I. The components of unsaponifiable matter of the algae *Scenedesmus, Agric. Biol. Chem.,* 25, 319, 1961.
41. **Kanazawa, A., Yoshioka, M., and Teshima, S.,** The occurrence of brassicasterol in the diatoms *Cyclotella nana* and *Nitzschia closterium, Bull. Jpn. Soc. Sci. Fish,* 37, 899, 1971.
42. **Knights, B. A.,** Sterols in *Ascophyllum nodosum, Phytochemistry,* 9, 903, 1970.
43. **Lenton, J. R., Goad, L. J., and Goodwin, T. W.,** Sterols of the mycobiont and phycobiont isolated from the lichen *Xanthoria parietina, Phytochemistry,* 12, 2249, 1973.
44. **Levin, E. Y. and Bloch, K.,** Absence of sterols in blue-green algae, *Nature (London),* 202, 90, 1964.
45. **Mercer, E. I., London, R. A., Kent, I. S. A., and Taylor, A. J.,** Sterols, sterol esters and fatty acids of *Botrydium granulatum, Tribonema aequale* and *Monodus subterraneus, Phytochemistry,* 13, 845, 1974.
46. **Meunier, H., Zelenski, S., and Warthen, L.,** Comparison of the sterol content of certain Rhodophyta, in *Proc. Food-Drugs from the Sea,* Youngken, H. W., Jr., Ed., Marine Technology Society, Washington, D.C., 1969, 319.
47. **Motzfeldt, A. M.,** Isolation of 24-oxocholoesterol from the marine brown algae *Pelvetia canaliculata* (Phaeophyceae), *Acta Chem. Scand.,* 24, 1846, 1970.
48. **Nadal, N. G. M.,** Sterols of *Spirulina maxima, Phytochemistry,* 10, 2537, 1971.
49. **Orcutt, D. M. and Richardson, B.,** Sterols of *Oocystis polymorpha,* a green alga, *Steroids,* 16, 429, 1970.
50. **Orcutt, D. M. and Patterson, G. W.,** Sterol, fatty acid and elemental composition of diatoms grown in chemically defined media, *Comp. Biochem. Physiol.,* 50B, 579, 1975.
51. **Paoletti, C., Pushparaj, B., Florenzano, G., Capella, P., and Lercker, G.,** Unsaponifiable matter of green and blue-green algae lipids as a factor of biochemical differentiation of their biomasses. II. Terpenic alcohol and sterol fractions, *Lipids,* 11, 266, 1976.
52. **Patterson, G. W.,** Sterols of Chlorella. II. The occurrence of an unusual sterol mixture in *Chlorella vulgaris, Plant Physiol.,* 42, 1457, 1967.
53. **Patterson, G. W.,** Sterols of *Laminaria, Comp. Biochem. Physiol.,* 24, 501, 1968.
54. **Patterson, G. W.,** Sterols of *Chlorella.* III. Species containing ergosterol, *Comp. Biochem. Physiol.,* 3, 391, 1969.
55. **Patterson, G. W.,** The distribution of sterols in algae, *Lipids,* 6, 120, 1971.
56. **Patterson, G. W.,** Sterols of *Nitella flexilis* and *Chara vulgaris, Phytochemistry,* 11, 3481, 1972.
57. **Patterson, G. W.,** Sterols of some green algae, *Comp. Biochem. Physiol.,* 47B, 453, 1974.
58. **Patterson, G. W. and Southall, A.,** unpublished.
59. **Patterson, G. W. and Krauss, R. W.,** Sterols of *Chlorella.* I. The naturally occurring sterols of *Chlorella vulgaris, Chlorella ellipsoidea, Plant Cell Physiol.,* 6, 211, 1965.
60. **Patterson, G. W., Thompson, M. J., and Dutky, S. R.,** Two new sterols from *Chlorella ellipsoidea, Phytochemistry,* 13, 191, 1974.
61. **Reiner, E., Topliff, J., and Wood, J. D.,** Hypocholestereolemic agents derived from sterols of marine algae, *Can. J. Biochem. Physiol.,* 40, 1401, 1962.
62. **Reitz, R. C. and Hamilton, J. G.,** The isolation and identification of two sterols from two species of blue-green algae, *Comp. Biochem. Physiol.,* 25, 401, 1968.
63. **Rohmer, M. and Brandt, R. D.,** Les sterols et leurs precursors chez *Astasia longa* Prengsheim. *Eur. J. Biochem.,* 36, 446, 1973.

64. **Rubenstein, I. and Goad, L. J.**, Sterols of the siphonous green alga, *Codium fragile, Phytochemistry,* 13, 481, 1974.
65. **Rubenstein, I. and Goad, L. J.**, Occurrence of (24*S*)-24-methyl-cholest-5, 22-E-dien-3β-ol in the diatom *Phaeodactylum tricornutum, Phytochemistry,* 13, 485, 1974.
66. **Safe, L. M., Wong, C. J., and Chandler, R. F.**, Sterols of marine algae, *J. Pharm. Sci.*, 63, 464, 1974.
67. **Saito, A. and Idler, D. R.**, Sterols in Irish moss *(Chondrus crispus), Can. J. Biochem.*, 44, 1195, 1966.
68. **Sechbach, J. and Ikan, R.**, Sterols and chloroplast structure of *Cyanidium caldarium, Plant Physiol.*, 49, 457, 1972.
69. **Shimadate, T., Rosenstein, F. U., and Kircher, H. W.**, Preparation of fucosterol from giant kelp, *Lipids,* 12, 241, 1977.
70. **Shimizu, Y., Alam, M., and Kobayashi, A.**, Dinosterol, the major sterol with a unique side chain in the toxic dinoflagelllate, *Gonyaulax tamarensis, J. Am. Chem. Soc.,* 98, 1059, 1976.
71. **Smith, L. L., Dhar, A. K., Gilchrist, J. L., and Lin, Y. Y.**, Sterols of the brown algae *Sargassum fluitans, Phytochemistry,* 12, 2727, 1973.
72. **Taylor, A.**, Sterol Composition of Two Haptophycean Algae, *Isochrysis galbana* and *Pavlova lutheri.* M.S. thesis, University of Maryland, College Park, 1977.
73. **Teshima, S. and Kanazawa, A.**, Occurrence of sterols in the blue-green alga, *Anabaena cylindrica, Bull. Jpn. Soc. Sci. Fish,* 38, 1197, 1972.
74. **Teshima, S. and Kanazawa, A.**, Biosynthesis of sterols in the red alga, *Porphyridium cruentum, Mem. Fac. Fish., Kagoshima Univ.,* 22, 1, 1973.
75. **Thornabene, T. G., Kates, M., and Volcani, B. E.**, Sterols, aliphatic hydrocarbons, and fatty acids of a nonphotosynthetic diatom, *Nitzschia alba, Lipids,* 9, 279, 1974.
76. **Tsuda, K. and Sakai, K.**, Steroids studies. XV. Sterols of green sea algae, *Chem. Pharm. Bull. (Tokyo),* 8, 554, 1960.
77. **Tsuda, K., Sakai, K., Tanabe, K., and Kishida, Y.**, Steroid studies. XVI. Isolation of 22-dehydrocholesterol from *Hypnea japonica, J. Am. Chem. Soc.,* 82, 1442, 1960.
78. **Tsuda, K., Hayatsu, R., Kishida, Y., and Akagi, S.**, Studies of the constitution of sargasterol, *J. Amer. Chem. Soc.,* 80, 921, 1958.
79. **Tsuda, K., Akagi, S., and Kishida, Y.**, Discovery of cholesterol in some red algae, *Science,* 126, 927, 1957.
80. **Tsuda, K., Akagi, S., and Kishida, Y.**, Cholesterol in some red algae, *Pharm. Bull. (Tokyo),* 6, 101, 1958.
81. **Tsuda, K., Akagi, S., Kishida, Y., Hayatsu, R., and Sakai, K.**, Sterols from ocean algae, *Pharm. Bull. (Tokyo),* 6, 724, 1958.
82. **Williams, B. L., Goodwin, T. W., and Ryley, J. F.**, The sterol content of some protozoa, *J. Protozool.,* 13, 227, 1966.
83. **Withers, N. W., Tuttle, R. C., Holz, G. G., Beach, D. H., Goad, L. J., and Goodwin, T. W.**, Dehydrodinosterol, dinosterone, and related sterols of a non-photosynthetic dinoflagellate *Cryptothecodium cohnii, Phytochemistry,* 17, p. 1987, 1978.
84. **Wright, D. C.**, Fatty Acid and Sterol Compositions of Six Green Algae Grown Autotrophically, Photoheterotrophically, and Heterotrophically, M.S. thesis, University of Maryland, College Park, 1978.

SYNTHESIS OF STEROIDS IN VASCULAR PLANTS

Waldemar Eichenberger

Several reviews covering the biosynthesis of plant sterols have been published in recent years.[1-6] Vascular plants (tracheophytes) comprise (a) pteridophytes, subdivided into Psilopsida, Lycopsida, Sphenopsida, and Filicopsida, and (b) spermatophytes, subdivided into gymnosperms and angiosperms. Sterols of spermatophytes have been investigated much more intensively than those of pteridophytes. There is good evidence that all tracheophytes, without exception, contain sterols.

STRUCTURE AND DISTRIBUTION

Plant sterols are characterized by a cyclopentanophenanthrene nucleus containing rings A to D, CH_3-groups at C-10 and C-13, a β-OH group at C-3, and a side chain at C-17 (Formula 1). In most cases, the nucleus contains a double bond localized in the B-ring. The side chain contains an additional C_1 or C_2 group at C-24, the configuration of which is α. A β-configuration at C-24 occurs only in 24-methylene compounds and $\Delta^{25(27)}$ compounds. Sterols with additional CH_3-groups at C-4 and C-14 also occur as minor components.

Quantitatively, sitosterol (XXI), stigmasterol (XXII), campesterol (XIII), and spinasterol (XX) are the most important sterols of higher plants. The first three sterols belong to the Δ^5-type, while the last one is a Δ^7-type.

The majority of vascular plants produce Δ^5-sterols. Usually, sitosterol, although also rarely stigmasterol,[7] predominate. A smaller group of plants comprising the Cucurbitaceae and Theacaea, as well as certain plants of the Chenopodiaceae and other families, contain Δ^7-sterols exclusively.

Particular plants contain ergosterol (24β-methylcholesta-5,7,22-trien-3β-ol), which predominates in *Lycopodium*.[8] Evidently, cholesterol (XII) is generally distributed, since it has been found in representatives of at least 35 families. Its relative amount in most cases accounts for less than 5% of the total sterols. Higher amounts seem to be exceptions.[9-11] Besides free sterols, steryl esters, glycosides, and monoacyl glycosides have been found regularly.[2,12]

Sterol content can vary considerably. The limits may be generalized as indicated by the following values: 0.01 to 0.1% of the fresh weight[1] or 0.1 to 1% of the dry weight.[1] Sterol compounds (including also esters, glycosides, acyl glycosides) may comprise 1 to 10% of the total lipid,[13] or 0.05 to 1% of the dry weight.[14,15]

BIOSYNTHESIS

The simplest structural precursor is acetate, and important intermediates are mevalonate (MVA), isopentenylpyrophosphate (IpPP) and squalene. The whole sequence may be divided into four steps: (1) formation of isopentenyl pyrophosphate (IpPP), (2) formation of squalene, (3) cyclization of squalene, and (4) secondary transformations. The most important steps leading to the C_5-precursor (IpPP) are described in Figure 1. MVA is phosphorylated twice to give MVA-5-P and MVA-5-PP, which after an intermediate (third) phosphorylation undergoes decarboxylation to produce IpPP.

The formation of squalene from IpPP is summarized in Figure 2. The chain elongation includes an isomerization of Δ^3-IpPP to Δ^2-IpPP and the intermediary formation of geranyl pyrophosphate (GePP) and farnesyl pyrophosphate (FaPP). The dimerization of FaPP leads to (cyclic) presqualene pyrophosphate[16] which undergoes reduction to form squalene.

R = H : sterol
R = acyl : steryl ester
R = hexose : steryl glycoside
R = 6-acylhexose : steryl acyl glycoside

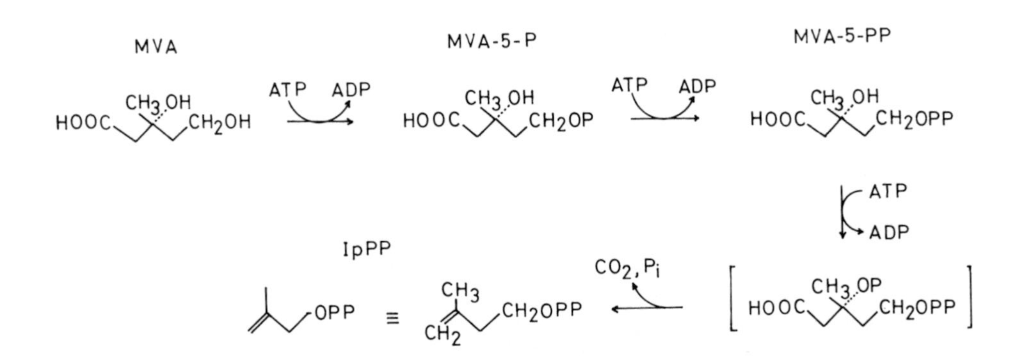

FIGURE 1. Formation of isopentenylpyrophosphate (IpPP) from MVA.

Squalene → Squalene-2,3-oxide → Cycloartenol (I)

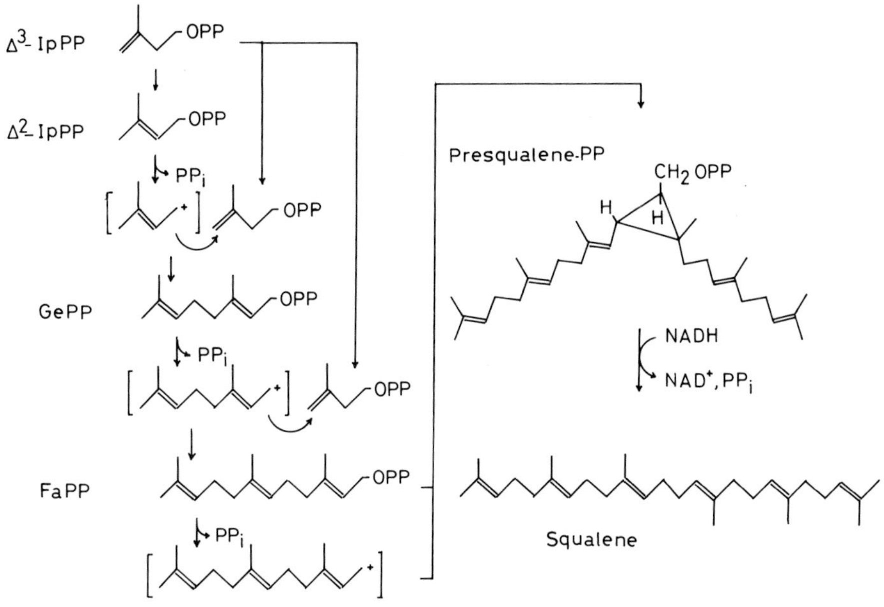

FIGURE 2. Formation of squalene.

The formation of IpPP and squalene proceed in an identical manner in both plants and animals. This was shown by the labeling pattern observed in squalene[17] and in sitosterol[18] formed from 2-^{14}C-MVA by plants, as compared to the labeling pattern of squalene and cholesterol formed by animal systems. The cyclization occurs with squalene-2,3-oxide as substrate, giving cycloartenol (I) as a product in all autotrophic organisms.[1,2]

Secondary transformations comprise (a) alkylation of the side chain, (b) elimination of methyl groups from C-4 and C-14, (c) opening of the cyclopropane ring and transposition of the double bond in the nucleus, and (d) hydrogenation and dehydrogenation of the side chain.

Since these steps may occur simultaneously or successively, many ways exist leading to particular end products. The most probable routes are summarized in Figure 3.

The alkylation of the side chain implies a preexisting $\Delta^{24(25)}$ double bond.[19] To give a 24-ethyl group, two successive methylation steps are necessary. The first methylation may take place as early as on cycloartenol (I) or on any of the subsequent intermediates (III, V, X). For the second methylation, VIII acts as the substrate. However, transformations of XII to XIII or XXII have also been reported.[20] As summarized in Figure 4, a 24-methylene sterol (XXIV) is needed for the formation of a 24 α-configuration, while a 25-methylene compound (XXV) is used for the formation of the rarer 24β-configuration. The elimination of methyl groups from the nucleus follows the sequence α-C-4, C-14, β-C-4.

The ring double bond is shifted from its primary Δ^8 position to the Δ^7 position, and from there to the Δ^5 position with a $\Delta^{5,7}$ diene (XIX, XXIII) as an intermediate. The introduction of the Δ^{22} double bond (if at all) is one of the last steps of the sequence.

FUNCTION

The function of sterols in vascular plants is not very well known and many findings are controversial. However, possible roles have been discussed[2,21] and two aspects are important:

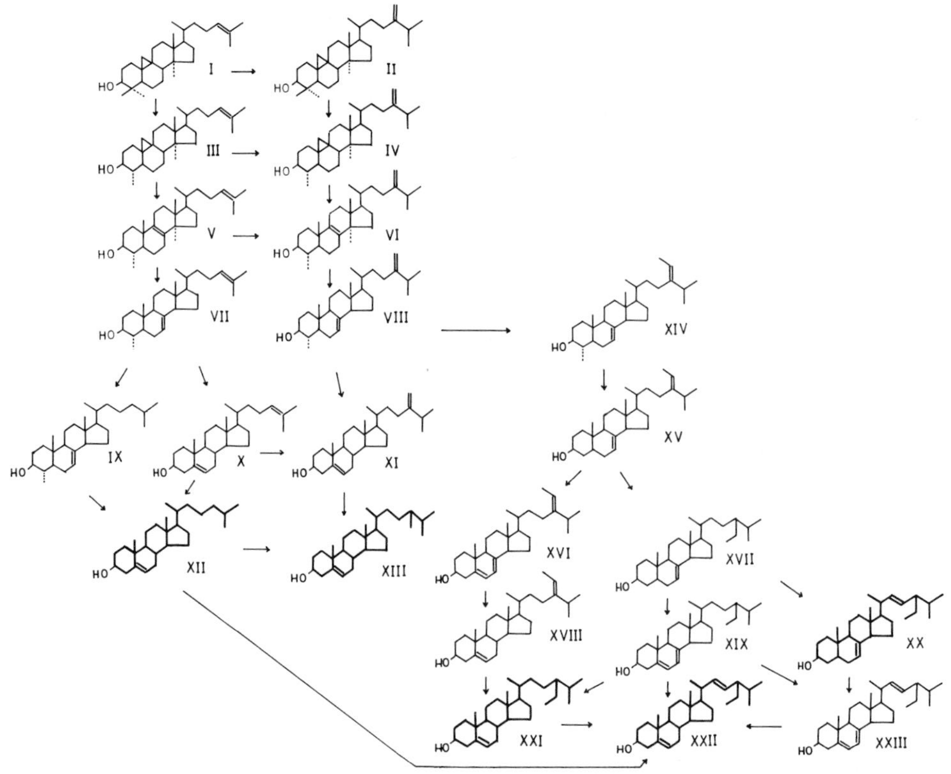

FIGURE 3. Formation of phytosterols from cycloartenol. I, cycloartenol; II, 24-methylenecycloartanol; III, 31-norcycloartenol; IV, cycloeucalenol; V, 31-norlanosterol; VI, obtusifoliol; VII, 24,25-dehydrolophenol; VIII, 24-methylenelophenol; IX, lophenol; X, desmosterol; XI, 24-methylenecholesterol; XII, cholesterol; XIII, campesterol; XIV, 24-ethylidenelophenol; XV, stigmasta-7,24(28)dien-3β-ol; XVI, stigmasta-5,7,24(28)trien-3β-ol; XVII, stigmasta-7-en-3β-ol; XVIII, avenasterol; XIX, stigmasta-5,7-dien-3β-ol; XX, spinasterol; XXI, sitosterol; XXII, stigmasterol; XXIII, stigmasta-5,7,22-trien-3β-ol.

1. As constituents of membranes. Evidence has been obtained mainly by alterations of permeability caused by free sterols, but not by esters or glycosides. Also, an increase in membrane material (as in seed germination) parallels the synthesis of sterols.
2. As hormones or precursors of hormones. Evidence has been obtained from the stimulation of growth by sterols and steryl glycosides.[22] Also sterols are transformed by a plant to other steroids, the biological significance of which has been confirmed.[21]

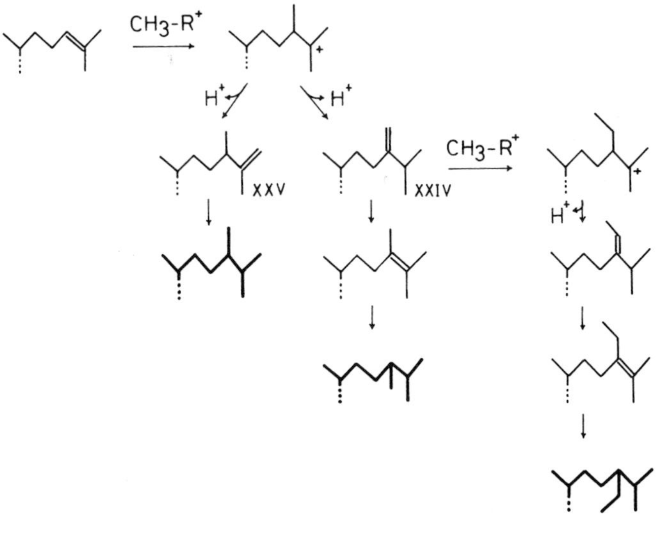

CH3-R+
H+
H+
XXV
XXIV
CH3-R+
H+

FIGURE 4. Alkylation of the sterol side chain.

REFERENCES

1. **Nes, W. R.,** The biochemistry of plant sterols, *Adv. Lipid Res.,* 15, 233, 1977.
2. **Grunwald, C.,** Plant sterols, *Annu. Rev. Plant Physiol.,* 26, 209, 1975.
3. **Goad, L. J.,** The biosynthesis of plant sterols, in *Lipids and Lipid Polymers in Higher Plants,* Tevini, M. and Lichtenthaler, H. K., Eds., Springer-Verlag, Berlin, 1977, chap. 8.
4. **Goad, L. J. and Goodwin, T. W.,** The biosynthesis of plant sterols, in *Progress in Phytochemistry,* Vol. 3, Reinhold, L. and Liwschitz, Y., Eds., Interscience, New York, 1972, 113.
5. **Bean, G. A.,** Phytosterols, *Adv. Lipid Res.,* 11, 193, 1973.
6. **Mulheirn, L. J. and Ramm, P. J.,** The biosynthesis of sterols, *Chem. Soc. Rev.,* 1, 259, 1972.
7. **Hillman, J. R., Knights, B. A., and McKail, R.,** Sterols of the adult and juvenile forms of ivy, *Hedera helix* L., *Lipids,* 10, 542, 1975.
8. **Nes, W. R., Krevitz, K., Behzadan, S., Patterson, G. W., Landrey, J. R., and Conner, R. L.,** The configuration of $\Delta^{5,7,22}$-sterols in a tracheophyte, *Biochem. Biophys. Res. Commun.,* 66, 1462, 1975.
9. **Cheng, A. L. S., Kasperbauer, M. J., and Rice, L. G.,** Sterol content and distribution in two *Nicotiana* species and their spontaneous-tumoring hybrid, *Phytochemistry,* 10, 1481, 1971.
10. **Ingram, D. S., Knights, B. A., McEvoy, I. J., and McKay, P.,** Studies in the Cruciferae. Changes in the composition of the sterol fraction following germination, *Phytochemistry,* 7, 1241, 1968.
11. **Eichenberger, W., Fürst, B., and Grob, E. C.,** Zum Vorkommen von Cholesterin in *Hosta undulata* (Liliaceen), *Chimia,* 26, 22, 1972.
12. **Eichenberger, W.,** Steryl glycosides and acylated steryl glycosides, in *Lipids and Lipid Polymers in Higher Plants,* Tevini, M. and Lichtenthaler, H. K., Eds., Springer-Verlag, Berlin, 1977, chap. 9.
13. **Eichenberger, W. and Grob, E. C.,** Ueber die quantitative Bestimmung von Sterinderivaten in Pflanzen und die intrazelluläre Verteilung der Steringlycoside in Blättern, *FEBS Lett.,* 11, 177, 1970.
14. **Méance, J. and Dupéron, R.,** Présence de glycosides stéroliques et de glycosides stéroliques acylés dans les semences. Evolution de ces substances au cours de la germination du radis. *C. R. Acad. Sci., Ser. D,* 277, 849, 1973.
15. **Katayama, M. and Katoh, M.,** Soybean sterols during maturation of seeds. I. Accumulation of sterols in the free form, fatty acid esters, acylated glucosides and non-acylated glucosides, *Plant Cell Physiol.,* 14, 681, 1973.
16. **Popjak, G., Edmond, J., and Wong, S.-M.,** Absolute configuration of presqualene alcohol, *J. Am. Chem. Soc.,* 95, 2713, 1973.
17. **Capstack, E., Rosin, N. L., Blondin, G. A., and Nes, W. R.,** Squalene in *Pisum sativum,* its cyclization to β-amyrin and labelling pattern, *J. Biol. Chem.,* 240, 3258, 1965.
18. **Battersby, A. R. and Parry, G. V.,** Tracer studies on β-sitosterol, *Tetrahedron Lett.,* 14, 787, 1964.

19. **Russell, P. T., van Aller, R. T., and Nes, W. R.,** The mechanism of introduction of alkyl groups at C-24 of sterols. II. The necessity of the Δ^{24} bond, *J. Biol. Chem.*, 242, 5802, 1967.
20. **Tso, T. S. and Cheng, A. L. S.,** Metabolism of cholesterol-4-^{14}C in *Nicotiana* plants, *Phytochemistry,* 10, 2133, 1971.
21. **Heftmann, E.,** Functions of steroids in plants, *Phytochemistry,* 14, 891, 1975.
22. **Smith, A. R. and Vonstaden, J.,** Biological activity of steryl glycosides in 3 *in vitro* plant bioassays, *Z. Pflanzenphysiol.*, 88, 147, 1978.

PRENYL VITAMINS OF VASCULAR PLANTS AND ALGAE

Hartmut K. Lichtenthaler

The green, brown, and red algae as well as mosses, ferns, and higher plants contain several fat-soluble vitamins which are regular constituents of their chloroplasts.[1,2] These are provitamin A, identical with β-carotene, the chromanol vitamin E, known as α-tocopherol, and the phytylnaphthoquinone vitamin K_1, known as phylloquinone. Vitamin A and vitamin D do not occur in plant tissues. The group of vitamin K_2-compounds, the menaquinones, represent methylnaphthoquinones with prenyl side chains of different lengths. The latter are components of many bacteria including intestinal bacteria such as *Escherichia coli.* The chemical structure of the various prenyl vitamins is shown in Figure 1. β-Carotene, vitamin A, and vitamin D are pure prenyllipids, their carbon skeleton derived from the isoprenoid (terpenoid) biosynthetic pathway. Vitamins E, K_1, and K_2 represent mixed prenyllipids with a prenyl chain bound to an aromatic nucleus (benzo- or naphthoquinone ring). Vitamin E is the cyclic form (chromanol) of the benzoquinone derivate, α-tocohydroquinone. Its oxidized form, β-tocoquinone, also occurs in green chlorophyll-containing plant tissue, though in lower concentration than vitamin E.

Various methods for the isolation and determination of prenyl vitamins have been described.[3-5] A new sophisticated method for their separation is high-pressure liquid chromatography (HPLC), which can also be applied for their quantitative determination (Table 1). In the case of plant lipid extracts, it is favorable to first separate the prenyl vitamins (β-carotene, vitamins E and K_1) from the bulk of the lipids (chlorophylls, xanthophylls, phospho- and galactolipids) by thin-layer chromatography before applying the HPLC method (for details see References 5 and 6). When working with prenyl vitamins, it should be noted that they are light-sensitive and quickly oxidize in air. Because of this property, vitamins E and β-carotene are often added as lipid antioxidants to food oils and pharmaceutical preparations.

Within the plant cell, provitamin A and vitamins E and K_1 are synthesized and accumulated in plastids. White, chlorophyll-free plant tissues, such as roots or tubers, which possess only leucoplasts or amyloplasts, are very poor in the three plant prenyl vitamins: vitamin E is always found; β-carotene and vitamin K_1, however, are sometimes not detectable. The yellowish leaves of dark-grown etiolated seedlings, which contain special plastid forms, the etioplasts (with prolamellar bodies and plastoglobuli), exhibit low amounts of β-carotene and vitamin K_1, but fair amounts of vitamin E. Etiolated tissue, especially of germinating wheat plants, has long been a major source for natural vitamin E (e.g., wheat germ oil). Green leaf tissue and certain ripening fruits, however, exhibit higher vitamin E levels. The individual levels of the three plant prenyl vitamins (in μg/g dry weight) in the various tissues of plants are summarized in Table 2.

β-Carotene, and vitamins K_1 and E are functional constituents of the photosynthetic membranes of chloroplasts (thylakoids). During the greening of plant tissues, they accumulate parallel to the light-induced formation of chlorophylls and thylakoids and are enriched on a dry weight basis. In fact, green plant tissues exhibit the highest levels of vitamin E, K_1, and provitamin A. As a general rule, the greener the tissue, the higher its β-carotene, vitamin K_1, and vitamin E content. In contrast to β-carotene and K_1, which within the chloroplasts are exclusively bound to thylakoids, vitamin E (α-tocopherol) is further accumulated even after the greening process of leaves. This excess amount of α-tocopherol is deposited in the osmiophilic plastoglobuli, the size and number of which rise with increasing age of leaves. Vitamin E accumulation is light-dependent and proceeds to a much higher extent in sun-exposed rather than shaded

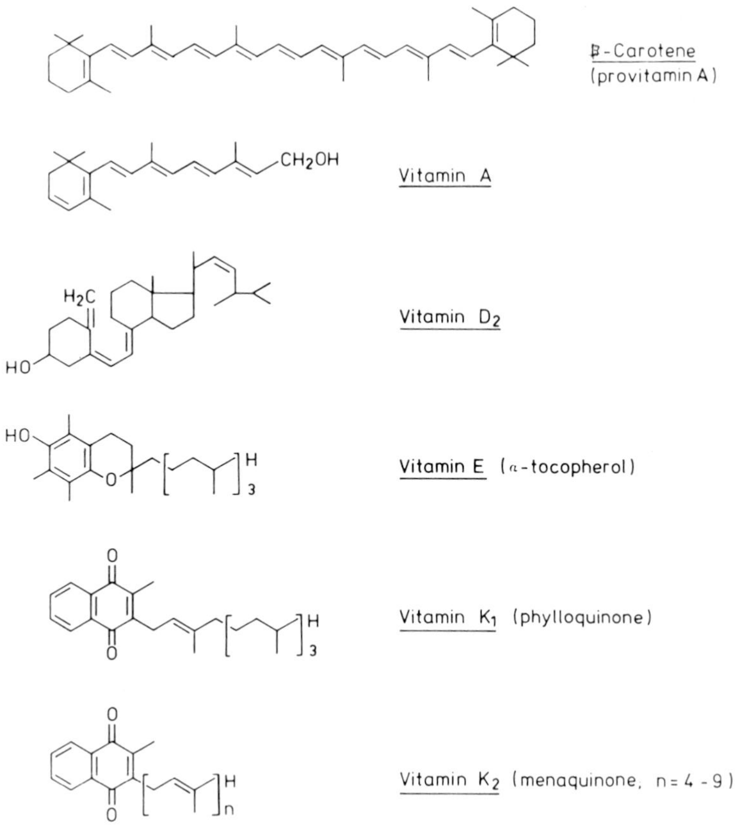

β-Carotene (provitamin A)

Vitamin A

Vitamin D_2

Vitamin E (α-tocopherol)

Vitamin K_1 (phylloquinone)

Vitamin K_2 (menaquinone, n = 4 - 9)

FIGURE 1. Chemical structure of prenyl vitamins (fat-soluble vitamins). Plant tissues contain provitamin A and the vitamins E and K_1.

leaves. This applies equally to the green leaves of herbs, vegetables, and trees, and to green fruits. Older green tissues, therefore, exhibit the highest vitamin E levels. The highest provitamin A levels of all plant tissues are found in the carrot root, which contains only β-carotene and no other carotenoids (Table 2). In ripening fruits as the chloroplasts develop into chromoplasts, the chlorophylls and thylakoids are gradually destroyed, the level of vitamin K_1 decreases, and often β-carotene does too. Vitamin E, however, is still accumulated during the chromoplast formation parallel to the appearance of secondary carotenoids. The α-tocopherol synthesis in developing chromoplasts is independent from the formation of the secondary carotenoids and also proceeds in mutants with little carotenoid formation (e.g., yellow, instead of red, fruits). However, when fruit ripening (color change from green to red) occurs while the fruits are still on the plant, the vitamin E formation is higher. In any case, the highest vitamin E levels are found in older red or yellow fruits and in older sun-exposed green leaves. These vitamin E levels are much higher than those detected in etiolated tissues.

Table 1
HIGH PRESSURE LIQUID CHROMATOGRAPHY OF PRENYL VITAMINS (REVERSED-PHASE HPLC)

Column			
Packing	Li Chrosorb-RP-8	5 μm, Merck®	
Dimensions	250 × 3mm		
Solvent	4.75% water in methanol	8% water in methanol	10% water in methanol
Pressure (bar)	155	160	180
Flow rate (mℓ/min)	1.3	1.2	1.2
Temperature (°C)	30	30	30
Detector	Spectral photometer PM 2 DLC, Zeiss®		

Compound	λ max (nm)[a]		t_R (min)[b]	
Vitamin A alcohol	250	1.7	2.1	2.8
Vitamin A aldehyde	250	1.7	2.3	3.0
Vitamin A palmitate	250	6.3	—	—
β-Carotene (provitamin A)	450	7.2	—	—
Vitamin D_2	265	2.5	4.1	6.6
Vitamin D_3	265	2.5	4.2	6.8
α-Tocopherol (Vitamin E)	290	3.0	6.0	9.0
Vitamin K_1 (phylloquinone)	260	4.2	8.7	—
Vitamin K_2 (menaquinone-4)	260	2.9	5.1	8.5
Vitamin K_3 (menadion)	260	1.3	1.5	1.6
Vitamin K_5	260	1.3	1.5	1.6

[a] λ max = wavelength of detection at absorbance maximum.
[b] t_R = retention time in minutes.

Table 2
AMOUNT OF PRENYLVITAMINS IN DIFFERENT GREEN AND NONGREEN PLANT TISSUES[a]

Type of plant tissue	Provitamin A	Vitamin E	Vitamin K_1
Green leaves			
Fagus sylvatica (beech)			
Sun leaves (August)	121	370	15
Shade leaves (August)	220	290	21
Spinacia oleracea (spinach)	580	413	30
Hordeum vulgare (barley)	515	256	56
Nicotiana tabacum	552	212	28
Etiolated leaves			
Hordeum vulgare (barley)	15	52	3
Phaseolus coccinaeus (bean)	30	98	6
Raphanus sativus	25	102	5
Fruit tissue			
Solanum lycopersicum			
Green tomato tissue	21	70	0.3
Red tomato tissue	74	118	0.1
Capsicum annuum (variety with red fruits)			

Green fruit tissue	25	86	0.5
Young red tissue	115	261	0.5
Full red tissue	103	330	0.2
Capsicum annuum (variety with yellow fruits)			
Green fruit tissue	32	36	3.6
Yellow-green tissue	8	51	1.9
Yellow fruit tissue	50	139	1.2
Chlorophyll-free tissue			
Solanum tuberosum (potato tubers)	—	0.6	0.03
Scorzonera hispanica (roots)	—	1.5	0.04
Allium cepa (white onion scales)	0.1	4.2	0.2
Brassica oleracea (cauliflower inflorescence)	0.6	3.0	0.8
Daucus carota (orange-red carrot roots)	614	32.0	0.6

[a] μg/g dry weight.

REFERENCES

1. **Lichtenthaler, H. K.,** Verbreitung und relative Konzentration der lipophilen Plastidenchinone in grünen Pflanzen, *Planta,* 81, 140, 1968.
2. **Lichtenthaler, H. K.,** Localization and functional concentrations of lipoquinones in chloroplasts, *Progress in Photosynthesis Research,* Vol. 1, 1969, 304.
3. Fat-soluble vitamins, in *International Encyclopedia of Food and Nutrition,* Vol. 9, Morton, R. A., Ed., Pergamon Press, New York, 1970.
4. *Fermente, Hormone, Vitamine,* Vol III(1), 3rd ed., Ammon, R. and Dirschel, W., Eds., Georg Thieme Verlag, Stuttgart, 1974.
5. **Lichtenthaler, H. K., Karunen, P., and Grumbach, K. H.,** Determination of prenylquinones in green photosynthetically active moss and liver moss tissue, *Physiol. Plant Pathol.,* 40, 105, 1977.
6. **Lichtenthaler, H. K.,** Separation of prenyllipids including chlorophylls, carotenoids, prenylquinones and fat-soluble vitamins, in *CRC Handbook Series in Chromatography,* Vol. 8, Mangold, H., Ed., in press.

HYDROCARBON COMPOSITION AND BIOSYNTHESIS IN PLANTS AND PHOTOSYNTHETIC BACTERIA

W. R. Finnerty

The ubiquitous distribution of hydrocarbons in plants has been extensively documented over the past four decades. Normal (n) — and branched — alkanes have been identified, wherein branched-alkanes comprised the smaller proportion of the hydrocarbon fraction.[1-3] Structural studies have established the presence of odd- and even-carbon n-alkanes and branched-alkanes as well as alkenes.[1-3] An extensive taxonomic study has documented structural variations within the alkane fraction along with changes in the types of alkanes that characterize specific plant tissue.[1,4]

The alkanes of plant cuticles differ qualitatively and quantitatively from noncuticular alkanes.[1,5-6] Shorter chain alkanes are characteristic of algae (Table 1) and photosynthetic bacteria (Table 1). The alkanes derived from ferns (Table 2), lichens (Table 2), and the seed plants (Table 3), consist of longer alkane carbon chains. Numerous alkenes are characteristic components of many algae (Table 1).

The expression of the genetic potential for alkane biosynthesis in plants appears to be regulated by a variety of intrinsic and extrinsic factors. Alkane composition differs between organs within a plant,[1-4,7] exhibits significant compositional change with age,[7-12] and undergoes drastic qualitative and quantitative changes depending on environmental conditions.[1-2,10,13] Laboratory studies with selected plants[9-11] have demonstrated a general decrease in total alkane content with age, whereas field studies have revealed both a general decrease in total alkane content with age, as well as qualitative changes in alkane composition, as reflected by increasing chain lengths.[1] Environmental variables, such as length of photoperiod, quality of light, site quality, moisture content, and temperature all predispose to and drastically affect the qualitative and quantitative composition of alkanes found in plants. The environmental conditions imposed on plants appear to represent overriding factors that negate or alter the intrinsic genetic potential of plants. Such variations in alkane composition have posed serious reservations as to the applicability of alkanes as a taxonomic aid.[1-3]

Mechanisms for alkane biosynthesis in plants involve condensation or elongation reactions of fatty acids.[1-3,14] The precise pathway remains indeterminant in context of intermediates, cofactors, and requisite enzymatic complement. Current interpretations of the available evidence favor the fatty acid elongation-decarboxylation mechanism.[14]

$$R{-}(CH_2)_x{-}CH_2{-}CH_2{-}COOH \xrightarrow{C_2\text{elongation}} R{-}(CH_2)_{x+2n}CH_2{-}CH_2{-}COOH$$

$$R{-}(CH_2)_{x+2n}CH_2{-}CH_2{-}COOH \xrightarrow{\text{decarboxylation}} R{-}(CH_2)_{x+2n}CH_2{-}CH_3 + CO_2$$

Synthesis of branched alkanes appears to occur through a elongation-decarboxylation mechanism involving the carbon skeleton of specific branched-chain amino acids.[14] The two classes of branched alkanes of biosynthetic origin are iso-alkanes and anteiso-alkanes.

$$(CH_3)_2CH{-}CH(NH_2){-}COOH \rightarrow (CH_3)_2CH{-}C(=O){-}C(=O){-}S{-}CoA \xrightarrow[C_2\text{elongation}]{\text{Iso-alkanes}} (CH_3)_2CH{-}(CH_2)_{x+2n}CH_2{-}CH_2{-}COOH$$

$$\begin{array}{c}CH_3\\ |\\ HC\\ |\\ CH_3\end{array}\!\!-\!(CH_2)_{x+2n}CH_2-CH_2-COOH \xrightarrow{\text{decarboxylation}} \begin{array}{c}CH_3\\ |\\ HC\\ |\\ CH_3\end{array}\!\!-\!(CH_2)_{x+2n}CH_2-CH_3 + CO_2$$

Anteiso-alkanes

$$\begin{array}{c}CH_3\\ |\\ CH_2\\ |\\ HC\\ |\\ CH_3\end{array}\!\!-\!\begin{array}{c}CH-COOH\\ |\\ NH_2\end{array} \rightarrow \begin{array}{c}CH_3\\ |\\ CH_2\\ |\\ HC\\ |\\ CH_3\end{array}\!\!-\!\underset{\displaystyle O}{\overset{}{\underset{\|}{C}}}-\overset{\displaystyle O}{\overset{\|}{C}}-S-CoA \xrightarrow{C_2\ \text{elongation}} \begin{array}{c}CH_3\\ |\\ CH_2\\ |\\ HC\\ |\\ CH_3\end{array}\!\!-\!(CH_2)_{x+2n}CH_2-CH_2-COOH$$

$$\begin{array}{c}CH_3\\ |\\ CH_2\\ |\\ HC\\ |\\ CH_3\end{array}\!\!-\!(CH_2)_{x+2n}CH_2-CH_2-COOH \xrightarrow{\text{decarboxylation}} \begin{array}{c}CH_3\\ |\\ CH_2\\ |\\ HC\\ |\\ CH_3\end{array}\!\!-\!(CH_2)_{x+2n}CH_2-CH_3 + CO_2$$

The hydrocarbon content of photosynthetic tissues represents less than 0.1% of the total dry weight (Tables 1 to 4). Variations in hydrocarbon composition occur intracellularly and extracellularly as well as being affected by age and environmental conditions.

ACKNOWLEDGMENTS

This paper was supported in part by a Department of Energy Contract EY76-S-09-0888 to W. R. Finnerty. Appreciation is extended to P. Kerr Falco in the preparation of this paper.

Table 1
COMPOSITION OF HYDROCARBONS FROM PHOTOSYNTHETIC MICROORGANISMS

Organism	% tissue dry weight	Range	Identities	Ref.
Chromatium	0.09	C_{15}—C_{20}	*n*-Alkanes	6
Rhodomicrobium vannielli		C_{15}—C_{30}	*n*-Alkanes	15
Rhodopseudomonas spheroides		C_{15}—C_{21}	*n*-Alkanes	15
Rhodospirillum rubrum		C_{15}—C_{21}	*n*-Alkanes	15
Chlorobium		C_{15}—C_{28}	*n*-Alkanes	15
	Cyanophyta (blue-green)			
Agmenellum quadruplicatum	0.005—0.12	C_{19}	*n*-Alkanes and alkenes	6
Anabaena variabilis	0.09	C_{17}—C_{18}		6
Trichedesmium erythaeum	0.05—0.12	C_{15}—C_{18}	*n*-Alkanes and alkenes	6
Microcoleus chthonoplastes	0.05—0.12	C_{15}—C_{18}	*n*-Alkanes	6
Plectonema terebrans	0.05—0.12	C_{15}—C_{18}	*n*-Alkanes and alkenes	6
Oscillatoria williamsii	0.015—0.12	C_{15}—C_{18}	*n*-Alkanes	6
Caccohloris elabens	0.015—0.12	C_{17}—C_{19}	*n*-Alkanes and alkenes	6

Table 1 (continued)
COMPOSITION OF HYDROCARBONS FROM PHOTOSYNTHETIC MICROORGANISMS

Organism	% tissue dry weight	Range	Identities	Ref.
Nostoc muscorum	0.035	C_{15}—C_{18}	*n*- and br-Alkanes	16
Nostoc sp.		C_{15}—C_{17}	*n*- and br-Alkanes	17
Anacystis nidulans	0.032	C_{15}—C_{18}	*n*-and br-Alkanes	16
A. cyanea		C_{17}—C_{29}	*n*- and br-Alkanes	17
A. montana		C_{17}—C_{29}	*n*- and br-Alkanes	17
Phormidium luridum	0.025	C_{17}	*n*- and br-Alkanes	16
Chlorogloea fritschii	0.038	C_{15}—C_{18}	*n*- and br-Alkanes	16
Spirulina paltensis		C_{15}—C_{17}	*n*- and br-Alkanes	17
Lyngbya aestuarii		C_{15}—C_{18}	*n*- and br-Alkanes	17
Chroococcus turgidus		C_{15}—C_{18}	*n*- and br-Alkanes	17
	Chlorophyta (green)			
Coelastrum microsporum		C_{17}	*n*-Alkanes	17
Scenedesmus quadricauda		C_{17}—C_{19}	*n*-Alkanes and alkenes	17
Tetraedron sp.		C_{15}—C_{27}	*n*-Alkanes	17
Chlorella pyrenoidosa		C_{17}	*n*-Alkanes and alkenes	17
Chlorella sp.	0.032	C_{15}—C_{19}	*n*-Alkanes and alkenes	6
C. vulgaris		C_{17}—C_{36}	*n*-Alkanes and alkenes	6
Chaetomorpha linum	0.00127	C_{17}	*n*-Alkanes	6
Spongomorpha arcta	0.02—0.14	C_{19}—C_{21}	*n*-Alkenes	6
Enteromorpha compressa	0.04	C_{15}—C_{17}	*n*-Alkanes and alkenes	6
Ulva lactuca	0.09	C_{17}—C_{21}	*n*-Alkenes	6
Cadium fragile	0.008—0.05	C_{15}—C_{19}	*n*-Alkanes and alkenes	6
	Phaeophyta (brown)			
Laminaria digitata	0.002	C_{15}—C_{21}	*n*- Alkanes and alkenes	6
L. saccharina	0.03	C_{15}—C_{24}	*n*-Alkanes and alkenes	6
Ascophyllum nodosum	0.01—0.06	C_{15}—C_{21}	*n*-Alkanes and alkenes	6
Fucus distichus	0.004—0.008	C_{15}—C_{21}	*n*-Alkanes and alkenes	6
Ecotocarpus fasiculatus	0.02	C_{15}—C_{21}	*n*-Alkanes and alkenes	6
	Rhodophyta (red)			
Porphyra leucosticta	0.04	C_{15}—C_{21}	*n*-Alkanes and alkenes	6
Chondrus crispus	0.07	C_{15}—C_{17}	*n*-Alkanes and alkenes	6
Polysiphonia urceolata	0.03	C_{15}—C_{19}	*n*-Alkanes and alkenes	6
	Chrysophyta (yellow-brown)			
Botryococcus braunii[a,b]	17.0	C_{27}—C_{31}	*n*-Alkanes and alkenes	10
Green (actively growing) colonies (3 weeks)				
Brown (nongrowing) colonies (6 weeks)	86.0	C_{34}	Botryococcene	10
Colony morphology lost (12 weeks)	< 0.01%			10
B. braunii grown in "natural" environment	76.0	C_{34}	br-Alkenes	18
Skeletonema costatum	0.12—0.135	C_{21}	*n*-Alkenes	6
Chaetoceros curirsetus	0.6	C_{21}	*n*-Alkenes	6
Cylindrotheca fusiformis	0.2	C_{21}	*n*-Alkenes	6
Lauderia borealis	0.7	C_{21}	*n*-Alkenes	6
Cyclotella nana	0.0004	C_{16}—C_{23}	*n*-Alkanes and alkenes	6
	Pyrophyta (dinoflagellates)			
Gonyaulax polyedra	0.004—0.2	C_{14}—C_{28}	*n*-Alkanes and alkenes	6

Table 1 (continued)
COMPOSITION OF HYDROCARBONS FROM PHOTOSYNTHETIC MICROORGANISMS

Organism	% tissue dry weight	Range	Identities	Ref.
Peridimium trochaideum	0.07	C_{14}—C_{25}	*n*-Alkanes and alkenes	6
Gymondinium splendens	0.06	C_{14}—C_{28}	*n*-Alkanes and alkenes	6

[a] Percent of total petrol extracts.
[b] Percent of fresh weight.

Table 2
COMPOSITION OF HYDROCARBONS FROM PHOTOSYNTHETIC NONVASCULAR PLANTS

Organism	% Tissue dry weight	Range	Identities	Ref.
Cetraria nivalis		C_{21}—C_{23}	*n*-Alkanes	19
C. crispa		C_{21}—C_{33}	*n*-Alkanes	19
Siphula ceratites		C_{17}—C_{33}	*n*- and br-Alkanes	19
Lobaria pulmonaria	0.05	C_{25}—C_{31}	*n*-Alkanes	20
Sphagnum fuscum (4 species)		C_{20}—C_{32}	*n*-Alkanes	21
Hypericum sp. (32 species)		C_{7}—C_{13}	*n*- and br-Alkanes	22
Oleandra distenta	0.12[a]		No alkanes	23
Nephrolepsis biserrata	0.06[a]	C_{27}—C_{33}	*n*-Alkanes	23
Ctenitis protensa	0.07[a]	C_{27}—C_{33}	*n*-Alkanes	23
Dryopteris felix-mas	0.08[a]	C_{27}—C_{33}	*n*-Alkanes	23
Lomariopsis palustris	0.05[a]	C_{29}—C_{33}	*n*-Alkanes	23
Asplenium adiantum-nigrum	0.07[a]	C_{27}—C_{33}	*n*-Alkanes	23
A. africanum	0.08[a]	C_{27}—C_{33}	*n*-Alkanes	23
A. trichomanes	0.02[a]	C_{24}—C_{33}	*n*-Alkanes	23
Phyllitis scolopendrium	0.03[a]	C_{25}—C_{33}	*n*-Alkanes	23
Blechnum spicant	0.02[a]	C_{25}—C_{35}	*n*-Alkanes	23
Microsorium punctatum	0.09[a]	C_{27}—C_{33}	*n*-Alkanes	23
Platycerium elephantotis	0.09[a]	C_{27}—C_{31}	*n*-Alkanes	23
Phymatodes scolopendria	0.19[a]	C_{31}—C_{33}	*n*-Alkanes	23
Adiantum vogelii	0.04[a]	C_{24}—C_{35}	*n*-Alkanes	23
Lygodium smithianum	0.04[a]	C_{33}—C_{35}	*n*-Alkanes	23
Marsilea diffusa	0.04[a]	C_{25}—C_{33}	*n*-Alkanes	23
M. polycarpa	0.04[a]	C_{24}—C_{31}	*n*-Alkanes	23
M. quadritolia	0.14[a]	C_{25}—C_{31}	*n*-Alkanes	23
Osmunda cinnamonea L.	0.013	C_{15}—C_{33}	*n*-Alkanes and alkenes	24
O. regalis var. *spectabilis*	0.024	C_{17}—C_{33}	*n*-Alkanes and alkenes	24
Lygodium japonicum	0.022	C_{19}—C_{33}	*n*-Alkanes and alkenes	24
Polypodium polypodioides	0.035	C_{35}—C_{33}	*n*-Alkanes and alkenes	24
Adiantum pedatum	0.032	C_{17}—C_{31}	*n*-Alkanes and alkenes	24
Asplenium sp.	0.025	C_{15}—C_{31}	*n*-Alkanes and alkenes	24
Pteridium aquilinum	0.003	C_{17}—C_{32}	*n*-Alkanes and alkenes	24
Dryopteris ludoirciana	0.008	C_{18}—C_{33}	*n*-Alkanes and alkenes	24
Thelypteris hexagonoptera	0.042	C_{17}—C_{33}	*n*-Alkanes and alkenes	24
Polystichum acrostichoides	0.027	C_{15}—C_{33}	*n*-Alkanes and alkenes	24

[a] Percent of total petrol extracts.

Table 3
COMPOSITION OF HYDROCARBONS FROM TISSUES OF VASCULAR PLANTS

Organism	% Tissue dry weight	Range	Identities	Ref.
Pinus jeffreyi		C_7, C_9, C_{11}	*n*-Alkanes	25
P. sabiniana		C_7—C_9,C_{11}	*n*-Alkanes	25
P. torreyana	3.1[a]	C_7—C_9, C_{11}	*n*-Alkanes	26
Pinus spp. (9 species)		C_{21}—C_{33}	Alkanes	27
Cupressus (8 species)		C_{21}—C_{35}	Alkanes	27
Juniperus (2 species)		C_{21}—C_{35}	Alkanes	27
Thuja occidentalis		C_{21}—C_{35}	Alkanes	27
Callitris (3 species)		C_{23}—C_{35}	Alkanes	27
Widdringtonia (4 species)		C_{21}—C_{35}		27
Nicotiana tabacum	0.15	C_{21}—C_{34}	*n*- and br-Alkanes	28
Cannabis sativa	0.1	C_{16}—C_{36}	*n*- and br-Alkanes	29
Trifolium repens	0.08	C_{21}—C_{33}	*n*-Alkanes	30
Artemisia ludoviciana	0.2	C_{19}—C_{33}	*n*-Alkanes	31
A. frigida	0.2	C_{19}—C_{31}	*n*-Alkanes	31
Rosmarinus officinalis		C_{16}—C_{36}	*n*- and br-Alkanes	32
Spinacia oleracea				
External lipid	0.06	C_{27}—C_{33}	*n*-Alkanes	5
Internal lipid	0.004	C_{27}—C_{33}	*n*-Alkanes	5
Brassica oleracea varieties				
Copenhagen	0.05	C_{18}—C_{31}		33
Round Dutch		C_{27}—C_{31}	*n*-Alkanes	34
Winnigstadt		C_{27}—C_{31}	*n*-Alkanes	35
Cassia obtusifolia		C_{16}—C_{40}	*n*- br-Alkanes	13
C. occidentalis		C_{16}—C_{36}	*n*- and br-Alkanes	13
Agave (21 species)		C_{21}—C_{33}	*n*-Alkanes	36
Kalanchoe (5 species)		C_{23}—C_{35}	*n*-Alkanes	36
Echeveria (5 species)		C_{23}—C_{35}	*n*- Alkanes	36
Crassula (4 species)		C_{23}—C_{35}	*n*- Alkanes	36
Sedum hispanicum		C_{25}—C_{35}	*n*- Alkanes	36
S. pachyphyllum		C_{27}—C_{33}	*n*- Alkanes	36
Eucalyptus (19 species)		C_{21}—C_{31}	*n*-Alkanes	36
Aloe (63 species)		C_{21}—C_{33}	*n*- and br-Alkanes, alkenes	4
Copernicia cerifera (3 samples)		C_{16}—C_{35}	*n*-Alkanes	1
Pholnix reclinata		C_{22}—C_{33}	*n*-Alkanes	1
P. Canariensis		C_{22}—C_{33}	*n*-Alkanes	1
Washingtonia filifera		C_{21}—C_{31}	*n*-Alkanes	1
Cocos plumosa		C_{22}—C_{33}	*n*-Alkanes	1
Sabal sp. (5 samples)		C_{22}—C_{33}	*n*-Alkanes	1
Hemerocallis aurantiaca		C_{21}—C_{33}	*n*-Alkanes and alkenes	1
Solandra grandiflora		C_{21}—C_{33}	*n*-Alkanes and alkenes	1
Allamandra cathartica		C_{23}—C_{33}	*n*-Alkanes	1
Monstera deliciosa		C_{21}—C_{33}	*n*-Alkanes	1
Thunbergia laurifolia		C_{25}—C_{35}	*n*-Alkanes	1
Hebe odora	3.9[b]	C_{23}—C_{33}	*n*-Alkanes	37
H. diosmifolia	3.2[b]	C_{27}—C_{33}	*n*- and br-Alkanes	37
H. parirflora	2.5[b]	C_{25}—C_{33}	*n*- and br-Alkanes	37
H. stricta	1.7[b]	C_{25}—C_{35}	*n*- and br-Alkanes	37
Gaultheria subcorymbosa	2.0[b]	C_{25}—C_{33}	*n*-Alkanes	37

Table 3 (continued)
COMPOSITION OF HYDROCARBONS FROM TISSUES OF VASCULAR PLANTS

Organism	% Tissue dry weight	Range	Identities	Ref.
G. antipoda	1.6[b]	C_{25}—C_{33}	*n*-Alkanes	37
Pimelea prostrata	1.3[b]	C_{25}—C_{31}	*n*- and br-Alkanes	37
Acaena anserinifolia	2.0[b]	C_{25}—C_{33}	*n*- and br-Alkanes	37
Arundo conspicua	2.5[b]	C_{23}—C_{33}	*n*-Alkanes	37
Aeonium (24 species)	0.5—12.0[b]	C_{25}—C_{35}	*n*- and br-Alkanes	38
Aichryson (3 species)	0.6—2.0[b]	C_{27}—C_{35}	*n*- and br-Alkanes	38
Greenovia (5 species)	1.0[b]	C_{27}—C_{35}	*n*- and br-Alkanes	38
Monanthes polyphylla	4.0[b]	C_{31}—C_{35}	*n*-Alkanes	38
M. anagensis	12.0[b]	C_{31}—C_{35}	*n*-Alkanes	38
Sedum anglicum		C_{27}—C_{33}	*n*-Alkanes	38
Dracaena draco		C_{25}-C_{33}	*n*-Alkanes	38
Lolium multiflora		C_{25}—C_{33}	*n*-Alkanes	38
Euphorbia peplus	1.0[b]	C_{25}—C_{33}	*n*- and br-Alkanes	38
E. balsimifera		C_{25}—C_{29}	*n*-Alkanes	38
E. atropurpurea		C_{25}—C_{33}	*n*- and br-Alkanes	38
E. regis-jubal	1.0[b]	C_{25}—C_{33}	*n*-Alkanes	38
E. aphylla		C_{25}—C_{33}	*n*-Alkanes	38
E. bourgaeanea		C_{27}—C_{31}	*n*-Alkanes	38
Marrubium vulglare		C_{17}—C_{35}	*n*- and br-Alkanes	39
Secale cereale	0.6	C_{23}—C_{35}	Hydrocarbon	40
Triticale hexaploide	0.6	C_{25}—C_{35}		40
Triticum durum cv. Stewart 63	0.4	C_{23}—C_{33}		40
T. aestivum varieties				
Demar 4	0.87—12.0	C_{20}—C_{33}	Alkanes	8
Selkirk	5.0—15.0[c]	C_{27}—C_{31}	*n*-Alkanes	7
Manitou	4.0—25.0[c]	C_{27}—C_{31}	*n*-Alkanes	7
Stewart	3.0—11.0[c]	C_{27}—C_{31}	*n*-Alkanes	7
Cortaderia toetoe	0.6[c]	C_{22}—C_{33}	*n*-Alkanes	41
C. atacamensis	0.3[c]	C_{24}—C_{32}	*n*-Alkanes	41
C. fulirda	0.3—0.6[c]	C_{25}—C_{31}	*n*-Alkanes	41
C. richardii	0.3—0.6[c]	C_{22}—C_{31}	*n*-Alkanes	41
C. selloana	0.3—0.6[c]	C_{25}—C_{32}	*n*-Alkanes	41
Asclepsis (6 species)		C_{23}—C_{33}	*n*-Alkanes and alkenes	42
Monarda (3 species)		C_{27}—C_{35}		43
Musca sapientum		C_{19}—C_{30}	*n*-Alkanes	44
Antirrhinum majus	0.18	C_{16}—C_{35}	*n*- and br-Alkanes	45
Citrus unshiu 6 cultivars	0.9—1.2	C_{20}—C_{35}	*n*-Alkanes and alkenes	46
5 cultivars	0.012—0.022	C_{20}—C_{35}	*n*-Alkanes and alkenes	47
C. sinensis 3 varieties	1.2[b]	C_{20}—C_{35}	*n*- and br-Alkanes and alkenes	48
C. paradisi 3 varieties	1.0	C_{20}—C_{35}	*n*- and br-Alkanes and alkenes	49
C. paradisi		C_{20}—C_{35}	*n*- and br-Alkanes and alkenes	50
C. aurantifolia	1.6[b]	C_{20}—C_{38}	*n*- and br-Alkanes	51
cv. Key		C_{20}—C_{35}	*n*- and br-Alkenes	
C. latifolia	1.2[b]	C_{20}—C_{38}	*n*- and br-Alkanes	51
cv. Persian		C_{20}—C_{33}	*n*- and br-Alkenes	
C. limettioides	1.2[b]	C_{20}—C_{38}	*n*- and br-Alkanes	51
cv. Columbia		C_{20}—C_{33}	*n*- and br-Alkenes	
C. limon	1.3[b]	C_{20}—C_{35}	*n*- and br-Alkanes	52
3 varieties			*n*- and br-Alkenes	
Anchusa azurea	≥5[d]	C_{15}—C_{29}	*n*- and br-Alkanes	53

Table 3 (continued)
COMPOSITION OF HYDROCARBONS FROM TISSUES OF VASCULAR PLANTS

Organism	% Tissue dry weight	Range	Identities	Ref.
Anthrisens cerefolium	≥5[d]	C_{17}—C_{29}	*n*- and br-Alkanes	53
Daucus carota	≥5[d]	C_{27}—C_{31}	*n*- and br-Alkanes	53
Helianthus annuus	≥5[d]	C_{16}—C_{31}	*n*-Alkanes	53
Myosotis scaroioides	≥5[d]	C_{18}—C_{29}	*n*- and br-Alkanes	53
Oenothera macrocarpa	≥5[d]	C_{20}—C_{31}	*n*-Alkanes	53
Orbignya oleifera	≥5[d]	C_{16}—C_{34}	*n*-Alkanes	53
Pastinaca sativa	≥5[d]	C_{18}—C_{31}	*n*-Alkanes	53
Shorea stenoptera	≥5[d]	C_{16}—C_{34}	*n*-Alkanes	53
Plantago ovata		C_{16}—C_{33}	*n*- and br-Alkanes	54
Balanites pedicellaris		C_{16}—C_{34}	*n*- and br-Alkanes	9
Salicarnia bigelovii		C_{22}—C_{33}	*n*- and br-Alkanes	12
Podocarpus nivalis	0.9[b]	C_{21}—C_{35}	*n*-Alkanes	55
P. acutifous	0.2	C_{21}—C_{35}	*n*-Alkanes	55
Balanites wilsoniana	0.001	C_{10}—C_{30}	*n*-Alkanes	56
Lophopetalum rigidum		C_{14}—C_{20}	*n*-Alkanes	57
Discorea deltoidea	0.001	C_{18}—C_{34}	*n*-Alkanes	58
Phormium tenax	1.0[b]	C_{24}—C_{32}	*n*- and br-Alkanes	37
Cordyline australis	1.3[b]	C_{25}—C_{31}	*n*-Alkanes	37
Cortaderia toetoe		C_{24}—C_{32}	*n*-Alkanes	41
Catharanthus longifolius		C_{27}—C_{33}	*n*-Alkanes	59
Salicornia bigelovii		C_{21}—C_{33}	*n*-Alkanes	12

[a] Percent of oleoresin.
[b] Percent of total petrol extracts.
[c] Percent of fresh weight.
[d] Total hydrocarbon content is comprised of alkanes, alkenes, etc., as well as terpenoid fractions.

Table 4
ALIPHATIC AND AROMATIC HYDROCARBONS FROM TWO MARSH PLANTS

Spartina cynosuroides[60,a]

Compound	% of oil	
Aliphatic hydrocarbons		
C_9H_{18}	0.2	
C_9H_{20}	T[b]	
$C_{10}H_{20}$	0.1	
n-Decane	T[b]	2.3% of oil;
$C_{10}H_{22}$	0.1	0.005% of weight
n-Tridecane	0.1	
$C_{14}H_{22}$	0.2	
n-Tetradecane	0.1	
n-Pentadecane	0.2	
n-Hexadecane	0.3	
n-Heptadecane	0.5	
2-Methylheptadecane	0.5	

Table 4 (continued)
ALIPHATIC AND AROMATIC HYDROCARBONS FROM TWO MARSH PLANTS

Compound	% of oil	
	Aromatic hydrocarbons	
Toluene	T[b]	
m-Xylene	0.2	
p-Xylene	0.1	
o-Methylstyrene	0.1	
1-Methyl-2-ethyltoluene	0.1	
Cumene	0.3	6.% of oil;
Trimethylbenzene	3.8	0.001% of weight
Naphthalene	0.1	
m-Diethylbenzene	0.3	
3-Ethyl-*o*-xylene	0.3	
Tetramethylbenzene	T[b]	
2-Methylnaphthalene	0.1	
1-Ethylindan	0.3	
Dimethylnaphthalene	T[b]	
1,1,4,5-Tetramethylindan	0.1	
	Juncus roemerianus[61,c]	
Dimethylbenzene	1.3	
Benzaldehyde	0.7	
Benzyl cyanide	1.2	
1-Methyl-4-ethylbenzene	0.8	
o-Tolualdehyde	0.6	
o-Methoxyphenol	0.4	
2-Methylpropenylbenzene	0.4	
Vinylbenzaldehyde	0.2	
1,3-Diethylbenzene	0.4	
1,3-Dimethyl-2-ethylbenzene	0.7	
1,2-Dimethyl-3-ethylbenzene	0.5	
Propoxyanisole	0.5	
1,2,3,4-Tetrachlorobenzene	1.2	
Naphthalene	1.4	
Decahydronaphthalene	1.4	
2-Methylnaphthalene	1.1	
1-Methylnapthalene	0.6	
1-Ethylindan	0.3	
1,1-Dimethylindan	0.3	
Dimethylnaphthalene	0.9	
2,6-Dimethylnapththalene	0.3	
Acetylnapththalene	0.3	
Acetylnaphthalene	0.1	
Phenanthrene	0.5	
4-Methylcyclohexanone	0.2	
Cyclohexane carboxaldehyde	0.1	
Methylcyclohexyl ketone	0.5	
2,2,6-Trimethylcyclohexanone	0.2	
2-Cyclohexylcyclohexanone	0.1	
(4-methyl-pent-3-enyl) Cyclohexadiene-1-carboxaldehyde	1.3	
(4-methyl-pent-3-enyl) Cyclohexane carboxyaldehyde	0.2	

[a] Oil recovered by distillation — yield 0.02% of weight.

[b] T = < 0.1%.

REFERENCES

1. **Herbin, G. A. and Robins, P. A.**, Patterns of variation and development in leaf wax alkanes, *Phytochemistry,* 8, 1985, 1969.
2. **Eglinton, G. and Hamilton, R. J.**, The distribution of alkanes, in *Chemical Plant Taxonomy,* Swain, T., Ed., Academic Press, New York, 1963, 187.
3. **Douglas, A. G. and Eglinton, G.**, The distribution of alkanes, in *Comparative Phytochemistry,* Swain, T., Ed., Academic Press, New York, 1966, 57.
4. **Herbin, G. A. and Robins, P. A.**, Studies on plant cuticular waxes. I. The chemotaxonomy of alkanes and alkenes of the genus *Aloe* (Liliaceae), *Phytochemistry,* 7, 239, 1968.
5. **Kaneda, T.**, Hydrocarbons in spinach: two distinctive carbon ranges of aliphatic hydrocarbons, *Phytochemistry,* 8, 2039, 1969.
6. **Tornabene, T. G.**, Microbial formation of hydrocarbons, in *Microbial Energy Conversion,* Schlegel, H. G. and Barnea, J., Eds., Erich Goltze K. G., Göttingen, 1976, 281.
7. **Tulloch, A. P.**, Composition of leaf surface waxes of *Triticum* species: variation with age and tissue, *Phytochemistry,* 12, 2225, 1973.
8. **Bianchi, G. and Corbellini, M.**, Epicuticular wax of *Triticum aestivum* Demar 4, *Phytochemistry,* 16, 943, 1977.
9. **Hardman, R., Wood, C. N., and Sofowora, E. A.**, Isolation and characterization of seed hydrocarbons from *Balanites aegyptiaca (B. Rofburghii)* and *B. pedicellaris, Phytochemistry,* 9, 1087, 1970.
10. **Brown, A. C. and Knight, B. A.**, Hydrocarbon content and its relationship to physiological state in the green alga *Botryococcus braunii, Phytochemistry,* 8, 543, 1969.
11. **Chang, S. Y. and Grunwald, C.**, Duvatrienediol, alkanes, and fatty acids in cuticular wax of tobacco leaves of various physiological maturity, *Phytochemistry,* 15, 961, 1976.
12. **Weete, J. D., Rivers, W. G., and Weber, D. J.**, Hydrocarbon and fatty acid distribution in the halophyte, *Salicornia bigelovii, Phytochemistry,* 9, 2041, 1970.
13. **Wilkinson, R. E.**, Sicklepod hydrocarbon response to photoperiod, *Phytochemistry,* 11, 1273, 1972.
14. **Kolattukudy, P. E.**, Biosynthesis of cuticular lipids, in *Annual Review of Plant Physiology,* Machlis, L., Ed., Annual Reviews, Palo Alto, Calif., 1970, 163.
15. **Han, J. and Calvin, M.**, Hydrocarbon distribution of algae and bacteria, and microbiological activity in sediments, *Proc. Natl. Acad. Sci., U.S.A.,* 64, 436, 1969.
16. **Han, J., McCarthy, E. D., and Calvin, M.**, Hydrocarbon constituents of the blue-green algae *Nostoc muscorum, Anacystis nidulans, Phormidium luridum* and *Chlorogloea fritschii, J. Chem. Soc.,* C, 275, 1968.
17. **Gelpi, E., Schneider, H., Mann, J., and Oro, J.**, Hydrocarbons of geochemical significance in microscopic algae, *Phytochemistry,* 9, 603, 1970.
18. **Maxwell, J. R., Douglas, A. G., Eglinton, G., and McCormick, A.**, The botryococcenes-hydrocarbons of novel structure from the alga *Botryococcus braunii,* Kutzing, *Phytochemistry,* 7, 2157, 1968.
19. **Gaskell, S. J., Eglinton, G., and Bruun, T.**, Hydrocarbon constituents of three species of Norwegian lichen: *Certraria nivalis, C. crispa,* and *Siphula ceratites, Phytochemistry,* 12, 1174, 1973.
20. **Catalano, S., Marsil, A., Morelli, I., and Pacchiani, M.**, Hydrocarbons, sterols and fatty acids of *Lobaria pulmonaria, Phytochemistry,* 15, 221, 1976.
21. **Corrigan, D., Kloos, C., O'Conner, C. S., and Timoney, R. F.**, Alkanes from four species of *Sphagnum* moss, *Phytochemistry,* 12, 213, 1973.
22. **Mathis, C. and Ourisson, G.**, Etude chimiotaxonomique du genre *Hypericum.* III. Repartition des carbures satures et des monoterpenes dans les huiles essentielles D'Hypericum, *Phytochemistry,* 3, 133, 1964.
23. **Bottari, F., Marsili, A., Morelli, I., and Pacchiani, M.**, Aliphatic and triterpenoid hydrocarbons from ferns, *Phytochemistry,* 11, 2519, 1972.
24. **Lytle, T. F., Lytle, J. S., and Caruso, A.**, Hydrocarbons and fatty acids of ferns, *Phytochemistry,* 15, 965, 1976.
25. **Mirov, N. T.**, U.S. Department of Agriculture Forest Service Tech. Bull. No. 1239, U.S. Department of Agriculture, Washington, D.C., 1961.
26. **Zavarin, E., Hathaway, W., Reicher, T., and Linhart, Y. B.**, Chemotaxanomic study of *Pinus torreyana* Parry turpentine, *Phytochemistry,* 6, 1019, 1967.
27. **Herbin, G. A. and Robins, P. A.**, Studies on plant cuticular waxes. III. The leaf wax alkanes and ω hydroxy acids of some members of the Cupressaceae and Pincaceae, *Phytochemistry,* 7, 1325, 1968.
28. **Weete, J. D., Venketeswaran, S., and Laseter, J. L.**, Two populations of aliphatic hydrocarbons of teratoma and habituated tissue culture of tobacco, *Phytochemistry,* 10, 939, 1971.
29. **Hendriks, H., Malingre, T. M., Batterman, S., and Bos, R.**, Alkanes of the essential oil of *Cannabis sativa, Phytochemistry,* 16, 719, 1977.

30. **Body, D. R.,** Neutral lipids of leaves and stems of *Trifolium* repens, *Phytochemistry,* 13, 1527, 1974.
31. **Bachelor, F. W., Paralikar, A. B., and Telang, S. A.,** Alkanes of three *Artemisia* species, *Phytochemistry,* 11, 442, 1972.
32. **Bieskorn, C. H. and Beck, H. R.,** Die Bohlenwasserstoffe des Blattwachses von *Rosmarinus officinalis, Phytochemistry,* 9, 1633, 1970.
33. **Hill, A. S. and Mattick, L. R.,** The n-alkanes of cabbage (var. Copenhagen) and sauerkraut, *Phytochemistry,* 5, 693, 1966.
34. **Laseter, J. L., Weber, D. J., and Oro, J.,** Characterization of cabbage leaf lipids: n-alkanes, ketone, and fatty acids, *Phytochemistry,* 7, 1005, 1968.
35. **Purdy, S. J. and Truter, E. V.,** Constitution of the surface lipids from the leaves of *Brassica oleracea* (var. capitata (Winnigstadt)). Isolation and quantitative fractionation, *Proc. Roy. Soc. London, Ser. B,* 158, 536, 1963.
36. **Herbin, G. A. and Robins, P. A.,** Studies on plant cuticular waxes. II. Alkanes from members of the genus *Agave* (Agavaceae), the genera *Kalanchoe, Echeveria, Crassula* and *Sedum* (Crassulaceae) and the genus *Eucalyptus* (Myrtaceae) with an examination of Hutchinson's sub-division of the angiosperms into herbaceae and lignosae, *Phytochemistry,* 7, 257, 1968.
37. **Eglinton, G., Hamilton, R. J., and Martin-Smith, M.,** The alkane constituents of some New Zealand plants and their possible taxonomic implications, *Phytochemistry,* 1, 137, 1962.
38. **Eglinton, G., Gonzalez, A. G., Hamilton, R. J., and Raphael, R. A.,** Hydrocarbon constituents of the wax coatings of plant leaves: a taxonomic survey, *Phytochemistry,* 1, 89, 1962.
39. **Brieskorn, C. H., and Feilner, K.,** Zum aufbau des Pflanzlichen abschlubgewebes: Die normalen und verzweigten alkane von *Marrubium vulgare* L., *Phytochemistry,* 7, 485, 1968.
40. **Tulloch, A. P. and Hoffman, L. L.,** Epicuticular waxes of *Secale cereale* and *Triticale hexaploide* leaves, *Phytochemistry,* 13, 2535, 1974.
41. **Martin-Smith, M., Subramanian, G., and Connor, H. E.,** Surface wax components of five species of *Cortaderia* (Gramineae) — a chemotaxonomic comparison, *Phytochemistry,* 6, 559, 1967.
42. **Pitak, D. M. and Eichmeier, L. S.,** Identification of the alkanes in six *Asclepias* species, *Phytochemistry,* 11, 436, 1972.
43. **Scora, R. W. and Tin, W.,** Isolation and identification of alkanes from three taxa of *Monarda, Phytochemistry,* 10, 462, 1971.
44. **Nagy, B., Modzeleski, V., and Murphy, Sister Mary T. J.,** Hydrocarbons in the banana leaf, *Musa sapientum, Phytochemistry,* 4, 945, 1965.
45. **Gulz, P. -G.,** Normale und verzweigte alkane in Chloroplastenpraparaten und Blattern von *Antirrhinum majus, Phytochemistry,* 7, 1009, 1968.
46. **Nordby, H. E. and Nagy, S.,** Saturated and mono-unsaturated long-chain hydrocarbon profiles from *Citrus unshiu* juice sacs, *Phytochemistry,* 14, 183, 1975.
47. **Nordby, H. E. and Nagy, S.,** Saturated and mono-unsaturated long-chain hydrocarbon profiles from mandarin juice sacs, *Phytochemistry,* 14, 1777, 1975.
48. **Nagy, S. and Nordby, H. E.,** Saturated and mono-unsaturated long-chain hydrocarbon profiles of sweet oranges, *Phytochemistry,* 12, 801, 1973.
49. **Nagy, S. and Nordby, H. E.,** Long-chain hydrocarbon profiles of grapefruit juice sacs, *Phytochemistry,* 11, 2789, 1972.
50. **Nagy, S., Nordby, H. E., and Lastinger, J. C.,** Variation of the long-chain hydrocarbon pattern in different tissues of duncan grapefruit, *Phytochemistry,* 14, 2443, 1975.
51. **Nagy, S. and Nordby, H. E.,** Saturated and mono-unsaturated long-chain hydrocarbons of lime juice sacs, *Phytochemistry,* 11, 2865, 1972.
52. **Nordby, H. E. and Nagy, S.,** Saturated and mono-unsaturated long-chain hydrocarbons from lemon juice sacs, *Phytochemistry,* 11, 3249, 1972.
53. **Brown, S. O., Hamilton, R. J., and Shaw, S.,** Hydrocarbons from suds, *Phytochemistry,* 14, 2726, 1975.
54. **Gelpi, E., Schneider, H., Doctor, V. M., Tennison, J., and Oro, J.,** Gas chromatographic-mass spectrometric identifications of the hydrocarbons and fatty acids of *Plantago ovata* seeds, *Phytochemistry,* 8, 2077, 1969.
55. **Bennett, C. R. and Cambie, R. C.,** Chemistry of the Podocarpaceae. XIII. Constituents of *Podocarpus nivalis* Hook, and *Podocarpus acutifolius* Kirk, *Phytochemistry,* 6, 883, 1967.
56. **Sofowora, E. A. and Hardman, R. A.,** Steroids, phthalyl esters and hydrocarbons from *Balanites wilsoniana* stem bark, *Phytochemistry,* 12, 403, 1973.
57. **Sainsbury, M. and Webb, B.,** Hydrocarbons and terpenoids from the bark of *Lophopetalum rigidum, Phytochemistry,* 11, 3541, 1972.
58. **Hardman, R. and Brain, K. R.,** Alkanes of tubers of *Dioscorea deltoidea, Phytochemistry,* 10, 115, 1971.

59. **Farnsworth, N. R., Pettler, F. H., Wagner, H., and Horhammer, L.,** Kurze mitteilung untersuchungen uber Catharanthus-arten. XVI. Uber alkanverbindungen der Wurzeln von *Catharanthus longifolius, Phytochemistry,* 7, 887, 1968.
60. **Mody, N. V., Bhattacharyya, J., and Miles, D. H.,** Survey of the essential oil in *Spartina cynosuroides, Phytochemistry,* 13, 1175, 1974.
61. **Miles, D. H., Mody, N. V., and Minyard, J. P.,** Constituents of marsh grass: Survey of the essential oils in *Juncus roemerianus, Phytochemistry,* 12, 1399, 1973.

TERPENOIDS OF ALGAE

William Fenical

The terpenoid constituents of algae have been poorly explored. This area of investigation has been largely hindered by the unicellular, microscopic forms of the majority of the algae, and hence, the unavailability of large unialgal collections of these organisms from nature. Of the 13 divisions of algae, 3 contain significant numbers of macroscopic forms which are mainly of marine origin.[1] The macroscopic green algae (Chlorophyta), the brown algae (Phaeophyta), and the red algae (Rhodophyta) are almost exclusively marine. These larger and more easily collected plants have been the subject of recent investigations which are summarized in this section.

The subject of algal terpenoids has not been extensively reviewed. There are, however, several more general discussions of the naturally occurring compounds from marine organisms which can be consulted, particularly the recent CRC publication "Handbook of Marine Science, Compounds from Marine Organisms".[2] A complete chemical discussion of algal sesquiterpenoids has appeared,[3] and the chemical features of the marine-derived diterpenoids have been described.[4]

In principle, terpenoids are considered all metabolites produced by the polymerization of isopentenyl pyrophosphate. From this viewpoint the steroids (modified triterpenoids produced via lanosterol and squalene) and the carotenoids (true tetraterpenoid pigments) should be considered part of this review; however, these substances are not covered. The diterpenoid component of chlorophyll, phytol, and its dehydration products, the phytadienes, also are excluded.

The emphasis of this section is on the terpenoid components of algae for energy resources, and hence a rigorous definition of the structural chemistry of the algal terpenoids, is not presented. Simply, these compounds are tabulated with respect to their origin, general structure types, and molecular composition. This organization permits the rapid identification of specific groups known for their terpenoid synthesis. It should be pointed out, in advance, that few algal species produce terpenoids, with the exceptions being brown algae of the family Dictyotaceae and several genera of the red algae, including *Chondrococcus, Plocamium,* and *Laurencia.*

Table 1
OXYGENATED TERPENOIDS FROM GREEN ALGAE

Source	Formula	Trivial name	Ref.
Caulerpa prolifera	$C_{21}H_{26}O_6$	Caulerpenyne	5
C. brownii	$C_{20}H_{34}O$	Caulerpol	6
C. flexilis	$C_{19}H_{28}O_4$	Flexilin	7
C. trifaria	$C_{24}H_{38}O_4$	Trifarin	7
Rhipocephalus phoenix	$C_{21}H_{28}O_6$	Rhipocephalin	8
	$C_{15}H^{20}O_3$	Rhipocephenal	8

TABLE 2
SESQUITERPENE HYDROCARBONS FROM BROWN ALGAE

Source	Formula	Trivial name	Ref.
Dictyopteris undulata	$C_{15}H_{24}$	Zonarene	9
Dilophus fasciola	$C_{15}H_{22}$	1-(*S*)trans-(−)-calamene	10
	$C_{15}H_{24}$	δ-Cadinene	10
	$C_{15}H_{24}$	1-Epibicyclosesquiphel-landrene	10
Dictyopteris divaricata	$C_{15}H_{24}$	(−)-Copaene	11
	$C_{15}H_{24}$	(−)-γ-Cadinene	11
	$C_{15}H_{18}$	Cadalene	11
	$C_{15}H_{24}$	(−)-β-Elemene	11

Table 3
OXYGENATED SESQUITERPENOIDS AND DITERPENOIDS FROM BROWN ALGAE

Source	Formula	Trivial name	Ref.
Dilophus fasciola	$C_{15}H_{26}O$	Cubenol	10
	$C_{15}H_{26}O$	4,10-Epoxymuurolane	10
	$C_{17}H_{26}O_2$	(2*S*,8*R*)-germacra-1(11)5(12)E6-trien-2-ol acetate	11
Dictyopteris divaricata	$C_{15}H_{26}O$	(−)-δ-Cadinol	12
	$C_{15}H_{24}O$	Dictyopterol	13
	$C_{15}H_{22}O$	Dictyopterone	13
Pachydictyon coriaceum	$C_{20}H_{32}O$	Pachydictyol A	14
Dictyota flabellata	$C_{20}H_{32}O_2$	Pachydictyol A epoxide	15
D. acutiloba	$C_{20}H_{32}O_2$	Dictyoxepin	16
	$C_{20}H_{32}O$	Dictyolene	16
Dilophus ligulatus	$C_{20}H_{32}O$	Dilophol	17
	$C_{20}H_{32}O_2$	Dictyol E	18,20
Dictyota dichotoma	$C_{20}H_{32}O_2$	Dictyotadiol	18
	$C_{22}H_{34}O_2$	Dictyol B acetate	18
	$C_{20}H_{32}O_2$	Pachydictyol A	18
	$C_{20}H_{32}O_2$	Dictyol C	18,20
	$C_{20}H_{30}O_2$	Dictyol A	19
	$C_{20}H_{32}O_2$	Dictyol B	19
	$C_{20}H_{32}O_2$	Dictyol D	20
Dilophus prolificans	$C_{20}H_{30}O_2$	Epoxydilophone	21
	$C_{20}H_{30}O_2$	Dilophone	21
	$C_{22}H_{32}O_4$	Epi-acetoxydilophone	21
	$C_{22}H_{32}O_4$	Acetoxydilophone	21
Cystoseira elegans	$C_{20}H_{32}O_2$	Eleganolone	22

Table 4
MONOTERPENE HYDROCARBONS AND HALOGENATED HYDROCARBONS FROM RED ALGAE

Source	Formula	Trivial name	Ref.
Chondrococcus hornemanni	$C_{10}H_{16}$	Myrcene	23
	$C_{10}H_{15}Cl$	7-Chloromyrcene	23
	$C_{10}H_{15}Br$	7-Bromomyrcene	23
	$C_{10}H_{15}Br$	(Z)-10-bromomyrcene	23
	$C_{10}H_{15}Br$	(Z)-10-Bromomyrcene	23
	$C_{10}H_{15}Br$	(E)-10-Bromomyrcene	23
	$C_{10}H_{14}BrCl$	(Z)-10-Bromo-7-chloromyrcene	23
	$C_{10}H_{14}BrCl$	(E)-10-Bromo-7-chloromyrcene	23
	$C_{10}H_{13}Br_2Cl$	3-Chloro-7,(Z)-10-bromomyrcene	23
	$C_{10}H_{13}Br_2Cl$	(Z)-10-Chloro-3,7-dibromomyrcene	23
	$C_{10}H_{14}BrCl$	3-Bromo-7-chloromyrcene	23
	$C_{10}H_{14}BrCl$	7-Bromo-10-chloromyrcene	23
	$C_{10}H_{15}Br_2Cl_3$		24
	$C_{10}H_{14}BrCl_3$		24
	$C_{10}H_{15}BrCl_2$		24
	$C_{10}H_{16}Br_2Cl_2$		24
	$C_{10}H_{16}Br_2Cl_2$		24
	$C_{10}H_{14}BrCl$		24
	$C_{10}H_{15}Br_2Cl_3$		24
	$C_{10}H_{15}Br_2Cl$		24
	$C_{10}H_{14}Br_2Cl_2$		24
Ochtodes secundiramea	$C_{10}H_{14}Br_2Cl_2$	Ochtodene	25
Plocamium cartilagineum	$C_{10}H_{12}Cl_4$		26
	$C_{10}H_{11}Cl_5$		26
	$C_{10}H_{11}Cl_5$		26
	$C_{10}H_{11}Cl_5$		26
	$C_{10}H_{12}Br_2Cl_2$		26
	$C_{10}H_{12}Br_2Cl_2$		26
	$C_{10}H_{10}BrCl_5$		26
	$C_{10}H_{10}BrCl_5$		26
	$C_{10}H_{10}BrCl_5$		26
	$C_{10}H_{11}Br_3Cl_2$		26
	$C_{10}H_{11}Br_3Cl_2$		26
	$C_{10}H_{13}BrCl_2$		27
	$C_{10}H_{14}Br_2Cl_2$		27
	$C_{10}H_{13}Br_2Cl_3$		27
	$C_{10}H_{14}Br_2Cl_2$		27
	$C_{10}H_{13}BrCl_2$		27
	$C_{10}H_{14}BrCl_3$		27
	$C_{10}H_{14}Br_2Cl_2$		28
	$C_{10}H_{14}Br_2Cl_2$		28
	$C_{10}H_{13}BrCl_2$	Plocamene D′	29
	$C_{10}H_{13}Cl_3$	Plocamene E	29
	$C_{10}H_{13}Cl_3$	Plocamene D	29
	$C_{10}H_{14}BrCl_3$		30
	$C_{10}H_{14}Br_2Cl_2$		30
P. mertensii	$C_{10}H_{14}BrCl_3$	Mertensene	30
P. violaceum	$C_{10}H_{13}BrCl_4$	Violacene	31,32
	$C_{10}H_{14}Br Cl_3$		33
	$C_{10}H_{13}Cl_3$	Plocamene B	34
	$C_{10}H_{15}Br_2Cl$		35
	$C_{10}H_{15}Cl_3$		35
Microcladia borealis	$C_{10}H_{13}BrCl_4$	Violacene	36
	$C_{10}H_{13}Cl_3$	Plocamene B	36
	$C_{10}H_{13}Cl_3$	Plocamene C	36
	$C_{10}H_{14}BrCl_3$	Plocamene D	36

Table 4 (continued)
MONOTERPENE HYDROCARBONS AND HALOGENATED HYDROCARBONS FROM RED ALGAE

Source	Formula	Trivial name	Ref.
M. californica	$C_{10}H_{13}BrCl_4$	Violacene	36
	$C_{10}H_{13}Cl_3$	Plocamene B	36
	$C_{10}H_{13}Cl_3$	Plocamene C	36
	$C_{10}H_{14}BrCl_3$	Plocamene D	36
M. coulteri	$C_{10}H_{13}BrCl_4$	Violacene	36
	$C_{10}H_{13}Cl_3$	Plocamene B	36
	$C_{10}H_{13}Cl_3$	Plocamene C	36
	$C_{10}H_{14}BrCl_3$	Plocamene D	36

Table 5
SESQUITERPENE HYDROCARBONS AND HALOGENATED HYDROCARBONS FROM RED ALGAE

Source	Formula	Trivial name	Ref.
Laurencia sp.-1	$C_{15}H_{18}$	Guaiazulene	37
Laurencia sp.-2	$C_{15}H_{23}Br_2Cl$		38
L. filiformis	$C_{15}H_{22}$	Dehydrolaurene	39
L. glandulifera	$C_{15}H_{20}$	Laurene	40
	$C_{15}H_{21}Br$	α-bromocuparane	41
	$C_{15}H_{21}Br$	α-Isobromocuparane	41
L. nidifica	$C_{15}H_{24}$	(+)-Selin-4,7(11)-diene	42
	$C_{15}H_{23}Br_2Cl$	Nidificene	43
	$C_{15}H_{22}BrCl$	Nidifidiene	43
L. nipponica	$C_{15}H_{20}$	Laurene	44
L. pacifica	$C_{15}H_{24}$	γ-Bisabolene	45
	$C_{15}H_{23}Br$	10-Bromo-α-chamigrene	46
	$C_{15}H_{23}Br$	10-Bromo-β-chamigrene	47
L. intricata	$C_{15}H_{24}BrCl$	Preintricatol	48
L. perforata	$C_{15}H_{20}BrCl$	Perforene	49

Table 6
OXYGENATED MONOTERPENOIDS FROM RED ALGAE

Source	Formula	Trivial name	Ref.
Chondrococcus hornemanni	$C_{10}H_{14}BrClO$	Chondrocole A	50,51
	$C_{10}H_{14}BrClO$	Chondrocole B	50
	$C_{10}H_{14}Br_2O$		24
	$C_{10}H_{12}BrClO_2$		51
Ochtodes secundiramea	$C_{10}H_{16}BrClO_2$	Ochtodiol	25
Plocamium cartilagineum	$C_{10}H_{11}Cl_3O$	Cartilagineal	52
P. costatum	$C_{10}H_{12}Br_2Cl_2O_2$	Costatone	53,54
	$C_9H_{10}Cl_2O_2$	Costatolide	53,54

Table 7
OXYGENATED SESQUITERPENOIDS FROM RED ALGAE

Source	Formula	Trivial name	Ref.
Laurencia caespitosa	$C_{15}H_{25}Br_2ClO_2$	Caespitol	55,56
	$C_{15}H_{25}Br_2ClO_2$	Isocaespitol	56
L. decidua	$C_{15}H_{19}BrO$	Aplysin	57
	$C_{15}H_{20}O$	Debromoaplysin	57
	$C_{15}H_{18}Br_2o$	Bromoaplysin	57
	$C_{15}H_{19}BrO_2$	Aplysinol	57
	$C_{15}H_{19}BrO$	Laurinterol	57
	$C_{15}H_{20}O$	Debromolaurinterol	57
	$C_{15}H_{18}Br_2O$	Bromolaurinterol	57
	$C_{17}H_{21}BrO_2$	Laurinterol acetate	57
	$C_{17}H_{22}O_2$	Debromolaurinterol acetate	57
	$C_{15}H_{19}BrO$	Isolaurinterol	57
	$C_{15}H_{20}O$	Debromoisolaurinterol	57
	$C_{17}H_{22}O_2$	Debromoisolaurinterol acetate	57
	$C_{15}H_{18}Br_2O$	Bromoisolaurinterol	57
L. elata	$C_{15}H_{22}BrClO$	Elatol	57
L. filiformis	$C_{15}H_{19}BrO$	10-Bromo-7-hydroxylaurene	39
	$C_{15}H_{19}BrO$	Filiformin	39
	$C_{15}H_{19}BrO_2$	Filiforminol	39
	$C_{15}H_{20}O$	7-Hydroxylaurene	39
L. glandulifera	$C_{15}H_{19}BrO$	Bromoether A	59
	$C_{15}H_{18}Br_2O$	Bromoether B	59
	$C_{15}H_{23}BrO$	Spirolaurenone	60
	$C_{15}H_{23}BrO$	10-Bromo-α-chamigrene-2-one	61
	$C_{15}H_{23}BrO$	10-Bromo-β-chamigrene-2-one	61
	$C_{15}H_{24}BrClO$	Glanduliferol	62
L. intricata	$C_{17}H_{27}Br_2ClO_3$	Acetoxyintricatol	48,63
	$C_{17}H_{26}BrClO_3$	Cyclodebromoacetoxyintricatol	48
L. johnstonii	$C_{15}H_{21}Br_2ClO_3$	Prepacifenol epoxide	64
L. majuscula	$C_{15}H_{20}Br_2O$		65
	$C_{15}H_{19}BrO$		65
L. nidifica	$C_{15}H_{22}BrClO$	Nidifidienol	66
	$C_{15}H_{24}O$		66
	$C_{15}H_{22}O$		66
	$C_{15}H_{22}BrClO$	Nidifocene	67,68
L. nipponica	$C_{15}H_{19}BrO_2$		69
	$C_{15}H_{19}BrO$	Laurenisol	70
L. obtusa	$C_{15}H_{23}Br_2ClO$	Obtusol	71
	$C_{15}H_{23}Br_2ClO$	Isoobtusol	71
	$C_{15}H_{24}BrClO$	Debromoobtusol	71
	$C_{15}H_{24}BrClO$	Debromoisoobtusol	71
	$C_{15}H_{24}BrO$	α-Snyderol	72
	$C_{15}H_{25}BrO$	3β-Bromo-8-epicaparrapi oxide	73
L. okamurai	$C_{15}H_{19}BrO$	Aplysin	74
	$C_{15}H_{20}O$	Debromoaplysin	74
	$C_{15}H_{19}BrO_2$	Aplysinol	74
	$C_{15}H_{19}BrO$	Laurinterol	74
	$C_{15}H_{20}O$	Debromolaurinterol	74
	$C_{15}H_{19}BrO$	Neolaurinterol	75
	$C_{15}H_{19}BrO$	Isoaplysin	75
	$C_{15}H_{20}O$		75
L. intermedia	$C_{15}H_{19}BrO$	Laurinterol	76,77
	$C_{15}H_{20}O$	Debromolaurinterol	76
	$C_{15}H_{19}BrO$	Isolaurinterol	77
L. pacifica	$C_{15}H_{19}BrO$	Aplysin	78
	$C_{15}H_{20}O$	Debromoaplysin	78
	$C_{15}H_{21}Br_2ClO_2$	Prepacifenol	79

	$C_{15}H_{23}Br_2ClO$	2,10-Dibromo-3-chloro-9-hydroxy-α-chamigrene	80
L. perforata	$C_{15}H_{20}Br_2$	Perforatone	81
	$C_{15}H_{22}O_2$	Perforenone A	81
	$C_{15}H_{21}ClO$	Perforenone B	81
L. poitei	$C_{15}H_{26}O_2$	Poitediol	82
	$C_{15}H_{26}O$	Dactylol	82
L. snyderae	$C_{15}H_{24}BrO$	β-Snyderol	72
L. synderae var. *guadalupensis*	$C_{15}H_{24}O$	Guadalupol	83
	$C_{15}H_{24}O$	Epiguadalupol	83
L. subopposita	$C_{15}H_{25}BrO$	Oppositol	84
	$C_{15}H_{20}O$	7-Hydroxylaurene	85
	$C_{15}H_{19}BrO$	10-Bromo-7-hydroxylaurene	85
	$C_{15}H_{26}O_2$	Opolopanone	85
	$C_{15}H_{24}O$		85
	$C_{15}H_{24}O$		85
	$C_{15}H_{26}O_2$		85
	$C_{15}H_{25}BrO_2$		85
L. tasmanica	$C_{15}H_{21}Br_2ClO_2$	Pacifenol	79
Laurencia sp.-1	$C_{15}H_{26}O$	Cycloeudesmol	86
Laurencia sp.-2	$C_{15}H_{25}BrO$	1-(S)-Bromo-4-(R)-hydroxy-(-)-selin-7-ene	37,87
Laurenciasp.-3	$C_{15}H_{23}Br_2ClO$	2,10-Dibromo-3-chloro-7,8-epoxychamigrene	38
Marginisporum aberrans	$C_{15}H_{19}BrO$	Aplysin	88
	$C_{15}H_{19}BrO_2$	Aplysinol	88
	$C_{15}H_{17}BrO_2$	Aplysinal	88
	$C_{15}H_{19}BrO$	Laurinterol	88
	$C_{15}H_{19}BrO$	Isolaurinterol	88

Table 8
OXYGENATED DITERPENOIDS FROM RED ALGAE

Source	Formula	Trivial name	Ref.
Laurencia concinna	$C_{20}H_{35}BrO_2$	Concinndiol	89
L. irieii	$C_{20}H_{32}Br_2O_2$	Iriediol	90
	$C_{20}H_{32}Br_2O_2$	Irieol A	90
	$C_{20}H_{32}Br_2O$	Irieol	91
	$C_{20}H_{34}Br_2O_3$	Irieol B	91
	$C_{22}H_{36}Br_2O_4$	Irieol C	91
	$C_{22}H_{34}Br_2O_4$	Irieol D	91
	$C_{20}H_{32}Br_2O_2$	Irieol E	91
	$C_{20}H_{34}Br_2O_4$	Irieol F	91
	$C_{22}H_{34}Br_2O_4$	Irieol G	91
L. obtusa	$C_{20}H_{32}Br_2O_2$	Obtusadiol	92
L. snyderae	$C_{20}H_{34}O_4$	Neoconcinndiol hydroperoxide	93
Sphaerococcus coronopifolius	$C_{20}H_{29}BrO_2$	Sphaerococcenol A	94
	$C_{20}H_{32}Br_2O$	Bromosphaerol	95
	$C_{20}H_{32}Br_2O_2$	Bromosphaerodiol	96

Table 9
TERPENOIDS OF MIXED OR UNUSUAL BIOGENESIS FROM ALGAE

Source	Formula	Trivial name	Ref.
Green algae			
Cymopolia barbata	$C_{16}H_{21}BrO_2$	Cymopol	97
	$C_{17}H_{23}BrO_2$	Cymopol methylether	97
	$C_{16}H_{20}Br_2O_2$	Cyclocymopol	97
	$C_{17}H_{22}Br_2O_2$	Cyclocymopol methylether	97
	$C_{16}H_{19}BrO_2$	Cymopochromenol	97
	$C_{16}H_{19}BrO_3$	Cymopolone	97
	$C_{16}H_{19}BrO_2$	Isocymopolone	97
Brown algae			
Taonia atomaria	$C_{27}H_{40}O_3$	Taondiol	98
	$C_{28}H_{42}O_4$	Atomaric acid	99
Stypopodium zonale	$C_{27}H_{38}O_4$	Stypoldione	100
	$C_{27}H_{40}O_4$	Stypotriol	100
Sargassum tortile	$C_{27}H_{40}O_3$		101
	$C_{27}H_{40}O_2$		101
Dictyopteris undulata	$C_{21}H_{30}O_3$	Zonarol	102
	$C_{21}H_{30}O_2$	Isozonarol	102
Cystoseira crinita	$C_{20}H_{34}O_2$	Crinitol	103
	$C_{14}H_{24}O_2$	Oxocrinol	103
Red algae			
Laurencia thysifera	$C_{30}H_{52}O_7Br$	Thysiferol	104

REFERENCES

1. **Round, F. E.,** *The Biology of the Algae,* 2nd Ed., St. Martins, New York, 1973, 231.
2. **Baker, J. T. and Murphy, V.,** *Handbook of Marine Science, Compounds from Marine Organisms,* Vol. I, CRC Press, Cleveland, 1976.
3. **Martin, J. D. and Darias, J.,** Algal sesquiterpenoids, in *Marine Natural Products; Chemical and Biological Perspectives,* Vol. 1, Scheuer, P. J., Ed., Academic Press, New York, 1978, chap. 3.
4. **Fenical, W.,** Diterpenoids, in *Marine Natural Products; Chemical and Biological Perspectives,* Vol. 2, Scheuer, P. J., Ed., Academic Press, New York, 1978.
5. **Amico, V., Oriente, G., Piattelli, M., Tringali, C., Fattorusso, E., Magno, S., and Mayol, L.,** Caulerpenyne, an unusual sesquiterpenoid from the green alga *Caulerpa prolifera, Tetrahedron Lett.,* p. 3593, 1978.
6. **Blackman, A. J. and Wells, R. J.,** Caulerpol, a diterpene alcohol related to vitamin A from *Caulerpa brownii, Tetrahedron Lett.,* p. 2729, 1976.
7. **Blackman, A. J. and Wells, R. J.,** Flexilin and trifarin, terpene 1,4-diacetoxybuta-1,3-dienes from two *Caulerpa* species (Chlorophyta), *Tetrahedron Lett.,* p. 3063, 1978.
8. **Sun, H. H. and Fenical, W.,** Rhipocephalin and rhipocephenal, toxic feeding deterrents from the tropical marine alga *Rhipocephalus phoenix, Tetrahedron Lett.,* p. 685, 1979.
9. **Fenical, W., Sims, J. J., Wing, R. M., and Radlick, P. C.,** Zonarene, a sesquiterpene from the brown seaweed *Dictyopteris zonarioides (undulata), Phytochemistry,* 11, 1161, 1972.
10. **Amico, V., Oriente, G., Piattelli, M., Tringali, C., Fattorusso, E., Magno, S., and Mayol, L.,** Sesquiterpenes based on the cadalane skeleton from the brown alga *Dilophus fasciola, Experientia,* 35, 450, 1979.
11. **Fattorusso, E., Magno, S., Mayol, L., Amico, V., Oriente, G., Piattelli, M., and Tringali, C.,** Isolation of (2S,8R)-germacra-1(11),5(12),E6-triene-2-ol acetate from the brown alga *Dilophus fasciola, Tetrahedron Lett.,* p. 4149, 1979.
12. **Irie, T., Yamamoto, K., and Masamune, T.,** Sesquiterpenes from *Dictyopteris divaricata,* I, *Bull. Chem. Soc. Jpn.,* 37, 1053, 1964.
13. **Kurosawa, E., Izawa, M., Yamamoto, K., Masamune, T., and Irie, T.,** Sesquiterpenes from *Dictyopteris divaricata,* II. Dictyopterol and dictyopterone, *Bull. Chem. Soc. Jpn.,* 39, 2509, 1966.

14. **Hirschfeld, D. R., Fenical, W., Lin, G. H. Y., Wing, R. M., Radlick, P., and Sims, J. J.,** Marine natural products. VIII. Pachydictyol A, an exceptional diterpene alcohol from the brown alga *Pachydictyon coriaceum, J. Am. Chem. Soc.,* 95, 4049, 1973.
15. **Robertson, K. J. and Fenical, W.,** Pachydictyol A epoxide, a new diterpene from the brown alga *Dictyota flabellata, Phytochemistry,* 16, 1071, 1977.
16. **Sun, H. H., Waraszkiewicz, S. M., Erickson, K. L., Finer, J., and Clardy, J.,** Dictyoxepin and dictyolene, two new diterpenes from the marine algo *Dictyota acutiloba* (Phaeophyta), *J. Am. Chem. Soc.,* 99, 3516, 1977.
17. **Amico, V., Oriente, G., Piatelli, M., Tringali, C., Fattorusso, E., Magno, S., and Mayol, L., Dilophol,** a new ten-membered ring diterpene alcohol from the brown alga *Dilophus ligulatus. J. Chem. Soc., Chem. Commun.,* 1024, 1976.
18. **Faulkner, D. J., Ravi, B. N., Finer, J., and Clardy, J.,** Diterpenes from *Dictyota dichotoma, Phytochemistry,* 16, 991, 1977.
19. **Fattorusso, E., Magno, S., Mayol, L., Santacroce, C., Sica, D., Amico, V., Oriente, G., Piattelli, M., and Tringali, C.,** Dictyol A and B, two novel diterpene alcohols from the brown alga *Dictyota dichotoma, J. Chem. Soc., Chem. Commun.,* 575, 1976.
20. **Danise, B., Minale, L., Riccio, R., Amico, V., Oriente, G., Piattelli, M., Tringali, C., Fattorusso, E., Magno, S., and Mayol, L.,** Further perhydroazulene diterpenes from marine organisms, *Experientia,* 33, 413, 1977.
21. **Kazlauskas, R., Murphy, P. T., Wells, R. J., and Blount, J. F.,** A series of novel bicyclic diterpenes from *Dilophus prolificans* (brown alga, Dictyotaceae), *Tetrahedron Lett.,* p. 4155, 1978.
22. **Francisco, C., Combaut, G., Teste, J., and Prost, M.,** Eleganolone nouveau cetol diterpenique lineaire de la pheophycee *Cystoseira elegans, Phytochemistry,* 17, 1003, 1978.
23. **Ichikawa, N., Naya, Y., and Enomoto, S.,** New halogenated monoterpenes from *Desmia (Chondrococcus) hornemanni, Chem. Lett. (Jpn.),* p. 1333, 1974.
24. **Burreson, B. J., Woolard, F. X., and Moore, R. E.,** Evidence for the biogenesis of halogenated myrcenes from the red alga *Chondrococcus hornemanni, Chem. Lett. (Jpn.),* p. 1111, 1975.
25. **McConnell, O. J. and Fenical, W.,** Ochtodene and ochtodiol, novel polyhalogenated cyclic monoterpenes from the red seaweed *Ochtodes secundiramea, J. Org. Chem.,* 43, 4238, 1978.
26. **Mynderse, J. S. and Faulkner, D. J.,** Polyhalogenated monoterpenes from the red alga *Plocamium cartilagineum, Tetrahedron,* 31, 1963, 1975.
27. **Higgs, M. D., Vanderah, D. J., and Faulkner, D. J.,** Polyhalogenated monoterpenes from *Plocamium cartilagineum* from the British Coast, *Tetrahedron,* 33, 2775, 1977.
28. **Gonzalez, A. G., Arteaga, J. M., Martin, J. D., Rodriquez, M. L., Fayos, J., and Martinez-Ripolls, M.,** Two new polyhalogenated monoterpenes from the red alga *Plocamium cartilagineum, Phytochemistry,* 17, 947, 1978.
29. **Crews, P., Kho-Wiseman, E., and Montana, P.,** Halogenated alicyclic monoterpenes from the red alga *Plocamium, J. Org. Chem.,* 43, 116, 1978.
30. **Norton, R. S., Warren, R. G., and Wells, R. J.,** Three new polyhalogenated monoterpenes from *Plocamium* species, *Tetrahedron Lett.,* p. 3905, 1977.
31. **Mynderse, J. S. and Faulkner, D. J.,** Violacene, a polyhalogenated monoterpene from the red alga *Plocamium violaceum, J. Am. Chem. Soc.,* 96, 6771, 1974.
32. **Van Engen, D., Clardy, J., Kho-Wiseman, E., Crews, P., Higgs, M., and Faulkner, D. J.,** Violacene, a reassignment of structure, *Tetrahedron Lett.,* p. 29, 1978.
33. **Mynderse, J. S., Faulkner, D. J., Finer, J., and Clardy, J.,** (1*R*,2*S*,4*S*,5*R*)-1-Bromo-trans-2-chlorovinyl-1,5-dimethylcyclohexane, a new monoterpene skeletal type from the red alga *Plocamium violaceum, Tetrahedron Lett.,* p. 2175, 1975.
34. **Crews, P. and Kho, E.,** Plocamene B, a new cyclic monoterpene skeleton from a red marine alga, *J. Org. Chem.,* 40, 2568, 1975.
35. **Crews, P. and Kho-wiseman, E.,** Acyclic polyhalogenated monoterpenes from the red alga *Plocamium violaceum, J. Org. Chem.,* 42, 2812, 1977.
36. **Crews, P., Ng, P., Kho-wiseman, E., and Pace, C.,** Halogenated monoterpene synthesis by the red alga *Microcladia, Phytochemistry,* 15, 1707, 1976.
37. **Howard, B. M. and Fenical, W.,** Structure, chemistry and absolute configuration of 1-*S*-bromo-4-*R*-hydroxy-(−)-selin-7-ene from a marine red alga *Laurencia* sp., *J. Org. Chem.,* 42, 2518, 1977.
38. **Howard, B. M. and Fenical, W.,** Structures and chemistry of two new hologen-containing chamigrene derivatives from *Laurencia, Tetrahedron Lett.,* p. 1687, 1975.
39. **Kazlauskas, R., Murphy, P. T., Quinn, R. J., and Wells, R. J.,** New laurene derivatives from *Laurencia filiformis, Aust. J. Chem.,* 29, 2533, 1976.
40. **Irie, T., Yasunari, Y., Suzuki, T., Imai, N., Kurosawa, E., and Masamune, T.,** A new sesquiterpene hydrocarbon from *Laurencia glandulifera, Tetrahedron Lett.,* p. 3619, 1965.
41. **Suzuki, T., Suzuki, M., and Kurosawa, E.,** α-Bromocuparane and α-isobromocuparane, new bromo compounds from *Laurencia* species, *Tetrahedron Lett.,* p. 3057, 1975.

42. **Sun, H. H. and Erickson, K. L.**, Sesquiterpenoids from the Hawaiian marine alga *Laurencia nidifica*, VIII. (+)-Selin-4,7(11)-diene, *J. Org. Chem.*, 43, 1613, 1978.
43. **Waraszkiewicz, S. M. and Erickson, K. L.**, Halogenated sesquiterpenoids from the Hawaiian marine alga *Laurencia nidifica:* nidificene and nidifidiene, *Tetrahedron Lett.*, p. 2003, 1974.
44. **Irie, T., Suzuki, T., Yasunari, Y., Kurosawa, E., and Masamune, T.**, Laurene, a sesquiterpene hydrocarbon from *Laurencia* species, *Tetrahedron*, 25, 459, 1969.
45. **Wolinsky, L. E.**, Biomimetic Approaches to Marine Natural Products, Doctoral dissertation, University of California, San Diego, 1976.
46. **Howard, B. M. and Fenical, W.**, 10-Bromo-α-chamigrene, *Tetrahedron Lett.*, p. 2519, 1976.
47. **Rose, A. F., Izac, R. R., and Sims, J. J.**, Applications of ^{13}C NMR to marine natural products, in *Marine Natural Products; Chemical and Biological Perspectives*, Vol. 2, Scheuer, P. J., Ed., Academic Press, New York, 1978.
48. **White, R. H. and Hager, L. P.**, A biogenetic sequence of halogenated sesquiterpenes from *Laurencia intricata*, in *Dahlem Workshop on the Nature of Seawater*, Goldberg, E. D., Ed., Dahlem Konferenzen, Berlin, 1975.
49. **Gonzalez, A. G., Aguilar, J. M., Martin, J. D., and Rodriguez, M. L.**, Perforene, a new halogenated sesquiterpene from the red alga *Laurencia perforata*, *Tetrahedron Lett.*, p. 205, 1976.
50. **Burreson, B. J., Woolard, F. X., and Moore, R. E.**, Chondrocole A and B, two halogenated dimethylhexahydrobenzofurans from the red alga *Chondrococcus hornemanni* (Mertens) Schmitz, *Tetrahedron Lett.*, p. 2155, 1975.
51. **Woolard, F. X., Moore, R. E., Van Engen, D., and Clardy, J.**, The isolation and absolute configuration of chondrocolactone, a halogenated monoterpene from the red alga *Chondrococcus hornemanni*, and a revised structure for chondrocole A, *Tetrahedron Lett.*, p. 2367, 1978.
52. **Crews, P. and Kho, E.**, Cartilagineal, an unusual monoterpene aldehyde from marine alga, *J. Org. Chem.*, 39, 3303, 1974.
53. **Stierle, D. B., Wing, R. M., and Sims, J. J.**, Marine natural products, XI. Costatone and cocsatolide, new halogenated monoterpenes from the red seaweed *Plocamium costatum*, *Tetrahedron Lett.*, p. 4455, 1976.
54. **Kazlauskas, R., Murphy, P. T., Quinn, R. J., Wells, R. J., and Schonholzer, P.**, Two polyhalogenated monoterpenes from the red alga *Plocamium costatum*, *Tetrahedron Lett.*, p. 4451, 1976.
55. **Gonzalez, A. G., Darias, J., and Martin, J. D.**, Caespitol, a new halogenated sesquiterpene from *Laurencia caespitosa*, *Tetrahedron Lett.*, p. 2381, 1973.
56. **Gonzalez, A. G., Darias, J., and Martin, J. D.**, Revised structure of caespitol and its correlation with isocaespitol, *Tetrahedron Lett.*, p. 1249, 1974.
57. **Caccamese, S. and Rinehart, K. L., Jr.**, New compounds and activities from *Laurencia* species, in *Drugs and Food from the Sea, Myth or Reality?* Kaul, P. N. and Sinderman, C. J., Eds., University of Oklahoma Press, Norman, 1978.
58. **Sims, J. J., Lin, G. H. Y., and Wing, R. M.**, Marine natural products. X. Elatol, a halogenated sesquiterpene alcohol from the red alga *Laurencia elata*, *Tetrahedron Lett.*, p. 3487, 1974.
59. **Suzuki, M. and Kurosawa, E.**, New bromo compounds from *Laurencia glandulifera* Kutzing, *Tetrahedron Lett.*, p. 4816, 1976.
60. **Suzuki, M., Kurosawa, E., and Irie, T.**, Spirolaurenone, a new sesquiterpenoid containing bromine from *Laurencia glandulifera* Kutzing, *Tetrahedron Lett.*, p. 4995, 1970.
61. **Suzuki, M., Kurosawa, E., and Irie, T.**, Three new sesquiterpenoids containing bromine, minor constituents of *Laurencia glandulifera* Kutzing, *Tetrahedron Lett.*, p. 821, 1974.
62. **Suzuki, M., Kurosawa, E., and Irie, T.**, Glanduliferol, a new halogenated sesquiterpenoid from *Laurencia glandulifera* Kutzing, *Tetrahedron Lett.*, p. 1807, 1974.
63. **McMillan, J. A., Paul, I. C., White, R. H., and Hager, L. P.**, Molecular structure of acetoxyintricatol: a new bromo compound from *Laurencia intricata*, *Tetrahedron Lett.*, p. 2039, 1974.
64. **Faulkner, D. J., Stallard, M. O., and Ireland, C.**, Prepacifenol epoxide, a halogenated sesquiterpene diepoxide, *Tetrahedron Lett.*, p. 1687, 1975.
65. **Suzuki, M. and Kurosawa, E.**, Two new halogenated sesquiterpenes from the red alga *Laurencia majuscula*, *Tetrahedron Lett.*, p. 4805, 1978.
66. **Sun, H. H., Waraszkiewicz, S. M., and Erickson, K. L.**, Sesquiterpenoid alcohols from the Hawaiian marine alga *Laurencia nidifica*. III, *Tetrahedron Lett.*, p. 585, 1976.
67. **Waraskiewicz, S. M. and Erickson, K. L.**, Halogenated sesquiterpenoids from the Hawaiian marine alga *Laurencia nidifica*. IV. Nidifocene, *Tetrahedron Lett.*, p. 1443, 1976.
68. **Waraskiewicz, S. M., Erickson, K. L., Finer, J., and Clardy, J.**, Nidifocene, a reassignment of structure, *Tetrahedron Lett.*, p. 2311, 1977.
69. **Suzuki, T., Furusaki, A., Hasiba, N., and Kurosawa, E.**, Novel skeletal bromo ether from the marine alga *Laurencia nipponica* Yamada, *Tetrahedron Lett.*, p. 3731, 1977.
70. **Irie, T., Fukuzawa, A., Izawa, M., and Kurosawa, E.**, Laurenisol, a new sesquiterpenoid containing bromine from *Laurencia nipponica* Yamada, *Tetrahedron Lett.*, p. 1341, 1969.

70. **Irie, T., Fukuzawa, A., Izawa, M., and Kurosawa, E.**, Laurenisol, a new sesquiterpenoid containing bromine from *Laurencia nipponica* Yamada, *Tetrahedron Lett.*, p. 1341, 1969.
71. **Gonzalez, A. G., Darias, J., Diaz, A., Fourneron, J. D., Martin, J. D., and Perez, C.**, Evidence for the biogenesis of halogenated chamigrenes from the red alga *Laurencia obtusa, Tetrahedron Lett.*, p. 3051, 1976.
72. **Howard, B. M. and Fenical, W.**, α-And β-synderol; new bromo-monocyclic sesquiterpenes from the seaweed *Laurencia, Tetrahedron Lett.*, p. 41, 1976.
73. **Faulkner, D. J.**, 3βBromo-8-epicaparrapi oxide, the major metabolite of *Laurencia obtusa, Phytochemistry*, 15, 1993, 1976.
74. **Irie, T., Suzuki, M., and Hayakawa, Y.**, Isolation of aplysin, debromoaplysin and aplysinol from *Laurencia okamurai, Bull. Chem. Soc. Jpn.*, 42, 843, 1969.
75. **Suzuki, M. and Kurosawa, E.**, New aromatic sesquiterpenoids from the red alga *Laurencia okamurai* Yamada, *Tetrahedron Lett.*, p. 2503, 1978.
76. **Irie, T., Suzuki, M., Ito, S., and Kurosawa, E.**, Laurinterol and debromolaurinterol, constituents from *Laurencia intermedia, Tetrahedron Lett.*, p. 1837, 1966.
77. **Irie, T., Suzuki, M., Kurosawa, E., and Masamune, T.**, Laurinterol, debromolaurinterol and isolaurinterol from *Laurencia intermedia* Yamada, *Tetrahedron*, 26, 3271, 1970.
78. **Sims, J. J., Fenical, W., Wing, R. M., and Radlick, P.**, Marine natural products. I. Pacifenol, a rare sesquiterpene containing bromine and chlorine from the red algal *Laurencia pacifica, J. Am. Chem. Soc.*, 93, 3447, 1971.
79. **Sims, J. J., Fenical, W., Wing, R. M., and Radlick, P.**, Marine natural products. IV. Prepacifenol, a halogenated epoxy sesquiterpene and precursor to pacifenol from the red alga *Laurencia filiformis, J. Am. Chem. Soc.*, 95, 972, 1973.
80. **Fenical, W.**, Chemical variation in a new bromo chamigrene derivative from the red seaweed *Laurencia pacifica, Phytochemistry*, 15, 511, 1976.
81. **Gonzalez, A. G., Aguilar, J. M., Martin, J. D., and Norte, M.**, Three new sesquiterpenoids from the marine alga *Laurencia perforata, Tetrahedron Lett.*, p. 2499, 1975.
82. **Fenical, W., Schulte, G. R., Finer, J., and Clardy, J.**, Poitediol, a new sesquiterpene diol from the marine alga *Laurencia poitei, J. Org. Chem.*, 43, 3628, 1978.
83. **Howard, B. M. and Fenical, W.**, Guadalupol and epiguadalupol, rearranged sesquiterpene alcohols from the red alga *Laurencia snyderae* var. *guadalupensis, Phytochemistry*, 18, 1224, 1979.
84. **Hall, S. S., Faulkner, D. J., Fayos, J., and Clardy, J.**, Oppositol, a brominated sesquiterpene alcohol of a new skeletal class from the red alga *Laurencia subopposita, J. Am. Chem. Soc.*, 95, 7187, 1973.
85. **Wratten, S. J. and Faulkner, D. J.**, Metabolites of the red alga *Laurencia subopposita, J. Org. Chem.*, 42, 3343, 1977.
86. **Fenical, W. and Sims, J. J.**, Cycloesdesmol, an antibiotic cyclopropane-containing sesquiterpene from the marine alga *Chondria oppositiclada Dawson, Tetrahedron Lett.*, p. 1137, 1974.
87. **Rose, A. F. and Sims, J. J.**, Marine natural products. XIV. 1-S-Bromo-4-(*R*)-hydroxyselin-7-ene, a metabolite of the marine alga *Laurencia* sp., *Tetrahedron Lett.*, p. 2935, 1977.
88. **Ohta, K. and Takagi, M.**, Halogenated sesquiterpenes from the marine red alga *Marginisporum aberrans, Phytochemistry*, 16, 1062, 1977.
89. **Sims, J. J., Lin, G. H. Y., Wing, R. M., and Fenical, W.**, Marine natural products, concinndiol, a bromo-diterpene alcohol from the red alga *Laurencia concinna, J. Chem. Soc., Chem. Commun.*, 470, 1973.
90. **Fenical, W., Howard, B. M., Gifkins, K. B., and Clardy, J.**, Irieol A and iriediol, dibromoditerpenes of a new skeletal class from *Laurencia, Tetrahedron Lett.*, p. 3983, 1975.
91. **Howard, B. M. and Fenical, W.**, Structure of the irieols, new dibromoditerpenoids of a unique skeletal class from the marine red alga *Laurencia irieii, J. Org. Chem.*, 43, 4401, 1978.
92. **Howard, B. M. and Fenical, W.**, Obtusadiol, a unique bromoditerpenoid from the marine red alga *Laurencia obtusa, Tetrahedron Lett.*, p. 2453, 1978.
93. **Howard, B. M., Fenical, W., Finer, J., Hirotsu, K., and Clardy, J.**, Neoconcinndiol hydroperoxide, a novel marine diterpenoid from the red alga *Laurencia, J. Am. Chem. Soc.*, 99, 6440, 1977.
94. **Fenical, W., Finer, J., and Clarfy, J.**, Sphaerococcenol A, a new rearranged bromo-diterpene from the red alga *Sphaerococcus coronopifolous, Tetrahedron Lett.*, p. 731, 1976.
95. **Fattorusso, E., Magno, S., Santacroce, C., Sica, D., DiBlasio, B., and Pedone, C.**, Bromosphaerol, a new bromine-containing diterpenoid from the red alga *Sphaerococcus coronopifolius, Gazz. Chim. Ital.*, 106, 779, 1976.
96. **Califieri, F., De Napoli, L., Fattorusso, E., Impellizzeri, G., Piattelli, M., and Scuito, S.**, Bromosphaerodiol, a minor bromo compound from the red alga *Sphaerococcus coronopifolius, Experientia*, 33, 1549, 1977.
97. **Hogberg, H. -E., Thomson, R. H., and King, T. J.**, The cymopols, a group of prenylated bromohydroquinones from the green calcareous alga *Cymopolia barbata, J. Chem. Soc.*, 1696, 1976.

98. **Gonzalez, A. G., Darias, J., and Martin, J. D.,** Taondiol, a new component from *Taonia atomaria, Tetrahedron Lett.,* p. 2729, 1971.
99. **Gonzalez, A. G., Darias, J., Martin, J. D., and Norte, M.,** Atomaric acid, a new component of *Taonia atomaria, Tetrahedron Lett.,* 3951, 1974.
100. **Gerwick, W. H., Fenical, W., Fritsch, N., and Clardy, J.,** Stypotriol and stypoldione, ichthyotoxins of mixed biogenesis from the marine alga *Stypopodium zonale, Tetrahedron Lett.,* p. 145, 1979.
101. **Kato, T., Kumanireng, A. S., Ichinose, I., Kitahara, Y., Kakanuma, Y., and Kato, Y.,** Structure and synthesis of active component from a marine alga *Sargassum tortile,* which induces the settling of swimming larvae of *Coryne uchidai, Chem. Lett. (Jpn.),* p. 335, 1975.
102. **Fenical, W., Sims, J. J., Squatrito, D., Wing, R. M., and Radlick, P.,** Zonarol and isozonarol, fungitoxic hydroquinones from the brown seaweed *Dictyopteris zonarioides (undulata), J. Org. Chem.,* 38, 2383, 1973.
103. **Fattorusso, E., Magno, S., Mayol, L., Santacroce, C., Sica, D., Amico, V., Oriente, G., Piattelli, M., and Tringali, C.,** Oxocrinol and crinitol, novel linear terpenoids from the brown alga *Cystoseira crinita, Tetrahedron Lett.,* p. 937, 1976.
104. **Blunt, J. W., Hartshorn, M. P., McLennan, T. J., Munro, M. H. G., Robinson, W. T., and Yorke, S. C.,** Thysiferol, a squalene-derived metabolite of *Laurencia thysifera, Tetrahedron Lett.,* p. 69, 1978.

SYNTHESIS OF PHENOLIC PLANT PRODUCTS

T. Higuchi

L-Phenylalanine and L-tyrosine are derived from prephenic acid diverged from chorismic acid in the shikimate pathway and are converted to phenylpropane derivatives, flavonoids, lignin, and tannins as secondary plant products by vascular plants. Details of the shikimate pathway have been reviewed by Davis,[1] Sprinson,[2] and Yoshida.[3]

PHENYLPROPANE DERIVATIVES

Formation of Cinnamic Acid

Tracer studies[4] have provided evidence that L-phenylalanine can be deaminated to cinnamic acid, which is converted to various phenylpropane derivatives via the cinnamate pathway, as shown in Figure 1. Deamination of L-phenylalanine to *trans*-cinnamate is catalyzed by phenylalanine ammonia lyase (PAL),[5] which is widely distributed in higher plants. An analogous enzyme, tyrosine ammonia lyase (TAL),[6] catalyzing the deamination of L-tyrosine to *trans-p*-coumarate, is found mainly in grasses, which is in accord with the tracer experiment in which ^{14}C- tyrosine is incorporated into lignin only in grasses.[4]

PAL is situated at an introductory point to secondary (phenolic) metabolism from primary (protein) metabolism, and it has been known that the PAL activity is high in lignifying tissues of various plant stems[7,8] but considerably lower in young, nondifferentiated, or completely lignified tissues. The enzyme is controlled by numerous internal and external factors such as substrate, product, inhibitors, hormones, light, wounding, or infection, which affect both the synthesis and activity of the lyase.[9]

PAL activity of pea, mustard, and radish seedlings germinated in the dark and of *Helianthus* tubers increases with short time illumination of red light. The induction effect is lost with a momentary illumination of near far-red light, suggesting the mediation of phytochrome for the induction. Blue light is observed as an additional effector for induction of PAL in *Helianthus* tuber and cucumber. Ethylene, produced by wounding or ripening of tissues, is known as another factor which induces PAL activity.[10] Control of PAL activity by isozymes has been suggested.[11]

Hydroxylation of Cinnamic Acid

Hydroxylation of cinnamate into *p*-coumarate at the next step in the cinnamate pathway is mediated by cinnamate-4-hydroxylase.[12,13] The hydroxylase has been studied extensively and is known to be located in microsome fraction of plant tissues. The enzyme is a monoxygenase of the P-450 type and requires cinnamate, NADPH, and O_2, as shown in Figure 2. The activity of cinnamate-4-hydroxylase also is induced by both ethylene and light.

An enzyme which mediates the hydroxylation of *p*-coumarate to caffeate has been isolated from spinach beet.[14] The hydroxylase is a phenolase and requires NADPH, ascorbate, or reduced pteridines as electron donors for the reaction (Figure 3). The microsomal fraction from the root of *Quercus* seedlings[15] is found to mediate the formation of caffeate from L-phenylalanine via *p*-coumarate, which suggests the occurence of *p*-coumarate-3-hydroxylase as a membrane-associated state.

Methylation of Hydroxycinnamic Acids

The conversion of caffeate to ferulate is catalyzed by *S*-adenosyl-L-methionine:catechol *O*-methyltransferases (OMT) which is widely distributed in higher

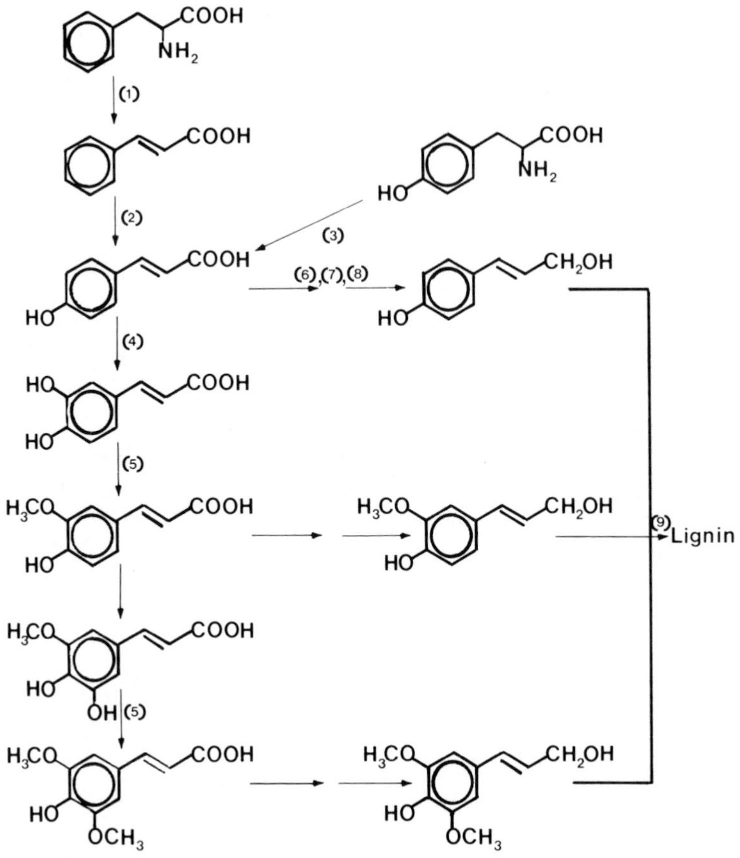

FIGURE 1. Cinnamate pathway in phenylpropanoids metabolism.

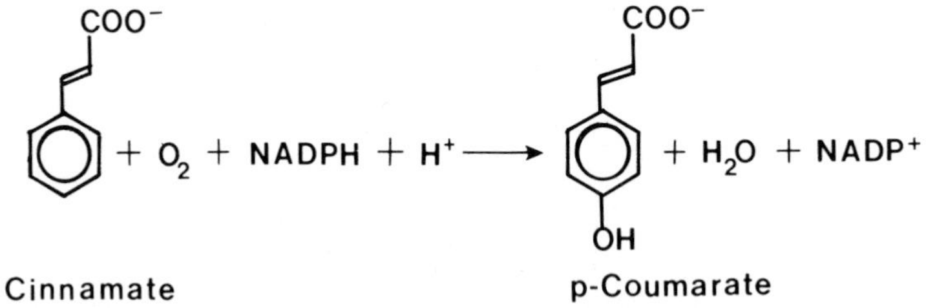

Cinnamate + O_2 + NADPH + H^+ → p-Coumarate + H_2O + $NADP^+$

FIGURE 2. Hydroxylation of cinnamate by cinnamate-4-hydroxylase.

plants. Ferulate thus formed may be hydroxylated to 5-hydroxyferulate, and 5-hydroxyferulate is methylated again to sinapate in angiosperms.[16,17] 5-Hydroxyferulate has not been found in nature, but it is presumed to be the intermediate between ferulate and sinapate by tracer experiment[18] (Figure 4).

In investigations on the substrate specificities of OMTs from gymnosperms[19] and angiosperms,[20] it is found that both caffeate and 5-hydroxyferulate are very good substrates, and 3,4,5-trihydroxycinnamate, 5-hydroxyvanillin, protocatechuic aldehyde, and chlorogenic acid also are fairly good substrates for angiosperm OMTs. The substrate specificity of the gymnosperm enzymes, on the other hand, is completely differ-

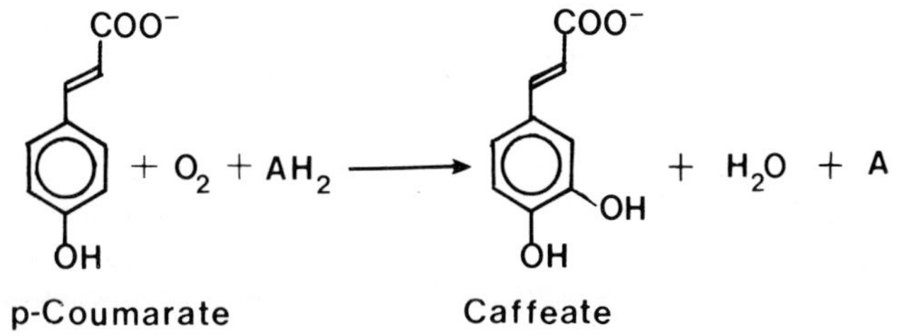

FIGURE 3. Hydroxylation of *p*-coumarate to caffeate by *p*-coumarate-3-hydroxylase (phenolase).

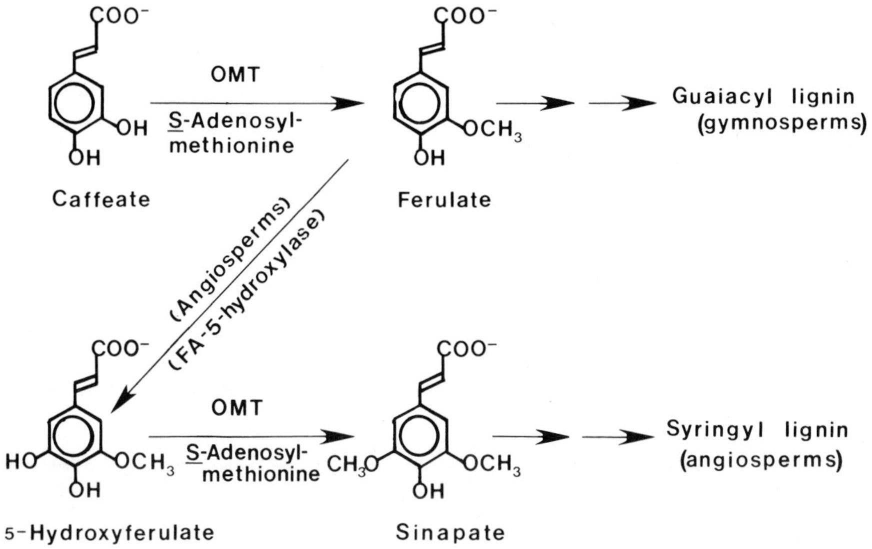

FIGURE 4. Formation of sinapate from ferulate in angiosperms.

ent. Caffeate is the most favorable substrate followed by protocatechuic aldehyde and 3,4-dihydroxyphenyl acetate; 5-hydroxyferulate is only slightly converted to sinapate (Table 1).

These results agree well with the distribution of guaiacyl lignin in gymnosperms and guaiacyl-syringyl lignins in angiosperms and indicate that the OMT is a main factor in controlling the formation of guaiacyl or syringyl lignin. It has been proposed that the enzymes of phenylalanine-cinnamate pathway occur as a membrane-associated multienzyme complex.[21]

Reduction of Hydroxycinnamic Acids

Ferulate and sinapate are reduced to the corresponding cinnamyl alcohols by successive mediation of three enzymes: hydroxycinnamate:CoA ligase, hydroxycinnamoyl-CoA reductase, and hydroxycinnamyl alcohol oxidoreductase, which recently were isolated from *Salix* and *Forsythia* by Mansell and co-workers[22] and from cell suspension culture of soybean by Ebel and Grisebach,[23] respectively. The same enzyme system has been found in *Brassica*[24] and many other plants.[25]

The first step in the reduction of ferulate is activation of the carboxyl group via CoA ester. Ferulate is converted to feruloyl adenylate and AMP in the presence of

Table 1
SUBSTRATE SPECIFICITY OF VARIOUS PLANT *O*-METHYLTRANSFERASES

Substrate	Relative methylation		
	Pine (%)	Poplar (%)	Bamboo (%)
Caffeic acid	100	100	100
5-Hydroxyferulic acid	10	320	124
3,4,5-Trihydroxycinnamic acid	25	60	50
Chlorogenic acid	10	46	3
iso-Ferulic acid	0	0	5
m-Coumaric acid	0	0	0
p-Coumaric acid	0	0	0
3,4-Dihydroxyphenyl acetic acid	54	0	0
3,4-Dihydroxymandelic acid	0	0	0
Protocatechuic aldehyde	68	46	45
5-Hydroxyvanillin	0	190	59
Protocatechuic acid	20	0	28
Gallic acid	0	0	0
Pyrocatechol	3	30	0
Pinosylvin	10	—	—
D-Catechin	0	0	0
Catechylglycerol-β-guaiacyl ether	0	—	—

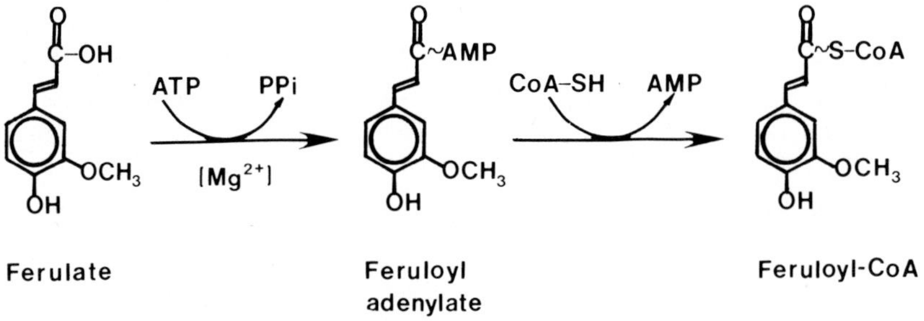

FIGURE 5. Activation of ferulate by hydroxycinnamate:CoA ligase.

ATP, and feruloyl adenylate is subsequently converted to feruloyl-CoA by CoA (Figure 5).

Cinnamate:CoA ligase which catalyzes this reaction is distributed in various higher plants, especially in young lignifying stems. The enzymes isolated from *Forsythia, Brassica* and soybean (ligase 2*) have similar substrate specificities, and hydroxycinnamates such as *p*-coumarate and ferulate are effective substrates. Methylated or glycosylated ferulate is not effective, and sinapate is also not effective[25] (Table 2).

Hydroxycinnamoyl-CoA is reduced to the corresponding aldehyde by cinnamoyl-CoA reductase.[23,26,27] The enzyme requires NADPH as hydrogen donor, and feruloyl-CoA is found to be the best substrate followed by *p*-coumaroyl, sinapoyl and 5-hydroxyferuloyl-CoAs[25] (Table 3).

* Two cinnamate:CoA ligases were isolated from cultured cells of soybean, and the ligase 1 was found to have a different substrate specificity from ligase 2 and preferentially acts on *p*-coumarate, sinapate, and methoxycinnamate.

Table 2
SUBSTRATE SPECIFICITY OF CINNAMATE:CoA LIGASES FROM HIGHER PLANTS[25]

Substrate	Relative activity (%)		
	Soybean (Isozyme 2)	*Brassica*	*Forsythia*
Cinnamate	24	0	0
o-Coumarate	—	15	26
m-Coumarate	—	21	54
p-Coumarate	104	137	104
Caffeate	90	93	64
Ferulate	100	100	100
Iso-Ferulate	104	83	87
3,4,5-Trihydroxycinnamate	—	—	43
Sinapate	0	2	0
p-Methoxycinnamate	0	0	0
3,4-Dimethoxycinnamate	0	0	0
3,4,5-Trimethoxycinnamate	0	0	0

Table 3
SUBSTRATE SPECIFICITY OF CINNAMOYL-CoA REDUCTASES FROM HIGHER PLANTS[25]

Substrate	Relative activity (%)		
	Forsythia	Soybean	Maple
Cinnamoyl-CoA	10	10	0
p-Coumaroyl-CoA	20	13	0
p-Methoxycinnamoyl-CoA	25	—	—
Caffeoyl-CoA	10	20	4
Feruloyl-CoA	100	100	100
3,4-Dimethoxycinnamoyl-CoA	40	—	34
5-Hydroxyferuloyl-CoA	20	25	—
Sinapoyl-CoA	20	30	8

The enzyme is specific to cinnamoyl-CoA; other aromatic or aliphatic CoA esters are not effective. Therefore, cinnamoyl-CoA:NADP oxydoreductase is proposed as a systematic name (Figure 6).

The last step in the formation of cinnamyl alcohol is the conversion of cinnamyl aldehyde to the corresponding alcohol mediated by alcohol dehydrogenase (Figure 7). Hydroxycinnamyl alcohol oxidoreductases from *Forsythia* and soybean have been investigated extensively.[28] The enzymes have rather broad substrate specificities for *p*-coumaraldehyde, coniferylaldehyde, sinapaldehyde, and their methyl derivatives and require NADPH as hydrogen donor.

The enzyme is distributed in several organs of various vascular plants, and is especially active in the cambial zone, which suggests that the enzyme is related to lignification.[25] The enzyme is specific for cinnamyl aldehydes; aliphatic aldehydes are not effective.

Cinnamyl alcohols thus formed are converted to various phenylpropane derivatives such as eugenol and related products, lignans, and lignin.

Many hydrocarbons, alcohols, acids, esters, aldehydes, ketones, phenols, and phenol ethers, which are related to phenylpropanoids and which may be derived from

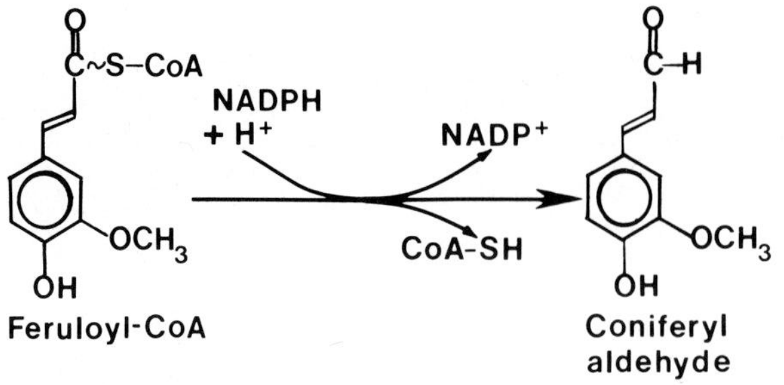

FIGURE 6. Reduction of feruloyl-CoA by cinnamoyl-CoA reductase.

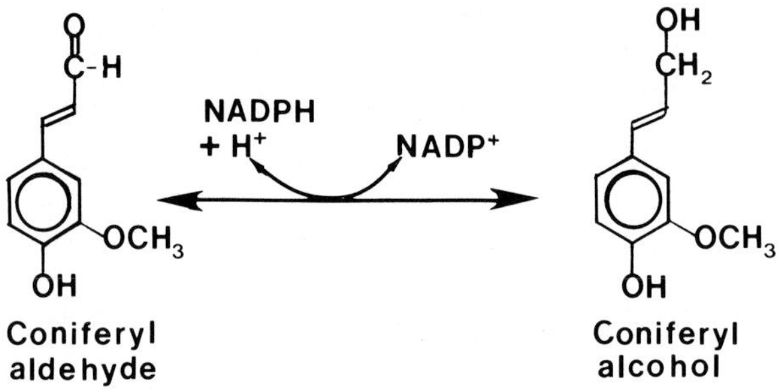

FIGURE 7. Reduction of coniferyl aldehyde by cinnamyl alcohol dehydrogenase.

shikimate-cinnamate pathway, have been isolated and identified.[29] Among them the biosynthesis of coumarin, chlorogenic acid, and eugenol has been investigated.

Biosynthesis of Coumarins

Coumarins generally occur as glucosides in plants of the Umbelliferae and Rutaceae. Umbelliferon (7-methoxycoumarin), esculetin (6,7-dihydroxycoumarin), and scopoletin (7-hydroxy-6-methoxycoumarin) correspond to *p*-coumaric, caffeic, and ferulic acids, respectively. Tracer experiments[30] provide evidence that coumarin can be formed, as shown in Figure 8.

The formation of both *p*-coumarate and *o*-coumarate from L-phenylalanine mediated by a microsomol fraction from potato tuber[31] indicates the occurrence of cinnamate-2-hydroxylase in addition to PAL and cinnamate-4-hydroxylase in the fraction. The cinnamate-2-hydroxylase requires NADPH. The enzyme is believed to be involved in coumarin biosynthesis and similar enzymes have been isolated from chloroplasts of *Melilotus alba* and *Hydrangea.*

Biosynthesis of Chlorogenic Acid

Chlorogenic acid, 3-caffeoylquinate, is widely distributed in fruits and leaves of angiosperms. The acid was found to be formed from caffeate and quinnate in the presence of CoA and Mg^{2+} by the enzyme isolated from cultured cells of *Nicotiana alata.*[32]

A recent investigation[33] has further shown that two enzymes, hydroxycinnamate:CoA ligase and hydroxycinnamoyl-CoA:quinnate hydroxycinnamoyltransferase, are involved in the biosynthesis of chlorogenic acid in tomato fruit. The ligase has a high affinity for *p*-coumarate followed by caffeate and ferulate, but not for cinnamate. The substrate affinity for the transferase is the following order, *p*-

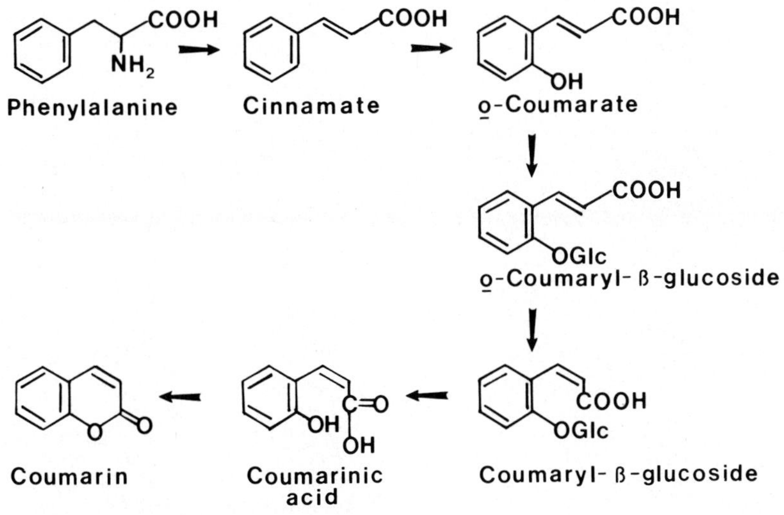

FIGURE 8. Biosynthetic pathway of coumarin in plants.

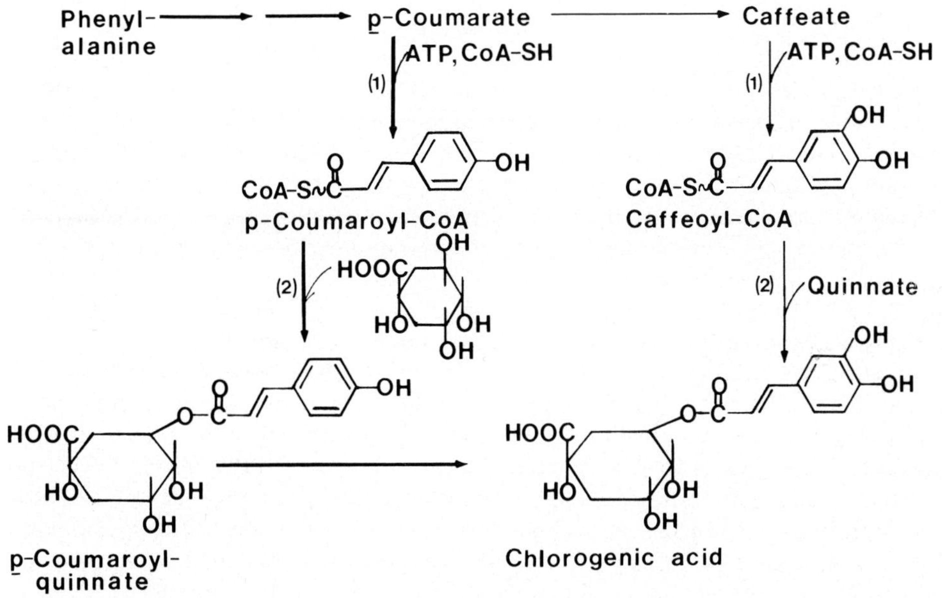

FIGURE 9. Biosynthesis of chlorogenic acid in tomato fruit.

coumaroyl-CoA>caffeoyl-CoA>feruloyl-CoA, which suggests that *p*-coumaroylquinnate formed from *p*-coumaroyl-CoA and quinate is hydroxylated to chlorogenic acid. Direct combination of caffeoyl-CoA with quinnate seems to be a minor pathway (Figure 9).

Tracer experiments[34] demonstrated that the allylphenols, eugenol and methyleugenol, are synthesized in *Ocimum basilicum* from ferulate and coniferyl alcohol, as shown in Figure 10.

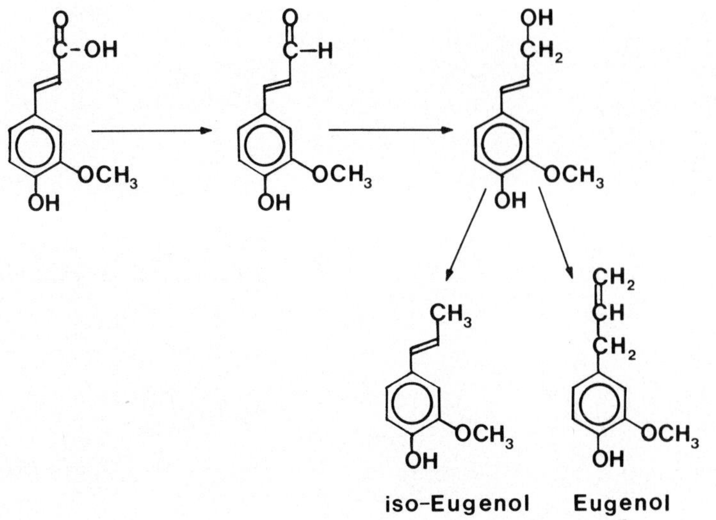

FIGURE 10. Biosynthesis of allylphenols in *Ocimum bacilicum*.

LIGNIN

Lignin, which makes up 17 to 33% of wood, is a complex aromatic polymer and plays a role in cementing polysaccharide components in cell walls. Estimated production of lignin is 13% of the total material produced by photosynthesis and that amounts to over 20 billion tons of lignin per year.[35] Lignins generally are classified into three major groups based on their structural monomer units: guaiacyl ligniOn in gymnosperms, guaiacyl-syringyl lignin in angiosperms, and guaiacyl-syringyl-*p*-hydroxyphenyl lignin in grasses.

Dehydrogenative Polymerization of Hydroxycinnamyl Alcohols

Freudenberg[36] found that a ligninlike dehydrogenation polymer (DHP) was produced in vitro by treating coniferyl alcohol with a crude mushroom phenol oxidase (laccase) under aerobic conditions. The DHP formed was closely related to spruce lignin in many aspects, such as functional groups, UV, IR, PMR, and ^{13}C-NMR spectra, and in degradation products in nitrobenzene oxidation, permanganate oxidation, and acidolysis. They further found that when coniferyl alcohol is dehydrogenated with laccase, or peroxidase, which was later elucidated[37] to be an actual enzyme involved in lignification, the alcohol results in mesomeric free radicals, as shown in Figure 11.

It was found by Gross et al.[38] that H_2O_2, as the substrate for peroxidase in the dehydrogenative polymerization of coniferyl alcohol, was produced by the peroxidase itself bound with cell walls via superoxide radical. The superoxide radical was suggested to be formed by the reduction of O_2 by NAD° which would be supplied by radical oxidation of NADH formed by malate dehydrogenase in cell walls (Figure 12).

The radicals of coniferyl alcohol formed can be coupled, nonenzymically, in a random fashion to give dimers, trimers, and higher oligomers as racemic mixtures. Figure 13 shows examples of the coupling of the radicals to give quinone methides which result in guaiacylglycerol-β-coniferyl ether, dehydrodiconiferyl alcohol, and *dl*-pinoresinol by the addition of water or intramolecular nucleophilic attack on the benzyl carbons by hydroxyl groups.

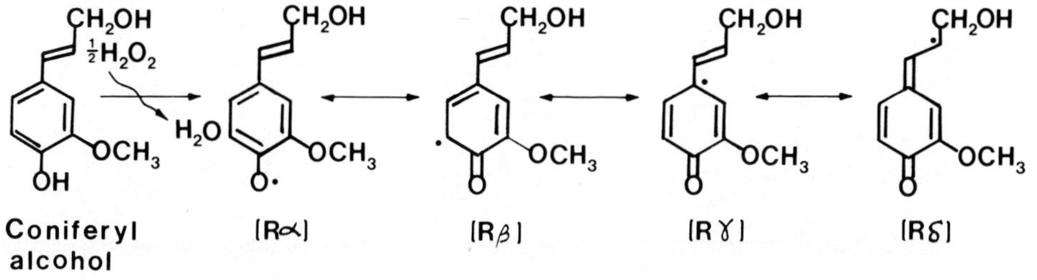

FIGURE 11. Dehydrogenation of coniferyl alcohol by peroxidase.

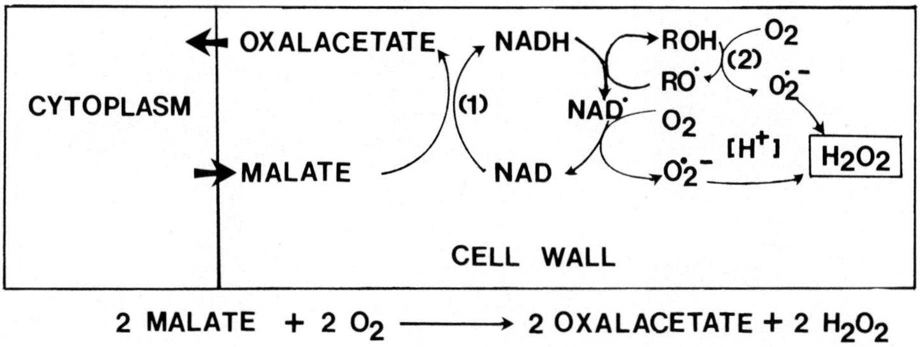

$$2 \text{ MALATE} + 2\, O_2 \longrightarrow 2 \text{ OXALACETATE} + 2\, H_2O_2$$

FIGURE 12. Hypothetical scheme on the formation of hydrogen peroxide in cell wall.[38]

It can be assumed that further dehydrogenation of these dimers and subsequent coupling of the arising radicals with the formation of biphenyl and diphenyl ether linkages results in the formation of lignin. The lignin formed by this process would be optically inactive as in natural lignins.

Figure 14 shows the structure of spruce lignin based on dehydrogenation experiments of coniferyl alcohol and on analytical and degradation investigations of spruce lignin.[39]

Freudenberg found [36] that a solution of coniferyl and sinapyl alcohols in approximately equal amounts gave a mixed dehydrogenation polymer by laccase treatment, and that the polymer was closely related to beech lignin in both analytical and chemical degradative features. It was accordingly concluded by Freudenberg that gymnosperm lignin was formed by the coupling of radicals formed by enzymic dehydrogenation of coniferyl alcohol, that angiosperm lignin is formed by the radical coupling of coniferyl and sinapyl alcohols, and that grass lignin is formed by the radical coupling of coniferyl, sinapyl, and *p*-coumaryl alcohol, respectively.

Biosynthesis of Syringyl Lignin

In the investigation on dehydrogenative polymerization of *p*-hydroxycinnamyl alcohols, Freudenberg and Neish[40] reported that sinapyl alcohol alone does not give a ligninlike polymer but yields mainly syringaresinol and dimethoxybenzoquinone. He suggested that no syringyl lignin occurs in nature. However, it was recently found that considerable amounts of DHP are formed from sinapyl alcohol alone[41] with peroxidase and H_2O_2. UV, IR, PMR, and ^{13}C-NMR spectra and functional group analysis and degradative studies of the polymer showed characteristic features of syringyl lignin, indicating that the phenoxy radicals of sinapyl alcohol formed enzymically are coupled not only by β-β bonds to form syringaresinol but also by β-O-4 bonds to produce

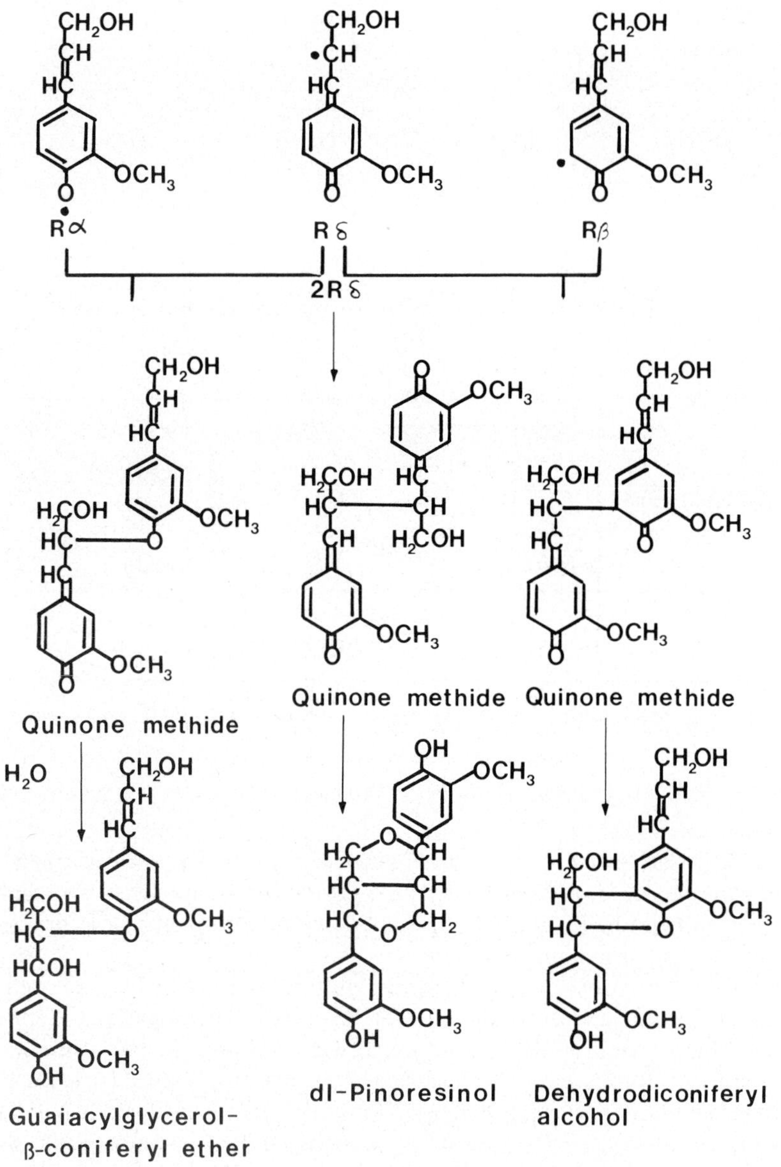

FIGURE 13. Formation of dilignols via quinone methides.

syringyl lignon as shown in Figure 15. Thus, guaiacyl lignin, which occurs in conifers, is mainly composed of a dehydrogenation polymer of coniferyl alcohol. Guaiacyl-syringyl lignin, which occurs in angiosperms, is composed of a mixed dehydrogenation polymer of coniferyl and sinapyl alcohols. Guaiacyl-syringyl-*p*-hydroxyphenyl lignin, which is found in grasses, is composed of a mixed dehydrogenation polymer of coniferyl, sinapyl, and *p*-coumaryl alcohols.

Differences between Gymnosperm and Angiosperm in the Formation of Lignin

Tracer and enzymic investigations provided evidence that the formation of guaiacyl lignin but not syringyl lignin in gymnosperms may be attributed to the following factors: absence of ferulate-5-hydroxylase, poor affinity of OMT towards 5-hydroxyfer-

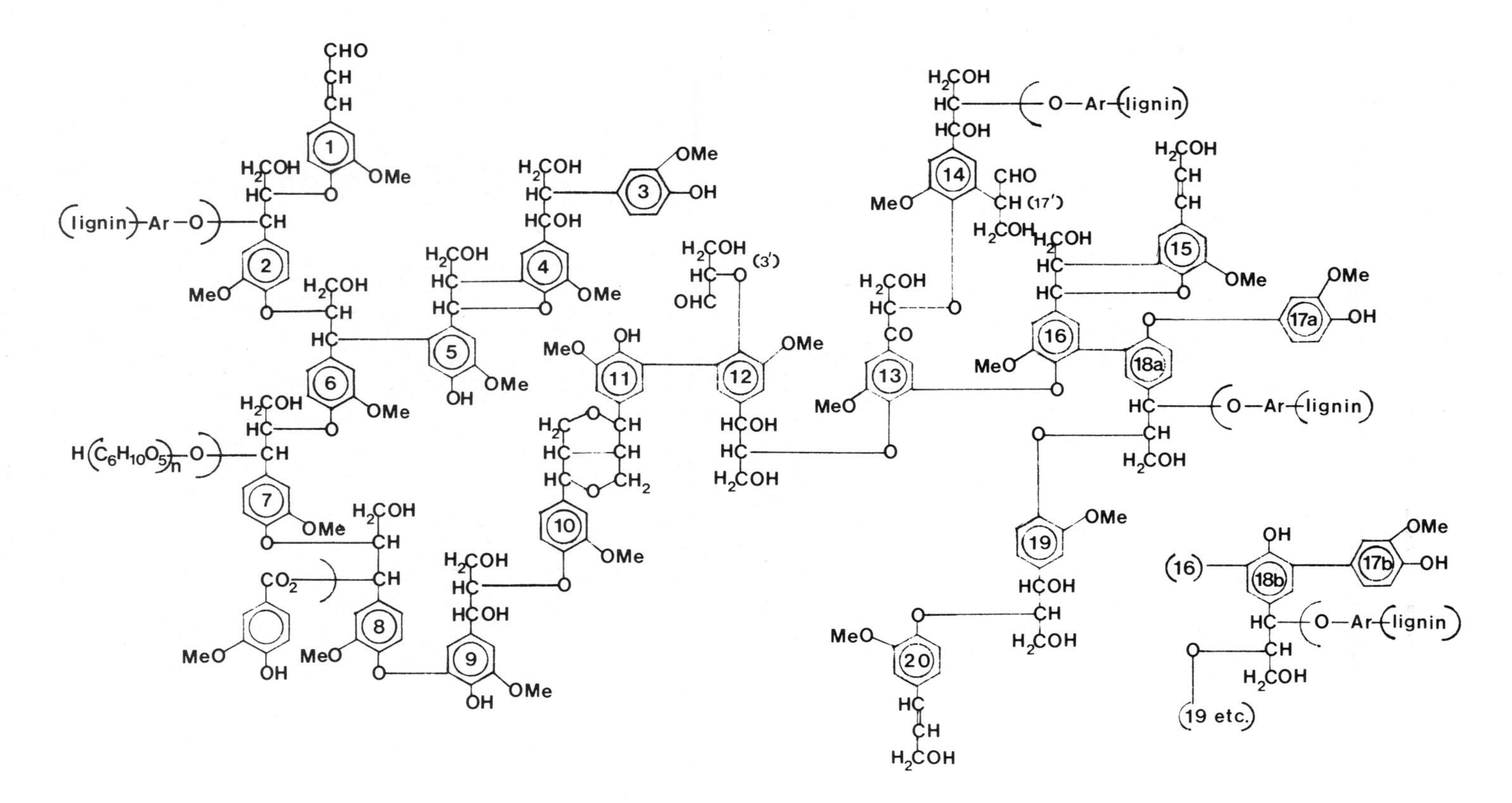

FIGURE 14. A schematic constitution of spruce lignin.[39]

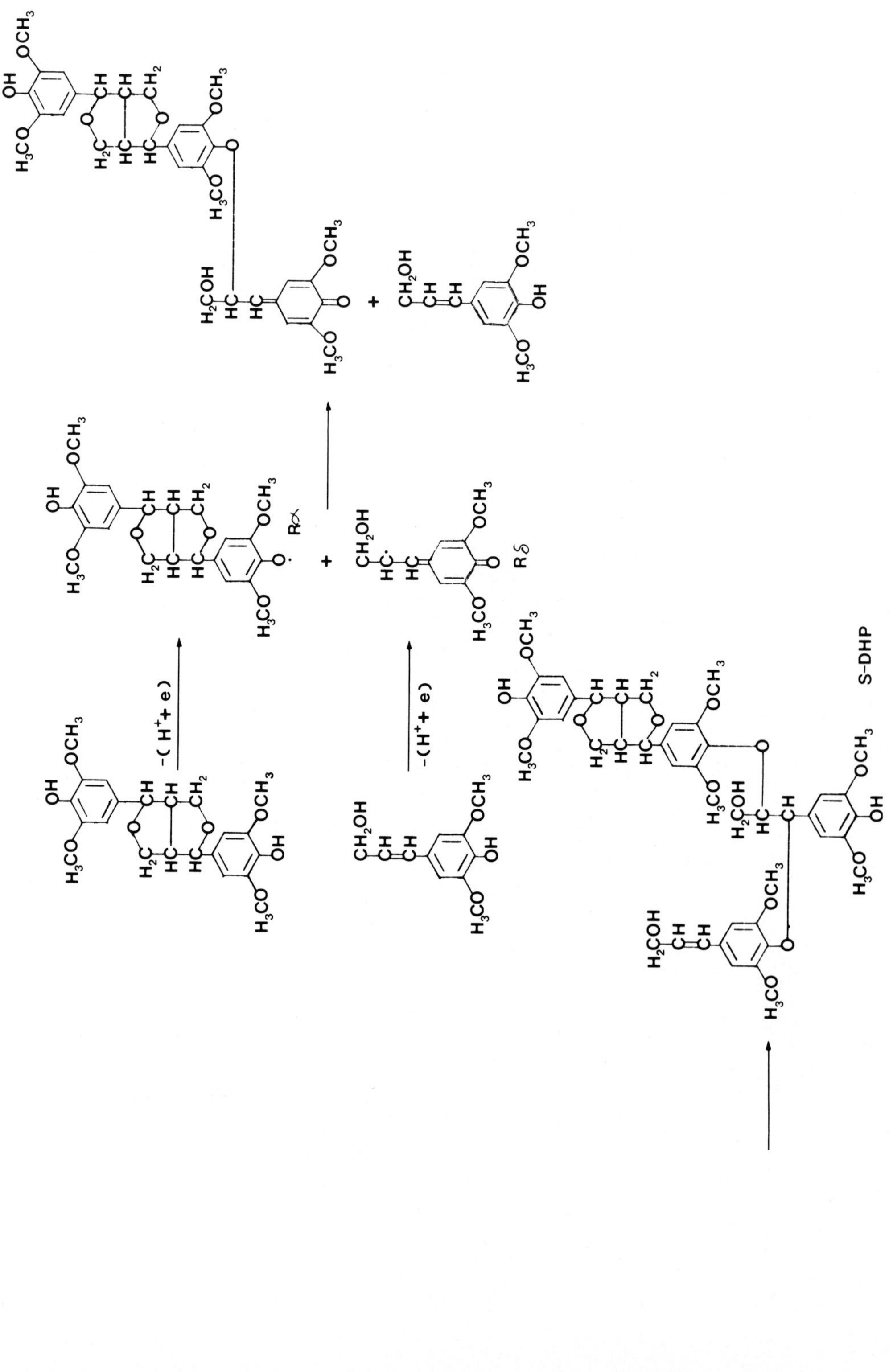

FIGURE 15. Hypothetical mechanism of biosynthesis of syringyl lignin.

ulate, and the lack of activation and/or reduction of sinapic acid. It was shown[42] that sinapaldehyde or sinapyl alcohol fed to conifer tissues is converted to syringyl lignin. Thus, differences in lignin metabolism between gymnosperms and angiosperms are as shown in Figure 16.

FLAVONOIDS

Types of Flavonoids

Flavonoids which consist of two aromatic rings joined by a 3-C unit are widely distributed in higher plants and are present in barks, heartwoods, and chromoplasts of flower petals. Flavonoids are classified according to the oxidation levels of the 3-C unit of the flavane nucleus (Figure 17), and are found many times as glycosides. A variety of flavones such as apigenin and luteolin, and flavonols, kaempferol, quercetin, and myricetin, and their glycosides are present in nature. Flavanone glycosides, hesperidin and naringin, are contained in large amounts in the outer pericarp of plants in Rutaceae. Chalcones, isomers of flavanones, are known to be present in equilibrium state with flavanones in plant tissues. Anthocyanins make an important contribution to red, pink, navy, and blue colors of flowers and leaves, and three main anthocyanidins: delphinidin, cyanidin, and pelargonidin, are found in nature. Catechin is contained in large amounts in tea leaves and barks of various trees.

Flavonoids are generally substituted at C5 and C7 of the A ring and C3′ and C4′ of the B ring with hydroxyl groups or methoxyl group, and they generally occur as 3 or 7 *O*-glycosides but sometimes as *C*-glycosides at C6 or C8.

Biosynthesis of Flavonoids

It was found that ^{14}C-acetate is preferentially incorporated into the A ring, whereas ^{14}C-labeled shikimate, phenylalanine, cinnamate and *p*-coumarate are efficiently incorporated into the B ring of quercetin[43] and aglycone of rutin in buckwheat. Similar results were obtained in the biosynthesis of cyanidin (red cabbage), phloretin (apple seedlings), catechin (tea seedlings *Chamaecyparis pisifera* and *Prunus* ssp.), and apigenin[10] (parsley).

Grisebach and Doerr[44] found that ^{14}C-labeled phenylalanine and acetate fed to a red clover are incorporated into hormononetin, isoflavonoid, and that the A ring is formed via acetate pathway while the C3 side chain and B ring are formed from phenylalanine. They[45] also found by using phenylanine labeled at C1, C2, and C3 that the phenyl group of the B ring at C2 is transfered to C3 in the formation of the isoflavone.

Grisebach and Hahlbrock[46] suggested that chalcone is the primary C6-C3-C6 compound formed via acetate and cinnamate pathways in the biosynthesis of flavonoids. Cinnamate-3-^{14}C was efficiently incorporated to chalcone in peas and the chalcone was converted to hormononetin by cell-free extracts of the same plant. It was also found that ^{14}C-chalcone glucoside was converted to quercetin and cyanidin in buckwheat and red cabbage, respectively. Chalcone-flavanone isomerases were later found in pea, beans, and parsley, and shown to mediate the reversible reaction between chalcone and flavanones.[46,47] It was assumed that the enzyme was responsible for the formation of flavanone. However, recent enzymic investigation by Kreuzaler and Hahlbrock[48] showed that naringenin, 4′, 5′, 7-trihydroxyflavanone, is formed from *p*-coumaroyl-CoA and malonyl-CoA, and that the former is incorporated into the B ring and the latter into the A ring. They[49] subsequently succeeded in isolating flavanone synthase not containing chalcone-flavanone isomerase, and in showing that the first enzymic product is flavanone. This biosynthetic pathway of flavonoids is shown in Figure 18.*

* Very recently Heller and Hahlbrook withdrew the proposal "flavanone as the first product," and demonstrated that the original product by this synthetase reaction is chalcone but not flavanone.[57]

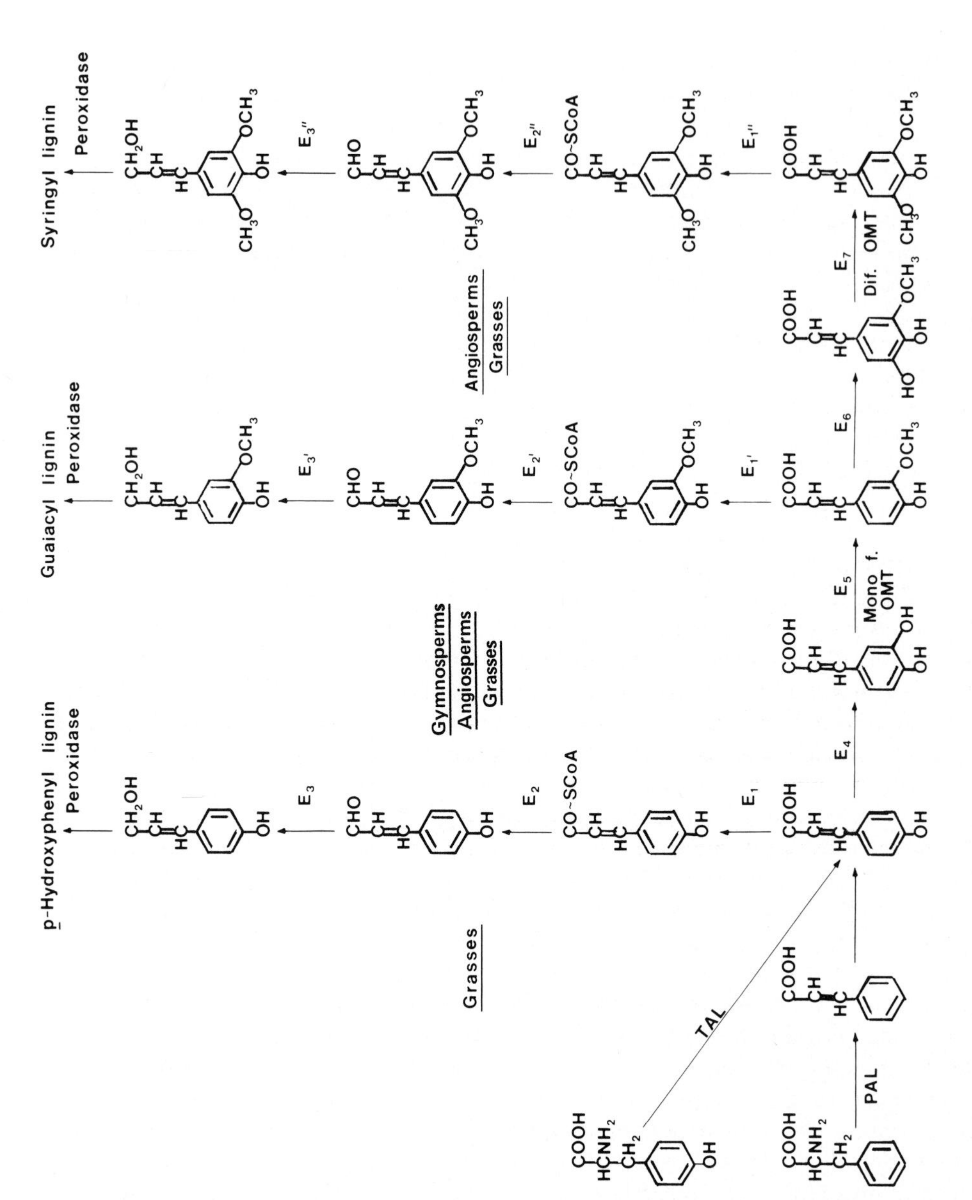

FIGURE 16. Differences in lignin biosynthesis between gymnosperms and angiosperms.

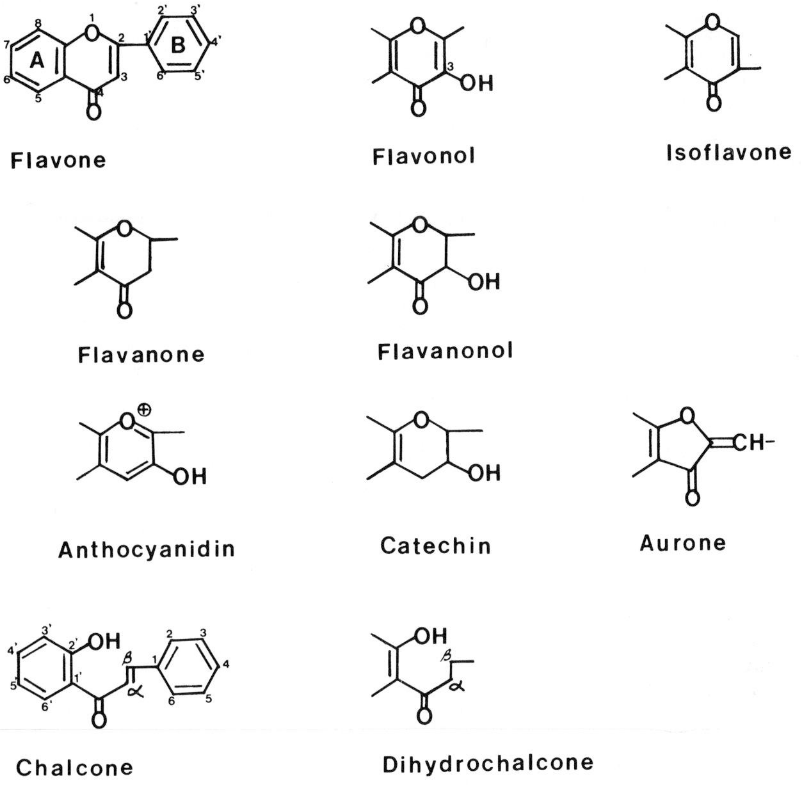

Flavone Flavonol Isoflavone

Flavanone Flavanonol

Anthocyanidin Catechin Aurone

Chalcone Dihydrochalcone

FIGURE 17. Chemical structures of flavonoids.

Flavanone is converted to flavone and flavanonol which then is converted to flavonol. Flavanone oxidase, which mediates the formation of apigenin (flavone) from naringenin (flavanone), was isolated from parsley.[46] It is assumed that anthocyanidin and catechin are formed from flavanonol, while isoflavone and auron are formed from chalcone, respectively. However, no enzymic evidence has been obtained.

Grisebach and Hahlbrock[46] found that 20 flavonoid glucosides such as apiin and 3′-methoxyapiin were formed when parsley suspension cultures were illuminated continuously with white light. They also found that UV light of wavelength less than 345 nm is effective in inducing the formation of flavonoids and that the enzymes in flavonoid biosynthesis are activated in two patterns with UV light. This is, the enzymes PAL, cinnamate-4-hydroxylase, *p*-coumarate:CoA ligase were activated rapidly within 2 to 2.5 hr illumination, reached maximum activities between 17 to 23 hr, and then the activity decreased. The enzymes, flavanone synthase, 7-*O*-glucosyltransferase, UDP-apiose synthase and malonyl-transferase were activated gradually after about 4 hr illumination, and reached maximum activities between 26 to 40 hr; the activities then gradually decreased.

The first group of enzymes is generally involved in phenylpropanoid metabolism but the second group is involved specifically in flavonoid synthesis. The results indicate that the induction of these two groups of enzymes is controlled by different mechanisms. The induction of the enzymes of flavonoid synthesis with light was also observed in cotyledon and leaves of germinated parsley and soybean suspension culture. Further, the suspension cultures of parsley and soybean in the light gave methoxylated flavonoids at the B ring, which was ascribed to the induction of *O*-methyltransferase

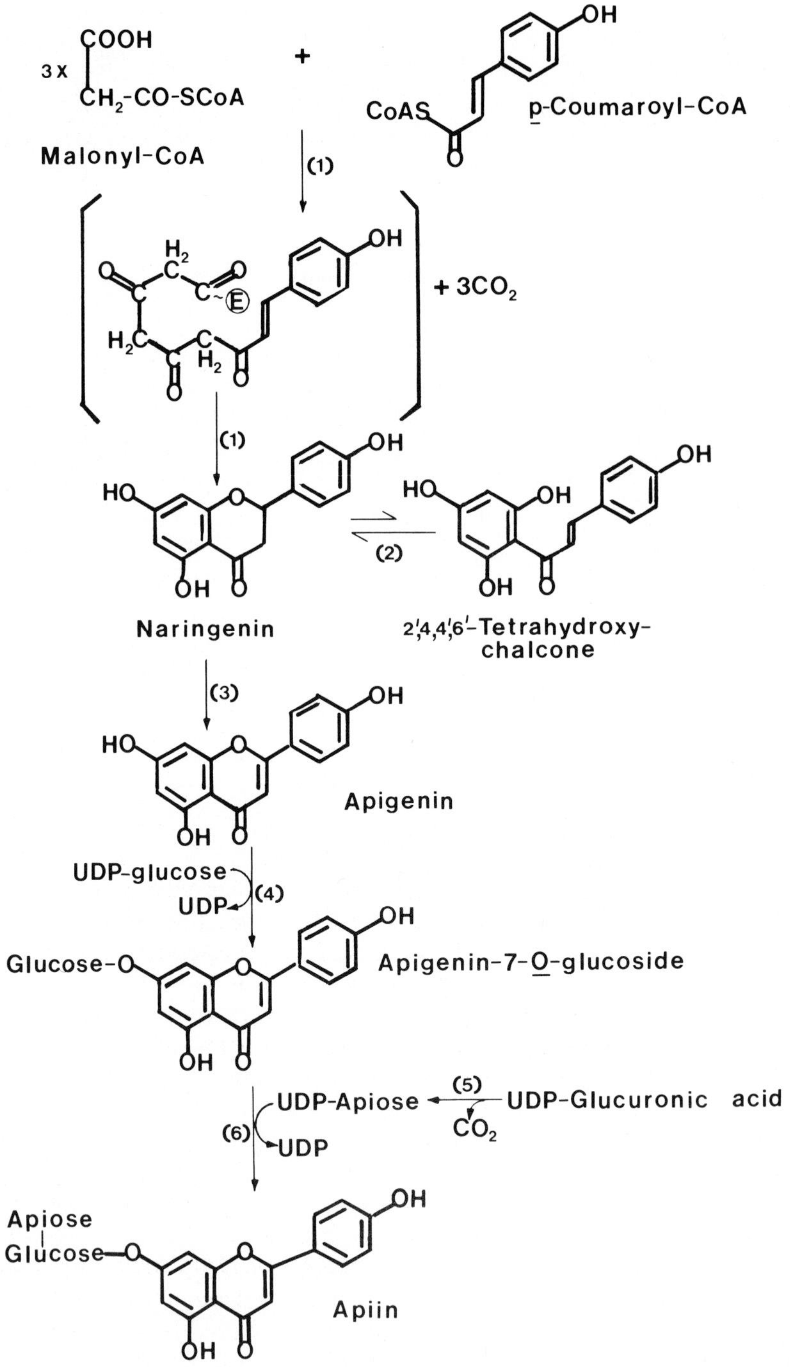

FIGURE 18. Biosynthetic pathway of flavonoids.

mediating the *m*-methylation of 3′, 4′-dihydroxyflavonoids as well as *p*-coumarate and caffeate.

TANNINS

Hydrolyzable Tannins

Tannins are generally classified into hydrolyzable and condensed tannins. Hydrolyzable tannins are composed of esters of polyalcohols, such as glucose with phenolic acids. A part of or all of the hydroxyl groups of glucose are esterified with gallic acid or related acids, which are easily hydrolyzed with alkali or tannase. Gallo- and ellagitannins are good representatives, but recent investigations have shown that both gallic and ellagic acids are present as mixtures in most of the hydrolyzable tannins.

It has been shown that 9 to 10 gallic acids are linked to one molecule of glucose and that half of the gallic acid is present as depside, *m*-digalloyl group, linked between *m*-hydroxyl groups and carboxyl groups. Tannins of the insect galls of *Rhus* and oak leaves are well known gallotannins which amount to 70% of the dry weight. Polyalcohol components (other than glucose), 1,5-anhydrosorbitol (maple tannin), hamamelose (hamamelitannin), and quinnic acid (tara-tannin) have been found.

Ellagitannins give ellagic acid by hydrolysis but it has been shown that the acid is not present in situ but formed from hexahydroxydiphenic acid esterified with glucose during hydrolysis. Corilagin is known as the simplest ellagitannin in myrobalan (*Terminalia* spp.), and eucalyptus. It is assumed that the hydrolyzable tannins are formed via esterification of sugar phosphate with galloyl CoA as in the formation of neutral lipids from glycerol and fatty acids. However, no firm information is available.[10]

Two pathways have been suggested for the biosynthesis of gallic acid by tracer studies. That is, ^{14}C-phenylalanine is efficiently incorporated into gallic acid in the leaves of *Geranium*,[50] suggesting the occurrence of β-oxidation of hydroxycinnamates. On the other hand, it was found that in a *Rhus*, shikimate is incorporated into gallic acid more efficiently than phenylalanine. These results indicate that both pathways, β-oxidation of cinnamate, and dehydrogenation of 5-dehydroshikimate, occur in higher plants (Figure 19).

Hexahydroxydiphenic acid should be formed via oxydative coupling of gallic acid as in the formation of lignin.

Condensed Tannins

Condensed tannins are polymers of flavonoids such as catechin and leucoanthocyanidins. As naturally occurring monomers of condensed tannins, *d*-catechin, *l*-epicatechin, and sevaral leucoanthocyanidins are present. It is known that tea leaves contain an ester of *l*-epicatechin with gallic acid. Since most tannins of plants contain catechin, the role of catechin in tannin formation has been noted. The distribution of leucoanthocyanidins in plants is wide, and their important contribution in tannin formation is also assumed.

Monomers of condensed tannin should be formed via condensation of a cinnamoyl-CoA and three molecules of malonyl-CoA as described in the section of flavonoid biosynthesis. Tracer investigations have supported the pathway, but no information is available on the branching points into catechin and leucoanthocyanidin in the pathway of flavonoid biosynthesis.

Polymerization of Catechins and Leucoanthocyanidins

It has been suggested that flavan-3-ols (catethin) and flavan-3,4-diols (leucoanthocyanidin) are polymerized to condensed tannins. Freudenberg and Weinges[51] found that 7 and 4′ hydroxylated flavan-3-ols and flavan-3,4-diols are easily polymerized by acid and suggested that condensed tannins are derived from these monomers contained in the barks of trees and heartwoods of *Acacia catechu* and *Schinopsis* species under

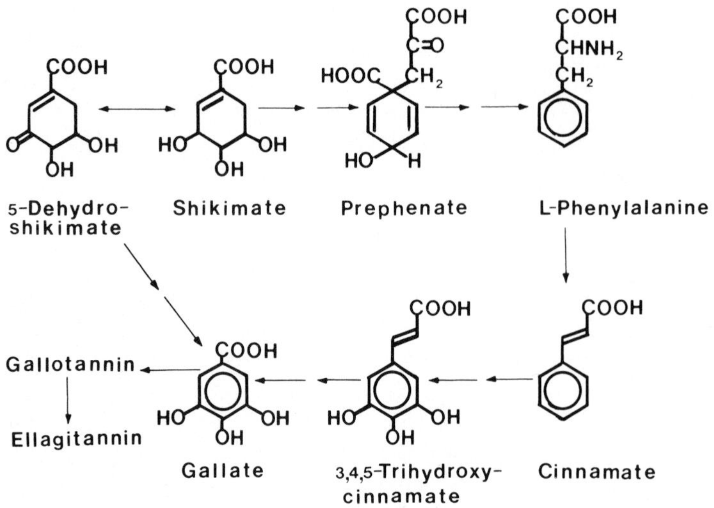

FIGURE 19. Biosynthetic pathway of gallic acid.

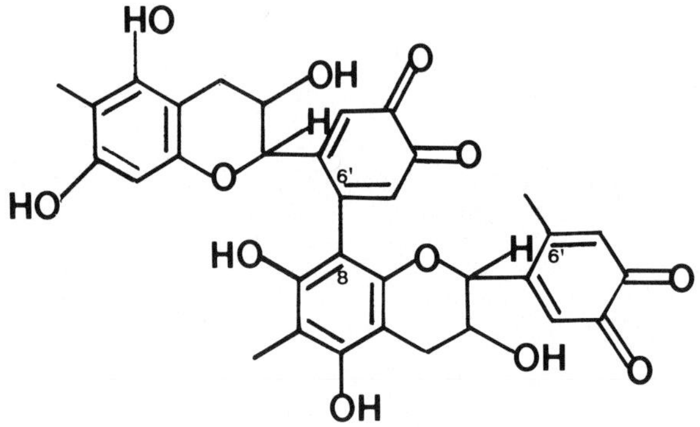

FIGURE 20. Oxidative polymerization of catechin.

the influence of plant acids. However, it is doubtful whether the intermolecular C–C linkages found in condensed tannins could be formed by the weak acidity in plant tissues.

Atmospheric or enzymic oxidation of catechins and leucoanthocynadins to condensed tannins has been presented as another more plausible explanation.[52] It was found that *d*-catechin and ℓ-epicatechin from the leaves of *Uncaria gambir* and heartwood of *Acacia catechu* are oxidized to dark-colored condensed tannins,[53] and that in catechin polymerization a product containing some head to tail linkage is formed (Figure 20). Further, enzymic oxidation of catechin by polyphenoloxidase[54] gives a polymer closely related to that produced by autooxidation, and the polymer has identical tannin properties and similar chemical properties to the main tannin extractives of *Uncaria gambir* leaves and *Acacia catechu* heartwood. It is thus probable that these condensed tannins arise by similar oxidations of catechin type precursors. Roux,[55] on the other hand, found that monomeric leucofisetinidin (ℓ-7, 3′, 4′-trihydroxyflavane-3,4-diol) in the sapwood of quebracho is transformed to various polymers present in the heartwood. Roux and Paulus[56] further found that some of the tannins are related to the monomeric leucorobinetinidin (7, 3′, 4′, 5′-tetrahydroxyflavan-3,4-diol) and leucofisetinidin, while mollisacacidin, (*d*-7, 3′, 4′-trihydroxyflavan-3, 4-diol) is related to black wattle heartwood tannins. Roux[55] concluded from these results that the polymeric leucoanthocyanidins of wattle and quebracho arise from the monomeric flavan-3,4-diols by enzymic condensation in vivo.

REFERENCES

1. **Davis, B. D.**, Biosynthesis of aromatic amino acids, *Symp. Amino Acid Metab.*, 799, 1955.
2. **Sprinson, D. B.**, The biosynthesis of aromatic compounds from D-glucose, *Adv. Carbohydr. Chem.*, 15, 235, 1960.
3. **Yoshida, S.**, Biosynthesis and conversion of aromatic amino acids in plants, *Ann. Rev. Plant Physiol.*, 20, 41, 1969.
4. **Brown, S. A.**, Lignins, *Ann. Rev. Plant Physiol.*, 17, 223, 1966.
5. **Koukol, J. and Conn, E. E.**, Purification and properties of the phenylalanine deaminase of *Hordeum vulgare*, *J. Biol. Chem.*, 236, 2692, 1961.
6. **Neish, A. C.**, Formation of *m*- and *p*-coumaric acids by enzymatic deamination of the corresponding isomers of tyrosine, *Phytochemistry*, 1, 1, 1961.
7. **Higuchi, T.**, Role of phenylalanine deaminase and tyrase in the lignification of bamboo, *Agric. Biol. Chem.*, 30, 667, 1966.
8. **Rubery, P. H. and Northcote, D. H.**, Site of phenylalanine ammonia-lyase activity and synthesis of lignin during xylem differentiation, *Nature (London)*, 219, 1230, 1968.
9. **Creasy, L. L. and Zucker, M.**, Phenylalanine ammonia-lyase and phenolic metabolism, *Recent Adv. in Phytochem.*, 8, 1, 1974.
10. **Yoshida, S. and Minamikawa, T.**, *Secondary Metabolism in Higher Plants (in Japanese)*, Tokyo-Diagaku Shuppan-Kai, 1977, 42.
11. **Boudet, A., Ranjeva, R., and Gadal, P.**, Proprietes allosteriques specifiques des deu isoenzymes de la phenylalanine-ammoniaque lyase chez *Quercus pedunculata*, *Phytochemistry*, 10, 997, 1971.
12. **Russel, D. W.**, The metabolism of aromatic compounds in higher plants, *J. Biol. Chem.*, 246, 3870, 1971.
13. **Hill, A. C. and Rhodes, M. J. C.**, The properties of cinnamic acid-4-hydroxylase of aged swede root disks, *Phytochemistry*, 1e, 2387, 1975.
14. **Vaughan, P. F. T. and Butt, V. S.**, The action of *o*-dihydric phenols in the hydroxylation of *p*-coumaric acid by a phenolase from leaves of spinach beet (*Beta vulgaris* L.), *Biochem. J.*, 119, 89, 1970.
15. **Alibert, G., Ranjeva, R., and Boudet, A.**, Recherches sur les enzymes catalysant la biosynthese des acides phenoliques chez *Quercus pedunculata*. III. Formation sequentielle, a partir de la phenylalanine, des acides cinnamique, *p*-coumarique et cafeique, par des organites cellulaires isoles, *Physiol. Plant.*, 27, 240, 1972.
16. **Higuchi, T., Shimada, M., Nakatsubo, F., and Tanahashi, M.**, Differences in biosyntheses of guaiacyl and syringyl lignins in woods, *Wood Sci. Technol.*, 11, 153, 1977.
17. **Shimada, M., Ohashi, H., and Higuchi, T.**, *O*-Methyltransferase involved in the biosynthesis of lignins, *Phytochemistry*, 9, 2463, 1970.
18. **Higuchi, T. and Brown, S. A.**, Studies of lignin biosynthesis using isotopic carbon. XII. The biosynthesis and metabolism of sinapic acid, *Can. J. Biochem. Physiol.*, 41, 614, 1963.
19. **Kuroda, H., Shimada, M., and Higuchi, T.**, Purification and properties of *O*-methyltransferase involved in the biosynthesis of gymnosperm lignin, *Phytochemistry*, 14, 1759, 1975.
20. **Shimada, M., Kuroda, H., and Higuchi, T.**, Evidence for the formation of methoxyl groups of ferulic and sinapic acids in *Bambusa* by the same *O*-methyltransferase, *Phytochemistry*, 12, 2873, 1973.
21. **Stafford, H. A.**, The metabolism of aromatic compounds, *Ann. Rev. Plant Physiol.*, 25, 459, 1974.
22. **Mansell, R. L., Stöckigt, T., and Zenk, M. H.**, Reduction of ferulic acid to coniferyl alcohol in a cell free system from a higher plant, *Z. Pflanzenphysiol.*, 68, 286, 1972.
23. **Ebel, J. and Grisebach, H.**, Reduction of cinnamic acids to cinnamyl alcohols with an enzyme preparation from cell suspension cultures of soybean *(Glycine max)*., *FEBS Lett.*, 30, 141, 1973.
24. **Rhodes, M. J. C. and Wooltorton, L. S. C.**, Reduction of the CoA thioesters of *p*-coumaric and ferulic acids by extracts of aged *Brassica napo-brassica* root tissue, *Phytochemistry*, 13, 107, 1974.
25. **Gross, G. G.**, Biosynthesis of lignin and related monomers, *Recent Adv. in Phytochem.*, 11, 141, 1977.

26. **Stöckigt, J., Mansell, R. L., Gross, G. G., and Zenk, M. H.**, Enzymic reduction of *p*-coumaric acid via *p*-coumaroyl CoA to *p*-coumaryl alcohol by a cell-free system from *Forsythia* sp., *Z. Pflanzenphysiol.*, 70, 305, 1973.
27. **Rhodes, M. J. C. and Wooltorton, L. S. C.**, Enzymes involved in the reduction of ferulic acid to coniferyl alcohol during the aging of disks of swede root tissues, *Phytochemistry*, 14, 1235, 1975.
28. **Mansell, R. L., Gross, G. G., Stockigt, J., Frank, H., and Zenk, M. H.**, Purification and properties of cinnamyl alcohol dehydrogenase from higher plants involved in lignin biosynthesis, *Phytochemistry*, 13, 2427, 1974.
29. **Stevens, G. and Nord, F. F.**, Natural phenylpropane derivatives, in *Modern Methods of Plant Analysis*, Paech, K. and Tracey, M. V., Eds., Springer-Verlag, Berlin, 1955, 392.
30. **Kosuge, T. and Conn, E. E.**, Metabolism of aromatic compounds in higher plants. III. β-Glucosides of *o*-coumaric, coumarinic and melilotic acids, *J. Biol. Chem.*, 236, 1617, 1961.
31. **Czichi, U. and Kindl, H.**, Formation of *p*-coumaric acid and *o*-coumaric acid from L-phenylalanine by microsomal membrane fractions from potato. Evidence of membrane-bound enzyme complexes, *Planta*, 125, 115, 1975.
32. **Stöckigt, J. and Zenk, M. H.**, Enzymatic synthesis of chlorogenic acid from caffeoyl coenzyme A and quinic acid, *FEBS Lett.*, 42, 131, 1974.
33. **Rhodes, M. J. C. and Wooltorton, L. S. C.**, The enzymatic conversion of hydroxycinnamic acids to *p*-coumarylquinic and chlorogenic acids in tomato fruits, *Phytochemistry*, 15, 947, 1976.
34. **Klischies, M., Stöckigt, J., and Zenk, M. H.**, Biosynthesis of the allylphenols eugenol and methyleugenol in *Ocimum basilicum* L., *J. Chem. Soc., Chem. Commun.*, 879, 1975.
35. **Kirk, T. K.**, Biodelignification research at the U.S. Forest Products Laboratory, Biological Delignification, Symp., Weyerhaeuser, 1976, 31.
36. **Freudenberg, K.**, Lignin: its constitution and formation from *p*-hydroxycinnamyl alcohols, *Science*, 148, 595, 1965.
37. **Higuchi, T.**, Studies on the biosynthesis of lignin, *Proc. 4th Int. Congr. Biochem.*, 2, 161, 1958.
38. **Gross, G. G., Janse, C., and Elstner, E. F.**, Involvement of malate, monophenols, and superoxide radical in hydrogen peroxide formation by isolated cell walls from horse radish (*Aromoracia lapathifolia* Gilib), *Planta*, 136, 271, 1977.
39. **Adler, E., Larsson, S., Lundquist, K., and Miksche, G. E.**, Acidolytic, alkaline, and oxidative degradation of lignin, Int. Wood Chem. Symp. Abstr., Seattle, Wash., 1969.
40. **Freudenberg, K. and Neish, A. C.**, Constitution and Biosynthesis of Lignin, Springer-Verlag, Berlin, 1968, 78.
41. **Yamasaki, T., Hata, K., and Higuchi, T.**, Dehydrogenation polymer of sinapyl alcohol by peroxidase and hydrogen peroxide, *Mokuzai Gakkaishi*, 22, 582, 1976.
42. **Nakamura, Y., Fushiki, H. and Higuchi, T.**, Metabolic differences between gymnosperms and angiosperms in the formation of syringyl lignin, *Phytochemistry*, 13, 1777, 1974.
43. **Underhill, E. W., Watkin, J. E. and Neish, A. C.**, Biosynthesis of quercetin in buckwheat. I.II., *Can. J. Biochem. Physiol.*, 35, 219, 229, 1957.
44. **Grisebach, H. and Doerr, N.**, Biogenesis of 7-hydroxy-4′-methoxyisoflavone *Naturwissenschaften*, 46, 514, 1959.
45. **Grisebach, H. and Doerr, N.**, Biogenesis of isoflavone. II. Mechanism of rearrangement, *Z. Naturforsch.*, 15b, 284, 1960.
46. **Grisebach, H. and Hahlbrock, K.**, Enzymology and regulation of flavonoid and lignin biosynthesis in plant and plant cell suspension cultures, *Recent Adv. in Phytochem.*, 8, 21, 1974.
47. **Wong, E. and Moustafa, E.**, Flavanone biosynthesis, *Tetrahedron Lett.*, p. 3021, 1966.
48. **Kreuzaler, F. and Hahlbrock, K.**, Flavonoid glycosides from illuminated cell suspension cultures of *Petroselinum hortense, Phytochemistry*, 12, 1149, 1973.
49. **Kreuzaler, F. and Hahlbrock, K.**, Enzymic synthesis of an aromatic ring from acetate units. Partial purification and some properties of flavanone synthase from cell suspension cultures of *Petroselium hortense, Eur. J. Biochem.*, 56, 205, 1975.
50. **El-Basyouni, S. Z., Chen, D., Ibrahim, R. K., Neish, A. C., and Towers, G. H. N.**, The biosynthesis of hydroxybenzoic acids in higher plants, *Phytochemistry*, 3, 485, 1964.
51. **Freudenberg, K. and Weinges, K.**, Catechins, other hydroxyflavans and hydroxyflavenes, *Fortschr. Chem. Org. Naturst.*, 16, 1, 1958.
52. **Hathway, D. E.**, Autooxidation of polyphenols. IV. Oxidative degradation of the catechin autooxidation polymer, *J. Chem. Soc.*, 520, 1958.
53. **Hathway, D. E. and Seakins, J. W. T.**, Autooxidation of catechin, *Nature (London)*, 176, 218, 1955.
54. **Hathway, D. E. and Seakins, J. W. T.**, Autooxidation of polyphenols. III. Autooxidation in neutral aqueous solution of flavans related to catechin, *J. Chem. Soc.*, 1562, 1957.
55. **Roux, D. G.**, Biogenesis of condensed tannins from leucoanthocyanins, *Nature (London)*, 181, 1454, 1958.

56. **Roux, D. G. and Paulus, E.,** Condensed tannins. XII. Polymeric leucofisetinidin tannins from the heartwood of *Acacia mearnsii, Biochem. J.*, 82, 320, 1962.
57. **Heller, W. and Hahlbrock, K.,** Highly purified "flavanone synthase" from parsley catalyzes the formation of naringenin chalcone, *Arch. Biochim. Biophys.*, 200, 617, 1980.

AMINO ACID SYNTHESIS AND CONVERSIONS: COMPARATIVE PATHWAYS IN BACTERIA AND PLANTS

John F. Thompson

The synthesis of amino acids is a major consideration in the contribution of autotrophic organisms to the maintenance of life systems on earth. The synthesis of amino acids by plants is especially important because monogastric animals must receive eight essential amino acids in their diets — tryptophan, lysine, isoleucine, leucine, valine, phenylalanine, methionine, and threonine. Animal proteins have 12 other amino acids that are considered nonessential including aspartic acid, glutamic acid, glycine, alanine, serine, cysteine, proline, asparagine, glutamine, tyrosine, arginine, and histidine. In this discussion, cystine is not considered separately since cystine is formed from cysteine by oxidation (either deliberately or incidentally). Cysteine and tyrosine are unusual nonessential amino acids because they are formed in animals only from the essential amino acids methionine and phenylalanine, respectively. The other nonessential amino acids need not be in the diet in any particular form or proportion because they can be interconverted. However, monogastric animals must have nonessential amino acids in their diet. Monogastric animals cannot reduce nitrate to ammonia or sulfate to sulfide, but they incorporate insignificantly small amounts of ammonia and sulfide into organic compounds. Hydroxylysine and hydroxyproline are found in animal proteins but the hydroxylation occurs after proline and lysine are incorporated into polypeptide chains. Therefore, the formation of hydroxyproline and hydroxylysine is not included. This section emphasizes the biosynthesis of the 20 amino acids found in plant and animal proteins because of the importance of these amino acids in nutrition.

Animals and plants contain over 200 amino acids that do not occur in proteins.[1] It is unrealistic to discuss these amino acids because we know little about their biosynthesis and their function. Some of the nonprotein amino acids (e.g., homoserine and cystathionine) will be discussed with respect to their role in the biosynthesis of protein amino acids. Nitrate and sulfate reduction are discussed elsewhere but their incorporation into organic compounds is considered. The synthesis of amino acids often involves the incorporation of an amino group into a keto acid (by aminotransferases) at the last step in amino acid biosynthesis. In many cases, the principal function of the keto acid is in the biosynthesis of an amino acid. The convention of tracing pathways of keto acid and amino acid synthesis from common compounds that are involved in several pathways (e.g., α-ketoglutaric acid and pyruvic acid) is adopted.

Protein amino acids are synthesized in plants and bacteria in the same way with the exceptions to be noted. Little is known about amino acid biosynthesis in photosynthetic bacteria, but it likely is the same as in heterotrophic bacteria. Most information will be presented in annotated figures. Amino acids are derived from nonnitrogenous glucose metabolites (Figure 1). Figure 1 also shows that several amino acids are formed from one nonnitrogenous precursor. The latter fact makes it logical and convenient to collate the biosynthetic pathways of amino acids derived from one compound into one diagram (Figures 2 to 7). Figure 1 also presents some of the abbreviations used in subsequent figures. Detailed information on amino acid synthesis has been given by Bryan,[2] in two chapters in "Metabolic Pathways",[3,4] and by Beevers.[5]

The synthesis of amino acids derived from α-ketoglutaric acid is presented first (Figure 2) because glutamic acid is the key intermediate in the incorporation of ammonia into organic compounds. Glutamic acid, glutamine, proline, and arginine are derived from α-ketoglutaric acid. Reaction 1 represents the reductive amination of α-ketoglu-

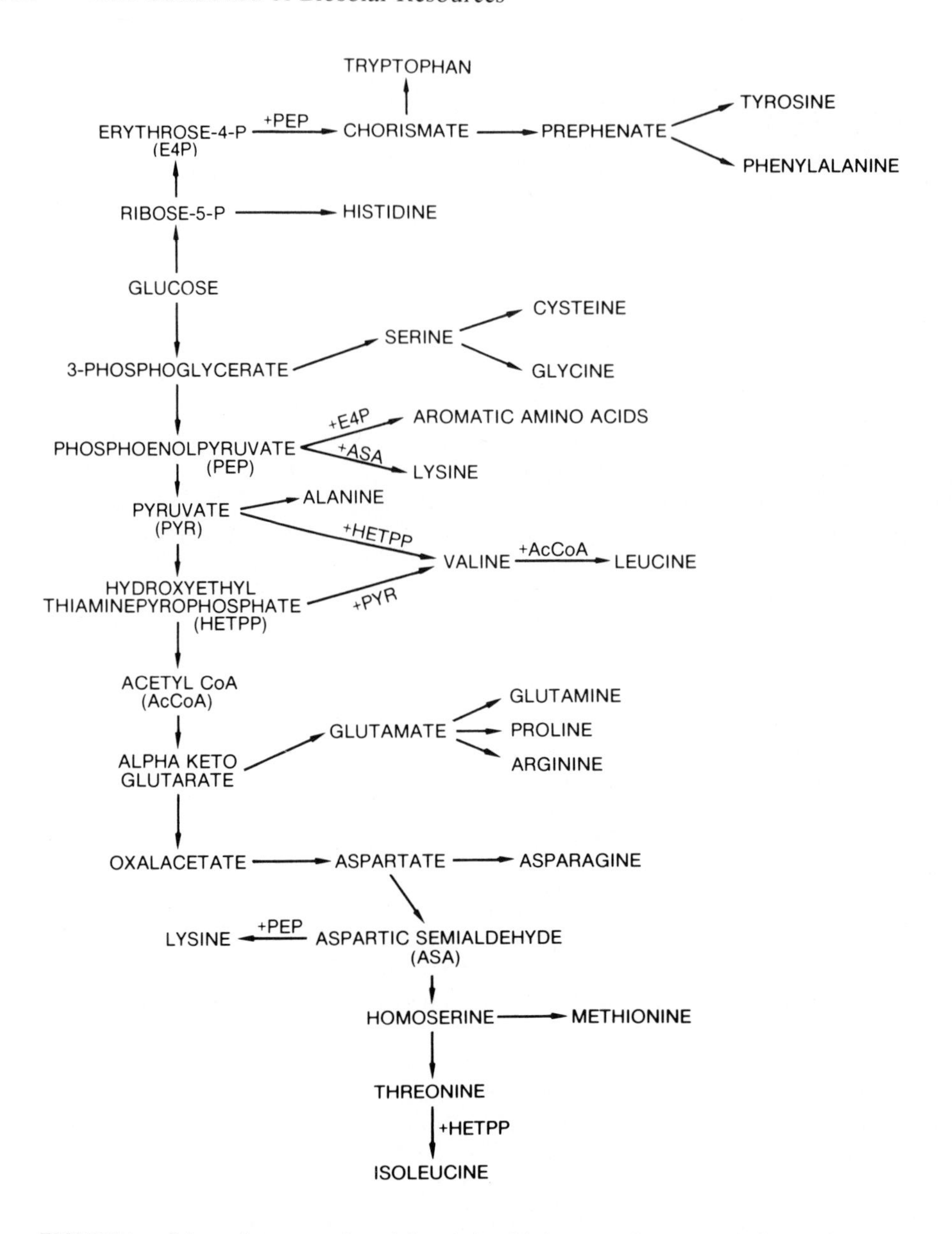

FIGURE 1. Schematic presentation of the relationship between glucose metabolism and amino acid synthesis. (Letters and numbers in parentheses are abbreviations for the compound named just above the parentheses).

taric acid catalyzed by glutamic dehydrogenase. The enzyme is responsible for the initial synthesis of glutamic acid although it now appears that the quantitatively important reaction for glutamic acid biosynthesis is glutamate synthetase (Reaction 3).[6] Since glutamine is a substrate for glutamate synthesis and glutamic acid is a substrate for glutamine synthesis, some glutamic acid must be formed as in Reaction 1 to start Reactions 2 and 3.

Glutamic acid is the ultimate source of amino nitrogen for all protein amino acids. Aminotransferase reactions utilize pyridoxal phosphate as a cofactor. Glutamine, proline, and arginine derive not only their alpha amino nitrogen from glutamic acid but also their carbons. Glutamine is not only incorporated into protein but its amide group

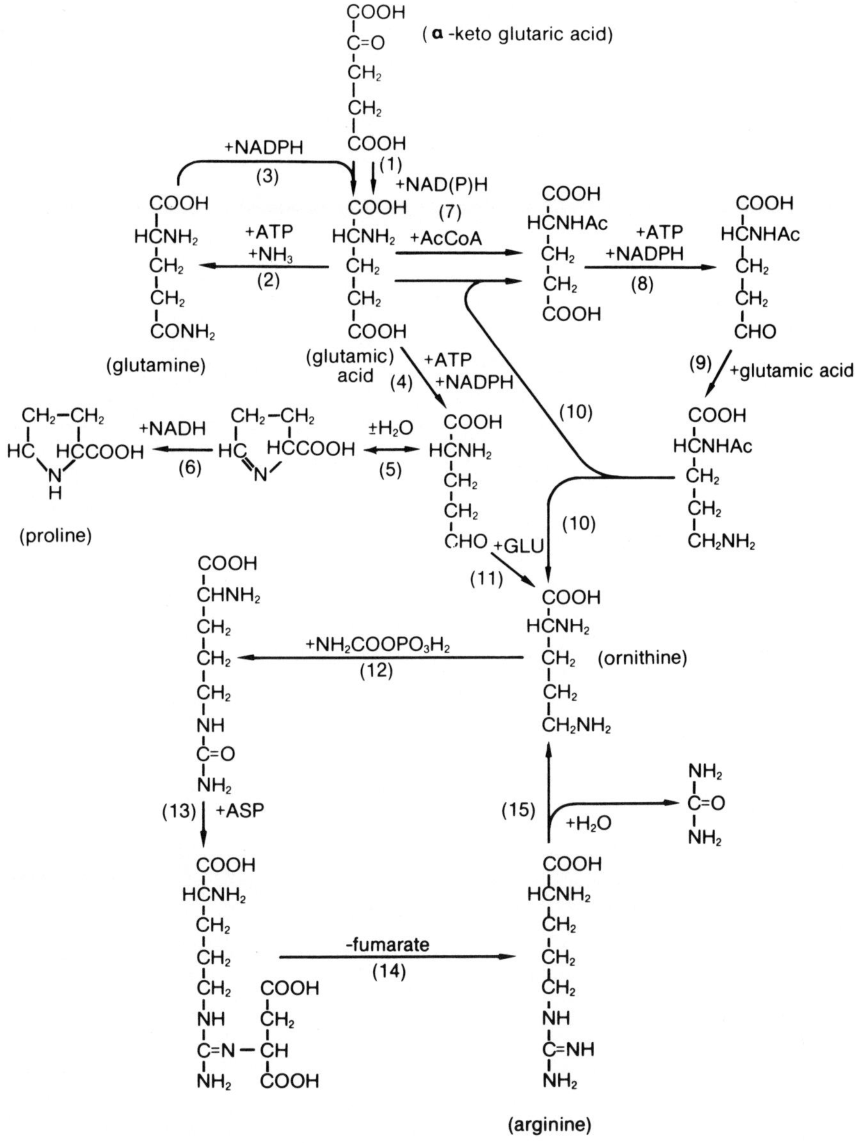

FIGURE 2. Biosynthesis of amino acids derived from α-ketoglutaric acid.

is an important amino donor in the biosynthesis of such compounds as carbamylphosphate (see Reaction 12), purines and sugar amines. Reaction 4 probably represents two reactions in which γ-glutamylphosphate is formed and reduced. The occurrence of γ-glutamylphosphate as an intermediate has not been demonstrated because γ-glutamylphosphate is unstable and must be enzyme bound. Two separate enzymes have not been demonstrated in photosynthetic plants and bacteria. The carboxyl group has to be activated by forming a mixed anhydride with phosphoric acid that can be reduced. Reaction 5 is a nonenzymatic reaction in which a cyclic Schiff base is formed from a straight chain compound and vice versa. Cyclization can be prevented by acetylation of the amino group as in Reaction 7. Probably much more ornithine is formed by Reaction 10 than by Reaction 11 because Reaction 5 is more rapid than Reaction 11. Reactions 7 to 10 provide a pathway for ornithine synthesis where reaction of the

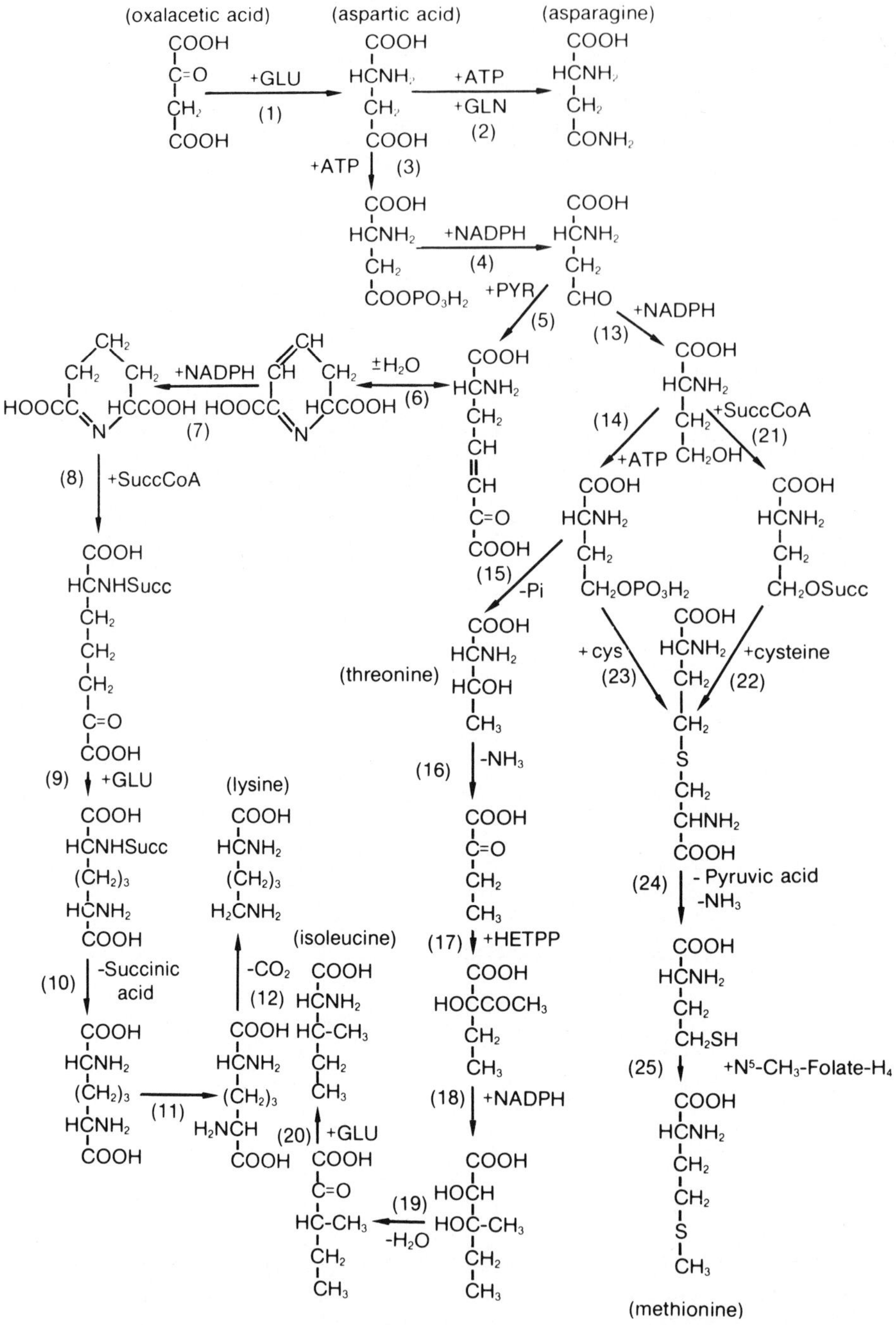

FIGURE 3. Biosynthesis of amino acids derived from oxalacetic acid.

amino group is prevented by acetylation. The remarks concerning Reaction 4 probably apply to reaction 8. Subsequently the acetyl group is recycled, as in Reaction 10, thus avoiding loss of energy involved in the formation of acetyl CoA (acetyl coenzyme A). Reactions 12 to 15 comprise the Krebs-Henseleit cycle or ornithine cycle whereby mammals convert amino nitrogen to urea. Plants utilize Reactions 12 to 14 to form arginine since they do not excrete urea and Reaction 15 is important when arginine is reutilized.

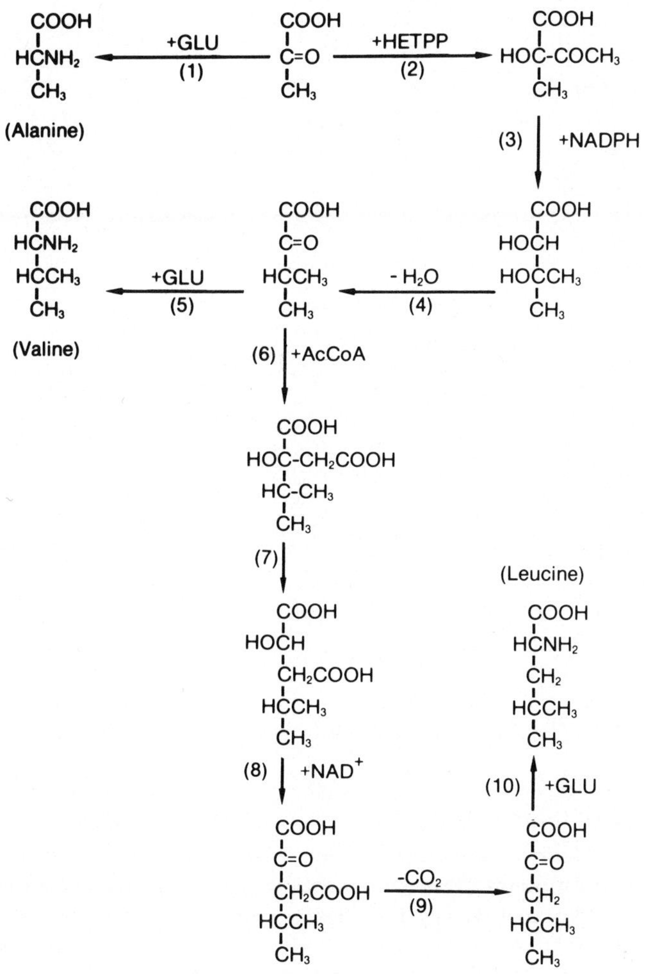

FIGURE 4. Biosynthesis of amino acids derived from pyruvic acid.

In plants, urea is hydrolyzed to CO_2 and ammonia prior to reutilization. Ornithine is converted to glutamic acid by reversal of Reactions 11 and 4.

Aspartic acid, asparagine, lysine, threonine, isoleucine, and methionine are derived from oxalacetic acid (Figure 3). Aspartic acid is the key compound and is formed from oxalacetic acid by an aminotransferase (Reaction 1). Asparagine is formed from aspartic acid by an amidation reaction (2) where the nitrogen is derived from the amide group of glutamine and the energy from ATP. This reaction is not analogous to the formation of glutamine (Reaction 2, Figure 2) because aspartyl adenylate is an intermediate. Asparagine is not an amino donor as is glutamine. The amide nitrogen of asparagine can be utilized after hydrolysis. Aspartic semialdehyde is formed by Reactions 3 and 4. In contrast to glutamic semialdehyde formation (Reaction 4, Figure 2), β-aspartylphosphate is a well-known intermediate. Aspartic semialdehyde does not spontaneously cyclize (as does glutamic semialdehyde) because a four-membered ring does not form readily. Aspartic semialdehyde can be condensed with pyruvic acid (Reaction 5) or reduced to homoserine (Reaction 13). In Reaction 5, water is eliminated to form a seven-carbon dicarboxylic acid (dihydrodipicolinic acid) that spontaneously cyclizes (Reaction 6). The cyclic compound is partially hydrogenated to tetrahydrodipicolinic acid (Reaction 7). The latter compound also represents a cyclic form of a

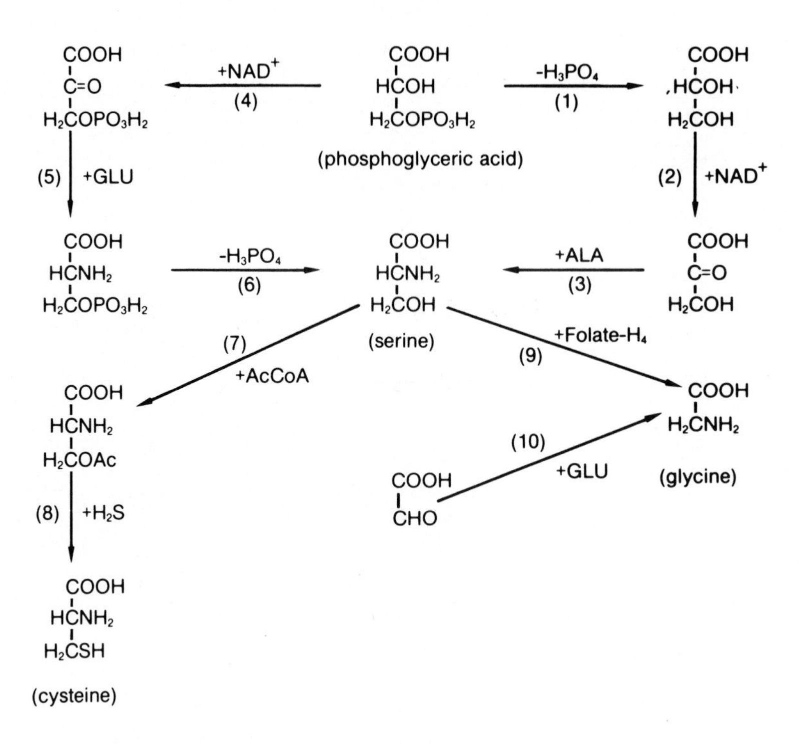

FIGURE 5. Biosynthesis of amino acids derived from phosphoglyceric acid.

straight chain compound. The reaction of the straight chain form with succinyl CoA forms a succinylated product that cannot cyclize. The latter compound is then transaminated to diaminopimelic acid which is decarboxylated to form lysine. Some fungi form lysine by an entirely different pathway where α-aminoadipic acid is an important intermediate. This pathway is not included here because it does not occur in autotrophs.

In higher plants and bacteria, the hydroxyl group of homoserine is phosphorylated (Reaction 14) to homoserine phosphate which is rearranged to threonine in a pyridoxal phosphate-mediated reaction. Threonine can be incorporated into protein or deaminated to α-ketobutyric acid in the first unique step in isoleucine biosynthesis (Reaction 16). Hydroxyethylthiamin pyrophosphate (HETPP) acetylates α-ketobutyric acid to form α-hydroxy, α-acetylbutyric acid. The latter compound is reductively rearranged to α,β-dihydroxy,β-methylbutyric acid. This compound is dehydrated to α-keto,β-methylvaleric acid that is transaminated to isoleucine.

The first unique step in methionine synthesis is the formation of cystathionine derived from homoserinephosphate and cysteine in higher plants and from *O*-succinylhomoserine and cysteine in bacteria. Cystathionine is split to form homosysteine which is methylated by N^5-methyltetrahydrofolic acid (N_5-CH_3-folate H_4) to form methionine. Methionine formation is the only reaction in which N^5-methyltetrahydrofolic acid acts as a methyl donor. In other methylation reactions, *S*-adenosylmethionine acts as methyl donor. In animals, the methylation of homocysteine to form methionine requires B_{12} cofactor. Plants do not use B_{12} cofactor in this reaction, do not make B_{12} cofactor, and do not require cobalt, a constituent of the B_{12} cofactor.

The great importance of pyridoxal phosphate in amino acid metabolism is illustrated by the fact that it is a cofactor in Reactions 1, 9, 11, 12, 15, 16, 20, and 22 to 24.

Alanine, valine, and leucine are derived from pyruvic acid (Figure 4). Alanine is formed directly by transamination with glutamic acid (Reaction 1). The last step in

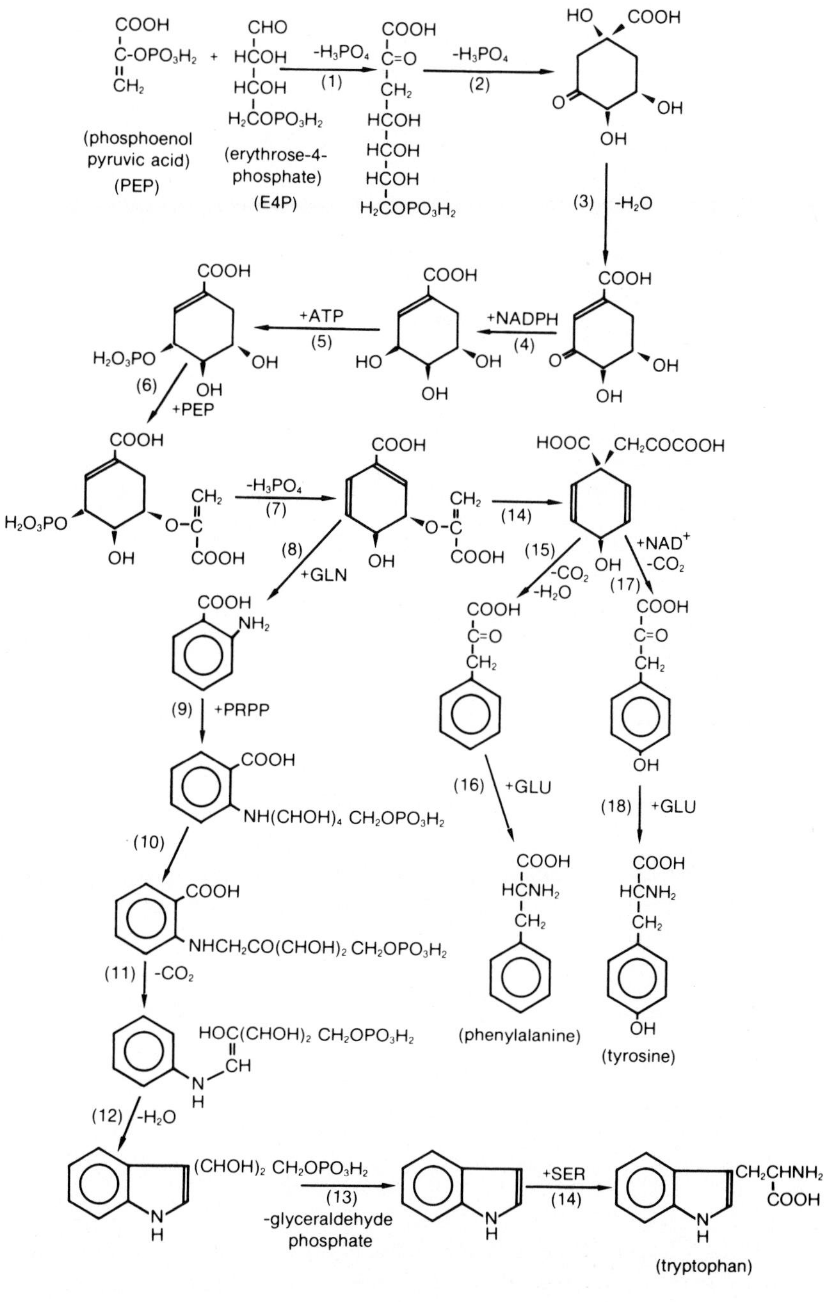

FIGURE 6. Biosynthesis of amino acids derived from phosphoenolpyruvic acid.

valine and leucine synthesis is also a transamination step (Reactions 5 and 10). In Reaction 2, pyruvic acid and a two-carbon fragment from hydroxyethylthiamine pyrophosphate are joined in a five-carbon keto acid (α-acetolactic acid). This acid is reductively rearranged to α,β-dihydroxy,β-methylbutyric acid. The latter acid is dehydrated to α-ketoisovaleric acid that is transaminated to form valine (Reaction 5) or coupled with acetyl-CoA to form α-isopropylmalic acid as the first unique step in leucine synthesis. Alpha-isopropylmalic acid is rearranged to β-isopropylmalic acid (Re-

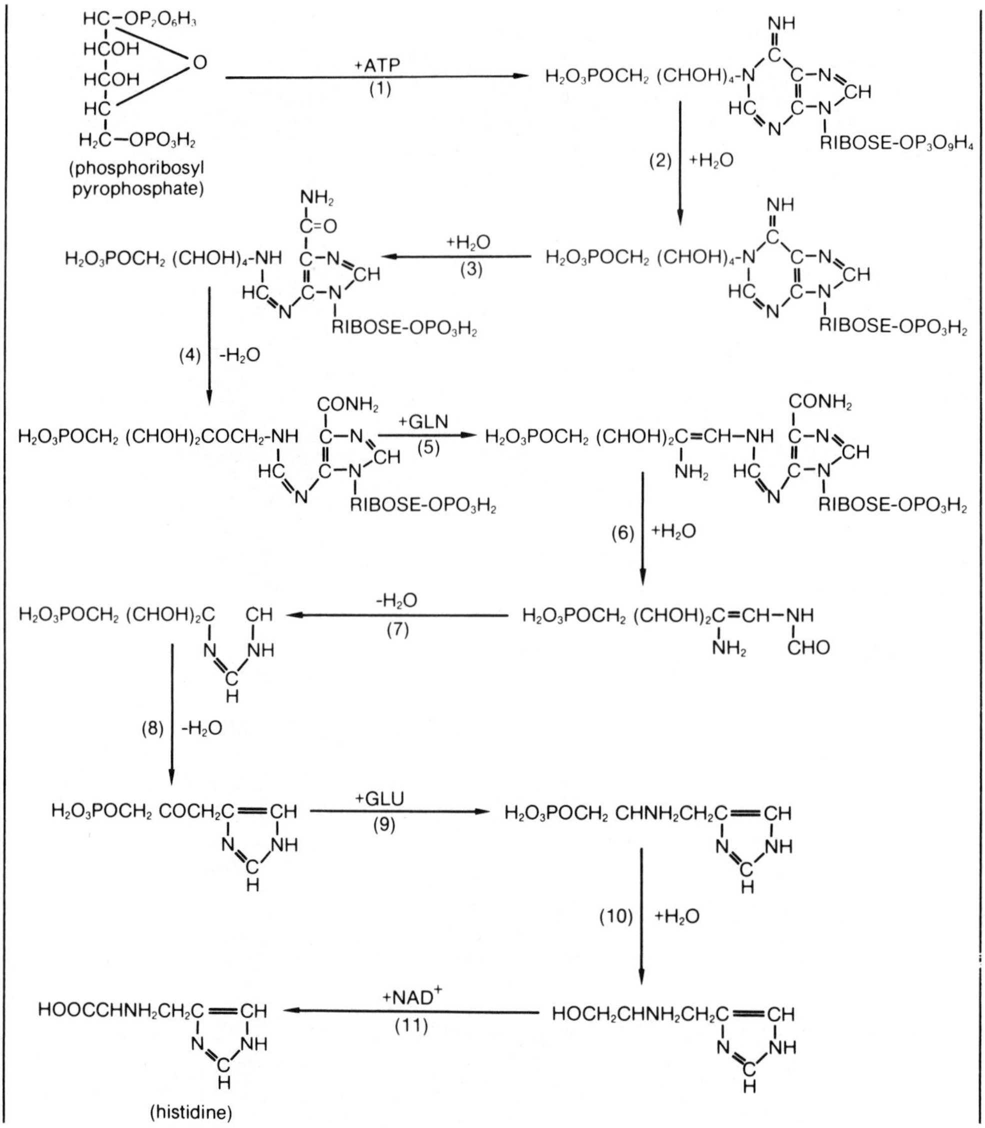

FIGURE 7. Biosynthesis of histidine from phosphoribosylpyrophosphate.

action 7). When β-isopropylmalic acid is dehydrogenated with NAD^+ (Reaction 8), the product spontaneously decarboxylates (Reaction 9) to α-ketoisocaproic acid, the keto analog of leucine. Reactions 2 to 5 are catalyzed by the same enzymes that promote the analogous reactions in isoleucine biosynthesis (Reactions 17 to 20 of Figure 3).

Serine, glycine, and cysteine are derived from 3-phosphoglyceric acid (Figure 5). The latter compound is an early product of CO_2 fixation in autotrophic organisms since phosphoglyceric acid is a product of ribulose bisphosphate carboxylase catalyzed reaction. Serine is formed from phosphoglyceric acid by a phosphorylated intermediate pathway (Reactions 4 to 6) or by a nonphosphorylated pathway (Reactions one to three). In either pathway, amination is accomplished by an amino transferase reaction (Reactions 3 and 5). Reaction 3 is unusual in that alanine rather than glutamic acid is the amino donor. The formation of glycine involves a tetrahydrofolate derivative and hydroxymethyl tetrahydrofolate is a product (Reaction 9). Glycine also is formed in plants by transamination of glyoxylic acid (Reaction 10). The glyoxylic acid is formed by oxidation of glycolic acid, a product of photorespiration. Two molecules of glycine can also form serine with the release of ammonia and CO_2 during photorespiration. Serine is the normal precursor of cysteine. Though serine will react with H_2S to form cysteine, *O*-acetylserine is much more active and must be the natural substrate for cysteine formation in plants and bacteria.

Phosphoenolpyruvic acid is a precursor of the aromatic amino acids (Figure 6). Phosphoenolpyruvic acid combines with erythrose-4-phosphate to form 3-deoxyarabino heptulosonic acid-7-phosphate (Reaction 1). The latter compound cyclizes (Reaction 2) to form a cyclohexane derivative (dehydroquinic acid) by a complex reaction requiring NAD^+ as catalyst, not a substrate. The dehydroquinic acid is dehydrated and reduced (Reactions 3 and 4) to shikimic acid. Phosphorylated shikimic acid reacts with phosphoenolpyruvic acid to form chorismic acid (Reaction 7) which is the branch point compound between the two pathways. Chorismic acid may react with glutamine to form anthranilic acid (Reaction 8). Phosphoribosylpyrophosphate combines with anthranilic acid to form a compound that undergoes a series of reactions (Reactions 9 to 13) to form indole. One Reaction (10) of this series involves an Amadori type of rearrangement. Indole combines with serine to form tryptophan (Reaction 14). Chorismic acid also undergoes an internal rearrangement (Reaction 14) to form prephenic acid that is the precursor of both phenylalanine and tyrosine. Prephenic acid is dehydrated and decarboxylated (Reaction 15) to form phenylpyruvic acid, the immediate precursor of phenylalanine. Prephenic acid may be oxidized and decarboxylated (Reaction 17) to *p*-hydroxy-phenylpyruvic acid, the precursor of tyrosine.

Phosphoribosylpyrophosphate and ATP furnish all the carbon atoms for histidine biosynthesis. Histidine is formed by a series of 10 reactions that are involved only in histidine biosynthesis. Figure 7 shows 11 reactions because Reaction 7 is spontaneous and is written separately to show the mechanism of imidazole ring formation. Nine enzymes catalyze the ten reactions because Reactions 8 and 10 are catalyzed by the same enzyme. Reaction 1 is unusual in that ATP is a substrate for its carbon and nitrogen atoms and not for phosphorylation. The adenine ring is hydrolytically opened (Reaction 3) and subsequently most of the adenine molecule is split off (Reaction 6). After the latter reaction, the resultant amino aldehyde spontaneously forms a Schiff base that is an unstable imidazole ring (Reaction 7). Reaction 5 illustrates one of the reactions where the amide group of glutamine supplies an amino group. Reaction 9 represents a transamination reaction where an α-keto acid is not the amino acceptor. The last step in histidine biosynthesis is the oxidation of an alcohol to an acid requiring the removal of four electrons.

REFERENCES

1. **Eichhorn, M. M.,** Data on the naturally occurring amino acids, in *CC Handbook of Biochemistry,* Sober, H. A., Ed., CRC Press, Boca Raton, Fla., 1970, B3.
2. **Bryan, J. K.,** Amino acid biosynthesis and its regulation, in *Plant Biochemistry,* 3rd Ed., Bonner, J. and Varner, J. E., Eds., Academic Press, New York, 1976, chap. 17.
3. **Greenberg, D. M.,** Biosynthesis of amino acids and related compounds, in *Metabolic Pathways,* Vol. III, 3rd Ed., Greenberg, D. M., Ed., Academic Press, New York, 1969, Chap. 16 (part 1).
4. **Rodwell, V. W.,** Biosynthesis of amino acids and related compounds, in *Metabolic Pathways,* Vol. III, 3rd Ed., Greenberg, D. M., Ed. Academic Press, New York, 1969, Chap. 16 (part 2).
5. **Beevers, L.,** *Nitrogen Metabolism in Plants,* Edward Arnold, London, 1976, chap. 2.
6. **Miflin, B. J. and Lea, P. J.,** Amino acid metabolism, *Ann. Rev. Plant Physiol.,* 28, 299, 1977.

Section 4
General Classification of Photosynthetic Organisms

TAXONOMIC CLASSIFICATION OF PHOTOSYNTHETIC BACTERIA (ANOXYPHOTOBACTERIA, PHOTOTROPHIC BACTERIA)

J. F. Imhoff and H. G. Trüper

Gibbons and Murray recently proposed a reorganization of higher bacterial taxa according to cell wall properties.[1] They also redefined phototrophic bacterial taxa above the family level. Accordingly, all phototrophic prokaryotes form the class of Photobacteria Gibbons and Murray.[1]

The first subclass Oxyphotobacteria Gibbons and Murray,[1] with the orders Cyanobacteriales Stanier[1] and Prochlorales Lewin,[2] includes all prokaryotes which contain plant-type chlorophylls and produce oxygen during photosynthesis using water as the electron donor. The second subclass, Anoxyphotobacteria Gibbons and Murray,[1] includes all prokaryotes which possess one or more of the different types of bacteriochlorophyll (a, b, c, d, or e). They do not evolve oxygen during photosynthesis, but carry out a phototrophic metabolism under anaerobic conditions with different organic and/or inorganic electron donors. The first order of this subclass, Rhodospirillales Pfennig and Trüper,[3] comprises the two families Rhodospirillaceae Pfennig and Trüper[3] and Chromatiaceae Bavendamm.[4] The second order, Chlorobiales Gibbons and Murray,[1] includes the families Chlorobiaceae Copeland[5] and Chloroflexaceae Trüper.[6]

Only the anoxygenic phototrophic bacteria, the Anoxyphotobacteria, which possess different types of bacteriochlorophyll, are described. The Oxyphotobacteria, which possess plant-type chlorophylls, are treated under "Algae". The Halobacteria, which can generate energy by a light-driven proton gradient, do not possess bacteriochlorophylls, and thus have no standing within the Photobacteria and are not considered.

The differentiation of the Anoxyphotobacteria down to the genus level is shown in Figures 1 and 2. The species of the different genera also are listed in these figures. They deal with our present knowledge about the taxonomy of Anoxyphotobacteria based upon the "Approved List of Names of Bacteria".[7] Thus the following genera and species, included in the eighth edition of *Bergey's Manual of Determinative Bacteriology,*[8] are not included here, because they either are not pure and have never been studied in pure cultures nor even in enrichment cultures, or as *Chloropseudomonas ethylica,* were discovered to be a syntrophic mixed culture:[9,10] the genus *Thiosarcina* and *T. rosea,* the genus *Chloropseudomonas* and *Chloropseudomonas ethylica,* the genus *Clathrochloris* and *C. sulfurica, Thiospirillum sanguineum* and *T. rosenbergii.* Compared with the data presented in the eighth edition of *Bergey's Manual of Determinative Bacteriology,* one new family has been created,[6] five new genera have been established,[11-15] and several new species have been described.

RHODOSPIRILLALES

The Rhodospirillales possess bacteriochlorophyll (bchl) *a* and, in four species, bchl *b* as the light-harvesting bchl and a great number of different carotenoids.[16] The photopigments are located in the cytoplasmic membrane, which forms more or less complex intracytoplasmic membrane systems such as vesticles, tubes, stacks, or only small intrusions. In most species, cell division occurs by binary fission and only exceptionally by budding.

Rhodospirillaceae

This family shows the greatest diversity of structure in intracytoplasmic membrane systems and comprises the most versatile organisms of the Anoxyphotobacteria. All

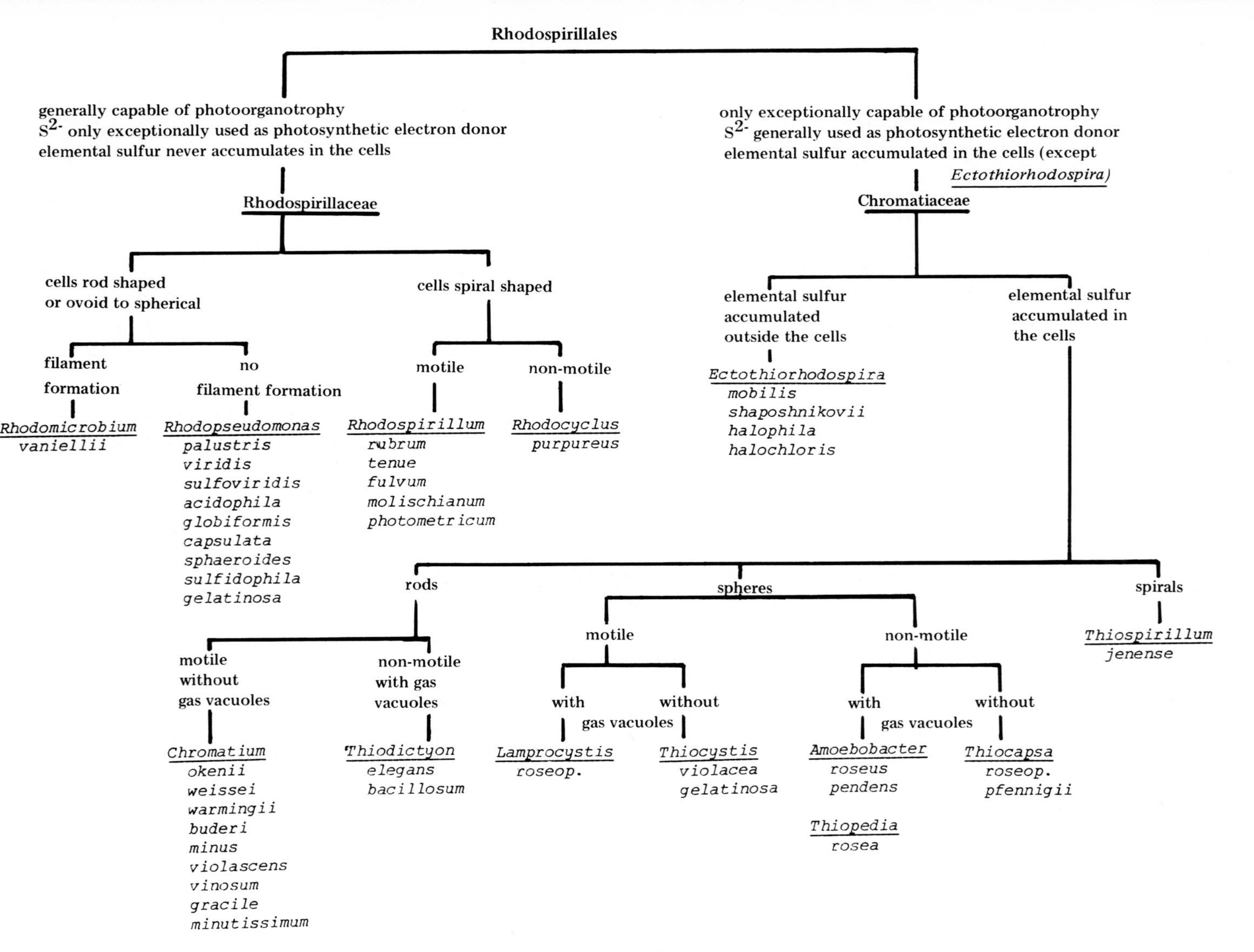

FIGURE 1. Classification of Anoxyphotobacteria through the first order.

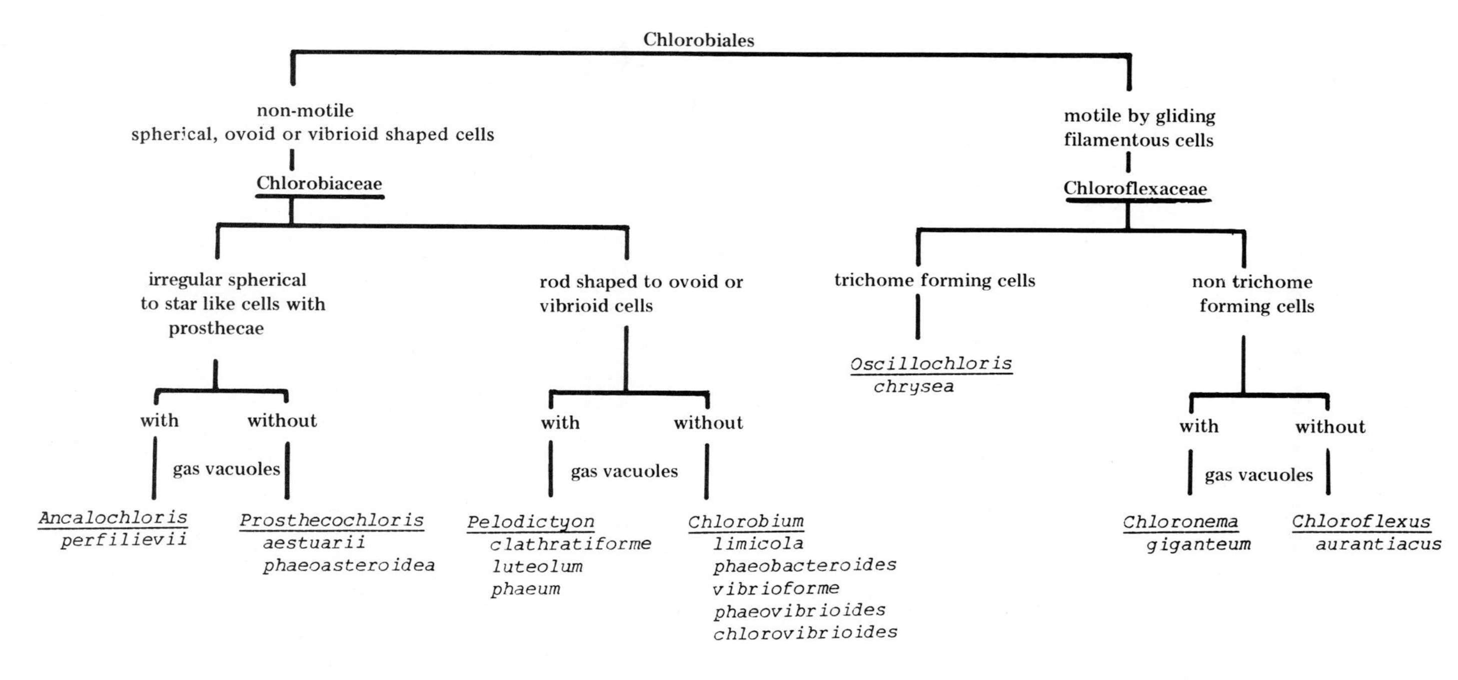

FIGURE 2. Classification of Anoxyphotobacteria through the second order.

species are able to use organic substances as photosynthetic electron donors. Most of them also are able to grow photoautotrophically with H_2 as electron donor and CO_2 as sole carbon source (except growth factors). Many species can grow chemoorganotrophically under aerobic dark conditions. Only a few species are known to oxidize hydrogen sulfide. If they do, they never accumulate elemental sulfur within the cells, as it is found in Chromatiaceae species. In the following description, the DNA base ratio in mol % G + C is given in brackets after each species name.

Rhodospirillum

The *Rhodospirillum* species can be clearly differentiated according to their spirilloid shape, cell-size, membrane structure, and pigment content. *R.rubrum* (63.8 to 65.8), the best-investigated *Rhodospirillum* species, contains bchl *a* and spirilloxanthin as the main carotenoid component. Cell size is 0.8 to 1.0 × 7 to 10 μm. Cultures are red-colored and photosynthetic membranes are present as vesicles. *R.tenue* (64.8) is the smallest species with 0.3 to 0.5 × 3 to 6 μm cell size. Violet and brown-colored variants exist due to different carotenoid content. Carotenoids of the rhodopinal series are present. The photosynthetic membranes occur as small tubular intrusions of the cytoplasmic membrane. The three brown-colored species, *R.fulvum* (64.3 to 65.3), *R.molischianum* (61.7 to 64.8), and *R.photometricum* (65.8), are rather sensitive to oxygen. The photosynthetic membranes occur as lamellar stacks which form a sharp angle with the cytoplasmic membrane. Main carotenoids are lycopene and rhodopin. The three species differ most obviously in their cell size, which is 0.5 to 0.7 × 3.5 μm for *R.fulvum,* 0.7 to 1.0 × 5.0 to 8.0 μm for *R.molischianum* and 1.2 to 1.5 × 7 to 10 μm for *R.photometricum.*

Rhodocyclus

The only nonmotile "spirillum" is *Rhodocyclus purpureus* (65.3).[15] Cultures are purple-violet, membranes occur as small tubular intrusions of the cytoplasmic membrane, and cell size is 0.6 to 0.7 × 2.7 to 5.0 μm. Individual cells of this bacterium are half-ring shaped before cell division. Carotenoids of the rhodopinal series are present.

Rhodomicrobium

Rhodomicrobium vannielii (61.3 to 63.8) is the only known species of this genus. Cells are ovoid to rod-shaped, 1.0 to 1.2 μm wide, 2 to 2.8 μm long, motile by peritrichous flagellation, and multiply by budding. Cultures are dark orange-red to brown. Photosynthetic membranes are present as parallel lamellae underlying the cytoplasmic membrane. The most characteristic feature of *Rhodomicrobium* is the formation of filaments (stalks) which may cause large cellular aggregates.

Rhodopseudomonas

Rod-shaped species that also multiply by budding and possess photosynthetic membranes as parallel lamellae underlying the cytoplasmic membrane belong to the genus *Rhodopseudomonas.* They are motile by subpolarly inserted flagella. The formation of rosettes in which the cells are attached to each other at their flagellated poles is characteristic of this group, which includes *R. palustris* (64.8 to 66.3), *R. acidophila* (62.2 to 66.8), *R. viridis* (66.3 to 71.4), and *R. sulfoviridis* (67.8 to 68.4).[17]

R. palustris cultures are red to dark brown-red. The cells have carotenoids of the spirilloxanthin series with mainly spirilloxanthin, and are 0.6 to 0.8 μm wide and 1.2 to 2.0 μm long. The pH-range for growth is 5.5 to 8.5; pH-optimum near 6.8. *R. palustris* is able to use $S_2O_3^{2-}$ and S^{2-} as photosynthetic electron donors. Cultures of *R. acidophila* are orange-brown to purple-red and grow in the pH-range from 4.8 to

7.0, with an optimum at pH 5.8. The size of the cells is 1.0 to 1.3 μm width and 2.0 to 6.0 μm length.

R. viridis and *R. sulfoviridis* are well characterized by their green to yellowish-brown coloration; their possession of bchl *b*. *R. sulfoviridis* differs from *R. viridis* in its dependence on reduced sulfur compounds for growth and the capability to grow with thiosulfate as electron donor in the presence of malate and yeast extract.

The other species of the genus *Rhodopseudomonas* have rod-shaped, ovoid, or spherical cells and multiply by binary fission.

R. gelatinosa (70.5 to 72.4) is able to liquify gelatine and to grow on it as sole carbon and nitrogen source. Cultures are pale brown to peach-colored and cells tend to clump together due to slime formation. Cells are 0.4 to 0.5 μm wide and 1 to 2 μm long. The photosynthetic membranes are present as small tubular intrusions of the cytoplasmic membrane.

The three species, *R. capsulata* (65.5 to 66.8), *R. sphaeroides* (68.4 to 69.9), and *R. sulfidophila*[18] (67.0 to 71.0), all have rod-shaped to ovoid cells, possess bchl *a* and carotenoids of the spheroidenone group located in vesicular-type intracytoplasmic membranes. The color of their cell suspensions is yellowish-brown to brown-red and purple-red in the presence of oxygen. All three species can be easily adapted to aerobic dark growth. While *R. sphaeroides* shows little tendency to chain formation, straight chains are formed in *R. sulfidophila* under proper culture conditions. Characteristic zig-zag arrangements of cells are observed in *R. capsulata*. Cell sizes are 0.5 to 1.2 × 2 to 2.5 μm for *R. capsulata*, 0.7 × 2.0 to 2.5 μm for *R. sphaeroides* and 0.6 to 0.9 × 0.9 to 2.0 μm for *R. sulfidophila*. *R. sulfidophila* tolerates rather high concentrations of sulfide and oxidizes it to sulfate without accumulation of elemental sulfur. Thiosulfate is also used as the photosynthetic electron donor.[18] Differentiation between *R. capsulata* and *R. sphaeroides* is possible due to the high penicillin sensitivity of *R. capsulata*. Growth of *R. capsulata* is completely inhibited by 0.1 unit penicillin G/mℓ. The tolerance of *R. sphaeroides* to penicillin is 10^3 to 10^5 times greater.[19] Genetic transfer mediated by "gene transfer agents" is limited to *R. capsulata* and does not extend to any other member of the Rhodospirillaceae.[20]

R. globiformis (66.3) is a motile sphere, 1.6 to 1.8 μm in diameter, unable to assimilate sulfate at higher concentrations. Thiosulfate or cysteine are preferred as sulfur compounds for growth. Cultures are purple-red and photosynthetic membranes are formed as vesicles.[21]

Detailed data on the utilization of carbon compounds by Rhodospirillaceae are given by Pfennig and Truper.[22]

Chromatiaceae

The general property of the species of this family is the capacity for photoautotrophic growth with sulfide or elemental sulfur as electron donors and CO_2 as sole carbon source. Elemental sulfur is generally deposited within cells during sulfide oxidation and can be detected easily by microscopic examination of growing cultures as highly refractile globules within the cells. The genus *Ectothiorhodospira* is the only exception here, depositing sulfur outside the cells. Cells multiply by binary fission and are, if at all, motile by polar flagella. The pigments, bchl *a* (only *Thiocapsa pfennigii* and *Ectothiorhodospira halochloris*[23] contain bchl *b*), and different carotenoids are located in intracytoplasmic membranes of the vesicular type. Exceptions are the genus *Ectothiorhodospira* with lamellar stacks similar to those in the brown-colored *Rhodospirillum* species and *T. pfennigii*, which contains bundles of tubes. Most species depend on the presence of reduced-sulfur compounds, and only a few species show a well-established capacity for photoorganotrophic growth. Only *T. roseopersicina* and *E.*

shaposhnikovii are capable of chemoorganotrophic growth. At limited partial pressure of oxygen, chemolithoautotrophic growth with bicarbonate and sulfide or thiosulfate in the dark has been shown for strains of *Chromatium vinosum, C. minus, C. gracile, C. violascens, Thiocystis violacea,*[24] *Thiocapsa roseopersicina,*[25] and *Amoebobacter roseus.*[26]

Ectothiorhodospira

Cells are vibrioid to spiral shaped and motile by polar flagella. The photosynthetic membranes are formed as lamellar stacks positioned at an angle to the cytoplasmic membrane. Elemental sulfur is accumulated extracellularly during oxidation of sulfide.

E. mobilis (67.3 to 69.9) cells are 0.7 to 1.0 μm wide and 2.0 to 2.6 μm long. Carotenoid pigments of the spirilloxanthin series are present. NaCl (2 to 8%) is required for growth. *E. shaposhnikovii* (62.3) cells are 0.8 to 0.9 μm wide and 1.5 to 2.5 μm long. Carotenoids of the spirilloxanthin series are present. Low NaCl concentrations are required for growth. *E. halophila* (68.4) cells are 0.8 μm wide and 2 to 5 μm long and contain carotenoids of the spirilloxanthin series. The organism is an extreme halophile and requires 11 to 22% total salts for optimal growth. *E. halochloris* (52.9) cells are 0.5 to 0.6 μm wide and 2.5 to 8.0 μm long.[23] They contain bchl *b* and small amounts of rhodopin carotenoids, mainly present as glucosides. Cultures are green-colored. *E. halochloris* is extremely halophilic. Salts (14 to 27% [w/v]) are required for optimal growth.[23]

Chromatium

Cells are rod-shaped (sometimes slightly curved), motile, and do not contain gas vacuoles. *C. okenii* (48 to 50) cells are 4.5 to 6 μm wide and 8 to 15 μm long and the flagella tuft is visible in the light microscope. Because of the presence of the carotenoid okenone single cells and cell suspensions are purple-red. *C. weissei* (48 to 50) cells are 3.5 to 4.5 μm wide and 7 to 9 μm long; otherwise *C. weissei* is as *C. okenii. C. warmingii* (55.1 to 60.2) cells are 3.5 to 4.0 μm wide and 5 to 11 μm long. The color of cell suspensions is purple-violet. Carotenoids of the rhodopinal series are present. Sulfur globules are located predominantly at the two poles of the cell. *C. buderi* (62.0 to 62.8) cells are 3.5 to 4.5 μm wide and 4.5 to 9.0 μm long. Cultures appear purple-violet. Carotenoids of the rhodopinal series are present. NaCl at 1 to 3% is required for growth. *C. minus* (52.0 to 62.2) cells are 2 μm wide and 2.5 to 6.0 μm long. Cultures appear purple-red. The carotenoid okenone is present. Thiosulfate is used as photosynthetic electron donor. *C. violascens* (61.8 to 64.3) cells are 2 μm wide and 2.5 to 6.0 μm long. Cultures are purple-violet. Carotenoids of the rhodopinal series are present. *C. vinosum* (61.3 to 66.3) cells are 2 μm wide and 2.5 to 6.0 μm long, cultures brown to brown-red. Carotenoids of the spirilloxanthin series are present. Photoorganotrophic growth is well-established. *C. gracile* (68.9 to 70.4) cells are 1 to 1.3 × 2 to 6 μm; otherwise *C. gracile* is as *C. vinosum. C. minutissimum* (63.7) cells are 1 to 1.2 × 2 μm; otherwise *C. minutissimum* is as *C. vinosum.*

Thiospirillum

Cells are spiral-shaped and motile by polar flagella. Only one species has been isolated as a pure culture. *Thiospirillum jenense* (45.5) cells are 2.5 to 4.0 μm wide and 30 to 40 μm long. Major carotenoids are rhodopin and lycopene. Cell suspensions are orange-brown. The organism does not grow in agar shake cultures.

Thiocystis

Cells are motile, spherical to ovoid-shaped, and do not contain gas vacuoles. *Thiocystis violacea* (62.8 to 67.9) cells are 2.5 to 3 μm in diameter. Cultures appear purple violet. Carotenoids of the rhodopinal series are present. Sulfur globules are evenly distributed throughout the cell. *T. gelatinosa* (61.3) cells are 3 μm in diameter and contain carotenoids of the okenone series. Cultures are purple-red. Globules of sulfur appear only in the peripheral part of the cytoplasm.

Thiodictyon

Cells are nonmotile, rod-shaped, and contain gas vacuoles. *Thiodictyon elegans* (65.3) cells are 1.5 to 2.0 μm wide and 3.0 to 8.0 μm long. They form aggregates in which they are arranged in an irregular net-like structure. Cultures are purple-violet and carotenoids of the rhodopinal series are present. *T. bacillosum* (66.3) cells are 1.5 to 2.0 μm wide and 3 to 6 μm long. Net-like aggregates are not formed. Cultures are purple-violet and carotenoids of the rhodopinal series are present.

Lamprocystis

Cells are motile, spherical to ovoid and contain gas vacuoles. *Lamprocystis roseopersicina* (63.8) cells are 3.0 to 3.5 μm in diameter. Cell suspensions are purple red. The carotenoids lycopenal and lycopenol are present.

Thiocapsa

Cells are nonmotile, spherical, and do not contain gas vacuoles. *Thiocapsa roseopersicina* (63.3 to 66.3) cells are 1.5 μm in diameter and cell suspensions are pink. Spirilloxanthin is present as the major carotenoid. Most strains produce a slime capsule. *T. pfennigii* (69.4 to 69.9) is 1.2 to 1.5 μm in diameter. Cultures are orange-brown. Photosynthetic pigments bchl *b* and tetrahydroxyspirilloxanthin are present, located in a tubular intracytoplasmic membrane system.

Amoebobacter

Cells are nonmotile, spherical, and contain gas vacuoles. *Amoebobacter roseus* (64.3) is 2 to 3 μm in diameter and contains carotenoids of the spirilloxanthin series. *A. pendens* (65.3) is 1.5 to 2.5 μm in diameter and contains carotenoids of the spirilloxanthin series.

Thiopedia

Cells are nonmotile, spherical to ovoid, and contain gas vacuoles. They are arranged in flat sheets with typical tetrads as structural units. *Thiopedia rosea* (at present not in pure culture) is 1 to 2 μm wide and 1.2 to 2.5 μm long and contains carotenoids of the rhodopinal series.

CHLOROBIALES

In contrast to the Rhodospirillales, within the Chlorobiales the light harvesting (lh) bacteriochlorophylls *c, d, e,* and carotenoids are located in *Chlorobium* vesicles (also called chlorosomes by Staehelin et al.[27]) which are present in and characteristic of all species of this order. Comparable structures have not been found in the Rhodospirillales. The photosynthetic reaction centers (rc) with small amounts of bchl *a* are incorporated in the cytoplasmic membranes to which the *Chlorobium* vesicles are closely attached.

Chlorobiaceae

All species of the family are nonmotile. They have restricted metabolic capacities which depend on anaerobic light conditions and the presence of reduced sulfur compounds (sulfide or thiosulfate) as photosynthetic electron donor and sulfur source and

CO_2 as carbon source. During sulfide oxidation to sulfate, elemental sulfur is excreted into the medium. Sulfur globules never appear inside the cells. During growth on sulfide and CO_2, only a limited number of organic compounds are photoassimilated, e.g., acetate, propionate, pyruvate, and fructose. Within three of the four genera, green and brown-colored species exist which possess morphologically and physiologically similar properties. The green species regularly contain bchl *c* or *d* (lh), a_p (rc) and the carotenoids chlorobactene and hydroxychlorobactene: the brown-colored species possess bchl *e* (lh), a_p (rc) and the carotenoids isorenieratene and β-isorenieratene.

Chlorobium

Rod- to ovoid-formed cells without gas vacuoles are typical for *C. limicola* (51.0 to 58.1) and its brown-colored counterpart *C. phaeobacteroides* (49.0 to 50.0). *C. limicola* is 0.7 to 1.1 × 0.9 to 1.5 μm in size and *C. phaeobacteroides* is 0.6 to 0.8 × 1.3 to 2.7 μm. The slightly curved to vibrio-shaped cells of the green-colored *C. vibrioforme* (52.0 to 57.1), *C. chlorovibrioides*,[28] and the brown-colored *C. phaeovibrioides* (52.0 to 53.0) are found in marine habitats and have a slight requirement for about 1% NaCl. They are 0.5 to 0.7 × 1.0 to 2.0 μm (*C. vibrioforme*) and 0.3 to 0.4 × 0.7 to 1.4 μm (*C. phaeovibrioides* and *C. chlorovibrioides*) in size. Green-colored species variants exist which are also able to use thiosulfate as the photosynthetic electron donor.

Pelodictyon

The species of this genus contain gas vacuoles. The green-colored *Pelodictyon clathratiforme* (48.5, at present not in pure culture) is rod-shaped, 0.7 to 1.2 μm wide and 1.5 to 2.5 μm long. Cells can undergo binary as well as ternary fission and thus form characteristic three-dimensional nets. Pfennig[29] reported the isolation of a brown-colored counterpart to this species; however, it was lost soon after isolation. The ovoid cells of the green *P. luteolum* (53.5 to 58.1) are 0.6 to 0.9 μm wide and 1.2 to 2.0 μm long. Cells of the respective brown species *P. phaeum* are 0.6 to 0.9 μm wide and 1 to 2 μm long.[30]

Prosthecochloris and Ancalochloris

Irregular spherical to star-like cells with prosthecae are characteristic of the genera *Ancalochloris,* which has gas vacuoles, and *Prosthecochloris,* which does not. *P. aestuarii* (50.0 to 56.1) is green colored, 0.5 to 0.7 μm wide, and 1.0 to 1.2 μm long. *P. phaeosteroidea*[31] is brown colored, 0.3 to 0.6 μm wide and 0.5 to 0.8 μm long. *A. perfilievii* (not in pure culture) is green colored and contains gas vacuoles.[11]

Chloroflexaceae

This family comprises those species of the Chlorobiales which have flexible filamentous cells and move by gliding.

Chloroflexus

In contrast to the restricted metabolism of the Chlorobiaceae, the metabolism of *Chloroflexus* is more versatile. Like Rhodospirillaceae, *Chloroflexus* grows primarily photoorganotrophically and can also adapt to chemoorganotrophic growth under aerobic dark conditions. Autotrophic growth with sulfide as electron donor and CO_2 as carbon source is also possible by some strains. The main oxidation product is elemental sulfur. *C. aurantiacus*[12] (53 to 55) is 0.6 to 0.7 μm wide. The length of the filaments varies from 30 to 300 μm. It is thermophilic with its optimum growth temperature from 52 to 60°C. The pH optimum is near 8. *C. auranticus* possesses bchl *a* and *c*, and the main carotenoids are β- and γ-carotene. Gorlenko[32] reported the isolation of some mesophilic strains of *Chloroflexus* which are quite similar to the thermophilic *C. aurantiacus.*

Chloronema

Among the filamentous green bacteria, species with gas vacuoles also exist.[13] *Chloronema giganteum* (at present not in pure culture) possesses bchl *d* and is phototrophic and facultatively aerobic like *Chloroflexus.*

Oscillochloris

Trichomes are formed which move by gliding. *Oscillochloris chrysea* (at present not in pure culture) has trichomes with 4.5- to 5.5-μm diameters. The trichome is septated. Most of the septa are incomplete. The *Chlorobium* vesicles are mainly located alongside the cell septa.[14]

REFERENCES

1. **Gibbons, N. E. and Murray, R. G. E.,** Proposals concerning the higher taxa of bacteria, *Int. J. Syst. Bacteriol.,* 28, 1, 1978.
2. **Lewin, R. A.,** Prochloron, type genus of the Prochlorophyta, *Phycologia,* 16, 217, 1977.
3. **Pfennig, N. and Trüper, H. G.,** Higher taxa of phototrophic bacteria, *Int. J. Syst. Bacteriol.,* 21, 17, 1971.
4. **Bavendamm, W.,** *Die farblosen und roten Schwefelbakterien des Süß- und Salzwassers,* Fischer Verlag, Jena, 1924.
5. **Copeland, H. F.,** *The Classification of Lower Organisms,* Pacific Books, Palo Alto, 1956.
6. **Trüper, H. G.,** Higher taxa of phototrophic bacteria: Chloroflexaceae fam. nov., a family for the gliding, filamentous, phototrophic "green" bacteria, *Int. J. Syst. Bacteriol.,* 26, 74, 1976.
7. International Commission of Systematic Bacteriology, Approved list of names of bacteria, *Int. J. Syst. Bacteriol.,* 30, 225, 1980.
8. **Pfennig, N. and Trüper, H. G.,** The phototrophic bacteria, in *Bergey's Manual of Determinative Bacteriology,* 8th ed., Buchanan, R. E. and Gibbons, N. E., Eds., Williams & Wilkins, Baltimore, 1974, 24.
9. **Gray, B. H., Fowler, C. F., Nugent, N. A., Rigopoulos, N., and Fuller, R. C.,** Reevaluation of *Chloropseudomonas ethylica* strain 2-K, *Int. J. Syst. Bacteriol.,* 23, 256, 1973.
10. **Biebl, H. and Pfennig, N.,** Growth yields of green sulfur bacteria in mixed cultures with sulfur and sulfate reducing bacteria, *Arch. Microbiol.,* 117, 9, 1978.
11. **Gorlenko, V. M. and Lebedeva, E. V.,** New green sulfur bacteria with apophyses, *Mikrobiologiya,* 40, 1035, 1971.
12. **Pierson, B. K. and Castenholz, R. W.,** A phototrophic gliding filamentous bacterium of hot springs, *Chloroflexus aurantiacus,* gen. and sp. nov., *Arch. Microbiol.,* 100, 5, 1974.
13. **Dubinina, G. A. and Gorlenko, V. M.,** New filamentous photosynthetic green bacteria with gas vacuoles, *Mikrobiologiya,* 44, 511, 1975.
14. **Gorlenko, V. M. and Pivovarova, T. A.,** On the belonging of blue-green alga *Oscillatoria coerulescens* Gicklhorn 1921, to a new genus of Chlorobacteria *Oscillochloris* nov. gen., *Izv. Akad. Nauk SSSR, Ser. Biol.,* 3, 396, 1977.
15. **Pfennig, N.,** *Rhodocyclus purpureus* gen. nov. and sp. nov., a ring-shaped, vitamin B_{12}-requiring member of the family Rhodospirillaceae, *Int. J. Syst. Bacteriol.,* 28, 283, 1978.
16. **Schmidt, K.,** Biosynthesis of carotenoids, in *The Photosynthetic Bacteria,* Clayton, R. K. and Sistrom, W. R., Eds., Plenum, New York, 1979, 729.
17. **Keppen, O. I. and Gorlenko, V. M.,** A new species of budding purple bacteria containing bacteriochlorophyll b, *Mikrobiologiya,* 44, 258, 1975.
18. **Hansen, T. A. and Veldkamp, H.,** *Rhodopseudomonas sulfidophila,* nov. spec., a new species of the purple nonsulfur bacteria, *Arch. Microbiol.,* 92, 45, 1973.
19. **Weaver, P. F., Wall, J. D., and Gest, H.,** Characterization of *Rhodopseudomonas capsulata, Arch. Microbiol.,* 105, 207, 1975.
20. **Wall, J. D., Weaver, P. F., and Gest, H.,** Gene transfer agents, bacteriophages, and bacteriocins of *Rhodopseudomonas capsulata, Arch. Microbiol.,* 105, 217, 1975.

21. **Pfennig, N.,** *Rhodopseudomonas globiformis,* sp. n., a new species of the Rhodospirillaceae, *Arch. Microbiol.,* 100, 197, 1974.
22. **Pfennig, N., Trüper, H. G.,** The Rhodospirillales, in *Handbook of Microbiology,* Vol. 1, 2nd ed., Laskin, A. I. and Lechevalier, H. A., Eds., CRC Press, Boca Raton, Fla., 1977, 119.
23. **Imhoff, J. F. and Trüper, H. G.,** *Ectothiorhodospira halochloris* sp. nov., a new extremely halophilic phototrophic bacterium containing bacteriochlorophyll b, *Arch. Microbiol.,* 114, 115, 1977.
24. **Kämpf, C.,** Untersuchungen über die Fähigkeit phototropher roter und grüner Schwefelbakterien zu chemotrophem Wachstum, Diploma Thesis, University of Göttingen, 1978.
25. **Kondratieva, E. N., Petushkova, Y. P., and Zhukov, V. G.,** Growth and oxidation of sulfur compounds by *Thiocapsa roseopersicina* in the darkness, *Mikrobiologiya,* 44, 389, 1975.
26. **Gorlenko, V. M.,** Oxidation of thiosulfate by *Amoebobacter roseus* in the darkness under microaerophilic conditions, *Mikrobiologiya,* 43, 729, 1974.
27. **Staehelin, L. A., Golecki, J. R., Fuller, R. C., and Drews, G.,** Visualization of the supramolecular architecture of Chlorosomes (Chlorobium type vesicles) in freeze-fractured cells of *Chloroflexus aurantiacus, Arch. Microbiol.,* 119, 269, 1978.
28. **Gorlenko, V. M., Chebotarev, E. N., and Kachalkin, V. I.,** Participation of microorganisms in sulfur turnover in Pomiaretzkoe Lake, *Mikrobiologiya,* 43, 908, 1974.
29. **Pfennig, N.,** Phototrophic green and purple bacteria: a comparative systematic survey, *Annu. Rev. Microbiol.,* 31, 275, 1977.
30. **Gorlenko, V. M.,** Phototrophic brown sulfur bacteria *Pelodictyon phaeum* nov. sp., *Mikrobiologiya,* 41, 370, 1972.
31. **Puchkova, N. N., Gorlenko, V. M.,** New brown colored chlorobacterium *Prosthecochloris phaeoasteroidea* nov. sp., *Mikrobiologiya,* 45, 655, 1976.
32. **Gorlenko, V. M.,** Characteristics of filamentous phototrophic bacteria from fresh-water lakes, *Mikrobiologiya,* 44, 756, 1975.

TAXONOMIC CLASSIFICATION OF ALGAE

Paul C. Silva

GENERAL CONSIDERATIONS

Taxonomy of algae has evolved, and will continue to evolve, through interaction of three processes: (1) progressive addition, modification, and abandonment of criteria suggested by information obtained by means of continually improving observational and analytical instruments and techniques; (2) discovery of organisms with combinations of characters that do not fit existing definitions of taxa; and (3) changing philosophical concepts. Taxonomy, like medicine, is an art based on scientific data. Its two main goals are to inventory (describe and name) the components of the biota and to arrange these components in a scheme that indicates interrelationships perceived to be natural. Since the nature of an organism is the sum of its structure, reproduction, functions, behavior, interaction with its environment and with sympatric organisms, genetic relationship with other living organisms, and phylogenetic relationship with extinct organisms, taxonomy is a synthesis of morphology, physiology, biochemistry, ecology, genetics, and paleontology. All information resulting from biological research is grist for the taxonomic mill. Realization of this fact alone should suffice to explain the bewildering fluidity of classification of a group of organisms as large and as morphologically and metabolically diverse as the algae, especially if it is appreciated that the large majority of algae are unicells with many diagnostic features that lay hidden until the advent of electron microscopy. Nonetheless, other factors will be discussed briefly to help explain why a nontaxonomist, when consulting a taxonomist, frequently receives equivocal and conflicting opinions.

Definition of Algae

Algae include all oxygen-evolving photosynthetic nonvascular organisms except bryophytes (liverworts and mosses). The latter are distinguished from algae by multicellular sex organs sheathed by sterile cells and by retention of the embryonic sporophyte within the female organ (archegonium). Many colorless organisms are also referable to the algae on the basis of their similarity to photosynthetic forms with respect to structure, life history, cell-wall composition, and storage products.

Four parts of the algal assemblage have equivocal boundaries. Cyanophyceae (blue-green algae) have been assigned to the bacteria on structural grounds, both groups lacking a membrane-bounded nucleus.[1] Despite this similarity and various biochemical similarities, the possession of chlorophyll *a* and the evolution of oxygen during photosynthesis support retention of the blue-greens among the algae. There are, however, colorless prokaryotic filamentous organisms (e.g., *Beggiatoa*) whose alignment with photosynthetic Cyanophyceae is equivocal. Several groups of algae present striking morphological similarities with corresponding groups of fungi. Rhodophyceae (red algae) are notably reminiscent of Ascomycetes, and the hypothesis that they are progenitors of higher fungi, proposed in 1874,[2] has received recent attention.[3,4] Classifications in which algae and fungi are intermixed, with the presence or absence of chlorophyll considered of secondary importance, are to be found in older literature.[2,5] Within the Chrysophyceae (golden algae), the spectrum of nutritional types, patterns of organelle ultrastructure, and the combination of amoeboid and flagellated stages lead to organisms that lie clearly within the province of the zoologist. Craspedomonads (choanoflagellates), for example, have been aligned with Chrysophyceae,[6-8] but recently obtained ultrastructural information supports their exclusion from the plant kingdom.

[9-11] The basic similarity among Chlorophyceae, Charophyceae, liverworts, mosses, ferns, and cycads led taxonomists as early as 1828 to place these organisms in a comprehensive group.[12] The scheme adopted by Chihara for this handbook extends the phylum (division) Chlorophyta to embrace all higher plants, including spermatophytes. A disadvantage of this treatment is that it sacrifices the expediency gained by demarcating higher plants from algae at the highest hierarchical level. Finally, there are assemblages of colorless flagellates (bodonids, hexamitids or distomatids, trichomonads, trypanosomatids) that have been held tenaciously by zoologists but nonetheless have been aligned with algae by various taxonomists since 1898.[13-15] Their taxonomic placement is immaterial for purposes of this handbook, however, since they are not photosynthetic.

According too much taxonomic importance to motility and not enough to photosynthesis, zoologists have traditionally considered phytoflagellates as protozoa.[16] Apportioning those major phyletic groups that include both monads and multicellular forms between two kingdoms, animal and plant, on the basis of somatic structure is so clearly artificial that numerous attempts, beginning at least as early as 1860, have been made to keep these groups together by recognizing a third kingdom, the protists. A classification that recognizes five kingdoms (prokaryotes, protists, fungi, animals, and plants) has been endorsed by Whittaker and Margulis,[17] who give an informative historical account of the rationale for this scheme. Cyanophyceae are placed in Monera (prokaryotes), Rhodophyceae and Phaeophyceae in Plantae, and the majority of algal groups in Protista. Separating Rhodophyceae and Phaeophyceae from other eukaryotic algae at such a high level is a questionable procedure, but the desirability of recognizing more than two kingdoms of organisms is unequivocal.

Diversity of Algae

In size and complexity algae range from single cells 1 to 2 μm in diameter to giant kelps often 30 m long. Unicells may be solitary or colonial, attached or free-living, with or without a protective cover, and motile (flagellated, amoeboid, gliding) or nonmotile. Colonies may be irregular or with a distinctive pattern, the latter type being flagellated or nonmotile. Multicellular algae may form packets, filaments, unistratose or multistratose laminae, or complex pseudoparenchymatous or parenchymatous thalli with organs resembling roots, stems, and leaves. Coenocytic algae range from microscopic spheres to thalli 10 m long with two levels of organization — gross morphological and anatomical. Life histories range from binary fission in some unicells to a sequence of four multicellular somatic phases in many red algae, involving highly specialized reproductive systems. Culture studies coupled with field observations have revealed that some algae are extremely plastic with regard to life history, being able to bypass one or another stage by recycling, apogamy, apospory, pedogenesis, parthenogenesis, and vegetative multiplication, and to produce special microscopic stages in response to seasonal, long-term, or sporadic changes in the environment.[18] Some species of benthic marine algae have different life histories in different parts of their geographic range,[19] the determining factors probably being temperature and light regimes.

Distinguishing between microalgae and macroalgae, while having procedural value, lacks a clear-cut morphological basis. In essence, microalgae are those forms (primarily unicells) that can be handled by microbiological methods. The goal is to obtain material of a particular genotype in sufficient quantity for experimentation. As culture techniques have improved markedly, so has the ability to obtain clonal material of structurally complex algae. No longer are culture collections restricted to unicellular or colonial forms: they now include a large number of red, brown, and green algae that by morphological standards would be considered macroscopic. In making comparative biochemical or physiological studies, the usual practice is to employ unicellular representatives of several major phyletic lines — for example, *Anacystis* (Cyanophyceae), *Porphyridium* (Rhodophyceae), *Monodus* (Xanthophyceae), *Ochromonas* (Chrysophyceae), *Hymenomonas* (Haptophyceae), *Hemiselmis* (Cryptophyceae), *Skeletonema* (Bacillariophyceae), *Gymnodinium* (Dinophyceae), *Chlorella* (Chlorophyceae), and *Euglena* (Euglenophyceae). The results of such studies tend to be kept separate from those obtained in experiments using macroalgae. Differences in methodology, however, should not be allowed to interfere with assimilation of information. Although structurally complex algae obviously engage in physiological activities unknown in microalgae (such as the transport of photosynthates from the blades to the holdfast of the giant kelp *Macrocystis*), basic processes should prove to be as homogeneous within a major phyletic line as is the biochemical and ultrastructural framework within which they operate.

As defined morphologically, macroalgae are to be found mostly among the benthic marine forms of Rhodophyceae, Phaeophyceae, and Chlorophyceae. In addition, all Charophyceae are macroscopic. In groups represented predominantly by unicellular or other microscopic forms, macroalgae include *Phaeosaccion* and *Hydrurus* (Chrysophyceae), *Tribonema* and *Vaucheria* (Xanthophyceae), exceptionally large filamentous blue-greens (e.g., *Lyngbya majuscula*), and some filamentous and foliose diatoms (e.g., *Navicula ulvacea*). Among Chlorophyceae, macroalgae are not restricted to the marine habitat, being represented in fresh water by such genera as *Cladophora, Draparnaldia, Hydrodictyon,* and *Stigeoclonium.*

It is impossible to state the number of biologically valid species of living algae, partly because the definition of a species varies from taxonomist to taxonomist and from group to group, and partly because of the paucity of monographic treatments. An estimate of 30,000 seems reasonable. Assuming that critical monographs (as distinguished from compilations) were available for all groups, the number of species recognized by the community of specialists would be subject to future change through the discovery of new organisms, continued discrimination within presently conceived species made possible by increased sophistication of instrumentation, a change of emphasis from individuals to populations facilitated by improved culture techniques, and changing philosophical concepts as one specialist supersedes another.

Recognition of Major Groups of Algae

References to algae as pond scum, plankton blooms, and beach wrack are to be found in the oldest extant writings. Linnaeus, whose works by international agreement constitute the starting point for the classification of all but a few kinds of organisms, treated "Algae" as a group of plants in which he placed liverworts, lichens, sponges, and some fungi in addition to algae as presently defined.[20] On the other hand, calcareous green and red algae and the only phytoflagellate known to him *(Volvox)* were aligned with worms and molluscs in the animal kingdom.[21] Among algae as presently defined, Linnaeus recognized 80 species in 10 genera representing about 47 modern genera in six major phyletic lines. In the last half of the 18th century, two important advances were made in algal taxonomy: the Linnaean definition of algae was purified (although calcareous algae were generally not included until 1819,[22] green flagellates until 1849,[23] and other phytoflagellates until 1890);[24] and hundreds of new species were described, creating a need for expansion and refinement of the Linnaean generic framework. During the first half of the 19th century, the inventory of algae greatly increased as voyages of exploration resulted in the discovery of diverse macroscopic forms in distant parts of the globe while intensified examination of water and soil revealed the wonders of the microscopic world. Accompanying the rapid increase in number of genera was an appreciation of the need to express intergeneric relationships (families) and interfamilial relationships (orders).

The concept of major phyletic lines (equivalent to modern classes and phyla or divisions) was the concern of only a few far-sighted workers and had its beginning in 1813 when Lamouroux[25] distinguished three primary groups of algae on the basis of color — reds, browns, and greens. Unknown to him, of course, were the biochemical and biophysical attributes of the pigments that are now deemed to be of fundamental phylogenetic significance. The morphological distinctness of diatoms, with the wall of each cell arranged like the two parts of a Petri dish, was recognized early in the 19th century and their golden brown color was attributed to a special pigment (diatomin) in 1849.[26] For many decades, however, desmids (unicellular nonmotile Chlorophyceae) were often aligned with diatoms because of their frequent possession of a median constriction that forms semicells superficially resembling the two halves of a naviculoid diatom as seen in valve view, despite the easily observable difference in pigmentation. Charophytes, with their unique vegetative structure and reproductive organs, were placed in a group coordinate with algae in 1824[27] after having been frequently misaligned with aquatic angiosperms. The scheme used herein accords Charophyceae a position between Chlorophyceae and bryophytes. Blue-green algae were dispersed among coccoid and filamentous members of other major groups until the first suprafamilial taxon was established for them in 1860.[28] At that time, their usual blue-green color was considered the primary character. Later, it was realized that the lack of a nucleus was of even greater importance.

Differences in flagellar structure deemed to be of primary phylogenetic significance were begun to be perceived during the last quarter of the 19th century, followed shortly by appreciation of the diagnostic importance of cell-wall structure and composition and of storage products. These new criteria, complementing those based on gross morphology and increasingly sophisticated pigment analysis, led to the separation of Xanthophyceae (heterokonts) from Chlorophyceae (isokonts and stephanokonts) during the period 1889 to 1899.[29-32] They also led to recognition of the fundamental distinctness of various groups of monads collectively termed Flagellata (dinoflagellates, chrysomonads, cryptomonads, chloromonads or raphidomonads, and euglenoids). Of the 20 classes of algae recognized herein, 11 had achieved general acceptance by 1927.[33] Of the remaining nine, charophytes were often retained in the Chlorophyceae and silicoflagellates in the Chrysophyceae. Prasinophytes were segregated from Chlorophyceae during the period 1941 to 1962[34-37] while haptophytes were segregated from Chrysophyceae in 1961-62,[37,38] both separations based on cytological and ultrastructural characters. Ebriids, formerly placed with or near silicoflagellates, were realigned adjacent to dinoflagellates while ellobiopsids and syndiniids were added as groups coordinate with dinoflagellates and ebriids during the period 1968 to 1976.[39-41] Eustigmatophytes were segregated from Xanthophyceae on cytological, ultrastructural, and biochemical criteria in 1970-71.[42,43] Finally, Prochlorophyceae was recognized in 1976-77[44,45] as a major group of prokaryotic photosynthetic organisms, differing from Cyanophyceae in their possession of typical chlorophycean pigments.

Level of Recognition of Major Phyletic Lines Within the Taxonomic Hierarchy

Despite the usual contention that scientific names are essential for accurate communication, there are many instances in which common names evoke clearer images. For example, diatoms are diatoms, but confusion reigns, even among taxonomists, as a result of this group of algae having scores of different scientific names, depending partly on the hierarchical level at which the group is recognized. The first scientific description of a species of diatoms *(Vibrio paxillifer)* was published in 1786,[46] followed by the first description of a genus of diatoms *(Bacillaria)* in 1791.[47] All diatoms known at the time were placed in a single family (Diatomeae) in 1822,[48] an order (Diatomeae) in 1824,[49] a class (Diatomophyceae) in 1864,[50] and a phylum (Diatomea) in 1866,[51] not to mention taxa at intermediate levels. We are thus burdened with a host of names (based on the stems Bacillari- and Diatom-), all referring to the same group of organisms. During the past few decades, there has been a tendency toward hierarchical inflation, with the unrealistic zoological tradition of treating each major group of phytoflagellates as an order of Protozoa[16] giving way to dubiously practical schemes that recognize numerous kingdoms of algae.[52] A workable compromise is to recognize only classes for most purposes, allowing individual taxonomists to juggle superclasses, subdivisions, divisions, superdivisions, subphyla, phyla, superphyla, subkingdoms, kingdoms, and superkingdoms as they deem appropriate. It should be noted, however, that in the phylogenetic scheme prepared by Chihara for this handbook, the major groups are shown at the level of phylum. A concordance of classes and phyla is given in Table 1. It should also be noted that the International Code of Botanical Nomenclature[53] sanctions the use of division but not phylum. Although most present-day taxonomists consider the two terms equivalent, there is precedent, especially in elaborate Germanic classifications, for treating phylum (Stamm) as a category above division (Abteilung).[54,55]

The vacuum created by pushing classes upward in the hierarchy has been filled by elevating orders to classes and families to orders. Thus, in recent schemes, 10 of the 20 classes recognized herein are treated as divisions or phyla, each comprising two to several classes.[56-75] The pendulum may well swing the other direction in the future.

Expression of Degree of Relationship among Major Groups of Algae

Corollary to the discrimination of groups of organisms is the expression of relationships among these groups. Degree of similarity (presumed phylogenetic affinity) among coordinate groups may be indicated in a scheme, whether in two or three dimensions, solely by position. There is a current trend, however, to express the varying degrees of similarity by bracketing two or more taxa in a taxon at the next highest rank. If this rank is preoccupied, a category of intermediate rank must be introduced. The upper part of the hierarchy can thus become excessively crowded. For example, if raphidomonads are treated as a class most closely related to xanthophytes and the two classes are thus bracketed in the phylum Xanthophyta, a superphylum must be used to indicate the phylogenetic distance between Xanthophyta and a cluster of groups characterized by the pigment complement chlorophyll *a*, chlorophyll *c*, and fucoxanthin that have similarly been elevated to the rank of phylum (Chrysophyta, Haptophyta, Bacillariophyta, Phaeophyta). Results of numerical analysis could be interpreted as suggesting that every linkage in a dendrogram represents a formal taxon, but the consequent classification would be so complex as to prove counterproductive. An opposite viewpoint was taken recently with regard to Phaeophyceae, the dendrogram being interpreted as supporting the demotion of certain taxa in order to indicate phylogenetic distances more accurately.[76] Manipulation of taxa upward and downward in particular groups, entailing a variable use of intermediate categories, is to be distinguished from general hierarchical inflation discussed in the preceding section.

Degree of Elaboration of Classification within Major Groups of Algae

The degree to which the taxonomic infrastructure of each major group of algae has been elaborated depends on the amount of diversity that has been perceived, the amount of hierarchical inflation, and tradition. Recently, diversity in ultrastructural details of organelles and cell processes has been perceived to such an extent by specialists as to evoke schemes of classification that cut across lines drawn by the application of usual morphological and biochemical criteria.[77] It must be kept in mind that tax-

Table 1
SYNOPSIS OF PHYLA (DIVISIONS), SUBPHYLA (SUBDIVISIONS), AND CLASSES OF ALGAE

Phylum Prochlorophyta
1. Class Prochlorophyceae
Phylum Cyanophyta
2. Class Cyanophyceae (Myxophyceae)
Phylum Rhodophyta
3. Class Rhodophyceae
Phylum Cryptophyta
4. Class Cryptophyceae
Phylum Chlorophyta
Subphylum Prasinophytina
5. Class Prasinophyceae
Subphylum Chlorophytina
6. Class Chlorophyceae
Subphylum Charophytina
7. Class Charophyceae
Subphylum Bryophytina }
Subphylum Pteridophytina } not algae
Subphylum Spermatophytina }
Phylum Euglenophyta
8. Class Euglenophyceae
Phylum Chromophyta
9. Class Chrysophyceae
10. Class Dictyochophyceae (Silicoflagellatophyceae)
11. Class Bacillariophyceae
12. Class Phaeophyceae
13. Class Xanthophyceae
14. Class Raphidophyceae (Chloromonadophyceae)
Phylum Eustigmatophyta
15. Class Eustigmatophyceae
Phylum Haptophyta
16. Class Haptophyceae
Phylum Dinophyta
17. Class Dinophyceae
18. Class Ebriophyceae
19. Class Ellobiophyceae
20. Class Syndiniophyceae

onomy is based on degree of similarity and dissimilarity and that relationships should be proposed and tested in accordance with the principle of correlation of multiple characters. Schemes based primarily on a single character or on very few characters, while useful in pointing out the need to reassess existing classifications, probably should be considered analytical tools rather than definitive statements. Diversity is taxonomically significant only against a background of homogeneity. Because the existence of homogeneity is often merely assumed or extrapolated, diversity within a particular character may prove to be the rule rather than the exception when comprehensive comparative studies have been made.

Tradition is an influential factor in taxonomy, varying from one group to another. Diatomists, for example, are conservative. As recently as 1924 all diatoms were placed in one family,[78] and some present-day workers recognize but one order to accommodate at least 15,000 species.[79] By contrast, dinoflagellate specialists, admittedly facing greater diversity, have recognized several segregate classes, at least 15 orders, and about 75 families for a much smaller number of species. Tradition is especially strong in deciding what constitutes a species and whether infraspecific variation is merely described or given formal taxonomic recognition. Among desmids and diatoms, recognition of infraspecific taxa (subspecies, varieties, subvarieties, forms, and subforms), many of which were described from individuals rather than populations, is excessive and almost totally devoid of demonstrated biological meaning.

Names of Taxa of Algae at Different Ranks

The formation and application of scientific names of algae is governed by the International Code of Botanical Nomenclature.[53] All names are treated as Latin regardless of their derivation. The name of a species is a binary combination of a generic name and a specific epithet. Additional epithets are used to indicate infraspecific taxa. For purposes of precision, the name of the author of the name of a species (or infraspecific taxon) is often appended. An epithet may be adjectival, in which case it agrees grammatically with the generic name, or it may be a substantive and thus remain undeclined. A generic name is a substantive in the singular number, or a word treated as such. The name of a family is a plural adjective used as a substantive; it is formed by adding -aceae to the stem of a legitimate name of an included genus. Names of subfamilies and tribes are similarly formed, with the endings -oideae and -inae, respectively. Names of taxa above the rank of family are of two kinds, descriptive and self-typified. A self-typified name is one that is based on the stem of the name of an included genus (the type genus). Self-typified names of orders end in -ales, suborders in -ineae. Descriptive names of orders and suborders have no prescribed form. Names of classes of algae, whether descriptive or self-typified, end in -phyceae, subclasses in -phycidae. Names of divisions, whether descriptive or self-typified, end in -phyta, subdivisions in -phytina. As noted previously, the category phylum, while not sanctioned by the botanical code, is often used instead of division, with the same ending. There are no rules governing names of such higher taxa as superclasses, superdivisions, subkingdoms, kingdoms, and superkingdoms.

CLASSES OF ALGAE

Primary diagnostic characters of the 17 classes of algae that include photosynthetic organisms are given in Table 2. General remarks concerning structural variation, distribution, reproduction, and taxonomy are provided for each class in the following synopsis. Within each basic classification, genera representative of the lowest included taxa (usually orders) are cited.

1. Class Prochlorophyceae

At the present time, only one species is known — *Prochloron didemni,* associated with didemnid ascidians in warm seas.[44,45,80] Establishing a new division (Prochlorophyta) rather than altering the circumscription of Cyanophyta to accommodate *Prochloron* has been criticized.[81]

2. Class Cyanophyceae (Myxophyceae, Schizophyceae)

Blue-green algae may be unicellular, colonial, filamentous, or pseudoparenchymatous. They range in size from spherical unicells 1 to 2 μm in diameter to filaments 10 cm long. Flagella are never produced. Specialized cells (heterocysts) in filamentous species have been shown to be the site of nitrogen fixation, but that process also occurs in some nonheterocystous forms. Gas vacuoles are common in planktonic species. Many filamentous forms are capable of a gliding movement. Almost all blue-greens are photosynthetic. Reproduction is by binary fission, various kinds of spores, and

Table 2
PRIMARY DIAGNOSTIC CHARACTERS OF ALGAL CLASSES

Class	Pigments	Plastid structure	Cell wall	Storage products	Flagella	Special cytological features
1. Prochlorophyceae (prochlorophytes)	chlorophyll *a* and *b*; β-carotene, xanthophylls unknown	no chloroplast, pigments in thylakoids lying free in cytoplasm	present, but details unknown; mucilaginous sheath absent	highly branched and unbranched α-1,4 glucans	none	no membrane-bounded organelles
2. Cyanophyceae (blue-green algae)	chlorophyll *a*; C-phycocyanin, C-phycoerythrin, allophycocyanin; β-carotene, xanthophylls (myxoxanthin, myxoxanthophyll, oscillaxanthin)	no chloroplast, thylakoids lying free in cytoplasm; phycobiliproteins in phycobilisomes on outer surface of thylakoids	usually 4-layered with mucopeptide as major component, hydrolyzing to muramic acid, glucosamine, alanine, glutamic and α,ε-diaminopimelic acids; surrounded by sheath composed of pectic acids and mucopolysaccharides	branched α-1,4 glucan; cyanophycin (protein containing only 2 amino acids, arginine and aspartic acid)	none	no membrane-bounded organelles
3. Rhodophyceae (red algae)	chlorophyll *a* (*d* in some Florideophycidae); R- and C-phycocyanin, R- and B-phycoerythrin, allophycocyanin; α- and β-carotene, xanthophylls (lutein, violaxanthin, zeaxanthin, antheraxanthin, neoxanthin)	stellate, discoid, or band-shaped, with 2 membranes; thylakoids single, not stacked, sometimes encircled by peripheral thylakoids; with or without pyrenoid	inner fibrillar layer (celluose in Florideophycidae, xylan in Bangiophycidae), outer nonfibrillar layer of mannan in Bangiophycidae; mucilaginous amorphous matrix of sulfated galactans; calcification in some genera	Floridean starch (α-1,4 glucan) in granules outside chloroplast; sugars, glycosides, polyols	none	pit-connections between cells in all Florideophycidae and some Bangiophycidae

4. Cryptophyceae (cryptomonads)	chlorophyll *a* and *c*; α-carotene dominant over β-carotene; acetylene-derived xanthophylls (alloxanthin, crocoxanthin, monadoxanthin)	usually 1 or 2 per cell, parietal, with 4 membranes; thylakoids loosely paired; phycobiliproteins in intrathylakoidal spaces, not in phycobilisomes; pyrenoid usually present	cell wall absent; periplast granular or fibrillar to the outside, layered to the inside	starch granules in perichloroplastic matrix near pyrenoid	2, equal or subequal, pleuronematic, arising from a ventral depression with a subapical opening	eyespot, when present, within chloroplast; ejectosomes common
5. Prasinophyceae (prasinophytes)	chlorophyll *a* and *b*; α- and β-carotene, xanthophylls (including at times siphonein and siphonaxanthin)	usually single, lobed; also numerous, discoid; with 2 membranes; 2 to 4 thylakoids per lamella; usually with pyrenoid	noncellulosic when present; cell body usually covered with organic scales or enclosed by organic theca	starch granules within chloroplast, sometimes sheathing pyrenoid	1, 2, or more (usually 4), equal, often arising from apical depression but also apical or subapical, usually covered with very small organic scales, sometimes with caducous hairs	trichocysts in some species
6. Chlorophyceae (green algae)	chlorophyll *a* and *b*; α-, β-, and γ-carotene, xanthophylls (lutein, violaxanthin, zeaxanthin, antheraxanthin, neoxanthin; siphonaxanthin and siphonein in Caulerpales)	highly variable in number and form, with 2 membranes; 2 to 5 thylakoids per lamella; usually with pyrenoid(s), although not always conspicuous	cellulose common; also hydroxyproline glycosides, xylans and mannans; calcified in some genera; lacking in some genera	starch granules within chloroplast, often sheathing pyrenoid	1 to 8, equal, apically inserted, acronematic; subapical crown of numerous flagella in some genera	eyespot, when present, within chloroplast; multinucleate cells and coenocytes common

Table 2 (continued)
PRIMARY DIAGNOSTIC CHARACTERS OF ALGAL CLASSES

Class	Pigments	Plastid structure	Cell wall	Storage products	Flagella	Special cytological features
7. Charophyceae (stoneworts)	chlorophyll *a* and *b;* β-carotene, xanthophylls as in Chlorophyceae	numerous, ellipsoid, in longitudinal rows surrounding central vacuole; varying number of thylakoids per lamella; pyrenoids lacking	cellulose; calcified in some species	starch granules within chloroplast	2, equal or subequal, subapically inserted, lacking hairs but covered with very small scales	internodal cells huge, multinucleate
8. Euglenophyceae (euglenoids)	chlorophyll *a* and *b;* β-carotene (sometimes also γ), xanthophylls (astaxanthin, antheraxanthin, neoxanthin, diadinoxanthin, diatoxanthin)	highly variable in number and form, with 3 membranes; usually 3 thylakoids per lamella; with or without pyrenoids	cell wall absent; pellicle composed of interlocking proteinaceous strips; pectic lorica in some genera	paramylon (β-1,3 glucan) granules in cytoplasm	usually 2, equal or unequal, stichonematic, thick (with paraflagellar rods), arising from subapical invagination, one often nonemergent; 3 or more in some parasitic forms	eyespot separate from chloroplast
9. Chrysophyceae (golden algae)	chlorophyll *a* (sometimes also *c*); β-carotene, xanthophylls (mostly fucoxanthin but also diadinoxanthin, diatoxanthin, echinenone)	usually 1 or 2, with 4 membranes; 3 thylakoids per lamella; with or without pyrenoids	cellulosic-pectic, sometimes silicified; wall often lacking; siliceous scales common; lorica common, cellulosic-pectic, sometimes silicified or calcified; statospores with 2 unequal silicified or calcified pieces	chrysolaminaran (β-1,3 glucan), usually in large posterior vesicle; lipids	usually 2, equal or unequal, apical, one (the longer) pantonematic, directed forward, the other acronematic, directed backward or laterally, often reduced or absent; flagellar scales in some genera	eyespot, when present, within chloroplast

10.	Dictyochophyceae (silicoflagellates)	chlorophyll *a* and *c* (presumably); carotenoids not determined	numerous, discoid; fine structure not determined	absent	unknown	1, arising from dense central region of protoplast, smooth (internal structure unknown)	internal skeleton of siliceous tubes
11.	Bacillariophyceae (diatoms)	chlorophyll *a* and *c;* β-carotene, xanthophylls (fucoxanthin, diatoxanthin, diadinoxanthin, neoxanthin)	variable in number and form, usually 1 or 2 laminate or numerous discoid, with 4 membranes; 3 thylakoids per lamella; often with simple internal pyrenoid	siliceous frustule of complex structure	chrysolaminaran (β-1, 3 glucan); oil droplets	only in male gametes of centric diatoms; 1, anterior, pleuronematic, lacking central pair of microtubules	
12.	Phaeophyceae (brown algae)	chlorophyll *a* and *c;* β-carotene, xanthophylls (mostly fucoxanthin but also zeaxanthin, antheraxanthin, violaxanthin)	variable in number and form, with 4 membranes; 2 to 6 (often 3) thylakoids per lamella; pyrenoid, when present, attached to chloroplast by short stalk	inner cellulose layer, outer mucilaginous layer; alginic acid often abundant; sulfated mucopolysaccharide (fucoidan)	laminaran (β-1, 3 glucan); polyols (mannitol); cyclitols (laminitol)	2, unequal, laterally inserted; one (usually the longer) directed forward, pleuronematic, the other directed backward, acronematic, vestigial in a few genera	eyespot (in male gametes) a reduced chloroplast located near insertion of flagella; physodes (vesicles lacking a membrane and containing polyphenols) common
13.	Xanthophyceae (yellow-green algae)	chlorophyll *a* and *c;* β-carotene, xanthophylls (diadinoxanthin, diatoxanthin, heteroxanthin, neoxanthin)	1 to many, laminate or discoid, with 4 membranes; 3 thylakoids per lamella; some with multilamellar internal pyrenoid	cellulose; some with 2 overlapping pieces (H-shaped in filaments); statospores with 2 equal or subequal pieces	lipids; possibly polysaccharides	2, unequal, apically or laterally inserted; one (usually the longer) directed forward, pleuronematic, the other directed backward, acronematic; compound zoospore of *Vaucheria* with numerous pairs of unequal smooth flagella	eyespot within chloroplast

Table 2 (continued)
PRIMARY DIAGNOSTIC CHARACTERS OF ALGAL CLASSES

Class	Pigments	Plastid structure	Cell wall	Storage products	Flagella	Special cytological features
14. Raphidophyceae (raphidomonads or chloromonads)	chlorophyll *a* and *c*; β-carotene; xanthophylls (fucoxanthin in marine forms)	numerous, discoid, with 4 membranes; 3 thylakoids per lamella; pyrenoids absent	absent	unknown	2, unequal, apically inserted; one directed forward, pleuronematic, the other trailing, acronematic	trichocysts common; eyespots lacking
15. Eustigmatophyceae (eustigmatophytes)	chlorophyll *a*; β-carotene, xanthophylls (violaxanthin, vaucheriaxanthin ester, diatoxanthin, heteroxanthin, neoxanthin)	usually 1, parietal, but also several, with 4 membranes; 3 thylakoids per lamella; stalked polyhedral or spherical pyrenoid on inner face of chloroplast (lacking in zoospores)	fibrillar-mucilaginous, composition not determined	lamellate vesicles surrounding pyrenoid or in cytoplasm, of unknown composition	usually 1, inserted subapically, directed forward, pantonematic; second flagellum sometimes emergent, directed backward or laterally, acronematic, but usually reduced to basal body	eyespot in zoospore, independent of chloroplast, composed of group of droplets without membrane
16. Haptophyceae (haptophytes)	chlorophyll *a* and *c*; α- and β-carotene, xanthophylls (fucoxanthin, diadinoxanthin, diatoxanthin)	usually 2, with 4 membranes; 3 thylakoids per lamella; 1 pyrenoid per chloroplast, often compound	absent; cell body covered with organic or calcified scales	paramylon-like substance in granules; water-soluble β-1,3 glucans	2, inserted apically or ventrally, equal or subequal, acronematic; in a few genera, flagella unequal, with fine hairs and/or small dense bodies	haptonema (filiform organelle) often present, arising between flagella, sometimes vestigial

17. Dinophyceae (dinoflagellates)	chlorophyll *a* and *c*; β-carotene, xanthophylls (peridinin, neoperidinin, dinoxanthin, neodinoxanthin, diadinoxanthin, diatoxanthin)	usually numerous, variable in form, with 3 membranes; 3 thylakoids per lamella; pyrenoids common, of variable form	cell covering (amphiesma) a series of concentric membranes, usually including a layer of thecal vesicles; thecal vesicles with or without cellulose thecal plates of variable thickness	starch granules in cytoplasm; sometimes oil	Desmophycidae: 2, inserted apically or subapically, heterodynamic, fine structure unknown; Dinophycidae: 2, arising from ventral depressions, one acronematic, directed backward, the other ribbonlike, beating in transverse groove	nucleus large, chromosomes condensed at all stages, centromeres and spindle absent; eyespots common, of varying form, some complex (ocelli); pusules; trichocysts; nematocysts (cnidocysts)

multicellular segments (hormogonia). Genetic recombination occurs by parasexual processes similar to those of bacteria.[82,83]

Blue-green algae are ubiquitous with respect to geography and habitat. They are especially abundant in the plankton of neutral or alkaline eutrophic fresh waters and tropical seas, often forming blooms. They also attach themselves to various substrates. Habitats include hot springs, snow and ice, soil, rocks, tree trunks, and buildings. They live symbiotically with a large variety of animals and plants, including sponges, diatoms, liverworts, cycads, the aquatic fern *Azolla,* and the angiosperm *Gunnera.* They constitute the phycobiont of many lichens.

The taxonomy of blue-green algae is in an especially unsteady state, being polarized by two diametrically opposed viewpoints. At one extreme is the Elenkin-Frémy-Geitler system, in which nearly every morphological variant is accorded taxonomic recognition. At the other extreme is the Drouet system, in which great morphological plasticity is attributed to each species. Geitler[84] recognized 4 orders, 22 families, 140 genera, and more than 1200 species. In four monographs covering all blue-greens except those with true branching (Stigonematales), Drouet[85-88] recognized no orders, 5 families, 20 genera, and 56 species. While Drouet is undoubtedly correct in contending that the number of names far exceeds the number of biologically valid taxa, his reductions have been too drastic, resulting in circumscriptions that are too broad to be meaningful. Culture studies must be combined with field observations to determine which of the innumerable distinctions perceived by previous taxonomists are biologically significant. Most specialists of Cyanophyceae steer a middle course.

Class Cyanophyceae

Subclass Coccogonophycidae: cells solitary, colonial, or in filaments without hormogonia or heterocysts; reproducing solely by spores (nannocytes, endospores, exospores)

- Order Chroococcales: cells solitary or in packets, pseudofilaments, or tubular colonies; reproducing by binary fission and nannocytes
  - Cells solitary or in packets: *Aphanothece* (incl. *Coccochloris*), *Chroococcus, Coelosphaerium, Eucapsis, Geitleribactron, Gloeocapsa, Gloeothece, Gomphosphaeria, Merismopedia* (incl. *Agmenellum*), *Microcystis* (incl. *Anacystis, Aphanocapsa*), *Synechococcus, Synechocystis*
  - Cells forming pseudofilaments: *Chlorogloea, Entophysalis*
  - Cells in gelatinous tubular colonies: *Johannesbaptistia, Tubiella, Wolskyella*
- Order Chamaesiphonales: cells solitary, in packets, or forming unbranched filaments; cells and filaments both with distinct morphological polarity; reproducing by exospores and endospores
  - Exospores: *Chamaesiphon, Cyanophanon, Siphononema*
  - Endospores: *Chroococcidiopsis, Clastidium, Cyanocystis* (incl. *Dermocarpella*), *Pascherinema (= Endonema)*
- Order Pleurocapsales: cells undergoing vegetative cell division (desmoschisis), in nonfilamentous colonies or forming prostrate and erect filaments, often pseudoparenchymatous; reproducing by endospores: *Dermocarpa* (incl. *Xenococcus*), *Hydrococcus* (incl. *Oncobyrsa*), *Hyella, Pleurocapsa* (incl. *Radaisia*)

Subclass Hormogonophycidae: filamentous, reproducing by hormogonia, with or without heterocysts

- Order Nostocales: filaments unbranched or falsely branched
  - Unbranched, trichome not attenuated to a hair: *Anabaena, Anabaenopsis, Aphanizomenon, Cylindrospermum, Gomontiella, Lyngbya* (incl. *Phormidium*), *Microchaete* (incl. *Fortiea*), *Microcoleus, Nodularia, Nostoc, Oscillatoria* (incl. *Arthrospira, Spirulina*), *Porphyrosiphon, Richelia, Schizothrix, Trichodesmium*

Unbranched or falsely branched, trichome attenuated to a hair: *Calothrix, Gloeotrichia, Homoeothrix, Rivularia*

Falsely branched: *Hydrocoryne, Plectonema, Scytonema, Tolypothrix*

Order Stigonematales: filaments with true branching, forming prostrate and upright systems: *Borzinema, Brachytrichia, Capsosira, Fischerella, Hapalosiphon, Loefgrenia, Loriella, Mastigocladopsis, Mastigocladus, Mastigocoleus, Nostochopsis, Pulvinularia, Stigonema, Westiella*

General and systematic treatments: References 60-62, 82,84-94

3. Class Rhodophyceae

Most red algae have multicellular thalli of microscopic or macroscopic size, including filaments (uniseriate or pluriseriate, unbranched or branched), laminae (monostromatic or distromatic), and complex plants of distinctive form produced by intergrowth of filamentous systems. In one subclass (Bangiophycidae) there are unicells (nonflagellate, but often capable of gliding or amoeboid movement) and mucilaginous colonies. Varying ratios of chlorophylls, carotenoids, and phycobilins provide a spectrum of color from blue-green through a live-green and reddish brown to bright pink. Rhodophyceae are chiefly marine, growing on various substrates along the shore and to depths limited by the availability of light. Of the few nonmarine forms, most are restricted to streams that are rapidly flowing, well aerated, and cold. Some, however, grow in quiet warm water while a few are subaerial. Calcification of cell walls occurs in various genera and is characteristic of the coralline algae (family Corallinaceae). Several genera produce polysaccharide hydrocolloids (e.g., carrageenan, agar) that are used extensively in industry and in microbiological laboratories. Most red algae are photosynthetic, but some are parasites or hemiparasites on other red algae. Asexual reproduction is by binary fission in unicellular forms and by fragmentation, special vegetative propagules (rare), and nonflagellate (but often somewhat amoeboid) spores in multicellular forms. Sexual reproduction is oogamous, involving colorless nonflagellate (but sometimes slightly amoeboid) male gametes (spermatia) and a specialized cell (carpogonium) whose contents (the egg) are fertilized *in situ*. The zygote produces diploid spores (carpospores) directly (in Bangiophycidae) or indirectly (in most orders of Florideophycidae). Carpospores, after being discharged, germinate to produce diploid asexual plants (sporophytes) which in turn produce meiospores, usually in tetrads (tetraspores). Meiospores germinate to produce male and female plants and thus complete an elaborate life history. The life cycle of most Florideophycidae entails a fourth somatic phase, the carposporophyte (gonimoblast), which is produced directly or indirectly by the zygote and remains an integral part of the female plant. Classification of the Bangiophycidae is based largely on vegetative and asexual reproductive features, while that of the Florideophycidae is based primarily on sexual reproductive features, secondarily on vegetative characters.

Class Rhodophyceae

Subclass Bangiophycidae: cells always uninucleate, chloroplast usually single, stellate, and axile; cell divisions usually intercalary; unicellular or multicellular; pit connections usually lacking; sexual reproduction rare

Order Porphyridiales: cells solitary or in irregular mucilaginous colonies: *Chroothece, Petrovanella, Porphyridium, Rhodella, Rhodosorus, Rhodospora, Vanhoeffenia*

Order Goniotrichales: thalli filamentous, branched or unbranched, usually epiphytic or epizoic; asexual reproduction by monospores, not involving special cell division: *Bangiopsis, Chroodactylon* (incl. *Asterocytis*), *Cyanoderma, Goniotrichopsis, Goniotrichum, Kneuckeria, Kyliniella, Neevea, Phragmonema*

Order Bangiales: thalli filamentous or parenchymatous (disc, blade, or sac); asexual reproduction by monospores, sometimes involving special cell division; sexual reproduction, where known, not involving a carposporophyte: *Bangia, Boldia, Erythrocladia, Erythrotrichia, Erythropeltis, Porphyra* (incl. *Porphyrella*), *Porphyropsis, Smithora*

Order Compsopogonales: polysiphonous filaments, with central cells surrounded by one or more layers of small cortical cells: *Compsopogon, Compsopogonopsis*

Order Rhodochaetales: uniseriate filaments with apical growth: *Rhodochaete*

Subclass Florideophycidae: cells often multinucleate, chloroplasts usually several to many per cell, peripheral; cell division commonly restricted to apical cells; multicellular; prominent pit connections present; sexual reproduction common, usually involving a carposporophyte

Order Acrochaetiales: uniseriate branched filaments, lacking cortication, usually forming prostrate and upright systems: *Acrochaetium, Audouinella, Kylinia, Rhodochorton, Rhodothamniella*

Order Nemaliales: thalli uniaxial or multiaxial; carposporophyte developing from zygote; meiosporophyte markedly different from sexual plants, often microscopic; meiospores of some species germinating *in situ: Atractophora, Batrachospermum, Cumagloia, Dermonema, Helminthocladia, Helminthora, Lemanea, Liagora, Naccaria, Nemalion, Nothocladus, Sirodotia, Thorea, Trichogloea, Tuomeya, Yamadaella*

Order Chaetangiales: thalli multiaxial; carposporophyte developing from zygote or cell subtending zygote, surrounded by ostiolate pericarp; tetrasporophyte microscopic or, if macroscopic, often differing in form from sexual plants: *Actinotrichia, Chaetangium, Galaxaura, Gloiophloea, Pseudogloiophloea, Pseudoscinaia, Scinaia*

Order Gelidiales: thalli uniaxial; carpogonium sessile; carposporophyte developing from multinucleate cell derived from zygote, gonimoblast filaments growing among nutritive filaments; adjoining carposporophytes often forming a compound cystocarp; cystocarp without pericarp; sexual plants and tetrasporophyte isomorphic: *Acanthopeltis, Beckerella, Gelidiella, Gelidium, Porphyroglossum, Pterocladia, Ptilophora, Suhria, Yatabella*

Order Bonnemaisoniales: thalli uniaxial; carpogonial branch three-celled; carposporophyte developing from zygote or cell subtending zygote, surrounded by pericarp; life history variable, with tetrasporophyte lacking, markedly different from sexual plants, or (yet to be confirmed) similar to sexual plants; gland cells usually present: *Asparagopsis, Bonnemaisonia, Delisea* (incl. *Ptilonia*)

Order Cryptonemiales: carposporophyte developing from special cell (generative auxiliary cell) of a special filament

Cell walls regularly calcified; thalli crustose or erect and articulated, calcified internodes alternating with noncalcified nodes (family Corallinaceae): *Alatocladia, Amphiroa, Arthrocardia, Bossiella, Calliarthron, Cheilosporum, Chiharaea, Clathromorphum, Corallina, Dermatolithon, Duthiophycus, Ezo, Fosliella, Goniolithon, Haliptylon, Hydrolithon, Jania, Litholepis, Lithophyllum, Lithoporella, Lithothamnium, Lithothrix, Marginisporum, Mastophora, Melobesia, Mesophyllum, Metagoniolithon, Metamasto-*

phora, Neogoniolithon, Pachyarthron, Phymatolithon, Porolithon, Serraticardia, Sporolithon, Tenarea, Yamadaea

Cell walls rarely calcified: *Acrosymphyton, Aeodes, Callophyllis* (incl. *Euthora*), *Carpopeltis, Chondrococcus, Choreocolax, Constantinea, Corynomorpha, Cryptonemia, Cryptosiphonia, Dermocorynus, Dilsea, Dudresnaya, Dumontia, Endocladia, Erythrophyllum, Farlowia, Glaphyrymenia, Gloiosiphonia, Grateloupia, Halymenia, Harveyella, Hildenbrandia, Hyalosiphonia, Kallymenia, Leptocladia, Ochtodes, Pachymenia, Peyssonnelia, Phyllymenia, Pikea, Polyides, Polyopes, Prionitis, Pugetia, Rhizophyllis, Rhodopeltis, Schimmelmannia, Thuretella, Tichocarpus, Weeksia*

Order Gigartinales: carposporophyte developing from special cell (generative auxiliary cell) of an ordinary vegetative filament: *Acrotylus, Agardhiella, Ahnfeltia, Areschougia, Calliblepharis, Callophycus, Calosiphonia, Catenella, Caulacanthus, Chondriella, Chondrus, Cruoria, Curdiea, Cystoclonium, Dicranema, Dicurella, Erythroclonium, Eucheuma, Furcellaria, Gelidiopsis, Gigartina, Gracilaria, Gymnogongrus, Halarachnion, Heringia, Hypnea, Iridaea, Melanthalia, Meristotheca, Mychodea, Mychodeophyllum, Nemastoma, Neurocaulon, Nizymenia, Opuntiella, Phacelocarpus, Phyllophora, Platoma, Plocamium, Predaea, Rhabdonia, Rhodoglossum, Rhodophyllis, Rissoella, Sarcodia, Sarcodiotheca, Sarconema, Schizymenia, Schmitzia, Sebdenia, Solieria, Sphaerococcus, Stenocladia, Stenogramme, Stictosporum, Titanophora, Trematocarpus, Turnerella*

Order Rhodymeniales: carposporophyte developing from generative auxiliary cell terminating two-celled filament produced by supporting cell of carpogonial branch before fertilization: *Binghamia, Botryocladia* (incl. *Gloiosaccion, Myrioglossa*), *Champia, Chrysymenia, Chylocladia, Coelarthrum, Coelothrix, Cryptarachne, Erythrocolon, Erythrymenia, Fauchea, Fryeella, Gastroclonium, Gloiocladia, Gloioderma, Halichrysis, Hymenocladia, Leptofauchaea, Lomentaria, Maripelta, Minium, Rhodymenia, Sciadophycus, Webervanbossea*

Order Palmariales:[95,183] carposporophyte lacking; carpogonia produced on minute female thalli, fertilized by spermatia from full-sized male thalli, zygote developing directly into tetrasporophyte; tetrasporangia cruciately divided, large, stalked: *Coriophyllum, Halosaccion, Leptosarca, Palmaria, Rhodophysema*

Order Ceramiales: carposporophyte developing from generative auxiliary cell produced by supporting cell of carpogonial branch after fertilization

Family Ceramiaceae: carposporophyte usually naked or with loose investment of involucral filaments; axial cells naked or corticated, pericentral cells not forming polysiphonous structure: *Anotrichium, Antarcticothamnion, Antithamnion, Antithamnionella, Ballia, Bornetia, Callithamnion, Carpoblepharis, Centroceras, Ceramium, Compsothamnion, Crouania, Dasyphila, Dasyptilon, Dohrniella, Falklandiella, Griffithsia, Gymnothamnion, Haloplegma, Halurus, Microcladia, Neoptilota, Platythamnion, Pleonosporium, Plumaria, Pterothamnion, Ptilota, Seirospora, Spermothamnion, Sphondylothamnion, Spyridia, Warrenia, Wrangelia*

Family Dasyaceae: carposporophyte enclosed by ostiolate pericarp; thallus with polysiphonous structure, axes sympodially developed; tetrasporangia borne in stichidia: *Dasya, Dictyurus, Heterosiphonia, Pogonophora, Rhodoptilum, Thuretia*

Family Delesseriaceae: thallus usually foliose, the blade developing from lateral pericentral cells: *Acrosorium, Botryoglossum, Caloglossa, Claudea, Cryptopleura, Delesseria, Grinnellia, Hemineura, Hymenena, Hypoglossum, Martensia, Membranoptera, Myriogramme, Nienburgia, Nitophyllum, Pantoneura, Phycodrys, Platysiphonia, Polyneura, Sarcomenia, Vanvoorstia*

Family Rhodomelaceae: thallus with polysiphonous structure, cylindrical or dorsiventrally flattened, axes monopodially developed; *Acanthophora, Amansia, Benzaitenia, Bostrychia, Brongniartella, Bryocladia, Bryothamnion, Chondria, Digenea, Euzonia, Halopitys, Herposiphonia, Janczewskia, Laurencia, Lenormandia, Leveillea, Levringiella, Lophosiphonia, Odonthalia, Picconiella, Placophora, Polysiphonia, Polyzonia, Pterosiphonia, Rhodomela, Rytiphlaea, Streblocladia, Vidalia*

General and systematic treatments: References 94, 96-98.

4. Class Cryptophyceae

This class comprises a relatively small group of biflagellate unicellular organisms (cryptomonads). The cells are asymmetric, flattened dorsiventrally, and bounded by a moderately flexible periplast. Varying ratios among chlorophylls, carotenoids, and phycobilins account for a spectrum of coloration, including green, olive-yellow, brown, red, and blue. Ejectosomes, analogous to the trichocysts of dinoflagellates, are usually present. Reproduction is by longitudinal binary fission. Biflagellate zoospores are formed in palmelloid genera. Cryptomonads occur in both fresh-water and marine habitats. Some are symbiotic with ciliates and invertebrates. A few osmotrophs are known.

Class Cryptophyceae

Order Cryptomonadales: flagellated unicells: *Chroomonas, Cryptaulax, Cryptomonas, Cyanomonas, Cyathomonas, Hemiselmis, Hillea, Isoselmis, Kathablepharis, Plagioselmis, Planonephros, Protochrysis* (photosynthetic); *Chilomonas, Leucocryptos* (colorless)

Order Tetragonidiales: cells in palmelloid or coccoid colonies, at times retaining or reforming flagella: *Bjornbergiella, Tetragonidium*

General and systematic treatments: References 75, 90, 91, 94, 99-103.

5. Class Prasinophyceae

As information from ultrastructural and biochemical studies accumulates, this class of scale-bearing phytoflagellates is achieving better definition. Its members were previously distributed among Chlorophyceae, Cryptophyceae, Chrysophyceae, and Xanthophyceae. Most prasinophytes are marine, constituting a component of the plankton whose significance has been largely unappreciated. Some are predominantly motile, but may form cysts, especially in temporary bodies of fresh water. Sessile coccoid forms that reproduce by zoospores are also known. In one genus (traditionally called *Prasinocladus*), nonmotile cells occupy the apices of a dendroid structure formed from old cell thecae, reproducing by zoospores which may become encysted. Some marine prasinophytes are best known from their spherical nonmotile stage, whose contents eventually cleave into numerous zoospores. Symbionts of a dinoflagellate *(Noctiluca)*, a radiolarian *(Thalassolampe)*, and a turbellarian worm *(Convoluta roscoffensis)* are known. Sexual reproduction has not been observed.

Class Prasinophyceae

Order Pedinomonadales: monads with single flagellum not arising from apical depression, without scales but with other prasinophycean ultrastructural characters: *Pedinomonas*

Order Nephroselmidales: uniflagellate or biflagellate monads, flagella not arising from apical depression; cell body and flagella with one layer of scales: *Dolichomastix, Mantoniella, Monomastix, Nephroselmis*

Order Pterospermatales: transitory or persistent nonmotile spherical stage with non-cellulosic wall, ornamented or with equatorial wings; zoospores with four flagella inserted eccentrically on flat or concave surface, directed backward when swimming; body scales three-layered, flagellar scales two-layered: *Pachysphaera, Pterosperma* (incl. *Coccopterum, Pterocystis, Pterosphaera*)

Order Pyramimonadales: motile cells with four flagella arising from apical depression, with body and flagellar scales of variable structure and arrangement; monads, some with a spherical nonmotile free-living stage, or sessile coccoid forms, the nonmotile stages forming *Pyramidomonas*-like zoospores: *Asteromonas, Halosphaera, Prasinochloris, Pyramimonas, Stephanoptera*

Order Prasinocladales: motile cells with noncellulosic theca and four flagella arising from apical depression; free-living monads or nonmotile cells in dendroid colony: *Prasinocladus, Tetraselmis* (incl. *Platymonas*)

General and systematic treatments: References 65, 66, 104-106, 184

6. Class Chlorophyceae

Almost all somatic types known for algae occur among green algae, the exceptions being rhizopodial unicells and complex multicellular thalli differentiated into macroscopic organs. Hundreds of genera and about 160 families have been proposed to accommodate this diversity. Green algae bridge the gap between monads and higher plants, and in the classification used in this handbook they are placed with charophytes, bryophytes, ferns, and seed plants in the phylum Chlorophyta (see preceding section, Definition of Algae). Five major lines of evolution are recognized: (1) volvocine, in which monads form motile colonies of definite structure (coenobia) with morphological and functional differentiation among the component cells; (2) tetrasporine, in which monads lose their flagella phylogenetically, forming sessile palmelloid colonies; (3) chlorococcine, in which nonmotile cells, phylogenetically derived from monads, occur singly or in colonies of definite or indefinite structure, with cell division of the type in which the walls of the progeny are entirely new and free from the parent wall (eleutheroschisis); (4) ulotrichine, with cell division of the type in which the parent wall forms part of the wall of the progeny (desmoschisis), thus leading to multicellular thalli; and (5) siphonous, in which macroscopic thalli are formed by the development of septate or nonseptate coenocytes.

Most Chlorophyceae occupy fresh-water, terrestrial, or subaerial habitats, but certain orders are predominantly marine. The majority of nonmarine forms are planktonic unicells or colonies, while the majority of marine forms are macrophytes growing on various substrates along the shore and to depths limited by the availability of light. Photosynthesis is the usual mode of nutrition, but facultative and obligate osmotrophs are common. A few parasites are known. Certain genera are phycobionts of lichens, while some species are symbionts of various animals. Calcification is common in marine siphonous forms, and the order Dasycladales has an excellent fossil record extending back as far as the Ordovician. Asexual reproduction is by binary fission of unicells, vegetative fragmentation of colonies or multicellular forms, and various kinds of spores. Sexual reproduction is common and ranges from isogamy through various degrees of anisogamy to oogamy. A particular type of sexual reproduction may be characteristic of a major subdivision (e.g. an order), but all types may be found within a single genus *(Chlamydomonas)*. The subclass Zygnematophycidae is characterized by nonflagellate amoeboid gametes.

Class Chlorophyceae

Subclass Chlorophycidae: morphologically variable, but not siphonous; motile cells usually with two to eight flagella, apically inserted

Order Volvocales: motile unicells or coenobia, some with a temporary sessile palmelloid stage: *Astrephomene, Brachiomonas, Carteria, Chlamydomonas, Chlorogonium, Chloromonas, Dunaliella, Dysmorphococcus, Eudorina, Gonium, Haematococcus, Pandorina, Phacotus, Platydorina, Pleodorina, Polytoma, Pteromonas, Pyrobotrys, Stephanosphaera, Volvox, Volvulina*

Order Tetrasporales: cells *Chlamydomonas*-like but lacking flagella in vegetative phase, solitary or colonial, with or without gelatinous matrix; sessile, neustonic, or planktonic; pseudoflagella present in some genera: *Apiocystis, Asterococcus, Chaetochloris, Chaetopeltis, Characiochloris, Chlorangiella, Chlorangiopsis, Gloeococcus, Gloeocystis, Hormotila, Hormotilopsis, Hypnomonas, Malleochloris, Nautococcus, Palmellopsis, Paulschulzia, Schizochlamys, Tetraspora*

Order Chlorococcales: nonmotile unicells or colonies of indefinite or definite structure, reproducing asexually by autospores, autocolonies, or zoospores, sexually by biflagellate gametes, rarely oogamous; products of cell division lacking walls or with walls completely free from parent wall

Suborder Chlorococcineae: cells uninucleate or multinucleate, cell division giving rise to zoospores and/or gametes: *Bracteacoccus, Characiosiphon, Characium, Chlorococcum, Hydrodictyon, Neochloris, Neospongiococcum, Pediastrum, Protosiphon, Sorastrum, Trebouxia*

Suborder Chlorellineae: cells uninucleate, not producing zoospores: *Ankistrodesmus, Chlorella, Coelastrum, Dictyosphaerium, Eremosphaera, Kirchneriella, Monoraphidium, Oocystis, Prototheca, Scenedesmus, Selenastrum*

Order Chlorosarcinales: cells uninucleate, undergoing desmoschisis to form diads, tetrads, or packets of similar cells that remain coherent for variable periods; reproduction by zoospores and flagellate gametes: *Axilosphaera, Borodinella, Borodinellopsis, Chlorosarcina, Chlorosarcinopsis, Planophila, Pseudotetracystis, Tetracystis*

Order Ulotrichales: unbranched uniseriate filaments with uninucleate cells; chloroplast variable, but often a parietal plate or cylinder: *Cylindrocapsa, Geminella, Koliella, Microspora, Radiofilum, Stichococcus, Ulothrix*

Order Ulvales: biseriate or pluriseriate filaments, monostromatic or distromatic blades, hollow tubes, or solid cylinders; cellular organization as in Ulotrichales: *Blidingia, Enteromorpha, Monostroma, Percursaria, Schizomeris, Ulva, Ulvaria*

Order Chaetophorales: branched filaments with uninucleate cells; thallus often differentiated into prostrate and upright systems; cellular organization as in Ulotrichales; unicellular colorless hairs, multicellular hairlike branches, or sheathed cytoplasmic processes often present: *Aphanochaete, Bolbocoleon, Chaetophora, Coleochaete, Draparnaldia, Draparnaldiopsis, Endophyton, Entocladia, Fritschiella, Gongrosira, Leptosira, Microthamnion, Pseudulvella, Stigeoclonium, Ulvella*

Order Trentepohliales: branched uniseriate filaments, erect or prostrate, with distinct sporocysts and gametocysts; subaerial; starch absent, hematochrome abundant; plasmodesmata present: *Cephaleuros, Phycopeltis, Stomatochroon, Trentepohlia*

Order Sphaeropleales: unbranched uniseriate filaments of very long multinucleate cells with cytoplasmic septa; chloroplast in each cell unit a parietal plate, often reticulate, with numerous pyrenoids; oogamous or anisogamous: *Sphaeroplea*

Order Prasiolales: packets, pluriseriate filaments, blades, or cylinders; chloroplast stellate, axile, with pyrenoid: *Prasiococcus, Prasiola, Prasiolopsis, Rosenvingiella, Schizogonium*

Order Cladophorales: unbranched or branched uniseriate filaments, cells usually multinucleate; chloroplast parietal, reticulate, one per cell or fragmented, with numerous pyrenoids: *Acrosiphonia, Anadyomene, Basicladia, Chaetomorpha, Cladophora, Microdictyon, Pithophora, Rhizoclonium, Spongomorpha, Urospora, Willeella*

Subclass Oedogoniophycidae: unbranched or branched uniseriate filaments of very long uninucleate cells; chloroplast a parietal reticulate cylinder with numerous pyrenoids; cell division highly specialized, resulting in an apical cap at the distal end of the parent cell; sexual reproduction oogamous, highly specialized, often involving dwarf male filaments derived from androspores; zoospores, androspores, and male gametes with apical crown of flagella (to 120); fresh-water.

Order Oedogoniales: *Bulbochaete, Oedocladium, Oedogonium*

Subclass Zygnematophycidae: nonmotile unicells or unbranched uniseriate filaments; cells uninucleate; flagellate reproductive cells absent; sexual reproduction by conjugation of amoeboid gametes; fresh-water or subaerial

Order Zygnematales

Suborder Zygnematineae: cell wall lacking pores or ornamentation

Family Zygnemataceae: filaments of cells not divided into semicells: *Debarya, Mougeotia, Mougeotiopsis, Pleurodiscus, Sirocladium, Sirogonium, Spirogyra, Temnogametum, Zygnema, Zygnemopsis, Zygogonium*

Family Mesotaeniaceae (saccoderm desmids): unicells resembling cells of Zygnemataceae, rarely forming filaments: *Ancylonema, Cylindrocystis, Mesotaenium, Netrium, Roya, Spirotaenia*

Suborder Desmidiineae (placoderm desmids): cell wall with pores and often ornamented; unicells composed of mirror-image semicells often demarcated by an equatorial constriction or incision; nucleus lying between semicells; sometimes colonial or filamentous

Family Desmidiaceae: semicells demarcated by an equatorial constriction; cells without zone of elongation: *Arthrodesmus, Bambusina, Cosmarium, Cosmocladium, Desmidium, Docidium, Euastrum, Hyalotheca, Micrasterias, Oocardium, Pleurotaenium, Sphaerozosma, Staurastrum, Staurodesmus, Tetmemorus, Xanthidium*

Family Peniaceae: semicells not demarcated by an equatorial constriction; cells with zone of elongation: *Closterium, Genicularia, Gonatozygon, Penium*

Subclass Siphonocladophycidae: coenocytes, septate to varying degrees; gametes biflagellate, zoospores with four apical flagella or subapical crown of numerous flagella; calcification common (especially in Caulerpales and Dasycladales)

Order Siphonocladales: septa formed by segregative division: *Boodlea, Chamaedoris, Cladophoropsis, Dictyosphaeria, Siphonocladus, Struvea, Valonia*

Order Caulerpales: simple or complex system of branched tubes generally lacking septa except for delimiting reproductive structures; septa formed by centripetal deposition of material: *Avrainvillea, Bryopsis, Caulerpa, Chlorodesmis, Codium, Derbesia, Dichotomosiphon, Halimeda, Penicillus, Trichosolen* (incl. *Pseudobryopsis*), *Tydemania, Udotea*

Order Dasycladales: nonseptate primary axis with whorls of laterals (radially symmetrical); single massive nucleus in rhizoid, fragmenting into numerous secondary nuclei just prior to reproduction; gametes usually released from operculate cysts formed in specialized gametangia: *Acetabularia, Batophora, Bornetella, Cymopolia, Dasycladus, Halicoryne, Neomeris*

General and systematic treatments: References 64-66, 91, 94, 107-114.

7. Class Charophyceae

Members of this class have macroscopic thalli to 30 cm (or more) high, with a regular succession of nodes and internodes. Each node bears a whorl of laterals of limited growth ("leaves"), while branches capable of unlimited growth may arise in the axils of the laterals. The plants are attached to the substrate by rhizoids. Sexual reproduction is oogamous and involves highly specialized structures. The oogonium is sheathed by sterile cells. Antheridia are borne in uniseriate filaments, several of which are surrounded by a common spherical envelope composed of eight cells. The egg is fertilized *in situ* and the resulting zygote secretes a thick wall. Fossil zygotes (gyrogonites) or their casts provide an elaborate phylogenetic record of charophytes extending back to the Devonian.[115]

Living charophytes are restricted to fresh and brackish water. Most grow in ponds and lakes with sandy or muddy bottoms, often forming extensive meadows. They thrive best in clear hard waters. The thalli of many species become encrusted with calcium carbonate, and their remains contribute to calcareous deposits on the bottom of the pond or lake.

Each internode is a single cell (with or without cortication), whose enormous size has made charophytes a favorite subject for biophysical research, especially on protoplasmic streaming and membrane physiology.

Class Charophyceae

Order Charales

Family Characeae: *Chara, Lamprothamnium, Lychnothamnus, Nitella, Nitellopsis, Tolypella*

General and systematic treatments: 66, 91, 94, 116-118.

8. Class Euglenophyceae

Most euglenoid algae are free-swimming unicells with two flagella (one of which may be nonemergent) arising from an anterior invagination. In two genera, the cells have three or more flagella. The cell body is bound by a proteinaceous pellicle. An external envelope (lorica) is present in some genera. One species of *Euglena* exists permanently in a palmelloid stage. *Colacium* forms a colony attached to various freshwater zooplankters, with each cell capable of metamorphosing into a zoospore. *Ascoglena* has flagellate cells affixed to a substrate. Sexual reproduction is unknown. Asexual reproduction is by binary fission, in some instances involving a temporary palmelloid stage. Thick-walled resting stages (cysts) are common in many genera.

Many euglenoids are colorless, with osmotrophic and/or phagotrophic nutrition. Photosynthetic forms may be facultatively osmotrophic. Euglenoids are most often found in bodies of fresh water rich in organic matter, frequently forming conspicuous blooms. They also occur on mud, in ice and snow, as endoparasites (of fresh-water flatworms, copepods, annelids, and tadpoles), and in various brackish and marine habitats.

Class Euglenophyceae

Order Eutreptiales: two flagella of same or different length, one directed forward and the other directed laterally or backward during swimming, highly mobile; cell nonrigid, but never phagotrophic; *Eutreptia, Eutreptiella* (green); *Distigma, Distigmopsis* (colorless)

Order Euglenales: two flagella, one emergent, the other nonemergent; phototrophic, osmotrophic, or phagotrophic but without special ingestion apparatus; *Ascoglena, Colacium, Euglena, Klebsiella, Lepocinclis, Phacus, Strombomonas, Trachelomonas* (green); *Astasia, Cyclidiopsis, Euglenopsis, Hyalophacus, Khawkinea* (colorless)

Order Rhabdomonadales: two flagella, one emergent and highly mobile throughout its length during swimming but usually held straight when cell is stationary; colorless and osmotrophic, never phagotrophic; cell rigid: *Gyropaigne, Menoidium, Parmidium, Rhabdomonas, Rhabdospira*

Order Sphenomonadales: two flagella, one or both emergent, one always directed forward, straight; colorless, osmotrophic and/or phagotrophic but without special ingestion apparatus; cell rigid or almost so, usually with keels or grooves: *Anisonema, Atraktomonas, Calycimonas, Notosolenus, Petalomonas, Sphenomonas, Tropidoscyphus*

Order Heteronematales: two flagella, one or both emergent, one directed forward during swimming, usually with coiling or flickering of just the tip; colorless, phagotrophic with special ingestion apparatus: *Dinematomonas (= Dinema), Entosiphon, Heteronema, Peranema, Peranemopsis, Urceolus*

Order Euglenamorphales: three or more flagella, all emergent and of same length, thickness, and activity; parasitic in digestive tract of tadpoles: *Euglenamorpha, Hegneria*

General and systematic treatments: References 90, 91, 94, 119-127

9. Class Chrysophyceae

Chrysophytes constitute a relatively large and diverse group of organisms. Freshwater forms outnumber marine forms and have a general preference for unpolluted cool or cold water. Most chrysophytes are photosynthetic, but osmotrophic and phagotrophic forms are known. Their somatic spectrum includes monads, free-living or attached, solitary or colonial; amoeboid or plasmodial forms; solitary or colonial coccoid cells; filaments and blades. A few genera can be considered macroalgae. The statospore, an internally formed resting stage or cyst, is a distinctive feature. The statospore is siliceous and comprises two pieces, a bottle-shaped portion and a plug. Chrysophycean cysts are abundant in the fossil record as far back as the Cretaceous. Although many chrysophytes have naked protoplasts, others have various cell coverings, including walls, scales, and loricas. Loricas may be siliceous, cellulosic-pectic, or (rarely) calcareous. Asexual reproduction is by binary fission, autospores, or zoospores in unicellular forms, and by fragmentation and zoospores in multicellular forms. Sexual reproduction, known for relatively few genera, is isogamous.

Classification of Chrysophyceae is especially fluid for several reasons. Pascher,[55,128-131] using as analogies the somatic spectra exhibited by Chlorophyceae and Xanthophyceae, accorded primary diagnostic value to gross morphology. As ultrastructural information has accumulated, the validity of this scheme has been questioned. Various chrysophytes have been realigned on ultrastructural features, but no satisfactory replacement for Pascher's scheme can be produced until much more information becomes available. Meanwhile, anomalous and equivocal taxonomic placements contribute to confusion as well as to clarification. The presence or absence, number, and structural details of flagella are now accorded major diagnostic value, but most genera have not been studied in culture and many life histories remain to be elucidated. Moreover, greatly reduced emergent and nonemergent flagella have been observed in certain forms. On ultrastructural evidence, many flagellate genera previously placed in this class (order Isochrysidales) have been realigned with Haptophyceae, while a few species have been transferred to Prasinophyceae. Certain nonmotile forms have been linked to life histories of motile haptophytes. Choanoflagellates (craspedomonads), formerly aligned with chrysophytes, are now considered unrelated and have been relinquished to zoologists.[9-11]

Class Chrysophyceae

Subclass Chrysophycidae: crown of tentacles absent from monads, flagellum not acting as contractile stalk

Order Chromulinales: cell wall lacking, but lorica sometimes present; monads or with palmelloid or pseudofilamentous organization; motile cells apparently uniflagellate: *Amphichrysis, Arthrogloea, Celloniella, Chromulina, Chrysapsis, Chrysocapsa, Chrysochaete, Chrysococcus, Chrysonebula, Chrysotilos, Heterochromulina, Hydrurus, Kephyrion, Lepochromulina, Monochrysis, Pascherella, Phaeaster, Sphaleromantis*

Order Chrysosphaerales: cell wall present; coccoid cells, solitary or colonial, free-living or attached, reproducing by autospores and apparently uniflagellate zoospores: *Aurosphaera, Chrysosphaera, Meringosphaera*

Order Rhizochrysidales: cell wall lacking, but lorica often present; rhizopodial forms, solitary or colonial, free-living or attached; zoospores uncommon, apparently uniflagellate, but at least at times with a vestigial second flagellum: *Bitrichia, Chrysamoeba* (incl. *Rhizochrysis*), *Chrysastridium (= Chrysidiastrum), Chrysothecopsis (= Stephanoporos), Heliapsis, Lagynion, Rhizaster, Stylococcus*

Order Chrysosaccales: cell wall lacking; gelatinous palmelloid colonies, flagellate cells unknown: *Chrysosaccus, Phaeosphaera*

Order Phaeoplacales: cell wall present; thallus filamentous, pseudoparenchymatous, or parenchymatous; flagellate cells unknown: *Chrysonema, Phaeoplaca, Sphaeridiothrix, Stichochrysis*

Order Ochromonadales: cells naked or with scales or lorica; monads, solitary or colonial, free-living or attached, or with palmelloid or rhizopodial organization; motile cells with two anteriorly inserted emergent flagella: *Anthophysa, Boekelovia, Chrysobotriella, Chrysoikos, Chrysolykos, Chrysosphaerella, Cyclonexis, Dendromonas, Dinobryon, Entodesmis, Epipyxis, Erkenia, Mallomonas, Microglena, Naegeliella, Ochromonas, Paraphysomonas, Poterioochromonas, Pseudokephyrion, Spumella, Syncrypta, Synura, Uroglena*

Order Chrysapiales: cell wall present; coccoid cells, solitary or colonial, reproducing by autospores or ochromonadoid zoospores: *Chrysapion, Stichogloea*

Order Phaeothamniales: cell wall present; uniseriate filaments, unbranched or branched, reproducing by ochromonadoid zoospores: *Apistonema, Nematochrysis, Phaeothamnion, Sphaeridothrix*

Order Sarcinochrysidales: monads, filaments, or a monostromatic sac; motile cells with two laterally inserted flagella: *Chrysomeris, Chrysowaernella, Giraudyopsis, Nematochrysopsis, Phaeosaccion, Sarcinochrysis*

Subclass Bicosoecophycidae: colorless monads, solitary or colonial, free-living or attached, with lorica formed of spiral or annular elements and one of the two flagella acting as a contractile stalk

Order Bicosoecales: *Bicosoeca* (incl. *Poteriodendron, Stephanocodon*), *Hydraeophysa*

Subclass Pedinellophycidae: radially symmetrical monads with up to six regularly arranged chloroplasts and a single apical flagellum, surrounded in most species by a ring of tentacles

Order Pedinellales: *Apedinella, Cyrtophora, Palatinella, Pedinella, Pseudopedinella*

General and systematic treatments: References 6-8, 67, 91, 94, 132-135.

10. Class Dictyochophyceae (Silicoflagellatophyceae)

Silicoflagellates constitute a relatively small group of marine planktonic monads. They are yellowish or greenish brown and photosynthetic. Each cell bears a single flagellum. The frothy cytoplasm flows around a radially symmetrical skeleton composed of siliceous tubules.[136] The dense central region contains the nucleus and perinuclear dictyosomes. Radiating from the central region are (1) globular or irregular processes connected by slender threads, and (2) fine pseudopodia. The peripheral cytoplasm contains numerous chloroplasts, mitochondria, and other organelles. The protoplast is covered by a delicate wall.[137] Reproduction is by binary fission, the skeleton of the daughter cell being a mirror image of that of its parent.

Although relatively common in the fossil record as far back as the Lower Createaceous, silicoflagellates seemingly are represented in modern seas by only a few taxa.[39] Generic distinctions are based on the structure of the skeleton. In clonal cultures, the skeleton has been found to be so variable as to cast doubt on the validity of these distinctions.[138]

Class Dictyochophyceae

Order Dictyochales

Family Dictyochaceae: *Dictyocha, Distephanus, Mesocena, Octactis*

General and systematic treatments: References 39, 139-141.

11. Class Bacillariophyceae

Diatoms are nonflagellated unicells encased in a silicified wall (frustule) composed of two large pieces (valves), one fitting over the other like the two parts of a Petri dish. Various small pieces (intercalary bands) lie in the region between the valves (the girdle). The valves are usually chambered and ornamented, with the various features as observed with light microscopy receiving such designations as puncta, striae, areolae, and costae. Electron microscopy (especially scanning E.M.) has revealed a complexity of frustule structure far greater than that known from light microscopy. Frustules fall into two major groups: those with radiately arranged structural features (centric diatoms) and those with bilaterally arranged structural features (pennate diatoms). Those pennate diatoms with a cleft (raphe) in at least one valve, through which mucilage and other substances can be exuded, are capable of a jerky or smooth gliding movement. A longitudinal clear space in a median or eccentric position on a valve is called a pseudoraphe.

Reproduction is usually asexual, by binary fission, with the smaller parent valve serving as the larger valve for one of the daughter protoplasts. Successive divisions thus progressively reduce the average size of the individuals of a population. Sexual reproduction in pennate diatoms is by amoeboid gametes, while in centric diatoms it is by oogamy, with the male gamete uniflagellate. The zygote rapidly increases in size and secretes a smooth or sculptured wall, at which stage it is called an auxospore. The auxospore then divides, producing typical diatoms of the maximum size for that species and thus counteracting the progressive diminution that results from successive binary fission. In some species, auxospore formation may be induced asexually. Diatoms produce various kinds of resting spores.

Most diatoms are free-floating, often dominating the plankton of marine, brackish, and fresh waters. They also occur in a variety of terrestrial and subaerial habitats, in sand and mud, and on other aquatic organisms. When growing on a substrate, they are attached by a gelatinous film, matrix, or stalk. They may be united in colonies of various forms, including macroscopic chains, branched pseudofilaments, and blades. Almost all diatoms are photosynthetic, although many marine littoral species require an external source of certain vitamins. A few osmotrophic free-living and attached species are known. The naked cells of two species of *Licmophora* are symbiotic in the tissue of a turbellarian worm *(Convoluta convoluta)*. Most genera of centric diatoms are strictly marine. Diatom frustules, well-known in the form of diatomaceous earth, are abundantly represented in the fossil record as far back as the Jurassic.

Diatoms are largely classified on structural features of the frustule, although other characters (such as chloroplast features) are of supplemental value.

Class Bacillariophyceae

Subclass Centrobacillariophycidae: structural features of valve arranged with reference to a central pole (centric valve) or to two or more poles (gonioid valve); raphe absent; chloroplasts commonly numerous; oogamous

Order Eupodiscales: cells discoid or short-cylindrical; valves radially symmetrical, usually circular but also semicircular, elliptical, or triangular, usually lacking special processes although margin may have spines or bristles: *Actinocyclus, Actinoptychus, Arachnoidiscus, Asterolampra, Asteromphalus, Aulacodiscus, Auliscus, Bacterosira, Charcotia, Coenobiodiscus, Corethron, Coscinodiscus, Cyclotella, Cylindropyxis, Cymatodiscus, Cymatotheca, Dactyliosolen, Detonula, Endictya, Ethmodiscus, Eupodiscus, Gossleriella, Groentvedia, Hemidiscus, Lauderia, Leptocylindrus, Melosira, Planktoniella, Porosira, Schimperiella, Schroederella, Skeletonema, Stephanodiscus, Stictodiscus, Thalassiosira* (incl. *Coscinosira*)

Order Rhizosoleniales: cells elongate, cylindrical or subcylindrical, circular or broadly elliptical in cross-section; girdle complex, with numerous intercalary bands; valve structure arranged with reference to eccentric pole: *Ditylum, Guinardia, Rhizosolenia*

Order Biddulphiales: cells box-shaped, pervalvar axis generally shorter, but sometimes slightly longer, than valvar axis; valves elliptical, semicircular, circular, or polygonal; valve structure arranged with reference to two or more poles, each pole represented by an angle, horn, or spine or by both angles and horns: *Anaulus, Bacteriastrum, Bellerochea, Biddulphia, Cerataulina, Cerataulus, Chaetoceros, Eucampia, Eunotogramma, Hemiaulus, Isthmia, Lithodesmium, Terpsinoe, Triceratium*

Order Rutilariales: valves boat-shaped (naviculoid), ornamentation radial or irregular: *Rutilaria*

Subclass Pennatibacillariophycidae: valves isobilateral, medianly zygomorphic, or less frequently dorsiventral, never centric; valve view mostly boat-shaped (naviculoid) or needle-shaped, with markings arranged pinnately with reference to a raphe or pseudoraphe; chloroplasts commonly few and large; isogamous amoeboid gametes

Order Fragilariales: cells rod-shaped to tabular, in valve view usually more or less linear, seldom clavate, in girdle view linear to rectangular; intercalary bands and septa frequent; chloroplasts usually small and numerous; valves without raphe, but usually each with a pseudoraphe: *Amphicampa, Asterionella, Auriculopsis, Campylosira, Catenula, Centronella, Climacosphenia, Cymatosira, Diatoma, Dimeregramma, Entopyla, Fragilaria, Fragilariopsis, Grammatophora, Hannaea, Licmophora, Meridion, Omphalopsis, Opephora, Plagiogramma, Podocystis, Protoraphis, Pseudohimantidium, Rhabdonema, Rhaphoneis, Sceptroneis, Semiorbis, Striatella, Subsilicea, Synedra, Tabellaria, Tetracyclus, Thalassionema, Thalassiothrix, Trachysphenia*

Order Eunotiales: valves arcuate, sublunate, hemispherical, or clavate; rudimentary raphe on one valve or both valves: *Actinella, Desmogonium, Eunotia, Peronia*

Order Achnanthales: valves elliptical or lanceolate, dissimilar; one valve with well-developed raphe, the other with pseudoraphe or at times with polar rudiments of a raphe: *Achnanthes, Campyloneis, Cocconeis, Peroniopsis, Pseudoperonia, Rhoicosphenia*

Order Naviculales: each valve with a well-developed raphe; raphe usually not enclosed in a keel or canal: *Amphipleura, Amphiprora, Amphora, Anomoeoneis, Auricula, Berkeleya, Brebissonia, Caloneis, Capartogramma, Cymbella, Diatomella, Diploneis, Entomoneis, Frustulia, Gomphocymbella, Gomphoneis, Gomphonema, Gyrosigma, Mastogloia, Nanoneis, Navicula, Neidium, Oestrupia, Okedenia, Phaeodactylum, Pinnularia, Plagiotropis, Pleurosigma, Scoliopleura, Scoliotropis, Scoresbya, Stauroneis, Stenoneis, Toxonidia, Trachyneis, Tropidoneis, Vanheurckia*

Order Epithemiales: each valve with a fully developed raphe; raphe enclosed in canal; costae present, between which are two rows of striae; cells solitary, free or epiphytic, valves arcuate or lunate, symmetrical on transapical axis, asymmetrical on apical axis: *Denticula, Epithemia, Rhopalodia*

Order Bacillariales (Nitzschiales): each valve with a canal-raphe enclosed in a keel: *Allonitzschia, Bacillaria, Chuniella, Cylindrotheca, Cymbellonitzschia, Gomphonitzschia, Hantzschia, Nitzschia, Pseudonitzschia*

Order Surirellales: valves elliptical, lanceolate, ovate, or subrectangular, with flat or twisted pseudoraphe; canal-raphe on winglike projection around margin of valve: *Campylodiscus, Cymatopleura, Surirella*

General and systematic treatments: References 7, 69, 70, 91, 94, 142-149

12. Class Phaeophyceae

The brown algae are notable in their lack of unicellular and colonial forms. The simplest thallus is one with prostrate and upright systems of filaments. Many brown algae of this type are minute epiphytes. In a large number of genera, the filaments are held together in varying degrees of firmness by mucilage and intergrowth, resulting in macroscopic thalli ranging from crusts to branched structures several meters long. Growth in filamentous forms is diffuse or localized at the base of certain filaments (trichothallic growth). In another large group of genera, the thallus is parenchymatous, with apical and intercalary growth resulting in the differentiation of structurally and functionally distinctive tissues and organs. The giant kelp *(Macrocystis)* is a striking example of this type, its huge thallus differentiated morphologically into a massive holdfast and innumerable stipes, blades, and air bladders. Anatomical differentiation results in tissues specialized for photosynthesis, transport, storage, initiation of growth, and reproduction. Almost all brown algae are marine, growing on various substrates along the shore and to depths limited by the availability of light. They constitute an enormous biomass in all regions, but exhibit maximum diversity in cool-temperate to subpolar waters. Only photosynthetic forms are known. Alginic acid (a polysaccharide), the major cell-wall constituent of brown algae, is used extensively in industry.

Asexual reproduction is by zoospores, aplanospores, special multicellular propagative structures, and fragmentation. Sexual reproduction is isogamous, anisogamous, or oogamous. The idealized brown-algal life history involves an alternation of a diploid sporophyte with similar or dissimilar haploid gametophytes. The sporophyte bears unilocular sporangia, the contents of each sporangium forming zoospores following meiosis and usually one to four subsequent mitotic divisions. Gametophytes bear plurilocular gametangia, each cell of which produces a single gamete. In simpler brown algae, bypassing one or the other phase is readily accomplished by the formation of mitospores in plurilocular organs on sporophytes and by the parthenogenetic development of gametes. Numerous other variations, including the formation of special interpolative phases, have been revealed in culture studies. In one major group of brown algae (subclass Fucophycidae) there is only one free-living phase, which is diploid.

Classification of Phaeophyceae has traditionally been based on differences in life history, manner of growth, vegetative structure, and sexual reproduction. Cytological characters have been considered of diagnostic value at the ordinal level by some taxonomists.[150,151] In the following synopsis, the phylogenetic distance between pairs of orders is highly variable, suggesting the utility and desirability of recognizing subclasses. The orders Dictyotales and Fucales are so distinctive that each has been accorded the rank of class by several authors. Difficulties are encountered, however, when an attempt is made to draw lines among the remaining orders.

Class Phaeophyceae

Subclass Phaeophycidae: haploid and diploid somatic phases; motile reproductive cells biflagellate

Order Ectocarpales: somatic phases essentially similar (sometimes slightly dissimilar), composed of uniseriate branched filaments with diffuse or somewhat localized growth; germlings with a loose-branching, creeping development; multiple chloroplasts per cell, pyrenoids present; isogamous (rarely anisogamous): *Ectocarpus, Feldmannia, Giffordia, Pilayella*

Order Ralfsiales: similar to Ectocarpales, but thallus crustose (at least in part); germlings forming discs of closely coherent filaments; usually a single platelike chloroplast per cell (several in *Pseudolithoderma*), pyrenoids absent; sexual reproduction (rarely observed) anisogamous: *Analipus, Heribaudiella, Lithoderma, Nemoderma, Petroderma, Pseudolithoderma, Ralfsia, Sorapion*

Order Chordariales: somatic phases dissimilar; gametophytes microscopic, filamentous; sporophyte usually macroscopic, pseudoparenchymatous; growth diffuse or localized; usually isogamous: *Acrothrix, Chordaria, Cladosiphon, Compsonema, Cylindrocarpus, Elachista, Eudesme, Haplogloia, Hecatonema, Leathesia, Leptonematella, Mesogloia, Myriactula, Myrionema, Spermatochnus, Stilophora, Tinocladia*

Order Sporochnales; somatic phases dissimilar; gametophytes microscopic, filamentous; sporophyte macroscopic, pseudoparenchymatous; growth trichothallic, involving tuft of hairs (multiaxial); oogamous: *Bellotia, Carpomitra, Encyothalia, Nereia, Perithalia, Sporochnus, Tomaculopsis*

Order Desmarestiales: somatic phases dissimlar gametophytes microscopic, filamentous; sporophyte macroscopic, pseudoparenchymatous; growth trichothallic, uniaxial; oogamous: *Arthrocladia, Desmarestia, Himantothallus, Phaeurus*

Order Cutleriales: somatic phases similar or dissimilar, composed of laterally coherent parenchymatous units, each derived from one of a marginal row of trichothallic hairs; gametophyte crustose, cylindrical, or foliose; sporophyte crustose; anisogamous: *Cutleria, Zanardinia*

Order Sphacelariales: somatic phases similar, with limited parenchymatous development, appearing filamentous or rarely crustose; primary growth initiated by prominent apical cells; anisogamous or oogamous: *Alethocladus, Battersia, Cladostephus, Halopteris, Sphacelaria*

Order Tilopteridales: somatic phases similar, with limited parenchymatous development, appearing filamentous; growth trichothallic; oogamous; asexual reproduction by globose monospores: *Haplospora, Tilopteris*

Order Dictyosiphonales: somatic phases dissimilar; gametophytes microscopic, filamentous; sporophyte parenchymatous, appearing filamentous, cylindrical, hollow-globose, or foliose; growth diffuse, sometimes accompanied by apical localization; multiple chloroplasts per cell, pyrenoids present; isogamous: *Asperococcus, Coilodesme, Dictyosiphon, Myriotrichia, Punctaria, Soranthera, Stictyosiphon, Striaria*

Order Scytosiphonales: somatic phases dissimilar; gametophytes parenchymatous, in the form of a blade, tube, net, or hollow hemisphere; sporophyte a pseudoparenchymatous crust; growth diffuse; single chloroplast with conspicuous pyrenoid in each vegetative cell; sexual reproduction isogamous or anisogamous, commonly bypassed: *Colpomenia, Hydroclathrus, Petalonia, Scytosiphon*

Order Laminariales (kelps): somatic phases dissimilar; gametophytes microscopic, filamentous; sporophytes parenchymatous, differentiated into structurally and functionally distinct tissues and organs; growth initiated in meristems; oogamous: *Agarum, Alaria, Arthrothamnus, Chorda, Costaria, Dictyoneurum, Ecklonia, Egregia, Eisenia, Hedophyllum, Laminaria, Lessonia, Lessoniopsis, Macrocystis, Nereocystis, Pelagophycus, Postelsia, Pterygophora, Saccorhiza, Thalassiophyllum, Undaria*

Subclass Dictyotophycidae: somatic phases similar, parenchymatous, foliose, usually erect (rarely procumbent); primary growth initiated by prominent apical cells; oogamous, male gamete with single emergent flagellum, the second flagellum vestigial; meiospores nonmotile, usually in sori

Order Dictyotales: *Dictyopteris, Dictyota, Dilophus, Glossophora, Lobophora* (incl. *Pocockiella*), *Pachydictyon, Padina, Spatoglossum, Stoechospermum, Stypopodium, Taonia, Zonaria*

Subclass Fucophycidae: diploid somatic phase only; parenchymatous, differentiated into structurally and functionally distinct tissues and organs; growth entirely diffuse or initiated by apical cells derived from trichothallic hairs or by an intercalary meristem; sexual reproduction oogamous (unknown in Ascoseirales), gametes produced in clusters within sunken chambers (conceptacles), motile gametes biflagellate

Order Ascoseirales: growth initiated by intercalary meristem; biflagellate cells developing into new thalli without fusing (probably to be interpreted as parthenogenesis in a life history in which isogamous or anisogamous sexual reproduction has been lost phylogenetically): *Ascoseira*

Order Fucales: growth initiated by apical cells derived from trichothallic hairs; oogamous: *Ascophyllum, Axillariella, Bifurcaria, Bifurcariopsis, Cystophora,*

Cystophyllum, Cystoseira, Fucus, Halidrys, Hesperophycus, Himanthalia, Hormophysa, Hormosira, Marginariella, Neoplatylobium, Pelvetia, Pelvetiopsis, Phyllospora, Platythalia, Sargassum, Scaberia, Scytothalia, Seirococcus, Turbinaria, Xiphophora

Order Durvillaeales:[152] growth diffuse, not initiated by intercalary meristem or by apical cells; oogamous: *Durvillaea*

General and systematic treatments: References 94, 153-155

13. Class Xanthophyceae

Although yellow-green algae include representatives of nearly every structural type known for algae, they are predominantly nonmotile unicells of fresh-water habitats. They also occur in brackish and marine waters, in and on soil and mud, in snow and ice, and on tree trunks and damp walls. Many are attached to aquatic plants. Colorless forms are rare. Unicells may be naked or walled; some have an outer envelope or lorica. Endogenous cysts, which are common, comprise two overlapping silicified pieces of wall material, usually of approximately equal size. The wall of the vegetative cell in many genera similarly comprises two overlapping pieces, which in filamentous forms are H-shaped. Germination of a cyst produces zoospores or amoeboid cells. Sexual reproduction (known for only a few genera) is isogamous or oogamous.

The following synopsis follows the Pascherian scheme of basing orders on the type of somatic organization.[55,156]

Class Xanthophyceae

Order Rhizochloridales: amoeboid stage dominant, unicellular or colonial; biflagellate zoospores: *Chlamydomyxa, Heterocalycina, Myxochloris, Rhizochloris, Rhizolekane, Stipitochloris, Stipitococcus, Stipitiporos*

Order Chloramoebales (Heterochloridales): naked biflagellate (rarely uniflagellate) monads, some capable of forming temporary palmelloid colonies: *Nephrochloris* (uniflagellate); *Ankylonoton, Chloramoeba, Heterochloris, Phacomonas, Polykyrtos* (biflagellate)

Order Heterogloeales: nonmotile coccoid forms, solitary or colonial, with or without a gelatinous matrix; zoospores with contractile vacuoles: *Characidiopsis, Helminthogloea, Heterogloea* (incl. *Gloeochloris*), *Malleodendron, Pelagocystis, Pleurochloridella*

Order Mischococcales: nonmotile coccoid forms, solitary or colonial, often attached to substrate by short mucilaginous stalk, not palmelloid; uniflagellate or biflagellate zoospores without contractile vacuoles: *Akanthochloris* (incl. *Groenlandiella*), *Arachnochloris* (incl. *Chlorarkys, Keriosphaera, Trachycystis*), *Asterogloea, Aulakochloris, Botrydiopsis* (incl. *Polychloris*), *Botryochloris* (incl. *Sphaerosorus*), *Bumilleriopsis, Centritractus, Characiopsis, Chlorallantus, Chlorapion, Chlorellidium* (incl. *Chlorellidiopsis*), *Chloridella, Chlorocloster, Chlorogibba, Chlorokoryne, Chloropedia, Chlororhabdion, Chlorosaccus, Clorothecium, Chytridiochloris, Diachros, Dichotomococcus, Dictyosphaeriopsis, Dioxys, Ducellieria, Ellipsoidion* (incl. *Monallantus*), *Endochloridion, Excentrochloris, Gaumiella, Gloeobotrys, Gloeopodium, Gloeoskene, Goniochloris, Hemisphaerella, Heterodesmus, Ilsteria, Leuvenia, Lutherella, Mischococcus, Monodus, Nephrodiella, Ophiocytium* (incl. *Sciadium*), *Perone, Peroniella, Pleurochloris* (see also Eustigmatophyceae), *Pleurogaster, Polyedriella, Prismatella, Pseudobumilleriopsis, Pseudopolyedriopsis, Pseudostaurastrum* (incl. *Isthmochloron*), *Pseudotetraedron, Radiosphaerella (=Radiosphaera), Raphidella,*

Rhomboidella, Sklerochlamys, Tetraedriella (incl. *Tetragoniella*), *Tetraktis, Tetraplektron, Trachychloron, Trachydiscus, Trypanochloris*

Order Tribonematales: simple or branched filaments with cross-walls (septa), uniflagellate or biflagellate zoospores: *Brachynematella (= Brachynema), Bumilleria, Chadefaudiothrix, Chaetopedia, Heterococcus* (incl. *Aeronema, Monocilia*), *Heteropedia* (incl. *Capitulariella*), *Heterotrichella, Neonema, Sphagnoikos (= Fremya), Tribonema, Xanthonema (= Heterothrix)*

Order Vaucheriales: multinucleate nonseptate vesicles or filaments (coenocytic); simple or compound zoospores: *Asterosiphon, Botrydium, Phyllosiphon, Phytophysa, Vaucheria*

General and systematic treatments: References 7, 73, 74, 94, 156-159.

14. Class Raphidophyceae (Chloromonadophyceae)

This class comprises a small group of biflagellate unicellular organisms (raphidomonads or chloromonads). The cells are flattened, dorsiventral or bilaterally symmetrical, and naked. Trichocysts are common; eyespots are unknown; contractile vacuoles are present in fresh-water forms. Temporary palmelloid stages are known. Most genera occur in fresh water (acidic to neutral), but there are marine forms that produce conspicuous blooms. Colorless raphidomonads have pseudopodia. Reproduction is by longitidinal binary fission.

Class Raphidophyceae

Order Raphidomonadales

Family Vacuolariaceae: photosynthetic: *Gonyostomum (= Raphidomonas), Merotrichia, Swirenkomonas, Trentonia, Vacuolaria* (fresh-water); *Chattonella* (incl. *Heterosigma, Hornellia*) (marine)

Family Thaumatomastigaceae: colorless, with pseudopodia: *Bodopsis, Colponema, Hyaloselene, Reckertia, Synoikomonas, Thaumatomastix, Thaumatomonas*

General and systematic treatments: References 75, 90, 91, 94, 101, 102, 160, 161

15. Class Eustigmatophyceae

This class was segregated from the Xanthophyceae on the basis of a suite of cytological, ultrastructural, and biochemical features. Some species previously assigned to the Chlorophyceae have recently been transferred to the group, giving a total of about a dozen eustigmatophytes. All are nonmotile unicells varying in shape from spherical to elongate and occurring mostly in fresh water. In one genus *(Pseudocharaciopsis)* the cell is attached by a stipe and a small discoid pad. Reproduction is by zoospores or autospores. The zoospore is especially diagnostic for the class. It is naked, amoeboid, elongate, and lacks both Golgi bodies and pyrenoids. Its eyespot is independent of the single chloroplast and consists of an irregular group of droplets not enclosed by a membrane. The pyrenoid (in vegetative cells) is unusual in being a polyhedral or spherical projection from the inner face of the chloroplast.

Because most eustigmatophytes are still assigned to genera whose type species are chlorophycean or xanthophycean, classification is nebulous. Only three generic names are definitely assignable here: *Chlorobotrys, Pseudocharaciopsis,* and *Vischeria. Pleurochloris* is a questionable fourth. The only available family name is Chlorobotryaceae, and at least one new family needs to be proposed. The disposition of *Ellipsoidion acuminatum, Monallantus salinus, Monodus subterraneus, Nannochloris oculata*, and *Polyedriella helvetica* remains for future determination, entailing the proposal of new genera.

General and systematic treatments: References 42, 43, 94, 158, 162, 163.

16. Class Haptophyceae (Prymnesiophyceae)

This assemblage of organisms has recently been segregated from the Chrysophyceae.[37] Motile cells of haptophytes typically have two equal or subequal acronematic flagella and sometimes a filiform organelle, the haptonema, arising between the flagella. Organic scales, which may be calcified, cover most motile cells and some nonmotile stages. Calcified scales complement or replace noncalcified scales. (By contrast, motile cells of chrysophytes have one acronematic and one pleuronematic flagellum, and their scales, when present, are siliceous.) The large majoriety of haptophytes are monads, but their somatic spectrum includes palmelloid colonies, cuboidal packets, coccoid unicells or clusters, and filaments. The life history of many haptophytes involves an alternation of two dissimilar vegetative phases, which in some instances have been shown to have a sporophyte/gametophyte relationship. Dissimilarity entails flagellar, haptonematal, and scale features of individual cells as well as differences in somatic expression. Culture studies have shown that benthic forms are highly variable in their morphology. Most haptophytes are marine, with coccolithophorids (unicells with calcified scales, or coccoliths) forming one of the three major components of phytoplankton, along with diatoms and dinoflagellates. All haptophytes are photosynthetic, although some are facultatively osmotrophic or phagotrophic.

Coccoliths are abundant in the fossil record as far back as the early Jurassic. Differences in coccolith morphology have been used as the basis for recognizing hundreds of genera and dozens of families of fossil coccolithophorids. There are relatively few genera of living coccolithophorids, for which about a dozen families have been proposed. Haptophytes were originally divided into two orders on the basis of the present or apparent absence of a haptonema.[37] The finding of vestigial haptonemata and the occurrence in the same life history of motile cells with and without a haptonema[164] cast doubt on the validity and utility of this primary distinction. The ability to produce calcified scales appears to be of greater taxonomic significance. A small group of problematic haptophytes has recently been set apart in their own order (Pavlovales), based on a combination of flagellar and haptonematal features.[165]

Class Haptophyceae

- Order Pavlovales: motile or palmelloid unicells; if nonmotile, producing motile swarmers; motile cells with two flagella that are unequal and bear fine hairs and/ or small dense bodies; shorter flagellum at times vestigial; haptonema present; scales lacking: *Corcontochrysis, Diacronema, Exanthemachrysis, Pavlova*
- Order Prymnesiales: monads or palmelloid or sarcinoid colonies; if nonmotile, producing motile swarmers; motile cells with two equal or subequal acronematic flagella; haptonema lacking, vestigial, or emergent; organic scales never calcified: *Chrysocampanula, Chrysochromulina, Chrysotila, Corymbellus, Dicrateria, Imantonia, Isochrysis, Phaeocystis, Platychrysis, Prymnesium*
- Order Coccolithophorales (Coccosphaerales): monads, with or without nonmotile stage; nonmotile stage represented by free-living unicells, benthic coccoid unicells and clusters, or benthic filaments; in one genus *(Ochrosphaera)* motile stage strictly reproductive, incapable of replication, giving rise to vegetative stage (irregular packets of cells); haptonema lacking, vestigial, or emergent; at least one motile cell-type in each species and some nonmotile cells covered by coccoliths in addition to or in place of organic uncalcified scales

Family Hymenomonadaceae: motile stage (coccolithophorid) alternating with benthic scale-covered filamentous stage: *Chrysonema, Hymenomonas, Pleurochrysis* (incl. *Cricosphaera*)

Family Ochrosphaeraceae: vegetative stage an irregular packet of cells, reproducing by coccolith-covered biflagellate cells: *Ochrosphaera*

Family Coccolithophoraceae: coccolithophorids without a benthic stage: *Acanthoica, Acanthosolenia, Algirosphaera, Anacanthoica, Anoplosolenia, Anthosphaera, Aspidiophora, Braarudosphaera, Calciarcus, Calcidiscus, Calcioconus, Calciopappus, Calciosolenia, Calyptrosphaera, Ceratolithus, Clavisphaera, Coccolithus, Corisphaera, Crystallolithus, Cyclococcolithus, Deutschlandia, Discosphaera, Gephyrocapsa* (incl. *Emiliania*), *Halopappus, Helicosphaera, Helladosphaera, Heyneckia, Homozygosphaera, Lohmannosphaera, Michaelsarsia, Najadea, Neosphaera, Ophiaster, Papposphaera, Pappomonas, Periphyllophora, Petalosphaera, Pontosphaera, Rhabdosphaera, Rhabdothorax, Scyphosphaera, Sphaerocalyptra, Syracolithus, Syracosphaera, Tergestiella, Thalassopappus, Thoracosphaera, Turrisphaera, Umbellosphaera, Umbilicosphaera, Wigwamma, Zygosphaera*

After more life histories of coccolithophorids have been elucidated, it may be feasible to recognize additional families.

General and systematic treatments: References 94, 104, 134, 139, 166-169.

17. Class Dinophyceae

Dinoflagellates exhibit a remarkable number of variations on a theme. The large majority are biflagellate monads in marine, brackish, and fresh waters, frequently producing blooms. Specialized habitats of free-living forms include moist sand at the edge of the sea, the surface of seaweeds, and snow. Amoeboid, palmelloid, colonial, coccoid (with or without a gelatinous matrix), and filamentous forms are also known. There is marked nutritional diversity. Most dinoflagellates are photosynthetic, although many marine species require an external source of certain vitamins. Photosynthesis may be accompanied by facultative osmotrophy or sporadic phagotrophy. Many dinoflagellates are symbionts of various marine protists and invertebrates. Both endoparasites and ectoparasites are known, the hosts including other dinoflagellates, diatoms, prasinophytes, coelenterates, molluscs, annelids, crustaceans, tunicates, and fish. Parasitic species are highly specialized or structurally reduced and their dinophycean affinity is shown only by features of their motile reproductive cells and of their nucleus.

There are two basic types of flagellar insertion. In one type, the two flagella are borne apically or subapically on a cell that is roughly bilaterally symmetrical. In the second type, an acronematic, posteriorly directed flagellum is located in a longitudinal groove (sulcus) and a ribbonlike flagellum is located in a transverse groove (cingulum). The cell covering (amphiesma) comprises a series of concentric membranes, one of which may be a layer of flattened vesicles. The vesicles, in turn, may contain cellulose plates of varying thickness. The number and the arrangement of these thecal plates are accorded taxonomic significance at various hierarchical levels. As observed with light microscopy, dinoflagellates were classified as thecate and nonthecate. The demonstration of previously unobserved plates through improved histochemical techniques and electron microscopy has necessitated certain taxonomic emendations. Dinoflagellates commonly reproduce only asexually, by binary fission. Sexual reproduction is isogamous or anisogamous. Many dinoflagellates have encysted stages, and the cysts are well represented in the fossil record as far back as the Triassic. Bioluminescence is well known in several genera *(Noctiluca, Pyrocystis, Pyrodinium, Gonyaulax).* Certain species of *Gonyaulax* and *Gymnodinium* produce toxic red tides.

Class Dinophyceae

Subclass Desmophycidae: motile cells with two apically or subapically inserted flagella, one directed forward and the second encircling the first, beating in a perpendicular plane

Order Desmomastigales: monads with apically inserted flagella; thecal layer in one piece or absent: *Adinomonas, Desmomastix, Haplodinium, Pleromonas*

Order Prorocentrales: monads with apically inserted flagella; theca divided longitudinally into two halves, sometimes with small flagellar pore plate: *Mesoporos, Prorocentrum* (incl. *Exuviaella*)

Order Protaspidales: colorless monads with subapically inserted flagella; amphiesma membranous: *Protaspis*

Order Desmocapsales: palmelloid; thecal layer absent; motile reproductive cells resembling *Desmomastix: Desmocapsa*

Subclass Dinophycidae: motile cells biflagellate, one flagellum arising in a longitudinal groove and extending backward, the other flagellum vibrating in a transverse groove

Order Blastodiniales: solitary or colonial ectoparasites or endoparasites; amphiesma nonthecate or possibly with thin plates; successive divisions in zoosporogenesis not separated by periods of rest: *Actinodinium, Amylodinium, Apodinium, Cachonella (= Diplomorpha), Duboscquodinium, Haplozoon, Oodinium, Parapodinium, Protoodinium*

Order Chytriodiniales: solitary ectoparasites; amphiesma apparently nonthecate; successive divisions in zoosporogenesis separated by periods of rest: *Chytriodinium, Myxodinium*

Order Gymnodiniales: monads, usually photosynthetic; amphiesma nonthecate, internal skeleton present in a few genera: *Achradina, Actiniscus, Adenoides, Amphidinium, Amphitholus (= Amphilothus), Aureodinium, Balechina, Bernardinium, Cochlodinium, Entomosigma, Erythropsidinium (= Erythropsis), Filodinium, Gymnodinium, Katodinium, Leucopsis, Monaster, Nematodinium, Nematopsides, Oxyrrhis, Polykrikos, Pronoctiluca, Proterythropsis, Protodinium, Protopsis, Schillingia, Spirodinium (= Gyrodinium), Torodinium*

Order Peridiniales: monads, photosynthetic or osmotrophic, sometimes phagotrophic; theca composed of many plates: *Acanthogonyaulax, Amphidiniopsis, Amphidoma, Archaeosphaerodiniopsis, Berghiella, Blepharocysta* (incl. *Lissodinium*), *Cachonina, Centrodinium, Ceratium, Ceratocorys, Cladopyxis, Congruentidium, Crypthecodinium, Dinosphaera, Diplopsalis* (incl. *Dissodium*), *Dolichodinium, Ensiculifera, Glenodiniopsis, Glenodinium, Goniodinium, Gonyaulax, Halophilodinium, Helgolandinium, Hemidinium, Heteraulacus* (incl. *Goniodoma*), *Heterodinium, Kolkwitziella, Kryptoperidinium, Lophodinium, Melanodinium, Minuscula, Ostreopsis, Oxytoxum, Pachydinium, Pavillardinium (= Murrayella), Peridiniella, Peridiniopsis, Peridinium, Podolampas, Protoceratium, Ptychodiscus, Pyrodinium, Pyrophacus, Roscoffia, Scrippsiella, Sphaerodinium, Spiraulax, Spiraulaxina, Thecadinium, Thompsodinium, Woloszynskia, Zygabikodinium (= Diplopeltopsis)*

Order Dinophysales: monads, photosynthetic or osmotrophic; amphiesma composed of 18 (rarely 19) plates, two of which are exceptionally large: *Amphisolenia, Citharistes, Dinoceras, Dinofurcula, Dinophysis* (incl. *Phalacroma = Prodinophysis*), *Heteroschisma (= Latifascia), Histioneis, Histiophysis, Me-*

tadinophysis, Metaphalacroma, Ornithocercus, Oxyphysis, Palaeophalacroma, Parahistioneis, Proheteroschisma, Pseudophalacroma, Thaumatodinium, Triposolenia

Order Noctilucales: phagotrophic monads; amphiesma nonthecate: *Cymbodinium, Kofoidinium, Leptodinium, Leptodiscus, Leptophyllus (= Abedinium), Noctiluca, Pavillardia, Petalodinium, Pomatodinium, Scaphodinium* (incl. *Leptospathium*), *Spatulodinium*

Order Gloeodiniales: nonmotile cells in packets surrounded by concentrically laminate gelatinous sheath, photosynthetic; reproductive cells hemidinoid: *Gloeodinium, Rufusiella*

Order Dinamoebales: free-living phagotrophic amoeboid unicells; cyst with nonthecate wall, releasing gymnodinoid zoospores upon germination: *Dinamoeba (= Dinamoebidium)*

Order Pyrocystales: nonmotile photosynthetic unicells, living free in plankton or hemiparasitic on copepod eggs, with thin smooth cellulosic wall, functioning as primary cysts within which secondary cysts (or vegetative cells of a second type) are formed; secondary cysts often lunate, producing peridinioid or gymnodinioid motile cells: *Diplodinium (= Dissodinium), Pyrocystis, Sporodinium*

Order Phytodiniales: nonmotile photosynthetic unicells, planktonic or attached; amphiesma nonthecate; reproduction by gymnodinioid motile cells or autospores: *Cystodinedria* (incl. *Phytodinedria*), *Cystodinium* (incl. *Gymnocystodinium*), *Dinastridium* (incl. *Bourrellyella*); *Dinococcus (= Raciborskia), Hypnodinium, Manchudinium, Phytodinium, Stylodinium* (incl. *Dinopodiella*), *Tetradinium, Thaurilens*

Order Zooxanthellales: coccoid unicells usually living symbiotically in various animals, capable of transforming contents into zoospores: *Endodinium, Zooxanthella* (incl. *Symbiodinium*)

Order Dinotrichales: branched uniseriate filaments; reproduction by gymnodinioid motile cells: *Dinoclonium, Dinothrix*

General and systematic treatments: References 40, 75, 90, 91, 94, 101, 102, 149, 170-175.

18. Class Ebriophyceae

Ebriids comprise a very small group of marine planktonic monads. The cells are colorless or tinged with yellow or rose, but without chloroplasts. They are phagotrophic, feeding principally on diatoms, but have also been reported to have pigmented endosymbionts. Each cell has an internal siliceous skeleton with tetraxial or triradial symmetry, suggesting an affinity with silicoflagellates, but the skeletal elements are solid rather than tubular. Moreover, the possession of two unequal flagella and a dinokaryon suggests a relationship with dinoflagellates. (A few dinoflagellates also have siliceous internal skeletons.) Reproduction is by binary fission, preceded by skeletal replication.

Although relatively common in the fossil record as far back as the Paleocene, ebriids seemingly are represented in modern seas by only a few species. Taxonomic distinctions are based on skeletal structure.

Class Ebriophyceae

Order Ebriales: *Ebria, Hermesinum* (incl. *Bosporella*)

General and systematic treatments: References 39, 139, 176.

19. Class Ellobiophyceae

This class was recently proposed to accommodate a small group of marine ectoparasites believed to have certain affinities with Dinophyceae.[40] Hosts include crustaceans (mostly pelagic) and polychaete annelids. The organism is composed of multinucleate trophic and reproductive segments and is bounded by a complex pellicle. Spores are poorly known, but have been reported to be uniflagellate *(Parallobiopsis),*[177] motile but apparently nonflagellate *(Ellobiopsis),*[178] and biflagellate *(Thalassomyces).*[179]

Class Ellobiophyceae

Order Thalassomycetales

Family Ellobiopsidaceae (Thalassomycetaceae): *Ellobiocystis, Ellobiopsis, Parallobiopsis, Rhizellobiopsis, Thalassomyces* (incl. *Amallocystis* = *Staphylocystis*)

General and systematic treatments: References 170, 180.

20. Class Syndiniophyceae

This class was recently proposed to accommodate a small group of marine endoparasites previously aligned with Dinophyceae.[41] Hosts include dinoflagellates, radiolarians, ciliates, coelenterates, and copepods. Syndiniids differ from free-living and ectoparasitic dinoflagellates in the following features: (1) chromosomes few (4 to 10), V-shaped, containing a sufficient quantity of basic proteins to be histochemically detectable; (2) centrioles associated with mitosis; (3) intracellular parasitism as their mode of nutrition; and (4) lack of cell covering containing thecal plates.

Class Syndiniophyceae

Order Syndiniales: *Amoebophrya, Atlanticellodinium, Dubosquella, Haematodinium, Hyalosaccus, Ichthyodinium, Keppenodinium, Merodinium, Solenodinium, Sphaeripara, Syndinium, Synhemidinium, Trypanodinium*

General and systematic treatments: References 41, 170, 171, 181, 182.

REFERENCES

1. **Stanier, R. Y., Kunisawa, R., Mandel, M., and Cohen-Bazire, F.,** Purification and properties of unicellular blue-green algae (order Chroococcales), *Bacteriol. Rev.,* 35, 171, 1971.
2. **Sachs, J.,** *Lehrbuch der Botanik,* Aufl. 4, Leipzig, 1874.
3. **Kohlmeyer, J.,** New clues to the possible origin of Ascomycetes, *BioScience,* 25, 86, 1975.
4. **Demoulin, V.,** The origin of Ascomycetes and Basidiomycetes: the case for a red algal ancestry, *Bot. Rev.,* 40, 315, 1975.
5. **Cohn, F.,** Conspectus familiarum cryptogamarum secundum methodum naturalem dispositarum, *Hedwigia,* 11, 17, 1872.
6. **Bourrelly, P.,** Recherches sur les Chrysophycées. Morphologie, phylogénie, systématique, *Rev. Algol., Mém. hors-sér.,* 1, 1957.
7. **Bourrelly, P.,** *Les Algues d'Eau Douce,* Tome II, Boubée, Paris, 1968.
8. **Loeblich, A. R. III and Loeblich, L. A.,** Division Chrysophyta, in *CRC Handbook of Microbiology,* Vol. 2, 2nd ed., Laskin, A. I. and Lechevalier, H. A., Eds., CRC Press, Boca Raton, Fla., 1979, 411.
9. **Leadbeater, B. S. C.,** Fine-structural observations on some marine choanoflagellates from the coast of Norway, *J. Mar. Biol. Assoc. U.K.,* 52, 67, 1972.
10. **Leadbeater, B. S. C. and Manton, I.,** Preliminary observations on the chemistry and biology of the lorica in a collared flagellate (*Stephanoeca diplocostata* Ellis), *J. Mar. Biol. Assoc. U.K.,* 54, 269, 1974.

11. **Hibberd, D. J.**, Observations on the ultrastructure of the choanoflagellate *Codosiga botrytis* (Ehr.) Saville-Kent with special reference to the flagellar apparatus, *J. Cell Sci.*, 17, 191, 1975.
12. **Reichenbach, H. G. L.**, *Conspectus Regni Vegetabilis*, Leipzig, 1828.
13. **Engler, A.**, *Syllabus der Pflanzenfamilien*, Ausg. 2, Berlin, 1898.
14. **Senn, G., Flagellata, in** *Die natürlichen Pflanzenfamilien*, Teil 1, Abt. 1a, Engler, A. and Prantl, K., Eds., Leipzig, 1900, 93.
15. **Papenfuss, G. F.**, Classification of the algae, in *A Century of Progress in the Natural Sciences, 1853-1953*, California Academy of Sciences, San Francisco, 1955, 115.
16. **Honigsberg, B. M.**, A revised classification of the phylum Protozoa, *J. Protozool.*, 11, 7, 1964.
17. **Whittaker, R. H. and Margulis, L.**, Protist classification and the kingdoms of organisms, *Bio-Systems*, 10, 3, 1978.
18. **Wynne, M. J. and Loiseaux, S.**, Recent advances in life history studies of the Phaeophyta, *Phycologia*, 15, 435, 1976.
19. **Dixon, P. S.**, Perennation, vegetative propagation and algal life histories, with special reference to *Asparagopsis* and other Rhodophyta, *Bot. Gothob. Acta. Univ. Gothob.*, 3, 67, 1965.
20. **Linnaeus, C.**, *Species Plantarum*, Tomus II, Stockholm, 1753.
21. **Linnaeus, C.**, *Systema Naturae*, Tomus I, Stockholm, 1758.
22. **Schweigger, A. F.**, *Beobachtungen auf naturhistorischen Reisen*, Berlin, 1819.
23. **Siebold, C. T. von**, Ueber einzellige Pflanzen und Thiere, *Z. Wiss. Zool., Abt. A*, 1, 270, 1849.
24. **Knoblauch, E.**, *Handbuch der systematischen Botanik von Dr. Eug. Warming*, Deutsche Ausgabe, Berlin, 1890.
25. **Lamouroux, J. V. F.**, Essai sur les genres de la famille des thalassiophytes non articulées, *Ann. Mus. Hist. Nat. (Paris)*, 20, 21, 1813.
26. **Nägeli, C.**, Gattungen einzelliger Algen, physiologisch und systematisch bearbeitet, *Neue Denkschr. Allg. Schweiz. Ges. Gesammten Naturwiss.*, 10(7), 1849.
27. **Greville, R. K.**, *Flora Edinensis*, Edinburgh, 1824.
28. **Stizenberger, E.**, *Dr. Ludwig Rabenhorst's Algen Sachsens resp. Mitteleuropa's Decade I-C. Systematisch Geordnet (mit Zugrundelegung eines Neuen Systems)*, Dresden, 1860.
29. **Borzì, A.**, *Botrydiopsis* nuovo genere di alghe verdi, *Boll. Soc. Ital. Microscop.*, 1, 60, 1889.
30. **Borzì, A.**, *Studi Algologici*, Fasc. 2, Palermo, 1895.
31. **Bohlin, K.**, Studier öfver några slägten af alggruppen Confervales Borzì, *Bih. Kongl. Svenska Vetensk.-Akad. Handl.*, 23(Afd. 3, 3), 1897.
32. **Luther, A.**, Ueber *Chlorosaccus* eine neue Gattung der Süsswasseralgen, nebst einigen Bemerkungen zur Systematik verwandter Algen, *Bih. Kongl. Svenska Vetensk.-Akad. Handl.*, 24(Afd. III, 13), 1899.
33. **West, G. S. and Fritsch, F. E.**, *A Treatise on the British Freshwater Algae, by the Late G. S. West, New Revised Edition*, Cambridge, 1927.
34. **Chadefaud, M.**, Sur l'organisation et la position systématique des Flagellés du genre *Pyramidomonas*, *Rev. Sci.*, 79, 113, 1941.
35. **Chadefaud, M.**, Études sur l'organisation de deux Volvocales sédentaires marines: *Prasinocladus lubricus* et *Chlorodendron subsalsum*, *Rev. Sci.*, 85, 862, 1947.
36. **Chadefaud, M.**, Les végétaux non vasculaires (cryptogamie), in *Traite de Botanique. Systematique*, Tome I, Chadefaud, M. and Emberger, L., Eds., Masson, Paris, 1960.
37. **Christensen, T., Alger**, in *Botanik*, Bind II, Nr. 2, Böcher, T. W., Lange, M., and Sørensen, T., Eds., Munksgaard, København, 1962.
38. **Parke, M.**, Some remarks concerning the class Chrysophyceae, *Brit. Phycol. J.*, 2, 47, 1961.
39. **Loeblich, A. R. III, Loeblich, L. A., Tappan, H., and Loeblich, A. R. Jr.**, Annotated index of fossil and recent silicoflagellates and ebridians with descriptions and illustrations of validly proposed taxa, *Geol. Soc. Amer. Mem.*, 106, 1968.
40. **Loeblich, A. R. III**, The amphiesma or dinoflagellate cell covering, in *Proc. N.A. Paleontological Convention*, Vol. 2, Allen, Lawrence, 1970, 867.
41. **Loeblich, A. R. III**, Dinoflagellate evolution: speculation and evidence, *J. Protozool.*, 23, 13, 1976.
42. **Hibberd, D. J., and Leedale, G. F.**, Eustigmatophyceae — a new algal class with unique organization of the motile cell, *Nature*, 225, 758, 1970.
43. **Hibberd, D. J. and Leedale, G. F.**, A new algal class — the Eustigmatophyceae, *Taxon*, 20, 523, 1971.
44. **Lewin, R. A.**, Prochlorophyta as a proposed new division of algae, *Nature*, 261, 697, 1976.
45. **Lewin, R. A.**, *Prochloron*, type genus of the Prochlorophyta, *Phycologia*, 16, 217, 1977.
46. **Müller, O. F.**, *Animalcula Infusoria Fluviatilia et Marina*, København, 1786.
47. **Gmelin, J. F.**, *Caroli à Linné . . . Systema Naturae, Ed. 13*, Tomus I, Pars VI, Leipzig, 1791.
48. **Dumortier, B. C.**, *Commentationes Botanicae*, Tournay, 1822.
49. **Agardh, C. A.**, *Systema Algarum*, Lund, 1824.

50. **Rabenhorst, L.**, *Flora Europaea Algarum Aquae Dulcis et Submarinae*, Sect. 1, Leipzig, 1864.
51. **Haeckel, E.**, *Generelle Morphologie der Organismen*, Band 2, Berlin, 1866.
52. **Leedale, G. F.**, How many are the kingdoms of organisms?, *Taxon*, 23, 261, 1974.
53. Stafleu, F. A. et al., Eds., International Code of Botanical Nomenclature adopted by the Twelfth International Botanical Congress, *Regnum Veg.*, 97, 1978.
54. **Wettstein, R. von**, *Handbuch der Systematischen Botanik*, Band 1, Leipzig 1901.
55. **Pascher, A.**, Systematische Übersicht über die mit Flagellaten in Zusammenhang stehenden Algenreihen und Versuch einer Einreihung dieser Algenstämme in die Stämme des Pflanzenreiches, *Beih. Bot. Centralbl.*, 48 (Abt. II, 2), 317, 1931.
56. **Wartenberg, A.**, *Systematik der Niederen Pflanzen*, Thieme, Stuttgart, 1972.
57. **Kornmann, P. and Sahling, P. H., Meeresalgen von Helgoland.** Benthische Grün-, Braun- und Rotalgen, *Helgol. Wiss. Meeresunters.*, 29, 1, 1977.
58. **Zinova, A. D.**, *Opredelitel' Zelenykh, Burykh i Krasnykh Vodoroslej Yuzhnykh Morej SSSR*, Nauka, Moskva-Leningrad, 1967.
59. **Round, F. E.**, *The Biology of the Algae*, 2nd ed., Edward Arnold, London, 1973.
60. **Gollerbakh [Hollerbach], M. M., Kosinskaya [Kossinskaja], E. K., and Polyanskij [Polyansky], V. I.**, Sinezelenye vodorosli, in *Opredelitel' Presnovodnykh Vodoroslej SSSR*, Vyp. 2, Sovetskaya Nauka, Moskva, 1953.
61. **Starmach, K., Cyanophyta-Sinice, Glaucophyta-Glaukofity**, in *Flora Slodkowodna Polski*, Tom 2, Starmach, K., Ed., Pânstwowe Wydawnictwo Naukowe, Warszawa, 1966.
62. **Kondrat'eva, N. V.**, Syn'o-zeleni vodorosti — Cyanophyta. Chastyna 2. Klas Hormohonievi - Hormogoniophyceae, in *Vyznachnykh Prisnovodnykh Vodorostej Ukrajns'koj RSR*, Vyp. 1, Chastyna 2, Naukova Dumka, Kyjv, 1968.
63. **Dixon, P. S. and Irvine, L. M., Rhodophyta**, in Parke, M. and Dixon, P. S., Eds., Check-list of British marine algae — third revision, *J. Mar. Biol. Assoc. U.K.*, 56, 532, 1976.
64. **Round, F. E.**, The taxonomy of the Chlorophyta, *Brit. Phycol. J.*, 2, 224, 1963.
65. **Round, F. E.**, The taxonomy of the Chlorophyta II, *Brit. Phycol. J.*, 6, 235, 1971.
66. **Bourrelly, P.**, *Les Algues d'Eau Douce. Tome I: Les Algues Vertes. Réimpression Revue et Augmentée*, Boubée, Paris, 1973.
67. **Matvienko, A. M.**, Zolotistye vodorosli, in *Opredelitel' Presnovodnykh Vodoroslej SSSR*, Vyp. 3, Sovetskaya Nauka, Moskva, 1954.
68. **Matvienko, O.[A.] M.**, Zolotisti vodorosti — Chrysophyta, in *Vyznachnykh Prisnovodnykh Vodorostej Ukrajns'koj RSR*, Vyp. III, Chastyna 1, Naukova Dumka, Kyjv, 1965.
69. **Zabelina, M. M., Kiselev, I. A., Proshkina-Lavrenko, A. I., and Sheshukova, V. S., Diatomovye vodorosli**, in *Opredelitel' Presnovodnykh Vodoroslej SSSR*, Vyp. 4, Sovetskaya Nauka, Moskva, 1951.
70. **Topachevs'kyj, O. V. and Oksiyuk, O. P.**, Diatomovi vodorosti — Bacillariophyta (Diatomeae), in *Vyznachnykh Prisnovodnykh Vodorostej Ukrajns'koj RSR*, Vyp. XI, Akad. Nauk URSR, Kyjv, 1960.
71. **Kylin, H.**, Uber die Entwicklungsgeschichte der Phaeophyceen, *Lunds Univ. Årsskr.*, 2, 28(8), 1933.
72. **Petrov, Yu. E.**, K sistematike klassa Cyclosporophyceae (Phaeophyta), *Nov. Sist. Nizsh. Rast. (Bot. Inst. Akad. Nauk SSSR)*, 1964, 146, 1964.
73. **Dedusenko-Shchegoleva, N. T. and Gollerbakh [Hollerbach], M. M.**, Zheltozelenye vodorosli — Xanthophyta, in *Opredelitel' Presnovodnykh Vodoroslej SSSR*, Vyp. 5, Akad. Nauk SSSR, Moskva-Leningrad, 1962.
74. **Matvienko, O.[A.] M. and Dogadina, T. V.**, Zhovtozeleni vodorosti — Xanthophyta, in *Vyznachnykh Prisnovodnykh Vodorostej Ukrajns'koj RSR*, Vyp. X, Naukova Dumka, Kyjv, 1978.
75. **Matvienko, O.[A.] M. and Litvinenko, R. M.**, Pirofitovi vodorosti — Pyrrophyta in *Vyznachnykh Prisnovodnykh Vodorostej Ukrajns'koj RSR*, Vyp. III, Chastyna 2, Naukova Dumka, Kyjv, 1977.
76. **Russell, G. and Fletcher, R. L.**, A numerical taxonomic study of the British Phaeophyta, *J. Mar. Biol. Assoc. U.K.*, 55, 763, 1975.
77. **Stewart, K. D. and Mattox, K. R.**, Comparative cytology, evolution and classification of the green algae with some consideration of the origin of other organisms with chlorophylls a and b, *Bot. Rev.*, 41, 104, 1975.
78. **Engler, A. and Gilg, E.**, *Syllabus der Pflanzenfamilien*, 9th und 10th Aufl. 9 & 10, Berlin, 1924.
79. **Hendey, N. I.**, A revised check-list of British marine diatoms, *J. Mar. Biol. Assoc. U.K.*, 54, 277, 1974.
80. **Lewin, R. A.**, A marine *Synechocystis* (Cyanophyta, Chroococcales) epizoic on ascidians, *Phycologia*, 14, 153, 1975.
81. **Antia, N. J.**, A critical appraisal of Lewin's Prochlorophyta, *Brit. Phycol. J.*, 12, 271, 1977.
82. **Fogg, G. E.**, Blue-green algae, in *CRC Handbook of Microbiology*, Vol. 2, 2nd ed., Laskin, A. I. and Lechevalier, H. A., Eds., CRC Press, Boca Raton, Fla., 1979, 347.

83. **Delaney, S. F., Herdman, M., and Carr, N. G.,** Genetics of blue-green algae, in *The Genetics of Algae,* Lewin, R. A., Ed., Blackwell, Oxford, 1976, 7.
84. **Geitler, L.,** Schizophyta: Klasse Schizophyceae, in *Die Natürlichen Pflanzenfamilien,* Aufl. 2, Band lb, Engler, A. and Prantl, K., Eds., Engelmann, Leipzig, 1942.
85. **Drouet, F. and Daily, W. A.,** Revision of the coccoid Myxophyceae, *Butler Univ. Bot. Stud.* 12, 1, 1956.
86. **Drouet, F .,** Revision of the classification of the Oscillatoriaceae, *Monogr. Acad. Nat. Sci. Philadelphia,* 15, 1968.
87. **Drouet, F.,** *Revision of the Nostocaceae with Cylindrical Trichomes (formerly Scytonemataceae and Rivulariaceae),* Hafner, New York, 1973.
88. **Drouet, F.,** Revision of the Nostocaceae with constricted trichomes, *Beih. Nova Hedwigia,* 57, 1, 1978.
89. **Desikachary, T. V.,** *Cyanophyta,* Indian Council of Agricultural Research, New Delhi, 1959.
90. **Bourrelly, P.,** *Les Algues d'Eau Douce,* Tome III, Boubee, Paris, 1970.
91. **Fott, B.,** *Algenkunde,* Aufl. 2, Fischer, Jena, 1971.
92. Carr, N. G. and Whitton, B. A., Eds., *The Biology of Blue-green Algae,* Blackwell, Oxford, 1973.
93. **Fogg, G. E., Stewart, W. D. P., Fay, P. and Walsby, A. E.,** *The Blue-green Algae,* Academic Press, New York, 1973.
94. **Bold, H. C. and Wynne, M. J.,** *Introduction to the Algae: Structure and Reproduction,* Prentice-Hall, Englewood Cliffs, N.J., 1978.
95. **Guiry, M. D.,** The importance of sporangia in the classification of the Florideophyceae, in *Modern Approaches to the Taxonomy of Red and Brown Algae,* Irvine, D. E. G. and Price, J. H., Eds., Academic Press, London, 1978, 111.
96. **Kylin, H.,** *Die Gattungen der Rhodophyceen,* Gleerup, Lund, 1956.
97. **Dixon, P. S.,** *Biology of the Rhodophyta,* Hafner, New York, 1973.
98. **Chapman, A. R. O., Rhodophyta,** in *CRC Handbook of Microbiology,* Vol. 2, 2nd ed., Laskin, A. I. and Lechevalier, H. A., Eds., CRC Press, Boca Raton, Fla., 1979, 503.
99. **Hollande, A.,** Classe des Cryptomonadines (Cryptomonadina Ehrenberg, 1832), in *Traite de Zoologie,* Tome I, Fasc. I, Grassé, P. P., Ed., Masson, Paris, 1952, 285.
100. **Butcher, R. W.,** An introductory account of the smaller algae of British coastal waters. Part IV: Cryptophyceae, *Gr. Brit., Minist. Agric. Fish. Food, Fish. Invest.,* Ser. IV [Part 4], 1967.
101. **Fott, B.,** Cryptophyceae, Chloromonadophyceae, Dinophyceae, in *Das Phytoplankton des Süsswassers,* Teil 3, Aufl. 2, Huber-Pestalozzi, G., Ed., Schweizerbart, Stuttgart, 1968.
102. **Starmach, K.,** Cryptophyceae — Kryptofity, Dinophyceae — Dinofity, Raphidophyceae — Rafidofity, in *Flora Slodkowodna Polski,* Tom 4, Starmach, K. and Siemińska, J., Eds., Pańśtwowe Wydawnictwo Naukowe, Warszawa, 1974.
103. **Dodge, J. D.,** The Cryptophyceae, in *CRC Handbook of Microbiology,* Vol. 2, 2nd ed., Laskin, A. I. and Lechevalier, H. A., Eds., CRC Press, Boca Raton, Fla., 1979, 513.
104. **Boney, A. D.,** Scale-bearing phytoflagellates: an interim review, *Oceanogr. Mar. Biol.,* 8, 251, 1970.
105. **Parke, M. and Green, J. C.,** Prasinophyceae, in Parke, M. and Dixon, P. S., Eds., Check-list of British marine algae — third revision, *J. Mar. Biol. Assoc. U. K.,* 56, 564, 1976.
106. **Chadefaud, M.,** Les Prasinophycées. Remarques historiques, critiques et phylogénétiques, *Bull. Soc. Phycol. France,* 22, 1, 1977.
107. **Randhawa, M. S.,** *Zygnemaceae,* Indian Council of Agricultural Research, New Delhi, 1959.
108. **Philipose, M. T.,** *Chlorococcales,* Indian Council of Agricultural Research, New Delhi, 1967.
109. **Fott, B.,** Chlorophyceae (Grünalgen). Ordnung: Tetrasporales, in *Das Phytoplankton des Süsswassers . . .* Teil 6, Huber-Pestalozzi, G., Ed., Schweizerbart, Stuttgart, 1972.
110. **Huber-Pestalozzi, G.,** Chlorophyceae (Grünalgen), Ordnung: Volvocales, in *Das Phytoplankton des Süsswassers,* Teil 5, Huber-Pestalozzi, G., Ed., Schweizerbart, Stuttgart, 1961.
111. **Mrozinska-Webb, T.,** Chlorophyta IV. Oedogoniales — Edogoniowe, in *Flora Słodkowodna Polski,* Tom 11, Starmach, K., Ed., Pańśtwowe Wydawnictwo Naukówe, Krakow, 1969.
112. **Kadlubowska, J. Z.,** Chlorophyta V. Conjugales. Zygnemaceae — Zrostnicowate, in Flora Słodkowodna Polski, Tom 12A, Starmach, K. and Siemińska, J., Eds., Pańśtwowe Wydawnictwo Naukowe, Kraków, 1972.
113. **Starmach, K.,** Chlorophyta, III. Zielenice nitkowate: Ulothrichales, Ulvales, Prasiolales, Sphaeropleales, Cladophorales, Chaetophorales, Trentepohliales, Siphonales, Dichotomosiphonales, in *Flora Słodkowodna Polski,* Tom 10, Starmach, K. and Siemińska, J., Eds., Pańśtwowe Wydawnictwo Naukowe, Kraków, 1972.
114. **Chapman, A. R. O.,** Chlorophyta, in *CRC Handbook of Microbiology,* Vol. 2, 2nd ed., Laskin, A. I. and Lechevalier, H. A., Eds., CRC Press, Boca Raton, Fla., 1979, 381.
115. **Grambast, L. J.,** Phylogeny of the Charophyta, *Taxon,* 23, 463, 1974.
116. **Smith, G. M.,** *The Fresh-water Algae of the United States,* 2nd ed., McGraw-Hill, New York, 1950.

117. **Pal, B. P., Kundu, B. C., Sundaralingam, V. S., and Venkataraman, G. S.,** *Charophyta,* Indian Council of Agricultural Research, New Delhi, 1962.
118. **Wood, R. D. and Imahori, K.,** *A Revision of the Characeae,* Cramer, Weinheim, 1974 (Vol. 2), 1975 (Vol. 1).
119. **Hollande, A.,** Classe des Eugléniens (Euglenoidina Bütschli, 1884), in *Traité de Zoologie,* Tome I, Fasc. I, Grassé, P. P., Ed., Masson, Paris, 1952, 238.
120. **Huber-Pestalozzi, G., Englenophyceen,** in *Das Phytoplankton des Süsswassers,* Teil 4, Huber-Pestalozzi, G., Ed., Schweizerbart, Stuttgart, 1955.
121. **Popova, T. G.,** Eglenovye vodorosli, in *Opredelitel' Presnovodnykh Vodoroslej SSSR,* Vyp. 7, Sovetskaya Nauka, Moskva, 1955.
122. **Leedale, G. F.,** *Euglenoid Flagellates,* Prentice-Hall, Englewood Cliffs, N. J., 1967.
123. **Wolken, J. J.,** *Euglena. An Experimental Organism for Biochemical and Biophysical Studies,* 2nd ed., Appleton-Century-Crofts, New York, 1967.
124. **Buetow, D. E., Ed.,** *The Biology of Euglena,* Vols. 1 and 2, Academic Press, New York, 1968.
125. **Leedale, G. F.,** *The Euglenoids,* Oxford University Press, London, 1971.
126. **Michajłow, W.,** *Euglenoidina Parasitic in Copepoda,* Państwowe Wydawnictwo Naukowe, Warszawa, 1972.
127. **Leedale, G. F.,** Euglenophyta, in *CRC Handbook of Microbiology,* Vol. 2, 2nd ed., Laskin, A. I. and Lechevalier, H. A., Eds., CRC Press, Boca Raton, Fla., 1979, 365.
128. **Pascher, A.,** Über Rhizopoden- und Palmellastadien bei Flagellaten (Chrysomonaden), nebst einer Übersicht über die braunen Flagellaten, *Arch. Protistenkd.,* 25, 153, 1912.
129. **Pascher, A.,** Chrysomonadinae, in *Die Süsswasser-Flora Deutschlands, Österreichs und der Schweiz,* Heft 2, Pascher, A., Ed., Fischer, Jena, 1913.
130. **Pascher, A.,** Über Flagellaten und Algen, *Ber. Dtsch. Bot. Ges.,* 32, 136, 1914.
131. **Pascher, A.,** Die braune Algenreihe der Chrysophyceen, *Arch. Protistenkd.,* 52, 489, 1925.
132. **Huber-Pestalozzi, G.,** Chrysophyceen, Farblose Flagellaten, Heterokonten, in *Das Phytoplankton des Süsswassers,* Teil 2, Hälfte 1, Huber-Pestalozzi, G., Ed., Schweizerbart, Stuttgart, 1941.
133. **Starmach, K.,** Chrysophyta I. Chrysophyceae — Złotowiciowce, in *Flora Słodkowodna Polski,* Tom 5, Starmach, K., Ed., Państwowe Wydawnictwo Naukowe, Warszawa, 1968.
134. **Hibberd, D. J.,** The ultrastructure and taxonomy of the Chrysophyceae and Prymnesiophyceae (Haptophyceae): a survey with some new observations on the ultrastructure of the Chrysophyceae, *Bot. J. Linn. Soc.,* 72, 55, 1976.
135. **Grassé, P. P. and Deflandre, G.,** Ordre des Biocoecidea (Bikoecidae S. Kent, 1880), in *Traité de Zoologie,* Tome I, Fasc. I, Grassé, P. P., Ed., Masson, Paris, 1952, 599.
136. **Van Valkenburg, S. D.,** Observations on the fine structure of *Dictyocha fibula* Ehrenberg. I. The skeleton, *J. Phycol.,* 7, 113, 1971.
137. **Van Valkenburg, S. D.,** Observations on the fine structure of *Dictyocha fibula* Ehrenberg. II. The protoplast, *J. Phycol.,* 7, 118, 1971.
138. **Van Valkenburg, S. D. and Norris, R. E.,** The growth and morphology of the silicoflagellate *Dictyocha fibula* Ehrenberg in culture, *J. Phycol.,* 6, 48, 1970.
139. **Gemeinhardt, K., Silicoflagellatae,** in *Dr. L. Rabenhorst's Kryptogamen-Flora von Deutschland, Österreich und der Schweiz,* Aufl. 2, Band 10, Abt. 2, Akademische Verlagsgesellschaft, Leipzig, 1930, 1.
140. **Deflandre, G.,** Classe des Silicoflagellidés (Silicoflagellata Borgert, 1891), in *Traité de Zoologie,* Tome I, Fasc. I, Grassé, P. P., Ed., Masson, Paris, 1952, 425.
141. **Glezer [Gleser], Z. [S.] I.,** Kremnevye zhgutikovye vodorosli (silikoflagellaty), in *Flora Sporovykh Rastenij SSSR,* Tom VII, Nauka, Moskva-Leningrad, 1966.
142. **Hustedt, F.,** Die Kieselalgen Deutschlands, Österreichs und der Schweiz unter Berücksichtigung der übrigen Länder Europas sowie der angrenzenden Meeresgebiete, in *Dr. L. Rabenhorst's Kryptogamen-Flora von Deutschland, Österreich und der Schweiz,* Aufl. 2, Band 7, Akademische Verlagsgesellschaft, Leipzig, 1927-1966.
143. **Fritsch, F. E.,** *The Structure and Reproduction of the Algae,* Vol. 1, University Press, Cambridge, Mass., 1935.
144. **Hendey, N. I.,** An introductory account of the smaller algae of British coastal waters. Part V: Bacillariophyceae (diatoms), *Gr. Brit., Minist. Agric. Fish. Food, Fish. Invest.,* Ser. IV [Part 5], 1964.
145. **Patrick, R . and Reimer, C. W.,** The diatoms of the United States exclusive of Alaska and Hawaii, *Monogr. Acad. Nat. Sci. Philadelphia,* 13, Vol. 1, 1966, Vol. 2, Part 1, 1975.
146. **Siemińska, J.,** Chrysophyta II. Bacillariophyceae — Okrzemki, in *Flora Słodkowodna Polski,* Tom 6, Starmach, K., Ed., Państwowe Wydawnictwo Naukowe, Warszawa, 1964.
147. Werner, D., Ed., *The Biology of Diatoms,* Blackwell, Oxford, 1977.
148. **Loeblich, A. R. III and Loeblich, L. A.,** Division Bacillariophyta, in *CRC Handbook of Microbiology,* Vol. 2, 2nd ed., Laskin, A. I. and Lechevalier, H. A., Eds., CRC Press, Boca Raton, Fla., 1979, 425.

149. **Drebes, G.,** *Marines Phytoplankton. Eine Auswahl der Helgoländer Planktonalgen (Diatomeen, Peridineen),* Thieme, Stuttgart, 1974.
150. **Feldmann, J.,** L'ordre des Scytosiphonales, in *Travaux botaniques dédiés à René Maire, Mém. hors-sér. Soc. Hist. Nat. Afr. Nord,* 2, 103, 1949.
151. **Nakamura, Y.,** A proposal on the classification of the Phaeophyta, in *Contributions to the Systematics of Benthic Marine Algae of the North Pacific,* Abbott, I. A. and Kurogi, M., Eds., Japanese Society of Phycology, Kobe, 1972, 147.
152. **Petrov, Yu. E.,** Sistematicheskoe polozhenie sem. Durvilleaceae i sistematika klassa Cyclosporophyceae (Phaeophyta), *Nov. Sist. Nizsh. Rast. (Bot. Inst. Akad. Nauk SSSR),* 1965, 70, 1965.
153. **Fritsch, F. E.,** *The Structure and Reproduction of the Algae,* Vol. 2, University Press, Cambridge, Mass., 1945.
154. **Papenfuss, G. F.,** Phaeophyta, in *Manual of Phycology,* Smith, G. M., Ed., Chronica Botanica, Waltham, 1951, 119.
155. **Chapman, A. R. O.,** Phaeophyta, in *CRC Handbook of Microbiology,* Vol. 2, 2nd ed., Laskin, A. I. and Lechevalier, H. A., Eds., CRC Press, Boca Raton, Fla., 1979, 401.
156. **Pascher, A., Heterokonten,** in *Dr. L. Rabenhorst's Kryptogamen-Flora von Deutschland, Österreich und der Schweiz,* Aufl. 2, Band II, Akademische Verlagsgesellschaft, Leipzig, 1937-1939.
157. **Starmach, K.,** Chrysophyta III. Xanthophyceae — Różnowiciowe, in *Flora Słodkowodna Polski,* Tom 7, Starmach, K., Ed., Państwowe Wydawnictwo Naukowe, Warszawa-Kraków, 1968.
158. **Silva, P. C.,** Review of the taxonomic history and nomenclature of the yellow-green algae, *Arch. Protistenkd.,* 121, 20, 1979.
159. **Loeblich, A. R. III and Loeblich, L. A.,** Division Xanthophyta, in *CRC Handbook of Microbiology,* Vol. 2, 2nd ed., Laskin, A. I. and Lechevalier, H. A., Eds., CRC Press, Boca Raton, Fla., 1979, 469.
160. **Hollande, A.,** Classe des Chloromonadines (Chloromonadina Klebs, 1892), in *Traité de Zoologie,* Tome I, Fasc. I, Grassé, P. P., Ed., Masson, Paris, 1952, 227.
161. **Loeblich, A. R. III and Loeblich, L. A.,** Division Chloromonadophyta, in *CRC Handbook of Microbiology,* Vol. 2, 2nd ed., Laskin, A. I. and Lechevalier, H. A., Eds., CRC Press, Boca Raton, Fla., 1979, 375.
162. **Hibberd, D. J. and Leedale, G. F.,** Observations on the cytology and ultrastructure of the new algal class, Eustigmatophyceae, *Ann. Bot. (London),* Ser. 2, 36, 49, 1972.
163. **Loeblich, A. R. III and Loeblich, L. A.,** Division Eustigmatophyta, in *CRC Handbook of Microbiology,* Vol. 2, 2nd ed., Laskin, A. I. and Lechevalier, H. A., Eds., CRC Press, Boca Raton, Fla., 1979, 481.
164. **Leadbeater, B. S. C.,** Preliminary observations on differences of scale morphology at various stages in the life cycle of *'Apistonema-Syracosphaera' sensu* von Stosch, *Brit. Phycol. J.,* 5, 57, 1970.
165. **Green, J. C.,** Notes on the flagellar apparatus and taxonomy of *Pavlova mesolychnon* van der Veer, and on the status of *Pavlova* Butcher and related genera within the Haptophyceae, *J. Mar. Biol. Assoc. U.K.,* 56, 595, 1976.
166. **Schiller, J.,** Coccolithinieae, in *Dr. L. Rabenhorst's Kryptogamen-Flora von Deutschland, Osterreich und der Schweiz,* Aufl. 2, Band 10, Abt. 2, Akademische Verlagsgesellschaft, Leipzig, 1930, 89.
167. **Stosch, H. A. von,** Haptophyceae, in *Handbuch der Pflanzenphysiologie,* Band 18, Ruhland, W., Ed., Springer-Verlag, Berlin, 1967, 646.
168. **Parke, M. and Green, J. C.,** Haptophyceae, in Parke, M. and Dixon, P. S., Eds., Check-list of British marine algae — third revision, *J. Mar. Biol. Assoc. U.K.,* 56, 551, 1976.
169. **Loeblich, A. R. III and Loeblich, L. A.,** Division Haptophyta, in *CRC Handbook of Microbiology,* Vol. 2, 2nd ed., Laskin, A. I. and Lechevalier, H. A., Eds., CRC Press, Boca Raton, Fla., 1979, 451.
170. **Schiller, J.,** Dinoflagellatae (Peridineae), in *Dr. L. Rabenhorst's Kryptogamen-Flora von Deutschland, Österreich und der Schweiz,* Aufl. 2, Band 10, Abt. 3, Akademische Verlagsgesellschaft, Leipzig, 1931-1937.
171. **Chatton, É.,** Classe des Dinoflagellés ou Péridiniens (Dinoflagellata Bütschli, 1885, Peridineae Ehrenberg, 1830, Peridiniales Schütt, 1896), in *Traité de Zoologie,* Tome I, Fasc. I, Grassé, P. P., Ed., Masson, Paris, 1952, 309.
172. **Kiselev, I. A.,** Pirofitovye vodorosli, in *Opredelitel' Presnovodnykh Vodoroslej SSSR,* Vyp. 6, Sovetskaya Nauka, Moskva, 1954.
173. **Loeblich, A. R., Jr., and Loeblich, A. R. III,** Index to the genera, subgenera, and sections of the Pyrrophyta, *Stud. Trop. Oceanogr.,* 3, 1966.
174. **Sarjeant, W. A. S.,** *Fossil and Living Dinoflagellates,* Academic Press, New York, 1974.
175. **Cox, E. R.,** Dinoflagellates, in *CRC Handbook of Microbiology,* Vol. 2, 2nd ed., Laskin, A. I. and Lechevalier, H. A., Eds., CRC Press, Boca Raton, Fla., 1979, 489.
176. **Deflandre, G.,** Classe des Ébriédiens (Ebriaceae Lemmermann 1900 emend. Ebriideae Deflandre 1936), in *Traité de Zoologie,* Tome I, Fasc. I, Grassé, P. P., Ed., Masson, Paris, 1952, 407.

177. **Hovasse, R.**, 'Parallobiopsis coutieri' Collin. Morphologie, cytologie, évolution. Affinités des Ellobiopsides, *Bull. Biol. Fr. Belg.*, 60, 409, 1926.
178. **Hovasse, R.**, Contribution à la connaissance biologique des Ellobiopsidae: la sporulation chez *Ellobiopsis fagei* sp. nov., *C. R. Acad. Sci.*, 233, 980, 1951.
179. **Galt, J. H. and Whisler, H. C.**, Differentiation of flagellated spores in *Thalassomyces* ellobiopsid parasite of marine Crustaceae, *Arch. Mikrobiol.*, 71, 295, 1970.
180. **Grassé, P. P.**, Les Ellobiopsidae (Ellobiopsidae Coutière, 1911), in *Traité de Zoologie*, Tome I, Fasc. I, Grassé, P. P., Ed., Masson, Paris, 1952, 1023.
181. **Ris, H. and Kubai, D. F.**, An unusual mitotic mechanism in the parasitic protozoan *Syndinium sp.*, *J. Cell Biol.*, 60, 702, 1974.
182. **Hollande, A.**, Étude comparée de la mitose syndinienne et de celle des Péridiniens libres et des Hypermastigines. Infrastructure et cycle évolutif des Syndinides parasites de Radiolaires, *Protistologica*, 10, 413, 1975.
183. **van der Meer, J. P. and Todd, E. R.**, The life history of *Palmaria palmata* in culture. A new type for the Rhodophyta, *Canad. J. Bot.*, 58, 1250, 1980.
184. **Norris, R. E.**, Prasinophytes, in *Phytoflagellates*, Cox, E. R., Ed., Elsevier/North-Holland, N.Y., 1980, 85.

PHYLOGENIC RELATIONSHIPS AMONG ALGAL PHYLA

Mitsuo Chihara

Phylogenetic relationships among the various algal phyla (divisions) can be deduced from the distribution of certain biochemical and structural characters. To be diagnostically significant at such a high level in the taxonomic hierarchy, a character must show minimal plasticity. Because photosynthesis, despite its complexity (or more likely because of its complexity), is homogeneous throughout the morphologically diverse assemblage of organisms that possess chlorophyll *a,* characters associated with this process are accorded the greatest value in delimiting phyla. These features include the presence of chlorophylls in addition to chlorophyll *a,* the presence of photosynthetically active accessory pigments, the type of storage product, and the structure of the photosynthetic unit (chloroplast). Of next greatest value in delimiting phyla are flagellar characters, including the presence or absence of flagella and the nature of their ornamentation (mastigonemes, scales).

The scheme proposed herein is given twice, first with respect to photosynthetic characters (Figure 1), and second with respect to flagellar morphology (Figure 2). It should be noted that phylum Prochlorophyta (prokaryotes with chlorophycean pigmentation) is not included in the scheme. Also, the presence or absence of a haptonema, a nonflagellar organelle, is taken into consideration along with flagellar characters.

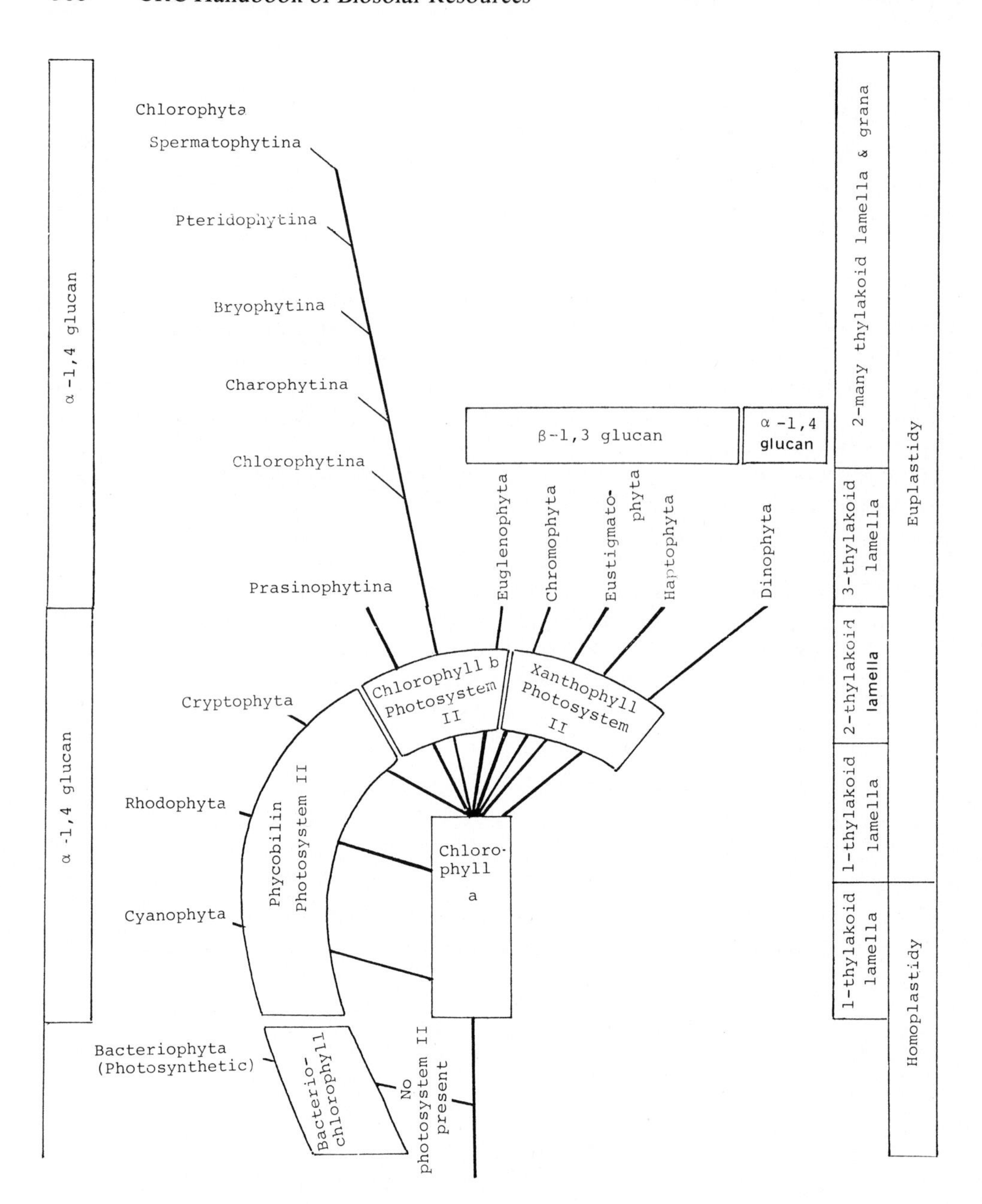

FIGURE 1. The relationships of photosynthetic characteristics of algal phyla. Photosynthetic characteristics: main pigments concerned with photosystem II, storage products, and chloroplast structure.

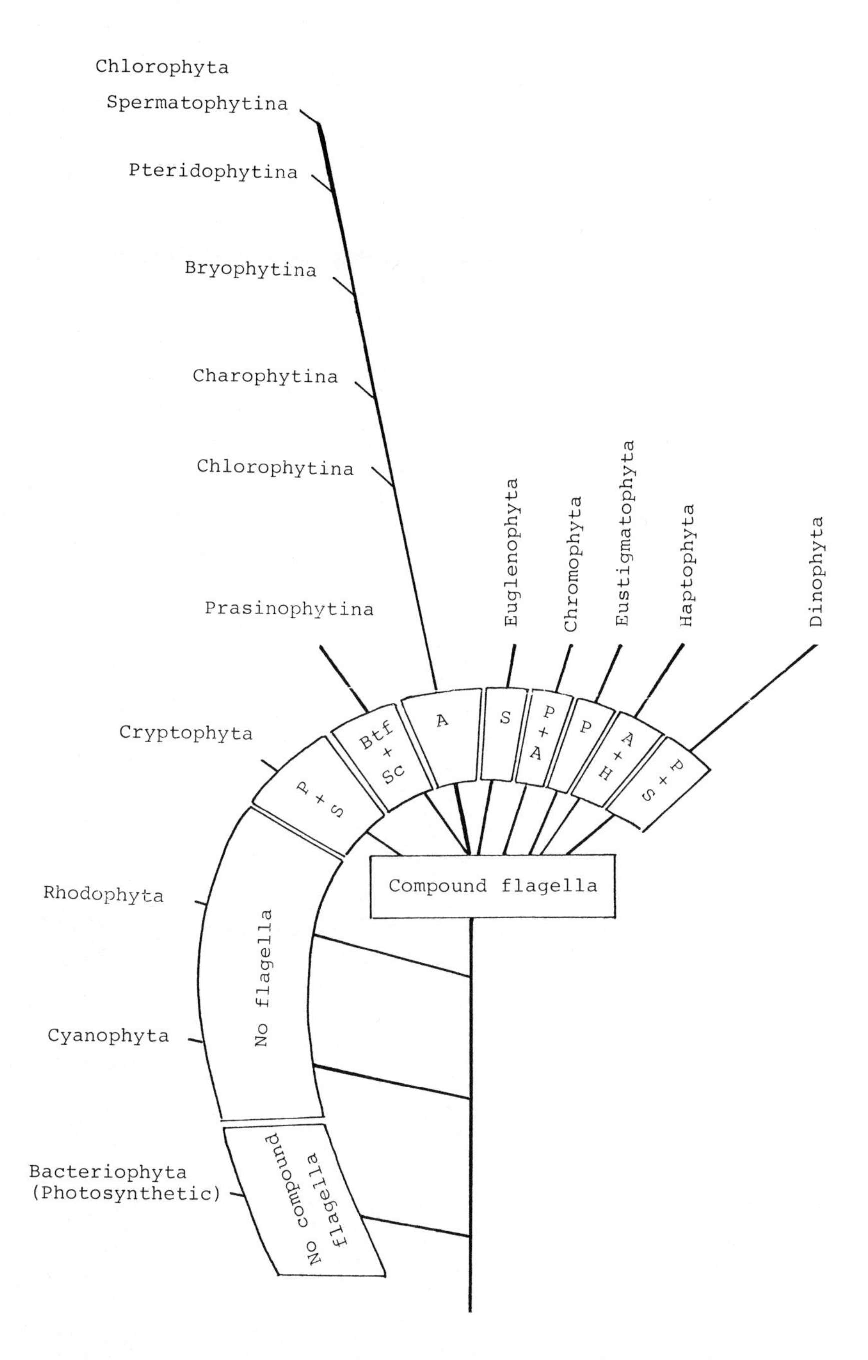

FIGURE 2. The relationships of flagellar morphology to algal phyla. Abbreviations used in figure: A, acronematic flagellum; S, stichonematic flagellum; P, pantonematic flagellum: H, haptonema; Btf + Sc, blunt tip flagellum with scales.

GENERAL CLASSIFICATION AND CHARACTERISTICS OF VASCULAR SEED PLANTS

David A. Young and David S. Seigler

INTRODUCTION

Extant vascular plants (Division Tracheophyta) represent the largest group of land plants with about 250,000 recognized species. Vascular plants can be classified into two major groups: nonseed plants (vascular cryptogams) and seed plants (phanerogams). Vascular cryptogams include ferns (subdivision Filicophytina), whisk ferns (subdivision Psilophytina), club mosses, quillworts and other "lower" vascular plants (subdivision Lycophytina), and horsetails (subdivision Sphenophytina) (Table 1). Extant seed plants consist of gymnosperms and angiosperms (subdivision Spermatophytina). This discussion is concerned primarily with the classification and general characteristics of gymnosperms and angiosperms.

Seed plants share several features in common. Their megaphyllous leaves are borne as lateral appendages on shoots or stems, and the leaves are flat and dorsiventral (i.e., flattened with a definite dorsal [adaxial] and ventral [abaxial] surface). Their vascular cylinder consists of internal xylem and external phloem, and is produced by a vascular cambium that typically forms a more or less continuous ring (absent in monocotyledonous angiosperms). They are heterosporous with separate microsporophylls (stamens) and megasporophylls, and at maturity (after pollination and fertilization) their seeds have an embryo consisting of cotyledons, a hypocotyl, and an epicotyl (plumule).

Angiosperms typically differ from "gymnosperms" (the term gymnosperm is used here in the broadest sense to include the classes Cycadopsida, Ginkgoöpsida, Coniferopsida and Gnetopsida) in several aspects. Angiosperm leaves have both secondary and tertiary venation which is produced by intercalary plate meristems between the midrib and leaf margin. In extant gymnosperms, a similar type of venation pattern is found only in *Gnetum* (Gnetopsida). It is generally considered that angiosperm stamens and carpels primitively are borne upon the same shoot (i.e., their flowers are bisexual) with the carpels uppermost. Unisexual flowers in angiosperms are considered by most workers to be secondarily derived. All extant gymnosperms have unisexual (monoecious or dioecious) reproductive structures (bisexual reproductive structures were, however, present in the extinct order Cycadeodales [= Bennettitales of some authors]).[1]

Another characteristic feature of angiosperms, and from which the name of the group is derived, is the closed carpel. In gymnosperms the ovules are always "naked", in the sense of not being enclosed in an ovary. There are, however, some angiosperms in which the carpel is not completely closed, of which probably the most noteworthy examples are *Degeneria* (Degeneriaceae) and some members of the Winteraceae.[2]

Some of the major differences between angiosperms and gymnosperms are in their reproductive cycles. In angiosperms there is considerable reduction of the gametophytes compared with those of most gymnosperms. For example, the immature male gametophyte (pollen grain) of angiosperms consists of two daughter nuclei and cells. The mature male gametophyte produces only three haploid nuclei. In extant gymnosperms the number of nuclei in the mature male gametophyte varies from four to six. The inner structure of most angiosperm pollen grains also is distinct from that of gymnosperm pollen. The majority of angiosperms have pollen grains that are tectate, with a nexine, columellae, and a tectum. This type of exine structure (i.e., with a middle layer of distinct well-defined columellae) is in sharp contrast to the spongy

Table 1
A GENERAL SYSTEM OF CLASSIFICATION OF EXTANT VASCULAR PLANTS

Division: Tracheophyta (Vascular plants)
  Subdivision: Psilophytina (Whisk ferns)
  Subdivision: Lycophytina (Lycophytes)
  Subdivision: Sphenophytina (Horsetails)
  Subdivision: Filicophytina (Ferns)
    Class: Filicopsida
    Class: Ophioglossopsida
    Class: Marattiopsida
  Subdivision: Spermatophytina (Seed plants)
    Class: Cycadopsida (Cycads)
    Class: Coniferopsida (Conifers)
    Class: Ginkgoöpsida *(Ginkgo)*
    Class: Gnetopsida *(Ephedra, Gnetum, Welwitschia)*
    Class: Angiospermopsida (Flowering plants)
      Subclass: Dicotyledonidae (Dicotyledons)
      Subclass: Monocotyledonidae (Monocotyledons)

(alveolar) or granular exine structure of gymnosperms, and is a major feature distinguishing the exine of angiosperm and gymnosperm pollen.[3-5] It should be mentioned, however, that some extant, presumably primitive angiosperms have pollen that lack columellae (i.e., are atectate).[6]

Similarly, the female gametophyte of angiosperms is reduced to an eight-nucleate (seven-celled) embryo sac. There are no archegonia as in most gymnosperms. Accompanying these reductions has been the evolution of the process of double fertilization, a phenomenon apparently unique to angiosperms. The second sperm of gymnosperms has no apparent function and degenerates. In addition, in angiosperms, unlike gymnosperms, pollination occurs some distance from the micropyle (on the stigma).

Embryo formation also differs in angiosperms. In gymnosperms (except *Gnetum* and *Welwitschia;* Gnetopsida) zygote formation is followed by a period of free nuclear divisions resulting in a multinucleate proembryo. In angiosperms, with the exception of peonies (*Paeonia*—Paeoniaceae),[7] the embryo develops directly from the zygote without a free nuclear stage. Free nuclear embryogeny in *Paeonia* probably is an example of convergent evolution in angiosperms and gymnosperms,[8] although Stebbins[9] considered it a vestigial character reflecting a relationship between angiosperms and certain extinct gymnosperm groups (Glossopteridales and Caytoniales).

Finally, the secondary phloem and xylem of angiosperms differ from those of gymnosperms. The sieve elements of angiosperm phloem consist of sieve tube members and companion cells, whereas gymnospermous phloem usually consists of sieve cells. In some gymnosperms the functional counterparts of companion cells are albuminous cells.[10] The secondary xylem of all gymnosperms, except *Ephedra, Gnetum* and *Welwitschia* (Gnetopsida), is vesselless, whereas vessels are present in the xylem of most angiosperms. The notable exceptions (i.e., vesselless angiosperms) are Amborellaceae, Tetracentraceae, Trochodendraceae, Winteraceae and *Sarcandra* of the Chloranthaceae. The major differences between angiosperms and gymnosperms are summarized in Table 2.

Before discussing the major gymnosperm and angiosperm groups, a comment on classification systems seems warranted. In the present discussion, we have elected to treat the angiosperms and various gymnosperm groups as classes (Angiospermopsida, Cycadopsida, Ginkgoopsida, Coniferopsida, and Gnetopsida) of the division Tracheophyta subdivision Spermatophytina, similar to the system of Bierhorst.[11] This differs

Table 2
COMPARISON OF THE MAJOR FEATURES DISTINGUISHING EXTANT GYMNOSPERMS AND ANGIOSPERMS

Feature	Gymnosperms	Angiosperms
Leaf venation	Primary	Primary, secondary and tertiary
Reproductive structures	Cones unisexual	Flowers primitively bisexual
Ovules	"Naked"	Enclosed in carpel
Gametophytes	Multinucleate, female typically with archegonia	Reduced number of nuclei, no archegonia
Pollen (exine)	Atectate (alveolar)	Mostly tectate
Pollination	At micropyle	On stigma, away from micropyle
Embryo formation	Delayed, with multinucleate proembryo	Direct from zygote, no multinucleate proembryo
Phloem	Sieve cells (sometimes albuminous cells)	Sieve tube members and companion cells
Xylem	Typically vesselless	Typically with vessels

from some authors[8,12,13] who have elevated each of these major groups to divisional status. In our opinion, the latter option seems to be an unwarranted "inflation" of rank.[14]

GYMNOSPERMS

Introduction

As mentioned above, extant gymnosperms are considered here to consist of four classes: Coniferopsida, Cycadopsida, Ginkgoöpsida, and Gnetopsida. In total, there are only a few more than 700 species and about 65 genera of extant gymnosperms (Tables 3 and 4), compared with over 200,000 species of extant angiosperms. The term "gymnosperm" (Gr. *gymnos,* naked, + Gr. *sperma,* seed) refers to the lack of enclosure of the seeds (ovules) in these plants. Taxonomic treatment of extant gymnosperms at higher categories is exceedingly variable.[8,11,15] However, among the gymnosperms there appear to be two major groups (or evolutionary lines), the cyadophytes (Cycadopsida) and the coniferophytes-gnetophytes (Coniferopsida, Ginkgoöpsida, Gnetopsida).[11] In the present treatment we basically have followed the classification system of Bierhorst.[11]

Cycadopsida

The Cycadopsida, or cycads, consists of 10 genera and approximately 100 species (Table 3). Cycads are palmlike plants mainly of tropical and subtropical regions of the world. Only a single species of Cycadopsida, *Zamia floridana* A. DC. (Zamiaceae), is native to the U.S. (Florida). They often have a distinct, cylindrical trunk which bears persistent leaf bases. Their stems are characterized by loosely arranged xylem and abundant parenchyma. The pith and cortex is extensive compared with the xylem and phloem. Their leaves are large, compound (usually pinnate), and fern- or palmlike. All cycads are dioecious and most produce terminal cones (except, for example, the megasporangia of *Cycas* spp.). Ovules are borne singly on leaves, in simple strobili (cones) on a megasporophyll or on a terminal conical axis. Female gametophytes produce archegonia and the sperms are flagellated.

Table 3
FAMILIES, GENERA (NUMBER OF SPECIES) AND GENERAL DISTRIBUTION (NATIVE) OF EXTANT TAXA OF THE CYCADOPSIDA, GINKGOÖPSIDA AND GNETOPSIDA

Class	Order	Family	Genus (number of species)	Distribution
Cycadopsida	Cycadales	Cycadaceae	*Cycas* (20)	Polynesia to Madagascar, north to Japan
		Stangeriaceae	*Stangeria* (1; *S. eriopus* [Kuntze] Nash)	South Africa
		Zamiaceae	*Lepidozamia* (2)	East Australia
			Macrozamia (14)	Temperate Australia
			Encephalartos (30)	Tropical and South Africa
			Dioön (3—5)	Mexico to Central America
			Microcycas (1)	Cuba
			Bowenia (2)	Northern Australia
			Zamia (ca. 40)	Tropical America, West Indies
			Ceratozamia (4)	Mexico
Ginkgoöpsida	Ginkgoales	Ginkgoaceae	*Ginkgo* (1; *G. biloba* L.)	Eastern China (?)
Gnetopsida	Ephedrales	Ephedraceae	*Ephedra* (35—40)	Warm temperate North and South America and Eurasia
	Gnetales	Gnetaceae	*Gnetum* (30—35)	Tropics of Old and New World
	Welwitschiales	Welwitschiaceae	*Welwitschia* (1; *W. mirabilis* Hook. f. = *W. bainesii* [Hook. f.] Carr.)	Southwest Africa

Table 4
FAMILIES, GENERA (NUMBER OF SPECIES) AND GENERAL DISTRIBUTION (NATIVE) OF EXTANT TAXA OF THE CONIFEROPSIDA

Order	Family	Genus (number of species)	Distribution
Coniferales	Pinaceae	*Abies* (50)	North temperate to Central America
		Cathaya (2)	China
		Cedrus (4)	Middle East, Algeria, Himalayas, Cyprus
		Keteleeria (ca. 8)	East Asia, Indochina
		Larix (12)	Eurasia, North America
		Picea (50)	North temperate regions
		Pinus (ca. 100)	Mostly north temperate
		Pseudolarix (2)	China
		Pseudotsuga (7)	Eastern Asia, western North America
		Tsuga (15)	Eastern Asia, North America
	Cephalotaxaceae	*Cephalotaxus* (ca. 7)	Eastern Asia
	Podocarpaceae	*Acmopyle* (3)	New Caledonia, Fiji
		Dacridium (ca. 25)	New Zealand to Indomalaysia, Chile
		Microcachrys (1)	Tasmania
		Microstrobus (2)	Australia, Tasmania
		Phyllocladus (7)	New Zealand, Tasmania to Philippines
	Podocarpaceae	*Podocarpus* (100)	Tropics, north to China and Japan
		Saxegothaea (1)	Argentina
	Araucariaceae	*Agathis* (20)	Indochina, Malaysia to New Zealand
		Araucaria (18)	New Zealand, Australia, New Caledonia, South America
	Cupressaceae		
	Subfamily: Sciadopitoideae	*Sciadopitys* (1)	Japan
	Subfamily: Cupressoideae	*Actinostrobus* (3)	Australia
		Austrocedrus (1)	Chile
		Callitris (16)	Australia
		Calocedrus (3)	North Pacific
		Chamaecyparus (6)	Northern hemisphere
		Cupressus (ca. 20)	Northern hemisphere
		Diselma (1)	Tasmania
		Fitzroya (1)	Chile
		Fokienia (2)	China
		Juniperus (60)	Northern hemisphere
		Libocedrus (9)	South Pacific
	Subfamily: Cupressoideae	*Neocallitropsis* (1)	New Caledonia
		Tetraclinis (1)	North Africa
		Thuja (5)	Northern hemisphere
		Thujopsis (1)	Japan
		Widdringtonia (3)	South Africa
		Athrotaxis (3)	Australia, Tasmania
		Cryptomeria (1)	Japan
		Cunninghamia (3)	China
		Glyptostrobus (1)	China
		Metasequoia (1)	China
		Sequoia (1)	California
		Sequoiadendron (1)	California
		Taiwania (1)	China, Formosa
		Taxodium (2)	North America
Taxales	Taxaceae	*Amentotaxus* (ca. 4)	China
		Austrotaxus (1)	New Caledonia
		Pseudotaxus (1)	China
		Taxus (10)	North temperate
		Torreya (6)	East Asia, Florida, California

Coniferopsida

Pines (*Pinus*) and related conifers (e.g., *Abies* and *Picea*) form extensive forests in various parts of the world. Economically, they are very important as sources of lumber, wood pulp (for the manufacture of paper), and various gums and resins. As treated here, extant conifers consist of two orders, the Coniferales and Taxales (Table 4), with approximately 55 genera and 580 species (the Taxales are sometimes regarded as a separate class, the Taxopsida[8]). Members of the Coniferopsida are all woody perennial plants with growth habits ranging from tall trees, such as *Sequoia sempervirens* of coastal California which attains heights up to nearly 120 m, to spreading trees and shrubs, to completely prostrate forms such as *Juniperus horizontalis*. Unlike cycads, the bulk of the stem of conifers is composed of xylem. The plants may be evergreen (as in *Pinus, Abies,* or *Picea*) or deciduous (as in *Larix* and *Taxodium*). Their leaves vary from scales, needles, linear to broad types, and are simple. All members of the Coniferopsida are either dioecious or monoecious. The female gametophytes of all conifers produce archegonia and may be polyembryonic. In conifers (and the Gnetopsida) the sperms are nonflagellated and are transported by the pollen tube directly to the archegonium. Recently, Eckenwalder[16] has argued rather convincingly for the merger of the Cupressaceae and Taxodiaceae.

Ginkgoöpsida

The Ginkgoöpsida consists of a single species, *Ginkgo biloba* L., the maidenhair tree. Although *Ginkgo* may or may not still be known in the wild, it is a commonly cultivated street tree in many cities of temperate regions of the world. One of the most distinctive features of *Ginkgo* is its fanshaped, deciduous leaves with their openly branched, dichotomous venation pattern. Like cycads, *Ginkgo* is dioecious, producing ovules and microsporangia on different individuals. Ovules are borne in pairs at the ends of short shoots. Microsporangia are clustered into conelike structures. The sperms of *Ginkgo* are flagellated. Fertilization may not occur until after the ovules have dropped from the tree, and embryos are formed during the later stages of development of the seeds. Seeds of *Ginkgo* have a rancid odor (because of the butyric acid in the seed coats) and often only male trees are cultivated. In some taxonomic treatments *Ginkgo* has been included as an order (Ginkgoales) of the Coniferopsida.[11]

Gnetopsida

The three orders, families, and genera comprising the Gnetopsida (Table 3) are indeed quite diverse and whether or not the Gnetopsida is a monophyletic taxon certainly is debatable.[17] Morphologically the three genera are quite distinct, and only *Ephedra* has species indigenous to the U.S. (semiarid regions of the Southwest). *Gnetum* (30 to 35 species) is a genus of tropical trees and vines with large leathery leaves, similar in general appearance to those of many dicotyledonous angiosperms. *Ephedra* (35 to 40 species) is a genus of small, much-branched shrubs with small, inconspicuous, scale-like leaves and more or less jointed stems. The single species of *Welwitschia* consists of a massive, woody, concave disc that produces just two long, strap-shaped leaves. Most of the plant is concealed underground. *Gnetum* and *Welwitschia* have several features which are otherwise known only in the angiosperms. Most notably, they lack archegonia (present in *Ephedra*), have vessels in their secondary xylem (as does *Ephedra*) and the broad reticulate veined leaves of *Gnetum* are very reminiscent of those of many dicotyledonous angiosperms. In the past, these features have led some authors to suggest that the Gnetopsida (especially *Gnetum*) may have been ancestral to the Angiospermopsida.[18-21] However, today such a suggestion is no longer considered very probable.[9,12-14,22]

ANGIOSPERMS

Introduction

The angiosperms, or flowering plants, are by far the largest group of vascular seed plants with over 200,000 species arranged in approximately 324 families (a conservative estimate!). Currently, there are six or seven major (modern) phylogenetic classification systems of the angiosperms. These are the systems of Cronquist,[12] Dahlgren,[23] Hutchinson,[24,25] Melchior and collaborators,[26] Takhtajan,[13] Stebbins,[9] and Thorne.[14] In the discussion that follows we have elected, for several reasons, to follow the system of Thorne.[14] We agree that the flowering plants are best regarded as a class, the Angiospermopsida, rather than a division (the Magnoliophyta or Anthophyta),[8,12] consisting of two subclasses, the Dicotyledonidae and Monocotyledonidae. Within each subclass, naturally related families are grouped together into superorders (e.g., Annoniflorae), which roughly are equivalent to the subclasses of the Cronquist[12] and Takhtajan[13] systems. The various orders of each superorder are further divided into suborders and families into subfamilies. This allows for the recognition of differences between related taxa without unnecessary multiplication of orders and families. Use of super- and subcategories avoids disintegration of natural groups while permitting emphasis on differences. A more detailed explanation of Thorne's system can be found elsewhere.[14] A complete listing of the classification is presented in Table 5, and has been presented pictorially elsewhere.[14,27] Becker[28] has presented a comparison of the major classification systems of flowering plants.

Subclass Dicotyledonidae

The general features of dicotyledonous angiosperms are summarized in Table 6. The Dicotyledonidae consists of 17 superorders, 39 orders, 58 suborders, 274 families, and 339 subfamilies. The phylogenetic relationships of these groups can be visualized as representing a "phylogenetic shrub.[14]" Central in this arrangement are the Annoniflorae, which are regarded as containing the most generalized ("primitive") of all extant angiosperm families. Large numbers of primitive features also are retained in the Hamamelidiflorae, Rosiflorae, and Rutiflorae. However, no one of these groups should be considered the ancestral stock of all other superorders, but each does retain relict species or primitive characteristics. In general, it is unrealistic to attempt to derive the extant members of a family or other higher category from the existing members of apparently closely related groups.

In preparing the following descriptions we have drawn heavily on the following reference works: Bailey,[29] Benson,[30] Cronquist,[12] Hill,[31] Lawrence,[32] Melchior,[26] Porter,[33] Smith,[34] Stebbins,[9] Takhtajan,[13] Thorne,[14,35,36] and Willis.[37]

Superorder Annoniflorae

The Annoniflorae is roughly equivalent to what has been called the Ranales or Ranalean "complex" in older literature. The Annoniflorae consists of 4 orders (Annonales, Nelumbonales, Berberidales, and Nymphaeales), 33 families, and probably more than 12,000 species.[36] A large proportion of the families of the Annoniflorae (e.g., Winteraceae) have retained numerous "primitive" features in their stems, leaves, flowers, and fruits. Some of these features include: numerous, distinct, spirally arranged floral parts; actinomorphic, bisexual flowers with undifferentiated perianth and apocarpous gynoecia; undifferentiated, partially open and styleless carpels; stamens which often are not differentiated into filament and anther; follicular fruits; monocolpate pollen, and vesselless xylem. These are all features generally considered to have been characteristic of the "proangiosperms", and the Annoniflorae certainly is one of the most "primitive" superorders of dicotyledons. In general, the superorder is character-

Table 5
PUTATIVELY PHYLOGENETIC CLASSIFICATION OF THE ANGIOSPERMOPSIDA

Subclass: Dicotyledonidae
- Superorder: Annoniflorae
  - Order: Annonales
    - Suborder: Winterineae
      - Winteraceae
    - Suborder: Illiciineae
      - Illiciaceae
      - Schisandraceae
    - Suborder: Annonineae
      - Magnoliaceae
      - Himantandraceae
      - Eupomatiaceae
      - Degeneriaceae
      - Annonaceae
      - Myristicaceae
      - Canellaceae
    - Suborder: Aristolochiineae
      - Aristolochiaceae
    - Suborder: Laurineae
      - Amborellaceae
      - Austrobaileyaceae
      - Trimeniaceae
      - Chloranthaceae
      - Lactoridaceae
      - Monimiaceae
      - Gomortegaceae
      - Calycanthaceae
      - Lauraceae
      - Hernandiaceae
    - Suborder: Piperineae
      - Saururaceae
      - Piperaceae
  - Order: Nelumbonales
    - Nelumbonaceae
  - Order: Berberidales
    - Suborder: Berberidineae
      - Lardizabalaceae
      - Sargentodoxaceae
      - Menispermaceae
      - Berberidaceae
      - Ranunculaceae
    - Suborder: Papaverineae
      - Papaveraceae (incl. Fumariaceae)
  - Order: Nymphaeales
    - Cabombaceae
    - Nymphaeaceae
    - Ceratophyllaceae
- Superorder: Theiflorae
  - Order: Theales
    - Suborder: Dilleniineae
      - Dilleniaceae
      - Paeoniaceae
    - Suborder: Theineae
      - Actinidiaceae
      - Stachyuraceae
      - Theaceae
      - Icacinaceae
      - Cardiopteridaceae
      - Aquifoliaceae
      - Phellinaceae
      - Oncothecaceae
      - Sphenostemonaceae
      - Parachryphiaceae
      - Marcgraviaceae
      - Caryocaraceae
      - Clethraceae
      - Cyrillaceae
      - Pentaphylacaceae
    - Suborder: Sarraceniineae
      - Sarraceniaceae
    - Suborder: Scytopetalineae
      - Ochnaceae
      - Quiinaceae
      - Scytopetalaceae
      - Sphaerosepalaceae
      - Medusagynaceae
      - Strasburgeriaceae
      - Dioncophyllaceae
    - Suborder: Nepenthineae
      - Nepenthaceae
    - Suborder: Hypericineae
      - Hypericaceae
      - Elatinaceae
    - Suborder: Lecythidineae
      - Lecythidaceae
  - Order: Ericales
    - Suborder: Ericineae
      - Ericaceae
      - Epacridaceae
    - Suborder: Empetrineae
      - Empetraceae
  - Order: Ebenales
    - Suborder: Sapotineae
      - Ebenaceae
      - Sapotaceae
    - Suborder: Styracineae
      - Symplocaceae
      - Lissocarpaceae
      - Styraceae
  - Order: Primulales
    - Suborder: Primulineae
      - Myrsinaceae
      - Theophrastaceae
    - Primulaceae
    - Suborder: Plumbagininеae
      - Plumbaginaceae
  - Order: Polygonales
    - Polygonaceae
- Superorder: Cistiflorae
  - Order: Cistales
    - Suborder: Cistineae
      - Flacourtiaceae
      - Dipentodontaceae
      - Peridiscaceae

Table 5 (continued)
PUTATIVELY PHYLOGENETIC CLASSIFICATION OF THE ANGIOSPERMOPSIDA

- Scyphostegiaceae
- Violaceae
- Cistaceae
- Suborder: Caricineae
  - Turneraceae
  - Malesherbiaceae
  - Passifloraceae
  - Achariaceae
  - Caricaceae
- Suborder: Cucurbitineae
  - Cucurbitaceae
- Suborder: Begoniineae
  - Begoniaceae
  - Datiscaceae
- Suborder: Loasineae
  - Loasaceae
- Order: Salicales
  - Salicaceae
- Order: Tamaricales
  - Tamaricaceae
  - Frankiniaceae
- Order: Capparales
  - Capparaceae
  - Moringaceae
  - Resedaceae
  - Brassicaceae
- Superorder: Malviflorae
  - Order: Malvales
    - Sterculiaceae
    - Huaceae
    - Elaecarpaceae
    - Tiliaceae
    - Dipterocarpaceae
    - Sarcolaenaceae
    - Bombacaceae
    - Bixaceae (incl. Cochlospermaceae)
    - Malvaceae
  - Order: Urticales
    - Ulmaceae
    - Cannabaceae
    - Urticaceae
    - Moraceae
  - Order: Rhamnales
    - Rhamnaceae
    - Elaeagnaceae
  - Order: Euphorbiales
    - Euphorbiaceae
    - Thymelaeaceae
    - Simmondsiaceae
    - Pandaceae
    - Aextoxicaceae
    - Didymelaceae
    - Dichapetalaceae
- Superorder: Santaliflorae
  - Order: Santalales
    - Suborder: Celastrineae
      - Medusandraceae
      - Celastraceae
      - Lophopyxidaceae
      - Stackhousiaceae
    - Suborder: Santalineae
      - Olacaceae
      - Santalaceae
      - Eremolepidaceae
      - Mysodendraceae
      - Loranthaceae
      - Viscaceae
    - Suborder: Balanophorineae
      - Balanophoraceae
      - Cynomoriaceae
- Superorder: Rafflesiiflorae
  - Order: Rafflesiales
    - Rafflesiaceae
    - Hydnoraceae
- Superorder: Geraniiflorae
  - Order: Geraniales
    - Suborder: Linineae
      - Linaceae
      - Ancistrocladaceae
      - Erthroxylaceae
      - Zygophyllaceae
    - Suborder: Geraniaceae
      - Oxalidaceae
      - Geraniaceae
      - Balsaminaceae
      - Tropaeolaceae
    - Suborder: Limnanthineae
      - Limnanthaceae
    - Suborder: Polygalineae
      - Malpighiaceae
      - Polygalaceae
      - Krameriaceae
      - Trigoniaceae
      - Vochysiaceae
- Superorder: Chenopodiiflorae
  - Order: Chenopodiales
    - Suborder: Chenopodiineae
      - Phytolaccaceae
      - Nyctaginaceae
      - Aizoaceae
      - Chenopodiaceae
      - Halophytaceae
      - Amaranthaceae
    - Suborder: Portulacineae
      - Portulacaceae
      - Basellaceae
      - Cactaceae
      - Didiereaceae
    - Suborder: Caryophyllineae
      - Molluginaceae
      - Caryophyllaceae
- Superorder: Hamamelidiflorae
  - Order: Hamamelidales
    - Suborder: Trochodendrineae

Table 5 (continued)
PUTATIVELY PHYLOGENETIC CLASSIFICATION OF THE ANGIOSPERMOPSIDA

- Trochodendraceae
- Tetracentraceae
- Eupteleaceae
- Cercidiphyllaceae
- Suborder: Eucommineae
  - Eucommiaceae
- Suborder: Hamamelidineae
  - Hamamelidaceae
  - Platanaceae
- Order: Casuarinales
  - Casuarinaceae
- Order: Fagales
  - Fagaceae
  - Betulaceae
- Superorder: Rutiflorae
  - Order: Rutales
    - Suborder: Rutineae
      - Rutaceae
      - Coriariaceae
      - Simaroubaceae
      - Meliaceae
      - Burseraceae
      - Anacardiaceae (incl. Julianiaceae)
    - Suborder: Fabineae
      - Fabaceae
      - Connaraceae
    - Suborder: Sapindineae
      - Sapindaceae
      - Surianaceae
      - Bataceae
      - Gyrostemonaceae
      - Sabiaceae
      - Melianthaceae
      - Akaniaceae
      - Aceraceae
      - Hippocastanaceae
      - Bretschneideraceae
    - Suborder: Juglandineae
      - Rhoiptelaceaceae
      - Juglandaceae
    - Suborder: Myricineae
      - Myricaceae
    - Suborder: Leitneriineae
      - Leitneriaceae
- Superorder: Rosiflorae
  - Order: Rosales
    - Suborder: Rosineae
      - Rosaceae
      - Chrysobalanaceae
      - Crossosomataceae
    - Suborder: Saxifragineae
      - Crassulaceae
      - Cephalotaceae
      - Saxifragraceae
      - Stylidiaceae
      - Droseraceae
      - Greyiaceae
      - Podostemaceae
      - Diapensiaceae
    - Suborder: Cunoniineae
      - Cunoniaceae
      - Davidsoniaceae
      - Brunelliaceae
      - Eucryphiaceae
      - Staphyleaceae
      - Corynocarpaceae
  - Order: Pittosporales
    - Suborder: Buxineae
      - Buxaceae
    - Suborder: Daphniphyllineae
      - Daphniphyllaceae
      - Balanopaceae
    - Suborder: Pittosporineae
      - Pittosporaceae
      - Byblidaceae
      - Tremandraceae
    - Suborder: Brunineae
      - Roridulaceae
      - Bruniaceae
      - Geissolomataceae
      - Grubbiaceae
      - Myrothamnaceae
      - Hydrostachyaceae
  - Order: Proteales
    - Proteaceae
- Superorder: Myrtiflorae
  - Order: Myrtales
    - Lythraceae
    - Trapaceae
    - Combretaceae
    - Oliniaceae
    - Penaeaceae
    - Myrtaceae
    - Melastomataceae
    - Onagraceae
- Superorder: Gentianiflorae
  - Order: Oleales
    - Salvadoraceae
    - Oleaceae
    - Barbeyaceae
  - Order: Gentianales
    - Loganiaceae
    - Buddlejaceae
    - Rubiaceae
    - Apocynaceae
    - Asclepiadaceae
    - Gentianaceae
    - Menyanthaceae
  - Order: Bignoniales
    - Bignoniaceae
    - Pedaliaceae
    - Martyniaceae
    - Myoporaceae
    - Schrophulariaceae

Table 5 (continued)
PUTATIVELY PHYLOGENETIC CLASSIFICATION OF THE ANGIOSPERMOPSIDA

Plantaginaceae
Lentibulariaceae
Acanthaceae
Gesneriaceae
Superorder: Solaniflorae
  Order: Solanales
    Suborder: Solanineae
      Solanaceae
      Convolvulaceae
    Suborder: Polemoniineae
      Polemoniaceae
    Suborder: Fouquieriineae
      Fouquieriaceae
  Order: Campanulales
    Pentaphragmataceae
    Campanulaceae
    Goodeniaceae
Superorder: Corniflorae
  Order: Cornales
    Suborder: Rhizophorineae
      Rhizophoraceae
    Suborder: Vitineae
      Vitaceae
    Suborder: Haloragineae
      Haloragaceae
      Gunneraceae
      Hippuridaceae
    Suborder: Cornineae
      Nyssaceae
      Cornaceae
      Alangiaceae
      Garryaceae
    Suborder: Araliineae
      Araliaceae
      Apiaceae
  Order: Dipsacales
    Caprifoliaceae
    Adoxaceae
    Valerianaceae
    Dipsacaceae
    Morinaceae
    Calyceraceae
Superorder: Lamiiflorae
  Order: Lamiales
    Suborder: Boraginineae
      Hydrophyllaceae
      Boraginaceae
      Lennoaceae
      Hoplestigmataceae
    Suborder: Lamiineae
      Verbenaceae
      Callitrichaceae
      Lamiaceae
Superorder: Asteriflorae
  Order: Asterales
    Asteraceae
      Subfamily: Cichorioideae
        Tribe: Mutisieae
          Cichorieae
          Vernonieae
          Cardueae
          Arctoteae
      Subfamily: Asteroideae
        Tribe: Eupatorieae
          Heliantheae
          Astereae
          Inuleae
          Ambrosieae
          Anthemideae
          Senecioneae
          Calenduleae
Subclass: Monocotyledonidae
  Superorder: Liliiflorae
    Order: Liliales
      Suborder: Liliineae
        Liliaceae
        Stemonaceae
        Dioscoreaceae
        Trichopodaceae
        Taccaceae
      Suborder: Iridineae
        Iridaceae
        Burmanniaceae
      Suborder: Orchidineae
        Orchidaceae
  Superorder: Alismatiflorae
    Order: Alismatales
      Butomaceae
      Alismataceae
      Hydrocharitaceae
    Order: Zosterales
      Suborder: Aponogetonineae
        Aponogetonaceae
      Suborder: Potamogetonineae
        Jucaginaceae
        Potamogetonaceae
        Posidoniaceae
        Zannichelliaceae
      Suborder: Zosterineae
        Zosteraceae
    Order: Najadales
      Najadaceae
    Order: Triuridales
      Triuridaceae
  Superorder: Ariflorae
    Order: Arales
      Suborder: Arineae
        Araceae
        Lemnaceae
      Suborder: Typhineae
        Typhaceae
  Superorder: Areciflorae
    Order: Arecales
      Arecaceae

Table 5 (continued)
PUTATIVELY PHYLOGENETIC CLASSIFICATION OF THE ANGIOSPERMOPSIDA

Order: Cyclanthales
  Cyclanthaceae
Order: Pandanales
  Pandanaceae
Superorder: Commeliniflorae
Order: Commelinales
  Suborder: Bromeliineae
    Bromeliaceae
    Rapateaceae
    Xyridaceae
  Suborder: Pontederiineae
    Pontederiaceae
    Philydraceae
  Suborder: Juncineae
    Juncaceae
    Cyperaceae
  Suborder: Commelinineae
    Commelinaceae
    Mayacaceae
  Suborder: Eriocaulineae
    Eriocaulaceae
  Suborder: Flagellariineae
    Flagellariaceae
    Restionaceae
    Ecdeiocoleaceae
    Centrolepidaceae
  Suborder: Poineae
    Poaceae
Order: Zingiberales
  Musaceae
  Strelitziaceae
  Heliconiaceae
  Lowiaceae
  Zingiberaceae
  Costaceae
  Cannaceae
  Marantaceae

Adapted and modified from Thorne, R. F., in *Evolutionary Biology*, Vol. 9, Hecht, M. K., Steere, W. C., and Wallace, B., Eds., Plenum Press, New York, 1976, 35, and personal communication. With permission.

ized by flowers which are often apocarpous, always polypetalous or apetalous, and frequently trimerous; stamens which are usually numerous and centripetal in development; and pollen which is binucleate and often monosulcate. Members of the Annonales are characterized by the presence of spherical secretory cells containing essential or ethereal oils in their parenchymatous tissues. Many members of the Annonales, Berberidales, and Nelumbonales are characterized by the presence of benzylisoquinoline alkaloids.[38] Some of the more familiar families of the Annoniflorae include the Magnoliaceae, Lauraceae, Berberidaceae, Ranunculaceae, Papaveraceae, and Nymphaeaceae. Economically important members of the Annoniflorae include *Myristica fragans* Houtl. (Myristicaceae; nutmeg and mace), *Piper nigrum* L. (Piperaceae; pepper), *Cinnamonum camphora* and *C. zeylanicum* (Lauraceae; camphor and cinnamon), *Persea americana* (Lauraceae; avocado), *Laurus nobilis* (Lauraceae; bay leaves), and *Papaver somniferum* (Papaveraceae; opium poppy). Several species of Magnoliaceae, Berberidaceae and Ranunculaceae are cultivated as ornamentals.

Superorder Theiflorae

The Theiflorae consists of 5 orders (Theales, Ericales, Ebenales, Primulales, and Polygonales), 42 families, and approximately 13,000 species. The superorder is a large diverse group primarily tropical in distribution. Because of this diversity, the Theiflorae is morphologically difficult to characterize as a whole. Generally, members of the Theiflorae tend to be woody plants with unspecialized xylem (although there is a trend toward herbaceousness, particularly in the Primulales and Polygonales) and usually simple, estipulate leaves. Their flowers are usually biseriate and pentamerous with a trend towards sympetaly. The stamens are as many as the petals or more and centrifugal in development. The calyx typically is imbricated in bud. The gynoecium typically is syncarpous with two too many carpels and the ovary is generally superior. Placen-

Table 6
MAJOR DIFFERENCES BETWEEN DICOTYLEDONS AND MONOCOTYLEDONS

Feature	Dicotyledons	Monocotyledons
Cotyledons	Two (rarely one)	One (embryo sometimes undifferentiated)
Floral parts	Typically 4- or 5-merous	Typically 3-merous
Pollen grains	Usually tricolpate (or a derived type; rarely monocolpate)	Usually monocolpate
Leaf venation	Typically netlike or reticulate (some exceptions)	Typically parallel (some exceptions)
Primary vascular bundles in stem	Arranged in a ring (i.e., eustele)	Scattered (i.e., atactostele)
True secondary growth from vascular cambium	Usually present	Absent
Mature root system	Either primary (tap root) or adventitious, or both	Totally adventitious

tation is often axile and the ovules are usually bitegmic and crassinucellate. Their seeds have copious to little or no endosperm. The Theiflorae appear to be closely allied with the Annoniflorae and this relationship is readily apparent in the apocarpous gynoecium of the Dilleniineae. The Primulales and Polygonales generally have been considered closely related to the Chenopodiiflorae (Centrospermae);[12] however, both orders lack many of the specialized features, other than free-central placentation (e.g., P-type sieve tube plastids[39] and betalain pigments), of the Chenopodiiflorae which excludes their placement in the Chenopodiiflorae.[40] Common families include the Theaceae, Aquifoliaceae, Sarraceniaceae, Ericaceae, Primulaceae, and Polygonaceae.

Superorder Cistiflorae

The Cistiflorae consists of 4 orders (Cistales, Salicales, Tamaricales, and Capparales), 22 families, and approximately 8500 species. The Cistiflorae basically is a tropical group of families related to the Theiflorae. They share in common with the Theiflorae features such as bisexual, actinomorphic, pentamerous, polypetalous flowers with imbricate parts. Stamens often are numerous and centrifugal in development. However, members of the Cistiflorae commonly are more herbaceous (often lianous) and have stipulate and often palmately veined or lobed leaves. Placentation is often parietal, and the fruits capsular or baccate. Within the Cistiflorae, the Capparales are distinctive in possessing secretory cells (myrosin cells) containing myrosinase and glucosinolates (mustard oil glucosides). Common families of the Cistiflorae include the Violaceae, Passifloraceae, Cucurbitaceae, Salicaceae, and Brassicaceae (= Cruciferae). The Brassicaceae and Cucurbitaceae are two economically very important families. Varieties of *Brassica oleracea* L. yield cabbage, kale, broccoli, cauliflower and brussels sprouts. *Raphanus sativus* L. is the radish; *Nasturtium officinale* R. Br. is watercress. From the Cucurbitaceae we get watermelon (*Citrullus vulgaris* Schrad.) and cucumber (*Cucumis sativus* L.), as well as pumpkin and various squashes from *Cucurbita* species.

Superorder Malviflorae

The Malviflorae consists of 4 orders (Malvales, Urticales, Rhamnales, and Euphorbiales), 22 families, and approximately 11,750 species. The Malviflorae are a diverse group of primarily tropical woody plants, which in general are characterized by their actinomorphic, pentamerous, polypetalous to sympetalous, more or less showy flowers. Their stamens often are numerous and connate, and centrifugal in development. The gynoecium is syncarpous with axile placentation, and the fruit typically is a capsule. Leaves generally are simple, alternate, palmately veined and often with stellate pubescence. Mucilage cells, canals or receptacles and cystoliths often are present. There are strong tendencies in members of the superorder toward sympetaly or apetaly, unisexuality, inferior ovaries, and herbaceousness. Most authors[9,12,13] include the Urticales in the Hamamelidiflorae, although as early as 1915 Bessey[41] recognized the close relationship of the Urticales with the Malvales, which recently have been discussed in detail by Thorne[42] and Berg.[43] Common families of the Malviflorae include the Malvaceae, Ulmaceae, Urticaceae, Moraceae, and Euphorbiaceae. Economically important members include *Gossypium* spp. (cotton, Malvaceae), *Theobroma cacao* L. (cocoa, Sterculiaceae), *Cola nitida* (Vent.) A. Chev. (cola, Sterculiaceae), *Cannabis sativa* L. (hemp, Cannabaceae), *Humulus lupulus* L. (hops, Cannabaceae), *Hevea brasiliensis* (Wild. ex A. Juss.) Müll.-Arg. (rubber, Euphorbiaceae), and *Ricinus communis* L. (castor bean, Euphorbiaceae).

Superorder Santaliflorae

The Santaliflorae consists of a single order (Santalales), 12 families, and approximately 2800 species. Because of its great evolutionary depth, the Santaliflorae is difficult to characterize as a whole. Basically, members of the superorder are tropical woody plants, many of which have become specialized (i.e., reduced in habit) to semi- or total parasites. Their actinomorphic flowers generally are much reduced in size and number of parts. Placentation is usually axile but tending toward free-central and basal. Their seeds are mostly with endosperm. Larger families include the Celastraceae, Olacaceae, Loranthaceae, and Viscaceae.

Superorder Rafflesiiflorae

The Rafflesiiflorae is a small group consisting of a single order (Rafflesiales) and only two families (Rafflesiaceae and Hydnoraceae) with approximately 70 species. All of the species are achlorophyllous root or stem parasites with extremely reduced vegetative bodies. The phylogenetic relationships of the Rafflesiiflorae are still unknown and the group is in need of thorough investigation.

Superorder Geraniiflorae

The Geraniiflorae, with a single order (Geraniales), consists of 14 families and approximately 5000 species. Members of the Geraniiflorae mostly are herbaceous plants (more or less woody in the Linineae) with bisexual, pentamerous, mostly actinomorphic flowers (tending toward zygomorphic in the Geraniineae and Limnanthineae, and strongly zygomorphic in many Polygalineae). Their stamens are often numerous and connate. The gynoecium is syncarpous and the ovary superior. Placentation is axile and their seeds have endosperm. Dehiscence of the fruit often is explosive. Common families include the Linaceae, Oxalidaceae, Geraniaceae, and Balsaminaceae. *Larrea divaricata* (creosote bush, Zygophyllaceae) is one of the most common plants of the deserts of North America. *Linum usitatissimum* L. (Linaceae) is the flax plant of commerce and *Pelargonium* spp. (Geraniaceae) is the florists' geranium.

Superorder Chenopodiiflorae

The Chenopodiiflorae (= Centrospermae of older literature) consists of a single order (Chenopodiales), 12 families, and approximately 6700 species. Members of the Chenopodiiflorae are characterized by flowers with a basically uniseriate perianth and often numerous, centrifugal stamens. Placentation is mainly free-central or basal, and the ovules are bitegmic and often campylotropous or amphitropous. Their seeds have a curved or coiled embryo and abundant perisperm but little or no endosperm. Pollen is trinucleate and tricolpate to polyporate. Anthocyanin pigments are replaced in all families, except the Caryophyllaceae and Molluginaceae, by betalain pigments in the flowering plants, (betalain pigments are known only from the Chenopodiiflorae). The P-type sieve tube plastids of members of the Chenopodiifloae accumulate protein as a single storage product or in addition to starch.[39] Many species have anomalous secondary thickenings (as in beets, *Beta vulgaris* L. [Chenopodiaceae]) produced by successive cambia with internal phloem. There is a strong tendency in many taxa toward succulence in leaves and stems (e.g., Cactaceae and Aizoaceae) and toward growing in saline habitats (e.g., Amaranthaceae and Chenopodiaceae). Several families (e.g., Bataceae, Gyrostemonaceae, and Theligonaceae) traditionally included in the Chenopodiiflorae recently have been removed because they lack many of the specialized features (such as betalain pigments and P-type sieve tube plastids) characteristic of the superorder.[44-46] Some economically important species include *Beta vulgaris* (sugar beets, common beets; Chenopodiaceae) and *Spinacia oleracea* L. (spinach, Chenopodiaceae).

Superorder Hamamelidiflorae

The Hamamelidiflorae is one of the smaller superorders of dicotyledons consisting of 3 orders (Hamamelidales, Casuarinales, and Fagales), 10 families, and approximately 1100 species. Some of the most common genera of the north temperate deciduous forests are members of this group: e.g., *Quercus* (oak, Fagaceae), *Fagus* (beech, Fagaceae), *Alnus* (alder, Betulaceae), *Betula* (birch, Betulaceae), and *Platanus* (sycamore, Platanaceae). The Hamamelidiflorae are all woody plants with simple, alternate, mostly deciduous leaves; they tend to be characterized by their highly reduced flowers. The perianth is often poorly developed or absent and the flowers typically are unisexual. In many of the families the flowers of one or both sexes are aggregated into dense catkinlike inflorescences (aments) and these families also tend to have unilocular, indehiscent, single-seeded fruits. There are strong tendencies toward wind dispersal of pollen and diaspores in the Hamamelidiflorae, and extreme floral reduction in the group appears to be associated with the shift from insect pollination to wind pollination. The Hamamelidiflorae includes the core families of the "Amentiferae" (e.g., Casuarinaceae, Fagaceae, and Betulaceae); however, the Urticales, Juglandales, Myricales, and Leitneriales are here removed and the Urticales are assigned to the Malviflorae and the remaining three orders to the Rutiflorae.[42,47]

Superorder Rutiflorae

The superorder Rutiflorae consists of a single order (Rutales), 22 families, and approximately 20,000 species, with over half of the species belonging to a single family, the Fabaceae (= Leguminosae; 12,000 to 13,000 species). Members of the Rutiflorae typically are tropical or subtropical woody plants with pinnately compound, alternate, estipulate leaves. Their flowers mostly are polypetalous, actinomorphic, and with imbricate perianth parts. There usually is a conspicuous disc and 8 to 10 stamens. The gynoecium typically is syncarpous and 3- to 5-carpelled with a superior ovary. There are strong tendencies toward anemophily, apetaly, unisexuality, and zygomorphic flowers. The number of locules and ovules in the ovary often are reduced to one. Their seeds often have no endosperm and a fairly large embryo. Representative families include the Rutaceae, Anacardiaceae, Aceraceae, Juglandaceae, and Fabaceae. Most authors[9,12,13] place the Fabaceae near the Rosales. However, Thorne and Dahlgren[56] are now convinced that the Fabaceae have their closest relationships with the families of the Rutiflorae. The Fabaceae is one of the most economically important of all families of angiosperms, yielding such crops as broad beans, clover, alfalfa, peas, common beans, lima beans, soybeans, lentils, and mung beans. *Citrus* species (Rutaceae) yield citrus fruits. Cashews, mangos, and pistachio nuts are obtained from members of the Anacardiaceae. *Swietenia mahogani* (L.) Jacq. (Meliaceae) is the mahogany tree.

Superorder Rosiflorae

The Rosiflorae consists of 3 orders (Rosales, Pittosporales, and Proteales), 30 families, and approximately 8500 species. Members of the Rosiflorae are extremely diverse in terms of habit, foliage, and anatomy, and have retained many "primitive" features similar to members of the Annoniflorae and Hamamelidiflorae. The group is difficult to characterize morphologically. However, in general, members of the Rosiflorae primitively are woody plants with mostly actinomorphic (sometimes zygomorphic), polypetalous or apetalous flowers. The perianth is often perigynous and imbricate in bud. Stamens are often numerous and centripetal in development. The gynoecium is sometimes apocarpous and the fruits follicular. When syncarpous the styles are usually distinct. These are strong tendencies toward the herbaceous and annual habit, zygomorphic flowers with fused and reduced numbers of parts, imperfect and epigynous flowers, and indehiscent one-seeded fruits. Representative families include the Rosaceae, Crassulaceae, Saxifragraceae *(sensu lato)*, and Droseraceae. A number of edible fruits, such as the strawberry, plum, cherry, apple, peach, pear, blackberry, and raspberry are obtained from members of the Rosaceae.

Superorder Myrtiflorae

The Myrtiflorae, with a single order (Myrtales), consists of 8 families and approximately 8000 species. Members of the Myrtiflorae most commonly are tropical woody plants (members of several families are mangrove taxa) with simple, entire, opposite, estipulate leaves. Their flowers generally are actinomorphic, bisexual, polypetalous, and often tetramerous, with the calyx tube perigynous or adnate to the ovary (ovary perigynous to epigynous). Stamens typically are numerous and centripetal in development. The gynoecium is syncarpous with axile or sometimes apical placentation. Seeds have a well-developed embryo and little or no endosperm. A characteristic feature of the superorder is the presence of internal phloem. Representative families include the Lythraceae, Myrtaceae, and Onagraceae. The Rhizophoraceae often are included in the Myrtales but have been removed by some authors[12,14] primarily because they lack internal phloem (see Corniflorae).

Superorder Gentianiflorae

The Gentianiflorae consists of 3 orders (Oleales, Gentianales, and Bignoniales), 19 families, and nearly 22,000 species. Members of the Gentianiflorae are characterized by their mostly bisexual, biseriate flowers with a sympetalous corolla. Their stamens are epipetalous and generally alternate with the corolla lobes. The gynoecium is syncarpous and mostly bicarpellate with a superior ovary. Ovules are unitegmic, and their seeds have a straight embryo embedded in fleshy endosperm and are often winged or comose. Their leaves are mostly opposite, simple, and entire. Internal phloem is often present. Major families are the Oleaceae, Apocynaceae, Asclepiadaceae, Bignoniaceae, Rubiaceae, Scrophulariaceae, Acanthaceae, and Gesneriaceae. *Coffea arabica* L. (Rubiaceae) is the source of coffee, *Olea europaea* L. (Oleaceae) is the olive, and *Sesamum indicum* L. (Pedaliaceae) is the sesame.

Superorder Solaniflorae

The Solaniflorae consists of 2 orders (Solanales and Campanulales), 7 families, and approximately 6500 species. Members of the Solaniflorae in general are characterized by their mostly actinomorphic, bisexual, pentamerous, hypogynous, sympetalous, showy flowers. Their stamens typically number five, are epipetalous and alternate with the corolla lobes. The gynoecium is syncarpous, often bicarpellate, and the ovary typically superior with axile placentation. The fruit typically is a capsule or baccate. The plants basically are woody with simple, alternate, estipulate leaves. Members of the Solanales are characterized by internal phloem. Representative families include the Solanaceae, Convolvulaceae, Polemoniaceae, and Campanulaceae. Economically important members of the superorder include *Capsicum* spp. (peppers; Solanaceae), *Lycopersicon esculentum* Mill. (tomato; Solanaceae); *Nicotiana tabacum* L. (tobacco; Solanaceae); *Solanum tuberosum* L. (potato; Solanaceae) and *Ipomoea batatas* (L.) Poir. (sweet potato; Convolvulaceae).

Superorder Corniflorae

The Corniflorae consists of 2 orders (Cornales and Dipsacales), 17 families, and approximately 4700 species. In general, members of the Corniflorae are woody plants with unspecialized xylem and estipulate leaves. Their flowers are mostly actinomorphic, bisexual, polypetalous, biseriate, and pentamerous (to tetramerous). The calyx tube often is adnate to the ovary. Flowers often are aggregated into umbels or dense heads. The stamens usually are equal in number to the number of petals and an intrastaminal disc often is present. The gynoecium is syncarpous with several to two carpels and the ovary inferior. Ovules are usually unitegmic and solitary in each locule of the ovary. Their seeds usually have a rudimentary embryo and abundant endosperm. Many members synthesize iridoid compounds.[48] Representative families include the Cornaceae, Apiaceae (= Umbelliferae), Caprifoliaceae, and Vitaceae. Economically important foodstuffs and spices such as dill, celery, caraway, coriander, cumin, carrot, parsnip, and anise are all obtained from members of the Apiaceae. *Vitis* spp. (Vitaceae) yield wine grapes.

Superorder Lamiiflorae

The Lamiiflorae, with a single order (Lamiales), consists of 7 families and approximately 8100 species. Members of the Lamiiflorae generally are characterized by their mostly bisexual, pentamerous flowers with sympetalous, tubular, actinomorphic, or more commonly bilabiate corollas. Their inflorescences are basically cymose. Stamens usually number four and are didynamous and epipetalous. The gynoecium is syncarpous, typically bicarpellate (often tetraloculate by false partitions), and the ovary superior. Their fruits are capsules, drupes (with four one-seeded pyrenes), or clusters of four one-seeded nutlets. Major families are the Hydrophyllaceae, Boraginaceae, Verbenaceae, and Lamiaceae (= Labiatae). Several species of Lamiaceae are of considerable economic importance as a source of aromatic oils used to flavor foods.

Superorder Asteriflorae

The Asteriflorae consists of a single family, the Asteraceae (= Compositae), but is the second largest family of flowering plants with nearly 20,000 species of subcosmopolitan distribution, and accounts for approximately 10% of all species of flowering plants. In their vegetative structures, composites are exceedingly variable, ranging from tiny annual herbs to shrubs and timber trees. Most composites readily are recognized by their characteristic inflorescence, the involucrate capitulum or head. Each head is composed of one to nearly 1000 individual flowers borne upon a conical or flattened receptacle. The five stamens usually are syngenesious. The ovary is inferior and contains a single basal ovule. The fruit is a cypsela usually with a pappus. Infrafamilial relationship of the tribes of the Asteraceae recently have been reviewed by several authors.[49-51] The traditional concept of placing the Cichorieae in one subfamily

(Cichorioideae, with all flowers of a head ligulate and plants with milky sap) and the remaining 12 tribes in another (Asteroideae) has been questioned by Carlquist,[50] who included the Mutisieae, Vernonieae, Cardueae, Arctoteae, and Eupatorieae in the Cichorioideae. The Heliantheae, Astereae, Inuleae, Anthemideae, Senecioneae, and Calenduleae were included in the Asteroideae. Wagenitz[51] has suggested that the Asteraceae consists of three groups: (1) the Cichorioideae, (2) the Vernonieae, Liabeae, Mutisieae, Cardueae, Echinopeae, and Arctotideae, and (3) the Eupatorieae, Heliantheae, Helenieae, Senecioneae, Calenduleae, Astereae, and Inuleae, but he did not give formal recognition to the latter two groups. Cronquist[49] regarded the classical classification as best expressing the pattern of diversity within the family, but suggested that perhaps three subfamilies could be recognized (similar to the system of Wagenitz[51]). Economically important plants in the Asteraceae include lettuce *(Lactuca),* chicory *(Cichorium),* artichoke *(Cynara),* sunflower *(Helianthus),* and ragweed *(Ambrosia).*

Subclass Monocotyledonidae

The Monocotyledonidae consists of 5 superorders, 11 orders, 15 suborders, 47 families, and 93 subfamilies. The general features of monocotyldeonous angiosperms are summarized in Table 6. Monocotyledons do not possess any single characteristic that is absent from all dicotyledons, and do show close affinities with several families of the Annoniflorae. The monocotyledons are a very diverse group, so diverse that it is very difficult to find connecting links between the five superorders. Most taxonomists[9,12-14,25] agree that the monocotyledons probably have been derived from some primitive dicotyledonous group, probably some time in the early cretaceous. The probable ancestral group of dicotyledons from which the monocotyledons have been derived has been much debated, and these theories recently have been reviewed by Huber.[52] He suggested that the differentiation of angiosperms into non-annonalean dicotyledons and annonalean dicotyledons (including the monocotyledons) took place prior to the differentiation of annonalean dicotyledons and monocotyledons! There also has been debate on the most primitive group of monocotyledons. Some authorities suggest that the Alismatiflorae are the most primitive monocotyledons,[12,13,24] whereas others argue in favor of the Liliiflorae.[14,53-55] It does appear that no single order of monocotyledons combines in itself all of the most primitive features of the group. In the present treatment the Liliiflorae are considered as the most "primitive" superorder of monocotyledons.

Superorder Liliiflorae

The Liliiflorae consists of a single order (Liliales), 8 families, and over 29,000 species, with over 20,000 of these members of probably the largest family of flowering plants, the Orchidaceae. Because of its great evolutionary depth it is difficult to characterize the group as a whole. Basically, members of the Liliiflorae have a biseriate perianth with both whorls usually petaloid. The ovary is syncarpous and usually tricarpellate. Stamens vary in number from 6 to 3 to 1, and pollen is binucleate. Their seeds either lack endosperm or when present contains food reserves of cellulose, fats, or protein, but seldom starch. Stomates typically are without subsidiary cells. Major families are the Liliaceae *(sensu lato)* with nearly 6900 species, the Iridaceae with some 1500 species, and the Orchidaceae. Economically important plants include *Allium* spp. (onion, leek, garlic, chives), *Asparagus officinalis* L. (asparagus), *Dioscorea alata* L. (yam, Dioscoreaceae), and *Vanilla planifolia* Andr. (vanilla, Orchidaceae).

Superorder Alismatiflorae

The Alismatiflorae consists of 4 orders (Alismatales, Zosterales, Najadales and Triuridales), 11 families, and only about 500 species. Members of the Alismatiflorae primarily are aquatic plants. Most are apocarpous, have trinucleate pollen, and, except for the Triuridales, lack endosperm in the mature seed. Nearly all have stomates with two subsidiary cells. Representative families include the Alismataceae, Potamogetonaceae, and Zosteraceae.

Superorder Ariflorae

The Ariflorae consists of a single order (Arales), 3 families (Araceae, Lemnaceae, and Typhaceae [including the Sparganiaceae]) and approximately 2100 species. Members of the Ariflorae are characterized by their minute flowers which often are clustered in a fleshy spike (spadix) subtended by a large bract (spathe). Their flowers are bisexual, trimerous, and with an undifferentiated perianth (sometimes absent). Their leaves often are without typical parallel venation. Plants are often aquatic or of moist, deeply shaded forests. They are of little economic importance, although a few species of *Alocasia, Xanthosoma,* and *Colocasia* are cultivated in the tropics and subtropics as root crops.

Superorder Areciflorae

The Areciflorae consists of three monofamilial orders — the Arecales (Arecaceae = Palmae), Cyclanthales (Cyclanthaceae), and Pandanales (Pandanaceae) — and approximately 4500 species. By far the largest group of the Areciflorae is the Arecaceae (palms), with nearly 3500 species. Palms are mostly woody plants with an unbranched trunk and a terminal cluster of large leaves. Their flowers are small, actinomorphic, trimerous, mostly unisexual (the plants typically monoecious), and borne in paniculate inflorescences that are subtended by one or more large, often woody spathes. The Cyclanthaceae are palmlike herbs and the Pandanaceae are woody plants (often with stilt roots) with long stiff (nonplicate) leaves. Economically the Arecaceae is extremely important, especially in tropical regions of the world, ranking second in importance only to the grasses (Poaceae). Palms provide food, shelter, and clothing to numerous peoples of the tropics. Economically important products obtained from palms include: coconut oil and meat (*Cocos nucifera* L.), other oils (e.g., *Elaeis guineensis* Jacq., the African Oil Palm), dates (*Phoenix dactylifera* L.), rattan cane (*Calamus* spp.), betel nuts (*Areca catechu* L.), and carnauba wax (*Copernicia cerifera* [Arr.] Mart.).

Superorder Commeliniflorae

The Commeliniflorae consists of 2 orders (Commelinales and Zingiberales), 22 families, and nearly 20,000 species, with over half of these members of a single family, the Poaceae (= Gramineae). Members of the Commeliniflorae generally are characterized by flowers either with a well-differentiated biseriate perianth or with the perianth much reduced. The gynoecium is mostly syncarpous. Pollen is binucleate or trinucleate. Endosperm is usually starchy. Stomates have two or more subsidiary cells. The plants rarely are aquatic. The largest families are the Bromeliaceae, Cyperaceae, Eriocaulaceae, Zingiberaceae, and Poaceae. The Poaceae, or grass family, is economically probably the most important of all flowering plant families. Economically important grass crops include rice, rye, barley, wheat, oats, sorghum, and corn, as well as numerous "grasses".

ADDENDUM

Since this paper was written (1979) several revisions have been made in the classification of flowering plants. The reader should consult the following references: Dahlgren,[57] Takhtajan,[58] Thorne[59].

REFERENCES

1. **Delevoryas, T.,** *Morphology and Evolution of Fossil Plants,* Holt, Rinehart & Winston, New York, 1962, 40.
2. **Carlquist, S.,** Toward acceptable evolutionary interpretation of floral anatomy, *Phytomorphology,* 19, 332, 1969.
3. **Doyle, J. A. and Hickey, L. J.,** Pollen and leaves from the mid-cretaceous potomac group and their bearing on early angiosperm evolution, in *Origin and Early Evolution of Angiosperms,* Beck, C. B., Ed., Columbia, New York, 1976, 139.
4. **Walker, J. W.,** Aperature evolution in the pollen of primitive angiosperms, *Am. J. Bot.,* 61, 1112, 1974.
5. **Van Campo, M.,** Précisions nouvelles sur les structures comparées des pollens de gymnospermes et d'angiospermes, *C. R. Acad. Sci. Ser. D,* 272, 2071, 1971.
6. **Walker, J. W. and Skvarla, J. J.,** Primitively columellaless pollen: a new concept in the evolutionary morphology of angiosperms, *Science,* 187, 445, 1975.
7. **Cave, M. S., Arnott, H. J., and Cook, S. A.,** Embryogeny in the California peonies with reference to their taxonomic position, *Am. J. Bot.,* 48, 397, 1961.
8. **Bold, H. C.,** *Morphology of Plants,* 3rd ed., Harper & Row, New York, 1973.
9. **Stebbins, G. L.,** *Flowering Plants Evolution above the Species Level.,* Belknap Press, Cambridge, Mass., 1974.
10. **Esau, K.,** *Anatomy of Seed Plants,* 2nd ed., John Wiley & Sons, New York, 1977, chap. 11.
11. **Bierhorst, D. W.,** *Morphology of Vascular Plants,* Macmillan, New York, 1971.
12. **Cronquist, A.,** *The Evolution and Classification of Flowering Plants,* Houghton Mifflin, Boston, 1968.
13. **Takhtajan, A.,** *Flowering Plants Origin and Dispersal,* Smithsonian Institution Press, Washington, D.C., 1969.
14. **Thorne, R. F.,** A phylogenetic classification of the Angiospermae, in *Evolutionary Biology,* Vol. 9, Hecht, M. K., Steere, W. C., and Wallace, B., Eds., Plenum Press, New York, 1976, 35.
15. **Arnold, C. A.,** Classification of the gymnosperms from the viewpont of paleobotany, *Bot. Gaz. (Chicago),* 110, 2, 1948.
16. **Eckenwalder, J. E.,** Re-evaluation of the Cupressaceae and Taxodiaceae: a proposed merger, *Madrono,* 23, 237, 1976.
17. **Eames, A. J.,** Relationships of the Ephedrales, *Phytomorphology,* 2, 79, 1952.
18. **Bailey, I. W.,** Origin of angiosperms: need for a broadened outlook, *J. Arnold Arbor. Harv. Univ.,* 30, 63, 1949.
19. **Hallier, H.,** Provisional scheme of the natural (phylogenetic) system of flowering plants, *New Phytol.,* 4, 151, 1905.
20. **Just, T.,** Gymnosperms and the origin of the angiosperms, *Bot. Gaz. (Chicago),* 110, 91, 1948.
21. **Pearson, H. W.,** *Gnetales,* Cambridge University Press, Cambridge, London, 1929.
22. **Thompson, W. P.,** The morphology and affinities of *Gnetum, Am. J. Bot.,* 3, 135, 1916.
23. **Dahlgren, R.,** A system of classification of the angiosperms to be used to demonstrate the distribution of characters, *Bot. Not.,* 128, 119, 1975.
24. **Hutchinson, J.,** *Evolution and Phylogeny of Flowering Plants,* Academic Press, New York, 1969.
25. **Hutchinson, J.,** *The Families of Flowering Plants,* Claredon Press, Oxford, 1973.
26. **Melchior, H., Ed.,** *A Engler's Syllabus der Pflanzenfamilien II Angiospermen,* 12th ed., Gebr. Borntraeger, Berlin, 1964.
27. **Jones, S. B., Jr., and Luchsinger, A. E.,** *Plant Systematics,* McGraw-Hill, New York, 1979.
28. **Becker, K. M.,** A comparison of angiosperm classification systems, *Taxon,* 22, 19, 1973.
29. **Bailey, L. H.,** *Manual of Cultivated Plants,* Macmillan, New York, 1969.
30. **Benson, L.,** *Plant Classification,* D. C. Heath, Boston, 1967.
31. **Hill, A. F.,** *Economic Botany,* McGraw-Hill, New York, 1952.
32. **Lawrence, G. H. M.,** *Taxonomy of Vascular Plants,* Macmillan, New York, 1951.
33. **Porter, C. L.,** *Taxonomy of Flowering Plants,* Freeman, San Francisco, 1967.
34. **Smith, J. P., Jr.,** *Vascular Plant Families,* Mad River Press, Eureka, Calif., 1977.
35. **Thorne, R. F.,** Synopsis of a putatively phylogenetic classification of the flowering plants, *Aliso,* 6, 57, 1968.
36. **Thorne, R. F.,** A phylogenetic classification of the Annoniflorae, *Aliso,* 8, 147, 1974.
37. **Willis, J. C.,** *A Dictionary of Flowering Plants and Ferns,* revised by Airy Shaw, H. K., Cambridge University Press, New York, 1973.
38. **Seigler, D. S.,** Plant systematics and alkaloids, in *The Alkaloids,* Vol. 16, Manske, R. H. F., Eds., Academic Press, New York, 1977, 1.

39. **Behnke, H.-D.,** The bases of angiosperm phylogeny: ultrastructure, *Ann. Mo. Bot., Gard.,* 62, 647, 1975.
40. **Mabry, T. J.,** The order Centrospermae, *Ann. Mo. Bot. Gard.,* 64, 210, 1977.
41. **Bessey, C. E.,** The phylogenetic taxonomy of flowering plants, *Ann. Mo. Bot. Gard.,* 2, 109, 1915.
42. **Thorne, R. F.,** The "Amentiferae" or Hamamelidae as an artificial group: a summary statement, *Brittonia,* 25, 395, 1973.
43. **Berg, C. C.,** Cecropiaceae: a new family of the Urticales, *Taxon,* 27, 39, 1978.
44. **Carlquist, S.,** Wood anatomy and relationships of Bataceae, Gyrostemonaceae, and Stylobasiaceae, *Allertonia,* 1, 297, 1978.
45. **Mabry, T. J., Eifert, I. J., Chang, C., Mabry, H., and Kidd, C.,** Theligonaceae: pigment and ultrastructural evidence which excludes it from the order Centrospermae, *Biochem. Syst. Ecol.,* 3, 53, 1975.
46. **Behnke, H.-D.,** Phloem ultrastructure and systematics position of Gyrostemonaceae, *Bot. Not.,* 130, 255, 1977.
47. **Hickey, L. J. and Wolfe, J. A.,** The bases of angiosperm phylogeny: vegetative morphology, *Ann. Mo. Bot. Gard.,* 62, 538, 1975.
48. **Jensen, S. R., Nielsen, B. J., and Dahlgren, R.,** Iridoid compounds, their occurrence and systematic importance in the angiosperms, *Bot. Not.,* 128, 148, 1975.
49. **Cronquist, A.,** The Compositae revisited, *Brittonia,* 29, 137, 1977.
50. **Carlquist, S.,** Tribal interrelationships and phylogeny of the Asteraceae, *Aliso,* 8, 465, 1976.
51. **Wagenitz, G.,** Systematics and phylogeny of the Compositae (Asteraceae), *Plant Syst. Evol.,* 125, 29, 1976.
52. **Huber, H.,** The treatment of the Monocotyledons in an evolutionary system of classification, in *Flowering Plants Evolution and Classification of Higher Categories,* Kubitzki, K., Ed., Springer-Verlag, New York, 1977, 285.
53. **Sargant, E.,** A theory of the origin of Monocotyledons founded on the structure of their seedlings, *Ann. Bot. (London),* 17, 1, 1903.
54. **Sargant, E.,** The evolution of Monocotyledons, *Bot. Gaz. (Chicago),* 37, 325, 1904.
55. **Lowe, J.,** The phylogeny of Monocotyledons, *New Phytol.,* 60, 355, 1961.
56. **Thorne, R. F. and Dahlgren, R.,** personal communication, 1980.
57. **Dahlgren, R.,** A revised system of classification of the angiosperms, *Bot. J. Linn. Soc.,* 80, 91, 1980.
58. **Takhtajan, A.,** Outline of the classification of flowering plants (Magnoliophyta), *Bot. Rev.,* 46, 225, 1980.
59. **Thorne, R. F.,** Phytochemistry and angiosperm phylogeny: a summary statement, in *Phytochemistry and Angiosperm Phylogeny,* Young, D. A. and Seigler, D. S., Eds., Praeger Scientific Press, N.Y.

SOURCES AND COLLECTIONS OF PHOTOSYNTHETIC BACTERIA (ANOXYPHOTOBACTERIA; PHOTOTROPHIC BACTERIA)

Norbert Pfennig

COLLECTIONS OF PHOTOSYNTHETIC ORGANISMS

Phototrophic Green and Purple Bacteria

Addresses of Culture Collections

ATCC:
American Type Culture Collection
12 301 Parklawn Drive
Rockville, Md., 20852, U.S.A.

DSM:
German Collection of Microorganisms
Grisebachstr. 8
D-3400 Göttingen, F.R.G.

NCIB:
National Collection of Industrial Bacteria
Torry Research Station
PO Box 31, 135 Abbey Road
Aberdeen, AB 9 8DG, Scotland, U.K.

Type and Neotype Strains of All Species of the Four Families of the Green and Purple Bacteria and Culture Collection Numbers

Chlorobiaceae: 7 species available.

Chlorobium limicola	DSM 245
Chlorobium vibrioforme	DSM 260
Chlorobium phaeobacteroides	DSM 266
Chlorobium phaeovibrioides	DSM 269
Pelodictyon luteolum	DSM 273
Pelodictyon phaeum	DSM 728
Prosthecochloris aestuarii	DSM 271

Chloroflexaceae: only one species described.

Chloroflexus aurantiacus	DSM 635

(Many strains of the thermophilic *Chloroflexus aurantiacus* were originally available from Dr. R. Castenholz, Department of Biology, University of Oregon, Eugene, Ore. 97403.)

Chromatiaceae: 22 species available.

Amoebobacter roseus	—	DSM 235
Amoebobacter pendens	—	DMS 236
Chromatium buderi	ATCC 25588	DSM 176
Chromatium gracile	—	DSM 203
Chromatium minus	—	DSM 178
Chromatium okenii	—	DSM 169
Chromatium vinosum	ATCC 17899	DSM 180

Chromatium violascens	ATCC 17096	DSM 198
Chromatium warmingii	ATCC 14959	DSM 173
Chromatium weissei	—	DSM 171
Ectothiorhodospira halochloris	—	DSM 1059
Ectothiorhodospira halophila	—	DSM 244
Ectothiorhodospira mobilis	—	DSM 237
Ectothiorhodospira shaposhnikovii	—	DSM 243
Lamprocystis roseopersicina	—	DSM 229
Thiocapsa pfennigii	—	DSM 1375
Thiocapsa roseopersicina	—	DSM 217
Thiocystis gelatinosa	—	DSM 215
Thiocystis violaceae	—	DSM 207
Thiodictyon bacillosum	—	DSM 234
Thiodictyon elegans	—	DSM 232
Thiospirillum jenense	—	DSM 216

Rhodospirillaceae: 16 species available.

Rhodocyclus purpureus		DSM 168
Rhodomicrobium vannielii	ATCC 17100	—
Rhodopseudomonas acidophila	ATCC 25092	DSM 137
Rhodopseudomonas capsulata	ATCC 11166	—
Rhodopseudomonas gelatinosa	ATCC 17011	—
Rhodopseudomonas globiformis	—	DSM 161
Rhodopseudomonas palustris	ATCC 17001	—
Rhodopseudomonas sphaeroides	ATCC 17023	—
Rhodopseudomonas sulfidophila	—	DSM 1374
Rhodopseudomonas sulfoviridis	—	DSM 729
Rhodopseudomonas viridis	ATCC 19567	—
Rhodospirillum fulvum	ATCC 15798	DSM 113
Rhodospirillum molischianum	ATCC 14031	—
Rhodospirillum photometricum	—	DSM 122
Rhodospirillum rubrum	ATCC 11170	NCIB 8355
Rhodospirillum tenue	ATCC 19137	DSM 109

Many species of the *Rhodospirillaceae* also are available in the NCIB.

SOURCES AND COLLECTIONS OF ALGAE (INCLUDING CYANOBACTERIA)

Paul C. Silva

Approximately 2,000 strains of algae are maintained in culture collections. As noted elsewhere in this handbook (Taxonomic Classification of Algae: Diversity of Algae), continually improving techniques are facilitating the culture of macroalgae, so that collections no longer are restricted to unicellular, colonial, and microscopic filamentous forms. Culture collections are of inestimable value in promoting research, and persons engaged in their maintenance have organized themselves into various national societies, capped by the World Federation for Culture Collections. An international directory has been published.[1] Most culture collections of microorganisms include few or no algae. Blue-green algae available in the most important algal collections have been summarized.[2] The following five collections have the largest number of algae representative of various classes:

1. The Culture Centre of Algae and Protozoa, 36 Storey's Way, Cambridge CB3 ODT, England
2. Die Sammlung von Algenkulturen, Pflanzenphysiologisches Institut der Universität, Nikolausberger Weg 18, D-3400 Göttingen, DBR (West Germany)
3. Culture Collection of Algae, Czechoslovak Academy of Sciences, Dukelskâ 145, Třeboň, Czechoslovakia
4. Culture Collection of Algae, Department of Botany, The University of Texas, Austin, Texas 78712, U.S.A.
5. Algal Collection of the Institute of Applied Microbiology, University of Tokyo, Bunkyo-ku, Tokyo 113, Japan

The collection at the University of Texas was recently transferred from Indiana University, the location given in the international directory. It should be noted that for some unexplained reason this collection was not included in the directory's systematic index. The major collections periodically issue catalogs and instructions for ordering. Data concerning the original isolation of each strain are usually available. A relatively small number of algae, for which such information is not available, may be obtained from commercial establishments. In the U.S. such sources include Carolina Biological Supply Company (2700 York Road, Burlington, N.C., 27215; Box 7, Gladstone, Ore. 97027), General Biological, Inc. [Turtox], (8200 South Hoyne Avenue, Chicago, Ill. 60620; 342 Western Avenue, Boston, Mass. 02135; 1945 Hoover Court, Birmingham, Ala. 35226), and Triarch Inc. (Box 98, Ripon, Wis. 54971).

In addition to the foregoing sources, collections are maintained by numerous individuals, universities, and industrial establishments chiefly for in-house use. While the keepers of these collections rarely have sufficient help to accommodate outside requests (and therefore do not wish to have their collections publicized), many are willing to provide cultures when approached on a personal basis.

REFERENCES

1. **Martin, S. M. and Skerman, V. B. D., Eds.,** *World Directory of Collections of Cultures of Microorganisms,* Wiley-Interscience, New York, 1972.
2. **Komârek, J.,** Culture collections, in *The Biology of Blue-green Algae,* Carr, N. G. and Whitton, B. A., Eds., Blackwell, Oxford, 1973, 519.

SOURCES AND COLLECTIONS OF PLANT GERMPLASM

George A. White

INTRODUCTION

Seeds and vegetative propagules for research purposes generally are available from commercial companies, state and federal experiment stations, crop germplasm collections, and plant scientists. Emphasis herein is on the sources of plant germplasm from collections held in the United States and how to obtain foreign materials. The National Plant Germplasm System is described in Program Aid Number 1188 of the U.S. Department of Agriculture (USDA).[1]

SOURCES OF PLANT GERMPLASM

Many plant species with a wide genetic base are available from germplasm working stock collections, including four Regional Plant Introduction Stations, and the National Seed Storage Laboratory. Since the germplasm collections are large, diverse, and much used, only small samples of seed or vegetative stocks can be provided. Samples are free, however, and available for research purposes in the U.S. and abroad.

When requesting materials from germplasm collections: (1) Clearly specify your needs, including any special characteristics; (2) Give the scientific and common names, the desired country or origin and, as appropriate, the cultivar name or other designation.

The locations of major germplasm collections are given in Table 1. In the crop listings, some emphasis is given to plants with high biomass productivity.

Other collections include cotton in Mississippi, Texas, and Arizona; soybeans in Illinois and Mississippi; sorghum in Texas; sugarcane in Florida; potatoes in Wisconsin; virus-free fruits in Washington; and subtropical and tropical fruits, vegetables, and ornamentals in Florida and Puerto Rico. Materials for research purposes are available from these and other germplasm collections. In addition, 12 fruit repositories are being planned.

Information about plant introductions has been published in Plant Inventories[2] continuously since 1898. Some specific examples of the use of plant introductions in crop improvement are described in the 1971 publication on The National Program for Conservation of Crop Germplasm.[3]

PROCEDURES FOR OBTAINING FOREIGN GERMPLASM

If a researcher needs plant materials which are not available in the U.S., the following procedures should be followed to obtain the desired materials: write the Plant Introduction Officer of the Germplasm Resources Laboratory for materials to come through the regular plant germplasm channels. Specify a source of the material if known. Contact the Permit Unit, Animal Plant Health Service (APHIS), USDA, Federal Building, Hyattsville, Md. 20782, for quarantine permits if you wish to import the material directly.

Strict quarantine regulations govern the importation of several plant species that have excellent biosolar potential. Vegetative stocks of sugarcane and bamboo are prohibited from all countries except for scientific purposes and then under departmental permit. Sugarcane quarantine is handled at the Beltsville Agricultural Research Center. Limited quantities of bamboo are given quarantine handling at the U.S. Plant Introduction Station, Glenn Dale, Md. This station also handles most of the virus indexing

Table 1
GERMPLASM WORKING COLLECTIONS OF VARIOUS CROPS[a]

Name and Address	Crop Germplasm
Germplasm Resources Laboratory USDA Agricultural Research Center Beltsville, Md. 20705	Wheat, barley, oats, rye, rice
Regional Plant Introduction Station University of Georgia Experiment, Ga. 30212	Sorghum, various warm-season grasses and legumes, kenaf, millet, cowpea, peanut, various vegetables
Regional Plant Introduction Station Iowa State University Ames, Iowa 50010	Corn, alfalfa, sunflower, tomatoes
Regional Plant Introduction Station New York Agricultural Experiment Station Geneva, N.Y. 14456	Red and white clover, timothy, pea, onion
Regional Plant Introduction Station Johnson Hall, Room 59 Washington State University Pullman, Wash. 99164	Safflower, fescue, ryegrass, wheatgrass, beans, cabbage
National Seed Storage Laboratory USDA-SEA Colorado State University Ft. Collins, Colo. 80521	Seeds of all crops (long-term storage)

[a] Each regional plant introduction station has collections of many unlisted agronomic and horticultural crops and related species.

of fruit and potato tuber importations. Corn, sorghum, and several related species require special quarantine permits for introduction from some countries. All plant material introduced for research purposes should enter the country through a quarantine facility for inspection. This provides a safeguard against introduction of plant pests. The Plant Germplasm Quarantine Center is located at the USDA Agricultural Research Center, Beltsville, Md. Other inspection stations are located at certain ports-of-entry into the U.S. The Plant Introduction Officer, in cooperation with APHIS officials, will assist in procuring needed plant materials from foreign sources.

REFERENCES

1. *Anon.,* The National Plant Germplasm System, U.S. Department of Agriculture Program Aid No. 1188, 1977.
2. USDA Plant Inventories, Nos. 1 to 184, 1898—1978.
3. *Anon.,* The National Program for Conservation of Crop Germplasm, University of Georgia Printing Department, Athens, Ga., 1971.

SOURCES AND COLLECTIONS OF WOODY FOREST PLANTS

Harry E. Sommer

INTRODUCTION

This section on sources and collections of woody forest plants is divided into two parts. The first part contains primarily lists of possible suppliers of seed or seedlings. The second part lists some collections of woody plants in arboretums and botanical gardens. The availability of research materials from the latter will vary with the institution's policy.

SOURCES OF SEEDS OR SEEDLINGS

The following are publications containing potential sources of seeds or seedlings:

Forest Tree Seed Directory, Food and Agricultural Organization of the United Nations, Rome, 1975, 283 pages; lists seed sources for most major and many minor species of trees along with data on germination, seed/kg, and seed treatment.

Seed and Planting Stock Dealers, USDA, Forest Service FS331, U.S. Department of Agriculture, Forest Service, Washington, D.C., 1979, 22 pages; lists names and addresses of 112 dealers; includes species handled, and checks on seed quality.

A Directory of Forest Tree Nurseries in the United States, U.S. Department of Agriculture, Forest Service, Washington, D.C., 1976, 33 pages; includes federal, state, other public, forest industry, and private nurseries; production information provided, but no information on species raised.

Forest Tree Seed Orchards, A Directory of Industry, State and Federal Forest Tree Seed Orchards in the United States, U.S. Department of Agriculture, Forest Service, Washington, D.C., 1974, 33 pages; lists species in orchard and seed production status.

A List of Seed in the Canadian Forestry Service Seed Bank, Information Report — PS-X-53, Wang, B. S. P. and Haddon, B. D., Eds., Environment Canada, Forestry Service, Ottawa, 1974, 26 pages; lists by species, year collected, place collected, and altitude.

Nursery Source Guide, a Handbook, McGrourty, F., Jr., Ed., in *Brooklyn Bot. Gard. Rec.,* 33, 2, 1977, 96 pages; lists 1200 trees and shrubs by ornamental use; list of nurseries included.

Forest Genetics Resources Information, Food and Agricultural Organization of the United Nations, Rome, issued irregularly. Part of the Forestry Occasional Papers series; includes recent collections of seed made, announcements of seed available, and seed distribution for provenance tests.

World Directory of Tree Seed Workers, Edwards, D. G. W., Ed., Environment Canada, Canadian Forest Service, Ottawa, 1976, 133 pages; indexed by species and discipline, so that a worker with a species or area of interest can be easily located.

COLLECTIONS OF WOODY FOREST SPECIES

Wyman[5] has compiled a listing of arboretums and botanical gardens of North America. In addition to details on size, purpose, and general description, featured collections are listed. Descriptions of these and other collections are sometimes available, e.g., *Tree Species and Hybrids in Nurseries and Aboreta* at the Ontario Tree Breeding Unit;[4] *The Douglas Fir Arboretum at Couichan Lake, Vancouver Island;*[2] *The Collection of Douglas Fir Inbreds at the Couichan Lake Experiment Station, Vancouver Island;*[3] and *A Study of the Genus Taxus,*[1] based in part on material at the Secrest Arboretum. Additional information on special collections, registration headquarters for specific genera and additional general information on gardens and arboretums can be found in *The Bulletin,* official publication of the American Association of Botanical Gardens and Arboreta, particularly Volume 8, No. 2, 1974.

REFERENCES

1. **Chadwick, L. C. and Keen, R. A.,** *A Study of the Genus Taxus, Research Bulletin 1086,* Ohio Agricultural Research and Development Center, Wooster, 1976.
2. **Orr-Ewing, A. L.,** *The Douglas Fir Arboretum at Couichan Lake, Vancouver Island, British Columbia Forest Service Research Note No. 57,* Research Division, British Columbia Forest Service, Victoria, 1973,
3. **Orr-Ewing, A. L.,** *The Collection of Douglas Fir Inbreds at the Couichan Lake Experiment Station, Vancouver, Island, British Columbia Forest Service Note No. 81,* Forest Service Research Division Ministry of Forests, Victoria, 1977.
4. *Tree Species and Hybrids in Nurseries and Arboreta,* Tree Breeding Unit, Research Branch, Ontario Department of Lands and Forests, Maple, 1971.
5. **Wyman, D.,** *The Arboretums and Botanical Gardens of North America,* Arnold Arboretum of Harvard University, Jamaica Plain, Mass., 1959.

LITERATURE GUIDE FOR PLANT TISSUE CULTURE

Harry E. Sommer

INTRODUCTION

This guide to plant tissue culture literature has two primary purposes. The first is to provide the investigator with references to aid in deciding if tissue culture methods are applicable to his problem. The second is to provide an introduction to the methods of plant tissue culture and the results obtained. Only a small portion of the literature is covered in this guide, and it is not intended as a substitute for a thorough literature search.

To our knowledge there are no type collections of plant tissue cultures. Therefore, an investigator must either obtain starter cultures from another laboratory or start his own cultures. Cultures of *Acer pseudoplatanus,* Paul's Scarlet rose, the WR-132 line of *Nicotiana tabacum* L., var. Xanthi, *Nicotiana tabacum,* var. Wisconsin 38, *Nicotiana glauca* × *N. langsdorfii,* and several lines of *Daucus carota* L. among others have been widely used. If one wishes to extend work done on the above systems, then these cultures should be obtained. However, one should be aware that the properties of the cultures may change with time and enviornment. In other instances it will be necessary to start one's own cultures.

For the purpose of this guide we have interpreted tissue culture to include cell, callus, organ, embryo, and protoplast culture. We have to include references useful for developmental biology, plant breeding, plant propagation, plant physiology, and biochemistry. All these areas contribute to the understanding of biological processes producing biosolar resources. For this presentation the literature has been broken up into 12 sections.

History

The first section, arranged in chronological order, consists of references that outline the history of plant tissue. This section is included primarily to allow the researcher to orient himself to the field and to find earlier works that may be pertinent to his interests. Haberlandt reported the first attempt to culture isolated higher plant cells, and pointed out the importance of such a technique. Krikorian and Berquam[1] published an English translation of the paper with comments on early work in these and related fields. Initial progress in achieving the successful indefinite culture of plant cells or tissue were slow. White[2,4,8,11] has reviewed the early work to 1945. Four important papers and two books appeared during that time. In 1934, White[3] reported the first successful long-term culture of tomato roots. Five years later Gautheret,[5] Nobécourt,[6] and White[7] reported success with long-term callus cultures. Further details of techniques and results are given in White's[8,9] and Gautheret's[10] books. During the first 50 years, much of the research effort had been directed to working out the nutritional requirements of plant tissues in culture. Emphasis was on finding the minimum requirements for maintenance of growth. White[12] and Gautheret[13] both reviewed the results of these efforts. This entire first phase in the development of plant tissue culture was organized and reviewed in detail by Gautheret[14] in his book *La Culture des tissus Végétaux: Techniques et Réalisation.* This book does contain a separate species index, so work on a particular species can be located quickly.

The second phase, with its emphasis on morphogenesis, could be said to have begun with the discovery of kinetin by Miller and Skoog[15] and their demonstration of its interaction with auxin in controlling organogenesis in tobacco tissue cultures. One of

Haberlandt's speculations was that it should be possible to obtain an embryo from a cell. Embryogenesis in carrot tissue culture was demonstrated by Steward[16,17] and Reinert[18] in 1958.

In 1960, Cocking[19] reported methods for the rapid isolation of higher plant protoplasts. The initial experiments were not done with tissue cultures, but the technique has had a profound impact on the recent course of plant tissue culture. Then Nitsch and Nitsch[20] reported a reliable method to obtain haploid tobacco plants. Shortly thereafter the fusion of protoplasts was observed,[21] and tobacco plants were regenerated from protoplasts.[22] The combination of the events since 1960 has enabled somatic cell genetics to be performed with plant cells. An excellent example is given by Melchers.[23,24]

There have been many other significant contributions to the development of plant tissue culture as a research tool, and examples of its application that could not be included in this section. Part of the purpose of the following sections is to provide direct or indirect reference to works not covered in the history. In addition some of the sections will provide more detailed references for some topics covered above.

General References

The general references consist of a list of books and a few review papers that provide general background information on plant tissue culture. Among the better current short surveys of the field is the book by Thomas and Davey;[34] unfortunately, literature citations are absent. For a detailed treatment of current work in plant tissue culture, Street's[33] book should be consulted. If one is particularly interested in determining whether a particular species has been cultured, in addition to Gautheret's[14] compendium, Sommer and Brown,[25] Murashige,[29] Reinert and Bajaj,[30] and de Fossard[43] are good starting places. Discussions of problems associated with morphogenesis in culture have been given in the reviews by Halperin,[28] Steward et al.,[31] and Street.[32] Gautheret[27] has described cases of differentiations and organogenesis particularly in the Jerusalem artichoke. The English translation of Butenko's[26] book serves as an introduction to the Soviet approach to plant tissue culture.

Symposia Proceedings

A number of symposia have been held on plant tissue culture. Since 1964, a major international symposium has been held every four years. The presentations of the plenary sessions have been published.[35,39,40,41] The most recent sessions have been under the sponsorship of the International Association for Plant Tissue Culture. Three other recent symposia, each aimed at a slightly different audience, should also be consulted at least for the current state of the art. They cover the range of interests from the propagator[36,38] to the applied researcher[37,38] to the basic researcher.[38]

Methods

In the methods section two sets of experiments from plant physiology laboratory manuals[45,48] have been included. These are provided as a starting place for those who have never done tissue culture work before. Street and Henshaw[46] and White[47] both describe general tissue culture methods, including the setup of the laboratory. Gamborg and Wetter[44] have assembled the methods used at the Prairie Regional Laboratory and associated organizations in their protoplast and tissue culture research. Durbin[42] contains some tissue culture methods using tobacco. Methods particularly for the propagation of many plants are contained in de Fossard.[43] Methods are also found in the sections on shoot tip and meristem culture, on protoplasts, on haploid cultures, and in some of the citations under General References.

Organ Culture

Organ cultures have been used in developmental studies such as Steeve's[53,54] studies of determination in leaf primordia, and Nitsch's[51] demonstration that excised ovaries could be grown in culture. The classical material for organ culture is the excised root. Butcher and Street[49] have reviewed studies on the culture of roots. Methods for excised root culture[52] and shoot and leaf culture[50] can be found in *Methods in Developmental Biology.*

Apical Meristem and Shoot Tip Cultures

These types of cultures have been applied to the propagation of orchids,[55,56] carnations,[59,60] and other crops.[58] The assumption has often been made incorrectly that disease-free plants are obtained (see Murashige and Jones,[97] Langhans et al.[96]). Meirstem cultures have also been used for physiological studies of meristem development.[57] Detailed description of methods are included in most of the references.[55,56,59,60]

Embryo Culture

Embryo culture has been reviewed several times.[62] Raghaven[63] not only discusses methods, uses of embryo culture, the development of embryos in vitro, but also covers somatic embryogenesis. Monnier[61] has made several improvements in the culture of immature embryos.

Protoplasts

Protoplasts, fusion of protoplasts, and plant regeneration from protoplasts have received much attention of late and several approaches have developed. Binding[65] was among the first to obtain regeneration from protoplasts, though he did not obtain true fusion. A reasonable idea of the range of uses for protoplasts can be obtained from the symposia listed.[71,73] Cocking has summarized information on the isolation of protoplasts and their regeneration as cells.[66] Cocking and Peberdy[67] have prepared a manual of experimental procedures for the isolation, fusion, and uptake of foreign particles by protoplasts. The research aims and procedures of the Prairie Regional Laboratory group have been covered by Constabel[68] and Gamborg et al.[69] Experimental methods can be found in their laboratory manual.[44] Smith's[72] papers contains illustrations that show some steps in the procedure. Keller and Melchers[70] have developed an alternate procedure to obtain fusion.

Haploid Cultures

Currently haploid cultures are used to produce haploid plants. The significance of haploid plants has been discussed by Sunderland[79] and Melchers and Labib.[76] A general idea of current activity can be obtained from Kasha[74] and Vasil and Nitsch.[80] Kasperbauer and Collins[75] have described the minimal conditions needed for anther culture. Sharp et al.[78] have described a nurse culture procedure for obtaining haploids from free microspores. Nitsch[77] has described a method to obtain haploid tobacco plantlets from microspores in a liquid culture. Pictures of the procedures are also contained. The female gametophyte has been used to obtain haploid cultures.[81]

Genetics and Plant Breeding

The genetic background of an explant often may influence its behavior in culture.[82,83,86] Since the development of the techniques for uptake of foreign particles by protoplasts, protoplast fusion, and production of haploid plants from pollen, there has been much discussion about the use of these techniques for plant breeding. Bottino,[84] Holl,[87] and Scowcroft[90] have presented reviews and discussion of these possibilities. More detailed coverage of experimental approaches and critical evaluation can be found in recent symposia volumes.[85,88,89] Giles[91] has been active in combining fungal and higher plant approaches.

Pathology

Some of the first tissue cultures were obtained from crown gall callus. Braun has been particularly active in investigating the autonomous growth by crown gall tissue[92] and its reversal.[93] Recent research[95] has shown a plasmid is involved in the crown gall response, opening up another possible way of modifying a plant genome. Tissue cultures have also been used for the production of disease-free plants,[96,97] study of virology,[98] and other areas of plant pathology.[94]

Natural Products from Tissue Cultures

Street[103] has reviewed the nutrition of plant tissue cultures from a physiological/biochemical viewpoint and more recently has published over 20 papers in large part dealing with the metabolism and related topics of *Acer pseudoplatanus* cultures. Nash and Davies[102] have presented a picture that is probably a fairly typical metabolic pattern in many culture systems. Tissue cultures have been considered as a potential source of rare and useful natural products. The control of biosynthesis in culture has been covered by Krikorian and Steward.[101] Further consideration of the potential for the production of natural products in culture can be found in the proceedings of two recent symposia.[99,100]

Photosynthesis and Plastid Development in Tissue Cultures

Commonly, plant tissue cultures are grown on a medium that contains sucrose or glucose as a carbon source, are often not green, and sometimes are grown in the dark. Exceptions are not commonly found in the references above, so the last section lists papers related to photosynthesis in plant tissue culture. Examples in conditions for autotrophic growth, plastid development, culture of C_4 plants, and selection for photosynthetic efficiency are given.

(SELECTED) TISSUE CULTURE LITERATURE

History

1. **Krikorian, A. D. and Berquam, D. L.**, Plant cell and tissue cultures: the role of Haberlandt, *Bot. Rev.*, 35, 59, 1969.
2. **White, P. R.**, Plant tissue culture, *Arch. Exp. Zellforsch.*, 10, 501, 1931.
3. **White, P. R.**, Potentially unlimited growth of excised tomato root tips in a liquid medium, *Plant Physiol.*, 9, 585, 1934.
4. **White, P. R.**, Plant tissue cultures, *Bot. Rev.*, 2, 419, 1936.
5. **Gautheret, R.**, Sur la possibilité de réaliser la culture indéfinie des tissus de Tubercules de carotte, *C. R. Acad. Sci.*, 208, 118, 1939.
6. **Nobécourt, P.**, Sur les proliférations spontanées de fragments de tubercules de carotte et leur culture sur milieu synthétique, *Bull. Soc. Bot. Fr.*, 85, 1, 1938.
7. **White, P. R.**, Potential unlimited growth of excised plant callus in an artificial nutrient, *Am. J. Bot.*, 26, 59, 1939.
8. **White, P. R.**, Plant tissue cultures, *Camb. Phil. Soc. Biol. Rev.*, 16, 34, 1941.
9. **White, P. R.**, *A Handbook of Plant Tissue Culture*, Jaques Cattell Press, Lancaster, Pa., 1943.
10. **Gautheret, R. J.**, *La Culture des Tissus*, Gallimard, Paris, 1945.
11. **White, P. R.**, Plant tissue cultures. II, *Bot. Rev.*, 12, 521, 1946.
12. **White, P. R.**, Nutritional requirements of isolated plant tissues and organs, *Annu. Rev. Plant Physiol.*, 2, 231, 1951.
13. **Gautheret, R. J.**, The nutrition of plant tissue cultures, *Annu. Rev. Plant Physiol.*, 6, 433, 1955.

14. **Gautheret, R. J.,** *La Culture des Tissus Végétaux: Techniques et Réalisation,* Masson & Cie, Paris, 1959.
15. **Skoog, F. and Miller, C. O.,** Chemical regulations of growth and organ formation in plant tissues cultured in vitro, *Symp. Soc. Exp. Biol.,* 11, 118, 1957.
16. **Steward, F. C., Mapes, M. O., and Smith, J.,** Growth and organized development of cultured cells. I. Growth and division of freely suspended cells, *Am. J. Bot.,* 45, 693, 1958.
17. **Steward, F. C., Mapes, M. O., and Mears, K.,** Growth and organized development of cultured cells, II. Organization in cultures grown from freely suspended cells, *Am. J. Bot.,* 45, 705, 1958.
18. **Reinert, J.,** Morphogenese und ihre Kontrolle an Gewebekulturen aus Carotten, *Naturwissenschaften,* 45, 344, 1958.
19. **Cocking, E. C.,** A method for the isolation of plant protoplasts and vacuoles, *Nature,* 187, 962, 1960.
20. **Nitsch, J. P. and Nitsch, C.,** Haploid plants from pollen grains, *Science,* 163, 85, 1969.
21. **Power, J. B., Cummins, S. E., and Cocking, E. C.,** Fusion of isolated plant protoplasts, *Nature,* 225, 1016, 1970.
22. **Takebe, I., Labib, G., and Melchers, G.,** Regeneration of whole plants from isolated mesophyll protoplasts of tobacco, *Naturwissenschaften,* 58, 318, 1971.
23. **Melchers, G.,** Kombination somatischer und konventioneller Genetik fur die Pflanzenzucktung, *Naturwissenschaften,* 64, 184, 1977.
24. **Melchers, G.,** The combination of somatic and conventional genetics in plant breeding, *Plant Res. Dev.,* 5, 86, 1977.

General References

25. **Brown, C. L. and Sommer, H. E.,** *An Atlas of Gymnosperms Cultivated in Vitro,* Georgia Forestry Research Council, Macon, 1975.
26. **Butenko, R. G., (Artmann, M., translator),** *Plant Tissue Culture and Plant Morphogenesis,* 1964 ed., Israel Program for Scientific Translations, Jerusalem, 1968.
27. **Gautheret, R. J.,** Factors affecting differentiation of plant tissues grown in vitro, in *Cell Differentiation and Morphogenesis,* Beerman, W., Gautheret, R., Nieuwkoop, P. D., Wardlaw, C. W., Wigglesworth, V. B., Wolf, E., and Zeevaart, J. A. D., Eds., North-Holland, Amsterdam, 1966, 55.
28. **Halperin, W.,** Morphogenesis in cell cultures, *Annu. Rev. Plant Physiol.,* 20, 395, 1969.
29. **Murashige, T.,** Plant propagation through tissue cultures, *Annu. Rev. Plant Physiol.,* 25, 135, 1974.
30. **Reinert, J. and Bajaj, Y. P. S.,** Eds., *Plant Cell, Tissue, and Organ Culture,* Springer-Verlag, Berlin, 1977.
31. **Steward, F. C., Mapes, M. O., and Ammerato, P. V.,** Growth and morphogenesis in tissue and free cell cultures, in *Plant Physiology, a Treatise, Vol. 5B, Analysis of Growth: the Responses of Cells and Tissues in Culture,* Steward, F. C., Ed., Academic Press, New York, 1969, 329.
32. **Street, H. E.,** Growth, differentiation and organogenesis in plant tissue and organ culture, in *Cells and Tissues in Culture, Methods, Biology, and Physiology,* Vol. 3, Willmer, E. N., Ed., Academic Press, New York, 1967, 631.
33. **Street, H. E., Ed.,** *Plant Tissue and Cell Culture,* 2nd ed., University of California Press, Berkeley, 1977.
34. **Thomas, E. and Davey, M. R.,** *From Single Cells to Plants,* Wykeham Publications Ltd., London, 1975.

Symposia Proceedings

35. **Anonymous,** Eds., *Les Cultures de Tissus de Planter, Colloques Internationaux du Centre National de la Recherche Scientifique No. 193,* Centre National de la Recherche Scientifique, Paris, 1971.
36. **Anonymous,** Eds., *Tissue Culture for Horiticultural Purposes, Acta Horticulture No. 78,* International Society for Horticultural Science, The Hague, 1977.
37. **Hughes, K. W., Ed.,** *Propagation of Higher Plants Through Tissue Culture: A Bridge Between Research and Application,* University of Tennessee Press, Nashville, 1979.
38. **Sharp, W. R., Larsen, P. O., Paddock, E. F., and Raghavan, V., Eds.,** *Plant Cell and Tissue Culture: Principle and Applications,* Ohio State University Press, Columbus, 1979.
39. **Street, H. E. Ed.,** *Tissue Culture and Plant Science,* Proc. 3rd Int. Congr. Plant Tissue Cell Culture, Academic Press, New York, 1974.
40. **Thorpe, T. A., Ed.,** *Plant Tissue, Cell, and Protoplast Research,* Proc. 4th Int. Congr. Plant Tissue Cell Culture, University of Calgary Press, Calgary, Canada, 1979.
41. **White, P. R. and Grove, A. R.,** *Proceedings of an International Conference on Plant Tissue Culture,* McCutchan Publishing Corp., Berkeley, 1965.

Methods

42. **Durbin, R. D., Ed.,** Nicotiana: procedures for experimental use, *U.S. Dep. Agric. Tech. Bull.,* 1586, 1979.
43. **de Fossard, R. A.,** *Tissue Culture for Plant Propagators,* University of New England Printery, Armidale, Australia, 1976.
44. **Gamborg, O. L. and Wetter, L. R.,** *Plant Tissue Culture Methods,* National Research Council of Canada, Prairie Regional Laboratory, Saskatoon, 1975.
45. **Skoog, F. and Armstrong, D. J.,** Tissue culture, in *Experimental Plant Physiology,* Pietro, A. S., Ed., C. V. Mosby, St. Louis, 1974, 110.
46. **Street, H. E. and Henshaw, G. G.,** Introduction and methods employed in plant tissue culture, in *Cells and Tissues in Culture, Methods, Biology, and Physiology,* Vol. 3, Willmer, E. N., Ed., Academic Press, New York, 1966, 459.
47. **White, P. R.,** The Cultivation of Animal and Plant Cells, 2nd ed., Ronald Press, New York, 1963.
48. **Witham, F. H., Blaydes, D. F., and Devlin, R. M.,** *Experiments in Plant Physiology,* Van Nostrand Reinholt, New York, 1971.

Organ Culture

49. **Butcher, D. N. and Street, H. E.,** Excised root culture, *Bot. Rev.,* 30, 513, 1964.
50. **Clutter, M. E., and Sussex, I. M.,** Shoot and leaf organ culture, in *Methods in Developmental Biology,* Wilt, H. and Wessells, N. K., Eds., Thomas Y. Crowell, New York, 1967, 435.
51. **Nitsch, J. P.,** Growth and development in vitro of excised ovaries, *Am. J. Bot.,* 38, 566, 1951.
52. **Street, H. E.,** Excised root culture, in *Methods in Developmental Biology,* Wilt, H. and Wessells, N. K., Eds., Thomas Y. Crowell, New York, 1967, 425.
53. **Steeves, T. A.,** Morphogenetic studies of fern leaves, *J. Linn. Soc. London, Bot.,* 58, 401, 1963.
54. **Steeves, T. A. and Sussex, I. M.,** Studies on the development of excised leaves in sterile culture, *Am. J. Bot.,* 44, 665, 1957.

Apical Meristem and Shoot Tip Cultures

55. **American Orchid Society,** *Meristem Tissue Culture, A Selection of Articles from the American Orchid Society Bulletin,* American Orchid Society, Cambridge, Mass.
56. **Arditti, J.,** Clonal propagation of orchids by means of tissue culture — a manual, in *Orchid Biology, Reviews and Perspectives 1,* Arditti, J., Ed., Comstock Publishing Associates, Ithaca, N.Y., 1977, 203.
57. **Ball, E.,** Sterile culture of the shoot apex of *Lupinus albus, Growth,* 24, 91, 1960.
58. **Morel, G. M.,** Morphogenesis of stem apical meristem cultivated in vitro: application to clonal propagation, *Phytomorphology,* 22, 265, 1972.
59. **Phillips, D. J.,** *Carnation Shoot Tip Cultures,* Technical Bulletin No. 102, Colorado State Experiment Station, Fort Collins, 1968.
60. **Stone, O. W.,** Factors affecting the growth of carnation plants from shoot apices, *Ann. Appl. Biol.,* 52, 199, 1963.

Embryo Culture

61. **Monnier, M.,** Culture in vitro de l'embryon immature de *Capsella bursa-pastorie* Moench, *Rev. Cyt. Biol. Végét.,* 39, 1, 1976.
62. **Narayanaswami, S. and Norstog, K.,** Plant embryo culture, *Bot. Rev.,* 30, 587, 1964.
63. **Raghaven, V.,** *Experimental Embryogenesis in Vascular Plants,* Academic Press, New York, 1977.
64. **Rappaport, J.,** In vitro culture of plant embryos and factors controlling their growth, *Bot. Rev.,* 20, 21, 1954.

Protoplasts

65. **Binding, H.,** Regeneration und Verschmelzung nackter Laubmoosprotoplasten, *Z. Pflanzenphysiol.,* 55, 305, 1966.
66. **Cocking, E. C.,** Plant cell protoplasts — isolation and development, *Annu. Rev. Plant Physiol.,* 23, 29, 1972.
67. **Cocking, E. C. and Peberdy, J. F., Eds.,** *The Use of Protoplasts from Fungi and Higher Plants as Genetic Systems, a Practical Handbook,* University of Nottingham, Nottingham, England, 1974.
68. **Constabel, F.,** Somatic hybridization in higher plants, *In Vitro,* 12, 743, 1976.

69. **Gamborg, O. L., Constabel, F., Fowke, L., Kao, K. A., Ohyama, K., Kartha, K., and Pelcher, L.,** Protoplast and cell culture methods in somatic hybridization in higher plants, *Can. J. Genet. Cytol.,* 16, 737, 1974.
70. **Keller, W. A. and Melchers, G.,** The effect of high pH and calcium on leaf protoplast fusion, *Z. Naturforsch.,* 28C, 737, 1973.
71. **Peberdy, J. F., Rose, H. J., Rogers, H. J., and Cocking, E. C.,** Eds., *Microbial and Plant Protoplasts,* Academic Press, New York, 1976.
72. **Smith, H. H., Kao, K. N., and Combatti, N. C.,** Interspecific hybridization by protoplast fusion in Nicotiana, confirmation and extension, *J. Hered.,* 67, 123, 1976.
73. **Tempe, J., Ed.,** Protoplastes et fusion de celles somatiques vegetales, *Colloq. Int. C. N. R. S.,* 212, 1973.

Haploid Cultures

74. **Kasha, K. J., Ed.,** *Haploids in Higher Plants, Advances and Potential,* Proc. 1st Int. Symp., University of Guelph, Guelph, Ontario 1974.
75. **Kasperbaner, M. J. and Collins, G. B.,** Anther-derived haploids in tobacco: evaluation of procedures, *Crop Sci.,* 14, 305, 1974.
76. **Melchers, G. and Labib, G.,** Die Bedentung Haploids hoherer Pflanzen fur Plfanzenphysiologie und Pflanzenzuchtung, die durch Antherenkultur erzeugten Haploiden, ein neuer Durchbruch fur die Pflanzenzuchtung, *Ber. Dtsch. Bot. Ges.,* 83, 129, 1970.
77. **Nitsch, C.,** La culture de pollen isole sur milieu synthetique, *C. R. Acad. Sci.,* 278D, 1031, 1974.
78. **Sharp, W. R., Raskin, R. S., and Sommer, H. E.,** The use of nurse culture in the development of haploid clones in tomato, *Planta,* 104, 357, 1972.
79. **Sunderland, N.,** Pollen plants and their significance, *New Sci.,* 47, 142, 1970.
80. **Vasil, I. K. and Nitsch, C.,** Experimental production of pollen haploids and their use, *Z. Pflanzenphysiol.,* 76, 191, 1975.
81. **Winton, L. and Huhtinen, O.,** Tissue culture of trees, in *Modern Methods in Forest Genetics,* Miksche, J. P., Ed., Springer-Verlag, Berlin, 1976, 243.

Genetics and Plant Breeding

82. **Baroncelli, S., Buiatti, M., and Bennici, A.,** Genetics of growth and differentiation "in vitro" of *Brassica oleracea var. botrytis.* I. Differences between 6 inbreed lines, *Z. Pflanzenzuecht.,* 70, 99, 1973.
83. **Bingham, E. T., Hurley, L. V., Kaatz, D. M., and Saunders, J. W.,** Breeding alfalfa which regenerates from callus tissue in culture, *Crop Sci.,* 15, 719, 1975.
84. **Bottino, P. J.,** The potential of genetic manipulation in plant cell cultures for plant breeding, *Radiat. Bot.,* 15, 1, 1975.
85. **Dudits, D., Farkas, G. L., and Maliga, P., Eds.,** *Cell Genetics in Higher Plants: Proceedings of an International Training Course,* Akademiai Kiodo, Budapest, 1976.
86. **Gresshoff, P. M. and Doy, C. H.,** Development and differentiation of haploid *Lycopersicon esculentum* (Tomato), *Planta,* 107, 161, 1972.
87. **Holl, F. B.,** Innovative approaches to genetics in agriculture, *Can. J. Genet. Cytol.,* 17, 517, 1975.
88. **Ledoux, L., Ed.,** *Genetic Manipulation with Plant Material,* Plenum Press, New York, 1975.
89. **Markham, R., Davies, D. R., Hopwood, D. A., and Horne, R. W., Eds.,** *Modification of the Information Content of Plant Cells,* Proc. 2nd John Innes Symp., North-Holland, Amsterdam, 1975.
90. **Scowcroft, W. R.,** Somatic cell genetics and plant improvement, *Adv. Agron.,* 29, 39, 1977.
91. **Vasil, I. K. and Giles, K. L.,** Induced transfer of higher plant chloroplasts into fungal protoplasts, *Science,* 190, 680, 1975.

Pathology

92. **Braun, A. C.,** A physiological basis for autonomous growth of the crown-gall tumor cell, *Proc. Natl. Acad. Sci. U.S.A.,* 44, 344, 1958.
93. **Braun, A. C.,** A demonstration of the recovery of the crown-gall tumor cell with the use of complex tumors of single cell origin, *Proc. Natl. Acad. Sci. U.S.A.,* 45, 932, 1959.
94. **Braun, A. C. and Lipetz, J.,** The use of tissue culture in phytopathology, in *Cells and Tissues in Culture, Methods, Biology and Physiology,* Vol. 3, Willmer, E. N., Ed., Academic Press, New York, 1966, 691.

95. **Chilton, M. D., Drummond, M. H., Merlo, D. J., Sciaky, D., Montoya, A. L., Gordon, M. P., and Nester, E. W.,** Stable incorporation of plasmid DNA into higher plant cells: the molecular basis of crow gall tumorigenesis, *Cell,* 11, 263, 1977.
96. **Langhans, R. W., Horst, R. K., and Earle, E. D.,** Disease-free plants via tissue culture propagation, *HortScience,* 12, 149, 1977.
97. **Murashige, T. and Jones, J. B.,** Cell and organ culture methods in virus disease therapy, in 3rd Int. Symp. Virus Diseases of Ornamental Plants, *Acta Hortic.,* 36, 207, 1974.
98. **Takebe, I.,** The use of protoplasts in plant virology, *Annu. Rev. Phytopathol.,* 13, 105, 1975.

Natural Products from Tissue Cultures

99. **Alferman, A. W., and Reinhard, E., Eds.,** *Production of Natural Compounds by Cell Culture Methods,* Gesellschaft für Strahlen und Umweltforchung, Munich, 1978.
100. **Barz, W., Reinhard, E., and Zenk, M. H., Eds.,** *Plant Tissue Culture and its Bio-technological Applications,* Proc. 1st Int. Congr. Medicinal Plant Research, Section B, Springer-Verlag, Berlin, 1977.
101. **Krikorian, A. D. and Steward, F. C.,** Biochemical differentiation: the biosynthetic potentialities of growing and quiescent tissue, in *Plant Physiology, a Treatise, Vol. 5B, Analysis of Growth: the Responses of Cells and Tissues in Culture,* Steward, F. C., Ed., Academic Press, New York, 1969, 227.
102. **Nash, D. T. and Davies, M. E.,** Some aspects of growth and metabolism of Paul's scarlet rose cell suspensions, *J. Exp. Bot.,* 23, 75, 1972.
103. **Street, H. E.,** The nutrition and metabolism of plant tissue and organ cultures, in *Cells and Tissues in Culture, Methods, Biology, and Physiology,* Vol. 3, Willmer, E. N., Ed., Academic Press, New York, 1966, 533.

Photosynthesis and Plastid Development in Tissue Culture

104. **Bergmann, L., and Bälz, A.,** Der Einfluss von Farblicht auf Wachstum und Zusammensetzung pflanzlicher Gewebekulturen. I. *Nicotiana tabacum* var. "Samsum", *Planta,* 70, 285, 1966.
105. **Bergmann, L.,** Wachstum grüner Suspensionskulturen von *Nicotiana tabacum* var. "Samsum" mit CO_2 als Kohlenstoffquelle, *Planta,* 74, 243, 1976.
106. **Berlyn, M. B. and Zelitch, I.,** Photoautotrophic growth and photosynthesis in tobacco callus culture, *Plant Physiol.,* 56, 752, 1975.
107. **Berlyn, M. B., Zelitch, I., and Beaudette, P. D.,** Photosynthetic characteristics of photoautotrophically grown tobacco callus cells, *Plant Physiol.,* 61, 606, 1978.
108. **Chandler, M. T., Marsac, N. T., and de Kouchkovsky, V.,** Photosynthetic growth of tobacco cells in liquid suspension, *Can. J. Bot.,* 50, 2265, 1972.
109. **Corduan, G.,** Autotrophe Gewebekulturen von *Ruta graveolens* und deren $^{14}CO_2$-Markierungsprodukte, *Planta,* 91, 291, 1970.
110. **Dalton, C. C. and Street, H. E.,** the role of the gas phase in the greening and growth of illuminated cell suspension of spinach *(Spinacia oleraceae L.), In Vitro,* 12, 485, 1976.
111. **Davey, M. R., Frearson, E. M., and Power, J. B.** Polyethylene glycol-induced transplantation of chloroplasts into protoplasts: an ultrastructural assessment, *Plant Sci. Lett.,* 7, 1976.
112. **Edelmann, J. and Hanson, A. D.,** Sucrose suppression of chlorophyll synthesis in carrot-tissue, *J. Exp. Bot.,* 23, 469, 1972.
113. **Hanson, A. D. and Edlemann, J.,** Photosynthesis by carrot tissue cultures, *Planta,* 102, 11, 1972.
114. **Heltne, J. and Bonnett, H. T.,** Chloroplast development in isolated roots of *Convolvulus arvensis* L., *Planta,* 92, 1, 1970.
115. **Husemann, W. and Barz, W.,** Photoautotropic growth and photosynthesis in cell suspension cultures of *Chenopodium rubrum, Plant. Physiol.,* 40, 77, 1977.
116. **Kaul, K. and Sabharwal, P. S.,** Effects of sucrose and kinetin on the growth and chlorophyll synthesis in tobacco tissue cultures, *Plant Physiol.,* 47, 691, 1971.
117. **Kennedy, R. A.,** Photorespiration in C_3 and C_4 tissue cultures: significance of Kranz anatomy to low photorespiration in C_4 plants, *Plant Physiol.,* 58, 573, 1976.
118. **Kennedy, R. A., Barnes, J. E., and Laetsch, W. A.,** Photosynthesis in C_4 plant tissue cultures: significance of Kranz anatomy to C_4 acid metabolism in C_4 plants, *Plant Physiol.,* 59, 600, 1977.
119. **Laetsch, W. M. and Stetler, D. A.,** Chloroplast structure and function in cultured tobacco tissue, *Am. J. Bot.,* 52, 798, 1965.
120. **de Marsac, N. T. and Péand-Lenoël, C.,** Exchanges of d'oxygene et assimilation de gaz caronique de tissue de tabac en culture photosynthêtique, *C. R. Acad. Sci.,* 274D, 3310, 1972.

121. **Melchers, G. and Labib, G.**, Somatic hybridization of plants by fusion of protoplasts. I. Selection of light resistant hybrids of "haploid" light sensitive varieties of tobacco, *Mol. Gen. Genet.*, 135, 277, 1974.
122. **Neuman, K-H. and Raafat, A.**, Further studies on the photosynthesis of carrot tissue cultures, *Plant Physiol.*, 51, 685, 1973.
123. **Pamplin, E. J. and Chapman, J. M.**, Sucrose suppression of chlorophyll synthesis in tissue culture: changes in the activity of the enzymes of the chlorophyll biosynthetic pathway, *J. Exp. Bot.*, 26, 212, 1975.
124. **Schieder, O.**, Hybridization experiments with protoplasts from chlorophyll-deficient mutants of some solanoceous species, *Planta*, 137, 253, 1977.
125. **Seyer, P., Marty, D., Lecure, A. M.,and Péaud-Lenoël**, Effect of cytokinin on chloroplastic differentiation in cultured tobacco cells, *Cell Differ.*, 4, 187, 1975.
126. **Siebertz, H. P., Heinz, E., and Bergman, L.**, Acyl lipids in photosynthetically active tissue cultures of tobacco, *Plant Sci. Lett.*, 12, 119, 1978.
127. **Stetler, D. A. and Laetsch, W. M.**, Kinetin-induced chloroplast maturation in cultures of tobacco tissue, *Science*, 149, 1387, 1965.
128. **Sunderland, N.**, Pigmented plant tissue culture. I. Auxins and pigmentation in chlorophyllous tissues, *Ann. Bot. (London)*, 30, 253, 1966.
129. **Sunderland, N. and Wells, B.**, Plastid structure and development in green callus tissues of *Oxalis dispar*, *Ann. Bot. (London)*, 32, 327, 1968.
130. **Yamaya, T. K., Ojima, K., and Ohira, K.**, Studies on the greening of cultured soybean and *Ruta* cells. II. Photosynthetic activities of the cultured green cells, *Soil Sci. Plant Nutr. (Tokyo)*, 23, 59, 1977.
131. **Yamada, Y. and Sato, F.**, The photoautotrophic culture of chlorophyllous cells, *Plant Cell Physiol.*, 19, 691, 1978.

Index

INDEX

A

B

C

D

E

F

G

H

I

J

K

L

M

N

O

P

Q

R

S

T

U

V

W

X

Z